AF834543

Novel Carbon Adsorbents

Novel Carbon Adsorbents

edited by
Juan M. D. Tascón
Instituto Nacional del Carbón, INCAR-CSIC, Oviedo, Spain

ELSEVIER

AMSTERDAM • BOSTON • HEIDELBERG • LONDON
NEW YORK • OXFORD • PARIS • SAN DIEGO
SAN FRANCISCO • SYDNEY • TOKYO

Elsevier
The Boulevard, Langford Lane, Kidlington, Oxford OX5 1GB, UK
Radarweg 29, PO Box 211, 1000 AE Amsterdam, The Netherlands
225 Wyman Street, Waltham, MA 02451, USA

First edition 2012

Copyright © 2012 Elsevier Ltd. All rights reserved

No part of this publication may be reproduced, stored in a retrieval system or transmitted in any form or by any means electronic, mechanical, photocopying, recording or otherwise without the prior written permission of the publisher

Permissions may be sought directly from Elsevier's Science & Technology Rights Department in Oxford, UK: phone (+44) (0) 1865 843830; fax (+44) (0) 1865 853333; email: permissions@elsevier.com. Alternatively you can submit your request online by visiting the Elsevier web site at http://elsevier.com/locate/permissions, and selecting Obtaining permission to use Elsevier material

Library of Congress Cataloging-in-Publication Data
A catalog record for this book is available from the Library of Congress

British Library Cataloguing in Publication Data
A catalogue record for this book is available from the British Library

ISBN: 978-0-08-097744-7

For information on all Elsevier publications
visit our web site at store.elsevier.com

Transferred to Digital Printing in 2012

Working together to grow
libraries in developing countries

www.elsevier.com | www.bookaid.org | www.sabre.org

ELSEVIER BOOK AID International Sabre Foundation

Contents

Preface vii
Foreword xi
Contributors xiii

Part I
Introduction

1. Novel Nanocarbons for Adsorption 3

Part II
Recent Developments in Theory

2. Accessibility of Gases and Liquids in Carbons 37
3. Virtual Porous Carbons 61

Part III
New Characterization Methodologies

4. Advanced Physical Adsorption Characterization of Nanoporous Carbons 107
5. Water Adsorption by Carbons. Hydrophobicity and Hydrophilicity 147
6. The Basicity of Carbons 173

Part IV
Adsorption by Novel Carbon Types

7. Adsorption by Carbon Gels 207
8. Adsorption by Phosphorus-Containing Carbons 245
9. Porous Carbide-Derived Carbons 269

10. Zeolite-Templated Carbon – Its Unique Characteristics and Applications 295
11. Adsorption by Soft-Templated Carbons 323
12. Hydrothermal Carbons: Synthesis, Characterization, and Applications 351
13. Porosity and Adsorption Properties of Single-Wall Carbon Nanohorn 401
14. Adsorption Behaviors of Graphene and Graphene-related Materials 435

Part V
Emerging Applications of Adsorption by Carbons

15. Porous Texture Versus Surface Chemistry in Applications of Adsorption by Carbons 471
16. Catalytic Removal of Water-Solved Aromatic Compounds By Carbon-Based Materials 499
17. Photochemical Behavior of Carbon Adsorbents 521
18. Carbon-based Catalyst Support in Fuel Cell Applications 549
19. Novel Carbon Materials for CO_2 Adsorption 583
20. Nonenvironmental Industrial Applications of Activated Carbon Adsorption 605
21. Biomedical Applications of Carbon Adsorbents 639

Index 671

Preface

It is well known that the entire realm of carbon materials science and technology has evolved at an incredible pace over the last 25 years. The impetus for these revolutionary changes has emanated from a number of discoveries that have led to the award of two Nobel Prizes (one for Chemistry in 1996, the other for Physics in 2010), an unprecedented enhancement of the impact of publications in this field and the increasing use of carbon-based commodities by our society.

One of the carbon-related areas that has benefited most from these achievements is that of carbon adsorbents. Suffice it to say that most of the key results dealt with in this book have been produced in the 21st Century, which explains why I have used the term "novel" in the book title, a curious paradox considering that the use of wood charcoal as adsorbent is even mentioned in the Old Testament. Much more recently, when one might have thought that porous (i.e. activated) carbons had reached their zenith, the advent of carbon nanoforms has led to the addition of a number of novel nanostructured materials to the set of already available carbon adsorbents; and, perhaps more significantly, has fostered the development of new concepts and methodologies for producing materials with novel and controlled architectures and functions. The success achieved so far has been due, in part, to the existence of an experienced carbon surface scientific community that has been prepared to assimilate and assume the challenges brought about by all these changes. As a matter of fact, even before the nano-revolution, we were already accustomed to dealing with pores in chars and activated carbons, the majority of which were sub-nanometric in width; but we simply referred to them as micropores, even though "micro" is three orders of magnitude larger than "nano".

In a sense, this book is a follow-up to *Adsorption by Carbons*, another book that I edited together with the late E.J. Bottani and that was published in 2008, also by Elsevier. It consists of 21 chapters dealing with topics not treated in the previous book, most of which concern "newer" subjects. The book is divided into five basic parts: introduction, recent developments in theory, the use of new characterization methodologies, adsorption by novel carbon types, and emerging applications of carbon adsorbents. An introductory chapter that provides an overview of nanocarbons within the framework of adsorption is followed by two chapters on new developments in theory, both of which are very much concerned with carbon porosity. Then a chapter on advanced physical adsorption methods (that has deep roots in theoretical work) leads on to a section dealing with the use of adsorption for characterizing carbon surface features such as hydrophilicity or basicity. The largest section of the book in terms of number of chapters describes the adsorption behavior of novel carbon materials, such as carbon gels, phosphorus-containing carbons, carbide-derived carbons, zeolite-templated carbons, soft-templated carbons, carbon nanohorns and, last but not least, graphene. The final part of the book begins with a challenging discussion on the relative roles of porous texture and surface chemistry in the applications of adsorption by carbons and is followed by a series of chapters on the emerging uses of carbon adsorbents in the areas of catalysis, photochemistry, fuel cells, carbon dioxide capture and in the industrial (excluding environment-related uses) and biomedical sectors. In summary, this book contains a set of 21 authoritative chapters that provide, I hope, a consistent and integrated body of knowledge revolving around the field of novel carbon adsorbents.

As in the case of *Adsorption by Carbons*, the strength of the present book mainly emanates from the stature of the contributing authors (none of whom contributed to the previous book—therefore the team is also new). The book has an unquestionably international flavor, as it includes authors with affiliations to no fewer than 16 different countries (this figure would be larger if one took into account the number of nationalities or mother languages involved). If the previous *Adsorption by Carbons* book was slightly unbalanced in favor of countries from the New World, that imbalance has now been redressed with the majority of the chapters coming from European countries and a significant proportion from Asia/Oceania. I am particularly proud of having succeeded in getting together an outstanding group of carbon scientists who managed to find the time to prepare their contributions when time is one of our scarcest and most precious resources. To convince them, I had recourse to my personal friendships, contacts established during collaboration projects and colleagues both in Spain and abroad. Once won over to the cause, they showed a willingness, an enthusiasm and a professionalism that knew no bounds. Rather than cite the participants' authors names here, which would make this text exceedingly long, I prefer to thank them collectively for their efforts. My thanks go out to the staff of Elsevier, particularly Ms. Louisa Hutchins, Editorial Project Manager, for her constant willingness to help me and even, on occasions, for spurring me on when, owing to my official duties as Director of INCAR, I was forced to slow down my pace as editor.

I also wish to thank Prof. John W. Patrick for contributing the Foreword to this book. From his privileged watch-tower position, first at Chesterfield, then at Loughborough and now at Nottingham, Prof. Patrick has surveyed for many years the progress made in the field of carbon-based materials. Our scientific community is much indebted to him for his almost lifelong commitment as Editor of *Fuel* and for his hard work in different fields of coal and carbon science and technology (I had the pleasure of working with him in a joint research project several years ago). Due to its relevance to the topics discussed here, I would like to make explicit mention of *Porosity in Carbons,* a book which he edited in 1995, and which is a fine predecessor for this book. My warmest thanks also go to Sendai, the workplace of Prof. Hirotomo Nishihara, who provided me with a highly esthetic, yet precise picture of the nanostructure of a zeolite-templated carbon that is featured on the cover of this book. This diagram beautifully illustrates the complexity and intricacy of the graphene sheet network that constitutes this type of adsorbent. *Nishihara–sensei, arigatō gozaimasu!*

Juan M.D. Tascón

Foreword

Carbon as an adsorbent is hardly a novel concept since there is evidence that this remarkable element was used for that purpose as early as 1550 BC and the published literature bears testimony to the wide range of such applications since then. Indeed numerous books on this subject have been produced over the years. So the question can be asked as to what is the motivation for another book on carbon as an adsorbent and what is novel about this ubiquitous element.

The answer is that despite its long history as an adsorbent, this fascinating element carbon still continues to amaze and intrigue us with an enigmatic character which provides it with almost unbelievable variation in structural form and adaptability in use.

When I started on my scientific career, I was taught that carbon always tended toward the formation of six-membered rings which stacked in parallel layers in an ABAB or ABCABC sequence. Now we know that five-membered rings are not uncommon along with spherical and tubular forms as alternative stable structures and I now have to contend with buckyballs, nanotubes and even single layered graphene as remarkably different carbon forms with distinctive properties.

The development of the computer has facilitated the development of modern analytical equipment for improved characterization of surfaces and the measurement of adsorptive capacity of materials and this has facilitated the development of modeling, thereby making a significant contribution to the understanding of the mechanisms involved in the adsorption of both gases and liquids by the different carbon forms.

At the same time the demand for adsorbents has increased remarkably through enlightened considerations of such topics as health and safety, alongside the ever increasing environmental constraints imposed on industries ranging from production of chemicals, water and air purification, to solvent recovery and even gold recovery. This demand allied to the gradual development of our understanding of both the nature of the various forms of carbon materials and the mechanisms involved in the adsorption and desorption processes provide another inducement to the writing of this book.

The twenty one chapters contributed by well-known experts in their field, form a most useful and timely contribution to our knowledge and understanding of the novel carbon materials now available to us and their application as adsorbents in a variety of processes.

John W. Patrick

Contributors

Gisèle Amaral-Labat, Institut Jean Lamour - UMR CNRS 7198, Université de Lorraine, ENSTIB, Épinal, France

Conchi O. Ania, Instituto Nacional del Carbón, INCAR-CSIC, Oviedo, Spain

Suresh K. Bhatia, School of Chemical Engineering, The University of Queensland, Brisbane, QLD, Australia

Eva Castillejos, Instituto de Catálisis y Petroleoquímica, CSIC, Campus Cantoblanco, Madrid, Spain

Alain Celzard, Institut Jean Lamour - UMR CNRS 7198, Université de Lorraine, ENSTIB, Épinal, France

Katie A. Cychosz, Quantachrome Instruments, Boynton Beach, FL, USA

Sheng Dai, Oak Ridge National Laboratory, Chemical Sciences Division, Oak Ridge, USA, Department of Chemistry, University of Tennessee, Knoxville, USA

Vanessa Fierro, Institut Jean Lamour - UMR CNRS 7198, Université de Lorraine, ENSTIB. Épinal, France

José Luís Figueiredo, Laboratório de Catálise e Materiais (LCM), Laboratório Associado LSRE/LCM, Departamento de Engenharia Química, Universidade do Porto, Porto, Portugal

Enrique Fuente, Instituto Nacional del Carbón, INCAR-CSIC, Oviedo, Spain

Sylwester Furmaniak, N. Copernicus University, Department of Chemistry, Physicochemistry of Carbon Materials Research Group, Toruń, Poland

Piotr A. Gauden, N. Copernicus University, Department of Chemistry, Physicochemistry of Carbon Materials Research Group, Toruń, Poland

Joanna Górka, Oak Ridge National Laboratory, Chemical Sciences Division, Oak Ridge, TN, USA

Meenakshi Goyal, University Institute of Chemical Engineering and Technology, Panjab University, Chandigarh, India

Eduardo Gracia-Espino, Advanced Materials Department, IPICYT, San Luis Potosí, Mexico

Antonio Guerrero-Ruiz, Unidad Asociada UNED-CSIC, "Grupo de Diseño y Aplicación de Catalizadores Heterogeneos", Spain, and Dpto. Química Inorgánica y Química Técnica, Facultad de Ciencias, UNED, Spain

Peter J.F. Harris, Centre for Advanced Microscopy, University of Reading, Whiteknights, UK

Carol A. Howell, University of Brighton, Brighton, UK

Denisa Hulicova-Jurcakova, ARC Centre of Excellence for Functional Nanomaterials, Australian Institute for Bioengineering and Nanotechnology, University of Queensland, St Lucia, QLD, Australia

Katsumi Kaneko, Research Center for Exotic Nanocarbons, Shinshu University, Nagano, Japan

Piotr Kowalczyk, Nanochemistry Research Institute, Department of Chemistry, Curtin University of Technology, Western Australia, Australia

Takashi Kyotani, Institute for Multidisciplinary Research for Advanced Materials, Tohoku University, Katahira, Sendai, Japan

Krisztina László, Department of Physical Chemistry and Materials Science, Budapest University of Technology and Economics, Budapest, Hungary

Seul-Yi Lee, Department of Chemistry, Inha University, Nam-gu, Incheon, South Korea

Ji Liang, ARC Centre of Excellence for Functional Nanomaterials, Australian Institute for Bioengineering and Nanotechnology, University of Queensland, St Lucia, QLD, Australia

Florentino López-Urías, Advanced Materials Department, IPICYT, San Luis Potosí S.L.P., Mexico, and Department of Physics, The Pennsylvania State University, University Park, PA, USA

Gao Qing (Max) Lu, ARC Centre of Excellence for Functional Nanomaterials, Australian Institute for Bioengineering and Nanotechnology, University of Queensland, St Lucia, QLD, Australia

J. Ángel Menéndez, Instituto Nacional del Carbón, INCAR-CSIC, Oviedo, Spain

Sergey V. Mikhalovsky, University of Brighton, Brighton, UK, and Nazarbayev University, Astana, Kazakhstan

Miguel A. Montes-Morán, Instituto Nacional del Carbón, INCAR-CSIC, Oviedo, Spain

Alexander V. Neimark, Rutgers University, Department of Chemical and Biochemical Engineering, Piscataway, NJ, USA

Vladimir G. Nikolaev, Kavetsky Institute of Experimental Pathology, Oncology and Radiobiology (IEPOR), Kiev, Ukraine

Hirotomo Nishihara, Institute for Multidisciplinary Research for Advanced Materials, Tohoku University, Katahira, Sendai, Japan

Tomonori Ohba, Graduate School of Science, Chiba University, Chiba, Japan

Soo-Jin Park, Department of Chemistry, Inha University, Nam-gu, Incheon, South Korea

Manuel Fernando R. Pereira, Laboratório de Catálise e Materiais (LCM), Laboratório Associado LSRE/LCM, Departamento de Engenharia Química, Universidade do Porto, Porto, Portugal

Gary J. Phillips, University of Brighton, Brighton, UK

Nadejda Popovska-Leipertz, Lehrstuhl für Chemische Reaktionstechnik, Universität Erlangen-Nürnberg, Erlangen, Germany

Alexander M. Puziy, Institute for Sorption and Problems of Endoecology, National Academy of Sciences of Ukraine, Kyiv, Ukraine

Shi Zhang Qiao, ARC Centre of Excellence for Functional Nanomaterials, Australian Institute for Bioengineering and Nanotechnology, University of Queensland, St Lucia, QLD, Australia

Inmaculada Rodríguez-Ramos, Instituto de Catálisis y Petroleoquímica, CSIC, Campus Cantoblanco, Madrid, Spain, and Unidad Asociada UNED-CSIC, Grupo de Diseño y Aplicación de Catalizadores Heterogeneos, Spain

Francisco Rodríguez-Reinoso, Instituto Universitario de Materiales, Departamento de Química Inorgánica, Universidad de Alicante, Alicante, Spain

Susan R. Sandeman, University of Brighton, Brighton, UK

Joaquín Silvestre-Albero, Instituto Universitario de Materiales, Departamento de Química Inorgánica, Universidad de Alicante, Alicante, Spain

Dimas Suárez, Departamento de Química Física y Analítica, Universidad de Oviedo, Oviedo, Spain

Hideki Tanaka, Department of Chemical Engineering, Kyoto University, Kyoto, Japan

Juan M.D. Tascón, Instituto Nacional del Carbón, INCAR-CSIC, Oviedo, Spain

Humberto Terrones, Department of Physics, The Pennsylvania State University, University Park, PA, USA, and Departamento de Física, Universidade Federal do Ceará, Fortaleza, CE, Brazil

Mauricio Terrones, Department of Physics, and Department of Materials Science and Engineering and Materials Research Institute, The Pennsylvania State University, University Park, PA, USA, and Research Center for Exotic Nanocarbons (JST), Shinshu University, Nagano, Japan

Artur P. Terzyk, N. Copernicus University, Department of Chemistry, Physicochemistry of Carbon Materials Research Group, Toruń, Poland

Matthias Thommes, Quantachrome Instruments, Boynton Beach, FL, USA

Maria-Magdalena Titirici, Max Planck Institute of Colloids and Interfaces, Potsdam, Germany

Ajna Tóth, Department of Physical Chemistry and Materials Science, Budapest University of Technology and Economics, Budapest, Hungary

Koki Urita, Faculty of Engineering, Nagasaki University, Nagasaki, Japan

Shigenori Utsumi, Faculty of Systems Engineering, Tokyo University of Science, Suwa, Japan

Teresa Valdés-Solís, Instituto Nacional del Carbón, INCAR-CSIC, Oviedo, Spain

Leticia F. Velasco, Instituto Nacional del Carbón, INCAR-CSIC, Oviedo, Spain

Chapter 1

Novel Nanocarbons for Adsorption

Eduardo Gracia-Espino
Advanced Materials Department, IPICYT, San Luis Potosí, Mexico

Florentino López-Urías
Advanced Materials Department, IPICYT, San Luis Potosí S.L.P., Mexico, and Department of Physics, The Pennsylvania State University, University Park, PA, USA

Humberto Terrones
Department of Physics, The Pennsylvania State University, University Park, PA, USA, and Departamento de Física, Universidade Federal do Ceará, Fortaleza, CE, Brazil

Mauricio Terrones
Department of Physics, and Department of Materials Science and Engineering and Materials Research Institute, The Pennsylvania State University, University Park, PA, USA, and Research Center for Exotic Nanocarbons (JST), Shinshu University, Nagano, Japan

Chapter Outline

1.1. Introduction 4
1.2. General Aspects of Carbon Nanostructures 4
1.3. Adsorption on Carbon Nanomaterials 9
1.3.1. Gases and Vapors Adsorbed on Carbon Nanotubes 9
1.3.2. Gas Adsorption on Graphene, Graphitic Nanoribbons and Other Carbon Materials 13
1.4. Biological Systems Adsorbed on Carbon Nanomaterials 16
1.4.1. Fullerenes 16
1.4.2. Carbon Nanotubes 19
1.4.3. Graphene 20
1.5. Adsorption of Heavy-Metals on Modified Carbon Nanomaterials 21
1.5.1. Lead Adsorption by Carbon Nanotubes 21
1.5.2. Chromium Adsorption by Carbon Nanotubes 22

Novel Carbon Adsorbents. DOI: 10.1016/B978-0-08-097744-7.00001-6
Copyright © 2012 Elsevier Ltd. All rights reserved

1.5.3. Zinc, Nickel, Copper and Other Materials Adsorption by Carbon Nanotubes 22
1.6. Carbon Dioxide Uptake on Carbon Nanostructures 24
1.7. Conclusions 26
References 26

1.1. INTRODUCTION

Carbon is a fascinating element of the periodic table since it can establish bonds with almost any element, thus resulting in a wide spectrum of compounds and allotropic forms. Adsorption is a phenomenon that occurs on surfaces, and materials with larger surface area are strong candidates to adsorb high concentrations of different chemical species. For a long time, activated carbons (porous materials with large surface area) have been extensively used for water treatment and gas adsorption. The new generation of carbon structures built at the nanoscale (e.g. nanofibers, nanotubes, fullerenes, nanocones, graphene, graphene nanoribbons, nanodiamonds, etc.) with relatively large surface area, and exhibiting novel electronic and chemical properties, provides new horizons for achieving enhanced adsorption, which could result in new applications. For example, some of these carbon nanostructures could be used in wastewater treatments, routes for clean energy generation, hydrogen storage devices, sensors, catalytic supports, virus inhibitors, etc. In this chapter, we review a wide spectrum of novel carbon nanostructures and their potential in the adsorption of heavy metals, gases, polymers, and biomolecules. We emphasize the role of surface modification of these nano-systems via chemical doping, defect engineering and chemical functionalization, and discuss their effects in the adsorption properties of carbon nanomaterials.

1.2. GENERAL ASPECTS OF CARBON NANOSTRUCTURES

Nowadays there are numerous carbon nanomaterials, some of which could be curved in order to form fascinating morphologies with novel physico-chemical properties (see Fig. 1.1). Graphite is the best example of a very flexible layered material, which is able to form a wide variety of shapes, ranging from zero-dimensional, such as fullerenes [1,2], cones [3,4] and toroids [5], to 1-, 2-, and 3-dimensional. In particular, one-dimensional (1D) systems include single- and multiwalled nanotubes [6,7], nanoribbons, and nanohelices. Two-dimensional (2D) structures such as graphene, antidot graphene and super-graphene, and Haeckelites [8]. Three-dimensional (3D) systems consist of graphite and Schwarzite-like structures [9–12]. In addition, Fullerenes and cones exhibit positive Gaussian curvature due to the presence of pentagonal carbon rings, and there are other structures containing heptagonal or higher carbon-membered

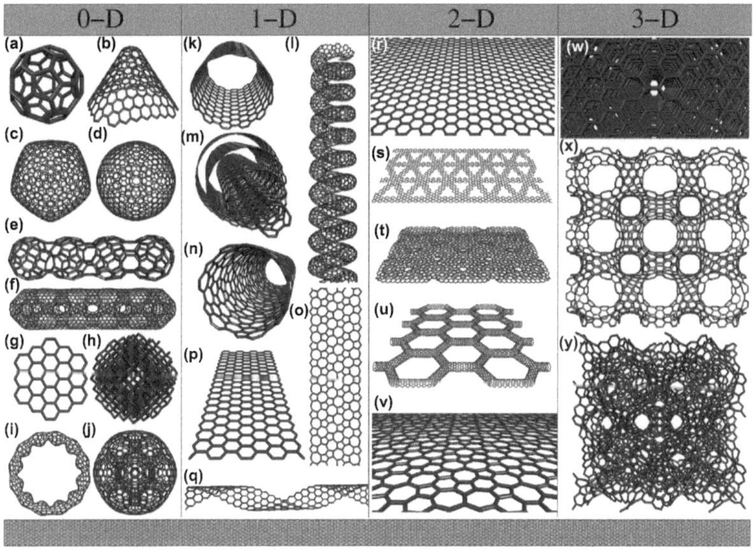

FIGURE 1.1 Molecular models of various types of hybridized carbon nanostructures with different dimensionalities, 0D, 1D, 2D, and 3D: (a) C_{60}: Buckminsterfullerene; (b) nanocone or nanohorn, (c) icosahedral giant fullerene, (d) spherical giant fullerene, (e) coalesced C_{60} molecules, (f) holey tube, (g) graphene cluster, (h) nanodiamond cluster, (i) nanotoroid of coalesced C_{60} molecules, (j) two interconnected fullerene surfaces with channels, (k) nanotube, (l) helicoidal nanotube, (m) multiwalled carbon nanotube, (n) Haeckelite nanotube (structure formed by only 5 and 7 carbon rings), (o) Haeckelite nanoribbon, (p) graphene nanoribbons, (q) helicoidal graphene nanoribbon, (r) graphene surface, (s) antidot graphene, (t) two interconnected graphene sheets, (u) super-graphene, (v) Haeckelite surface, (w) graphite, (x) Schwarzite P688 crystals, and (y) Schwarzite D688 crystals.

rings exhibiting negative Gaussian curvature, such as Schwarzites. More generally, the concept of curvature in graphitic nanostructures can be understood by introducing different sizes of carbon rings within the hexagonal framework (e.g. 4-, 5-, 7-, and 8- membered rings). The introduction of non-hexagonal rings modifies the Gaussian curvature of the structure, thus creating unexpected properties exactly where the defect is located; these defective sites could then be attractive for enhanced adsorption properties (see below).

Carbon allotropes (e.g. graphene, diamond, graphite, and fullerenes) exhibit different electronic properties. For example, graphite is a semimetal, whereas diamond is a semiconductor with large electronic bandgap. Furthermore, carbon nanotubes formed by a rolled single graphene layer could be either metallic or a semiconductor depending, on the way they are rolled up [13,14,15]. In particular, nanotubes with armchair chirality exhibit metallic characteristics whereas zigzag tubes, a semiconducting or semimetal behavior. By unzipping a carbon nanotube, graphene nanoribbons could be formed exhibiting a semiconducting behavior (ribbons having armchair edges) or metallic, if spin polarization is not considered (ribbons with zigzag edges); three different families of nanoribbbons could exist according to their width [16,17,18]. When spin polarization is considered, the most stable state for zigzag nanoribbons is semiconducting. In addition, antiferromagnetic and ferromagnetic couplings along the zigzag edges could be possible [18].

The chemical activity of these novel carbon nanomaterials could be correlated to their morphology. For example, flat systems (zero Gaussian curvature) containing all saturated bonds as in graphite, do not interact strongly with other chemical species, unless defects or doping are present. Therefore, the chemical activity of carbon nanosurfaces could be studied as a function of their Gaussian curvature and carbon bond hybridization (sp, sp^2, or sp^3). In this context, there are novel carbon nanostructures that exhibit different types of Gaussian curvature and consequently they display unusual chemical activity along their surface.

A selective surface modification of materials could result in significant changes in their chemical, electronic, magnetic, and mechanical properties. The addition of foreign atoms (doping), the creation of defects, or the addition of functional groups might result in strongly modified nanocarbons that change dramatically their interactions with the surrounding environment. *Doping* is a very useful technique able to tune the electronic and chemical properties of different carbon nanomaterials, including fullerenes [19]; carbon nanotubes [20,22,23]; and graphene nanoribbons [25–29]. Different atoms, such as nitrogen, boron, phosphorus, sulfur, and silicon, have been introduced into the graphene-like lattice of the carbon nanomaterials (see Fig. 1.2). *Defect generation* of the graphitic surface

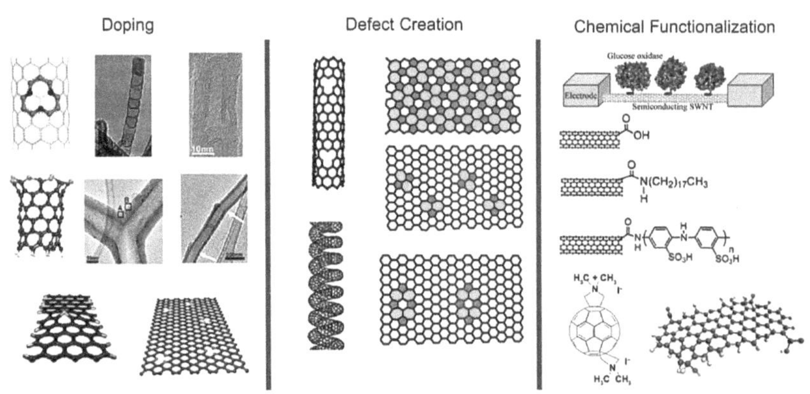

FIGURE 1.2 Different strategies to modify the electronic and chemical properties of carbon nanomaterials. Doping by introducing atoms such as nitrogen [20], phosphorous [30], and sulfur [31] among others has been widely used. Vacancies and bond rotations are typically present on carbon nanomaterials. Chemical functionalization involves the adsorption of functional groups such as carboxylic acid or polymers and biomolecules on the graphene lattice [32,33].

constitutes an alternative approach to modify the electronic and chemical properties of carbon nanomaterials. Defects such as vacancies (from monovacancy, divacancy up to antidot systems [34,35], bond rotations such as Thrower-Stone-Wales defects [36,37], and the introduction of non-hexagonal carbon rings (4-,5-,7-, and 8- membered rings that modify the Gaussian curvature [3,5]) could significantly alter the local properties of carbon nanostructures [38] (see Fig. 1.2). The *chemical functionalization* consists of adding diverse functional groups to the surface of different nanomaterials (see Fig. 1.2). The functionalization could be performed by the introduction of small functional groups, such as carboxylic and amine groups [39,40], or large molecules such as polymers [41] and biomolecules [42].

One could also construct 2D and 3D arrays of nanocarbons by assembling different nanoscale-building blocks (e.g. nanotubes, nanoribbons, etc). Along this line, 3D carbon nanomaterials exhibiting large surface area (e.g. 3600 m^2/g) could be constructed by interconnecting covalently carbon nanotubes [43]. In particular, Romero-Herrera et al. [43] proposed different types of covalent nanotube junctions using tubes of different chiralities (see Fig. 1.3). These authors demonstrated that novel systems also exhibit outstanding shock absorbing properties. 3-D carbon nanotube networks have been synthesized by Lepró et al. [44]. In particular, these researchers synthesized branched carbon nanotubes and networks that are believed to show enhanced chemically active sites when compared to straight 1-D carbon nanotubes. Different types of heterojunctions with "Y" and "T" morphologies were identified (see Fig. 1.4). In this direction, experimental work related to branched carbon nanotubes was reported by Romo-Herrera et al. [23,31]. The authors showed SEM and HRTEM images, as well as elemental characterization of various Y junctions generated by pyrolyzing thiophene and nickelocene in an inert atmosphere. Sulfur atoms were mainly located in regions of the networks that exhibited the largest curvature. They also demonstrated that sulfur plays an important role in the formation of branched nanotube networks with stacked-cone morphologies (see Fig. 1.5). The growing branches possessed minute amounts of sulfur that are sufficient to promote the formation of heptagons (negative curvature) and pentagons (positive curvature). However, further experimental and theoretical work is still needed along this line, and more control in the fabrication of 3-D nanotube networks with enhanced adsorption properties is awaiting to be conducted.

Various research groups have demonstrated that pristine carbon nanomaterials could exhibit low chemical reactivity, and therefore low adsorption properties. For this reason, their surface modification has been a useful route to improve the interaction of carbon nanomaterials with their surrounding environment including different molecules, clusters, and compounds.

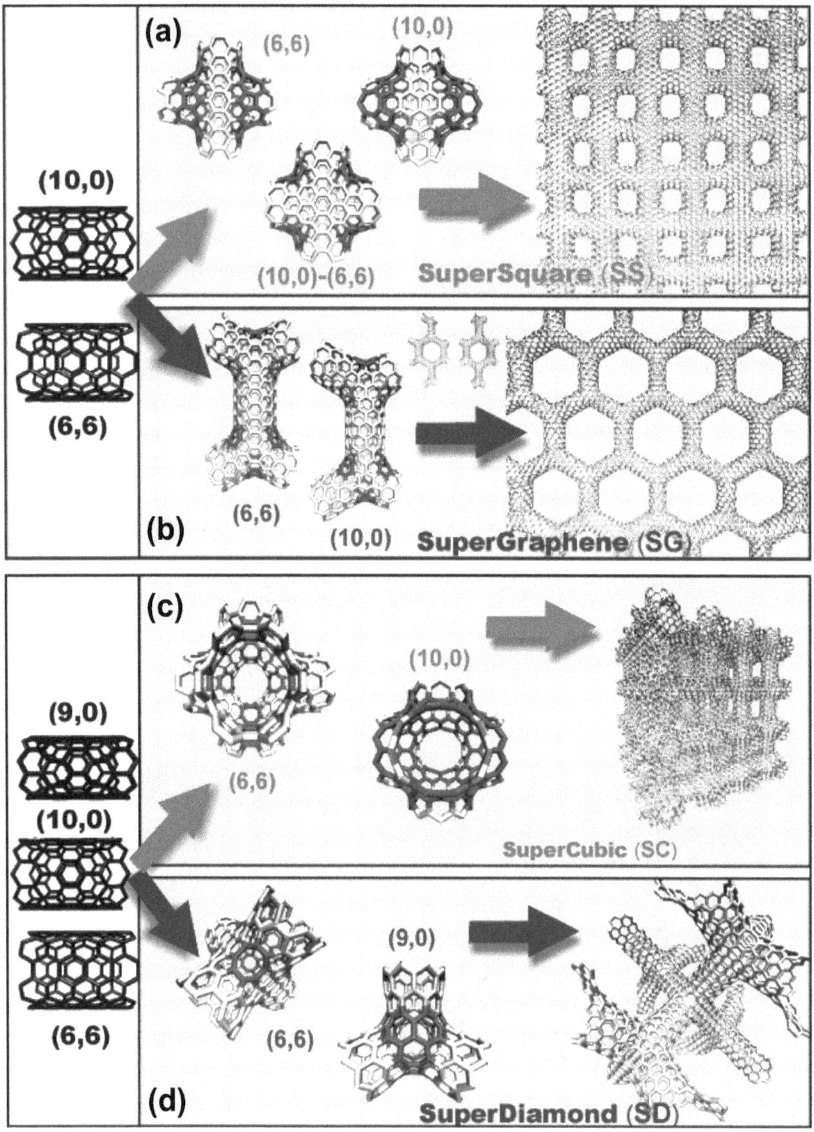

FIGURE 1.3 Model for four different architectures of ordered networks based on carbon nanotubes (1D nanostructures). (a and b) Super-square and super-graphene correspond to 2D networks, whereas (c and d) super-cubic and super-diamond represent 3D network examples. The four families are constructed from either armchair or zigzag carbon nanotubes in order to study the chirality effects. The red rings point out the non-hexagonal carbon rings in each node [43].

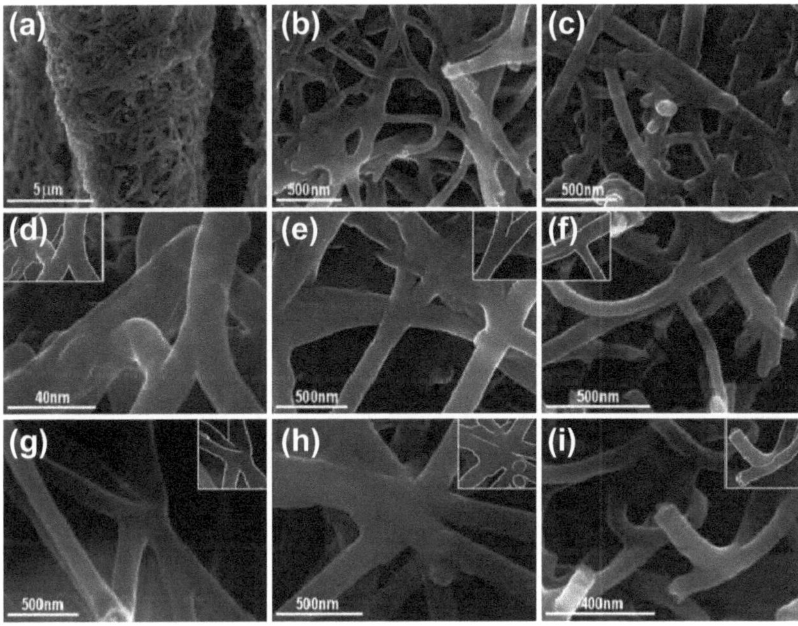

FIGURE 1.4 Scanning electron microscopy images of nanotube junctions: (a) overall morphology of the CNx@CNT networks; (b) and (c) higher magnification images showing different ways in which the nanotubes are joined; (d) and (e) are representative images of "Y"-junctions; (f) a "T"-junction; (g) and (h) show multibranched structures; and (i) exhibits one nanocable embedded over another, denoted as an "on-junction" [44].

1.3. ADSORPTION ON CARBON NANOMATERIALS

1.3.1. Gases and Vapors Adsorbed on Carbon Nanotubes

Carbon nanotubes have proved to be excellent candidates for developing sensing applications [45,46]. In this context, multi-walled carbon nanotube (MWCNT)-based sensors are sensitive to a variety of gases such as NH_3, NO, NO_2, H_2, SF_6, and Cl_2 [47–49]. In addition, chemically functionalized CNTs have been used as an active component in sensors via the addition of functional groups such as carboxylic acid on SWCNTs or MWCNTs; the modified tubes revealed enhanced sensitivity to ethanol [50] and NO gas [51]. Other approaches involve the incorporation of carbon nanotubes into polymers [52,53], such as polypyrrole that showed an enhanced sensitivity. SWNTs with covalently attached poly- (*m*-amino benzene sulfonic acid) have also shown improved sensing for the detection of NH_3, thus allowing the detection of this gas at concentrations as low as 5 ppm. Doping has been also used to improve the gas selectivity and sensitivity of sensors. For example, nitrogen-doped multi-walled carbon nanotubes (CN_X-MWNTs) have shown to be excellent for gas sensing applications [54,55]. Villalpando-Paez et al. [54] fabricated a pellet

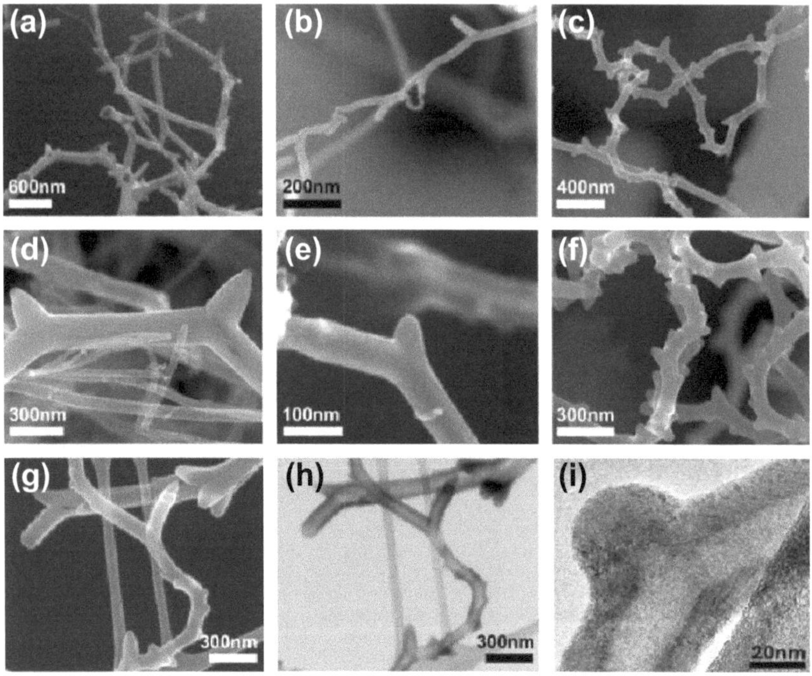

FIGURE 1.5 General morphology of carbon nanostructure junctions. (a) and (d) SEM images of thick junctions (diameters ca. 200 nm), (b) and (e) thinner junctions(diameters ca. 50 nm), (c) and (f) networks with a higher density of branching points, (g) SEM and (h) bright field STEM, showing the hollow interior of the junctions, and (i) TEM image, showing the general structure of an opening junction [31].

sensor based on aligned carbon nanotubes doped with nitrogen, which efficiently detected different gas vapors such as ethanol, gasoline, acetone, ammonia, etc (see Fig. 1.6). Recently, Rebollo-Plata et al. [55] demonstrated that CNx-MWNT films could be also used as efficient gas sensors. Other systems consisting of nanotubes decorated with metal or metal-oxide nanoparticles have been proposed as active components in the fabrication of gas sensors. These hybrid nanoparticles–nanotube systems have attracted much attention because they displayed enhanced sensitivity towards gaseous species such as H_2, H_2S [56,57], NO_2 [58,59], NH_3, and CO [60], and biological materials [61–63] when compared to pure carbon nanotube systems [64]. More recently, hetero-doped carbon nanotubes with nitrogen and phosphorus have been synthesized [22,30,65]. Density functional calculations predict that substitutional phosphorus atoms strongly modify the chemical properties of the surface of nanotubes, thus creating highly localized sites with specific affinity towards acceptor molecules [24]. Phosphorus–nitrogen co-dopants have a similar effect on acceptor molecules, but the P–N bond could also accept

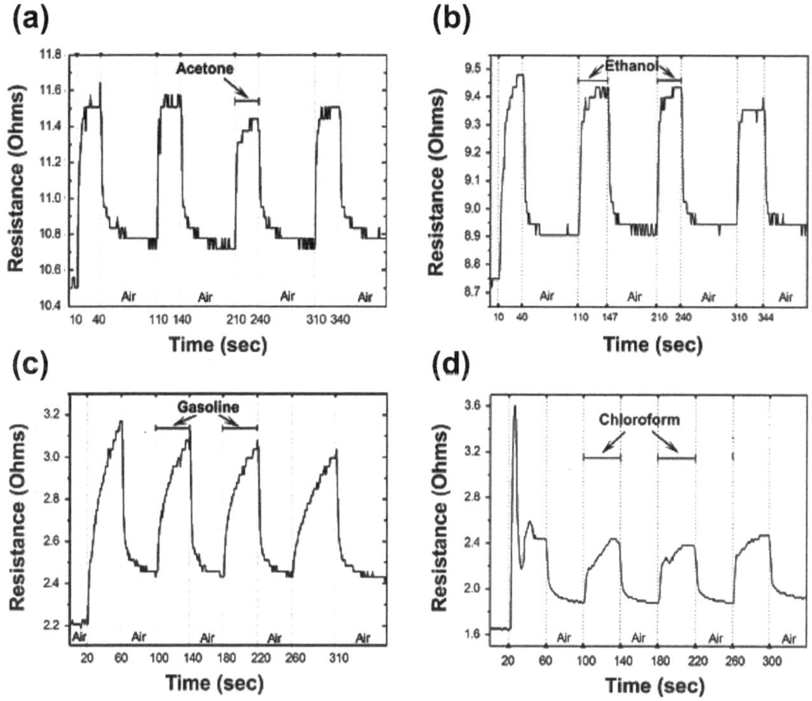

FIGURE 1.6 Resistance vs. time for nitrogen-doped multiwalled carbon nanotubes (CNx-MWNTs) films subjected to different vapors: (a) acetone, (b) ethanol, (c) gasoline, and (d) chloroform. The films are formed by aligned carbon nanotubes [54]. The authors demonstrated that CNx-MWNTs could be potentially more efficient in detecting gaseous hazardous species due to the presence of active sites localized around the nitrogen atoms.

charge, thus resulting in an affinity to donor molecules. This molecular selectivity is illustrated for CO and NH_3 adsorbed onto PN-doped nanotubes, O_2 on P-doped nanotubes, and NO_2 and SO_2 on both P- and PN-doped nanotubes [24].

Along this line, we have theoretically studied four different functional groups that have been anchored to a (10,0) semiconductor SWCNT. In particular, carboxylic acid (COOH-), amide (NH_2-), a repeating unit of poly(ethylene glycol), $C_2H_5O_2$, and poly(amino-benzene sulfonic) acid strongly interacted with the (10,0) semiconducting tube. The relaxed geometries are shown in Fig. 1.7. In all cases, the nanotube geometries are not significantly modified by the addition of the functional group, only the carbon atom containing the additional group exhibits an sp^3 hybridization. The electronic density of states is mainly located in the functional groups. The oxygen and nitrogen atoms of the different functional groups exhibit the highest electronic density. Therefore, these reactive sites could enhance the interaction with some molecules, suitable for further chemical reactions or for sensing

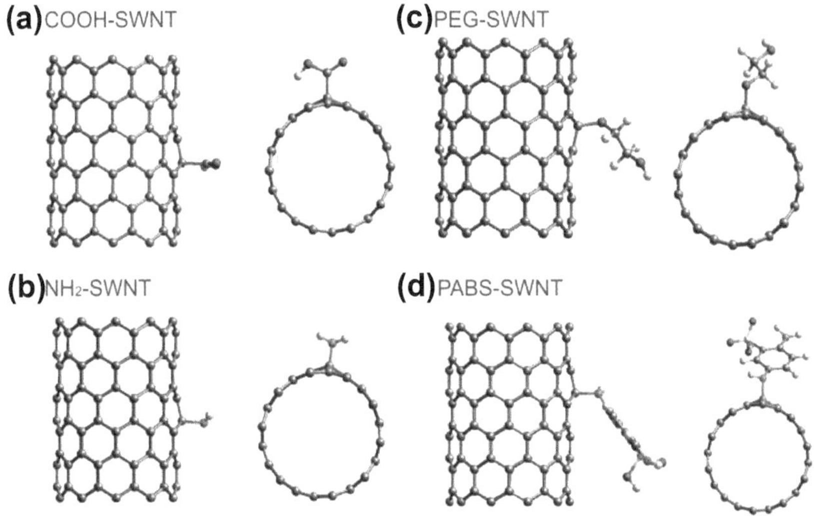

FIGURE 1.7 Density functional calculations results different molecules adsorbed onto a single walled (10,0) carbon nanotube (SWNT). All systems were relaxed using the conjugated gradient method. The different images correspond to the top and side views of: (a) Carboxyl (COOH), (b) amide (NH_2), (c) ethylene glycol (PEG), and (d) amino benzene sulfonic (PABS). Note that the molecules were adsorbed by the carbon nanotubes via a covalent bond. Our results were obtained using the Siesta code [66].

applications with improved selectivity. Figure 1.8 depicts the band structure of (b) carboxyl-, (c) amide-, (d) ethylene glycol-, and (e) amino-benzene sulfonic acid-SWCNT. Note that, the band structure of (10,0)-SWCNTs exhibit a bandgap characteristic of semiconductor nanotubes (0.754 eV). All the functionalized nanotubes possess a highly localized and partially filled band passing through the Fermi level. These results are in agreement with previous publications, in which all the monovalent functionalization generates an impurity state at the Fermi level. PEG is a well-known insulating material, and our results demonstrate that the PEG chain does not exhibit states near the Fermi level within the PEG-SWCNT system, thus, suggesting that thin films of PEG-SWCNTs could not have a good electrical conduction if only PEG-polymer chains are in contact. However, the PABS chain exhibits states near the Fermi level, suggesting that it could act as a conductor, enhancing the conductivity through the nanotube-polymer system. These results are in good agreement with our experimental observations, in which the PABS-SWNT composite exhibited enhanced electrical conductivity when compared to the other functionalized carbon nanotubes [67].

Recently, arsenic, lead, and cadmium, in an aqueous environment, have been detected using CN_X-MWNTs [68–70]. Due to their enhanced reactivity, CN_X-MWNTs have demonstrated to be an enhanced adsorbent material when

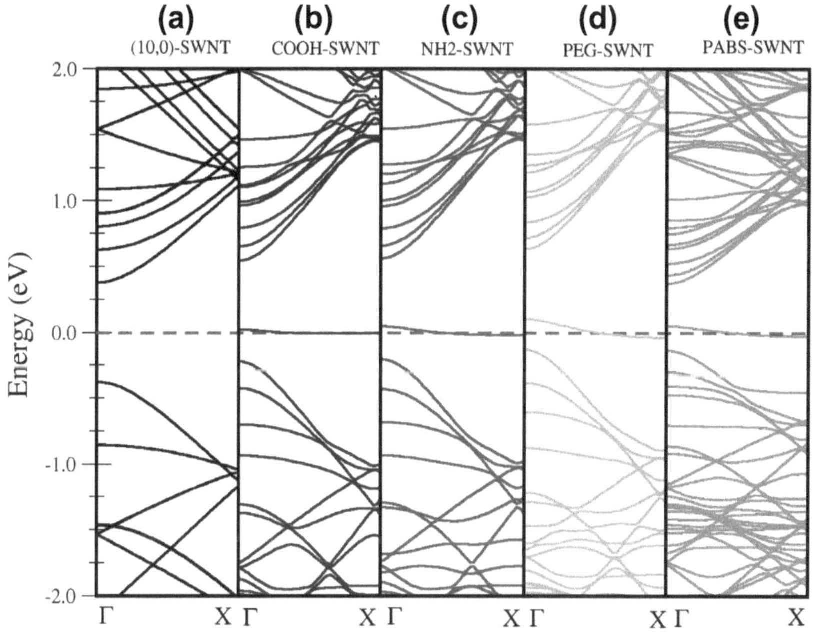

FIGURE 1.8 Band structure calculations for the (10,0) single walled carbon nanotubes (SWNTs) with different molecules adsorbed onto the SWNT waist. In (a), the pristine or non-functionalized case is shown. (b) Carboxyl- SWCNT, (c) amide-SWCNT (d) (ethylene glycol)-SWCNT, (e) (amino benzene sulfonic)-SWCNT. The entire functionalized nanotubes exhibit highly localized bands (quasi-flat bands) around the Fermi level (dashed red line at zero energy). The corresponding relaxed structures can be seen in Fig. 1.6. Our results were obtained by Siesta code [66].

compared to pure (undoped) MWNTs [71]. Density functional theory (DFT) calculations of a Cd atom on the surface of a N-doped single walled carbon nanotubes are shown in Figs. 1.9 and 1.10. From these calculations, it is clear that the Cd atom is strongly adsorbed onto the nitrogenated site within the tube, and the bandgap is reduced by the presence of the metal atom. Therefore, the adsorption of metal atoms or clusters onto carbon nanotubes (doped or modified) offers a variety of possibilities able to efficiently remove metallic nanoparticles from aqueous media.

1.3.2. Gas Adsorption on Graphene, Graphitic Nanoribbons and Other Carbon Materials

Since the first experiments related to the isolation of graphene and few-layered graphene [72], their sensing properties started to be investigated by different research groups worldwide. The two-dimensional material and its exceptional properties make graphene an attractive material for sensing applications, thus

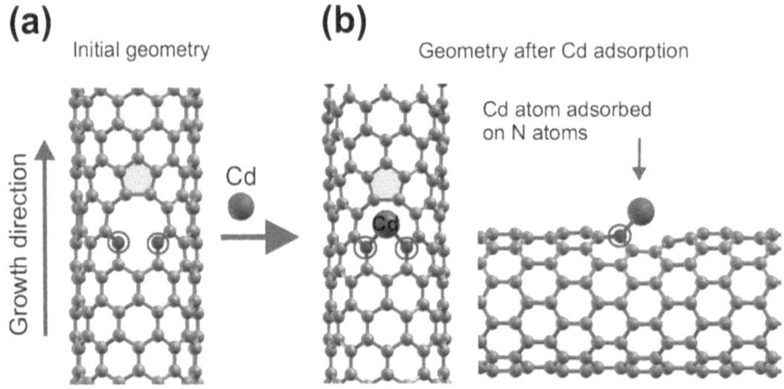

FIGURE 1.9 First-principles calculations of adsorbed cadmium atom on the surface of a nitrogen-doped single walled (10,0) carbon nanotube (SWNT). The left side depicts the starting geometry in our conjugated gradient geometry relaxation method. The nitrogen atoms are closed by a circle. After geometry relaxation, the cadmium atom is adsorbed by the carbon nanotube via the nitrogen atoms. The structure of the carbon nanotube was obtained by removing a carbon atom (vacancy creation) and by replacing two carbon atoms by nitrogen around the vacancy, after geometry optimization, a pentagonal carbon ring appears [71]. Our results were obtained using the Siesta code [66].

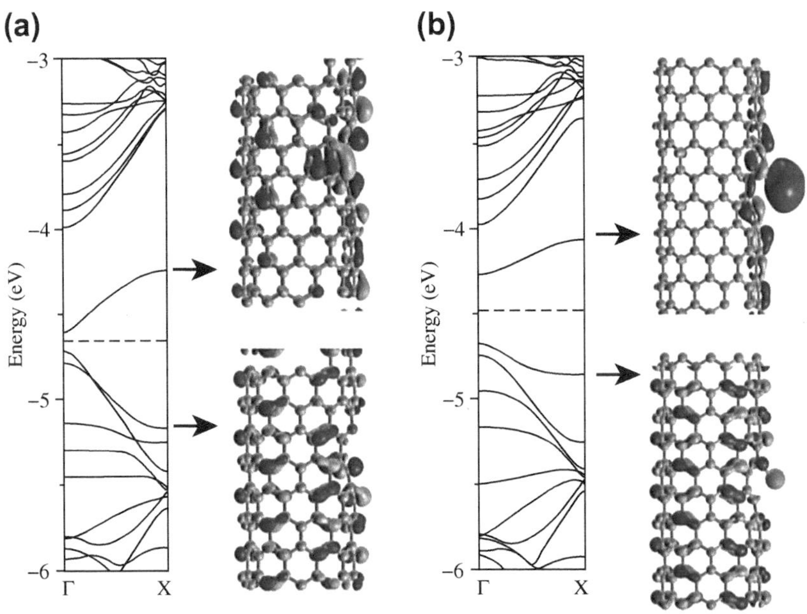

FIGURE 1.10 Band structures and the highest occupied molecular orbital HOMO (bottom) and the lowest unoccupied molecular orbital LUMO (up) wavefunctions at the gamma point. (a) Results only N-doped carbon nanotubes and (b) N-doped nanotube with cadmium adsorbed. Note that the electronic bandgap is increased when the cadmium is adsorbed onto the surface of the carbon nanotubes; the optimized structure can be also seen in Fig. 1.8. Our results were obtained by the Siesta code [66].

increasing the sensitivity for detecting individual molecules. In this context, Schedin et al. [73] demonstrated that mechanically exfoliated graphene flakes were able to detect a single NO_2 molecule. These authors fabricated micrometer-size sensors made from graphene; the adsorbed molecules changed the local carrier concentration in graphene by one electron, which lead to step-like changes in resistance. The high sensitivity that was achieved is due to the fact that graphene is an exceptionally low-noise material (electronically), thus suggesting that it could be an ideal candidate not only for chemical detectors but also for other local sensing probe applications involving external charges, magnetic fields, or mechanical strains. Large arrays of graphene-based sensors could also increase the uptake area [74], thus allowing higher sensitivity for short-time exposures and the detection of active (toxic) gases in extremely low concentrations.

Fowler et al. [75] reported the fabrication of graphene-based chemical sensors for NO_2, NH_3, and 2,4-dinitrotoluene. These authors deposited graphene using spin coating in order to create single-layer films on inter-digitized electrode arrays. The time response was relatively slow at room temperature but this could be accelerated at higher temperatures at the expense of sensitivity. In addition, Joshi et al. [76] reported gas sensing properties of graphene films and ribbons toward O_2, CO, and NO_2; they determined that graphene films exhibited an enhanced sensing response when compared to graphene ribbons. Pt/graphene/SiC devices have also proved to be excellent sensors for detecting hydrogen gas. These experiments also showed that hydrogen gas is readily desorbed from the sensor (after exposure) [77]. More recently, the adsorption of chemical vapor deposition (CVD)-grown graphitic nanoribbons was investigated for carbon dioxide and water by Asai et al. [78]. The authors found an irreversible adsorption behavior at room temperature that is different from the behavior of microporous carbon and carbon black. The authors claimed that the irreversible adsorptions of both systems (CO_2 and H_2O) are due to the large number of sp^3 hybridized carbon atoms located at the edges (terrace-like edges; see Fig. 1.11).

On the other hand, Kowalczyk et al. [86] investigated, using Grand canonical Monte Carlo simulations, the hydrogen storage in the primitive, gyroid, diamond, and quasi-periodic icosahedral nanoporous carbon materials, and in carbon nanotubes (see Fig. 1.12). They found that quasi-periodic icosahedral nanoporous carbon material could reach 6 wt % of storage capacity at 3.8 MPa and 77 K, but the volumetric density does not exceed 24 kg H_2 m^{-3}. Bundles of single-walled carbon nanotubes could store up to 4.5 wt %, but with a high volumetric density of 42 kg H_2 m^{-3}. All investigated microporous carbon materials are not effective against compression above 20 MPa at 77 K, because the adsorbed density approaches the density of the bulk fluid. It follows from this work that the geometry of carbon surfaces could enhance the storage capacity only to a limited extent. Thus, only a combination of the most effective structures with appropriate additives (metals or other functionalities) could

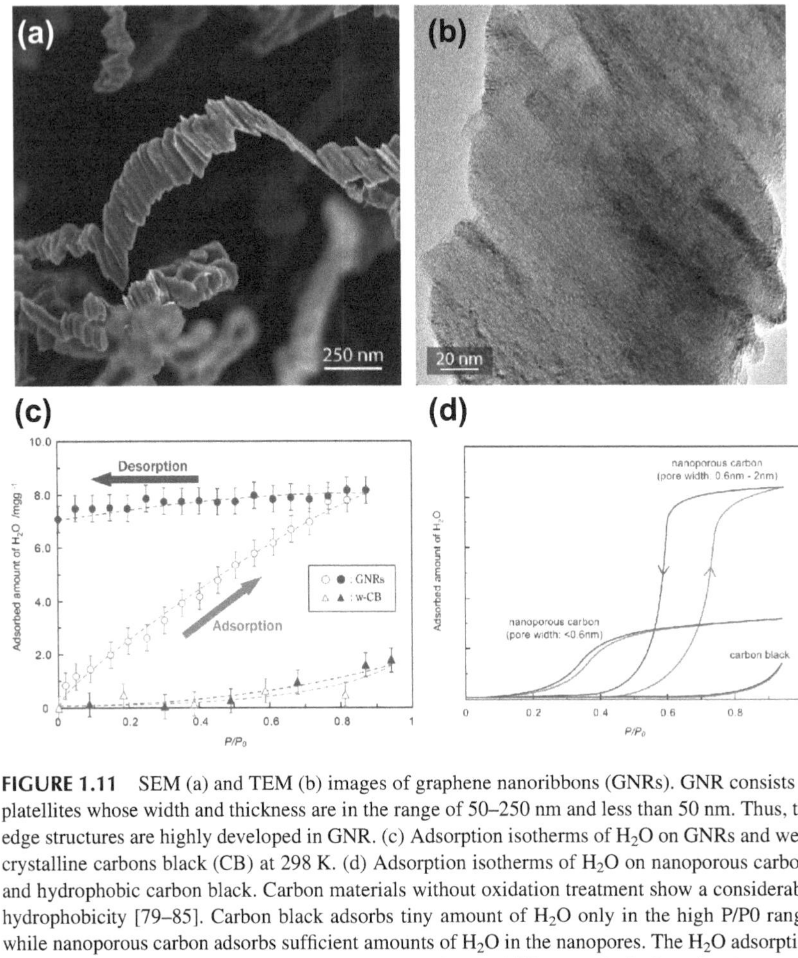

FIGURE 1.11 SEM (a) and TEM (b) images of graphene nanoribbons (GNRs). GNR consists of platellites whose width and thickness are in the range of 50–250 nm and less than 50 nm. Thus, the edge structures are highly developed in GNR. (c) Adsorption isotherms of H_2O on GNRs and well-crystalline carbons black (CB) at 298 K. (d) Adsorption isotherms of H_2O on nanoporous carbons and hydrophobic carbon black. Carbon materials without oxidation treatment show a considerable hydrophobicity [79–85]. Carbon black adsorbs tiny amount of H_2O only in the high P/P0 range, while nanoporous carbon adsorbs sufficient amounts of H_2O in the nanopores. The H_2O adsorption isotherm of nanoporous carbon having pores of 0.6–2 nm exhibits a marked adsorption hysteresis. On the contrary, the H_2O adsorption isotherm of nanoporous carbon with width less than 0.6 nm has almost no adsorption hysteresis. Results from [86].

provide an efficient storage medium for hydrogen in the quest of a source of "clean" energy.

1.4. BIOLOGICAL SYSTEMS ADSORBED ON CARBON NANOMATERIALS

1.4.1. Fullerenes

Fullerenes have attracted considerable attention in different fields since their discovery in 1985 [1]. Investigations of their physical, chemical, and biological

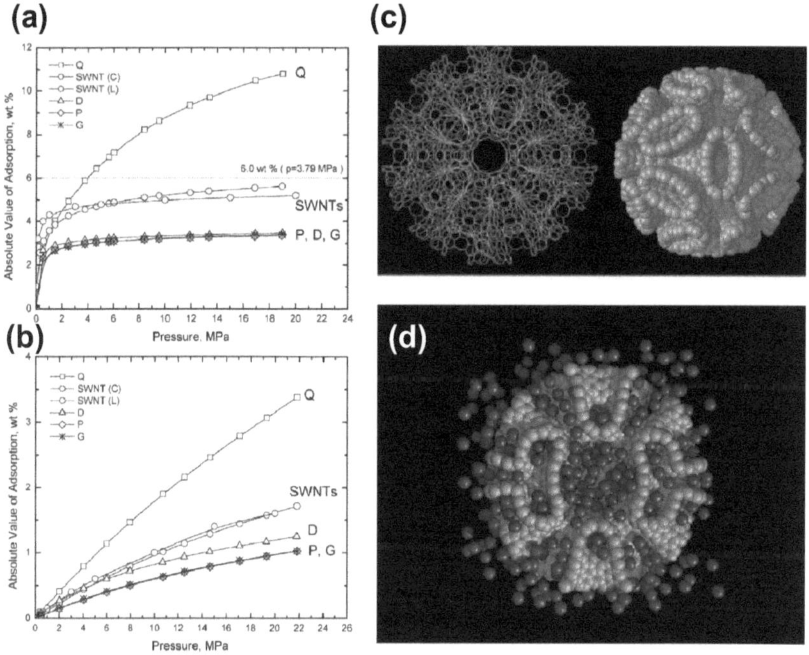

FIGURE 1.12 (a) Simulated absolute value of adsorption (hydrogen gravimetric weight percent) at 77 K from the computer simulations. Abbreviations: Q – quasi-periodic icosahedral nanoporous carbon material, hexagonal bundle of SWNTs (C), SWNT (L) – simulation results of hydrogen storage in SWNTs (single wall carbon nanotubes) taken from Levesque et al. [87] for comparison with these results, D – diamond nanoporous carbon material, P – primitive nanoporous carbon material, and G – gyroid nanoporous carbon material. (b) The same at 303 K. (c) The quasi-periodic icosahedral nanoporous carbon material. The left figure shows carbon rings on the surface and the right figure shows eight unit cells of the structure decorated by carbon atoms. (d) A snapshot of hydrogen adsorbed in the quasi-periodic icosahedral nanoporous carbon material at 3.8 MPa and 77 K [86].

properties have thrown up valuable information (see Fig. 1.13). It is inferred that size, hydrophobicity, dimensionality, and electronic properties make fullerenes an appealing material in the chemistry field, especially in medicine. Their unique cage-like structure coupled with the possibility of derivatization make fullerenes a potential therapeutic agent. The study of biological applications has attracted increasing attention despite the low solubility of fullerenes in physiological media.

In particular, C_{60} is insoluble in aqueous media and the cages tend to aggregate easily [93]. Fullerene derivatives cover practically all known classes of chemical compounds, demonstrating both high chemical activity and a broad versatility of chemical reactions. Fullerene compounds with antiviral activity have potential in medicine, with strong implications for the treatment of human

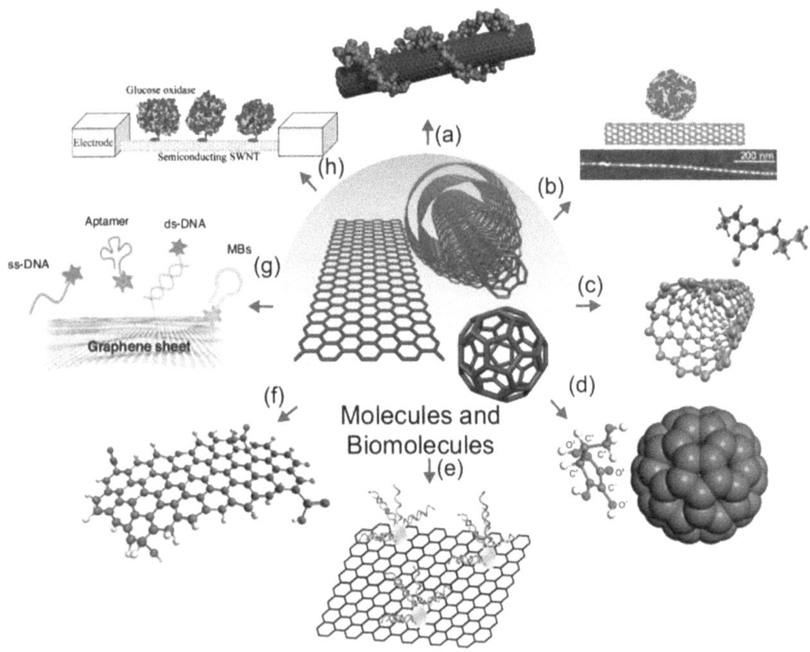

FIGURE 1.13 Adsorption of different molecules and biomolecules on carbon nanostructures. (a) A single walled carbon nanotube (SWNT) coated by a single-stranded DNA via π–π stacking [88]. (b) Schematic illustration of globular protein adsorption onto a nanotube [89]. (c) Atrazine pesticide molecule adsorbed onto a nitrogen-doped SWCNT. (d) Adsorption of ascorbic acid onto the C_{60} fullerene [90]. (e) Detection of DNA by adsorption onto functionalized graphene [91]. (f) Graphene oxide with carbonyl and carboxyl groups adsorbed onto its surface [33]. (g) ssDNA, aptamers, and MBs can be adsorbed onto the surfaces of graphene or graphene derivates [92]. (h) Glucose oxidase coated-SWCNT as biosensor [32].

immunodeficiency virus (HIV) [93]. However, further studies still need to be conducted in order to find the optimum C_{60} system able to inhibit the activity of viruses including HIV.

Results published in 1999 have shown that fullerenes offer potential as biological antioxidants [94]. The antioxidant property is based on the fact that fullerenes possess large amounts of conjugated double bonds and low-lying lowest unoccupied molecular orbital (LUMO), which can easily take up an electron, thus making possible an attack of radical species. It has been reported that up to 34 methyl radicals have been added onto a single C_{60} molecule [95]. The fullerene could then react with many superoxides without being consumed. Due to this feature, fullerenes are considered to be the world's most efficient radical scavenger and are described as radical sponges. The adsorption properties of crystals formed by C_{60} molecules have been also investigated by Schur et al. [96]. These authors used different experimental methods for fullerite hydrogenation; however, they do not find the maximum hydrogen content. Thus, more investigations are needed in this direction.

A large number of metal clusters (such as Au, Ag and Pt [97,98]) could be potentially used as carriers for cellular delivery because of their versatile properties, including selective targeted delivery and controlled release of the carried drugs. Fullerenes could be considered as inorganic nanoparticles and show wide availability due to their small size (~1 nm) and biological activity. The activity of this carbon molecule rests upon the properties of both, the fullerene core and its chemical modification. The fullerene core is highly hydrophobic, whereas the functional groups attached to the core add further complexity to the behavior of the fullerene molecule. By anchoring hydrophilic moieties, fullerenes become water soluble and are capable of carrying drugs and genes for the cellular delivery. Derivatized fullerene could cross the cell membrane and bind to the mitochondria, as demonstrated by Foley et al. [99].

Moreover, DNA-functionalized fullerenes are able to enter the COS-1 cells, and experiments have showed a comparable or even better efficiency when compared to that of commercially available lipid-based vectors [100,101].

1.4.2. Carbon Nanotubes

As grown, pristine carbon nanotubes have highly hydrophobic surfaces, and are not soluble in aqueous solutions. For biomedical applications, surface chemistry or functionalization is required to solubilize CNTs, and to render biocompatibility and low toxicity [102]. The functionalization strategies presented above suggest that both single- and multi-walled carbon nanotubes may act as scaffolds for biomolecule immobilization, thus allowing subsequent applications for biosensing. The remarkable physical properties of carbon nanotubes, including high surface area, semiconducting behavior, and strong Raman scattering spectra, make them useful to obtain data sensitive to adsorbed biomolecule interactions occurring along the carbon nanotube sidewalls (see Fig. 1.13).

For molecular recognition applications, using carbon nanotubes, ligand–target affinity must be maintained. Therefore, the non-specific immobilization of analyte proteins onto the hydrophobic nanotube sidewall is not desirable. Small drug molecules could be covalently conjugated to CNTs for *in vitro* delivery. Fluorescent dyes and drug cargoes were simultaneously linked to 1,3-dipolar cycloaddition functionalized CNTs via amide bonds for the delivery of an anti-cancer drug [103] or an antifungal drug [104] into cells. Targeting ligands including folic acid [88,105], peptides [106,107], and antibodies [108–111] have been used in conjunction with CNTs in order to specific types of cells *in vitro*, or to tumors *in vivo*.

Unlike various small drug molecules that are able to diffuse into cells, biomacromolecules including proteins, DNA, and RNA rarely cross cell membranes. Intracellular delivery is thus required in order to use these molecules for therapeutic applications. Proteins can be either conjugated or noncovalently absorbed onto nanotubes for intracellular delivery [112,113]. It is also noteworthy that the hydrophobic surface of SWNTs allows non-specific binding of proteins [114].

1.4.3. Graphene

Graphene has been employed as a substrate to establish interfaces with various biomolecules and cells. The biological modification of graphene in order to improve its biocompatibility, solubility, and selectivity has recently attracted great interest. Several studies have focused on graphene modification and functionalization [115].

The large 2D aromatic surface of graphene makes it an ideal substrate for adsorption of specific biomolecules. Importantly, single-stranded DNA (ssDNA) could be strongly adsorbed onto the graphene surface. In particular, the tethering process of ssDNA on graphene is preferential on relatively thick sheets ($\sim$8 nm), and on wrinkled rather than flat surfaces of chemically modified graphene. Patterning ssDNA on graphene layers increases the hole density in the graphene layer in the order of $5.61 \times 10^{12}/\text{cm}^2$. The resultant ssDNA–graphene biointerface has been used in a field-effect transistor (FET) for label-free, reversible detection of complementary ssDNA [116]. Through a self-assembly process under strong ultrasonication, monolayers of ssDNA molecules have been adsorbed onto both sides of graphene sheets by p–p stacking [117]. In another example, graphene adsorbed with ssDNA has been used directly for the study of surface-enhanced laser desorption ionization time-of-flight mass spectrometry (SELDI-TOF-MS) [118]. Graphene efficiently preconcentrated the ssDNA and served as a direct platform for SELDI-TOF-MS analysis. Another elegant approach of graphene tethering with DNA includes the formation of nanoparticles (NPs) decorated with graphene biosystems. Thiolated ssDNA is first adsorbed onto graphene oxide (GO) sheets. The resultant thiol-DNA–GO sheet is used as a 2D bionanointerface for the assembly of Au NPs. Furthermore, the Au NPs retain their optical properties after assembly [119].

Chemically modified graphene has been used in biosensors. For example, immunoglobulin E (IgE) aptamers with an approximate height of 3 nm were successfully immobilized on graphene. The aptamer-modified graphene field-effect (G-FET) transistor showed selective electrical detection of the IgE protein, and exhibited good affinity [120]. Hou et al. [121] found that graphene modified with ethylenediamine triacetic acid (EDTA) could be used as an ideal electrode material to fabricate biosensors. These authors fabricated a dopamine electrochemical biosensor with a high selectivity and sensitivity, which might allow its potential use in the investigation and diagnosis of dopamine-related diseases.

1.5. ADSORPTION OF HEAVY-METALS ON MODIFIED CARBON NANOMATERIALS

Trace metals are one of the major pollutants in the environment. The major sources of these heavy metals are soils and dusts from traffic, industry, and weathered materials. Trace metals are useful indicators of pollution in surface soil environments resulting from man-made emission sources.

In general, carbon nanotubes show excellent performance when adsorbing different divalent metal ions in aqueous solution (Cd^{2+}, Cu^{2+}, Ni^{2+}, Pb^{2+}, and Zn^{2+}). It is noteworthy that these ions exhibit different affinity; for example, at pH = 9 the affinity order of the metal ions toward CNTs is Cu(II) > Pb(II) > Co(II) > Zn(II) > Mn(II) [68–70,122]. Different research groups have shown that the adsorption capacities of CNTs increased substantially after oxidizing them with NaOCl, HNO_3, and $KMnO_4$ solutions, and such surface oxidized CNTs reveal great potential as superior adsorbents for environmental protection applications. An important fact is that the recovery of metal ions and the regeneration of CNTs can be achieved using acid elution with little effect on the CNT performance; the adsorption mechanisms appear mainly attributable to chemical interactions established between the metal ions and the surface functional groups of the CNTs [123].

1.5.1. Lead Adsorption by Carbon Nanotubes

Yan-Hui et al. demonstrated that MWCNTs exhibited exceptional adsorption capability and high adsorption efficiency for Pb(II) removal from water. The adsorption efficiency was significantly influenced by the pH of the solution, and the nanotube functionalization [124].

Other groups have used additives to improve the metal uptake by the carbon nanomaterials. For example, using oxidized multi-walled carbon nanotubes (Ox-MWCNTs), Jiaxing et al. [125] studied different types of surfactants, sodium dodecylbenzene sulfonate (SDBS), octyl-phenol-ethoxylate (TX-100), and benzalkonium chloride (BKC), in order to improve the Pb adsorption capabilities. The adsorption of Pb(II) was enhanced by the presence of SDBS, but slightly inhibited by the presence of BKC. The authors determined that the adsorption process in the Pb–surfactant–Ox-MWCNT ternary system was attributed to electrostatic, hydrophobic, and π–π interactions. In addition, polymer composites have been also used as adsorbent materials for Pb in aqueous systems. Shubin Yang et al. [126] used polyacrylamide (PAAM)-grafted MWNTs (MWCNTs/PAAM) and used a natural organic macromolecular compound (humic acid, HA) as additive, in order to improve the lead adsorption. In this case, the presence of HA led to a strong increase in Pb(II) adsorption at low pH; this effect is attributed to the modification of the adsorbent surface properties and the partial complexation of Pb(II) with the adsorbed HA. The addition of nanoparticles on CNTs is also a novel strategy to improve lead adsorption. In this case, Shu-Guang et al. [127] used CNTs decorated with MnO_2 nanoparticles (MnO_2/MWCNTs) to remove Pb(II) from aqueous solutions. The MnO_2/MWCNTs composite exhibited an increment in Pb(II) adsorption when compared to pure carbon nanotubes (i.e. no anchored nanoparticles). In this case, the Pb(II) adsorption by MnO_2/MWCNTs occurred within the first 15 min of contact.

1.5.2. Chromium Adsorption by Carbon Nanotubes

The removal of hexavalent chromium Cr(VI) from aqueous solution when using Ox-MWCNTs has been studied by Jun Hu et al. [128]. These authors found that the Cr(VI) removal was favored at low pH values, and demonstrated the best performance at $pH = 2$. Jun Hu et al. also determined the adsorption mechanism. They also demonstrated that Ox-MWCNTs can effectively remove Cr(VI) from aqueous solution under a wide range of experimental conditions, without significant Cr(III) release. An interesting fact is that the adsorbed Cr(VI) could also be desorbed readily from the Ox-MWCNTs surface at high pH, thus allowing to recycle the nanotubes.

Later, Pillay K. et al. [129] studied the Cr(VI) adsorption capabilities on MWNTs and compared the results with those obtained with activated carbons. Their studies revealed that MWCNTs showed superior adsorption capabilities when compared to activated carbons. In addition, the removal of Cr(VI) was higher at lower pH values; they determined that the uptake of Cr(VI) was hindered in the presence of competing anions such as Cl^- and SO_4^{2-}. The principal adsorption mechanisms were identified as an ion exchange mechanism, intraparticle diffusion, and electrostatic interactions.

Other strategies such as nanoparticle decoration of CNTs have been also used for chromium removal. For example, Ze-Chao Di et al. [130] used ceria nanoparticles supported on aligned carbon nanotubes (CeO_2/MWCNTs) as adsorbent for Cr(VI) from drinking water. The best Cr(VI) adsorption effect of CeO_2/MWCNTs was observed at a pH range of 3.0–7.4, which is closer to the real pH values usually found in wastewaters; thus CeO_2/MWCNTs appear to have great potential in environmental protection.

1.5.3. Zinc, Nickel, Copper and Other Materials Adsorption by Carbon Nanotubes

Chungsying Lu et al., [131] studied the Zn(II) adsorption capabilities of functionalized SWCNTs and MWCNTs in aqueous solutions. The authors compared their results with commercially activated carbons, and they demonstrated that both types of CNTs exhibit enhanced performance. Lu et al. also determined that Zn(II) ions could be easily removed from the surface of SWCNTs and MWCNTs by the addition of an acid solution, thus maintaining the adsorption capacity up to 10 cycles of the adsorption/desorption process. This suggests that both CNTs could be reused through many cycles of water treatment. More recently, ZnO nanoparticles (ZnO-NPs) anchored on the surface of CN_X-MWNTs were successfully synthesized by Gracia-Espino et al. [132]. The authors used a self-assembly chemical method involving thiophene as a passivated agent. By varying the thiophene concentration, they could control the ZnO-NP size. SEM and HRTEM studies revealed that the ZnO-NPs were homogeneously distributed along the nanotube surface as depicted in Fig. 1.14.

Studies concerning Ni(II) adsorption onto Ox-MWCNTs have been carried out by Shitong Yang et al. In this case, the authors used polyacrylic acid (PAA) as additive. Their results indicated that the adsorption of Ni(II) on Ox-MWCNTs increased up to zero to 99% at pH 2–9, and the high adsorption level was then maintained with increasing pH. A positive effect of PAA on Ni(II) adsorption was found at pH < 8. The adsorption of Ni(II) on Ox-MWCNTs may be mainly attributed to surface complexion and ion exchange [133].

MWCNTs have been also used as adsorbent materials for Cu. In this context, Sheng et al. [134] studied the adsorption of Cu(II) on MWCNTs using humic acid (HA) or fulvic acid (FA) as additives for enhancing the adsorption activity. As in most adsorption processes, the adsorption is also strongly dependent on the pH but independent of ionic strength. Sheng et al. observed a positive effect of HA/FA on Cu(II) adsorption at pH <7.5. The authors found that adsorption was spontaneous and enhanced at higher temperature.

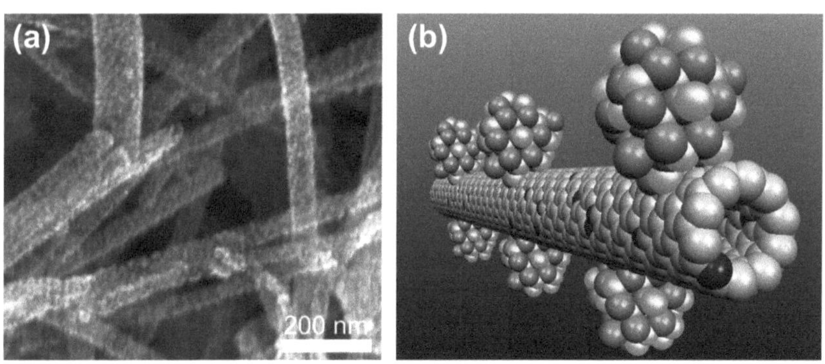

FIGURE 1.14 (a) Scanning electron microscopy (SEM) image: ZnO nanoparticles (ZnO-NPs) anchored on the surface of nitrogen-doped multiwalled carbon nanotubes (CN_X-MWNTs). The anchoring process consists of a self-assembly method involving the mixing of CN_X-MWNTs (previously synthesized using chemical vapor deposition method) in a solution with dimethylformamide, zinc acetylacetonate, and thiophene. The thiophene solvent was used as a capping agent for controlling the size and distribution of ZnO-NPs on the tubes. (b) Molecular model showing a single-walled carbon nanotube doped with nitrogen and with ZnO nanoparticles attached to its surface [132].

Finally, CNTs have been also used to remove radioactive materials such as Th(IV). Chen C.L. et al. [135] studied the adsorption capabilities of Th(IV) onto Ox-MWCNTs, and their results showed that Th(IV) adsorption strongly depended on the Ox-MWCNT concentration and pH; the suggested adsorption mechanism might be surface complexation. Uranium was also adsorbed onto MWCNTs, thus showing the adsorption of U(VI) onto Ox-MWCNTs. The removal of U(VI) by Ox-MWCNTs was strongly dependent on pH and ionic strength. The presence of humic acid or fulvic acid enhanced U(VI) removal at low pH [136]. For both, thorium and uranium, Ox-MWCNTs exhibited enhanced performance when compared to pristine MWNTs.

The employment of carbon nanotubes and other carbon nanomaterials for water and wastewater treatments, especially containing divalent metal ions, is gaining more attention. The method appears to be simple and effective in pollution control. Therefore, modified nanocarbons including MWCNTs and SWNTs may be excellent candidates for the removal, solidification, and pre-concentration of heavy metals in aqueous solutions.

1.6. CARBON DIOXIDE UPTAKE ON CARBON NANOSTRUCTURES

For fullerenes, experimental evidence related to the CO_2 adsorption was reported in 1994 by Nagano et al. [137]. The authors determined that solid C_{60} adsorbs a large amount of CO_2 under a supercritical CO_2 treatment. The absorption kinetics demonstrates that the process is significantly slower than the normal physical adsorptions and could be accelerated by increasing temperature; their results also suggested a strong interaction established between CO_2 and C_{60}. More recently, some theoretical experiments have been performed in order to study the CO_2 adsorption onto the C_{60} fullerene. Martínez-Alonso et al. [138] determined Ar and CO_2 adsorption onto a crystallized sample of fullerene C_{60} by means of Monte Carlo computer simulations. Analysis of the structure of the adsorbed film indicated that both gases are adsorbed in the voids of the C_{60} structure in a solid state with densities slightly larger than the bulk solid. Both gases exhibited a behavior similar to nitrogen. Subsequently, Gao et al. [139] studied theoretically the CO_2 and N_2O adsorption on pristine and calcium (Ca)-decorated fullerenes (C_{60}) by density functional theory (DFT). They evaluated the potential application of the (Ca)-decorated C_{60} for the capture of CO_2 and N_2O. Their results indicated that the adsorption of CO_2 and N_2O molecules on the pristine C_{60} is considerably weak and accompanied by neglectable charge transfer. However, when C_{60} is decorated with Ca atoms the CO_2 and N_2O adsorptions are greatly enhanced and up to five CO_2 molecules can be adsorbed onto the Ca-C_{60} system due to the electrostatic interaction. Moreover, when four Ca atoms are decorated on the surface of C_{60}, the maximum number of the adsorbed CO_2 molecules was 16. Their results might be useful to provide an effective method to capture or transform greenhouse gases.

Functionalized SWNTs have also been used as a superior adsorbent material. Cinke et al. [140] determined the CO_2 adsorption capabilities of purified SWNTs at different temperatures. The authors found that the main interaction process was dominated by physisorption, and purified SWNTs adsorb nearly twice the volume of CO_2 compared to activated carbons. In this way, Dillon et al. [141] used fluorinated single-wall carbon nanotubes (F-SWNTs) that were functionalized with polyethyleneimine (PEI). The resulting polymeric composite (PEI-SWNTs) adsorbed as much as 9.2% w/w for PEI-SWNT at 27 °C, while the CO_2 desorption was accomplished by heating under argon atmosphere at 75 °C. In this case, the adsorption process generates carbamate as a consequence of the reversible CO_2 absorption on the primary amine of the PEI.

As previously pointed out, polymeric nanocomposites have shown excellent behavior for gas adsorption, used in gas sensing and gas storage. Lu et al. [142,143] compared three different materials, CNTs, granular activated carbon (GAC), and zeolites; each material was previously functionalized with 3-aminopropyl-triethoxysilane (APTS). First, they observed that after functionalization, all adsorbent materials improved their storage CO_2 capacity. Then, under the same conditions, it was determined that modified CNTs possess the greatest adsorption capacity, followed by the modified zeolites, and then the modified GAC. The mechanism of CO_2 adsorption onto these adsorbents appears mainly attributable to physical force, which makes feasible low temperature regenerations. Their results suggested the APTS-CNTs nanocomposite was an efficient adsorbent with potential applications for CO_2 capture from gas streams. Further studies on this composite, determined that cyclic CO_2 adsorption onto APTS-CNTs showed that the adsorbed CO_2 could be effectively desorbed via thermal treatment and via vacuum suction, where the best results were obtained at 120 °C and 0.145 atm, in just 5 min. The adsorption capacities of APTS-CNTs were preserved during 20 cycles of adsorption and regeneration [144]. Additional improvements in the APTS-CNTs synthesis were accomplished, and the adsorption capacities and physicochemical properties of APTS-CNTs were preserved through 100 adsorption–desorption cycles, displaying the stability of APTS-CNTs during a prolonged cyclic operation [145].

Theoretical studies on the adsorption of small molecules such as N_2, O_2, CO, CO_2, H_2O, and HF in model bundles of CNTs have been performed [146]. Their results show that the nature and the stability of the adsorption sites are strongly dependent on the molecular species and the SWNT diameter. Improving the size homogeneity of the ropes and judiciously calibrating their diameter would provide an efficient route for selecting molecular species.

Graphene from exfoliation of graphitic oxide has exhibited good hydrogen uptake at 100 atm and 298 K (3 wt %). Equally interestingly, the graphene samples show high uptake of CO_2, the value reaching up to 35 wt % at 1 atm and 195 K. First-principles calculations performed by the same group showed that CO_2 molecules sit alternatively in a parallel fashion on the six-membered rings of the graphene, offering a maximum uptake of 37.93 wt % in single-layer graphene [147]. In addition, theoretical studies have been carried out, and DFT calculations were performed in order to investigate the adsorption of CO_2 on defective graphite (0001). Cabrera-Sanfelix et al. [148] studied the adsorption mechanism and obtained a dissociation of CO_2 via the formation of epoxy groups, followed by oxygen recombination and desorption of O_2. The global minimum ("O_2 desorbed + graphene sheet") entails an energy release of about 3.4 eV with respect to the initial state. Huang et al. [149] studied the adsorption of CO, NO, NO_2, O_2, N_2, CO_2, and NH_3 on graphene nanoribbons (GNRs) using first-principles methods. Their results demonstrate that the electronic and transport properties of armchair GNRs are sensitive to the adsorption of NH_3, and the system exhibits n-type semiconducting behavior after NH_3 adsorption. Other gas molecules have little effect on the conductance of GNRs.

1.7. CONCLUSIONS

It is important to note that different types of nanocarbons would adsorb chemical species differently, depending on their local curvature and nature of their surface. A relevant aspect in the adsorption of metals and molecules is based on the accessible surface area of the material, and its reactivity. One could enhance the surface area of different carbon nanomaterials by creating reactive pores, by introducing dopants, defects and chemical groups. It has been demonstrated that the adsorption properties of different nanoscale carbons could be tailored, and further experimental and theoretical work is still needed in order to elucidate the adsorption mechanisms to generate the ultimate materials for cleaning polluted waters, sensing diverse molecules, and storing different gases.

ACKNOWLEDGMENTS

This work was supported in part by CONACYT-México grant: 60218-F1 (FLU). MT thanks JST-Japan for funding the Research Center for Exotic NanoCarbons, under the Japanese Regional Innovation Strategy Program by the Excellence. H.T. acknowledges support of CAPES, Brazil, through its foreign scientist invited program. MT is also grateful to the Penn State Center for Nanoscale Science for a seed grant on layered systems.

REFERENCES

[1] H.W. Kroto, J.R. Heath, S.C. O'Brien, R.F. Curl, R.E. Smalley, C_{60}: buckminsterfullerene, Nature 318 (1985) 162–163.

[2] M. Terrones, G. Terrones, H. Terrones, Structure, chirality, and formation of giant icosahedral fullerenes and spherical graphitic onions, Structural Chemistry 13 (2002) 373–384.
[3] H. Terrones, Curved graphite and its mathematical transformations, Journal of Mathematical Chemistry 15 (1994) 143.
[4] H. Terrones, T. Hayashi, M. Muñoz-Navia, M. Terrones, Y.A. Kim, N. Grobert, R. Kamalakaran, J. Dorantes-Dávila, R. Escudero, M.S. Dresselhaus, M. Endo, Graphitic cones in palladium catalysed carbon nanofibres, Chemical Physics Letters 343 (2001) 241–250.
[5] M. Terrones, W.K. Hsu, J.P. Hare, D.R.M. Walton, H.W. Kroto, H. Terrones, Graphitic structures: from planar to spheres, toroids and helices, Philosophical Transactions of the Royal Society London A 354 (1996) 2025–2054.
[6] S. Iijima, Helical microtubules of graphitic carbon, Nature 354 (1991) 56–58.
[7] A. Oberlin, M. Endo, T. Koyama, Filamentous growth of carbon through benzene decomposition, Journal of Crystal Growth 32 (1976) 335–349.
[8] H. Terrones, M. Terrones, E. Hernández, N. Grobert, J.C. Charlier, P.M. Ajayan, New metallic allotropes of planar and tubular carbon, Physical Review Letters 84 (2000) 1716.
[9] A.L. Mackay, H. Terrones, Diamond from graphite, Nature 352 (1991) 762.
[10] H. Terrones, A.L. Mackay, The geometry of hypothetical curved graphite structures, Carbon 30 (1992) 1251.
[11] T. Lenosky, X. Gonze, M. Teter, V. Elser, Energetics of negatively curved graphitic carbon, Nature 355 (1992) 333–335.
[12] A.L. Mackay, H. Terrones, Hypothetical graphite structures with negative gaussian curvature, Philosophical Transactions of the Royal Society of London A 343 (1993) 113–127.
[13] M.S. Dresselhaus, G. Dresselhaus, R. Saito, Physical Review B 45 6234 (1992).
[14] S. Niyogi, M.A. Hamon, H. Hu, B. Zhao, P. Bhowmik, R. Sen, M.E. Itkis, R.C. Haddon, Chemistry of single-walled carbon nanotubes, Accounts of Chemical Research 35 (2002) 1105–1113.
[15] R. Saito, G. Dresselhaus, M.S. Dresselhaus, Physical Properties of Carbon Nanotubes, Imperial College Press, 1998.
[16] K. Nakada, M. Fujita, G. Dresselhaus, M.S. Dresselhaus, Edge state in graphene ribbons: nanometer size effect and edge shape dependence, Physical Review B 54 (1996) 17954–17961.
[17] Y.-W. Son, M.L. Cohen, S.G. Louie, Halft-metallic graphene nanoribbons, Nature 444 (2006a) 347–349.
[18] Y.W. Son, M.L. Cohen, S. Louie, Energy gaps in graphene nanoribbons, Physical Review Letters 97 (2006b) 216803.
[19] A.F. Hebard, M.J. Rosseinky, R.C. Haddon, D.W. Murphy, S.H. Glarum, T.T.M. Palstra, A.P. Ramirez, A.R. Karton, Potassium-doped C_{60}, Nature 350 (1991) 600–601.
[20] R. Czerw, M. Terrones, J.C. Charlier, X. Blase, B. Foley, R. Kamalakaran, N. Grobert, H. Terrones, D. Tekleab, P.M. Ajayan, W. Blau, M. Rühle, D.L. Carroll, Identification of electron donor states in N-Doped carbon nanotubes, Nano Letters 1 (2001) 457.
[21] F. Villalpando-Paez, A. Zamudio, A.L. Elias, H. Son, E.B. Barros, S.G. Chou, Y.A. Kim, H. Muramatsu, T. Hayashi, J. Kong, et al., Synthesis and characterization of long strands of nitrogen-doped single-walled carbon nanotubes, Chemical Physics Letters 424 (2006) 345–352.
[22] I.O. Maciel, J. Campos-Delgado, E. Cruz-Silva, M.A. Pimenta, B.G. Sumpter, V. Meunier, F. López-Urías, E. Muñoz-Sandoval, H. Terrones, M. Terrones, A. Jorio, Synthesis, electronic structure, and raman scattering of phosphorus-doped single-wall carbon nanotubes, Nano Letters 9 (2009) 2267–2272.

[23] J.M. Romo-Herrera, D.A. Cullen, E. Cruz-Silva, D. Ramírez, B.G. Sumpter, V. Meunier, H. Terrones, D.J. Smith, M. Terrones, The role of sulfur in the synthesis of novel carbon morphologies: from covalent Y-Junctions to sea-urchin-like structures, Advanced Functional Materials 19 (2009) 1193–1199.
[24] E. Cruz-Silva, F. Lopez-Urias, E. Muñoz-Sandoval, B.G. Sumpter, H. Terrones, J.–C. Charlier, V. Meunier, M. Terrones, Phosphorus and phosphorus-nitrogen doped carbon nanotubes for ultrasensitive and selective molecular detection, Nanoscale 3 (2011) 1008.
[25] W. Zhiyong, H. Huifang, Z. Hui, The electronic properties of graphene nanoribbons with boron/nitrogen codoping, Applied Physics Letters 96 (2010) 243110.
[26] D. Wei, Y. Liu, Y. Wang, H. Zhang, L. Huang, G. Yu, Synthesis of N-doped graphene by chemical vapor deposition and its electrical properties, Nano Letters 9 (2009) 1752.
[27] S. Yu, W. Zheng, Q. Wen, Q. Jiang, First principle calculations of the electronic properties of nitrogen-doped carbon nanoribbons with zigzag edges, Carbon 46 (2008) 537.
[28] J. Xiaoting, J. Campos-Delgado, M. Terrones, V. Meunier, M.S. Dresselhaus, Graphene edges: a review of their fabrication and characterization, Nanoscale 3 (2011) 86–95.
[29] M. Terrones, A.R. Botello-Méndez, J. Campos-Delgado, F. López-Urías, Y. Vega-Cantú, F.J. Rodríguez-Macías, A.L. Elias, E. Muñoz-Sandoval, A.G. Cano-Márquez, J.-C. Charlier, H. Terrones, Graphene and graphite nanoribbons: morphology, properties, synthesis, defects and applications, Nanotoday 5 (2010) 351–372.
[30] E. Cruz-Silva, D.A. Cullen, L. Gu, J.M. Romo-Herrera, E. Muñoz-Sandoval, F. López-Urías, B.G. Sumpter, V. Meunier, J.C. Charlier, D.J. Smith, H. Terrones, M. Terrones, Heterodoped nanotubes: theory, synthesis, and characterization of phosphorus–nitrogen doped multiwalled carbon nanotubes, ACS Nano 2 (2008) 441.
[31] J.M. Romo-Herrera, B.G. Sumpter, D.A. Cullen, H. Terrones, E. Cruz-Silva, D.J. Smith, V. Meunier, M. Terrones, An atomistic branching mechanism for carbon nanotubes: sulfur as the triggering agent, Angewandte Chemie International Edition 47 (2008) 2948.
[32] K. Besteman, J.-O. Lee, F.G.M. Wiertz, H.A. Heering, C. Dekker, Enzyme-coated carbon nanotubes as single-molecule biosensors, Nano Letters 3 (2003) 727–730.
[33] V.C. Tung, M.J. Allen, Y. Yang, R.B. Kaner, High-throughput solution processing of large-scale graphene, Nature Nanotechnology 4 (2009) 25–29.
[34] L. Rosales, M. Pacheco, Z. Barticevic, A. León, A. Latgé, P.A. Orellana, Transport properties of antidot superlattices of graphene nanoribbons, Physical Review B 80 (2009). 073402.
[35] T.G. Pedersen, C. Flindt, J. Pedersen, N.A. Mortensen, A.-P. Jauho, K. Pedersen, Graphene antidot lattices: designed defects and spin qubits, Physical Review Letters 100 (2008) 136804.
[36] A. Stone, D. Wales, Theoretical studies of icosahedron C_{60} and some related species, Chemical Physics Letters 128 (1986) 501–503.
[37] P.A. Thrower, The study of defects in graphite by transmission electron microscopy, Chemistry and Physics of Carbon 5 (1969) 217–319.
[38] O.V. Yazyev, S.G. Louie, Topological defects in graphene: dislocations and grain boundaries, Physical Review B 81 (2010) 195420.
[39] Y.–P. Sun, K. Fu, Y. Lin, W. Huang, Functionalized carbon nanotubes: properties and applications, Accounts of Chemical Research 35 (2002) 1096–1104.
[40] J. Liu, A.G. Rinzler, H. Dai, J.H. Hafner, R.K. Bradley, P.J. Boul, A. Lu, T. Iverson, K. Shelimov, C.B. Huffman, F. Rodriguez-Macias, Y.-S. Shon, T.R. Lee, D.T. Colbert, R.E. Smalley, Fullerene pipes, Science 280 (1998) 1253–1256.

[41] B. Zhao, H. Hu, R.C. Haddon, Synthesis and properties of water-soluble single-walled carbon nanotube-poly(aminobenzene sulfonic acid) graft copolymer, Advanced Functional Materials 14 (2004) 71–76.
[42] M. Prato, K. Kostarelos, A. Bianco, Functionalized carbon nanotubes in drug design and discovery, Accounts of Chemical Research 41 (2008) 60–68.
[43] J.M. Romo-Herrera, M. Terrones, H. Terrones, S. Dag, V. Meunier, Covalent 2D and 3D networks form 1D nanostructures: designing new materials, Nano Letters 7 (2007) 570–576.
[44] X. Lepro, Y. Vega-Cantu, F.J. Rodriguez-Macias, Y. Bando, D.M. Golberg, M. Terrones, Production and characterization of coaxial nanotube junctions and networks of CNx/CNT, Nano Letters 7 (8) (2007) 2220.
[45] C. Li, E.T. Thostenson, T.–W. Chou, Sensors and actuators based on carbon nanotubes and their composites: a review, Composites Science and Technology 68 (2008) 1227–1249.
[46] J. Wang, Carbon-nanotube based electrochemical biosensors: a review, Electroanalysis 17 (2005) 7–14.
[47] D.R. Kauffman, A. Star, Carbon nanotubes gas and vapor sensors, Angewandte Chemie International Edition 47 (2008) 6550–6570.
[48] N. Sinha, J. Ma, J.T.W. Yeow, Carbon nanotubes-based sensors, Journal of Nanoscience and Nanotechnology 6 (2006) 573–590.
[49] O.K. Varghese, P.D. Kichambre, D. Gong, K.G. Ong, E.C. Dickey, C.A. Grimes, Gas sensing characteristics of multi-wall carbon nanotubes, Sensors and Actuators B 81 (2001) 32–41.
[50] H.-L. Hsu, J.-M. Jehng, Y. Sung, L.-C. Wang, S.-R. Yang, The synthesis, characterization of oxidized multi-walled carbon nanotubes, and application to surface acoustic wave quartz crystal gas sensor, Material Chemistry and Physics 109 (2008) 148–155.
[51] J. Maklin, T. Mustonen, K. Kordas, S. Saukko, G. Toth, J. Vahakangas, Nitric oxide gas sensors with functionalized carbon nanotubes, Physica Status Solidi B 244 (2007) 4298–4302.
[52] K.H. An, S.Y. Jeong, H.R. Hwang, Y.H. Lee, Enhanced sensitivity of a gas sensor incorporating single-walled carbon nanotube-polypyrrole nanocomposites, Advanced Mater 16 (2004) 1005–1009.
[53] E. Bekyarova, M. Davis, T. Burch, M.E. Itkis, B. Zhao, S. Sunshine, R.C. Haddon, Chemically functionalized single-walled carbon nanotubes as ammonia sensors, Journal of Physical Chemistry B 108 (2004) 19717–19720.
[54] F. Villalpando-Paez, A.H. Romero, E. Munoz-Sandoval, L.M. Martinez, H. Terrones, M. Terrones, Chemical Physics Letters 386 (2004) 137–143.
[55] B. Rebollo-Plata, E. Muñoz-Sandoval, F. López-Urías, E.L. Hernández-Cortina, H. Terrones, M. Terrones, Efficient vapor sensors using foils of dispersed nitrogen-doped and pure carbon multiwalled nanotubes, Journal of Nanoscience and Nanotechnology 10 (2010) 3965.
[56] A. Star, V. Joshi, S. Skarupo, D. Thomas, J.-C.P. Gabriel, Gas sensor array based on metal-decorated carbon nanotubes, Journal of Physical Chemistry B 110 (2006) 21014–21020.
[57] S. Mubeen, T. Zhang, N. Chartuprayoon, Y. Rheem, A. Mulchandani, N.V. Myung, M.A. Deshusses, Sensitive detection of H_2S using gold nanoparticle decorated single-walled carbon nanotubes, Analytical Chemistry 82 (2010) 250–257.
[58] M. Penza, R. Rossi, M. Alvisi, G. Cassano, E. Serra, Functional characterization of carbon nanotube networked films functionalized with tuned loading of Au nanoclusters for gas sensing applications, Sensors and Actuators B: Chemical 140 (2009) 176–184.

[59] C. Bittencourt, A. Felten, E.H. Espinosa, R. Ionescu, E. Llobet, X. Correig, J.-J. Pireaux, WO_3 films modified with functionalised multi-wall carbon nanotubes: morphological, compositional and gas response studies, Sensors and Actuators B: Chemical 115 (2006) 33–41.

[60] D.R. Kauffman, D.C. Sorescu, D.P. Schofield, B.L. Allen, K.D. Jordan, A. Star, Understanding the sensor response of metal-decorated carbon nanotubes, Nano Letters 10 (2010) 958–963.

[61] Y.-C. Tsai, P.-C. Hsua, Y.-W. Lin, T.-M. Wu, Silver nanoparticles in multiwalled carbon nanotube–nafion for surface-enhanced Raman scattering chemical sensor, Sensors and Actuators B: Chemical 138 (2009) 5–8.

[62] M. Sanles-Sobrido, L. Rodríguez-Lorenzo, S. Lorenzo-Abalde, A. Gonzalez-Fernandez, M.A. Correa-Duarte, R.A. Alvarez-Puebla, L.M. Liz-Marzán, Label-free SERS detection of relevant bioanalytes on silver-coated carbon nanotubes: the case of cocaine, Nanoscale 1 (2009) 153–158.

[63] W. Zhao, H. Wang, X. Qin, X. Wang, Z. Zhao, Z. Miao, L. Chen, M. Shan, Y. Fang, Q. Chen, A novel nonenzymatic hydrogen peroxide sensor based on multi-wall carbon nanotube/silver nanoparticle nanohybrids modified gold electrode, Talanta 80 (2009) 1029–1033.

[64] C.S. Yeung, L.V. Liu, Y.A. Wang, Adsorption of small gas molecules onto Pt-doped single-walled carbon nanotubes, Journal of Physical Chemistry C 112 (2008) 7401–7411.

[65] E. Cruz-Silva, F. López-Urías, E. Muñoz-Sandoval, B.G. Sumpter, H. Terrones, J.–C. Charlier, V. Meunier, M. Terrones, Electronic transport and mechanical properties of phosphorus- and phosphorus–nitrogen-doped carbon nanotubes, ACS Nano 3 (2009) 1913–1921.

[66] J. Junquera, O. Paz, D. Sánchez-Portal, E. Artacho, Numerical atomic orbitals for linear-scaling calculations, Physical Review B 64 (2001) 235111.

[67] E. Gracia-Espino, G. Sala, F. Pino, N. Halonen, J. Luomahaara, J. Mäklin, G. Toth, K. Kordas, H. Jantunen, M. Terrones, P. Helistö, H. Seppä, P.M. Ajayan, R. Vajtai, Electrical transport and field-effect transistors using inkjet-printed swcnt films having different functional side groups, ACS Nano 4 (2010) 3318–3324.

[68] G. Andrade-Espinosa, E. Munoz-Sandoval, M. Terrones, M. Endo, H. Terrones, J.R. Rangel-Mendez, Acid modified bamboo-type carbon nanotubes and cup-stacked-type carbon nanofibres as adsorbent materials: cadmium removal from aqueous solution, Journal of Chemical Technology and Biotechnology 84 (2009) 519–524.

[69] N.V. Perez-Aguilar, E. Muñoz-Sandoval, P.E. Diaz-Flores, J.R. Rangel-Mendez, Adsorption of cadmium and lead onto oxidized nitrogen-doped multiwall carbon nanotubes in aqueous solution: equilibrium and kinetics, Journal of Nanoparticle Research 12 (2010) 467–480.

[70] P.E. Diaz-Flores, F. López-Urías, M. Terrones, J.R. Rangel-Méndez, Efficient adsorption of Cd^{2+} and phenol groups on modified N-doped carbon nanotubes: experimental and DFT studies, Journal of Colloid and Interface Science 334 (2009) 124–131.

[71] E. Gracia-Espino, F. Lopez-Urias, H. Terrones, M. Terrones, Doping (10,0)-semiconductor nanotubes with nitrogen and vacancy defects, Materials Express 1 (2011) 127–135.

[72] K.S. Novoselov, A.K. Geim, S.V. Morozov, D. Jiang, Y. Zhang, S.V. Dubonos, I.V. Grigorieva, A.A. Firsov, Electric field effect in atomically thin carbon films, Science 306 (2004) 666–669.

[73] F. Schedin, A.K. Geim, S.V. Morozov, E.W. Hill, P. Blake, M.I. Katsnelson, K.S. Novoselov, Detection of individual gas molecules adsorbed on graphene, Nature Materials 6 (2007) 652–655.

[74] P.E. Sheehan, L.J. Whitman, Detection limits for nanoscale biosensors, Nano Letters 5 (2005) 803–807.
[75] J.D. Fowler, M.J. Allen, V.C. Tung, Y. Yang, R.B. Kaner, B.H. Weiller, Practical chemical sensors from chemically derived graphene, ACS Nano 3 (2009) 301–306.
[76] R.K. Joshi, H. Gomez, F. Alvi, A. Kumar, Graphene films and ribbons for sensing of O_2, and 100 ppm of CO and NO_2 in practical conditions, Journal of Physical Chemistry C 114 (2010) 6610–6613.
[77] M. Shafiei, P.G. Spizzirri, R. Arsat, J. Yu, J. du Plessis, S. Dubin, R.B. Kaner, K. Kalantar-zadeh, W. Wlodarski, Platinum/graphene nanosheet/SiC contacts and their application for hydrogen gas sensing, Journal of Physical Chemistry C 114 (2010) 13796–13801.
[78] M. Asai, T. Ohba, T. Iwanga, H. Kanoh, M. Endo, J. Campos-Delgado, M. Terrones, K. Nakai, K. Kaneko, Marked adsorption irreversibility of graphitic nanoribbons for CO_2 and H_2O, Journal of American Chemical Society 133 (2011) 14880.
[79] M. Nakamura, T. Ohba, P. Branton, H. Kanoh, K. Kaneko, Equilibration-time and pore-width dependent hysteresis of water adsorption isotherm on hydrophobic microporous carbons, Carbon 48 (2010) 305.
[80] J. Miyawaki, T. Kanda, K. Kaneko, Hysteresis-associated pressure-shift-induced water adsorption in carbon micropores, Langmuir 17 (2001) 664.
[81] L.R. Radovic, Chemistry and Physics of Carbon, vol. 28, Marcel Dekker, New York, 2003, p. 229.
[82] R.C. Bansal, J.B. Donnet, F. Stoeckli, Active Carbon, Marcel Dekker, New York, 1988, p. 191.
[83] E.A. Muller, L.F. Rull, L.F. Vega, K.E. Gubbins, Adsorption of water on activated carbons: a molecular simulation study, Journal of Physical Chemistry 100 (1996) 1189.
[84] P. Lodewyckx, The effect of water uptake in ultramicropores on the adsorption of water vapour in activated carbon, Carbon 48 (2010) 2549.
[85] S.J. Gregg, K.S.W. Sing, Adsorption, Surface Area, and Porosity, second ed., Academic Press, New York, 1982, p 262.
[86] P. Kowalczyk, R. Holyst, M. Terrones, H. Terrones, Hydrogen stoarage in nanoporous carbon materials: myth and facts, Physical Chemistry Chemical Physics 9 (2007) 1786.
[87] D. Levesque, A. Gicquel, F.L. Darkrim, S.B. Kayiran, Monte Carlo simulations of hydrogen storage in carbón nanotubes, Journal of Physics: Condensed Matter 14 (2002) 9285.
[88] N.W.S. Kam, M. O'Connell, J.A. Wisdom, H. Dai, Carbon nanotubes as multifunctional biological transporters and near-infrared agents for selective cancer cell destruction, Proceedings of the National Academy of Science 102 (2005) 11600–11605.
[89] R.J. Chen, S. Bangsaruntip, K.A. Drouvalakis, N.W.S. Kam, M. Shim, Y. Li, W. Kim, P.J. Utz, H. Dai, Noncovalent functionalization of carbon nanotubes for highly specific electronic biosensors, Proceedings of the National Academy of Science 100 (2003) 4984–4989.
[90] S.G. Santos, J.V. Santana, F.F. Maia Jr., V. Lemos, V.N. Freire, E.W.S. Caetano, B.S. Cavada, E.L. Albuquerque, Adsorption of ascorbic acid on the C fullerene, Journal of Physical Chemistry B 112 (2008) 14267–14272.
[91] X. Dong, Y. Shi, W. Huang, P. Chen, L.J. Li, Electrical detection of DNA hybridization with single-base specificity using transistors based on CVD-grown graphene sheets, Advanced Materials 22 (2010) 1649–1653.
[92] Y. Wang, Z. Li, J. Wang, J. Li, Y. Lin, Graphene and graphene oxide: biofunctionalization and applications in biotechnology, Trends in Biotechnology 29 (2011) 205–212.
[93] M. Prato, Fullerene chemistry for materials science applications, Journal of Materials Chemistry 7 (1997) 1097–1099.

[94] I.C. Wang, L.A. Tai, D.D. Lee, P.P. Kanakamman, C.K.-F. Shen, T.-Y. Luh, C.H. Cheng, K.C. Hwang, C_{60} and water-soluble fullerene derivatives and antioxidants against radical-initiated lipid peroxidation, Journal of Medicinal Chemistry 42 (22) (1999) 4614–4620.

[95] P.J. Krusic, E. Wasserman, P.N. Keizer, Radical reactions of C_{60}, Science 254 (1991) 1183.

[96] D.V. Schur, B.P. Tarasov, Y.M. Shulga, S.Y. Zaginaichenko, Z.A. Matysina, A.P. Pomytkin, Hydrogen in fullerites, Carbon 41 (2003) 1331–1342.

[97] P.K. Jain, X. Huang, I.H. El-Sayed, M.A. El-Sayed, Noble metals on the nanoscale: optical and photothermal properties and some applications in imaging, sensing, biology, and medicine, Accounts of Chemical Research 41 (2008) 1578–1586.

[98] H.M. Chen, R.S. Liu, Architecture of metallic nanostructures: synthesis strategy and specific applications, Journal of Physical Chemistry C 115 (2011) 3513–3527.

[99] S. Foley, C. Crowley, M. Smaihi, Cellular localisation of a water-soluble fullerene derivative, Biochemical and Biophysical Research Communications 294 (2002) 116–119.

[100] H. Isobe, N. Tomita, S. Jinno, et al., Synthesis and transfection capability of multi-functionalized fullerene polyamine, Chemical Letters 12 (2001) 1214.

[101] E. Nakamura, H. Isobe, Functionalized fullerenes in water, the first 10 years of their chemistry, biology, and nanoscience, Accounts of Chemical Research 36 (2003) 807.

[102] A.B. Castle, E. Gracia-Espino, C. Nieto-Delgado, H. Terrones, M. Terrones, S. Hussain, Hydroxyl-functionalized and N-doped multiwalled carbon nanotubes decorated with silver nanoparticles preserve cellular function, ACS Nano 5 (2011) 2458–2466.

[103] G. Pastorin, W. Wu, S. Wieckowski, J.P. Briand, K. Kostarelos, M. Prato, A. Bianco, Double functionalisation of carbon nanotubes for multimodal drug delivery, Chemical Communications (2006) 1182–1184.

[104] W. Wu, S. Wieckowski, G. Pastorin, M. Benincasa, C. Klumpp, J.P. Briand, R. Gennaro, M. Prato, A. Bianco, Targeted delivery of amphotericin B to cells by using functionalized carbon nanotubes, Angewandte Chemie International Edition 44 (2005) 6358–6362.

[105] S. Dhar, Z. Liu, J. Thomale, H. Dai, S.J. Lippard, Targeted single-wall carbon nanotube-mediated Pt (IV) prodrug delivery using folate as a homing device, Journal of American Chemical Society 130 (2008) 11467–11476.

[106] Z. Liu, X. Sun, N. Nakayama, H. Dai, Supramolecular chemistry on water-soluble carbon nanotubes for drug loading and delivery, ACS Nano 1 (2007) 50–56.

[107] Z. Liu, W.B. Cai, L.N. He, N. Nakayama, K. Chen, X.M. Sun, X.Y. Chen, H.J. Dai, *In vivo* biodistribution and highly efficient tumour targeting of carbon nanotubes in mice, Nature Nanotechnology 2 (2007) 47–52.

[108] K. Welsher, Z.D.D. Liu, H. Dai, Selective probing and imaging of cells with single walled carbon nanotubes as near-infrared fluorescent molecules, Nano Letters 8 (2008) 586–590.

[109] Z. Liu, X. Li, S.M. Tabakman, K. Jiang, S. Fan, H. Dai, Multiplexed multi-color Raman imaging of live cells with isotopically modified single walled carbon nanotubes, Journal of American Chemical Society 130 (2008) 13540–13541.

[110] P. Chakravarty, R. Marches, N.S. Zimmerman, A.D. Swafford, P. Bajaj, I.H. Musselman, P. Pantano, R.K. Draper, E.S. Vitetta, Thermal ablation of tumor cells with antibody-functionalized single-walled carbon nanotubes, Proceedings of the National Academy of Science USA 105 (2008) 8697–8702.

[111] M.R. McDevitt, D. Chattopadhyay, B.J. Kappel, J.S. Jaggi, S.R. Schiffman, C. Antczak, J.T. Njardarson, R. Brentjens, D.A. Scheinberg, Tumor targeting with antibody-functionalized, radiolabeled carbon nanotubes, Journal of Nuclear Medicine 48 (2007) 1180–1189.

[112] N.W.S. Kam, T.C. Jessop, P.A. Wender, H.J. Dai, Nanotube molecular transporters: internalization of carbon nanotube-protein conjugates into mammalian cells, Journal of American Chemical Society 126 (2004) 6850–6851.

[113] N.W.S. Kam, H. Dai, Carbon nanotubes as intracellular protein transporters: generality and biological functionality, Journal of American Chemical Society 127 (2005) 6021–6026.

[114] Z. Liu, S. Tabakman, K. Welsher, H. Dai, Carbon nanotubes in biology and medicine: in vitro and in vivo detection, imaging and drug delivery, Nano Research 2 (2009) 85–120.

[115] M.J. Allen, et al., Honeycomb carbon: a review of graphene, Chemical Review 110 (2010) 132–145.

[116] N. Mohanty, V. Berry, Graphene-based single-bacterium resolution biodevice and DNA transistor: interfacing graphene derivatives with nanoscale and microscale biocomponents, Nano Letters 8 (2008) 4469–4476.

[117] A.J. Patil, et al., Aqueous stabilization and self-assembly of graphene sheets into layered bio-nanocomposites using DNA, Advanced Materials 21 (2009) 3159–3164.

[118] L.A.L. Tang, et al., Graphene-based SELDI probe with ultrahigh extraction and sensitivity for DNA oligomer, Journal of American Chemical Society 132 (2010) 10976–10977.

[119] J.B. Liu, et al., Noncovalent DNA decorations of graphene oxide and reduced graphene oxide toward water-soluble metal-carbon Irbid nanostructures via self-assembly, Journal of Material Chemistry 20 (2010) 900–906.

[120] O. Yasuhide, M. Kenzo, M. Kazuhiko, Label-free biosensors based on aptamer-modified graphene field-effect transistors, Journal of American Chemical Society 132 (2010) 18012–18013.

[121] S. Hou, M.L. Kasner, S. Su, K. Patel, R. Cuellari, Highly sensitive and selective dopamine biosensor fabricated with silanized graphene, Journal of Physical Chemistry C 114 (2010) 14915.

[122] A. Stafiej, K. Pyrzynska, Adsorption of heavy metal ions with carbon nanotubes, Separation and Purification Technology 58 (2007) 49–52.

[123] G.P. Rao, C. Lu, F. Su, Sorption of divalent metal ions from aqueous solution by carbon nanotubes: a review, Separation and Purification Technology 58 (2007) 224–231.

[124] Y.-H. Li, S. Wang, J. Wei, X. Zhang, C. Xu, Z. Luan, D. Wu, B. Wei, Lead adsorption on carbon nanotubes, Chemical Physics Letters 357 (2002) 263–266.

[125] J. Li, S. Chen, G. Sheng, J. Hu, X. Tan, X. Wang, Effect of surfactants on Pb(II) adsorption from aqueous solutions using oxidized multiwall carbon nanotubes, Chemical Engineering Journal 166 (2011) 551–558.

[126] S. Yang, J. Hu, C. Chen, D. Shao, X. Wang, Mutual effects of Pb(II) and humic acid adsorption on multiwalled carbon nanotubes/polyacrylamide composites from aqueous solutions, Environmental Science & Technology 45 (2011) 3621–3627.

[127] S.G. Wang, W.X. Gong, X.W. Liu, Y.W. Yao, B.Y. Gao, Q. Yue, Removal of lead(II) from aqueous solution by adsorption onto manganese oxide-coated carbon nanotubes, Separation and Purification Technology 58 (2007) 17–23.

[128] J. Hu, C. Chen, X. Zhu, X. Wang, Removal of chromium from aqueous solution by using oxidized multiwalled carbon nanotubes, Journal of Hazardous Materials 162 (2009) 1542–1550.

[129] K. Pillay, E.M. Cukrowska, N.J. Coville, Multi-walled carbon nanotubes as adsorbents for the removal of parts per billion levels of hexavalent chromium from aqueous solution, Journal of Hazardous Materials 166 (2009) 1067–1075.

[130] Z.C. Di, J. Ding, X.J. Peng, Y.H. Li, Z.K. Luan, J. Liang, Chromium adsorption by aligned carbon nanotubes supported ceria nanoparticles, Chemosphere 62 (2006) 861–865.

[131] C. Lu, H. Chiu, C. Liu, Removal of zinc(II) from aqueous solution by purified carbon nanotubes: kinetics and equilibrium studies, Industrial and Engineering Chemical Research, 45 (2006) 2850–2855.

[132] E. Gracia-Espino, F. Lopez-Urias, H. Terrones, M. Terrones, to be published.

[133] S. Yang, J. Li, D. Shao, J. Hu, X. Wang, Adsorption of Ni(II) on oxidized multi-walled carbon nanotubes: effect of contact time, pH, foreign ions and PAA, Journal of Hazardous Materials 166 (2009) 109–116.

[134] G. Sheng, J. Li, D. Shao, J. Hu, C. Chen, Y. Chen, X. Wang, Adsorption of copper(II) on multiwalled carbon nanotubes in the absence and presence of humic or fulvic acids, Journal of Hazardous Materials 178 (2010) 333–340.

[135] C.L. Chen, X.L. Li, X.K. Wang, Application of oxidized multi-wall carbon nanotubes for Th(IV) adsorption, Radiochimica Acta 95 (2007) 261–266.

[136] Y. Sun, S. Yang, G. Sheng, Z. Guo, X. Wang, The removal of U(VI) from aqueous solution by oxidized multiwalled carbon nanotubes, Journal of Environmental Radioactivity 105 (2012) 40–47.

[137] Y. Nagano, T. Kiyobayashi, T. Nitta, CO_2 absorption in C60 solid, Chemical Physics Letters 217 (1994) 186–190.

[138] A. Martínez-Alonso, J.M.D. Tascón, E.J. Bottani, Physical adsorption of Ar and CO_2 on C_{60} fullerene, Journal of Physical Chemistry B 105 (2001) 135–139.

[139] B. Gao, J.X. Zhao, Q.H. Cai, X.G. Wang, X.Z. Wang, Doping of calcium in C_{60} fullerene for enhancing CO_2 capture and N_2O transformation: a theoretical study, Journal of Physical Chemistry A 115 (2011) 9969–9976.

[140] M. Cinke, J. Li, C.W. Bauschlicher Jr., A. Ricc, M. Meyyappan, CO_2 adsorption in single-walled carbon nanotubes, Chemical Physics Letters 376 (2003) 761–766.

[141] E.P. Dillon, C.A. Crouse, A.R. Barron, Synthesis, characterization, and carbon dioxide adsorption of covalently attached polyethyleneimine-functionalized single-wall carbon nanotubes, ACS Nano 2 (2008) 156–164.

[142] C. Lu, H. Bai, B. Wu, F. Su, J.F. Hwang, Comparative study of CO_2 capture by carbon nanotubes, activated carbons, and zeolites, Energy Fuels 22 (2008) 3050–3056.

[143] F. Sua, C. Lua, W. Cnena, H. Baib, J.F. Hwangc, Capture of CO_2 from flue gas via multiwalled carbon nanotubes, Science of the Total Environment 407 (2009) 3017–3023.

[144] S.C. Hsu, C. Lu, F. Su, W. Zeng, W. Chen, Thermodynamics and regeneration studies of CO_2 adsorption on multiwalled carbon nanotubes, Chemical Engineering Science 65 (2010) 1354–1361.

[145] F. Su, C. Lu, H.S. Chen, Adsorption, desorption, and thermodynamic studies of CO_2 with high-amine-loaded multiwalled carbon nanotubes, Langmuir 27 (2011) 8090–8098.

[146] M. Arab, F. Picaud, M. Devel, C. Ramseyer, C. Girardet, Molecular selectivity due to adsorption properties in nanotubes, Physical Review B 69 (2004) 165401.

[147] A. Ghosh, K.S. Subrahmanyam, K.S. Krishna, S. Datta, A. Govindaraj, S.K. Pati, C.N.R. Rao, Uptake of H_2 and CO_2 by Graphene, Journal of Physical Chemistry C 112 (2008) 15704–15707.

[148] P. Cabrera-Sanfelix, Adsorption and reactivity of CO_2 on defective graphene sheets, Journal of Physical Chemistry A 113 (2) (2009) 493–498.

[149] B. Huang, Z. Li, Z. Liu, G. Zhou, S. Hao, J. Wu, B.L. Gu, W. Duan, Adsorption of gas molecules on graphene nanoribbons and its implication for nano-scale molecule sensor, Journal of Physical Chemistry C 112 (2008) 13442–13446.

Chapter 2

Accessibility of Gases and Liquids in Carbons

Suresh K. Bhatia
School of Chemical Engineering, The University of Queensland, Brisbane, QLD, Australia

Chapter Outline

2.1. Introduction 37
2.2. Experimental Background 39
2.2.1. Characterization of Carbons by Low-Temperature Adsorption 39
2.2.2. Prediction of High-Temperature Gas Adsorption Based on Low-Temperature Characterization 42
2.3. Percolation Theory-Based View of Accessibility 47
2.4. Atomistic Modeling of Accessibility 51
2.4.1. Cluster Analysis of the Adsorbed Phase 51
2.4.2. Transition State Theory 53
2.5. Open Loop Hysteresis 54
2.6. Conclusions 57
References 58

2.1. INTRODUCTION

Microporous carbons have a multitude of uses in adsorption and gas separation [1], as electrode materials in batteries and supercapacitors [2], in water purification, and as catalyst supports [3]. In the last two decades, their use as media for gas storage, particularly of methane and hydrogen, has also received considerable attention [4], creating enhanced interest in their adsorption properties. The features of nanoporous carbons that make them attractive for these applications are their high internal surface area and pore volume (mainly micropores and mesopores), which enable high equilibrium capacity for fluids in their structure [5]. However, for any of these applications to effectively take advantage of this large surface area and pore volume, the internal structure of the carbon must be accessible to the infiltrating gas or liquid. Thus, fluid

Novel Carbon Adsorbents. DOI: 10.1016/B978-0-08-097744-7.00002-8
Copyright © 2012 Elsevier Ltd. All rights reserved.

molecules entering the pores from the external surface of the particle must be able to percolate throughout the internal pore structure, and not face significant restrictions to their entry into the nanopores offering the large surface area for adsorption.

Nanoporous carbons possess complex internal structure, whose understanding has been the subject of much research for several decades [6–9]. At the fundamental level the internal morphology of most carbons comprises a turbostratic structure, but displays much long-range disorder, suggestive of random aggregation of nanoscale graphitic crystallites [10]. However, the graphene layers in the turbostratic structure possess numerous defects and distortions, and present a tortuous surface of complex topology. This leads to constrictions and bottlenecks in the structure, which can then be viewed as comprising nanoscale pore spaces or cages connected by narrow necks. These constrictions can present energy barriers to infiltrating fluid molecules [11,12], and inhibit access to the pore space that is only approachable through these constrictions or pore necks.

The existence of energy barriers to the entry of fluids into the internal pore spaces of carbons has long been known, and was revealed as early as 1952 in a seminal paper by Maggs [13] who showed an increase in apparent nitrogen adsorption capacity of some coals with increase in temperature, and that this capacity had a kinetic character – i.e. equilibration was extremely slow and was not attained in practical time scales. The difficulty in equilibration of nitrogen and some other gases in carbons has been subsequently reported in other studies [14,15], while an increase in argon adsorption of some coals with increase in temperature has recently also been reported [16] – both suggestive of internal energy barriers. However, the traditional interpretation of accessibility has been that it is essentially pore and molecular size dependent, i.e. molecules smaller than the pore size can freely enter a given pore while those that are larger cannot [17,18] – overlooking energy barriers. Percolation theory provides the most useful framework through which to predict this molecular size-dependent accessibility of the structure, and thereby the variation of accessible pore volume with molecular size. Nevertheless, such an approach fails to explain the temperature dependence of the apparent capacity, and its kinetic nature. These anomalous features have since been explained by interpretations based on transition state theory [11,12], and simulation methods have been developed [11] that can predict if a specified pore space is accessible at a given temperature. Besides adsorption, this view of accessibility has been found applicable to carbon oxidation kinetics, where the large increase in surface area at low (3–5%) conversion has been attributed to increase in accessibility of gases used in adsorption characterization due to the preferential reaction at defect sites at pore mouths and their opening up by the oxidation [19,20].

This chapter reviews the above developments, starting with the experimental background that has shaped our view of accessibility. The percolation view of accessibility is then discussed, and subsequently the simulation and transition state theory techniques recently developed to predict the accessibility are reviewed. Finally, we discuss open loop hysteresis and the permanent trapping of adsorbed molecules.

2.2. EXPERIMENTAL BACKGROUND

2.2.1. Characterization of Carbons by Low-Temperature Adsorption

Nitrogen adsorption at 77 K has been the most widely used method for characterization of porous material structure [5], and was the focus of early nonlocal density functional theory methods [21,22] solving for a slit pore-size distribution, $f(H_{in})$, by inverting the generalized adsorption integral (GAI)

$$\Gamma_{ex}(P) = \int [\hat{\rho}(P, H_{in}) - \rho_b] f(H_{in}) dH_{in} \tag{2.1}$$

for the excess adsorption $\Gamma_{ex}(P)$. Here $H_{in} = (H_{cc} - \sigma_c)$ is the geometrical pore width, H_{cc} is the physical pore width, σ_c is the effective diameter of the carbon atom, and ρ_b is the bulk density. $\hat{\rho}(P, H_{in})$ is the average density in a pore of width H_{in} at a given pressure P, obtained using density functional theory. However, the use of nitrogen for nanoporous carbons has long been considered controversial because of accessibility issues and the difficulty of attaining its adsorption equilibrium at 77 K in such materials. This leads to unrealistically low surface areas and pore volumes, recently confirmed by Scherdel and Reichenauer [23], who showed significantly larger pore volumes using small-angle X-ray scattering compared to nitrogen adsorption. As an early example, in studies of nitrogen and argon adsorption on bituminous coals, Lecky, Hall, and Anderson [24] reported surface areas in the range of 0.4–11 m^2/g, much lower than those obtained using methyl alcohol adsorption, in the range of 40–200 m^2/g. While the low values were confirmed by Zwietering, Oele, and van Krevelen [25], their explanation that the polar nature of methyl alcohol resulted in peptizating action and creation of a new surface not permeable to inert gases was challenged by Maggs [26] and Dryden [27] who argued that the higher surface area and porosity were permanent. Shortly thereafter, in a classic study which provided direction for the modern understanding of accessibility, Maggs [13] showed that nitrogen adsorption in a Welsh coking coal had an unusual temperature dependence, with an increase in amount adsorbed with increase in temperature between 110 °C and 175 °C. Maggs' results showed also the kinetically governed nature of the capacity at any temperature, and that equilibrium was not attained even after 4 h of exposure at each temperature.

Although Maggs did not exclude the possibility of a physical change such as thermal expansion leading to greater accessibility above 110 °C, and the increase in adsorption, the alternate currently accepted view is that of existence of constrictions and narrow pore necks presenting energy barriers that are more easily overcome at higher temperatures due to increase in kinetic energy of the adsorbate. Thus, as temperature is increased more pores become accessible as their entry barriers are overcome, and this leads to an increase in adsorption. At sufficiently high temperature, all the pores become accessible and the temperature dependence becomes normal again, with a decrease in adsorption with increased temperature. This existence of energy barriers also explains the slow approach to equilibrium observed by Maggs, and is supported by the well-known activated nature of the diffusivity in microporous materials [28]. Similar anomalous behavior of the adsorption capacity of nitrogen upon temperature increase has since also been reported for carbonized almond shell by Rodriguez-Reinoso et al. [14], who showed that this unusual behavior was eliminated upon small amount of activation (>10% burn-off). These authors found the adsorption kinetics to be very slow at 77 K, but much faster at 90 K, leading to higher capacities at 90 K in the carbonized material. They concluded that constrictions of size comparable to the N_2 molecule existed which were reacted away upon activation, thereby removing the energy barriers to entry of the N_2 into the micropores. Such a viewpoint also explains the large reported increase in the surface area with small amounts of gasification of 3–5% reported by other workers, for coals measured by N_2 adsorption at 77 K [29], and of coal char [19] or cokes [20] measured by Ar adsorption at 87 K – since defect sites at pore entrances and constrictions will comprise the most reactive parts of the structure and gasify early, increasing the accessibility of the pore structure.

While the above problem of low accessibility of N_2 at 77 K has received much attention, it is now clear that even Ar adsorption at 87 K can suffer from this problem. Besides the large increase in surface area of coal char with very small amount of burn-off reported by Feng and Bhatia [19], more direct confirmation of energy barriers is evident from the results of Bae et al. [16], depicted in Fig. 2.1 showing a very large increase in adsorption capacity of coals for argon when the temperature increases from 87 K to 313 K – a direct indication of energy barriers inhibiting access of argon to a large portion of the microporosity at 87 K.

Interestingly, the adsorption at 313 K was reduced on heat treatment at 573 K and 673 K, but increased on heat treatment at 773 K, suggesting that volatiles produced on heat treatment were condensing in narrow regions and creating constrictions at the lower heat treatment temperatures [30]. Nevertheless, despite the above findings it appears that Ar at 87 K faces less inaccessibility issues than N_2 at 77 K, and is preferable for characterization. Figure 2.2 [12] compares the pore-size distributions (PSDs) obtained by nonlocal density functional theory with a slit pore model, using the adsorption of both Ar at 87 K and N_2 at 77 K, in BPL- and PCB-activated carbon.

For the BPL carbon the PSDs depicted in Fig. 2.2a clearly indicate greater accessibility to Ar at 87 K, which is able to sample smaller micropore widths compared to N_2 at 77 K. Above about 0.6-nm pore width, the PSDs and accessibility in both cases appear similar. Interestingly, the area under the first

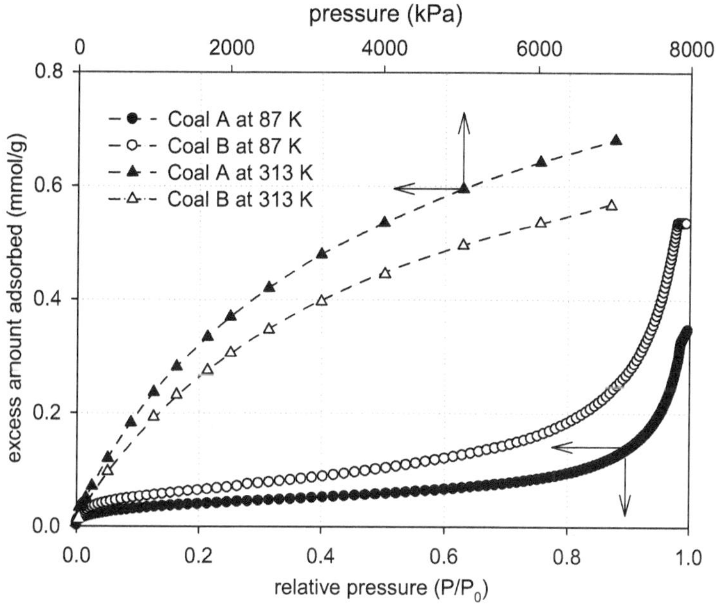

FIGURE 2.1 Argon adsorption isotherms on two coals at 87 K and 313 K. *(Reprinted with Permission from Bae, J.-S. and Bhatia, S.K. High Pressure Adsorption of Methane and Carbon Dioxide on Coal.* Energy & Fuels, ***20****, 2599–2607. Copyright 2006 American Chemical Society)*

peak of the N_2-based PSD compares well with that of the Ar-based PSD, suggesting that as the adsorption continues the smallest micropores are accessed by N_2 but at a higher than equilibrium pressure. This leads to the shift in the position of the first peak for N_2 to larger pore width. A somewhat similar situation exists for the PCB carbon, whose PSD is depicted in Fig. 2.2b although the difference in accessibility between the two gases is much smaller in this case, and the first peaks are much closer.

As a result of such accessibility issues at the low temperatures, the alternative of CO_2 adsorption at 195 K was proposed by Marsh [31], despite showing partial inaccessibility to CO_2 in some carbons derived from polymeric sources even at this higher temperature. Marsh also proposed the alternative of CO_2 adsorption at 273 K and 1 bar pressure when such inaccessibility occurs. Nevertheless, at this temperature the vapor pressure of CO_2 is 34.66 bar, and at 1 bar pressure only pores smaller than 1 nm show significant adsorption. This has led others to suggest a combination of CO_2 adsorption at 273 K and N_2 adsorption at 77 K for more complete characterization [32,33]. However, work by Rios et al. of CO_2 adsorption in Takeda 3A carbon molecular sieve (CMS) has demonstrated that care must be exercised to ensure sufficiently long equilibration times even at 273 K, when pore openings are comparable to the CO_2 molecular size [15]. Independent work in the author's group has indeed

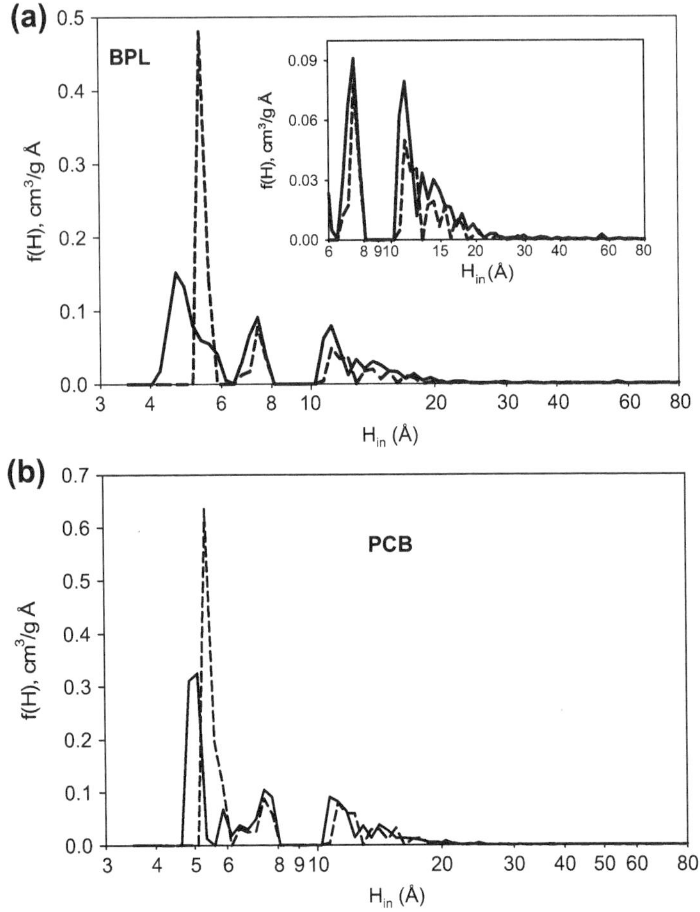

FIGURE 2.2 The results of the pore-size distribution of (a) the BPL and (b) PCB carbon, determined from experimental adsorption isotherms of Ar at 87 K (solid line) and N_2 at 77 K (dashed line). The inset shows the PSD in the large pore-size range ($H_{in} > 6$ Å). H_{in} is internal pore width. *(Reprinted with Permission from Nguyen, T.X. and Bhatia, S.K. Kinetic Restriction of Simple Gases in Porous Carbons: Transition State Theory Study.* Langmuir, ***24***, *146–154. Copyright 2008 American Chemical Society)*

demonstrated the absence of hysteresis for CO_2 in the Takeda 3A CMS at 273 K [12], confirming its equilibration, while N_2 at 77 K displayed hysteresis and lack of equilibration, as depicted in Fig. 2.3.

2.2.2. Prediction of High-Temperature Gas Adsorption Based on Low-Temperature Characterization

A key goal of the low-temperature characterization using N_2 at 77 K, Ar at 87 K, or CO_2 at 273 K, discussed above, is to enable prediction of the

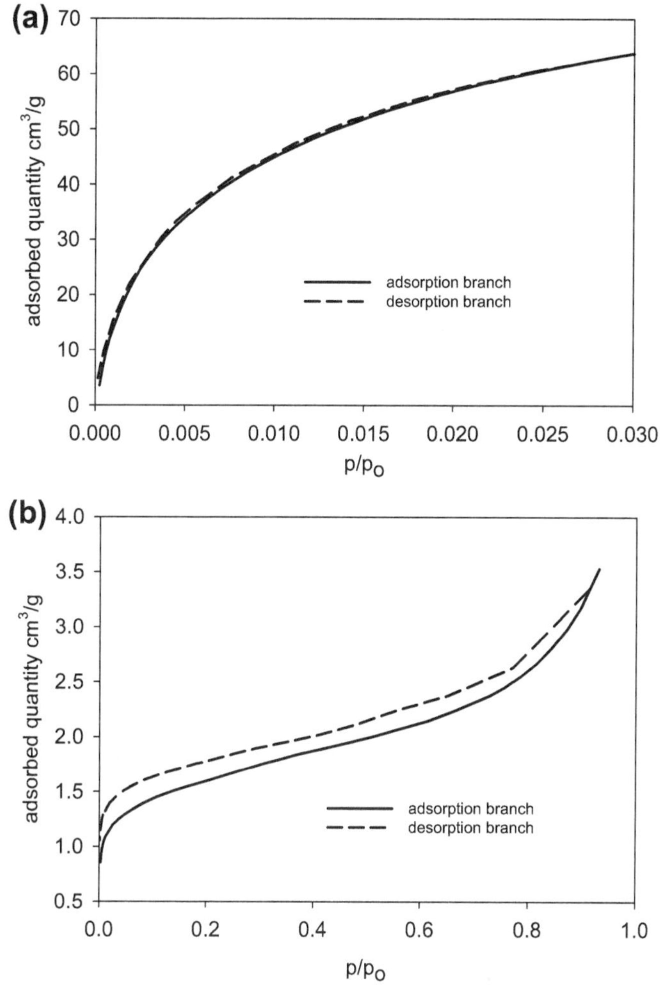

FIGURE 2.3 Experimental sub-atmospheric pressure adsorption isotherms of (a) CO_2 at 273 K and (b) N_2 at 77 K, in carbon molecular sieve Takeda 3 Å. *(Reprinted with Permission from Nguyen, T.X. and Bhatia, S.K. Kinetic Restriction of Simple Gases in Porous Carbons: Transition State Theory Study.* Langmuir*, **24**, 146–154. Copyright 2008 American Chemical Society)*

adsorption of other gases using Eqn (2.1) with the pore-size distribution $f(H_{in})$ obtained from the characterization. While this is often successful, complications can arise when there is a difference in accessibility between the conditions of the characterization and those of the prediction. Under these circumstances, it is not possible to accurately predict the high-temperature adsorption isotherm of another gas, forcing process design to rely on more empirical procedures. Figure 2.4 [12] depicts experimental and predicted isotherms of CO_2 at various

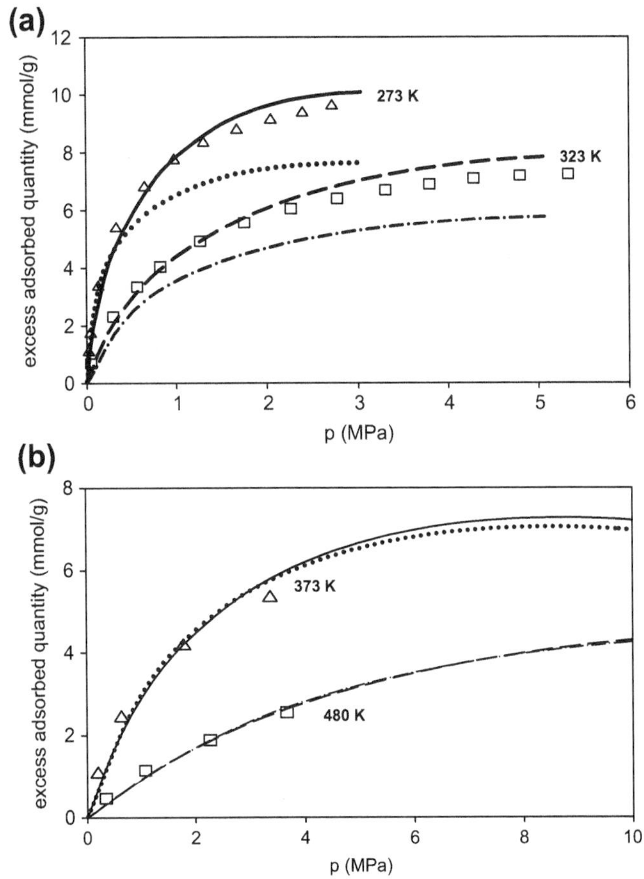

FIGURE 2.4 Experimental adsorption isotherms of CO_2 in (a) BPL carbon [36] at 273 K (empty triangle) and 323 K (empty square) and (b) PCB carbon [37] at 373 K (empty triangle) and 480 K (empty square), and model predictions. Solid and dashed lines depict the predicted CO_2 isotherms (a) at 273 K and 323 K and (b) at 373 K and 480 K respectively, using the structural parameters determined from Ar adsorption at 87 K. Similarly, dotted and dashed-dotted lines illustrate the predicted CO_2 isotherms (a) at 273 K and 323 K and (b) at 373 K and 480 K respectively, using the structural parameters determined from N_2 adsorption at 77 K. *(Reprinted with Permission from Nguyen, T.X. and Bhatia, S.K. Kinetic Restriction of Simple Gases in Porous Carbons: Transition State Theory Study.* Langmuir*; **24**, 146–154. Copyright 2008 American Chemical Society)*

temperatures in BPL- and PCB-activated carbons, using the Ar and N_2 PSDs depicted in Fig. 2.2.

It is evident that the predictions using the Ar-based PSD are in good agreement with the experimental isotherm for both the BPL and PCB carbons, indicating that Ar at 87 K is able to access the same porosity as CO_2 at the higher temperatures in both carbons. For the PCB carbon even the N_2 PSD-based predictions for CO_2 are accurate and match those based on the Ar-based

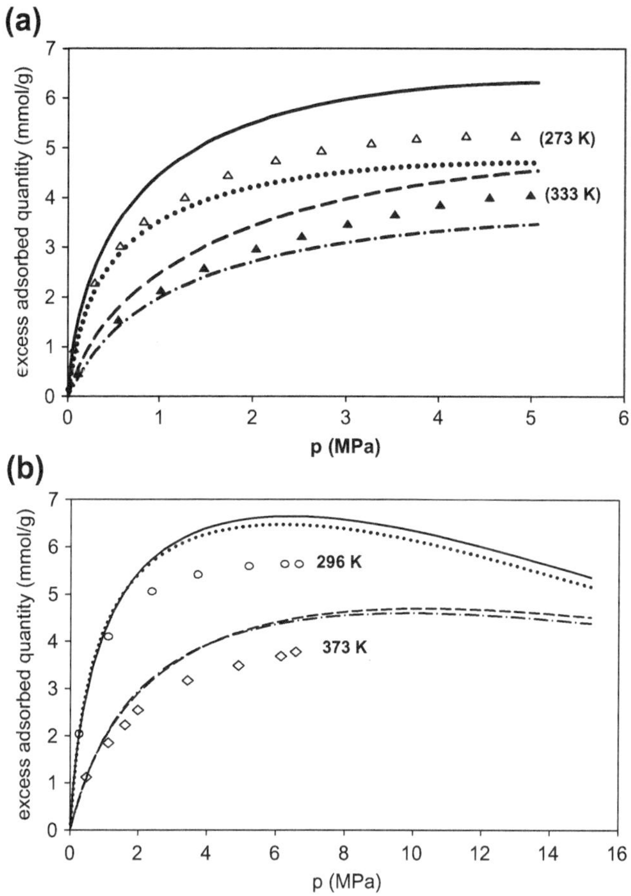

FIGURE 2.5 Experimental adsorption isotherms of CH_4 in (a) BPL carbon [36] at 273 K (empty triangle) and 333 K (empty square), and (b) PCB carbon [37] at 296 K (empty triangle) and 373 K (empty square), and model predictions. Solid and dashed lines depict the predicted CH_4 isotherms (a) at 273 K and 333 K and (b) at 296 K and 373 K respectively, using the structural parameters, determined from Ar adsorption at 87 K. Similarly, dotted and dashed-dotted lines illustrate the predicted CH_4 isotherms (a) at 273 K and 333 K and (b) at 296 K and 373 K respectively, using the structural parameters determined from N_2 adsorption at 77 K. *(Reprinted with Permission from Nguyen, T.X. and Bhatia, S.K. Kinetic Restriction of Simple Gases in Porous Carbons: Transition State Theory Study.* Langmuir, ***24****, 146–154. Copyright 2008 American Chemical Society)*

PSD, consistent with the small difference in accessibility evident in Fig. 2.2. On the other hand, the lower accessibility of N_2 at 77 K in the BPL carbon, evident in Fig. 2.2, leads to underprediction of the CO_2 isotherm in Fig. 2.4a. Similar predictions for methane in these carbons [12] yielded the results in Fig. 2.5, showing that the N_2-based PSD underpredicts the isotherm of even the larger CH_4 molecule in BPL carbon at 273 K and 333 K. However, the Ar PSD-based results show overprediction of the isotherm compared to the experimental values. Thus, the accessibility of Ar in the BPL micropore structure exceeds that of CH_4 at the higher temperatures, while that of N_2 is lower. On the other hand, Fig. 2.5b shows that both Ar and N_2 at 87 K and 77 K respectively show greater accessibility than CH_4 at the higher temperatures in PCB carbon. Thus, it is clear that differences of the PSDs of different carbons can lead to contrasting pictures of differences between accessibilities of different gases, and there is no universal characterization method that can be used as a basis for predictions. Independent validation of accessibility and therefore accuracy of predictions is therefore necessary before they can be reliably used in process design.

It is evident from the above discussion that CO_2 will in general have a higher accessibility than N_2 as well as methane. Consistent with this, in recent work Narkiewicz and Mathews [34] have shown CO_2 to have 12.4% higher accessibility than methane in a molecular level representation of coal. In an attempt to account for differences in accessibility between the characterization gas and other adsorptives, Davies and Seaton [35] modified Eqn (2.1) to

$$\Gamma_{ex}(P) = A\int[\hat{\rho}(P,H_{in}) - \rho_b]f(H_{in})dH_{in} \tag{2.2}$$

where A is a correction factor representing the fraction of pore volume accessible to the species for which the prediction is being made. When this approach was employed by Nguyen and Bhatia [12] to the results in Fig. 2.5a for methane in BPL carbon, using the PSD based on argon adsorption at 87 K, it was found that the factor A had the value of 0.86 at 273 K and 0.91 at 333 K, consistent with the expectation of increase in methane accessibility with increase in temperature. However, the value of A needs to be predicted *a priori*, and as yet no simple method is available, as the accessibility must depend on the size of the pore entry or constriction at the pore mouth, and this is not addressed in the simple slit pore model underlying the PSD.

While the different accessibilities of various gases are related to constrictions and pore entry barriers, and are therefore temperature dependent, they do have a correlation with molecular size – as is to be expected. Figure 2.6 depicts the pore volumes of Norit R1 Extra activated carbon accessible to different gases, determined by Nguyen et al. [38] based on analysis of the adsorption data of Herbst and Harting [39].

A decrease in accessible volume (denoted adsorbent volume in Fig. 2.6) with increase in molecular size is evident. In principle, this should also be a function of temperature, as discussed above, but this was not considered in the data interpretation. In their own analysis of the adsorption data of the different gases in Fig. 2.6, Herbst and Harting [39] did use a temperature-dependent pore volume, assumed to be the same as the helium pore volume. This volume did increase with temperature, as expected, but may not be meaningful for interpreting the data of the other gases having different accessibility.

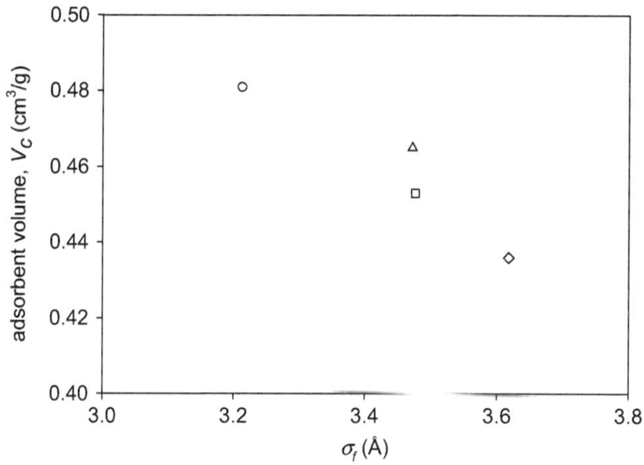

FIGURE 2.6 Variation in the value of adsorbent volume with molecular size, determined by different probing gases. *(Reprinted with Permission from Nguyen, T.X., Bhatia, S.K. and Nicholson, D. Prediction of High-Pressure Adsorption Equilibrium of Supercritical Gases Using Density Functional Theory.* Langmuir*, **21**, 3187–3197. Copyright 2005 American Chemical Society)*

2.3. PERCOLATION THEORY-BASED VIEW OF ACCESSIBILITY

The molecular size dependence of the accessible pore volume suggests a percolation view of accessibility, in which the accessible porosity comprises not only pores smaller than the molecular size but also larger pores that are accessed only through such small pores. Figure 2.7 [18] depicts an illustration of this concept.

This concept was proposed by Lopez-Ramon et al. [17], in connection with the characterization of a microporous carbon, Carbosieve G, using molecular probes of different sizes. Figure 2.8 depicts the pore-size distributions obtained based on the density functional theory approach in Eqn (2.1) with isotherms of CH_4, CF_4, and SF_6, showing the smallest adsorptive, CH_4, yielding the narrowest pores.

To interpret the isotherm data considering the different accessibilities of each adsorptive, Lopez-Ramon et al. [17] proposed the percolation theory approach in which the pore network is characterized by a coordination number, Z, representing the number of pores connected at an intersection. For any adsorbate of molecular size d_c, the accessible pore volume is then obtained as

$$V_{acc} = \frac{\Omega^a(Z,\Omega)}{\Omega} V_{tot} \int_{d_c}^{\infty} f(H)dH \qquad (2.3)$$

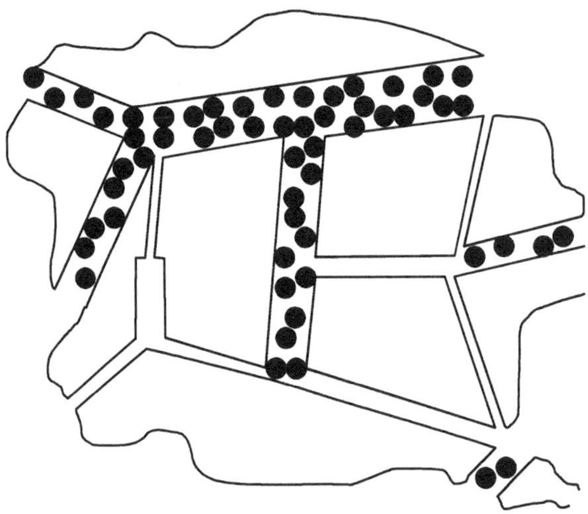

FIGURE 2.7 Schematic illustration of pore network connectivity. *(Reprinted with Permission from Ismadji, S. and Bhatia, S.K. Investigation of Network Connectivity in Activated Carbons by Liquid Phase Adsorption.* Langmuir, ***16****, 9303–9313. Copyright 2000 American Chemical Society)*

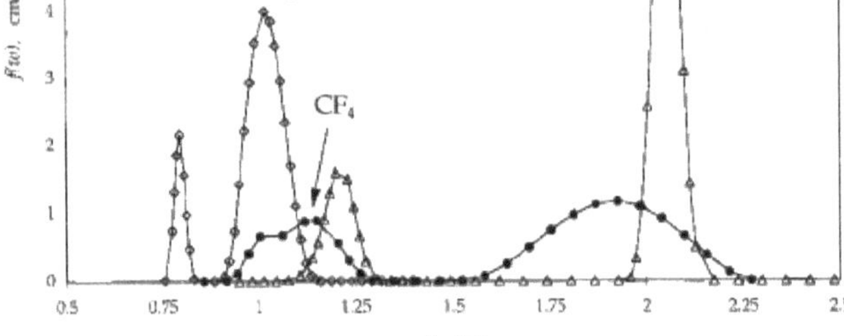

FIGURE 2.8 Pore-size distribution of Carbosieve G, obtained using CH_4 at 295 K, and CF_4, SF_6 at 296 K. *(Reprinted with Permission from Lopez-Ramon, M.V., Jagiello, J., Bandosz, T.J. and Seaton, N.A. Determination of Pore Size Distribution and Network Connectivity in Microporous Solids by Adsorption Measurements and Monte Carlo Simulation.* Langmuir, ***13****, 4435–4445. Copyright 1997 American Chemical Society)*

where V_{tot} is the actual total pore volume, $f(H)$ is the slit pore volume distribution commonly employed for carbons, Ω is the number fraction of available pores, given by

$$\Omega = \int_{d_c}^{\infty} \frac{f(H)}{H}\,\mathrm{d}H \bigg/ \int_{0}^{\infty} \frac{f(H)}{H}\,\mathrm{d}H \tag{2.4}$$

and $\Omega^a(Z,\Omega)$ is the fraction of accessible pores. This fraction is given by percolation theory, based on an assumed network model. For a random network of uniform coordination number, an expression for the function $\Omega^a(Z,\Omega)$ has been provided by Lopez-Ramon et al. [17]. Figure 2.9 depicts the overall pore-size distributions obtained by these authors from isotherms of the three adsorptives at three temperatures, based on the use of Eqn (2.2) with the accessible pore volume fraction A calculated from the accessible pore volume V_{acc}.

It is evident that the pore-size distribution is a function of temperature, with the higher temperature-based PSD showing a shift in the second peak to smaller pore sizes. This is most likely because at higher temperatures the pore mouth barriers of these pores are overcome at lower pressures, leading to lower estimated pore size. Thus, it is clear that the percolation theory approach is missing the temperature effect as it is based purely on a slit-pore distribution with pore access related to molecular size, and overlooks the entry constrictions and energy barriers.

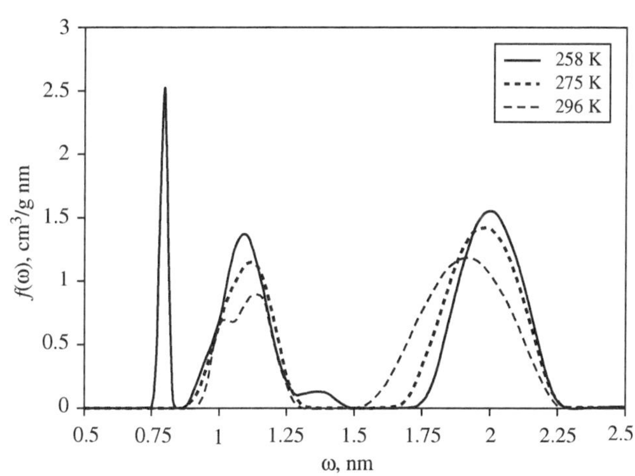

FIGURE 2.9 Overall pore-size distribution of Carbosieve G, obtained at different temperatures. *(Reprinted with Permission from Lopez-Ramon, M.V., Jagiello, J., Bandosz, T.J. and Seaton, N.A. Determination of Pore Size Distribution and Network Connectivity in Microporous Solids by Adsorption Measurements and Monte Carlo Simulation.* Langmuir, ***13****, 4435–4445. Copyright 1997 American Chemical Society)*

The above approach has also been used by Ismadji and Bhatia [18,40,41] for liquid-phase adsorption of three esters (ethyl propionate, ethyl butyrate, and ethyl isovalerate) from aqueous solutions onto three activated carbons (Filtrasorb F-400, Norit ROW 0.8, and Norit ROX 0.8). By fitting the isotherms, the saturation capacities of different esters were determined, and from these the accessible pore volume for each ester was estimated assuming the adsorbed phase density corresponding to that of the liquid. Interpretation of the variation of accessible volume with molecular size using Eqns (2.3) and (2.4) and the expressions for $\Omega^a(Z,\Omega)$ provided by Lopez-Ramon et al. [17] yielded the coordination number Z for each carbon. This was found to be about 3 for the Filtrasorb F-400 and 4 for the Norit carbons. These values are slightly higher than the value of about 2.3 reported by Lopez-Ramon et al. for Carbosieve G. Figure 2.10 [18] depicts the variation of accessible pore volume with ester molecular size for Norit ROX 0.8, showing the fit of the percolation model to the experimentally determined results.

Ismadji and Bhatia [41] extended the above concept to binary systems of esters, considering pores inaccessible to both molecules, pores where only the smaller molecule penetrates, and large pores accessed by both molecules. It was found that experimental binary isotherms could be well predicted using the ideal adsorbed solution theory [42] based on the pure component isotherms, over a range of temperature, despite the percolation theory neglecting temperature effects on the accessibility.

The above success of the percolation theory may be traced to the fact that in practice observation of the kinetic nature of the accessibility is very much dependent upon the difference between the experimental time scale and that of activated diffusion of adsorbate molecules in the porous solid. The diffusion time scale is dictated by the ratio of activation energy to kinetic energy,

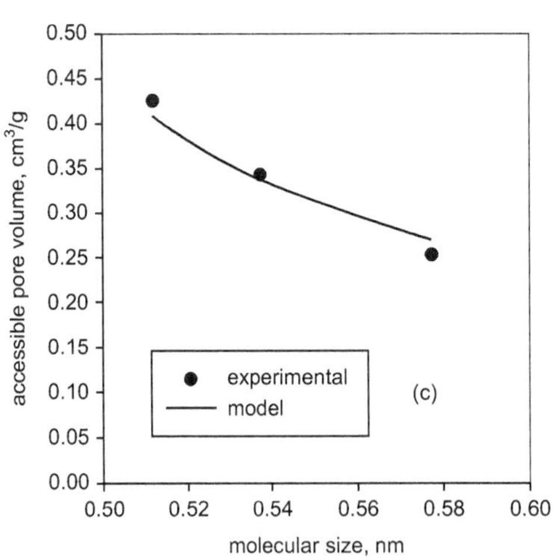

FIGURE 2.10 Correlation between accessible pore volume and molecular size for Norit ROX 0.8 activated carbon. *(Reprinted with Permission from Ismadji, S. and Bhatia, S.K. Investigation of Network Connectivity in Activated Carbons by Liquid Phase Adsorption.* Langmuir, ***16****, 9303–9313. Copyright 2000 American Chemical Society)*

i.e. E_a/k_BT. Since large adsorbate molecules interact strongly with the adsorbent, the activation energy may be significantly larger than the maximum kinetic energy in the experimental temperature range. This results in accessibility being only weakly temperature dependent, since pores with constricted entries remain inaccessible over a wide temperature range. On the other hand, since small adsorbate molecules interact weakly, the time scale of activated diffusion through the pore mouth is very sensitive to temperature, and the accessibility strongly temperature dependent. Thus, one expects the percolation theory approach to work satisfactorily for liquids and other large molecule adsorptives, but not for light gases.

2.4. ATOMISTIC MODELING OF ACCESSIBILITY

2.4.1. Cluster Analysis of the Adsorbed Phase

Given the importance of the pore mouth entry barrier, it is clear that a correct picture of accessibility can only be obtained through an atomistic model of the carbon structure, such as that obtained through reverse Monte Carlo simulation. Once the carbon structure is known, it is possible to identify inaccessible pore spaces in the structure, and methods for this have been developed in the author's laboratory [11,12]. Monte Carlo simulation is one of these, although traditional use of this method cannot discriminate between closed and open pore spaces. This is because of the random insertion of adsorbate molecules in the structure during the simulation, and this can occur even in the inaccessible pore space. Nguyen and Bhatia [11] developed a procedure involving cluster analysis of the adsorbate to analyze the connectivity of the fluid phase in the carbon structure. Since temperature (i.e. kinetic energy) is the main factor responsible for pores being kinetically closed, the analysis is based on the persistence of this effect even at high densities corresponding to that of the liquid phase, i.e. when the average intermolecular distance, $\overline{d}_a$, is approximately equal to the equilibrium separation, $2^{1/6}\sigma_f$, based on the Lennard–Jones (LJ) pair potential. Here σ_f is the fluid LJ size parameter or collision diameter.

In the technique all the pore spaces in the solid, including closed pores, are first filled up with fluid molecules to form a closed-packed phase. Subsequently, the structure of the fluid phase is analyzed to determine discrete clusters, if any. Such discrete clusters are indicative of closed pores, and their absence indicates continuous accessible pore space. The algorithm comprises three key steps, as described below.

1. The first step is straightforward and comprises the filling up of the structure using conventional grand canonical Monte Carlo simulation. To distinguish physically closed and kinetically closed pores, the adsorption simulations may be performed at a reasonably high temperature at which kinetically closed pores are absent. To ensure that detection of closed pores is

independent of the microstates, the configuration of the adsorbed phase is collected every 10^6 steps. The total number of configurations used for every GCMC run is 60 million.

2. In the second step all adsorbate clusters in the solid structure are determined. In order to determine if nearest neighbor particles lie in the same cluster or in different clusters, it is necessary to assign the maximum intermolecular distance, $RC_{[i][i]}$, between two nearest neighbor adsorbate molecules in the same adsorbed cluster i, and the minimum distance $RC_{[i][j]}$ between molecules in different clusters. For a close-packed adsorbate phase, these distances are given by [11]

$$RC_{[i][i]} = 1.2225\sigma_f \tag{2.5}$$

$$RC_{[i][j]} = 2\sigma_{sf} \tag{2.6}$$

In addition, the largest molecular size of the adsorbate, which ensures complete resolution of two clusters, is given as

$$\sigma_f^{\max} = 4.495\sigma_s \tag{2.7}$$

Using the above values of $RC_{[i][i]}$ and $RC_{[i][j]}$, all clusters in the close-packed adsorbate phase are then determined, considering the structure of this phase as a graph in which the adsorbate molecules are vertices. Each vertex connects to its neighboring vertices determined using $RC_{[i][i]}$. The *breadth-first search* algorithm [43] is employed to traverse all distinct pathways connected to two arbitrary vertices. During this traversal visited vertices are marked to guarantee that each pathway is not repeatedly visited. Accordingly, two molecules are in the same cluster if there exists at least one pathway connecting them. In this way, all the clusters are determined.

3. The third step determines the continuity of each cluster throughout the structure. This is determined by examining the clusters in a periodic supercell, constructed from the primitive unit cell by superimposing mirror images. Figure 2.11 illustrates the unit cell and the corresponding supercell respectively, for a saccharose char whose hybrid reverse Monte Carlo-based structure has been determined by Nguyen et al. [9].

The portions with right slanted black stripes represent the solid phase, and balls depict adsorbate molecules of the close-packed adsorbed phase. It is evident in these figures that *the number of adsorbate molecules in a cluster associated with a closed pore is invariant from the periodic unit cell to the periodic supercell, while that associated with open space is doubled from the periodic unit cell to the periodic supercell.* This observation is key to determining pore network connectivity in the carbon structure model. The degree of connectivity of a pore space, N_f, is then given as

$$N_f = \frac{N_s^{\{i\}}}{N_u^{\{i\}}} - 1 \tag{2.8}$$

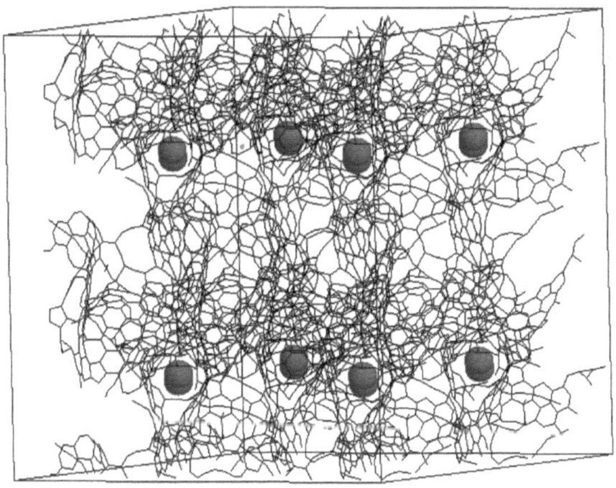

FIGURE 2.11 Illustration of supercell, with kinetically closed pores at 77 K that can accommodate 2 nitrogen molecules. *(Reprinted with Permission from Nguyen, T.X. and Bhatia, S.K. Determination of Pore Accessibility in Disordered Nanoporous Materials.* J. Phys. Chem C. ***111****, 2212–2222. Copyright 2007 American Chemical Society)*

where $N_u^{\{i\}}$ and $N_s^{\{i\}}$ are the number of adsorbate molecules in the ith cluster in the periodic unit cell and in the periodic supercell respectively. Thus, a cluster is associated with a closed pore if N_f is equal to zero, and otherwise, to an open pore, for which N_f varies between unity and $(2^d - 1)$. Here d is the number of dimensions in which periodic boundary conditions are imposed. Thus, for a 3-dimensional system, N_f can be as high as 7.

Nguyen and Bhatia [11] applied the above procedure to determine the pore connectivity of the saccharose char using nitrogen as the probing molecule. At 77 K it was found that there exists one closed pore space that accommodates up to 2 nitrogen molecules. On the other hand at 300 K no closed pore space was found. This clearly indicates the presence of a kinetically closed space, rather than a physically closed space at 77 K. This was confirmed by molecular dynamics simulations, which showed that nitrogen could enter and leave this space at 300 K but not at 77 K.

2.4.2. Transition State Theory

Transition state theory (TST) offers a convenient route to identifying kinetically closed pores in the atomistic structure of a carbon [11]. According to TST, the mean crossing time between cages A and B is given as

$$\tau_{A\rightarrow B} = \frac{1}{k_{A\rightarrow B}} \tag{2.9}$$

where $k_{A\rightarrow B}$ is a rate constant, given as

$$k_{A\rightarrow B} = \kappa\sqrt{\frac{k_B T}{2\pi m}}\,\frac{\int\limits_{DS} e^{-\beta\phi_{sf}(\mathbf{r})}d^2\mathbf{r}}{\int\limits_{V_{\text{cage A}}} e^{-\beta\phi_{sf}(\mathbf{r})}d^3\mathbf{r}} \tag{2.10}$$

where the integral in the numerator is taken over the dividing surface of maximum potential energy, between cages A and B. Here κ is a transmission coefficient shown to be nearly unity, k_B is the Boltzmann constant, T is the temperature, and m is the mass of the particle. ϕ_{sf} is the interaction potential between the adsorbate particle i at position $\mathbf{r}$ and all solid atoms of the adsorbent phase, given as

$$\phi_{sf}(\mathbf{r}) = \sum_{j=1} u(|\mathbf{r}_j - \mathbf{r}|) \tag{2.11}$$

where u is the solid–fluid pair potential, taken to be the 12–6 Lennard–Jones potential. Details on evaluation of the integrals in Eqn (2.10) are available elsewhere [11]. As an example of the use of this method, Fig. 2.12a [9] depicts the temperature dependence of the crossing time of Ar and N_2 between cages A and B, depicted in Fig. 2.12b in the atomistic model of saccharose char.

Figure 2.12a shows that the estimated crossing time, or adsorption time, of N_2 at 77 K from cage A to B is extremely large (about 5 h to ~3.7 million years), for both choices of LJ parameters for N_2 [22,44], while that of argon at 87 K is of the order of minutes. This suggests an accessibility problem for N_2 at 77 K, but not for Ar at 87 K. This is supported by the results shown in Fig. 2.4, where the experimental adsorption data for CO_2 in BPL carbon at temperatures of 273 K and 323 K (triangles and squares respectively) are correctly predicted by the use of structural parameters derived from experimental Ar adsorption data at 87 K (solid and dashed lines), but significantly underpredicted by the use of parameters derived from experimental N_2 adsorption data at 77 K (dotted and dashed dotted) lines. Thus, it is clear from both experiment and theory that Ar at 87 K is a better adsorptive for characterization than N_2 at 77 K, and suffers lesser problems of accessibility.

The above technique has been used by Nguyen and Bhatia [12] to determine the crossing times for CO_2 and CH_4 between cages A and B of the saccharose char, illustrated in Fig. 2.12b. Table 2.1 depicts these results, showing the difference between the crossing times at various temperatures.

It is clear from the results in Table 2.1 that the crossing time of CO_2 is very rapid, and its adsorption/desorption equilibrium is very rapid. On the other hand, the crossing time of CH_4 is very large, particularly at low temperatures, which may be expected to lead to an apparent inaccessibility in practical time scales. This suggests excellent molecular sieving ability for CO_2/CH_4 mixtures in carbons having pore mouths of sufficiently narrow size, which has indeed been experimentally shown [45].

2.5. OPEN LOOP HYSTERESIS

For long, hysteresis in porous materials has been of much interest to adsorption scientists. In general, experimentally observed hysteresis phenomena can be grouped into two classes. The first class of hysteresis, or closed-loop hysteresis,

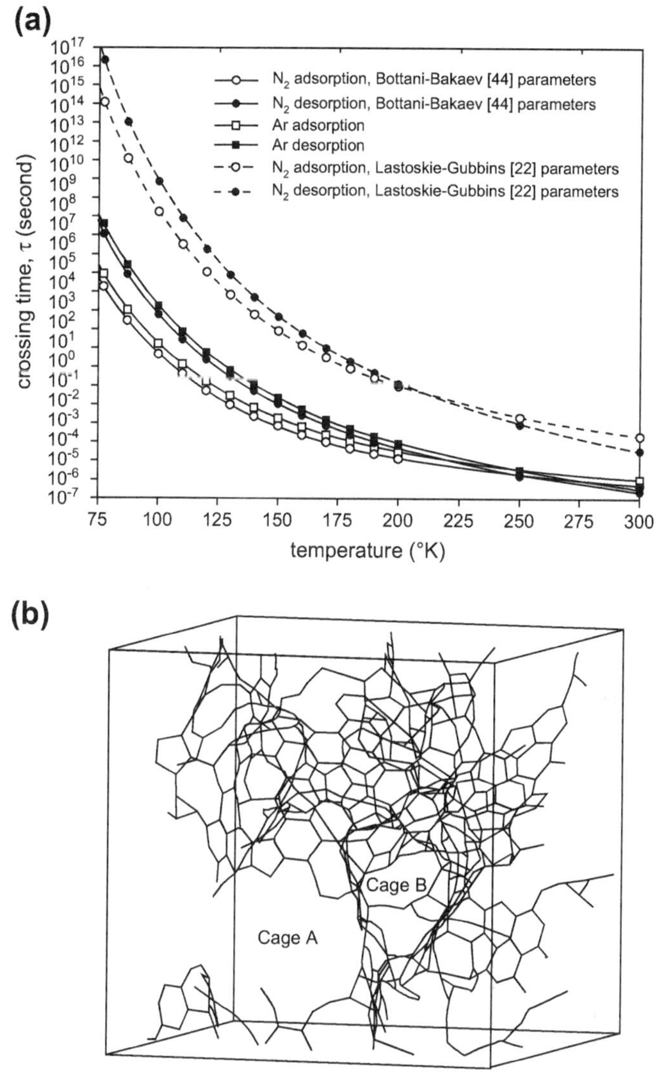

FIGURE 2.12 (a) Temperature variation of the estimated crossing time of a single nitrogen molecule (circles) and argon (squares) between cages A and B and (b) illustration of cages A and B. *(Reprinted with Permission from Nguyen, T.X. and Bhatia, S.K. Determination of Pore Accessibility in Disordered Nanoporous Materials.* J. Phys. Chem C. ***111****, 2212–2222. Copyright 2007 American Chemical Society)*

involves adsorbate condensation in mesoporous materials [46], while the other class of hysteresis, or open-loop hysteresis, is normally observed in ultramicroporous carbons such as coals and molecular sieve carbons. In particular, closed-loop hysteresis shows irreversible adsorption only in the vicinity of the condensation region, but is reversible outside this region. Such irreversibility is

TABLE 2.1 Calculated Results of the Crossing Time of CO_2 and CH_4 between Cages A and B, for the HRMC Based Structural Model of Saccharose Char, Using the TST Approach

CO_2			CH_4		
Temp. (K)	$\tau_{ads(A \to B)}$ (sec)	$\tau_{des(B \to A)}$ (sec)	Temp. (K)	$\tau_{ads(A \to B)}$ (sec)	$\tau_{des(B \to A)}$ (sec)
273	5.9209×10^{-5}	1.7996×10^{-4}	200	4.4559×10^{7}	5.3373×10^{7}
283	3.7958×10^{-5}	9.5426×10^{-5}	253	8.2091×10^{3}	3.8807×10^{3}
293	2.529×10^{-5}	5.2853×10^{-5}	263	2.4836×10^{3}	9.9048×10^{2}
303	1.7468×10^{-5}	3.0498×10^{-5}	273	8.2799×10^{2}	2.7938×10^{2}
313	1.2408×10^{-5}	1.8187×10^{-5}	283	3.0135×10^{2}	8.6321×10^{1}
323	9.0849×10^{-6}	1.1234×10^{-5}	323	1.0488×10^{1}	1.6270×10^{0}

Reprinted with Permission from Nguyen, T.X. and Bhatia, S.K. Kinetic Restriction of Simple Gases in Porous Carbons: Transition State Theory Study. *Langmuir*, **24**, 146–154. Copyright 2008 American Chemical Society.

related to the high activation energy of adsorbate droplet growth [46] resulting from the confinement. In contrast, open-loop hysteresis shows the occurrence of irreversible adsorption down to the lowest experimental pressure. It can be seen in Fig. 2.12a that the crossing time of N_2 and Ar at low temperatures ($77\ K < T < 100\ K$) from cage A to cage B, or adsorption time ($4\ s < \tau_{ads} < 5\ h$), is within the experimental adsorption time scale, while their crossing time from cage B to cage A, or desorption time ($588\ s < \tau_{des} < 134$ days), is far beyond the experimental time scale. This suggests that the open-loop hysteresis is related to the high activation energy arising from constriction of the pore mouth. As another example, it is evident from Table 2.1 that the crossing time of carbon dioxide at 273 K between cages A and B is very short, leading its instantaneous equilibrium under practical adsorption experimental conditions, i.e. no open-loop hysteresis is expected to be observed. This prediction is consistent with experimental observation of adsorption hysteresis of N_2 at 77 K in Takeda 3 Å, but not for CO_2 at 273 K in the same carbon, seen in Fig. 2.3a and b. On the other hand, open-loop hysteresis has been recently reported by Bae et al. [30] for both CH_4 and CO_2 at 313 K in an Australian coal, suggesting the presence of ultrafine microporous constrictions that act as molecular sieves for even CO_2, restricting its desorption from the pore bodies.

While the above discussion of irreversibility of the adsorption at low pressure relates to adsorption/desorption processes in microporous carbons, the apparent trapping of molecules on heterogeneous surfaces due to slower desorption has been predicted by Seri-Levi and Avnir [47] based on dynamic Monte Carlo simulations. Thus, even micropores in the form of surface irregularities and heterogeneities can lead to the open-loop hysteresis effect. Their simulations demonstrated the quasi-equilibrium nature of the adsorption isotherm, which was found to lead to hysteresis according to the surface geometry. A dynamic population of permanently trapped molecules was demonstrated.

Bhatia et al. [48] have demonstrated the irreversible adsorption of iodine with open-loop hysteresis in a microporous carbon. They showed that the reversibly adsorbed iodine followed a modified Langmuirian isotherm

$$C_a(P^*) = C_t(P^*) - C_o = \frac{K^* P^* C_{as}}{[1 + (K^* - 1)P^*]} \tag{2.12}$$

at all temperatures, suggesting this reversible iodine to be adsorbed in mesopores. Here $C_t(P^*)$ is the total amount adsorbed at relative pressure P^*, on the desorption or re-adsorption branch, $C_a(P^*)$ the amount of reversibly adsorbed iodine, and C_o the amount of trapped iodine $[=C_t(0)]$. Further C_{as} is the amount of reversibly adsorbed iodine at $P^* = 1$. The kinetics of the adsorption–desorption of this reversibly adsorbed part was also found to be Langmuirian, with the desorption rate constant being larger at a bulk relative pressure of 1.0 compared to that at $P^* = 0.6$. This suggests that the larger amount of iodine adsorbed at the higher pressure leads to a "crowding out" effect, due to the reduced degrees of freedom and the lower entropy under these conditions. Such an effect is also apparent in the desorption crossing times of CH_4 in Table 2.1, which are smaller than the adsorption time at temperatures above 283 K.

2.6. CONCLUSIONS

The understanding of the accessibility of fluids in nanoporous materials and carbons is critical for meaningful characterization by low-temperature adsorption, as well as for the behavior of adsorption processes – particularly for molecular sieving, gas and electrochemical storage, and gas separation. It is found from experimental and theoretical considerations that nitrogen at 77 K has poor accessibility in micropores, and argon adsorption at 87 K provides more reliable characterization. This has been confirmed by comparison of experimental data and predictions of the adsorption of other gases such as CO_2 and CH_4 in the carbon based on pore-size distributions obtained by density functional theory interpretation of argon and nitrogen isotherms. For large molecules the accessibility has been found to be predominantly a function of the molecular size, and is relatively independent of temperature, and can be predicted using percolation theory concepts. On the other hand for small molecules, the presence of a large activation barrier at the pore mouth plays a critical role. Simulation as well as transition state theory methods can be used to determine the existence of physically or kinetically closed pores in an atomistic carbon structure. Open loop hysteresis occurs when desorption time in the pores is larger than the adsorption time, leading to an apparent permanent trapping of adsorbed molecules.

REFERENCES

[1] R.T. Yang, Gas Separation by Adsorption Processes. Butterworth (1986).

[2] B.E. Conway, Electrochemical Supercapacitors: Scientific Fundamentals and Technological Applications, Springer, 1999.

[3] L. Radovic, in: H. Marsh, E.A. Heintz, F. Rodriguez-Reinoso (Eds.), Introduction to Carbon Technologies, University of Alicante, 1997.

[4] S.K. Bhatia, A.L. Myers, Optimum conditions for adsorptive storage, Langmuir 22 (2006) 1688–1700.

[5] S.J. Greg, K.S. Sing, Adsorption, Surface Area and Porosity, Academic Press, 1982.

[6] T.J. Bandosz, M.J. Biggs, K.E. Gubbins, Y. Hattori, T. Liyama, K. Kaneko, J. Pikunic, K.T. Thomson, Molecular models of porous carbons, Chem. Phys. Carbon 28 (2003) 41–228.

[7] J. Pikunic, C. Clinard, N. Cohaout, K.E. Gubbins, J.-M. Guet, R.J.-M. Pellenq, I. Rannou, J.-N. Rouzaud, Structural modeling of porous carbons: constrained reverse Monte Carlo method, Langmuir 19 (2003) 8565–8582.

[8] T.X. Nguyen, S.K. Bhatia, S.K. Jain, K.E. Gubbins, Structure of saccharose-based carbon and transport of confined fluids: hybrid reverse Monte Carlo reconstruction and simulation studies, Mol. Simul. 32 (2006) 567–577.

[9] T.X. Nguyen, N. Cohaut, J.-S. Bae, S.K. Bhatia, New method for atomistic modeling of the microstructure of activated carbons using hybrid reverse Monte Carlo simulation, Langmuir 24 (2008) 7912–7922.

[10] R.E. Franklin, The structure of graphitic carbons, Acta Crystallogr. 4 (1951) 253–261.

[11] T.X. Nguyen, S.K. Bhatia, Determination of pore accessibility in disordered nanoporous materials, J. Phys. Chem. C 111 (2007) 2212–2222.

[12] T.X. Nguyen, S.K. Bhatia, Kinetic restriction of simple gases in porous carbons: transition state theory study, Langmuir 24 (2008) 146–154.

[13] F.A.P. Maggs, Reversal of temperature dependence for physical adsorption of nitrogen, Research correspondence 6 (1953) 13S–14S.

[14] F. Rodriguez-Reinoso, J. de D. Lopez-Gonzalez, C. Berenguer, Activated carbons from almond shells – 1. Preparation and characterization by nitrogen adsorption, Carbon 20 (1982) 513–518.

[15] R.V.R. Rios, J. Silvestre-Albero, A. Sepulveda-Escribano, M. Molina-Sabio, F. Rodriguez-Reinoso, Kinetic restrictions in the characterization of narrow microporosity in carbon materials, J. Phys. Chem. C 111 (2007) 3803–3805.

[16] J.-S. Bae, S.K. Bhatia, High pressure adsorption of methane and carbon dioxide on coal, Energy Fuels 20 (2006) 2599–2607.

[17] M.V. Lopez-Ramon, J. Jagiello, T.J. Bandosz, N.A. Seaton, Determination of pore size distribution and network connectivity in microporous solids by adsorption measurements and Monte Carlo simulation, Langmuir 13 (1997) 4435–4445.

[18] S. Ismadji, S.K. Bhatia, Investigation of network connectivity in activated carbons by liquid phase adsorption, Langmuir 16 (2000) 9303–9313.

[19] B. Feng, S.K. Bhatia, Variation of the pore structure of coal chars during gasification, Carbon 41 (2003) 507–523.

[20] K.N. Tran, S.K. Bhatia, A. Tomsett, Air Reactivity of petroleum cokes: role of inaccessible porosity, Ind. Eng. Chem. Res. 46 (2007) 3265–3274.

[21] N.A. Seaton, J.P.R.B. Walton, N. Quirke, A New analysis method for the determination of the pore-size distribution of porous carbons from nitrogen adsorption measurements, Carbon 27 (1989) 853–861.

[22] C. Lastoskie, K.E. Gubbins, N. Quirke, Pore-size distribution analysis of microporous carbons – a density-functional theory approach, J. Phys. Chem. B 97 (1993) 4786–4796.
[23] C. Scherdel, G. Reicheauer, Microstructure and morphology of porous carbons derived from sucrose, Carbon 47 (2009) 1102–1111.
[24] J.A. Lecky, W.K. Hall, R.B. Anderson, Adsorption of water and methanol on coal, Nature 168 (1951) 124–125.
[25] P. Zwietering, A.P. Oele, D.W. van Krevelen, Pore structure and internal surface of coal, Fuel 30 (1951) 203–204.
[26] F.A.P. Maggs, Accessible surface areas of coals, Nature 169 (1952) 269.
[27] I.G.C. Dryden, Accessible surface areas of coals, Nature 169 (1952) 269–270.
[28] J. Kärger, D.M. Ruthven, Diffusion in Zeolites and Microporous Solids, Wiley, New York, 1992.
[29] B. Kawahata, P.L. Walker, Mode of porosity development in activated anthracite. Proc. 5th Carbon Conference, 2 (1962) 251–263.
[30] J.-S. Bae, S.K. Bhatia, V. Rudolph, P. Massarotto, Pore accessibility of methane and carbon dioxide in coals, Energy Fuels 23 (2009) 3319–3327.
[31] H. Marsh, W.F.K. Wynne-Jones, The surface properties of carbon-1. The effect of activated diffusion in the measurement of surface area, Carbon 1 (1964) 269–279.
[32] J. Garrido, A. Linares-Solano, J.M. Martin-Martinez, M. Molina-Sabio, F. Rodriguez-Reinoso, R. Torregrosa, Use of N_2 vs. CO_2 in the characterization of activated carbons, Langmuir 3 (1987) 76–81.
[33] D. Lozano-Castello, D. Cazorla-Amoros, A. Linares-Solano, Microporous solid characterization: use of classical and "new" techniques, Chem. Eng. Technol. 26 (2003) 852–853.
[34] M.R. Nartkiewicz, J.P. Mathews, Visual representation of carbon dioxide adsorption in a low-volatile bituminous coal molecular model, Energy Fuels 23 (2009) 5236–5246.
[35] G.M. Davies, N.A. Seaton, Development and validation of pore structure models for adsorption in activated carbons, Langmuir 15 (1999) 6263–6276.
[36] S. Himeno, T. Komatsu, S. Fujita, High-Pressure adsorption equilibria of methane and carbon dioxide on several activated carbons, J. Chem. Eng. Data 50 (2005) 369–376.
[37] J.A. Ritter, R.T. Yang, Equilibrium adsorption of multicomponent gas-mixtures at elevated pressures, Ind. Eng. Chem. Res. 26 (1987) 1679–1686.
[38] T.X. Nguyen, S.K. Bhatia, D. Nicholson, Prediction of high-pressure adsorption equilibrium of supercritical gases using density functional theory, Langmuir 21 (2005) 3187–3197.
[39] A. Herbst, P. Harting, Thermodynamic description of excess isotherms in high-pressure adsorption of methane, argon and nitrogen, Adsorption 8 (2002) 111–123.
[40] S. Ismadji, S.K. Bhatia, Characterization of activated carbons by liquid phase adsorption, Carbon 39 (2001) 1237–1250.
[41] S. Ismadji, S.K. Bhatia, Effect of pore network connectivity on multicomponent adsorption of large molecules, AIChE J. 49 (2003) 65–81.
[42] A.L. Myers, J.M. Prausnitz, Thermodynamics of mixed-gas adsorption, AIChE J. 11 (1965) 121–127.
[43] T.H. Cormen, C.E. Leierson, R.L. Rivest, C. Stein, Introduction to Algorithms, second ed., MIT Press, Cambridge, 2001.
[44] E.J. Bottani, V.A. Bakaev, The grand-canonical ensemble Monte-Carlo simulation of nitrogen on graphite, Langmuir 10 (1994) 1550–1555.
[45] D. Lozano-Castello, J. Alcaniz-Monge, D. Cazorla-Amorós, A. Linares-Solano, W. Zhu, F. Kapteijn, J.A. Moulijn, Adsorption properties of carbon molecular sieves prepared from an activated carbon by pitch pyrolysis, Carbon 43 (2005) 1643–1651.

[46] R. Valiullin, S. Naumov, P. Galvosas, J. Kärger, H.-J. Woo, F. Porcheron, P.A. Monson, Exploration of molecular dynamics during transient sorption of fluids in mesoporous materials, Nature 443 (2006) 965–968.

[47] A. Seri-Levy, D. Avnir, Effects of heterogeneous surface geometry on adsorption, Langmuir 9 (1993) 3067–3076.

[48] S.K. Bhatia, F. Liu, G. Arvind, Effect of pore blockage on adsorption isotherms and dynamics: anomalous adsorption of iodine on activated carbon, Langmuir 16 (2000) 4001–4008.

Chapter 3

Virtual Porous Carbons

Artur P. Terzyk, Sylwester Furmaniak and Piotr A. Gauden
N. Copernicus University, Department of Chemistry, Physicochemistry of Carbon Materials Research Group, Toruń, Poland

Peter J.F. Harris
Centre for Advanced Microscopy, University of Reading, Whiteknights, UK

Piotr Kowalczyk
Nanochemistry Research Institute, Department of Chemistry, Curtin University of Technology, Western Australia, Australia

Chapter Outline

3.1. Introduction **61**
3.2. VPC Models **63**
3.2.1. Virtual Oxidation 64
3.2.2. VPC Absolute (Geometric) Porosity, and the Modification of the Bhattacharya and Gubbins Method 66
3.3. Simulation Details **69**
3.3.1. Grand Canonical Monte Carlo 70
3.3.2. Molecular Dynamics 73
3.4. Results **74**
3.4.1. The Reality of Studied VPC Models 74
3.4.2. Examples of VPC Application 76
3.5. Summary and Perspectives **95**
References **96**

3.1. INTRODUCTION

Due to progress in the development of computers, molecular simulation techniques have become the basic tool of current theoretical chemistry. That is why using a computer, one can try to build a model of carbon and then perform molecular simulations to study different phenomena, among them adsorption processes [1–9].

We know exactly the atomistic models of a single-carbon slit-like pore, carbon nanotube, etc. However, in the case of activated carbons, existing models

Novel Carbon Adsorbents. DOI: 10.1016/B978-0-08-097744-7.00003-X
Copyright © 2012 Elsevier Ltd. All rights reserved.

are still far from reality. Therefore, we have tried to build a realistic model mimicking the properties of real carbon adsorbents. In this case, a very important role has been played by the so-called virtual porous carbons (VPCs).

As stated by Biggs and Buts [10] "*we use the term Virtual Porous Carbon to describe computer-based molecular models of nanoporous carbons that go beyond the ubiquitous slit pore model and seek to engage with the geometric, topological and chemical heterogeneity that characterizes almost every form of nanoporous carbon*". Of course, VPC models should mimic the properties of a real carbon as far as possible. That is why different research groups try to use the results of experimental observations obtained for example from the application of diffraction methods, adsorption calorimetry, etc. [5–49]. In this way, using the reverse Monte Carlo (RMC) simulation method (or its more sophisticated version – where also the energy is minimized during fitting of theoretical RDFs to experimental ones; i.e. the hybrid RMC – HRMC), one can, for example, obtain a more or less realistic replica of a studied carbon sample (see for example [22–37]). This method leads to a realistic carbon model. However, this model is not "*flexible*" i.e. it is hardly able to change, for example, porosity of the model in a systematic way. Moreover, the obtained replica is not unique.

In contrast, using the realistic VPC model one can easily change the porosity as well as the chemical composition of the carbon surface layer. Simultaneously, one knows exactly the chemical composition of carbon as well as its pore-size distribution (PSD) curve. Therefore, VPC from the very beginning provides information unknown for each real porous carbon. For real carbons, we still have problems with determination of chemical composition of surface and/or porosity. More precisely, some experimental methods are available, but the application of those methods needs the application of models (for example we must assume the pore geometry) and/or often experimentally available methods do not provide satisfactory qualitative and quantitative information. For example, the XPS technique applied for the determination of the chemical composition of the carbon surface layer leads only to a statement about the type of groups, but does not provide the amount of them (moreover only the surface layer is detected and the method does not penetrate the internal space of the carbon material [50]).

Summing up, using the VPC we know exactly what type of surface groups are present and the number and location of all surface functionalities. Moreover, we can introduce these groups in an ordered or random way. We can also choose the place of location (for example attach them to the edge atoms or to the atoms inside a carbon fragment).

As mentioned above, we also know exactly the PSD of the carbon. The question arises as to how to perform VPC chemical modification, and how to determine the absolute VPC porosity. This is described below for two VPC models, the model of "hard" carbon proposed by Harris et al. [9,40–46,48,49,51–59] and the simple model of "soft" carbon proposed recently in our group [47,60].

3.2. VPC MODELS

Several years ago, a model of nongraphitizing carbon which incorporated fullerene-related elements was proposed by Harris et al. [51–59]. This model was based on detailed studies of typical nongraphitizing microporous carbons using high-resolution transmission electron microscopy (HRTEM). These studies showed that carbons heated to high temperatures (up to 2800 °C) often contained fullerene-related structures including closed nanoparticles and saddle points. On the basis of these findings, and of observations of the "fresh" carbons, a new model was put forward in which the structure consists of discrete fragments of curved carbon sheets in which pentagons and heptagons are dispersed randomly throughout networks of hexagons. An illustration of this model is given in Fig. 3.1(a) where we also show some of the individual fragments and the complete (3-D) structure. This fullerene-like model explains many properties of so-called "hard" synthetic activated carbons, especially their porosity and resistance to graphitization. The presence of fullerene-like structures has been supported by both experimental [34,51–59,63–65] and theoretical [34,63,64,66,67] studies. In addition, the reality of this model was recently confirmed by quenched MD simulations of different carbon structures by Powles et al. [20].

The original model proposed by Harris based on HRTEM analysis can be modified. To do this, we can simply cut some fragments from the internal part of the initial box (such a procedure was applied in [41]). In this way, models

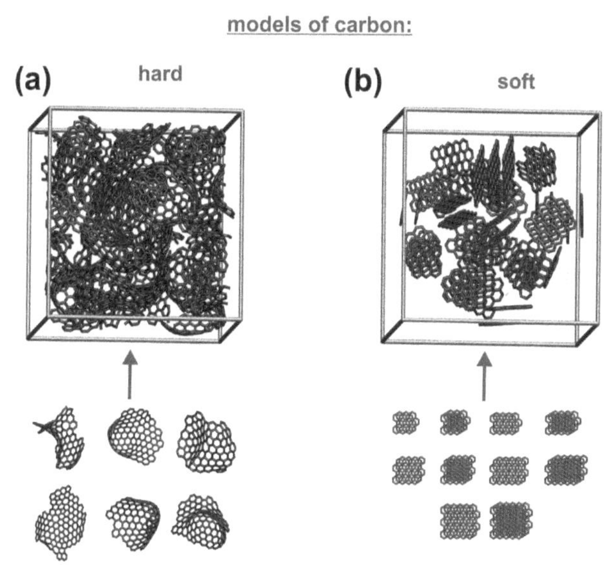

FIGURE 3.1 Graphical representation of two studied carbon models with the individual fullerene-like fragments and the basis of microcrystallites. Note that this figure and others (i.e. Figs 3.1–3.4, and 3.6) were created using the VMD program [61,62].

differing in porosity are obtained. However, if we wish to study the effect of PSD on adsorption, we must use a series of VPCs having more systematic and controlled changes in porosity. They can be obtained if one applies the second (proposed by us) procedure of VPC construction. Using the initial microporous structure (for example the one named S0 - see [40]) and the Monte Carlo program, a fragment containing a small number of carbon atoms (for example around 50) can be placed randomly in the arbitrarily chosen starting structure (for example S0 – 2704 carbon atoms). To do this, first, the position of the center of mass of the fragment is sampled; next, the code samples the angular orientation. The fragment is accepted in the structure if an overlap does not occur, i.e. if the distance between pairs of carbon atoms in adjacent fragments is larger than 0.34 nm. The configurations of each structure are recorded by the program after the addition of each fragment. The next carbon structures can be obtained using the same method but with a fragment approximately half the size.

To study the properties of graphitizable (i.e. soft) carbon, we proposed a simple model (shown in Fig. 3.1(b)) of activated carbon composed of a system of slit-like pores. It is widely accepted that this type of activated carbon contains microcrystallites [6,7,9,68–70]. Therefore, the proposed procedure inserts randomly double- and triple-layered graphite microcrystallites into a simulation box in such a way that a microporous carbon model is created. The carbon–carbon bond length was assumed as equal to 0.142 nm and the interlayer spacing between layers was fixed as equal to 0.335 nm (since we try to prepare the model of "soft" carbon, both values are the same as observed for graphite). Also like in graphite, the centers of benzene rings in each layer are shifted by 1/3 (we assumed ABC stacking of graphene layers). Figure 3.1(b) shows the applied basis of microcrystallites. The program generating the carbon structure randomly uses one of 10 microcrystallites and introduces it inside the simulation box at a random position and with random angular orientation. This insertion is accepted if there are no overlaps between fragments (see above). To perform molecular simulations, usually one arbitrarily chosen structure (created using this method) is used. For example, the one applied previously for the simulation of the immersion process had a radius equal to 2.3 nm (this means that all 1844 carbon atoms forming this structure are closed in a sphere having this radius) and contained 24 microcrystallites. In this case, as well as in the case of the "hard" carbon model, one can easily expand the structure to obtain a VPC series with systematic changes in porosity. To do this, the structures with systematically shifted fragments can be prepared from the initial models. These VPCs were obtained by proportional displacement of the centers of microcrystallites (for example by factors of 1.125, 1.25, and 1.5 [47,60]), forming the initial structure – Fig. 3.2.

3.2.1. Virtual Oxidation

To introduce surface carbon functionalities, we proposed a so-called "virtual oxidation" procedure [9,43–45,47–49,60,71,72]. This procedure is a useful tool

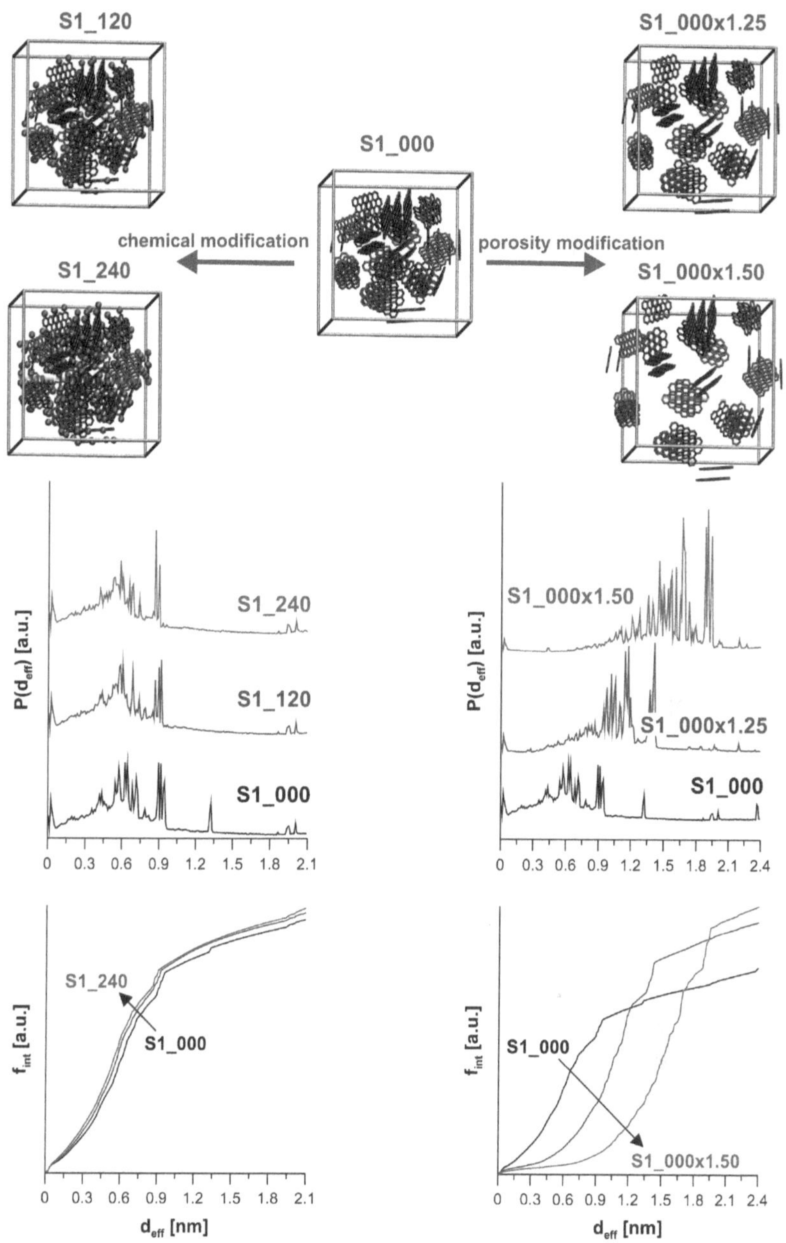

FIGURE 3.2 Schematic representation of two procedures: virtual oxidation and the method leading to systematic changes in VPC porosity. Below we present typical BG PSD curves.

to study the effects caused by gradual changes in the composition of carbon surface layer. Two important assumptions were made by us while this method was developed. First, we assumed that after modification of the VPC, surface the carbon skeleton remains unchanged, i.e. the modification procedure does not remove carbon from the skeleton (this can be easily achieved in a real experiment if the modification conditions are not drastic). The second assumption is that carbon oxygen functionalities are attached to the most reactive atoms located on the edges of the structure. This is in fact observed in many experiments, as was mentioned by Marsh and Rodriguez-Reinoso [7] and by Zawadzki et al. [73–75]. This location of surface groups also makes it possible to explain the experimental changes in adsorption isotherms determined for adsorption of nonelectrolytes from aqueous solutions [76–79].

Since the simplest groups to model are carbonyls, we started our study using this type of functionalities. The procedure works as follows. First we choose, as the starting structure imitating the real carbon, one of the VPC structures. To generate the oxidized VPC models, a program was developed. This program starts by detecting carbon atoms having a valence equal to two, i.e. being the most reactive and acting as potential centers for bonding oxygen atoms. The crucial part of virtual oxidation i.e. attachment of oxygen is performed in such a way that the oxygen atom and the carbon atom attaching this oxygen (and two nearest neighbors of this carbon atom) lie in the same plane, and the axis of the carbon–oxygen bond is determined by the secant of the C–C–C angle (to the middle C atom the oxygen atom is attached). This oxygen atom is placed in such a way that the length of the carbon–oxygen bond is equal to 0.1233 nm [80] and that there are no overlaps between this atom and the remaining fragments of the structure. In this way, we attach the maximum number of surface carbonyl groups which may be introduced in the studied structure (for example in our first study of this type it was around 180; see Table 3.1 in [43]). Performing this virtual oxidation one should remember that the number of attached oxygen atoms should lead to realistic (i.e. as observed in experiment) mass content of oxygen. From oxidized structures obtained in such a way, subsequent structures were generated by the random removal of an assumed percentage (for example 20%) of the carbonyls from the starting model (see Fig. 3.2). In this way, a number of different structures can be obtained, starting from one almost totally saturated with carbonyls and finishing on the initial structure without oxygen. In this way, by performing molecular simulations on a series of structures, one can study the influence of systematic changes in carbon surface composition.

3.2.2. VPC Absolute (Geometric) Porosity, and the Modification of the Bhattacharya and Gubbins Method

In the case of crystalline adsorbents, such as zeolites, the diameters of pores (cages, windows etc.) are known exactly. In the case of real activated carbons that are amorphous solids, the methods of PSD function determination are still not perfect. In this case, the PSD function is usually obtained from the solution of the so-called global adsorption isotherm (GAI) equation using the density functional theory (DFT) (and/or grand canonical Monte Carlo (GCMC)) local isotherms simulated for assumed pore geometry (usually perfect infinite slits) [5–7,81–87]. In this way, we assume the ideal shape of pores neglecting pore connectivity, the presence of heteroatoms or defects in structure etc.

In the case of VPC models, the "geometric" pore-size distribution can be calculated and this is the major advantage of virtual carbons. The method of VPC porosity determination was published few years ago. In their pioneering work, Bhattacharya and Gubbins [88] proposed a new method of PSD function determination. This method was tested for solids with known pore diameters (model structures), for example SBA-15 and MCF and the authors observed good agreement between simulation and experiment.

In the method proposed by Bhattacharya and Gubbins (called the BG method [88]) to determine the absolute (geometric) PSD curve, a uniform grid of points is generated in the simulation box (in our studies $100 \times 100 \times 100$), and for each such point (located in a pore) the largest sphere containing this point (and situated in the pore) is found (Fig. 3.3). The diameter of this sphere is equal to the diameter of the pore, and the collection of histograms of the diameters of pores for each such a point of the grid makes it possible to plot the histogram of pore diameters (related to the PSD curve). The program, prepared in our group, works in an iterative way [40,41]. For each node of the grid located in the structure of carbon, the center of the testing sphere is placed randomly (at the start the coordinates of the node are assumed as the center of the sphere; the diameter of the sphere is limited by the distance to the nearest fragment of carbon structure). Each displacement of the center of the sphere (the sphere must contain the test point) leading to a rise in the diameter is accepted and the next attempts of displacement are made. The program stops calculations for a given point when after 1000 subsequent iterations a sphere with a larger diameter is not found. We assume that the radii of C and O are equal to 0.17 and 0.15 nm, respectively. This method leads to the PSD function of any structure placed in a simulation box. Note that we tested the reality of BG method results using the model of ideal and defected carbon slit-like pores [41].

The pioneering idea of the BG method can be implemented for the study of adsorption phenomena and can be used, after minor modification, for determination of the diameters of pores containing molecules during the adsorption process. We have developed this procedure [44,60] in such a way that additionally the PSD functions for adsorbed molecules can also be calculated. For example, if adsorption of phenol on VPC was studied, the nodes in the structures (used in the original BG method) were replaced by the centers of benzene rings of adsorbed phenol molecules in equilibrated simulation boxes (see Fig. 3.4). Cumulative PSD curves obtained in this way provide information

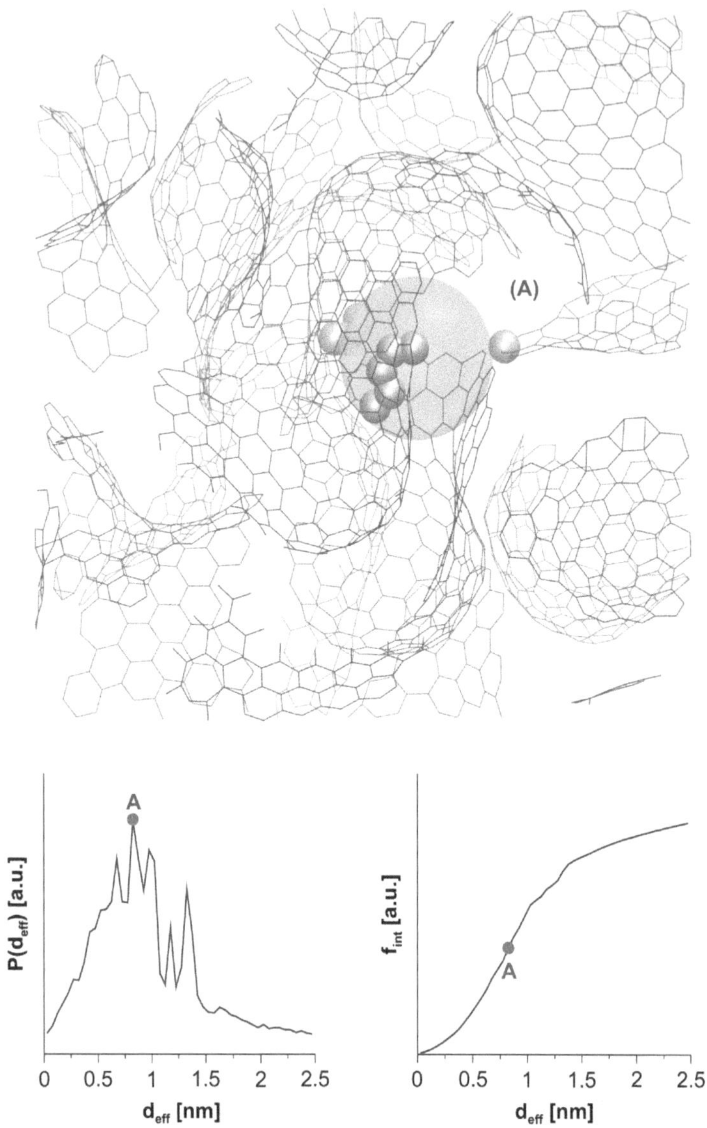

FIGURE 3.3 The graphical representation of original approach proposed by Bhattacharya and Gubbins [88]. We show the BG detection sphere, carbon skeleton atoms (small balls and lines), and the point obtained on differential and integral PSD curves.

about diameters of pores filled by adsorbed molecules, and allow one to conclude what is the influence of, for example, porosity as well as carbon surface oxidation on the distribution of adsorbed molecules among pores with different diameters.

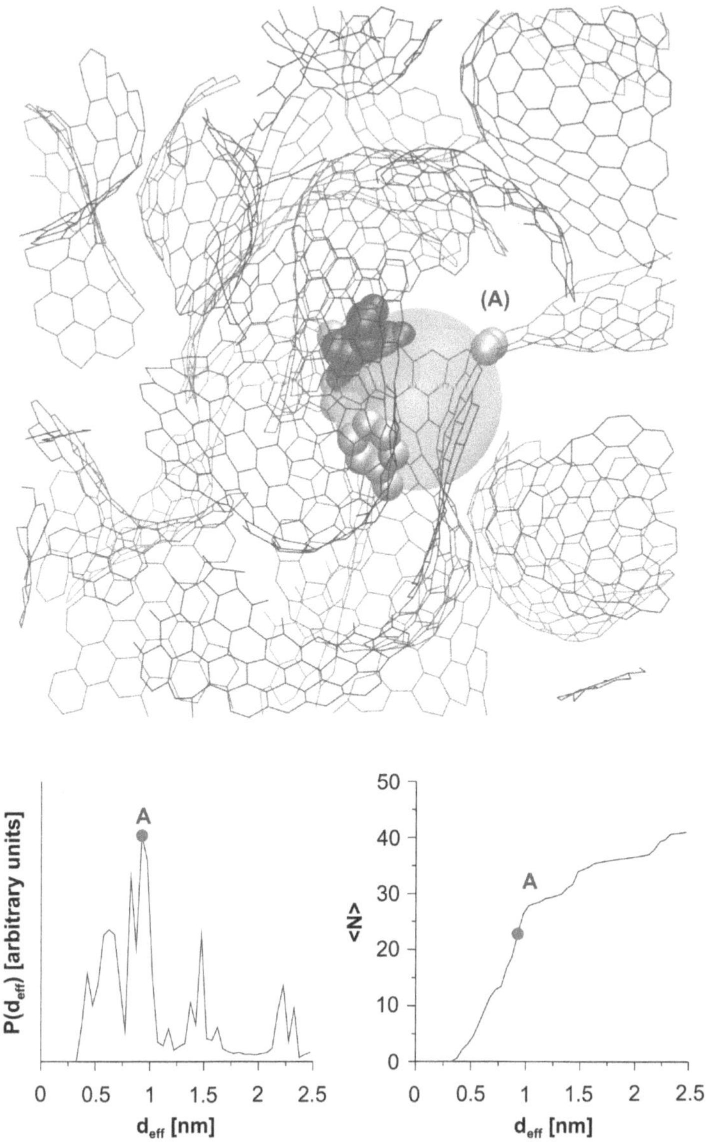

FIGURE 3.4 The application of the modified BG–phenol molecule (black balls) in a pore and the sphere detecting pore diameter at the adsorption center [44,60].

3.3. SIMULATION DETAILS

We use two fundamental methods of molecular simulations i.e. the grand canonical Monte Carlo and molecular dynamics (MD). The first method was applied for the simulation of adsorption from gaseous phase, while the second one to study adsorption from solutions.

3.3.1. Grand Canonical Monte Carlo

Adsorption isotherms for the assumed temperature were simulated using the classical GCMC method. Usually, a cubicoid simulation box with a solid model placed in a center having assumed dimensions was applied. Periodic boundary conditions were used in all three directions. For all systems for each adsorption point, 2.5×10^7 iterations were performed during the equilibration, and next 2.5×10^7 equilibrium ones, applied for the calculation of the averages (one iteration = an attempt to change the state of the system by creation, annihilation, displacement, or rotation (only for molecules)).

The probability of attempts of changing a state of a system by creation, annihilation, and rotation and displacement (the latter one is connected with the change in angular orientation) were equal to 1/3, 1/3, 1/6, and 1/6.

Noble gases were modeled as Lennard–Jones (LJ) centers. Other adsorbates were modeled as the complex of a different number of LJ centers and/or point charges. Carbonaceous structures and surface groups were also modeled as a combination of LJ centers and point charges. The geometry of adsorbate molecules and surface groups is characterized in Table 3.1. Tables 3.2 and 3.3 show the values of the parameters applied (we used the Lorentz–Berthelot mixing rules).

Since the majority of results discussed in this book were obtained using the GCMC technique, which is in fact a computer program, the values of the applied parameters (the so-called force field) are crucial. Below we discuss briefly the applied methodology.

Generally, the energy of interactions between a pair of atoms/molecules at a distance r (for the case of multiatomic molecules, this energy depends also on the angular position - ω_1 and ω_2) can be written as

$$U_{ff}(r, \omega_1, \omega_2) = \sum_{i=1}^{N_{LJ}} \sum_{j=1}^{N_{LJ}} U_{LJ}^{ij}(r_{ij}) + U_{\mathrm{electr}}(r, \omega_1, \omega_2) \tag{3.1}$$

where N_{LJ} denotes the number of LJ centers and $U_{\mathrm{LJ}}^{ij}(r_{ij})$ is the energy of dispersion interactions between a pair of centers i and j being placed at the distance r_{ij} and calculated from the truncated LJ potential:

$$U_{\mathrm{LJ}}^{ij}(r_{ij}) = \begin{cases} 4\varepsilon_{ij}\left[\left(\dfrac{\sigma_{ij}}{r_{ij}}\right)^{12} - \left(\dfrac{\sigma_{ij}}{r_{ij}}\right)^{6}\right] & r_{ij} < r_{\mathrm{cut},ij} \\ 0 & r_{ij} \geq r_{\mathrm{cut},ij} \end{cases} \tag{3.2}$$

where σ_{ij} and ε_{ij} are the collision diameter and the depth of the potential energy well for interactions between centers i and j. The second term in Eqn (3.1)

TABLE 3.1 Geometric Characterization of Adsorbate Molecules and Surface Groups Introduced in Structures

Molecule/ Group	Bond Type	Length of the Bond [nm]	Angle Type	Value of the Angle [deg]	References
Adsorbate Molecules					
N_2	N–N	0.1100		–	[89]
CO_2	C=O	0.1160		–	[89]
CCl_4	C–Cl	0.1766	Cl–C–Cl	109.5	[90]
C_6H_6	UA^a–UA^a	0.1806		–	[91]
CH_4	C–H	0.1090	H–C–H	109.5	[92]
SF_6	S–F	0.1564	F–S–F	90.0	[93]
Surface Groups					
Carbonyl	C=O	0.1233		–	[80]
Hydroxyl	C^b–O	0.1364	C^b–O–H	110.5	
	O–H	0.0960			
Carboxyl	C^b–C	0.1520	C^b–C–O	111.0	
	C=O	0.1214	O=C–O	123.0	
	C–O	0.1364	C–O–H	107.0	
	O–H	0.0970			

[a] *An united atom*
[b] *An atom of the carbon structure bounding the groups*

($U_{electr}(r, \omega_1, \omega_2)$) is the energy of electrostatic interactions between a pair of molecules which can be written as

$$U_{electr}(r) = \begin{cases} \dfrac{1}{4\pi\varepsilon_0} \displaystyle\sum_{i=1}^{N_{PC}} \sum_{j=1}^{N_{PC}} \dfrac{q_i q_j}{r_{ij}} & r < r_{cut,C} \\ 0 & r \geq r_{cut,C} \end{cases} \tag{3.3}$$

where N_{PC} is the number of point charges in the molecule, q_i and q_j denote the values of the charges of the centers, and ε_0 ($8.8543 \times 10^{-12}\,C^2\,J^{-1}\,m^{-1}$) is the dielectric permittivity of free space. We use the cutoffs for electrostatic interactions for the whole molecules [91] and if the centers of mass of two molecules are located at a distance smaller than $r_{cut,C}$, the sum of the energy of

TABLE 3.2 The Values of LJ Potential Parameters and Point Charges Applied in Simulations for Considered Adsorbates

	Number of						
Adsorbate	LJ centers (N_{LJ})	Point charges (N_{PC})	Center	σ [nm]	ε/k_B [K]	q/e	References
Ne	1	0	Ne	0.2780	34.900	–[a]	[94]
Ar	1	0	Ar	0.3400	120.000	–[a]	[95]
Kr	1	0	Kr	0.3685	164.410	–[a]	[94]
Xe	1	0	Xe	0.4055	229.000	–[a]	[94]
N_2	2	3	N	0.3310	36.000	−0.482	[89]
			COM[b]	–[c]	–[c]	+0.964	
CO_2	3	3	C	0.2800	27.000	+0.700	[89]
			O	0.3050	79.000	−0.350	
CCl_4	5	0	C	0.4600	39.000	–[a]	[90]
			Cl	0.3500	105.000	–[a]	
C_6H_6	6	0	UA[d]	0.3246	89.400	–[a]	[91]
CH_4	5	5	C	0.3400	55.055	−0.660	[92]
			H	0.2650	7.901	+0.165	
			C–H[e]	0.3025	30.600	–	
SF_6	7	7	S	0.3228	165.140	+0.660	[93]
			F	0.2947	27.020	−0.110	

[a]*LJ-type center without charge*
[b]*A center of mass*
[c]*A center treated only as a point charge and not as a LJ center*
[d]*An united atom*
[e]*LJ parameters between different centers in methane molecules*

interactions between all pairs of charges occurring in the molecules is calculated, otherwise the electrostatic interactions are neglected. The cutoff for electrostatic interactions ($r_{\mathrm{cut},C}$) was assumed at 1.5 nm. At this distance the energy of interactions of a pair of molecules is negligibly small (the energy of electrostatic interactions for the studied molecules (having quadrupole or higher moments) decreases proportionally to the fifth or higher power of the distance [96]).

TABLE 3.3 The Values of LJ Potential Parameters and Point Charges Applied in Simulations for Carbonaceous Structures and Surface Groups [80]

Group	Center	σ [nm]	ε/k_B [K]	q/e
C[a]	C[a]	0.340	28.0	–[b]
Carbonyl	C	0.340	28.0	+0.50
	O	0.296	105.8	−0.50
Hydroxyl	C[c]	0.340	28.0	+0.20
	O	0.307	78.2	−0.64
	H	–[d]	–[d]	+0.44
Carboxyl	C[c]	0.340	28.0	+0.08
	C	0.375	52.8	+0.55
	=O	0.296	105.7	−0.50
	O	0.300	85.6	−0.58
	H	–[d]	–[d]	+0.45

[a]*An atom of carbon structure nonbounded with hydroxyl and/or carboxyl*
[b]*LJ-type center without charge*
[c]*An atom of carbon structure bounding a group*
[d]*A center treated only as a point charge and not as a LJ center*

The interactions between the studied molecules and the structure of the adsorbent can be written as

$$U_{sf} = \sum_{i=1}^{N_{CS}} \left(\sum_{j=1}^{N_{LJ}} U_{LJ}^{ij}(r_{ij}) + \frac{1}{4\pi\varepsilon_0} \sum_{j=1}^{N_{PC}} \frac{q_i q_j}{r_{ij}} \right) \tag{3.4}$$

where N_{CS} is total number of atoms building carbonaceous structure and surface groups in simulation box. For all LJ-type interactions, the cutoff is assumed as equal to $r_{cut,ij} = 5 \times \sigma_{ij}$.

3.3.2. Molecular Dynamics

All simulations were performed using Gromacs molecular dynamics package [44,47,60,71,72,97–100] in the isobaric–isothermal ensemble with a Berendsen thermostat (assuming a temperature studied) and barostat. For arbitrarily chosen systems, we also perform simulations for different temperatures to check the temperature dependence of studied effects (for example pore-blocking mechanism [72]). Water was modeled using the TIP4P model and dispersion interactions were taken into account using the Lennard–Jones (LJ) potential with respective parameters for carbon [44,47,60,71,72]. Periodic boundary conditions were applied in all three directions. The method for calculating electrostatics and van der Waals interactions was based on cutoffs, and these cutoffs were located at 0.90 nm. Usually, the starting configuration of the cubicoid box containing the carbon model and water was prepared using the Genbox procedure, i.e. the standard procedure adopted in the Gromacs package for preparation of simulation boxes. After calculations the density of water far from the structure was always close to the density of bulk water. Usually, depending on a system simulations were performed for long times (around 10 000 ps). All equilibrium properties were calculated from the last 1500 ps.

3.4. RESULTS

3.4.1. The Reality of Studied VPC Models

Before discussing the advantages of VPC models, we will demonstrate the results showing that VPC models studied by us behave similarly to real systems. One should remember that a model of activated carbon recovering **all** physical and chemical properties of this material is still not available. The answer to the question: will we find a model imitating all carbon properties (?) is simple: no we will not. We can find a model recovering, for example, some adsorption properties, but we must remember that activated carbon is a really complicated amorphous solid. This amorphous nature of carbon is a major difficulty if we try to construct a model. But this is unfortunately not all. This amorphous solid contains different Lewis acidic and basic centers, heteroatoms (oxygen, nitrogen, chloride etc.) forming different groups (often with unknown structures), pores with different and unknown shapes connecting in an unknown way etc. Moreover, it is obvious that the acid–base properties of surface oxygen groups depend on the properties of the carbon center attaching this group. Recently, a very interesting study was [101] published showing how the length of a carbon nanotube, and the place where surface carboxylic group is attached, changes the acidity of this group. Summing up, we can propose a VPC model for application in the field of adsorption, but the same model used for the study of (for example) acid–base properties of carbon is probably completely unrealistic. The same can be stated about the models obtained from application of the most advanced HRMC methods.

Now we will show a few examples of realistic "adsorption" behavior of studied VPCs. Correlations referred below were obtained based on the results of GCMC simulations of adsorption on two VPC models [42]. We studied the adsorption of noble gases (Ar, Kr, Ne, and Xe), benzene, and carbon tetrachloride. Both the VPC models studied by us represent strictly microporous-activated carbons; however, one of them has dispersed microporosity (i.e. pores up to 2 nm) while the second has only small micropores (the BG distribution shows that pores having diameters larger than ca. 0.95 nm are not present [42]).

The first example of realistic VPC adsorption behavior is the confirmation of Chen and Yang's [102,103] results. These authors, while deriving the Dubinin–Astakhov (DA) equation, showed that the characteristic energy of adsorption ($E_0\beta$) is directly related to the mean potential in the pore. Therefore, accepting this derivation, one should expect a relationship between the isosteric enthalpy of adsorption at the zero coverage limit (this enthalpy, if we neglect the small contribution from translational energy, is in fact the potential energy of solid–fluid interaction) and the value of ($E_0\beta$). This situation is in fact observed in Fig. 3.5(a) for data obtained from GCMC simulations on a highly

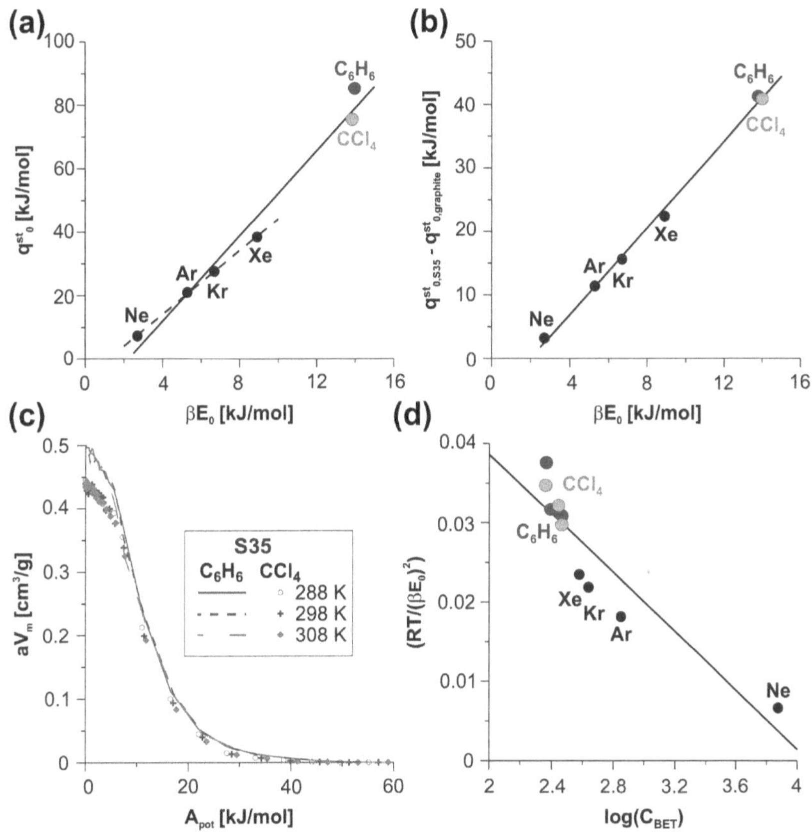

FIGURE 3.5 (a) The correlation between βE_0 and the isosteric enthalpy at zero coverage (q_0^{st}) for the data on S35 structure; CCl_4 and C_6H_6 at 298 K, dashed line – only noble gases; solid line – all adsorbates. (b) Recovering the correlation of Stoeckli and Morell from GCMC simulations on structure S35 and on graphite. (c) Adsorption isotherms of CCl_4 and C_6H_6 converted into "characteristic curves". (d) Recovering of the Tsunoda's correlation from simulation data on S35 structure.

microporous VPC structure. Our results confirm the validity of the postulate of Chen and Yang, showing that the parameter ($E_0\beta$) is proportional to the mean potential in the pores.

The next example of realistic VPC behavior is the recovery of Stoeckli and Morell's experimental correlation [104]. These authors using chromatographic data concluded, for a series of adsorbates, that there is a linear relationship between the difference in energy of adsorption in micropores (at zero coverage) and the same energy on graphite, and ($E_0\beta$) values. Therefore, they showed that ($E_0\beta$) represents an average excess of adsorption energy in the micropores, with respect to an open graphitic surface. To check the validity of this relation for VPCs, we plotted the values obtained from GCMC simulations on carbon models and on graphite (Fig. 3.5(b)). One can observe a very good recovery of this correlation observed on the experimental data of those authors. In this figure, we show that a more realistic plot (i.e. approaching (0,0)) is observed if only adsorption of noble gases is considered.

The next example is the way in which the VPC models predict the temperature dependence of the so-called "characteristic" curve of adsorption. It is well known from different experimental measurements that for strictly microporous carbons, and if nonpolar molecules are adsorbed, the "characteristic curve" is temperature independent. Figure 3.5(c) shows benzene and carbon tetrachloride adsorption isotherms simulated for three temperatures and converted into a "characteristic" curve. As one can observe, this curve is in fact temperature independent as is observed for experimental data on real microporous carbon samples.

Finally, we checked whether the VPC models obey the relations proposed by Tsunoda [105] who studied many empirical systems. He observed the relationships between the C constant of the BET model and ($E_0\beta$) from the DR equation. Figure 3.5(d) shows that this correlation can be recovered from the results of our simulation study. In fact, Rudziński and Everett [106] derived this relation from the theory of adsorption on heterogeneous surfaces.

Summing up this part, one can see that the VPC models applied by us behave similarly to real activated carbons as far as their "adsorption" behavior is concerned. Note that previous studies showed that the density of the studied VPC also approaches that observed for real carbons [42].

3.4.2. Examples of VPC Application

Molecular simulation results using VPC can be applied for different purposes. In this section, we discuss some examples (dividing the obtained results into five groups).

3.4.2.1. Reality of Carbon PSD Determination Methods Based on Adsorption Data of Noble Gases

In this section, we discuss the results showing the reality of the most popular methods of carbon porosity determination. To do this, we have proposed the application of very simple methodology. It is well known that carbon PSD determination is a very difficult task [6,7,81–88,107–109]. We do not know the shapes of activated carbon pores, or the way they are connected. However, using the VPC model and the above-described BG method, we can find the geometric (i.e. absolute) PSD of our model solid [88]. Therefore for the VPC, we know the exact (geometric) pore-size distribution.

A majority of PSD determination methods uses adsorption isotherm data as the left-hand side (*lhs*) of the GAI equation. Mathematically, this equation is the so-called Fredholm integral equation of the first kind and it is "ill posed". However, this problem can be solved using advanced computer methods [81–84,107–109]). In the GAI equation, one must propose the kernel. This kernel of the *rhs* of GAI is usually obtained using molecular simulations. One assumes the shapes of pores present in a solid (for example a slit-like shape), and using simulation methods (usually GCMC or DFT) simulates a series of local adsorption isotherms. Advanced computer procedures are used to equalize the *lhs* with the *rhs* of the GAI equation, and, as a final result, one obtains the PSD function.

In commercial adsorption apparatus, the slit-like carbon pore model is assumed [110,111]. Therefore, the methodology used by us is simple. We simulated the adsorption isotherm of different gases on a VPC model and next, we treated the isotherms as experimental ones (i.e. as the *lhs* of the GAI equation). Since the absolute PSD is known, we have an opportunity to compare the PSD obtained from the simulated global isotherm with the real ones.

Figure 3.6 shows three model VPC structures and the PSD calculated using the BG model [41]. As one can observe, the analysis of PSD curves for the initial carbon structure (called HO) showed the presence of a micro-mesoporous system of pores. The PSD of the next studied structure called HC1 was in fact similar to HO but the ratio of micropores to mesopores was larger. Finally, the third studied structure (named HC2) was strictly microporous. Note that there was excellent compatibility between the PSDs and the shapes of simulated low temperature (87 K) Ar adsorption isotherms. Thus, we have observed that the development of mesoporosity (from HC2 structure to HO) leads to a conversion of simulated adsorption isotherms from type I of the IUPAC classification to type IV i.e. the shape changes from one typical for microporous solids to one typical for mesoporous ones.

The PSDs obtained from simulated Ar adsorption were calculated using the following methods: Horvath–Kawazoe (HK) [112–114] (we limit the results from this method to a discussion of PSDs from the micropore range), nonlocal DFT (with the local isotherms simulated for the slit-like carbon pores) [85,86,110,111,115–117], Nguyen and Do (ND) (also with a system of local isotherms obtained for carbon slits) [81,82,84,107–109,118,119], and the Barrett–Joyner–Halenda (BJH) [120,121] (this method was applied for mesoporous sample). For the BJH model, two approaches were applied, namely the diameters of the pores were calculated by applying the classical equation in which it was assumed that desorption takes place from the meniscus having cylindrical and/or spherical shapes. The parameters were taken from the paper by Kruk and Jaroniec [121]. We also introduced the correction term (0.438 nm) suggested by them [121]. This correction term was estimated from the analysis of Ar adsorption data on MCM-41 materials [121,122]. The volume of the adsorbate in the monolayer was calculated from the standard Ar adsorption isotherm measured at 87 K on a nongraphitizing carbon black Cabot BP280.

We concluded that the HK method shows the presence of distributed microporosity for all structures, with a strongly marked maximum at 1 nm (Fig. 3.7(a)). It was also interesting that the average micropore diameter was

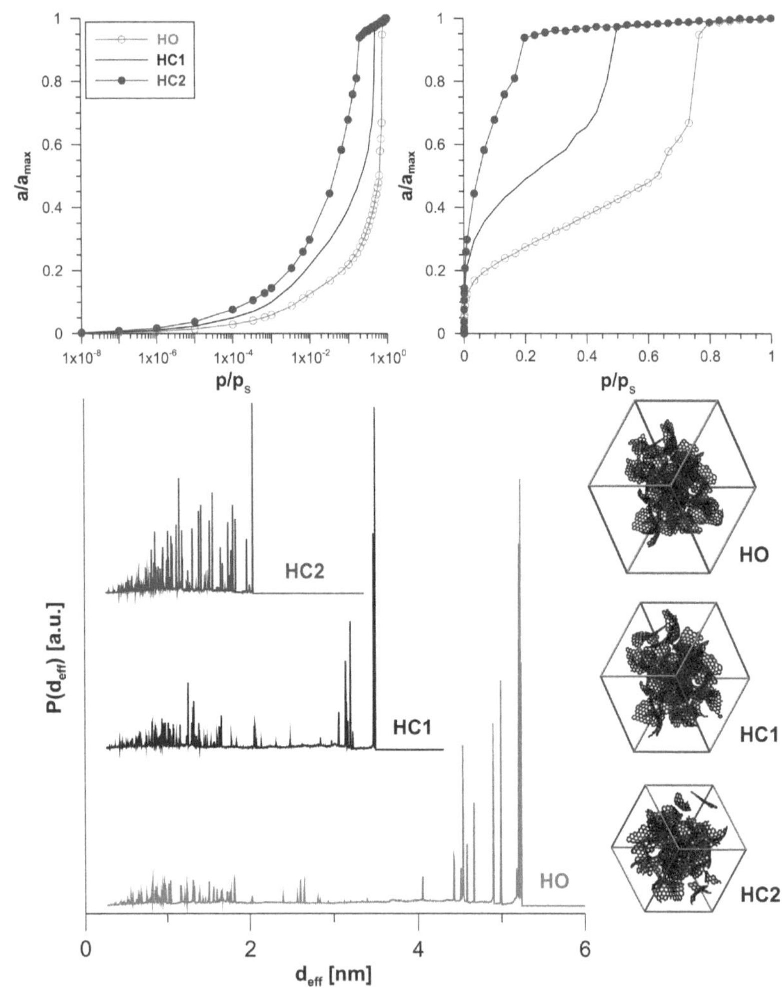

FIGURE 3.6 Ar adsorption isotherms (87 K) simulated by the GCMC method for three studied VPC structures differing in porosity. The PSDs from BG method for all studied structures.

recovered relatively well. This was probably caused by the quasi-Gaussian PSDs of micropores in the studied systems, since the HK method usually leads to a single peak in the PSD covering the whole range of micropore diameters.

The comparison of ND and NLDFT PSDs with geometric ones leads to conclusion that NLDFT gives slightly better results than the ND method. For instance, the true PSD for one of the studied structures (namely for HC1) has a large spike about 3.5 nm; the NLDFT PSD peak is much closer to this spike. The same was observed for HC2. The most important finding is the detection of a bimodal microporous structure for all three carbon models. This is usually observed on the PSDs of real microporous carbons calculated from DFT (ND) software connected with sorption apparatus. In fact from Fig. 3.7(b) and (c), one can observe that the BG method in the range of microporosity shows quasi-Gaussian PSD curves. Therefore, using the slit-like model of micropores and neglecting pore connectivity lead to an absence of pores around ca. 1 nm on the PSD curve (note that the origin of this effect was discussed by us in [81,84]). Both methods show the presence of mesopores, and the diameters of that type of pores are recovered relatively well, especially for the HO structure.

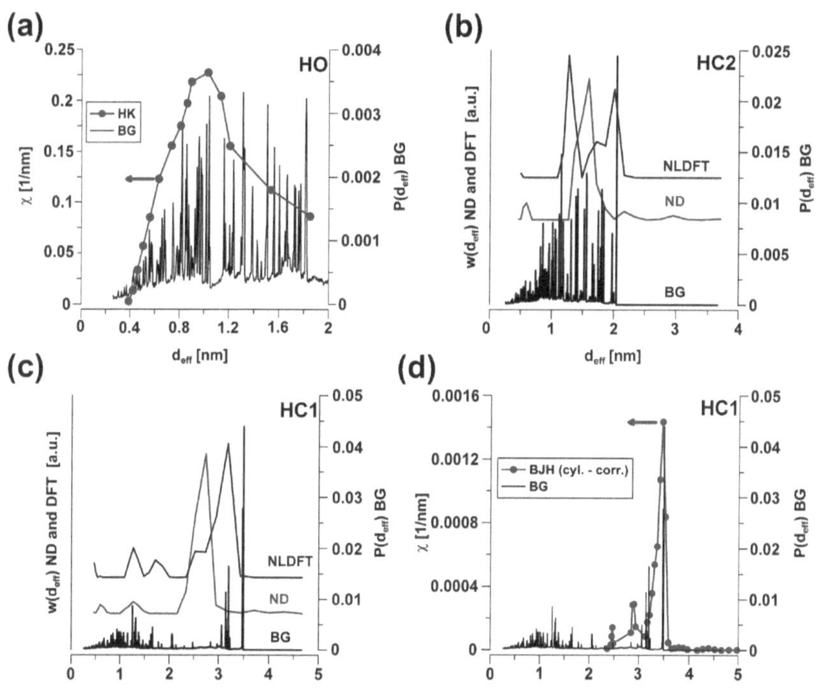

FIGURE 3.7 Selected PSD curves for the studied carbon structures calculated using BG, HK, DFT, ND, and BJH methods from adsorption data simulated by GCMC method. Note that the BG method provides absolute (i.e. geometric) PSD curve.

The results from the BJH method (Fig. 3.7(d)) showed that the assumption of a cylindrical meniscus and the correction proposed by Kruk and others [121,122] leads to the most reasonable results. As one can see, PSDs in the range of mesopores for two of three studied structures (HO and HC1) were reasonable.

In carbon science, the models based on the potential theory of adsorption are still very popular [83,114,123–128]. Many of them are strictly modifications of Dubinin's approach, i.e. the new isotherm equation can be developed using Dubinin's isotherm as a local one in the GAI equation [83,106,114,123–128]. Then assuming adsorption energy distribution, by simple integration, a new equation can be easily derived. Two models derived in such a way, namely Dubinin–Radushkevich–Stoeckli (DRS) [83,106,114,126,127,129] and Jaroniec–Choma (JCh) [83,114,126,128,129] were analyzed by us. Using a simple fitting procedure based on genetic algorithms, we fitted the both models to a series of simulated adsorption isotherms [42]. The results are shown in Fig. 3.8. For the structure having narrow micropores (named S35), both models describe the data with good

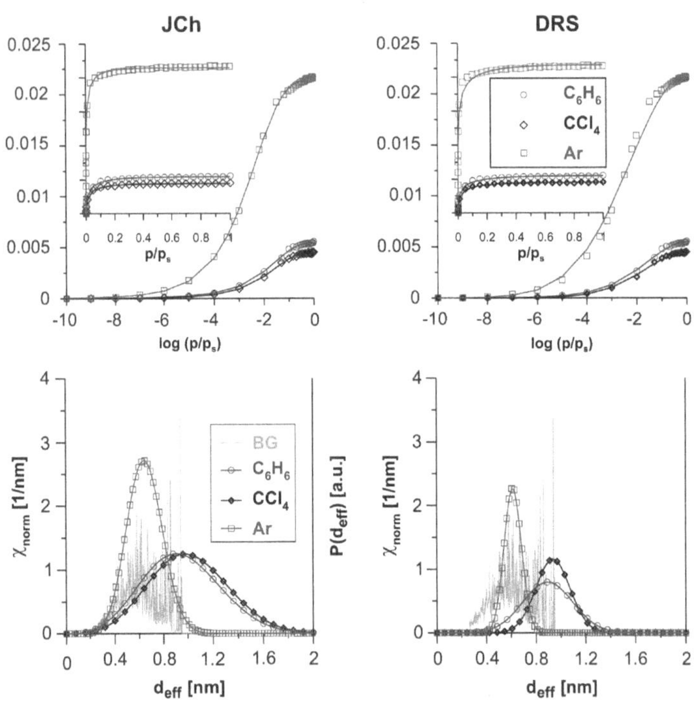

FIGURE 3.8 Typical fits of JCh and DRS models to adsorption isotherms simulated for S35 structure (CCl_4 and C_6H_6 at 298 K). The comparison between the absolute pore-size distributions (from BG method) and those obtained from DRS and JCh models.

accuracy; however, for the structure possessing wider micropore dispersion (S0) and for the adsorption of benzene and carbon tetrachloride, an insufficient fit was recorded for the JCh model. One can observe good recovery of the absolute BG pore-size distribution. What is more important is that the distribution recovered by the models depends on the type of the adsorbate. For the S0 structure, the JCh model works less well than the DRS if one compares the shapes of PSDs. Moreover, more realistic shapes and locations of PSDs are shown for the DRS model applied for all studied adsorbates excepting benzene (here the PSD is shifted toward larger micropores). For this structure, the JCh distribution is too wide, due to the applied gamma-type model. On the other hand, for the S35 structure the JCh model seems to be more realistic than the DRS (for adsorption of noble gases) and for benzene and carbon tetrachloride the distribution is too wide. The latter effect is also seen on DRS distribution calculated for this structure.

These results are crucial from an experimental point of view. One can conclude that if there is a wide distribution of micropores in activated carbon, the DRS model leads to the most realistic PSD, and noble gases or carbon tetrachloride adsorption data should be applied to obtain reliable pore characteristics. On the other hand, if the studied sample is highly microporous, the JCh model should be preferred and one should apply the data of adsorption of noble gases to obtain the most realistic PSD curve.

Finally, the question arises about the most important parameter of the structure of activated carbons, i.e. the average micropore diameter. Figure 3.9 compares the values calculated from the both models for both studied structures, with the average pore diameter from the BG method. For the structure having distributed microporosity (S0), the DRS model predicts the average micropore diameter well, and the best results are recorded for Ar, while for the S35 structure both models lead to almost the same average micropore

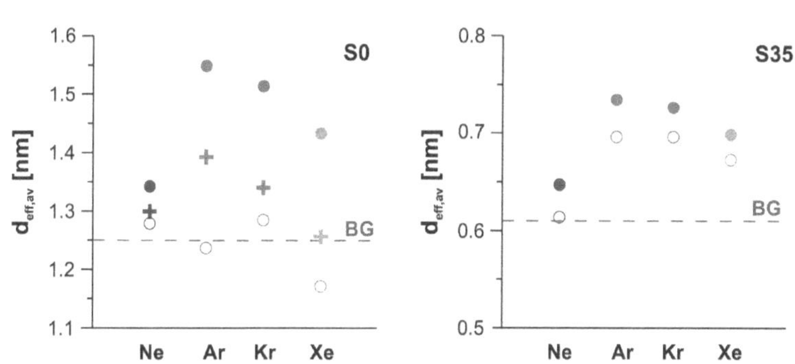

FIGURE 3.9 The comparison of the average micropore diameters from BG method (dashed line) with those calculated from the JCh model (filled circles), DRS (open circles), and DI equations (crosses) for two VPC structures.

diameters, while slightly better results are observed for the DRS model, and for Ne adsorption.

3.4.2.2. Reality of Carbon PSD Determination Methods Based on Adsorption Data of Gases Having Permanent Electric Moments

In this section, we report results checking the applicability of experimentally used gases having permanent electric moments for determination the PSD of carbons.

It is well established that there are two factors determining the adsorption properties of activated carbons, namely the PSD and the chemical nature of surface functionalities [5–9,73–79,85–87]. That is why we use, say, water adsorption data, to check the carbon surface polarity and, for example, Ar adsorption data to check porosity. For this reason, it is rather risky to use adsorption data of polar molecules for detecting PSD and it is also risky to apply nonpolar gas adsorption data to probe the polarity of the carbon surface. Nevertheless, the application of CO_2 adsorption data measured at ambient temperature has been proposed for the purpose of PSD determination for activated carbons [6,7,85,87]. The reasons for CO_2 application are as follows: the method is fast and, moreover, selected experimental findings [130] show that some pores can be inaccessible for nitrogen due to diffusion restriction. In fact, simulation results of Sweatman and Quirke [131] showed that nitrogen is diffusionally limited in microporous carbons and that pore connectivity can significantly affect the adsorption of nitrogen at 77 K in carbons. This has important consequences for characterization studies of microporous carbons that employ nitrogen at this temperature. The authors also claimed that measurements of nitrogen adsorption above 90 K and CO_2 above 260 K (but below their respective bulk critical temperatures and up to their respective saturation pressures) will provide more accurate characterization of carbon porosity.

In [43] we used a VPC model of strictly microporous carbon (called S0). Carbonyl surface functionalities were introduced *via* the procedure of virtual oxidation, described above. It was shown using the BG method that after introduction of surface oxygen functionalities, the PSD of the VPC remains practically unchanged. We simulated, using the GCMC method, adsorption isotherms of Ar, N_2, and CO_2 (see above for the details of the applied models). These isotherms were treated as the *lhs* of the GAI equation. On the *rhs*, we used local isotherms obtained from simulation in carbon slits. Next we compared the geometric PSDs with those obtained from GAI (Fig. 3.10).

The most important practical conclusions from this study were as follows. Mild oxidation of the carbon structure, if nondestructive for the carbon skeleton, does not change absolute carbon porosity in a drastic way. For carbon without surface oxides, we observe systematic shifts of the pore-size

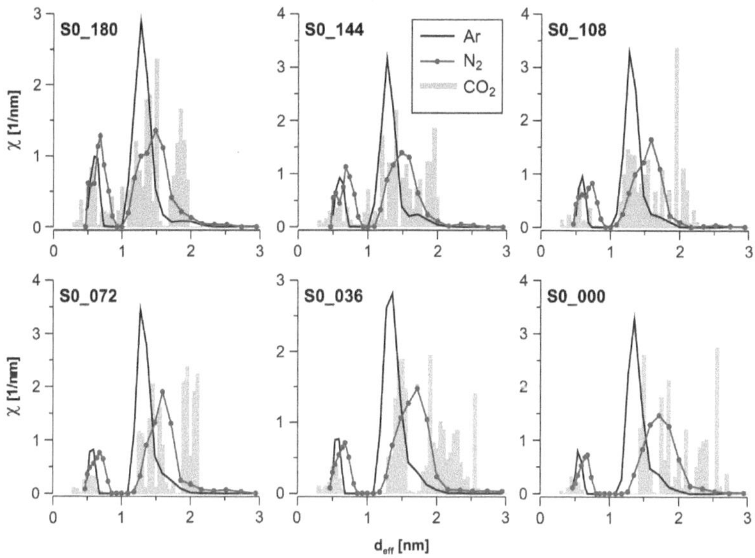

FIGURE 3.10 The comparison of the PSDs obtained from simulated isotherms of adsorption on VPC described by local isotherms generated for slit-like pores (Ar and N_2: ND method, CO_2: local isotherms were calculated by GCMC and fitted to the global isotherm).

distributions toward larger diameters with the rise in the value of electric moments of studied molecules (Fig. 3.11). Moreover, for CO_2 some artificial peaks in the PSD are seen, especially in the range of larger pores. This effect is strongly pronounced after introduction of surface oxygen groups. Our results also showed that Ar is a "safe" molecule for determination of PSD and the average pore diameter of activated carbons. Moreover, for this molecule the approach of the porosity of real carbon by the system of slit-like pores can be applied as a rough approximation, and the values of approximate average pore diameters can be calculated even for carbons having large numbers of surface oxides. In contrast, the results of PSD calculation from CO_2 adsorption isotherms are questionable even for carbons without surface oxygen groups.

Since in the above-cited study [43] only one carbon structure was studied, we performed additional simulations for more microporous carbons and for different surface groups from carbonyls. The results were reported in [45]. In this case since in activated carbons different types of surface groups are usually present, we considered a more realistic situation than studied previously, i.e. we take into account the presence of three of the most frequently occurring groups (carbonyl, hydroxyl, and carboxyl groups). Moreover, adsorbents containing a mixture of all three groups were also considered. To study GCMC results quantitatively isotherms were described by the DA model. The results obtained lead to the conclusion that the presence of surface carboxylic groups mainly

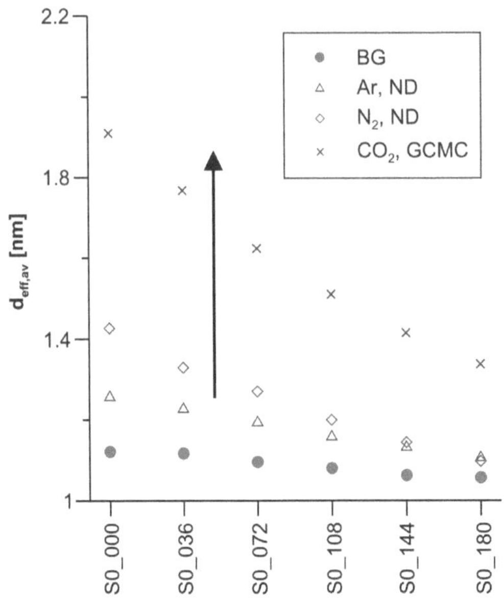

FIGURE 3.11 The comparison of absolute average pore diameters (BG method) with those calculated from adsorption data simulated for adsorption of different adsorbates on VPC models (containing different amount of oxygen groups). One can see that the difference between absolute (BG) average pore diameters and the diameters calculated from simulated adsorption data increases with a rise in electric moment of an adsorbate (from Ar up to CO_2).

affects the parameters obtained from fitting the DA model to simulated isotherms. For carbon models with different pore-size distributions, as well as containing three types of the most frequently occurring surface groups (and their mixtures), we observe the remarkable influence of oxygen on the characteristic energy of adsorption and the heterogeneity parameter of the DA model, while the maximum adsorption is affected to a smaller degree.

Since the largest influence is observed for surface carboxylic groups (being the most frequently observed in commercial carbons), the results of porosity parameters calculated based on fitting of the DA model to experimental CO_2 adsorption data are questionable.

Similar conclusions were obtained in [49]. In this study using GCMC simulation on VPC models, we showed, for the first time, the influence of carbon porosity and surface oxidation on the parameters of the Dubinin–Astakhov adsorption isotherm equation fitted to SF_6 adsorption data. It was proved that both the characteristic energy of adsorption and the parameter n of the DA model are in this case affected by the number of oxygen groups present on the carbon surface. Therefore, as in the case of nitrogen and/or carbon dioxide adsorption, it can be concluded that SF_6 adsorption data on microporous carbon, described using the DA model, cannot be applied for microporosity characterization if the carbon contains oxygen surface functionalities. On the other hand, for carbons not containing oxygen, E_0 is related to the micropore diameter. Therefore, the characteristic adsorption energy of the DA model can be applied for calculation of the average micropore diameter, but only if the new relation between those two values is developed, since the parameter n of this equation is also correlated with the average micropore diameter. Therefore, the application of a simple inverse relationship between the characteristic energy and pore diameter seems to be questionable.

3.4.2.3. Average Micropore Diameter and Reality of the BET Surface Area

To characterize the texture of activated carbons, two values are usually used, namely average micropore diameter and the surface area (called sometimes "apparent" surface area).

Pore diameter is usually calculated based on old Dubinin's experimental proof showing that one of the parameters calculated from the fitting of Dubinin–Astakhov model to experimental results, called the characteristic energy of adsorption, is related to the half-width of activated carbon slit-like pore [132]. Dubinin's result stimulated great interest, and in the 80s and 90s of the last century many relationships were proposed for example by Stoeckli, McEnaney, Jaroniec and Choma, and other authors [114,132]. It is striking that if all pore diameters obtained from those relationships are plotted on the same figure, large differences (even around 100 %) are observed. Of course, the differences are due to the nature of studied carbon samples, applied measurement techniques etc. For example, if we consider two small micropores having exactly the same diameters but possessing different number of graphite layers forming pore walls, we will observe experimentally differences between E_0 values, and subsequently differences between calculated micropore diameters (but in fact they are the same!). However, this is completely ignored and the influence of the origin of carbon samples on obtained E_0 vs. $d_{eff,av}$ relationships has not been studied so far.

Using molecular simulations and VPC models, one can easily check the validity of obtained $d_{eff,av}$ values and compare them with the geometric ones [40]. The procedure in this case is very simple and analogous to that described above. Using a series of VPC models, created using the procedure of addition of fragments to the initial structure (see above), we can apply the BG method and calculate the absolute PSD. From this PSD, we can easily evaluate the absolute $d_{eff,av}$ value. Next using GCMC simulation, a series of, say Ar, adsorption isotherms is simulated. Isotherms obtained in this way are next described by the DA model. In this case, to assure that the fit of the DA model to simulated isotherms will be good enough, and will be observed for the wide pressure range, the VPC models should possess small pores. In this way, we avoid the problem of pressure range choice during the fitting procedure.

The problem of average micropore diameter calculation using the $d_{eff,av} = f(1/E_0)$ relationships was studied in [40]. We created 36 VPC models with progressively changed PSDs using the method described above of carbon fragments insertion. Next, Ar adsorption isotherms were simulated and described by the DA model. We also studied PSDs obtained by application of different models, as in [41]. It is very interesting that simulation results on a series of VPC models lead to a hyperbolic shape of the $d_{eff,av} = f(E_0)$ plot. It is also important to note that the shape of this plot is almost identical to those obtained from the two simplest experimentally derived relations proposed by Stoeckli [133]. On the other hand, the assumption of the inverse linear relation between the parameter n and the pore diameter is not correct. It was additionally concluded that the DA adsorption isotherm equation correctly predicts the maximum amount of adsorption if the studied system is strictly microporous.

The next important problem in adsorption science is the validity of BET surface area. From an analysis of published results, it can be seen that for different types of adsorbents, an overestimation, an underestimation, and/or a similarity of the surface area calculated by the BET method (in comparison with the geometrical considerations) can be observed [23,134–144]. Therefore, it is very difficult to correlate the differences between both types of surface areas with that of an adsorbent, the structural heterogeneity, and porosity. This subject was studied by us for models of carbon nanotube bundles [144] and for VPC models [46] using a similar procedure, i.e. the application of VPC and molecular simulations. Here the application of BET model is used as a standard tool recommended by IUPAC [6,7,85,87]. In this case, the BET model is fitted to simulated isotherms and we have used the Rouqueroll [145] procedure to avoid negative C values. The values of absolute geometric areas have been calculated using the Vega ZZ package [146,147], and the details were described by us previously [144]. The reality of Vega ZZ results was tested for a slit-like carbon pore [46] and the geometric surface area of this slit was calculated by the method analogous to this proposed by Kaneko et al. [148].

To compare all three values of surface area (calculated by the above-mentioned methods i.e. geometric, calculated from VEGA ZZ and from BET model fitted to simulated data), Ar adsorption isotherms (87 K) were simulated by the classical GCMC method. Additionally from those isotherms, the active surface for adsorption (ASA) was calculated by the method proposed by Pellenq et al. [139] and applied in a recent study [144].

The values obtained from application of the VEGA ZZ package were in very good agreement with the geometric surface areas calculated for slit-like pore. Simultaneously, it was shown that for the case of ideal carbon slit-like pore, large values of surface areas, as suggested by Kaneko et al., are possible [148]. In this way, using molecular simulations and geometric considerations we confirm the validity of the results obtained by the mentioned authors.

Finally we showed that one can modify the ASA method, taking into account the multiplication of the specific surface area of adsorbed molecules. This leads to a better agreement between the geometric and "adsorption" types of surface areas. For realistic VPC models, the BET-type methods provide, generally, too small values of surface areas, and a better agreement is observed for the results obtained from the linearization method proposed by Rouquerol et al. [145] than from a typical BET procedure [6,7,85,87]. The original ASA method was modified leading to the values of surface areas being more realistic; however, the presented approach is not perfect yet. Figure 3.12 shows the comparison of different surface area values calculated for ten VPC models.

3.4.2.4. Searching the Meaning of Empirical Parameters

In adsorption science, different interesting empirical correlations are observed; however, the origin of these correlations is often difficult to explain, even based

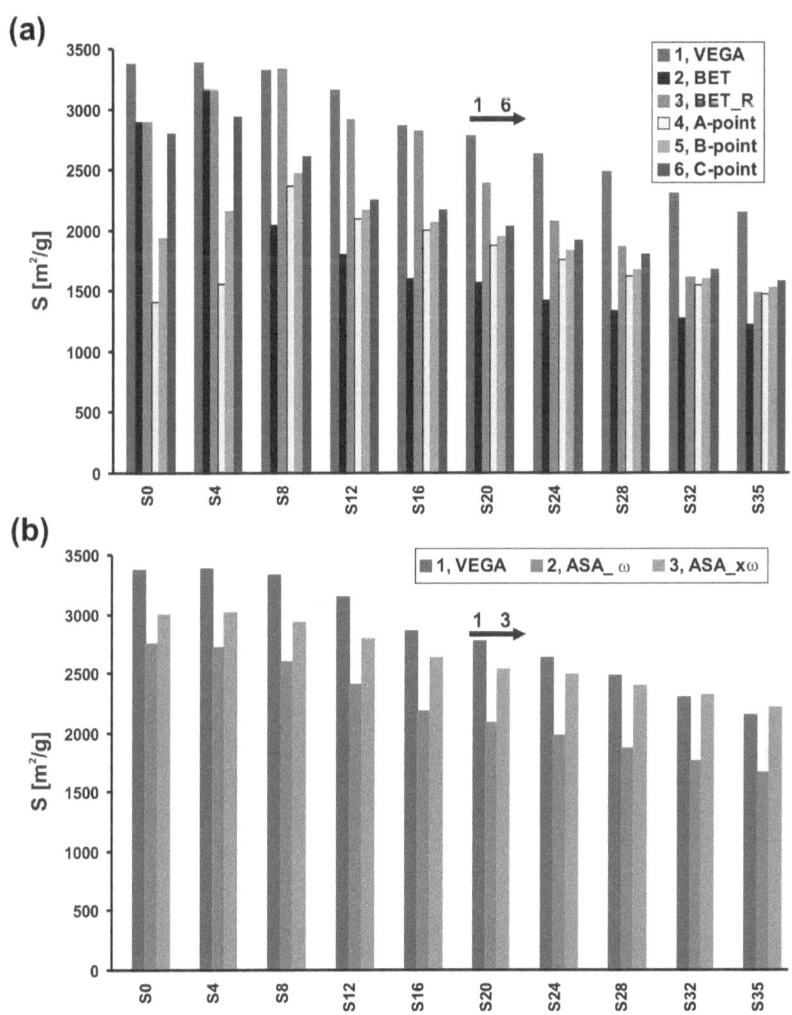

FIGURE 3.12 (a) The comparison of surface area values calculated for VPCs models (using different methods based on simulated adsorption data) with the true (i.e. obtained from the VEGA package) surface area values. Bottom figure (b) shows the results of ASA method modification.

on advanced theoretical models. In this case, molecular simulations offer excellent possibilities since we can observe the mechanisms of processes on molecular level.

Thus, for example, for application of the DA model to describe experimental methane supercritical adsorption data, one must calculate the "apparent" saturation pressure value since we are above a critical point, and liquid does not exist at those conditions. To do this, different more or less empirical advanced equations have been proposed. Among those relations the most general was proposed by Amankwah and Schwarz [149]; however, this relationship contains the values of critical temperature and pressure respectively, and the empirical parameter k. As was stated by the authors, k is the parameter specific for the system studied. In their paper, Amankwah and Schwarz showed [149] that for experimental data measured for different activated carbons the value of this parameter changes in the range 2.10–2.73 (methane adsorption) or 2.72–4.16 (hydrogen adsorption). Note that in the case of $k = 2$, equation proposed by them reduces to the relation proposed by Dubinin [125]; however, this original relation not always leads to the single characteristic curve as pointed out by Ozawa et al. [150].

The meaning of the k parameter, as well as the relation between different constants (for example the constants of Toth and/or Sips equations [151,152]) and the parameters of carbon structure can be easily established from a series of data obtained using GCMC simulation on VPC models [48]. The methodology is simple. Using a series of VPC having progressively changed porosity (and for a fixed PSD progressively changed oxygen content), one can easily simulate methane adsorption isotherms. Next these isotherms are described using theoretical models, and values of best-fit parameters obtained from this description are correlated with the average micropore diameters or oxygen contents (known for VPC model exactly).

The results shown in Fig. 3.13 are in our opinion the most spectacular example of simulation on VPC model application for determination of the meaning of empirical parameters [48]. It is clearly seen that the values of the parameter k occurring in the empirical relation proposed by Amankwah and Schwarz [149] are strictly related, by an inverse relation, to activated carbon micropore diameter. Moreover, the k value increases with the contents of surface oxygen. Interestingly, also with the rise in the oxygen content we observe the rise in the value of Toth's equation "heterogeneity parameter" t.

3.4.2.5. Molecular Simulations as a Tool Explaining Experimental Regularities

In this section we will discuss the application of MD simulations for the explanation of possible adsorption mechanisms and some experimental regularities. A good example here is the explanation of phenol adsorption mechanism on carbons, and the exact role of oxygen groups in this mechanism.

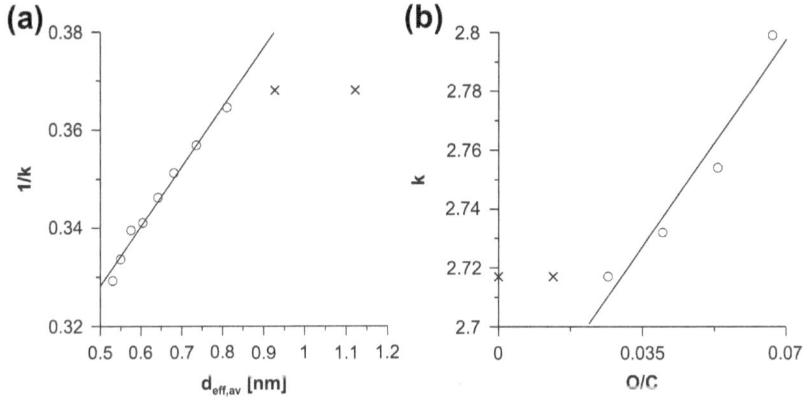

FIGURE 3.13 (a) The correlation between average diameters of pores calculated from BG method and the parameter *k* occurring in the empirical relation proposed by Amankwah and Schwarz [149]. (b) The similar relationship but plotted for the ratio of oxygen to carbon contents in VPCs. In the case of parameter *k*, we excluded from linear regression those systems where the *k* value reaches the lowest physical range (points marked as crosses).

Adsorption of phenol – a common water pollutant – is widely studied because activated carbons are used in water plant stations, filters etc. [153–155]. It has been proved and widely documented by experimental results that carbon surface oxidation, causing the introduction of oxygen-containing surface groups, reduces the adsorption of phenol at small equilibrium concentrations [6]. However, there is a long-standing debate in the literature about the mechanism of this reduction. Two major explanations of this effect were proposed during discussion of the experimental data [76,156]. The first postulates that the decrease in phenol adsorption after carbon oxidation is caused by the reduction in the energy of π–π interactions between carbon matrix and benzene ring of phenol molecule. This effect is due to changes in the electron density of carbon after the appearance of surface oxygen groups. The second proposed effect is the "pore blocking" postulated by Coughlin and Ezra [157], who suggested that surface oxygen complexes increase the affinity of water to carbon, and therefore solvent molecules can effectively block some micropores for phenol.

To check how the PSD of carbon and the contents of oxygen functionalities determine phenol adsorption, we performed a very simple MD simulation [44]. For a series of VPC models (having progressively changed porosity and oxygen contents), we introduced water and a fixed number of randomly placed (at the vicinity of VPC structure) phenol molecules. Next using the Gromacs MD package [97–100], we fill the simulation boxes with water molecules and the system is equilibrated at a chosen temperature. Then, we used a modified BG method [44,60] to determine the distribution of pores filled with adsorbed phenol molecules. In this way, one can easily find the optimal carbon pore diameter for phenol adsorption as well as to check how the addition of surface oxygen groups changes the accessibility of molecules to pores, and what is the mechanism of adsorption process. We can also check the number of hydrogen bonds formed between adsorbed phenol and water molecules and surface oxygen functionalities.

Figure 3.14 shows the average number of phenol molecules in the carbon structure after equilibration of simulation boxes, plotted as the function of the average micropore diameter of pores adsorbing those molecules. This diameter was calculated using the above-described, modified BG method. Simulation results show that with the rise in microporosity, a larger number of phenol molecules is adsorbed in smaller micropores; however, total adsorption in micropores decreases due to the decrease in the pore volumes (since we use the periodic boundary conditions, it can be stated that the remaining molecules, adsorbed on the external surface of the structures, are adsorbed in large mesopores created between periodic images of the box). An important conclusion obtained from MD simulation results is that the optimal pores for phenol adsorption possess diameters close to the diameter of the benzene ring (ca. 0.6 nm). From Fig. 3.14, it can also be concluded that a larger influence on phenol adsorption is caused by carbon surface chemical composition and this is in accordance with experimental reports, suggesting that this factor has a larger influence than porosity on the adsorption of phenol (and generally organics) from aqueous solutions. Therefore, in the case of carbon without surface functionalities, the "solvent effect" does not play an important role. This is in accordance with the results of GCMC simulations of phenol adsorption from the gaseous phase on carbon models published by Bertoncini et al. [158]. On the other hand, if those groups are present on the carbon surface, one can observe gradually decreasing maximum adsorption with the rise in their contents (see Fig. 3.14). Mild oxidation of carbon (our simulation data show that up to ca. 1.5% of oxygen) does not, in practice, reduce the number of adsorbed molecules

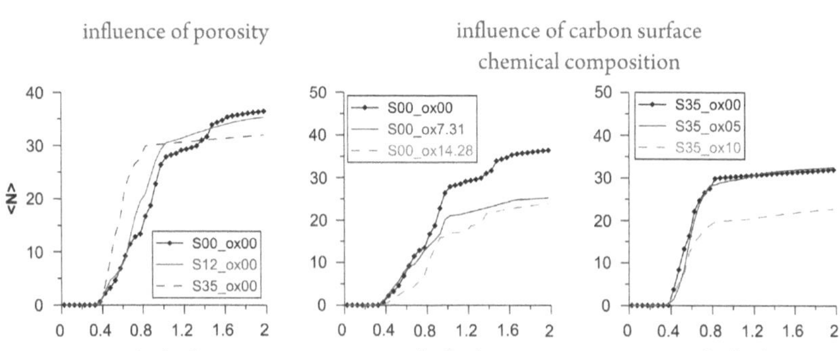

FIGURE 3.14 The influence of porosity and chemical composition of carbon surface layer on phenol adsorption on microporous VPC models.

in all studied structures. However, if carbon is strongly oxidized, a drastic decrease in adsorption of phenol in micropores is observed. To explain this effect, we calculated the hydrogen-bond statistics using the procedure proposed by Gordillo and Martí [159], concluding that the average number of phenol molecules interacting with carbonyl groups *via* hydrogen bonds is small (depending on structure from 2 up to 6 molecules only). In contrast, the number of water molecules forming hydrogen bonds with carbonyls almost linearly increases with the number of surface oxygen groups in the structures. Therefore, it was concluded that the pore-blocking effect is responsible for decreasing adsorption of phenol with the rise in contents of surface oxygen groups.

This effect was also observed on animations of phenol in equilibrated structures. Water-saturating carbonyl groups block some parts of the structure making them inaccessible for phenol molecules and decreasing in this way adsorption on oxidized structures. This conclusion is in agreement with the results of DFT calculations [160], showing that phenol interaction with surface functional groups is characterized by low energy and, for most carbon surface oxygen groups, is thermodynamically unstable at room temperature. In the case of adsorption from aqueous solution, this kind of interaction is negligible due to competitive and much stronger interactions of water molecules with polar adsorption sites.

Deeper insights into the pore-blocking effect was achieved from a detailed analysis of the snapshots [44] and it was also shown that interaction of water molecules with surface oxygen groups leads to the creation of hydrogen-bonded clusters of water with larger number of hydrogen bonds than in structures without oxygen. This effect causes the pore blocking for phenol molecules, and the experimentally observed decrease in adsorption at low coverages after carbon oxidation. In summary, we have reported the first MD simulation results of phenol adsorption on low coverages on VPC models [44]. An optimal microporous carbon for phenol adsorption should possess the largest possible volume of micropores with diameters around 0.6 nm. Pore blocking causes a decrease in phenol adsorption. The presented results support the idea of Coughlin and Ezra [157] showing that highly hydrogen-bounded water clusters around surface oxygen groups enable the penetration of some micropores by subsequent phenol molecules. The number of phenol molecules interacting with carbon oxygen groups is small at low coverages and this is in agreement with the DFT calculations [160].

Since in the studies referred to above only phenol adsorption was investigated, we decided to expand our study and to simulate (using MD) adsorption from aqueous solution of other organic molecules i.e. benzene and paracetamol [60]. As in the case of phenol adsorption, we studied the effect of porosity and carbon surface oxidation. In the latter case, we also studied the influence of surface groups location, concluding that there is almost no influence of location of surface groups on benzene accessibility to small micropores (this can be seen only for large number of groups), and a decrease in adsorption after surface oxidation observed for the two remaining molecules. For the initial carbon (i.e. not containing oxygen), adsorption decreases from benzene down to paracetamol. This is in accordance with experimental data [76,79]. The location of surface groups has an influence on accessibility of molecules to the smallest pores, whereas this effect is the most strongly pronounced for phenol and paracetamol only if carbon contains larger amount of surface oxygen.

The comparative plots of probability of adsorption (in micropores i.e. in pores with diameter smaller than 2 nm) vs. oxygen contents, and average micropore diameters revealed very interesting regularities. Considering the changes in porous structure due to the rise in distance between the microcrystallites (i.e. the rise in micropore diameter), a progressive decrease in adsorption was observed with the rise in distance between microrystallites forming the carbon structure. We also observed that the effect of carbon surface chemical groups vanishes with the rise in micropore diameters. Generally, it was concluded that a progressive rise in micropore diameter (for carbons without oxygen) has small influence on probability of benzene adsorption. In this case, the strongly marked decrease in adsorption is observed for the VPC series having the largest micropore diameter. However, considering the series containing oxygen the decrease in benzene adsorption was strongly pronounced if the number of groups increased. A similar situation occurred for the two remaining molecules.

Simultaneously we also recorded a decrease in adsorption with a rise in the number of surface oxygen groups. Therefore, a nice interrelation between two effects was seen, i.e. pore accessibility strongly depends on the concentration of surface groups; however, this effect vanishes for larger micropores. In this case, the probability of adsorption of all molecules is almost the same and surface oxygen almost does not play a role in adsorption.

The average micropore diameter calculated from BG method for adsorbed molecules was correlated with the BG pore diameter calculated for the structures. We observed that the average micropore diameter calculated for adsorbed molecules was similar for all three adsorbates. For the structure without oxygen groups, we observed that the smallest pores are accessible for benzene and this is in agreement with Gurvitch's rule. Moreover, one could also clearly observe how the distribution of surface groups influences accessibility of molecules to pores. It was interesting that for the structure with larger micropores those differences in accessibility vanished and we observed the same accessibility for all three molecules and for each oxygen contents. This result, in our opinion, clearly demonstrates that for larger micropores the differences in adsorption are not caused by the collision diameters of adsorbed molecules but that those differences in accessibility to micropores are resulted by the pore-blocking effect.

In the case of adsorption of organics next interesting application of simulations can be demonstrated. Namely, in the same study we were interested in the calculation of the number of water molecules displaced by a single adsorbed molecule. This ratio is necessary if one wishes to fit advanced models to experimental data. In all advanced models, one must take into account the correction on displacement effect, i.e. if the one solvent molecule is displaced by one solute molecule, the ratio $r = 1$. It means that in this case monolayer adsorption value does not need corrections. However, since solute molecules are usually larger than water molecules, one must correct monolayer capacity by taking r into account in the equation describing maximum adsorption (see for example [161]).

Since it is impossible to determine this value experimentally r is usually calculated as the ratio of molar volumes of solute and water (if water is applied as solvent). In our previous study [162], the ratio calculated in this way for phenol/water system was equal to 4.91 and for paracetamol/water to 5.81. Therefore, we calculated r values for the above-described systems and obtained from simulation data results are shown in Fig. 3.15 for VPC structure where water has full accessibility to the internal part of carbon. As one can observe, r value increases with the rise in molar volume of a solute molecule. Moreover, the results showed in Fig. 3.15 clearly indicate that r depends on the chemical composition of carbon layer! This means that the r value is not constant and depends also on the type of studied carbon.

Another example of the application of molecular simulation for explanation of experimental regularities is the use of MD technique for explanation of the mechanisms of carbon immersion in water. It is well known from the 1950s that the more or less perfect, but linear correlation between the surface oxygen content and the value of carbon enthalpy of immersion in water is observed [6,7,87,163–166].

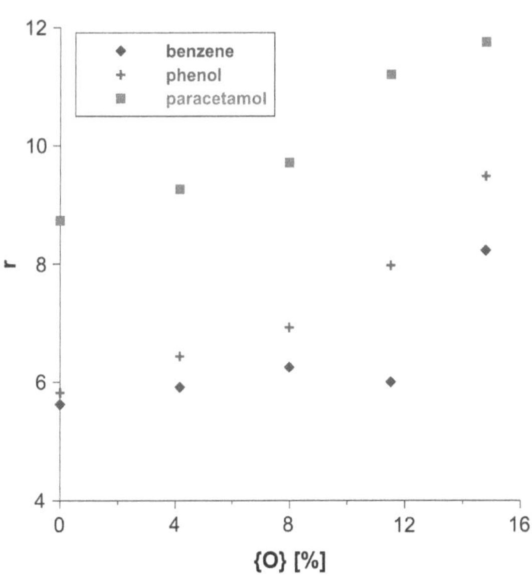

FIGURE 3.15 A correlation between the number of water molecules displaced per one organics molecule (r) and the contents of oxygen in VPC.

However, since the value of this enthalpy, particularly in this solvent, depends on oxygen content, it is hard to determine the influence of carbon pore diameter on water enthalpy of immersion. Using a series of VPC models having (as usually) progressively changed porosity as well as the contents of surface oxygen, and placing those models inside a simulation box, one can easily (using a barostat) equilibrate the system to imitate the process of immersion.

The results were reported by us recently [47]. We checked the dependency between the enthalpy of carbon immersion in water value and the oxygen content. Also, the influence of porosity was studied. Using a series of "soft" VPC models obtained according to the procedure described in Sect. 2 and the "virtual oxidation" procedure (see Sect. 2.1), it was shown that there is a small influence of the distribution of surface groups on the enthalpy of the immersion value. A linear relation between this enthalpy and the concentration of surface groups was recorded, independently of micropore size distribution and this observation is in accordance with experimental findings [6,7,87,163–166]. The obtained enthalpy of immersion was also a function of polar surface area, and the ratio of the polar/nonpolar surface area. The latter finding is in accordance with the correlations between XPS results and enthalpy of immersion [167].

The mechanism of the immersional process depends on carbon porosity and the concentration of oxygen on the surface. For unoxidized carbons, the enthalpy of immersion can even be positive if the micropores are small; however, it increases (i.e. is more exothermic) with the rise in the average micropore diameter. The slopes of the relations between enthalpy value and oxygen content are almost the same for carbons with different porosity; therefore, the shifts between them are caused by the differences in the energy of the water interaction with starting carbon structure. That is why enthalpy increases with the rise of average micropore diameter.

Using VPC and MD simulations, we also concluded that generally the behavior of carbon in immersion process follows the Pratt–Chandler theory of hydrophobicity [168–170]. We observed the creation of a liquid–vapor interface in the case of strongly hydrophobic systems. This interface, however, vanishes with the rise in the number of carbon oxygen functional groups. The density of water approached the density of bulk with the rise in average micropore diameter. This is also in accordance with the Pratt–Chandler mechanism, because with the rise in micropore diameter the microcrystallites forming VPC start to behave more independently. For such a case, a solute can be treated as a system of small microcrystallites and the water molecules' behavior approaches the behavior of bulk water. Analysis of proximal distributions suggested a slightly larger accessibility of water to carbonyl groups located on the edges of structures. Also, the restriction of accessibility to the surface groups by some water molecules interacting with neighboring groups was observed. We also observed the effect of limited accessibility to carbon atoms due to the presence of carbonyls. The average number of hydrogen bonds calculated per single carbonyl group increased with the average micropore diameter. Therefore, in microporous carbons the formation of hydrogen bonds between water and surface groups is strongly restricted. Moreover, the number of created hydrogen bonds between water and carbonyl groups decreased with the rise in the number of those groups. This is caused by the above-mentioned restriction of accessibility to the surface groups by water molecules interacting with neighboring surface functionalities.

3.5. SUMMARY AND PERSPECTIVES

It is obvious that molecular simulations are now one of the most important tools in science. There are different more or less spectacular examples of application of simulations. In carbon adsorption science, we also know the examples of practical application of simulation results, and one of the most popular is in determination of the PSD curve.

However, the application of VPC linked with molecular simulations opens completely new perspectives. Major advantages offered by those models are: known composition and porosity of the VPC model, known number and composition of surface groups, possibility of easy manipulation providing systematic changes in porosity, and/or surface chemical composition.

However, there are also disadvantages and they can be divided into two major groups. First are the limitations due to application of simulation methods, where the results depend on the applied force field i.e. on the sets of used parameters. The second group of limitations originates from the reality of the applied carbon model. We still do not have enough information about the composition of carbon surface layers, pore connectivity, occurrence of defects, impurities, heteroatoms, acid–base centers etc. We also still do not know how the fragments forming the carbon structure are connected.

In our opinion the future of VPC depends on the development of our knowledge about real carbon. We still have serious limitation in modeling the chemical composition of carbon surfaces. It is well established that heteroatoms located on a carbon surface must change the electron density of this surface but application of *ab inbitio* methods to calculate those changes is still limited only to small systems. We also must develop the methods of carbon structure investigation. New possibilities are offered for example by HRTEM techniques [39,59] and we believe that in the 21st century we will be able to say more about carbon, the so-called "*old but new material*".

ACKNOWLEDGMENTS

We acknowledge professor Juan Tascon for his invitation. The authors acknowledge the use of the computer cluster at Poznań Supercomputing and Networking Center and the Information and Communication Technology Center of the Nicolaus Copernicus University (Toruń, Poland).

REFERENCES

[1] P. Kowalczyk, R. Hołyst, M. Terrones, H. Terrones, Hydrogen storage in nanoporous carbon materials: myth and facts, Phys. Chem. Chem. Phys. 9 (2007) 1786–1792.

[2] G.K. Dimitrakakis, E. Tylianakis, G.E. Froudakis, Pillared graphene: a new 3-D network nanostructure for enhanced hydrogen storage, Nano Lett. 8 (2008) 3166–3170.

[3] R.P. Wesołowski, A.P. Terzyk, Pillared graphene as a gas separation membrane, Phys. Chem. Chem. Phys. 13 (2011) 17027–17029.

[4] A.P. Terzyk, P.A. Gauden, P. Kowalczyk, Carbon Materials - Theory and Practice, Research Signpost, 2008.

[5] T.J. Bandosz, M.J. Biggs, K.E. Gubbins, Y. Hattori, T. Iiyama, K. Kaneko, J. Pikunic, K.T. Thomson, Molecular models of porous carbons, in: Chemistry and Physics of Carbon, vol. 28, Marcel Dekker Inc, 2003.

[6] R.C. Bansal, M. Goyal, Activated Carbon Adsorption, CRC Press, 2005.

[7] T.H. Marsh, F. Rodriguez-Reinoso, Activated Carbon, Elsevier, 2006.

[8] H. Bock, K.E. Gubbins, J. Pikunic, Models of porous carbons, in: Adsorption by Carbons, Elsevier, 2008.

[9] A.P. Terzyk, S. Furmaniak, P.A. Gauden, P.J.F. Harris, R.P. Wesołowski, P. Kowalczyk, Virtual porous carbon (VPC) models - application in study of fundamental activated carbon properties by molecular simulations, in: Activated Carbon: Classification, Properties and Applications, Nova Science Publishers, 2011.

[10] M.J. Biggs, A. Buts, Virtual porous carbons: what they are and what they can be used for, Mol. Sim. 32 (2006) 579–593.

[11] T. Roussel, A. Didion, R.J.-M. Pellenq, R. Gadiou, C. Bichara, C. Vix-Guterl, Experimental and atomistic simulation study of the structural and adsorption properties of faujasite zeolite - templated nanostructured carbon materials, J. Phys. Chem. C 111 (2007) 15863–15876.

[12] T. Roussel, R.J.-M. Pellenq, C. Bichara, Structure of narrow-diameter single-wall carbon nanotubes grown in $AlPO_4$-5 zeolite, Phys. Rev. B 76 (2007) 235418.

[13] Q. Cai, A. Buts, M.J. Biggs, N.A. Seaton, Evaluation of methods for determining the pore size distribution and pore-network connectivity of porous carbons, Langmuir 23 (2007) 8430–8440.

[14] D.D. Do, D. Nicholson, H.D. Do, Heat of adsorption and density distribution in slit pores with defective walls: GCMC simulation studies and comparison with experimental data, Appl. Surf. Sci. 253 (2007) 5580–5586.

[15] G.R. Birkett, D.D. Do, On the physical adsorption of gases on carbon materials from molecular simulation, Adsorption 13 (2007) 407–424.

[16] D.D. Do, H.D. Do, Modeling of adsorption on nongraphitized carbon surface: GCMC simulation studies and comparison with experimental data, J. Phys. Chem. B 110 (2006) 17531–17538.

[17] A. Wongkoblap, D.D. Do, Characterization of Cabot non-graphitized carbon blacks with a defective surface model: adsorption of argon and nitrogen, Carbon 45 (2007) 1527–1534.

[18] S. Furmaniak, A.P. Terzyk, P.A. Gauden, N.A. Marks, R.C. Powles, P. Kowalczyk, Simulating the changes in carbon structure during the burn-off process, J. Coll. Interf. Sci. 360 (2011) 211–219.

[19] Y. Shi, A mimetic porous carbon model by quench molecular dynamics simulation, J. Chem. Phys. 128 (2008) 234707.

[20] R.C. Powles, N.A. Marks, D.W.M. Lau, Self-assembly of sp^2-bonded carbon nanostructures from amorphous precursors, Phys. Rev. B 79 (2009) 075430.
[21] J.C. Palmer, A. Llobet, S.-H. Yeon, J.E. Fischer, Y. Shi, Y. Gogotsi, K.E. Gubbins, Modeling the structural evolution of carbide-derived carbons using quenched molecular dynamics, Carbon 48 (2010) 1116–1123.
[22] J.K. Brennan, T.J. Bandosz, K.T. Thomson, K.E. Gubbins, Water in porous carbons, Coll. Surf. A: Physicochem. Eng. Asp. 187–188 (2001) 539–568.
[23] K.T. Thomson, K.E. Gubbins, Modeling structural morphology of microporous carbons by reverse Monte Carlo, Langmuir 16 (2000) 5761–5773.
[24] J. Pikunic, C. Clinard, N. Cohaut, K.E. Gubbins, J.-M. Guet, R.J.-M. Pellenq, I. Rannou, J.-N. Rouzaud, Structural modeling of porous carbons: constrained reverse Monte Carlo method, Langmuir 19 (2003) 8565–8582.
[25] J. Pikunic, K.E. Gubbins, R.J.-M. Pellenq, N. Cohaut, I. Rannou, J.-M. Guet, C. Clinard, J.-N. Rouzaud, Realistic molecular models for saccharose-based carbons, Appl. Surf. Sci. 196 (2002) 98–104.
[26] J. Pikunic, R.J.-M. Pellenq, K.T. Thomson, J.-N. Rouzaud, P. Levitz, K.E. Gubbins, Improved molecular models for porous carbons, Stud. Surf. Sci. Catal. 132 (2001) 647–652.
[27] J. Pikunic, P. Llewellyn, R.J.-M. Pellenq, K.E. Gubbins, Argon and nitrogen adsorption in disordered nanoporous carbons: simulation and experiment, Langmuir 21 (2005) 4431–4440.
[28] S.K. Jain, J. Fuhr, R.J.-M. Pellenq, J. Pikunic, C. Bichara, K.E. Gubbins, Stability of porous carbon structures obtained from reverse Monte Carlo using tight binding and bond order hamiltonians, Stud. Surf. Sci. Catal. 160 (2006) 169–176.
[29] S.K. Jain, J. Pikunic, R.J.-M. Pellenq, K.E. Gubbins, Effects of activation on the structure and adsorption properties of a nanoporous carbon using molecular simulation, Adsorption 11 (2005) 355–360.
[30] S.K. Jain, R.J.-M. Pellenq, J.P. Pikunic, K.E. Gubbins, Molecular modeling of porous carbons using the hybrid reverse Monte Carlo method, Langmuir 22 (2006) 9942–9948.
[31] T.X. Nguyen, S.K. Bhatia, Pore accessibility of N_2 and Ar in disordered nanoporous solids: theory and experiment, Adsorption 13 (2007) 307–314.
[32] T.X. Nguyen, N. Cohaut, J.-S. Bae, S.K. Bhatia, New method for atomistic modeling of the microstructure of activated carbons using hybrid reverse Monte Carlo simulation, Langmuir 24 (2008) 7912–7922.
[33] P. Zetterström, S. Urbonaite, F. Lindberg, R.G. Delaplane, J. Leis, G. Svensson, Reverse Monte Carlo studies of nanoporous carbon from TiC, J. Phys: Condens. Matt. 17 (2005) 3509–3524.
[34] T.C. Petersen, I.K. Snook, I. Yarovsky, D.G. McCulloch, B. O'Malley, Curved-surface atomic modeling of nanoporous carbon, J. Phys. Chem. C 111 (2007) 802–812.
[35] G. Opletal, T.C. Petersen, D.G. McCulloch, I.K. Snook, I. Yarovsky, The structure of disordered carbon solids studied using a hybrid reverse Monte Carlo algorithm, J. Phys: Condens. Matt. 17 (2005) 2605–2616.
[36] G. Opletal, T. Petersen, B. O'Malley, I. Snook, D.G. McCulloch, N.A. Marks, I. Yarovsky, Hybrid approach for generating realistic amorphous carbon structure using metropolis and reverse Monte Carlo, Mol. Sim. 28 (2002) 927–938.
[37] V. Petkov, R.G. Difrancesco, S.J.L. Billinge, M. Acharya, H.C. Foley, Local structure of nanoporous carbons, Philosoph. Magazine B 79 (1999) 1519–1530.
[38] J.-M. Leyssale, J.-P. Da Costa, Ch. Germain, P. Weisbecker, G.L. Vignoles, An image-guided atomistic reconstruction of pyrolytic carbons, App. Phys. Lett. 95 (2009) 231912.

[39] V. Fernandez-Alos, J.K. Watson, R. vander Wal, J.P. Mathews, Soot and char molecular representations generated directly from HRTEM lattice fringe images using Fringe3D, Combust. Flame 158 (2011) 1807–1813.
[40] A.P. Terzyk, S. Furmaniak, P.A. Gauden, P.J.F. Harris, J. Włoch, P. Kowalczyk, Hyper parallel tempering Monte Carlo simulations of Ar adsorption in new models of microporous non-graphitizing activated carbon - effect of microporosity, J. Phys: Condens. Matt. 19 (2007) 406208.
[41] A.P. Terzyk, S. Furmaniak, P.J.F. Harris, P.A. Gauden, J. Włoch, P. Kowalczyk, G. Rychlicki, How realistic is the pore size distribution calculated from adsorption isotherms if activated carbon is composed of fullerene like fragments? Phys. Chem. Chem. Phys. 9 (2007) 5919–5927.
[42] A.P. Terzyk, S. Furmaniak, P.A. Gauden, P.J.F. Harris, J. Włoch, Testing isotherm models and recovering empirical relationships for adsorption in microporous carbons using virtual carbon models and GCMC simulations, J. Phys: Condens. Matt. 20 (2008) 385212.
[43] S. Furmaniak, A.P. Terzyk, P.A. Gauden, P. Kowalczyk, P.J.F. Harris, Can carbon surface oxidation shift the pore size distribution curve calculated from Ar, N_2 and CO_2 adsorption isotherms? Simulation results for realistic carbon model, J. Phys: Condens. Matt. 21 (2009) 315005.
[44] A.P. Terzyk, P.A. Gauden, S. Furmaniak, R.P. Wesołowski, P.J.F. Harris, Molecular dynamics simulation insight into the mechanism of phenol adsorption at low coverages from aqueous solutions on microporous carbons, Phys. Chem. Chem. Phys. 12 (2010) 812–817.
[45] S. Furmaniak, A.P. Terzyk, P.A. Gauden, P.J.F. Harris, P. Kowalczyk, The influence of carbon surface oxygen groups on Dubinin-Astakhov equation parameters calculated from CO_2 adsorption isotherm, J. Phys: Condens. Matt. 22 (2010) 085003.
[46] P.A. Gauden, A.P. Terzyk, S. Furmaniak, P.J.F. Harris, P. Kowalczyk, BET surface area of carbonaceous adsorbents - verification using geometric considerations and GCMC simulations on virtual porous carbon models, Appl. Surf. Sci. 256 (2010) 5204–5209.
[47] A.P. Terzyk, P.A. Gauden, S. Furmaniak, R.P. Wesołowski, P. Kowalczyk, Activated carbon immersed in water - the origin of linear correlation between enthalpy of immersion and oxygen content studied by molecular dynamics simulation, Phys. Chem. Chem. Phys. 12 (2010) 10701–10713.
[48] A.P. Terzyk, S. Furmaniak, R.P. Wesołowski, P.A. Gauden, P.J.F. Harris, Methane storage in microporous carbons - effect of porosity and surface chemical composition tested on realistic carbon model, in: Advances in Adsorption Technology, Nova Science Publishers, 2010.
[49] S. Furmaniak, A.P. Terzyk, P.A. Gauden, P. Kowalczyk, P.J.F. Harris, The influence of carbon surface chemical composition on Dubinin-Astakhov equation parameters calculated from SF_6 adsorption data - GCMC simulation, J. Phys: Condens. Matt. 23 (2011) 395005.
[50] J.P. Boudou, A. Martinez-Alonzo, J.M.D. Tascon, Introduction of acidic groups at the surface of activated carbon by microwave-induced oxygen plasma at low pressure, Carbon 38 (2000) 1021–1029.
[51] P.J.F. Harris, S.C. Tsang, High-resolution electron microscopy studies of non-graphitizing carbons, Philosoph. Magazine A 76 (1997) 667–677.
[52] P.J.F. Harris, Structure of non-graphitising carbons, Int. Mater. Rev. 42 (1997) 206–218.
[53] P.J.F. Harris, A. Burian, S. Duber, High-resolution electron microscopy of a microporous carbon, Philosoph. Magazine Lett. 80 (2000) 381–386.

[54] P.J.F. Harris, New perspectives on the structure of graphitic carbons, Crit. Rev. Solid State Mater. Sci. 30 (2005) 235–253.
[55] S.C. Tsang, P.J.F. Harris, J.B. Claridge, M.L.H. Green, A microporous carbon produced by arc-evaporation, J. Chem. Soc., Chem. Commun. 90 (1993) 1519–1522.
[56] P.J.F. Harris, Impact of the discovery of fullerenes on carbon science, in: Chemistry and Physics of Carbon, vol. 28, Marcel Dekker Inc, 2003.
[57] P.J.F. Harris, Fullerene-related structure of non-graphitizing carbons, in: Carbon Materials - Theory and Practice, Research Signpost, 2008.
[58] P.J.F. Harris, Non-graphitizing carbons, in: The Encyclopedia of Materials: Science and Technology, Elsevier, 2001.
[59] P.J.F. Harris, Z. Liu, K. Suenaga, Imaging the atomic structure of activated carbon, J. Phys: Condens. Matt. 20 (2008) 362201.
[60] A.P. Terzyk, P.A. Gauden, W. Zieliński, S. Furmaniak, R.P. Wesołowski, K.K. Klimek, First molecular dynamics simulation insight into the mechanism of organics adsorption from aqueous solutions on microporous carbons. Chem. Phys. Lett. 515 (2011) 102–108.
[61] W. Humphrey, A. Dalke, K. Schulten, VMD: visual molecular dynamics, J. Mol. Graphics 14 (1996) 33–38.
[62] http://www.ks.uiuc.edu/research/vmd/.
[63] M. Acharya, M.S. Strano, J.P. Mathews, S.J.L. Billinge, V. Petkov, S. Subramoneyk, H.C. Foley, Simulation of nanoporous carbons: a chemically constrained structure, Philosoph. Magazine B 79 (1999) 1499–1518.
[64] M.A. Smith, H.C. Foley, R.F. Lobo, A simple model describes the PDF of a non-graphitizing carbon, Carbon 42 (2004) 2041–2048.
[65] A. Goel, P.V. Hebgen, J.B. Vander Sande, J.B. Howard, Combustion synthesis of fullerenes and fullerenic nanostructures, Carbon 40 (2002) 177–182.
[66] A. Kumar, R.F. Lobo, N.J. Wagner, Porous amorphous carbon models from periodic Gaussian chains of amorphous polymers, Carbon 43 (2005) 3099–3111.
[67] F. Cataldo, The impact of a fullerene-like concept in carbon black science, Carbon 40 (2002) 157–162.
[68] H. Marsh, F. Rodriguez-Reinoso, Science of Carbon Materials, Universidad de Alicante, 2000.
[69] R. Setton, P. Bernie, S. Lefrant, Carbon Molecules and Materials, Taylor and Francis, 2002.
[70] P.A. Gauden, A.P. Terzyk, S. Furmaniak, Models of activated carbons: past, the present and future, Wiadomości Chemiczne 62 (2008) 403–447. in Polish.
[71] A.P. Terzyk, P.A. Gauden, S. Furmaniak, R.P. Wesołowski, P.J.F. Harris, P. Kowalczyk, Adsorption from aqueous solutions on opened carbon nanotubes - organic compounds speed up delivery of water from inside, Phys. Chem. Chem. Phys. 11 (2009) 9341–9345.
[72] P.A. Gauden, A.P. Terzyk, R. Pieńkowski, S. Furmaniak, R.P. Wesołowski, P. Kowalczyk, Molecular dynamics of zigzag single walled carbon nanotube immersion in water, Phys. Chem. Chem. Phys. 13 (2011) 5621–5629.
[73] J. Zawadzki, Infrared spectroscopy in surface chemistry of carbons, in: Chemistry and Physics of Carbon, vol. 21, Marcel Dekker Inc, 1988.
[74] J. Zawadzki, M. Wiśniewski, J. Weber, O. Heintz, B. Azambre, IR study of adsorption and decomposition of propan-2-ol on carbon and carbon-supported catalyst, Carbon 39 (2001) 187–192.
[75] J. Zawadzki, IR spectral studies on carbonization process of cellulose, Pol. J. Chem. 52 (1978) 2157–2166.

[76] A.P. Terzyk, Molecular properties and intermolecular forces-factors balancing the effect of carbon surface chemistry in adsorption of organics from dilute aqueous solutions, J. Coll. Interf. Sci. 275 (2004) 9–29.
[77] A.P. Terzyk, G. Rychlicki, S. Biniak, J.P. Łukaszewicz, The new correlations between the compositions of carbon surface layer of and its physicochemical properties exposed while paracetamol is adsorbed at different temperatures and pH, J. Coll. Interf. Sci. 257 (2003) 13–30.
[78] A.P. Terzyk, Adsorption of biologically active compounds from aqeous solutions on to commercial unmodified activated carbons. Part IV. Do the properties of amphoteric carbon surface layer influence paracetamol adsorption at the acidic pH? Adsorpt. Sci. Technol. 21 (2003) 93–123.
[79] A.P. Terzyk, G. Rychlicki, M.S. Ćwiertnia, P.A. Gauden, P. Kowalczyk, The chemical composition of carbon surface layer and benzene adsorption from vapour and from dilute aqueous solutions, Langmuir 21 (2005) 12257–12267.
[80] M. Jorge, C. Schumacher, N.A. Seaton, Simulation study of the effect of the chemical heterogeneity of activated carbon on water adsorption, Langmuir 18 (2002) 9296–9306.
[81] P.A. Gauden, A.P. Terzyk, M. Jaroniec, P. Kowalczyk, Bimodal pore size distributions for carbons: experimental results and computational studies, J. Coll. and Interf. Sci. 310 (2007) 205–216.
[82] P. Kowalczyk, A.P. Terzyk, P.A. Gauden, R. Leboda, E. Szmechtig-Gauden, G. Rychlicki, Z. Ryu, H. Rong, Estimation of the pore size distribution from nitrogen adsorption isotherm. The comparison of density functional theory and the method of Do and co-workers, Carbon 41 (2003) 1113–1125.
[83] P.A. Gauden, A.P. Terzyk, The normalisation of the micropore-size distribution function in the Polanyi-Dubinin type of adsorption isotherm equations, J. Coll. Interf. Sci. 227 (2000) 482–494.
[84] P.A. Gauden, A.P. Terzyk, P. Kowalczyk, Some remarks on the calculation of the pore size distribution function of activated carbons, J. Coll. Interf. Sci. 300 (2006) 453–474.
[85] S. Lowell, J.E. Shields, M.A. Thomas, M. Thommes, Characterization of Porous Solids and Powders: Surface Area, Pore Size and Density, Springer, 2004.
[86] P.A. Webb, C. Orr, Analytical Methods in Fine Particles Technology, Micromeritics Instrument, 1997.
[87] F. Rouquerol, J. Rouquerol, K.S.W. Sing, Adsorption by Powders and Porous Solids, Academic Press, 1999.
[88] S. Bhattacharya, K.E. Gubbins, Fast method for computing pore size distributions of model materials, Langmuir 22 (2006) 7726–7731.
[89] J.J. Potoff, J.I. Siepmann, Vapor-liquid equilibria of mixtures containing alkanes, carbon dioxide, and nitrogen, AIChE J. 47 (2001) 1676–1682.
[90] D.D. Do, H.D. Do, Adsorption of carbon tetrachloride on graphitized thermal carbon black and in slit graphitic pores: five-site versus one-site potential models, J. Phys. Chem. B 110 (2006) 9520–9528.
[91] P. Ungerer, B. Travitian, A. Boutin, Application of Molecular Simulation in the Oil and Gas Industry, Editions Technip, 2005.
[92] A.P. Terzyk, S. Furmaniak, P.A. Gauden, P. Kowalczyk, Fullerene intercalated graphene nanocontainers - the mechanism of Ar adsorption and the test of high pressure CH_4 and CO_2 storage capacities, Adsorpt. Sci. Technol. 27 (2009) 281–296.
[93] D. Dellis, J. Samios, Molecular force field investigation for sulfur hexafluoride: a computer simulation study, Fluid Phase Equilibria 291 (2010) 81–89.

[94] D.D. Do, H.D. Do, Effects of potential models in the vapor - liquid equilibria and adsorption of simple gases on graphitized thermal carbon black, Fluid Phase Equilibria 236 (2005) 169–177.
[95] M. Miyahara, T. Yoshika, M. Okazaki, Determination of adsorption equilibria in pores by molecular dynamics in a unit cell with imaginary gas phase, J. Chem. Phys. 106 (1997) 8124–8134.
[96] C.G. Gray, K.E. Gubbins, Theory of Molecular Fluids, vol. 1, Clarendon Press, 1984.
[97] D. van Der Spoel, E. Lindahl, B. Hess, G. Groenhof, A.E. Mark, H.J. Berendsen, GROMACS: fast, flexible, and free, J. Comput. Chem. 26 (2005) 1701–1718.
[98] H.J.C. Berendsen, D. van der Spoel, R. van Drunen, GROMACS: a message-passing parallel molecular dynamics implementation, Comput. Phys. Commun. 91 (1995) 43–56.
[99] E. Lindahl, B. Hess, D. van der Spoel, GROMACS 3.0: A package for molecular simulation and trajectory analysis, J. Mol. Model. 7 (2001) 306–317.
[100] B. Hess, C. Kutzner, D. van der Spoel, E. Lindahl, GROMACS 4: algorithms for highly efficient, load-balanced, and scalable molecular simulation, J. Chem. Theory Comput. 4 (2008) 435–447.
[101] T. Kar, S. Scheiner, A.K. Roy, The effect on acidity of size and shape of carboxylated single-wall carbon nanotubes. A DFT-SLDB study, Chem. Phys. Lett. 460 (2008) 225–229.
[102] S.G. Chen, R.T. Yang, Theoretical investigation of relationships between characteristic energy and pore size for adsorption in micropores, J. Coll. Interf. Sci. 177 (1996) 298–306.
[103] S.G. Chen, R.T. Yang, Theoretical basis for the potential theory adsorption isotherms. The Dubinin-Radushkevich and Dubinin-Astakhov equations, Langmuir 10 (1994) 4244–4249.
[104] H.F. Stoeckli, D. Morel, On the physical meaning of parameters E_0 and β in Dubinin's theory, Chimia 34 (1980) 502–503.
[105] R. Tsunoda, Relationships between adsorption parameters of the BET and the Dubinin-Radushkevich equations in nitrogen adsorption on active carbons, J. Coll. Interf. Sci. 117 (1987) 291–292.
[106] W. Rudziński, D.H. Everett, Adsorption of Gases on Heterogeneous Surfaces, Academic Press, 1992.
[107] P. Kowalczyk, A.P. Terzyk, P.A. Gauden, The application of CONTIN package for the evaluation of micropore size distribution function, Langmuir 18 (2002) 5406–5413.
[108] A.P. Terzyk, P.A. Gauden, P. Kowalczyk, What kind of pore size distribution assumes the Dubinin-Astakhov adsorption isotherm equation? Carbon 40 (2002) 2879–2886.
[109] P. Kowalczyk, L. Solarz, A.P. Terzyk, P.A. Gauden, V.M. Gun'ko, Solving the unstable linear Fredholm integral equation of the first kind by means of a new stochastic algorithm (ASA), Shedae Informaticae 11 (2002) 75–98.
[110] http://www.micromeritics.com/Library/DFT-NLDFT-Density-Functional-Theory.aspx.
[111] http://www.quantachrome.com/technical/dft.html.
[112] G. Horvath, K. Kawazoe, Method for the calculation of effective pore size distribution in molecular sieve carbon, J. Chem. Eng. Jpn. 16 (1983) 470–475.
[113] A.P. Terzyk, P.A. Gauden, G. Rychlicki, R. Wojsz, The new relationships between the characteristic energy of nitrogen adsorption (at 77.5 K) and the average effective diameter of carbon slit-like micropores, Carbon 36 (1998) 1703–1706 (Erratum (1999). Carbon, 37, 539).
[114] P.A. Gauden, A.P. Terzyk, Theory of Adsorption in Micropores of Carbonaceous Materials, WICHiR, 2002. in Polish.
[115] P. Tarazona, R. Evans, A simple density functional theory for inhomogeneous liquids. Wetting by gas at a solid-liquid interface, Mol. Phys. 52 (1984) 847–857.

[116] L.D. Gelb, K.E. Gubbins, R. Radhakrishnan, M. Śliwińska-Bartkowiak, Phase separation in confined systems, Rep. Progr. Phys. 62 (1999) 1573–1659.

[117] M. Borówko, Adsorption on heterogeneous surfaces, in: Adsorption: Theory, Modeling and Analysis, Marcel Dekker Inc, 2002.

[118] C. Nguyen, D.D. Do, A new method for the characterization of porous materials, Langmuir 15 (1999) 3608–3615.

[119] D.D. Do, C. Nguyen, H.D. Do, Characterization of micro-mesoporous carbon media, Coll. Surf. A: Physicochem. Eng. Asp. 187 (2001) 51–71.

[120] E.P. Barrett, L.G. Joyner, P.P. Halenda, The determination of pore volume and area distributions in porous substances. I. Computations from nitrogen isotherms, J. Am. Chem. Soc. 73 (1951) 373–380.

[121] M. Kruk, M. Jaroniec, Accurate method for calculating mesopore size distributions from argon adsorption data at 87 K developed using model MCM-41 materials, Chem. Mater. 12 (2000) 222–230.

[122] M. Kruk, M. Jaroniec, A. Sayari, Application of large pore MCM-41 molecular sieves to improve pore size analysis using nitrogen adsorption measurements, Langmuir 13 (1997) 6267–6273.

[123] M.M. Dubinin, A.V. Astakhov, The theory of volume filling of micropores for vapours and gases adsorption on microporous solids, Izvestiya Akademii Nauk SSSR, Seriya Khimicheskaya 5 (1971) 125–134.

[124] M.M. Dubinin, Fundamentals of the theory of adsorption in micropores of carbon adsorbents: characteristics of their adsorption properties and microporous structures, Carbon 27 (1989) 457–467.

[125] M.M. Dubinin, The potential theory of adsorption of gases and vapors for adsorbents with energetically nonuniform surfaces, Chem. Rev. 60 (1960) 235–241.

[126] M. Jaroniec, R. Madey, Physical Adsorption on Heterogeneous Solids, Elsevier, 1988.

[127] H.F. Stoeckli, A generalization of the Dubinin-Radushkevich equation for the filling of heterogeneous micropore systems, J. Coll. Interf. Sci. 59 (1977) 184–185.

[128] M. Jaroniec, J. Choma, Characterization of heterogeneity of activated carbons by utilizing the benzene adsorption data, Mater. Chem. Phys. 15 (1986) 521–536.

[129] P. Kowalczyk, P.A. Gauden, A.P. Terzyk, D.D. Do, G. Rychlicki, Characterising porous carbonaceous materials using the new methods. An attempt for estimation of micropore size distribution function from adsorption isotherm of nitrogen and benzene, Annales Universitatis Mariae Curie-Skłodowska Sectio AA Chemia 58 (2002) 46–74.

[130] D. Cazorla-Amoros, J. Alcaniz-Monge, A. Linares-Solano, Characterization of activated carbon fibers by CO_2 adsorption, Langmuir 12 (1996) 2820–2824.

[131] M.B. Sweatman, N. Quirke, Characterization of porous materials by gas adsorption: comparison of nitrogen at 77 K and carbon dioxide at 298 K for activated carbon, Langmuir 17 (2000) 5011–5020.

[132] P.A. Gauden, A.P. Terzyk, G. Rychlicki, P. Kowalczyk, M.S. Ćwiertnia, J.K. Garbacz, Estimating the pore size distribution of activated carbons from adsorption data of different adsorbates by various methods, J. Coll. Interf. Sci. 273 (2004) 39–63.

[133] H.F. Stoeckli, F. Kraehenbuehl, The external surface of microporous carbons, derived from adsorption and immersion studies, Carbon 22 (1986) 297–299.

[134] A. Seri-Levy, D. Avnir, The Brunauer-Emmett-Teller equation and the effects of lateral interactions. A simulation study, Langmuir 9 (1993) 2523–2529.

[135] A. Seri-Levy, D. Avnir, Effects of heterogeneous surface geometry on adsorption, Langmuir 9 (1993) 3067–3076.

[136] G.F. Cerofolini, N. Re, The mathematical theory of adsorption on non-ideal surfaces, Rivista del Nuovo Cimento della Societa Italiana di Fisica 16 (1993) 1–52.

[137] B. Coasne, A. Grosman, C. Ortega, R.J.-M. Pellenq, Physisorption in nanopores of various sizes and shapes: a grand canonical Monte Carlo simulation study, Stud. Surf. Sci. Catal. 144 (2002) 35–42.

[138] B. Coasne, K.E. Gubbins, R.J.-M. Pellenq, A grand canonical Monte Carlo study of adsorption and capillary phenomena in nanopores of various morphologies and topologies: testing the BET and BJH characterization methods, Part. Part. Syst. Character. 21 (2004) 149–160.

[139] R.J.-M. Pellenq, B. Coasne, P.E. Levitz, Adsorption/condensation of xenon in mesopores having a microporous texture or a surface roughness, in: Adsorption and Transport at the Nanoscale, CRC Press, 2006.

[140] B. Coasne, F. Di Renzo, A. Galarneau, R.J.M. Pellenq, Adsorption of simple fluid on silica surface and nanopore: effect of surface chemistry and pore shape, Langmuir 24 (2008) 7285–7293.

[141] L.D. Gelb, K.E. Gubbins, Characterization of porous glasses: simulation models, adsorption isotherms, and the Brunauer-Emmett-Teller analysis method, Langmuir 14 (1998) 2097–2111.

[142] L.F. Herrera, D.D. Do, G.R. Birkett, Characterization of virtual nano-structures through the use of Monte Carlo integration, Adv. Mater. Res. 32 (2008) 275–278.

[143] K.S. Walton, R.Q. Snurr, Applicability of the BET method for determining surface areas of microporous metal-organic frameworks, J. Am. Chem. Soc. 129 (2007) 8552–8556.

[144] S. Furmaniak, A.P. Terzyk, P.A. Gauden, R.P. Wesołowski, P. Kowalczyk, Ar, CCl_4 and C_6H_6 adsorption outside and inside of the bundles of multi-walled carbon nanotubes - simulation study, Phys. Chem. Chem. Phys. 11 (2009) 4982–4995.

[145] J. Rouquerol, P. Llewellyn, F. Rouquerol, Is the BET equation applicable to microporous adsorbents? Stud. Surf. Sci. Catal. 160 (2007) 49–56.

[146] A. Pedretti, L. Villa, G. Vistoli, VEGA - an open platform to develop chemo-bio-informatics applications, using plug-in architecture and script programming, J. Comp.-Aid. Mol. Des. 18 (2004) 167–173.

[147] http://www.vegazz.net/.

[148] K. Kaneko, C. Ishii, M. Ruike, H. Kuwabara, Origin of superhigh surface area and microcrystalline graphitic structures of activated carbons, Carbon 30 (1992) 1075–1088.

[149] K.A.G. Amankwah, J.A. Schwarz, A modified approach for estimating pseudo-vapor pressures in the application of the Dubinin-Astakhov equation, Carbon 33 (1995) 1313–1319.

[150] S. Ozawa, S. Kusumi, Y. Ogino, Physical adsorption of gases at high pressure. IV. An improvement of the Dubinin-Astakhov adsorption equation, J. Coll. Interf. Sci. 56 (1976) 83–91.

[151] B.-U. Choi, D.-K. Choi, Y.-W. Lee, B.-K. Lee, S.-H. Kim, Adsorption equilibria of methane, ethane, ethylene, nitrogen, and hydrogen onto activated carbon, J. Chem. Eng. Data 48 (2003) 603–607.

[152] D.D. Do, Adsorption Analysis: Equilibria and Kinetics, Imperial College Press, 1998.

[153] A. Dąbrowski, P. Podkościelny, Z. Hubicki, M. Barczak, Adsorption of phenolic compounds by activated carbon - a critical review, Chemosphere 58 (2005) 1049–1070.

[154] L.R. Radovic, C. Moreno-Castilla, J. Rivera-Utrilla, Carbon materials as adsorbents in aqueous solutions, in: Chemistry and Physics of Carbon, vol. 27, Marcel Dekker Inc, 2001.

[155] C. Moreno-Castilla, Adsorption of organic solutes from dilute aqueous solutions, in: Adsorption by Carbons, Elsevier, 2008.

[156] A.P. Terzyk, Further insights into the role of carbon surface functionalities in the mechanism of phenol adsorption, J. Coll. Interf. Sci. 268 (2003) 301–329.

[157] R.W. Coughlin, F.S. Ezra, Role of surface acidity in the adsorption of organic pollutants on the surface of carbon, Environ. Sci. Technol. 2 (1968) 291–297.

[158] C. Bertoncini, J. Raffaelli, L. Fassino, H.S. Odetti, E.J. Bottani, Phenol adsorption on porous and non-porous carbons, Carbon 41 (2003) 1101–1111.

[159] M.C. Gordillo, J. Martí, Hydrogen bond structure of liquid water confined in nanotubes, Chem. Phys. Lett. 329 (2000) 341–345.

[160] I. Efremenko, M. Sheintuch, Predicting solute adsorption on activated carbon: phenol, Langmuir 22 (2006) 3614–3621.

[161] J.K. Garbacz, S. Furmaniak, A.P. Terzyk, M. Grabiec, New model describing adsorption from liquid binary mixtures of nonelectrolytes with limited and unlimited miscibility of components, J. Coll. Interf. Sci. 359 (2011) 512–519.

[162] A.P. Terzyk, J. Chatłas, P.A. Gauden, G. Rychlicki, P. Kowalczyk, Developing the solution analogue of the Toth adsorption isotherm equation, J. Coll. Interf. Sci. 266 (2003) 473–476.

[163] F.H. Healey, Y.F. Yu, J.J. Chessick, The detection of hydrophilic heterogeneities on a carbon surface, J. Phys. Chem. 59 (1955) 399–402.

[164] H.F. Stoeckli, K. Kraehenbuehl, V. Morel, The adsorption of water by active carbons, in relation to the enthalpy of immersion, Carbon 21 (1983) 589–591.

[165] F. Carrasco-Marin, A. Mueden, T.A. Centeno, H.F. Stoeckli, C. Moreno-Castilla, Water adsorption on activated carbons with different degrees of oxidation, J. Chem. Soc., Faraday Trans. 93 (1997) 2211–2215.

[166] A.P. Terzyk, G. Rychlicki, The influence of activated carbon surface chemical composition on the adsorption of acetaminophen (paracetamol) in vitro: the temperature dependence of adsorption at the neutral pH, Coll. Surf. A: Physicochem. Eng. Asp. 163 (2000) 135–150.

[167] A.P. Terzyk, The influence of activated carbon surface chemical composition on the adsorption of acetaminophen (paracetamol) in vitro: part II. TG, FTIR, and XPS analysis of carbons and the temperature dependence of adsorption kinetics at the neutral pH, Coll. Surf. A: Physicochem. Eng. Asp. 177 (2001) 23–45.

[168] L.R. Pratt, D. Chandler, Theory of the hydrophobic effect, J. Chem. Phys. 67 (1977) 3683–3704.

[169] L. Maibaum, D. Chandler, Segue between favorable and unfavorable solvation, J. Phys. Chem. B 111 (2007) 9025–9030.

[170] G. Hummer, S. Garde, A.E. Garcia, A. Pohorille, L.R. Pratt, An information theory model of hydrophobic interactions, Proc. Natl. Acad. Sci. United States of America 93 (1996) 8951–8955.

Chapter 4

Advanced Physical Adsorption Characterization of Nanoporous Carbons

Matthias Thommes and Katie A. Cychosz
Quantachrome Instruments, Boynton Beach, FL, USA

Alexander V. Neimark
Rutgers University, Department of Chemical and Biochemical Engineering, Piscataway, NJ, USA

Chapter Outline

4.1. Introduction **107**
4.2. Experimental Aspects **109**
4.2.1. General Comments 109
4.2.2. Choice of Adsorptive 111
4.3. Adsorption Mechanism **116**
4.4. Aspects of Surface Area Assessment **119**
4.5. Pore Size and Porosity Analysis **122**
4.5.1. General Comments 122
4.5.2. Classical, Macroscopic Thermodynamic and Empirical Methods for Pore Size and Porosity Analysis 123
4.5.3. Modern, Microscopic Methods for Pore Size and Porosity Analysis: Density Functional Theory 126
4.5.4. Pore-Size Characterization of Micro–Mesoporous Carbon Materials – Selected Application Examples 130
4.6. Conclusions **138**
References **139**

4.1. INTRODUCTION

A proper characterization of the textural properties of nanoporous carbons (e.g. activated carbon, carbon aerogels, ordered mesoporous carbons, nanotubes, and

Novel Carbon Adsorbents. DOI: 10.1016/B978-0-08-097744-7.00004-1
Copyright © 2012 Elsevier Ltd. All rights reserved.

nanohorns) is crucial for a better understanding of their behavior in important industrial applications including separation, gas and energy storage, batteries, and many others [1–5]. Major advancements in materials characterization and practical utilization are related to the progress made in the development of various experimental techniques such as gas adsorption, X-ray diffraction (XRD), small-angle X-ray and neutron scattering (SAXS and SANS), mercury porosimetry, electron microscopy (scanning and transmission), thermoporometry, NMR methods, and others. Each method has a limited length scale of applicability for pore-size analysis and an overview of different methods for pore-size characterization and their application range was given by the International Union of Pure and Applied Chemistry (IUPAC) [6]. However, among all these methods, gas adsorption (i.e. physical adsorption) is still the most popular one because it allows for assessing a wide range of pore sizes (from 0.35 nm up to 100 nm). Furthermore, gas adsorption techniques are convenient to use and less cost intensive than some of the other methods.

Physisorption (physical adsorption) occurs whenever a gas (the adsorptive) is brought into contact with a solid surface (the adsorbent). The matter in the adsorbed state is known as the adsorbate, as distinct from the adsorptive, which is the gas or vapor to be adsorbed. The forces involved in physisorption always include the long-range London dispersion forces and the short-range intermolecular repulsion. These combined forces give rise to nonspecific molecular interactions. Physisorption in porous materials is governed by the interplay between the strength of fluid–wall and fluid–fluid interactions as well as the effects of confined pore space on the state and thermodynamic stability of fluids in narrow pores [7–12]. This is also reflected in the shape or type of the adsorption isotherm. IUPAC has published a classification of six types of adsorption isotherms and proposed to classify pores by their internal pore width (the pore width defined as the diameter in the case of cylindrical and spherical pores and as the distance between opposite walls in case of slit pores) [13]. According to their diameter, the pores are divided into three groups: micropores of width less than 2 nm, mesopores of width between 2 and 50 nm, and macropores of width greater than 50 nm. Micro- and mesopores are often collectively referred to as nanopores. The micropore range is often further subdivided into ultramicropores (pores smaller than 0.7 nm) and supermicropores (pores from 0.7 to 2 nm). As indicated, the pore size is generally specified as the internal pore width, although for mesopores it has been customary to refer to the pore radius (or pore diameter). Of course, such terms have a precise meaning only when the pore shape is properly defined. Furthermore, the internal or effective pore width must not be confused with the distance between the centers of surface atoms (the outer atoms of solid in the opposite walls of a pore), which determines the pore geometry in molecular level theries and simulations. According to [1], the pore morphology describes geometrical shape and structure of the pores including pore width and volume as well as the nature of the pore walls. On the other hand, pore topology describes how the pores are arranged and the connectivity of pores. It should be noted that a new IUPAC task group has been formed in 2010 which focuses on revising the 1985 recommendations (for more information, see http://www.iupac.org/web/ins/2010-009-1-100).

The recent advances in physical adsorption characterization (for reviews, see for instance [10–12]) are, to a large extent, related to (i) discovery of novel ordered mesoporous materials (e.g. M41S materials), which exhibit a uniform pore size with periodically ordered structure and can therefore be used as model adsorbents to test the theories of gas adsorption; (ii) carefully performed adsorption experiments coupled with complimentary experimental techniques; (iii) development and application of microscopic approaches such as the density functional theory (DFT) of inhomogeneous fluids, as well as methods based on molecular simulation. These modern theoretical and computational methods, which are based on the statistical mechanics of nanophases, describe the configuration of adsorbed fluid (the adsorbate) on a molecular level. It has been shown that the application of DFT methods allows one to obtain reliable pore-size distributions over the complete range of micro- and mesopores. The methods for pore-size analysis based on DFT and molecular simulation are now widely used; they are featured in ISO standard ISO 15901-3 and are commercially available for many important adsorptive/adsorbent systems. The DFT methods include hybrid methods, which assume different pore geometries for the micro- and mesopore size ranges, as found in materials with hierarchical pore structures. More recent advances include the development of quenched solid density functional theory (QSDFT), which quantitatively takes into account the surface geometrical inhomogeneity characterized by a roughness parameter. It has been demonstrated that this novel method significantly improves the accuracy of the pore-size distribution calculations for many micro- and mesoporous carbon materials [14–16].

Within this context, this chapter reviews some important aspects of applying gas adsorption for the textural characterization of micro- and mesoporous carbons, but it should not be considered a comprehensive review. Rather, it discusses important aspects of surface area and pore-size analysis of porous carbons. We address briefly some experimental aspects, focus on adsorption mechanisms in micro- and mesoporous carbons, and then discuss the methods of pore-size/volume characterization on selected examples.

4.2. EXPERIMENTAL ASPECTS

4.2.1. General Comments

The adsorption isotherm, which represents the adsorbed gas amount as a function of pressure at constant temperature, can be obtained by volumetric (manometric) and gravimetric methods, carrier gas and calorimetric techniques, nuclear resonance, as well as by a combination of calorimetric and impedance spectroscopic measurements (for an overview, see [7–9]). However, the most frequently used methods are the volumetric (manometric) and gravimetric methods. The gravimetric method is based on a sensitive microbalance and a pressure gauge. The adsorbed amount can be measured directly, but a pressure-dependent buoyancy correction is necessary. The gravimetric method is convenient to use for the study of adsorption not too far from room temperature. The volumetric method is recommended to measure adsorption of nitrogen, argon, and krypton at cryogenic temperatures, i.e. the temperatures of liquid nitrogen (77 K) and argon (87 K), which are primarily used for surface area and pore-size characterization [13]; an important advantage of the volumetric method is that the adsorbent is in direct contact with the thermostat and the adsorbent temperature is well defined. The volumetric method is based on calibrated volumes and pressure measurements, and it uses the general gas equation of state. The adsorbed amount is calculated by determining the difference of the total amount of gas admitted to the sample cell with the adsorbent and the amount of gas in the dead space. In this scenario, the dead volume needs to be very accurately determined.

Physical adsorption in micropores occurs at relative pressures substantially lower (very often down to relative pressures below 10^{-5}) than in the case of adsorption in mesopores and special requirements are necessary to obtain accurate adsorption isotherms on microporous samples. The most important point is that physical adsorption in micro- and mesoporous adsorbents can span a broad spectrum of pressures (up to seven orders of magnitude). In order to study the adsorption of gases such as nitrogen and argon (at their boiling temperatures) within the relative pressure range from $10^{-7} \leq P/P_0 \leq 1$ with sufficiently high accuracy, it is desirable to use a combination of different transducers within different pressure ranges. In addition, one has to assure that the sample cell and the manifold can be evacuated to very low pressures, which requires a proper high vacuum pumping system, i.e. a turbomolecular pump. Further it is also highly recommended to monitor the saturation pressure, P_0, throughout the entire analysis that can be achieved by means of a dedicated saturation pressure transducer.

In order to obtain correct data, it is required to remove all physically adsorbed material from the adsorbent surface to ensure a reproducible initial state. In principle, this can be accomplished by vacuum pumping or purging with an inert gas at elevated temperatures. Outgassing under vacuum is desirable because it prepares the surface under the same conditions (evacuated sample cell) that are required to start a static volumetric adsorption experiment. This is particularly important for carbons with very narrow micropores; here, the adsorption measurements often start at relative pressures as low as 10^{-7}, and the samples require an outgassing at pressures below 0.01 Pa. This can be achieved by using a turbomolecular pump which, if coupled with a diaphragm roughing pump, allows the sample to be outgassed even in a completely oil-free system.

4.2.2. Choice of Adsorptive

4.2.2.1. Standard Adsorptives for Carbon Characterization

The question of which adsorptive to choose for the physical adsorption characterization is important in particular with regard to the fact that for nitrogen adsorption at 77 K, micropore filling of carbon pores smaller than 1 nm (which are present in many carbon materials) occurs at very low pressures. The rate of diffusion and adsorption equilibration is extremely slow at relative pressures below 10^{-5}. Furthermore, a pore width of 0.7 nm corresponds to the thickness of the bilayer formed by nitrogen molecules. Pre-adsorbed N_2 molecules near the entry of a narrow micropore may block further adsorption. This leads to time-consuming measurements and may cause under-equilibration of measured adsorption isotherms, giving erroneous results for the analysis. For characterization of microporous materials, it may be necessary to utilize other gases (such as argon or carbon dioxide) to accurately determine pore-size distributions in some carbons. Although the kinetic diameters of nitrogen (0.36 nm), argon (0.34 nm), and carbon dioxide (0.33 nm) are similar, the adsorption behavior of these three adsorptives is quite different.

Argon at 87 K shows nearly perfect adsorption behavior because, compared to nitrogen and carbon dioxide, argon does not give rise to specific interactions with a variety of surface functional groups, which can lead to enhanced adsorption/specific interactions caused by quadrupole moment characteristic to non-symmetric molecules. As a consequence of this factor and a slightly higher temperature, argon at 87 K fills micropores of dimensions 0.5–1 nm at higher relative pressures compared to nitrogen at 77 K, leading to accelerated diffusion and faster equilibration times. This effect is much more pronounced, for instance, for zeolites and metal–organic frameworks than for carbons. The pore-filling pressures of argon (87 K) are often shifted 1–1.5 decades in relative pressure as compared to nitrogen [17,18]. A comparison between argon (87 K) and nitrogen (77 K) adsorption on a physically activated carbon (obtained from Norit) is shown in Fig. 4.1 and is also discussed in detail in [19].

The question arises whether the advantages of argon adsorption at 87 K will also be present at liquid nitrogen temperature (77 K). Unfortunately, argon adsorption at 77 K is less than optimal for both pore size and surface area analysis because (i) a combined and complete micro- and mesopore size analysis with argon is not possible at 77 K (which is ca. 6.5 K below the triple point temperature of bulk argon) and is limited to pore diameters smaller than 16 nm (for details see [20]); (ii) although the argon adsorption isotherm at 77 K is still shifted to higher relative pressures as compared to the appropriate nitrogen adsorption, the shift toward higher absolute pressure is less pronounced as compared to argon adsorption at 87 K. This is due to the low saturation pressure of argon at 77 K (205 torr for solid argon and 230 torr for

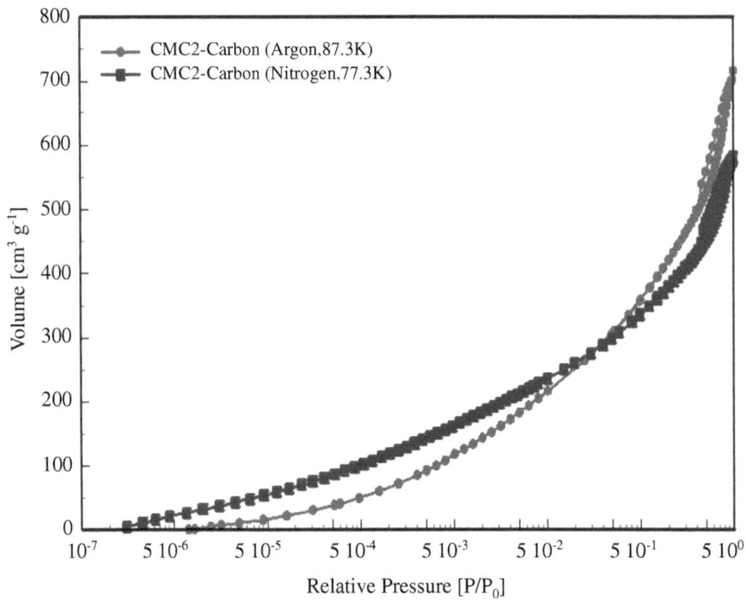

FIGURE 4.1 Nitrogen (77 K) and argon (87 K) adsorption isotherms on activated carbon.

supercooled liquid argon). Hence, argon adsorption at 77 K does not offer the same "experimental benefits" as argon adsorption at 87 K.

Despite the advantages which argon adsorption at 87 K offers, pore filling of ultramicropores still occurs at very low pressures. Associated with the low pressures is the well-known problem of restricted diffusion, which prevents nitrogen and also argon molecules from entering the narrowest micropores – pores of width <0.45 nm. In order to address this, the use of CO_2 as adsorptive at temperatures close to room temperature has been suggested. At 273 K for instance, CO_2 is still ca. 32 K below its critical temperature and because the saturation pressure is very high (26,200 torr), the relative pressure measurements necessary for the micropore analysis are achieved in the range of moderate absolute pressures (1–760 torr). Hence, due to the relatively high absolute temperatures and pressures compared with nitrogen and argon adsorption at cryogenic temperatures, it has been established that diffusion limitations of nitrogen in micropores can be eliminated by the use of CO_2 [21–23]. CO_2 molecules are able to easily access the ultramicropores despite the fact that the dimensions of N_2, Ar, and CO_2 are similar. Furthermore, the experimental setup needed to perform micropore analysis with CO_2 is much simpler than the corresponding experimental setup for micropore analysis by argon or nitrogen adsorption, because a turbomolecular pump and low-pressure transducers are not necessary. Because of these experimental advantages, CO_2 adsorption has become a standard tool for the assessment of microporous carbons. Consequently, advanced DFT and grand canonical Monte Carlo (GCMC) methods have been developed for calculating the pore-size distribution from the CO_2 adsorption isotherm (more details found below) [24,25].

One of the main advantages for performing CO_2 adsorption experiments at 273 K is that, because of the enhanced kinetics, a micropore characterization with CO_2 can be performed in few hours – much faster than adsorption experiments at cryogenic temperatures. However, it should be noted that in particular for carbons with very narrow pores, as found in some carbon molecular sieves, the adsorption is sensitive to the chosen experimental equilibration parameter [26]. The kinetics of CO_2 adsorption in carbon micropores has also been investigated (for carbon aerogels) by combining adsorption experiments with small-angle X-ray scattering [27]. Additionally, similar to nitrogen, CO_2 has a quadrupole moment, which affects the adsorption behavior on carbons containing oxygen surface functionalities, complicating the pore-size analysis by CO_2 for such carbons [28].

A nice example which demonstrates the complementary use of various adsorptives along with the proper use of DFT for calculation of pore-size distribution (discussed in detail below) is shown in Fig. 4.2. Nitrogen (77 K), argon (87 K), and carbon dioxide (273 K) were used to characterize an ultramicroporous carbon (microwave-exfoliated graphene oxide) useful for supercapacitor applications [29]. Figure 4.2a shows the nitrogen, argon, and carbon dioxide isotherms on this carbon (plotted in a linear scale) and Fig. 4.2b shows the three isotherms plotted in a semi-logarithmic scale. The semi-logarithmic plots illustrate that carbon dioxide fills even the narrowest micropores at much higher relative pressures than both nitrogen and argon. In the case of this carbon, the NLDFT pore-size distribution calculated from the carbon dioxide adsorption isotherm clearly shows the presence of pores smaller than 0.7 nm (Fig. 4.2c). Only with the combination of carbon dioxide and nitrogen adsorption can complementary and complete pore-size and pore-volume information be determined. It is also worthwhile noting that the NLDFT cumulative pore-volume data (Fig. 4.2c) allows one to differentiate between micro- and mesopore volume in a straightforward manner.

4.2.2.2. Other Adsorptives for Carbon Characterization

With the recent growing interest in hydrogen storage and fuel cell applications, there has been increased research on hydrogen adsorption on carbon materials and pore-size determination using hydrogen at 77 K. Although highly supercritical at this temperature, it has been shown that by using hydrogen adsorption at 77 K coupled with a proper NLDFT approach, pore-size analysis can be extended to pore sizes smaller than those that can be characterized with other adsorptives, including CO_2 [30,31]. However, the use of hydrogen as an adsorptive for pore-size analysis is limited to a pore width range smaller than 0.5 nm.

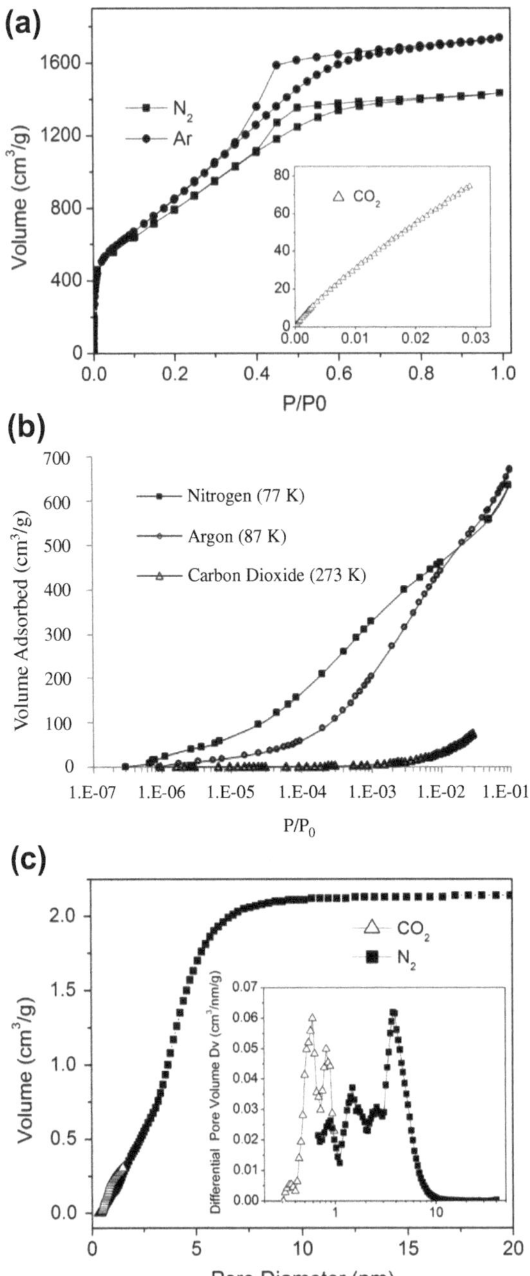

FIGURE 4.2 (a) Nitrogen (77 K) and argon (87 K) isotherms on a microwave-exfoliated graphene oxide material. Inset is the carbon dioxide (273 K) isotherm on the same material, (b) semi-logarithmic display of the nitrogen (77 K), argon (87 K), and carbon dioxide (273 K) isotherms highlighting the low-pressure adsorption data up to a relative pressure of 10^{-2}, and (c) cumulative pore-volume and (inset) pore-size distribution calculated using DFT methods. *(From [26]. Reprinted with permission from AAAS)*

The method of nonane preadsorption has been suggested many decades ago [7] for the characterization of complex micro- and mesoporous carbons. Here, one assesses microporosity by the retention of *n*-nonane in narrow pores. In this method, the narrow micropores are filled with nonane, which can only be desorbed at elevated temperatures – the narrower the micropore, the higher the temperature to remove the nonane. However, this method is not always straightforward, and in carbons which exhibit a wide pore-size distribution, nonane adsorbed in narrow micropores may block the entrances to the wider pores. Recently, *n*-nonane preadsorption has been successfully applied to lignocellulosic-derived activated carbons [32] in order to elucidate the details of their complex micro–mesopore network.

The use of water as a probe for surface chemistry and pore structure analysis has attracted a great deal of interest [33–47]. Water is an appealing adsorbtive because it can be adsorbed at room temperature and has a very small kinetic diameter (0.28 nm), which allows for probing the pores that are not accessible to nitrogen or carbon dioxide [38]. A comparison of water (at 298 K) and nitrogen (at 77 K) adsorption on purely microporous physically activated carbon TE80 (from Pica Corporation) is shown in Fig. 4.3 [48]. The isotherms are of completely different shape, type I for nitrogen adsorption (based on the IUPAC classification) but type V for water adsorption. Whereas in the case of nitrogen the micropores are already completely filled at a relative pressure <0.1, micropore filling with water occurs at relative pressures >0.4 due to the hydrophobic nature of the carbon surface. Up to this pressure, almost no

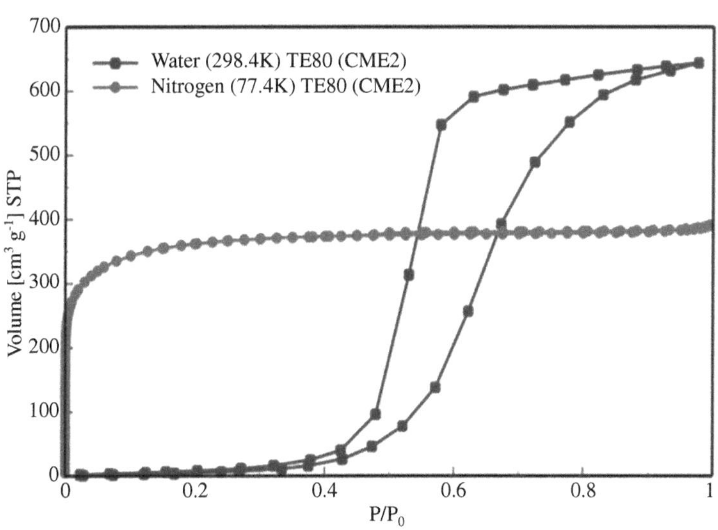

FIGURE 4.3 Comparison of nitrogen (77 K) and water (298 K) adsorption isotherms on a physically activated carbon (Picactif TE80 from Pica) illustrating the adsorption hysteresis observed for the micropores in the water isotherm. *(From [48]. Reprinted with Permission)*

adsorption was observed. Further, the water sorption isotherms show hysteresis, due to different mechanisms of adsorption/pore filling (*via* clustering) and desorption (*via* molecular evaporation), as discussed in the literature [34,46]. It should be noted that the occurrence of hysteresis in the range where nitrogen and argon isotherms are reversible can potentially be useful for textural characterization.

However, interpretation of water isotherms is difficult because of the competing effects of pore structure and surface chemistry. In a recent study, these effects were investigated for three commercially available activated carbons [48]. Water adsorption was coupled with textural analysis *via* nitrogen and argon adsorption, as well as surface characterization by TPD-MS and XPS. Although all three carbons studied had essentially identical micropore structures (as evidenced by a combination of Ar (87 K), N_2 (77 K), and CO_2 (273 K) adsorption), differences in the water adsorption isotherms could be clearly associated with differences in surface chemistry, indicating the potential of water adsorption for surface chemistry assessments.

In order to assess the smallest pore dimensions in microporous materials, Kaneko et al. suggested helium which is the smallest inert molecule [49,50]. Of course, the experiments have to be performed below helium's critical temperature (ca. 5.2 K) and Kaneko et al. measured helium adsorption at 4.2 K on activated carbon fibers. From these data, micropore volumes 20–50% greater than those obtained from nitrogen adsorption at 77 K were found. The excess amount of helium adsorption was ascribed to the presence of ultramicropores, which cannot be accessed by nitrogen molecules.

4.3. ADSORPTION MECHANISM

In order to obtain surface area, pore-size distribution, pore volume, and other structural information from the analysis of gas adsorption isotherms, it is necessary to understand the underlying adsorption mechanisms. The mechanisms of micro- and mesopore filling differ substantially. The filling of micropores is, in most cases, a continuous process. The filling of ultramicropores occurs at low relative pressures ($P/P_0 < 0.01$) and is entirely governed by the enhanced gas–solid interactions. However, in addition to the strong adsorption potential, a cooperative mechanism may play a role in the pore-filling process of supermicropores ($P/P_0 \approx 0.01$–0.2). The relative pressure, at which the micropore filling occurs, depends on a number of factors, including the size and nature of adsorptive molecules, pore shape, and effective pore width. The pore-filling capacity depends essentially on the accessibility of the pores for the probe molecules, which is determined by the molecular size and chosen experimental conditions. A basic understanding of the micropore-filling process has already been achieved by Dubinin and co-workers based on the pioneering work of Polanyi [51,52].

By contrast, mesopores fill *via* capillary condensation, which represents a gas–liquid phase transition. The sorption behavior in mesopores depends not only on the fluid–wall attraction, but also significantly on the attractive interactions between the adsorbate molecules. This leads to the occurrence of multilayer adsorption and capillary condensation. Capillary condensation represents a phenomenon whereby gas condenses to a liquid-like phase in pores at a pressure less than the saturation pressure, P_0, of the bulk fluid. It represents an example of a shifted bulk transition under the influence of the attractive fluid–wall interactions. Capillary condensation is very often accompanied by hysteresis, and it is widely accepted that there is a correlation between the shape of the hysteresis loop and the texture of the adsorbent. An empirical classification of hysteresis loops was given by IUPAC [13].

Traditionally, pore condensation has been described within the Kelvin approach [see Refs. 7–9 and references therein]. For pores of uniform shape and width (ideal slit-like or cylindrical mesopores), the shift of the gas–liquid-phase transition of a confined fluid from bulk coexistence is expressed in terms of the macroscopic quantities like the surface tension of the bulk fluid, the densities of the coexistent liquid, and the contact angle of the liquid meniscus against the pore wall. However, the Kelvin concept fails to correctly describe the peculiarities of the critical region, confinement-induced shifts of the phase diagram of condensed fluid, and hysteresis behavior, specifically the disappearance of hysteresis with decreasing pore size (at given temperature) or increasing temperature (for a given pore size) (for a review see [12,53]).

By contrast, it has been shown that modern, microscopic approaches such as DFT (NLDFT and QSDFT) and molecular simulation (grand canonical Monte Carlo (GCMC) simulation) are capable of qualitatively and quantitatively predicting capillary condensation and hysteresis behavior of fluids in highly ordered model materials. Two principal factors determine the hysteretic behavior: hysteresis on the level of a single pore of given shape and the cooperative effects due to the specifics of pore network connectivity. On the pore level, adsorption hysteresis is considered as an intrinsic property of the vapor–liquid phase transition in a finite volume system. A classical scenario of capillary condensation implies that the vapor–liquid transition is delayed due to the existence of metastable adsorption films and hindered nucleation of liquid bridges. In open, uniform cylindrical pores of finite length, metastable states occur only on the adsorption branch. In an open pore filled by liquid-like condensate, the liquid–vapor interface is already present and evaporation occurs without nucleation, *via* a receding meniscus. That is, the desorption process is associated with the equilibrium vapor–liquid transition. This mechanism of adsorption hysteresis is present in ordered mesoporous materials with uniform cylindrical pores (e.g. MCM-41, SBA-15). In order to describe hysteresis phenomena in more disordered adsorbents, network models have been developed [54–60]. The network models attribute hysteresis to pore-blocking/percolation effects.

In the case of pore blocking, in a material with an ink-bottle shape pore, desorption is delayed and the wide body of the pore remains filled until the neck evaporates at a lower vapor pressure. Additionally, the desorption branch of the hysteresis loop is significantly steeper than the adsorption branch, resulting in a hysteresis loop of type H2 according to the IUPAC classification. Adsorption/desorption mechanisms were recently studied with the aid of model materials containing well-defined ink-bottle pores (e.g. SBA-16, hierarchically ordered porous materials such as KLE) [61–67]. These theoretical and experimental studies revealed that if the pore neck diameter is smaller than a certain critical size at a given temperature and adsorptive, desorption occurs *via* cavitation, or spontaneous nucleation of a bubble in the pore. In this case, the pore body empties, while the pore neck remains filled. Figure 4.4 illustrates that for a given temperature and adsorptive, the neck size dictates the desorption mechanism. In the case of pore blocking (neck width larger than the critical width of ca. 5–6 nm for nitrogen and argon adsorption at 77 and 87 K, respectively), evaporation occurs at the equilibrium pressure of the corresponding meniscus and, therefore, information about the neck (pore entrance size) can be obtained from the desorption branch of the isotherm. In the case where the neck is smaller than the critical size, desorption from the pore body occurs *via* cavitation. Studies of the temperature dependence of the hysteresis loop in such materials showed a transition between cavitation and pore-blocking regimes, suggesting a possibility of

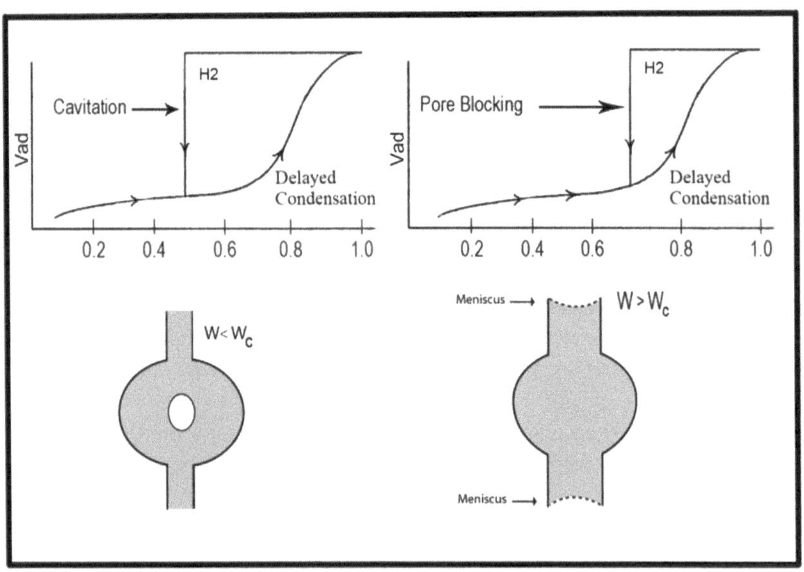

FIGURE 4.4 Schematic of pore-blocking and cavitation phenomena. W is the size of the pore neck and W_c is the critical neck size below which cavitation is the desorption mechanism. *(Reprinted with Permission from [69]. Copyright 2006 American Chemical Society)*

estimating the neck-size distribution from the desorption isotherm by tuning the experimental conditions [64]. Another possibility is to use various probe molecules and it has been demonstrated that adsorption experiments performed with two adsorptives (e.g. nitrogen and argon at 77 K and 87 K, respectively) allow for detecting and separating the effects of pore-blocking/percolation and cavitation [67].

Cavitation-induced evaporation appears to be important for many micro- and mesoporous solids and is responsible for the often observed characteristic stepdown in the desorption isotherm at relative pressure ranges from 0.4 to 0.5 (giving rise to type H3 and H4 hysteresis). In the past, this characteristic stepdown was discussed within the framework of the tensile strength hypothesis [7,68] which was associated with the lower closure point of hysteresis. In this classical approach, it was supposed that the tensile stress limit of condensed fluid, which is indicated by cavitation, does not depend on the nature and pore structure of the adsorbent, but is a universal feature of the adsorptive. However, the observed cavitation-induced stepwise desorption in ink-bottle pores occurs at appreciably larger pressures (up to relative pressures of 0.5) than the "universal" lower closure point of hysteresis mentioned above, which for nitrogen (at 77 K) and argon (at 87 K) was assumed to be at relative pressures of 0.42 and 0.38, respectively. Although there is a weak dependency of the cavitation pressure on the pore size for pore diameters smaller than ca. 12 nm (for spherical pores) [69], the position of the cavitation transition is mainly determined by the state of the pore fluid. Hence, pore-size analysis from the desorption isotherm would lead to artificial spikes around a pore size of 4 nm (for reviews, see [11,12]).

4.4. ASPECTS OF SURFACE AREA ASSESSMENT

Surface area is the crucial parameter for optimized porous carbons for many applications. However, due to the complex nature of micro- and mesoporous materials, no single experimental technique can be expected to provide an evaluation of the "absolute" surface area. Surface area values are of relative nature and should always be related to the method, conditions, and probe molecules used in the experimental work. Despite the well-known limitations, the Brunauer–Emmett–Teller (BET) method continues to be widely used for the evaluation of surface areas of micro- and mesoporous adsorbents. The BET equation is applicable for the surface area estimates of nonporous, macro-, and mesoporous materials (pore width > 4 nm), but its extension to microporous materials contradicts the basic physical mechanism of micropore volume filling rather than monolayer coverage implied by the BET theory [8,9]. In the case of microporous materials, the surface area obtained using the BET method should be considered as an "equivalent" surface area. Despite this limitation, the BET method is the most frequently used method for surface area determination of porous carbons.

In order to determine the BET surface area, one must first transform the measured isotherm into a BET plot from which the monolayer capacity, n_m, is derived. Once the monolayer capacity has been determined, the surface area can be calculated using the molecular cross-sectional area. The monolayer capacity, n_m, is calculated using the BET equation:

$$\frac{p/p_0}{n_a[1-(p/p_0)]} = \frac{1}{n_m C} + \frac{C-1}{n_m C} \cdot \frac{p}{p_0}$$

where n_a is the adsorbed amount and C is an empirical constant that gives an indication of the attractive adsorbent–adsorbate interactions. Emmett and Teller originally found that various adsorbents gave linear BET plots over the relative pressure range $P/P_0 = 0.05$–0.3.

In microporous materials, micropores fill at relative pressures below 0.1 and it is difficult to separate monolayer formation from micropore filling. The linear BET range is often shifted to significantly lower relative pressures, but the problem of which relative pressure range should be chosen for BET analysis is of prominent importance because of the possibility for significant overestimation of the monolayer capacity. In order to address the problem, Rouquerol et al. have determined a procedure for finding the linear BET range for a microporous material [70,117]. Based on an alternative plotting of the BET equation first proposed in 1961 [71], they suggest that (a) C must be positive (i.e. any negative intercept on the BET plot indicates that one is outside the valid range of the BET equation) and (b) application of the BET equation must be limited to the range where the term $[V(1-P/P_0)]$ continuously increases with P/P_0. This method has recently been adopted by the International Organization for Standardization (ISO) (ISO FDIS 9277, Determination of the Specific Surface Area of Solids by Gas Adsorption BET Method, 2009) and is successful at accurately determining the linear BET range for microporous materials. This method is illustrated with the example of a microporous activated carbon fiber (ACF) sample in Fig. 4.5. If the traditional linear BET pressure range is applied to the data, the resulting BET plot (Fig. 4.5a) is nonlinear, indicating an incorrect BET plot/range. Additionally, the BET plot intercept is negative, meaning that the C-constant is negative as well, which is unphysical. In this case, the calculated BET area is 1008 m^2/g. If, instead, the method of Rouquerol et al. is applied, a plot such as the one shown in Fig. 4.5b results. The maximum of this plot indicates the highest relative pressure in the linear BET range. This information is used to construct the linear BET plot in Fig. 4.5c. As such, a linear BET range of 0.026–0.071, well below the range recommended for mesoporous materials, is found, corresponding to the "equivalent" BET area of 1167 m^2/g, over 100 m^2/g more than if the traditional BET range is used.

In order to address problems associated with applying the BET equation to microporous materials, it was also suggested to treat the physisorption data of nitrogen by a modified BET equation, which includes the micropore volume,

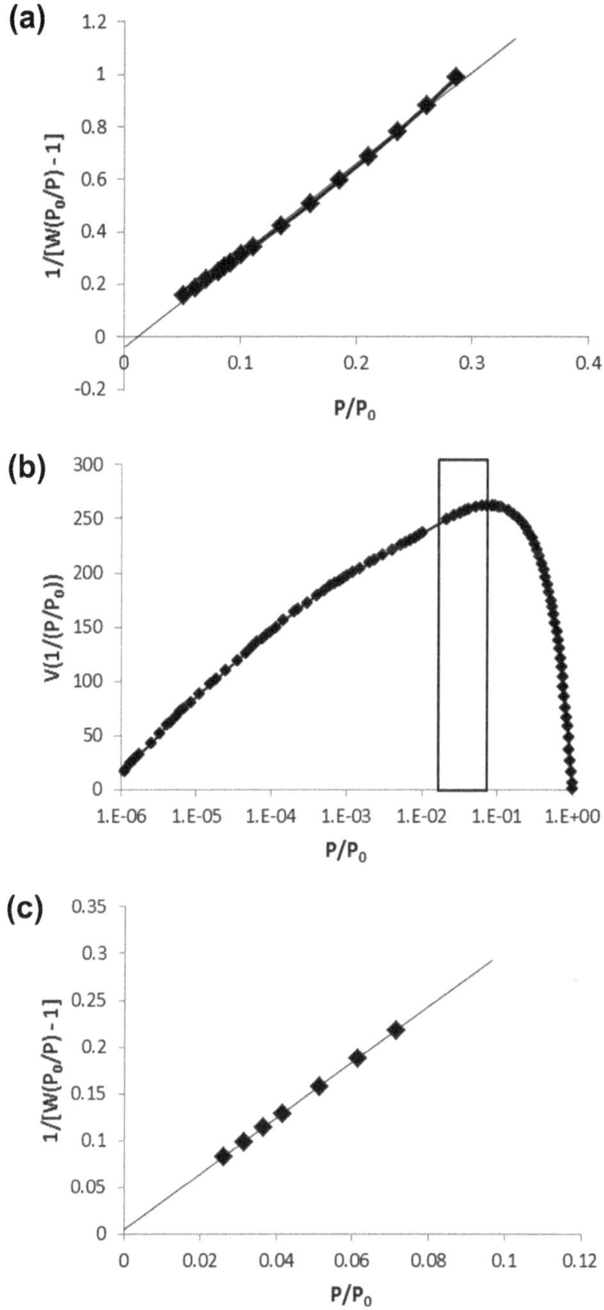

FIGURE 4.5 (a) BET plot for an activated carbon fiber sample using the classical relative pressure range ($P/P_0 - 0.05$–0.3), (b) plot of P/P_0 vs. $V(1 - P/P_0)$ used to determine the upper limit (graph maximum) for the BET range, and (c) BET plot in the relative pressure range determined using the method proposed by Rouquerol et al.

V_{micro}. The suggested approach allows one to extract the V_{micro} value from the adsorption data and consequently to determine the value of C_{BET} and specific surface area of the nonmicroporous part of the sample [72]. Other methods to obtain a specific surface area from gas adsorption are based on the application of the standard isotherm concept, e.g. *t*- and alpha-s methods which allow one to obtain the external (nonmicroporous) surface area [7–9]. By subtracting the external surface area from the BET area, one can determine a micropore surface area. A micropore surface area can also be obtained by methods related to the Dubinin approach (i.e. the Dubinin–Kaganer method) [7–9].

Another alternative to obtain the surface area of microporous solids and to differentiate between micro- and external surface area is the application of the DFT method, which calculates the specific cumulative surface area (specific surface area as a function of pore size) over the complete range of micro- and mesopores. However, accurate results are obtained only if the given experimental adsorptive/adsorbent system is compatible with one of the available DFT methods. Details of the DFT methods are discussed below.

A recent study of 190 different porous carbons confirmed that the application of the BET method is problematic for assessing surface areas [73]. It was suggested that the application of methods derived from the Dubinin-Radushkevitch approach, methods based on density functional theory (NLDFT and QSDFT), and the application of immersion calorimetry are useful concerning the surface area determination of microporous carbons.

4.5. PORE SIZE AND POROSITY ANALYSIS

4.5.1. General Comments

As indicated before, over the last two decades tremendous progress has been achieved with regard to the understanding of sorption phenomena in narrow pores, which in turn led to major progress in physical adsorption characterization. This progress is greatly associated with the development and application of microscopic methods, such as the density functional theory (DFT) of inhomogeneous fluids (e.g. nonlocal density functional theory, NLDFT) or computer simulation methods (e.g. Monte Carlo (MC) and molecular dynamic (MD) simulations) [53,74–83]. These molecular-level methods, which are based on statistical mechanics, describe the specifics of distribution of adsorbed molecules in pores, in contrast to the classical methods which are based on macroscopic, thermodynamic assumptions (e.g. Dubinin–Radushkevich (DR), Horvath–Kawazoe (HK), comparison plot methods for the analysis of micropores, or Kelvin equation-based methods such as the Barrett–Joyner–Halenda (BJH) or Broekhoff–de Boer (BDB) methods for analysis of mesopores) [7–9]. Each of these classical methods is based on the assumption of a certain adsorption mechanism within a pore width range (i.e. micro- or mesopores) and, consequently, one needs to combine at least two methods for characterization of porous carbons which contain both micro- and mesopores. On the other hand, application of a DFT-based method allows one to obtain a reliable pore-size analysis over the complete micro- and mesopore-size range.

4.5.2. Classical, Macroscopic Thermodynamic and Empirical Methods for Pore Size and Porosity Analysis

4.5.2.1. Micropore Analysis

In the absence of mesoporosity, the physisorption isotherm is of type I (according to the IUPAC classification) with a plateau which is virtually horizontal. In this case, one can correlate the adsorbed amount in the plateau region directly with the micropore volume by applying the Gurvitsch rule [84]. The Gurvitsch rule implies that the pores are filled with a homogeneous, bulk-like liquid, an assumption that does not account for the fact that the degree of molecular packing in small pores depends on both the pore size and shape. This problem has been addressed by applying microscopic methods (e.g. molecular simulation or DFT), which reflect the degree of molecular packing in the details of the density profile.

In order to assess the micropore volume (and average pore width) of nanoporous carbons, methods based on the Dubinin–Radushkevich (DR) approach [52,85] are frequently applied; the DR method is still considered a standard tool for obtaining micropore volume. The DR equation is written as follows:

$$V_{\mathrm{a}} = V_{\mathrm{micro}} \exp\left[- \left\{ \left(\frac{RT}{\beta E_o} \right) \ln \frac{p_0}{p} \right\}^2 \right]$$

where V_a is the volume of liquid adsorbate, V_{micro} is the micropore volume, β is the affinity coefficient (e.g. 0.33 for nitrogen), and E_0 is the characteristic adsorption energy. A plot of $\log V_a$ vs. $[\log(P_0/P)]^2$ should give a straight line, which is very often found at relative pressures between 10^{-4} and 10^{-2}. The volume of mesopores, V_{meso}, is then estimated by subtracting V_{micro} from the total pore volume, which is obtained from the Gurvitsch rule at a relative pressure where all pores are filled (typically at $P/P_0 = 0.95$).

The DR equation often fails to linearize the data when the microporous adsorbent is very heterogeneous with regard to surface chemistry and texture. To overcome such deficiencies in the original DR equation, a more general equation, known as the DA equation, was proposed by Dubinin and Astakhov [86]. Another generalization of the DR equation has been introduced by Stoeckli et al. [87,88]. Based on empirical studies, Dubinin and Stoeckli proposed that E_0 can be related to the average micropore width, W, for a carbonaceous adsorbent by the following simple formula: $W = 26$ (kJ nm/mol)/E_0. Later, Stoeckli et al. [89,90] suggested a slightly modified relationship: $W = 13.7$ (kJ nm/mol)/$(E_0 - 9.7)$. More recent work focused on the influence of carbon surface oxygen groups on the DA equation parameters for CO_2 adsorption on carbons [29].

It is also worthwhile to note that within this context the relationship between the pore volumes and surface areas derived from nitrogen adsorption isotherms using several different methods (DR, BET, and t-plot) has been recently investigated [91]. Carbon aerogels exposed to temperatures varying from 1000 °C to 2500 °C were used as model materials and a linear correlation was observed between the BET surface area and the DR pore volume.

Furthermore, the application of the standard and comparison isotherm concept allows one to determine micropore volume, (external) surface area, and, in principle, information about the average pore size. The t-plot method proposed by Lippens and de Boer [92] compares the test isotherm on a microporous material with a reference type II isotherm composed of averaging the experimental data on a number of nonporous adsorbents with a BET C constant similar to that of the microporous sample being tested. The experimental test isotherm is then redrawn as a *t*-curve, i.e. a plot of the volume of gas adsorbed as a function of *t*, which is the standard multilayer thickness on the reference nonporous material at the corresponding P/P_0. Another method for estimating micropore volume is the alpha-s method, developed by Sing et al. [93,94]. An advantage compared to the *t*-method is that the construction of the alpha-s plot does not require the monolayer capacity and allows, therefore, for a more direct comparison between the test isotherm and the reference isotherm (for details, see [7]). The reference isotherm in the alpha-s method is a plot of the amount of gas adsorbed, normalized by the amount of gas adsorbed at a fixed relative pressure vs. P/P_0. The reference relative pressure is usually taken as $P/P_0 = 0.4$, and the normalized $V_{ads}/V_{ads}^{0.4}$ is plotted. Principally, the alpha-s method allows one to (i) identify the physisorption mechanisms; (ii) check the validity of the BET nitrogen area; (iii) evaluate the total surface area of nonporous or mesoporous solids; and (iv) assess the effective micropore volume and external area of an ultramicroporous or supermicroporous solid. It is also worthwhile mentioning that Kaneko et al. [95] introduced the high-resolution alpha-s analysis, based on a high-resolution standard isotherm. The high-resolution method makes particular use of the alpha-s plot below $P/P_0 = 0.4$ and allows one to obtain more information concerning micro- and mesoporosity in the adsorbent.

A more advanced method, which allows for the determination of the micropore-size distribution of microporous carbons up to a pore width of 2 nm, was put forward by Horvath and Kawazoe (HK) and is based on the work of Everett and Powl [96,97]. The HK method allows for the calculation of the micropore-volume distribution from a low-pressure nitrogen adsorption isotherm obtained on a slit-pore carbon. Saito and Foley extended this approach to argon (87 K) adsorption in zeolitic pores with cylindrical pore geometry [98]. Contrary to the classical, Dubinin-related approaches, the HK approach employs the adsorption potential in slit carbon pores and provides a detailed account for the enhancement of the adsorption potential as the pore-size decreases. The advantage of the HK method is that it is specific with respect to the pore shape and adsorbate–adsorbent interaction potential. However, a disadvantage is that it does not give a realistic description of the micropore filling because it neglects the inhomogeneity of the adsorbed molecules in the micropores, leading to inaccuracies in the pore size and volume determination. Some improvements of the HK method have been considered [99], but treatments such as DFT or methods of molecular simulation are superior and provide a much more accurate approach for pore-size analysis over the complete range of micro- and mesopores.

4.5.2.2. Mesopore Analysis

The Kelvin equation serves as the basis for many methods of mesopore analysis, including the widely used Barrett–Joyner–Halenda (BJH) method. The Kelvin equation provides a relationship between the pore diameter and the pore condensation pressure and predicts that pore condensation shifts to a higher relative pressure with increasing pore diameter and temperature. In order to account for the multilayer adsorbed film, the Kelvin equation is combined with a standard isotherm (i.e. *t*-curve), which usually refers to the adsorption on a reference nonporous solid. The Kelvin equation is written as

$$\ln(P/P_0) = -2\gamma\cos\theta/\mathrm{RT}\Delta\rho(r_\mathrm{p} - t_\mathrm{c})$$

where R is the universal gas constant, r_p the pore radius, and t_c is the thickness of an adsorbed multilayer film, which is formed prior to the capillary condensation. For nitrogen and argon adsorption at their boiling temperatures (77.4 K and 87.3 K, respectively), the contact angle, θ, of the adsorbate adsorbed on the pore walls is zero (i.e. situation of complete wetting).

The multilayer adsorbed film is assessed by the statistical (mean) thickness of an adsorbed film on a nonporous solid of a surface similar to that of the sample under consideration. However, it is obvious that assuming the situation of a nonporous, planar surface for the evaluation of the adsorbed film thickness cannot be valid in the case of narrow pores, where the curvature of the pore walls and their adsorption potential has a pronounced effect on the film thickness and its interfacial tension. The validity of the Kelvin equation also becomes questionable because it does not allow for the effect of the adsorption forces on the pore condensation. Furthermore, the macroscopic, thermodynamic concept of a smooth liquid–vapor interface and bulk-like core fluid cannot be realistically applied to condensation in narrow mesopores. The development of model mesoporous molecular sieves (e.g. M41S materials) allowed for the first time direct experimental tests of the validity of the modified Kelvin equation and the BJH method. Because of the high degree of order, the pore diameter of such model substances can be derived by independent methods

(X-ray-diffraction, high-resolution transmission electron microscopy, etc.). It was found that the BJH method and related procedures underestimate the pore size by up to 20–30 % for pores smaller than 10 nm if not properly corrected. Some improvement was achieved by modifying the classical Broekhoff–de Boer approach and by calibrating the Kelvin equation using a series of mesoporous molecular sieves (MCM-silicas) of known pore diameter [118,119]. However, it needs to be stressed that approaches based on calibration are strictly only valid over a limited pore-size range (for reviews, see [11,12]). On the contrary, as discussed in the following section, advanced methods based on density functional theory (DFT) and molecular simulation allow one to obtain an accurate pore-size analysis and combined micro- and mesopore analysis. This is also demonstrated in Fig. 4.6 which shows the pore-filling pressure for nitrogen (77 K) in slit-pore carbons as predicted by DFT, Horvath Kawazoe (HK), and modified Kelvin equation (MK) for a series of pore widths [75]. For a given pore-filling pressure, the corresponding micropore width, as predicted by the HK method (micropore range) and modified Kelvin equation (mesopore range), is underestimated compared to the prediction by DFT.

4.5.3. Modern, Microscopic Methods for Pore Size and Porosity Analysis: Density Functional Theory

DFT and computer simulation methods (such as molecular dynamics and Monte Carlo simulation) are powerful techniques that describe sorption and phase

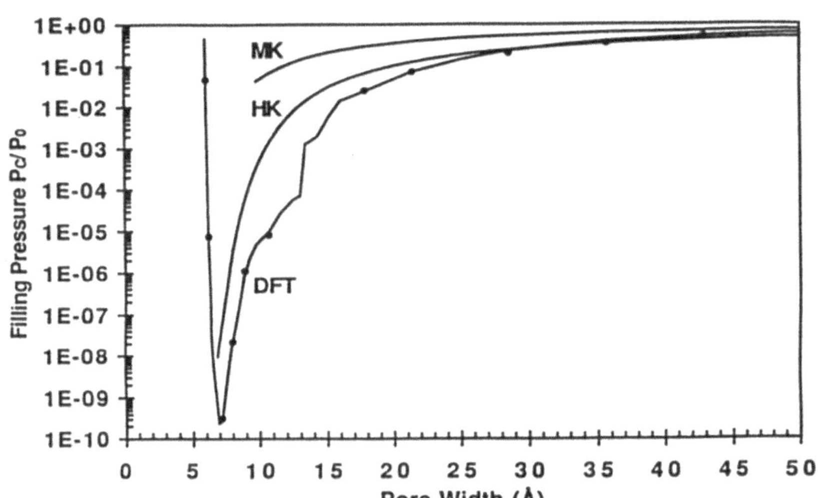

FIGURE 4.6 Pore-filling pressures for nitrogen (77 K) in slit-pore carbons as a function of pore width as predicted by DFT, HK, and modified Kelvin (MK) equations. *(Reprinted with Permission from [75]. Copyright 1993 American Chemical Society)*

behavior of inhomogeneous fluids confined in nanoporous materials [74–83]. These approaches describe the distribution of adsorbed molecules in pores and thus provide detailed information about the local fluid structure near curved solid walls as compared to the bulk fluid structure. It is generally assumed that at experimental conditions the adsorbate in a pore is in thermodynamic equilibrium with a bulk gas phase. The local density, $\rho(r)$, of the pore fluid is therefore determined by minimization of the grand thermodynamic potential, which depends on the intermolecular potentials of fluid–fluid (adsorbate–adsorbate) and fluid–solid (adsorbate–adsorbent) interactions, whereby the latter depends on the pore geometry. The parameters of the fluid–fluid interactions are usually determined in a way that they reproduce the bulk fluid properties (e.g. surface tension and gas–liquid coexistence curve). The parameters of the solid–fluid potential are chosen to fit the standard adsorption isotherms on well-defined nonporous carbon material. Once $\rho(r)$ is known, other thermodynamic properties, such as the adsorption isotherm, energy of adsorption, and positions of phase transitions, can be calculated.

Pioneering studies on DFT and the adsorption and phase behavior of fluids in pores were performed by Evans and Tarazona [100,101]. Seaton et al. [81] were the first to apply DFT for the calculation of the pore-size distribution (PSD) in both the meso- and micropore ranges. In this first attempt, the pore-size distribution analysis based on the local version of DFT was used. Although the local DFT provides a qualitatively reasonable description of adsorption in pores, it is quantitatively inaccurate especially in the range of micropores. A significant improvement in accuracy was obtained with the nonlocal density functional theory (NLDFT), which was first reported for the pore-size analysis of microporous carbons in 1993 [75]. Since then, NLDFT has been frequently used for the pore-size analysis of micro- and mesoporous materials (for a review, see [10] and references therein) and the NLDFT method is now commercially available for many adsorbent/adsorptive systems.

A drawback of the NLDFT method for the pore-size characterization of disordered carbon materials is that the solid surface is treated as molecularly smooth and its predictions imply pronounced layering steps on adsorption isotherms, which are not observed in experimental adsorption isotherms on amorphous carbons. This can cause prominent artifacts in the NLDFT PSDs, such as the gap at ca. 1 nm, which is characteristic of many porous-activated carbons [24,102]. Several approaches have been suggested to account for the heterogeneity of carbon materials including the development of advanced structural models of porous carbons by reverse Monte Carlo techniques [1,78,80] which, however, are still too complex to be implemented for routine pore-size analyses. Others suggested modeling porous carbons with a mixed geometry model [103]. Molecular simulations have also demonstrated that the surface roughness and defects significantly affect the shape of adsorption isotherms on heterogeneous surfaces [35,104–106].

Within the framework of the standard slit-pore model of carbons, variability of pore-wall thickness was introduced [107,108], but it led to just a marginal improvement over the standard NLDFT approach. The drawbacks of the conventional NLDFT model were also addressed by the introduction of two-dimensional DFT approaches [109].

In order to account quantitatively for the effects of surface heterogeneity in a practical way, the quenched solid density functional theory (QSDFT) was introduced. In contrast to the conventional NLDFT models that assume structureless graphitic pore walls, the solid is modeled using the distribution of solid atoms rather than the source of the external potential field. Within the framework of QSDFT, the grand potential of both solid and fluid is considered. In fact, QSDFT is a multicomponent DFT, in which the solid is treated as one of the components of the adsorbate–adsorbent system. The surface heterogeneity in the QSDFT model is characterized by a single roughness parameter that represents the characteristic width of molecular level surface corrugations. It has been demonstrated that the QSDFT model significantly improves the method of adsorption porosimetry [15,19]. The QSDFT method for carbons [15] has originally been developed assuming slit-shaped pores, which are typical model pores in activated microporous carbons. However, due to the emergence of novel materials with predesigned pore morphology obtained by synthesis routes which make use of structure directing agents or hard templates, the QSDFT method has recently been extended to micro–mesoporous carbons with cage-like and channel-like pore geometries [16].

In this chapter, we do not focus on the theoretical and mathematical details of the NLDFT and QSDFT methods, which were described in detail in [75,110] for NLDFT and in [15,16] for QSDFT. The calculation of the pore-size distribution is based on a solution of the *generalized adsorption equation* (GAE) which is sometimes also called *integral adsorption equation*, as given in the equation below, which correlates the kernel of theoretical adsorption (or desorption) isotherms with the experimental adsorption (desorption) isotherm:

$$N_{\mathrm{exp}}(P/P_0) = \int_{D_{\mathrm{min}}}^{D_{\mathrm{max}}} N_{\mathrm{QSDFT}}(P/P_0, D) f(D) \mathrm{d}D$$

Here, D_{min} and D_{max} are the minimum and maximum pore sizes in the kernel. The kernel $N_{\mathrm{QSDFT}}(P/P_0, D)$ represents a set of theoretical isotherms in model pores for a series of pore diameters, which cover the whole range of micro- and mesopores accessible in the adsorption experiment.

The kernel can be regarded as a theoretical reference for a given class of adsorbent/adsorptive systems. A typical QSDFT kernel of theoretical isotherms of nitrogen (77 K) adsorption in cylindrical carbon pores is shown in Fig. 4.7 [16].

The experimental isotherm is represented as the convolution of the DFT kernel (set of the theoretical isotherms) in a series of pores within a given range of pore sizes, D, and the sought pore-size distribution function *f*(D). The procedure is illustrated schematically in Fig. 4.8.

Solution of the GAE equation is obtained using the quick non-negative least-squares method [110]. In this method, the equation is represented as a matrix equation, which is solved using the discrete Tikhonov regularization method combined with the non-negative least-squares algorithm.

As mentioned before, the standard NLDFT approaches do not sufficiently take into account the chemical and geometrical heterogeneity of the pore walls and a structureless, molecularly smooth pore-wall model is assumed. This mismatch between the theoretical NLDFT assumptions and the experimental systems causes the theoretical adsorption isotherms to exhibit multiple steps associated with layering transitions related to the formation of a monolayer, second adsorbed layer, and so on. These layering steps on the theoretical isotherms can cause artificial gaps on the calculated pore-size distributions. Figure 4.9 illustrates this problem for nitrogen (77 K) adsorption on activated carbon fiber (ACF-15). The prominent step at $P/P_0 = 0.3 \times 10^{-3}$, which is characteristic of the theoretical NLDFT isotherms and is due to the monolayer transition on the smooth graphite surface, is completely eliminated in the QSDFT isotherm. As a consequence, a sharp minimum in the NLDFT pore-size

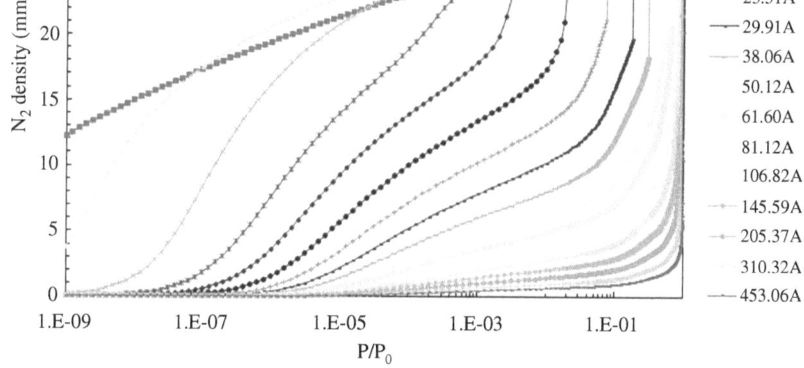

FIGURE 4.7 Kernel of adsorption isotherms of nitrogen at 77 K in cylindrical pores with molecularly rough walls. The pore widths are given in the right panel. *(Reprinted from [16], with Permission from Elsevier)*

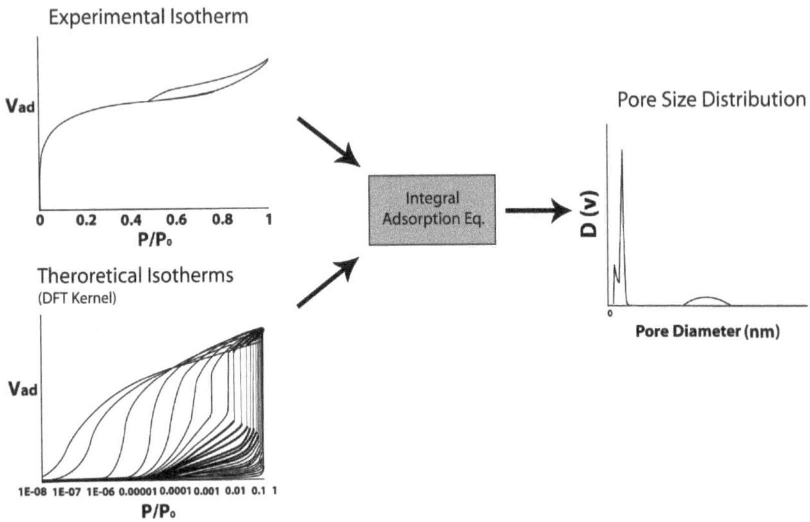

FIGURE 4.8 Schematic representation of the procedure applied to calculate a DFT pore-size distribution from experimental adsorption/desorption isotherms.

distribution curve at ~1 nm, which is typical to the NLDFT pore-size distribution curves for many microporous carbons, does not appear in QSDFT calculations. This confirms that the minimum on the differential NLDFT pore-size distribution is indeed an artifact caused by the monolayer step in the NLDFT approach. The pore-size distribution curves in Fig. 4.9b show that QSDFT and NLDFT agree above and below the regions where artificial gaps were observed with NLDFT; this is in line with the observation that the cumulative pore-volume curves calculated by NLDFT and QSDFT (Fig. 4.9c) are also in good agreement [15] outside the 1-nm pore width range.

A similar scenario is observed for argon (87 K) adsorption. Pore-size distributions calculated using the argon QSDFT methods likewise do not exhibit the layering transition characteristic of NLDFT and any false minima in the pore-size distribution, as shown for a Calgon F400 activated carbon in Fig. 4.10.

4.5.4. Pore-Size Characterization of Micro–Mesoporous Carbon Materials – Selected Application Examples

In order to validate and test recently developed QSDFT methods for cylindrical carbon pores, ordered mesoporous carbon CMK-3, which is an inverse replica of SBA-15 silica [111], has been employed. The complete micro- and meso-pore-size distribution of ordered CMK-3 was calculated from both the adsorption and desorption branches of the nitrogen (77 K) isotherm (Fig. 4.11a) by applying a N_2 (77 K) cylindrical pore carbon QSDFT method to both the

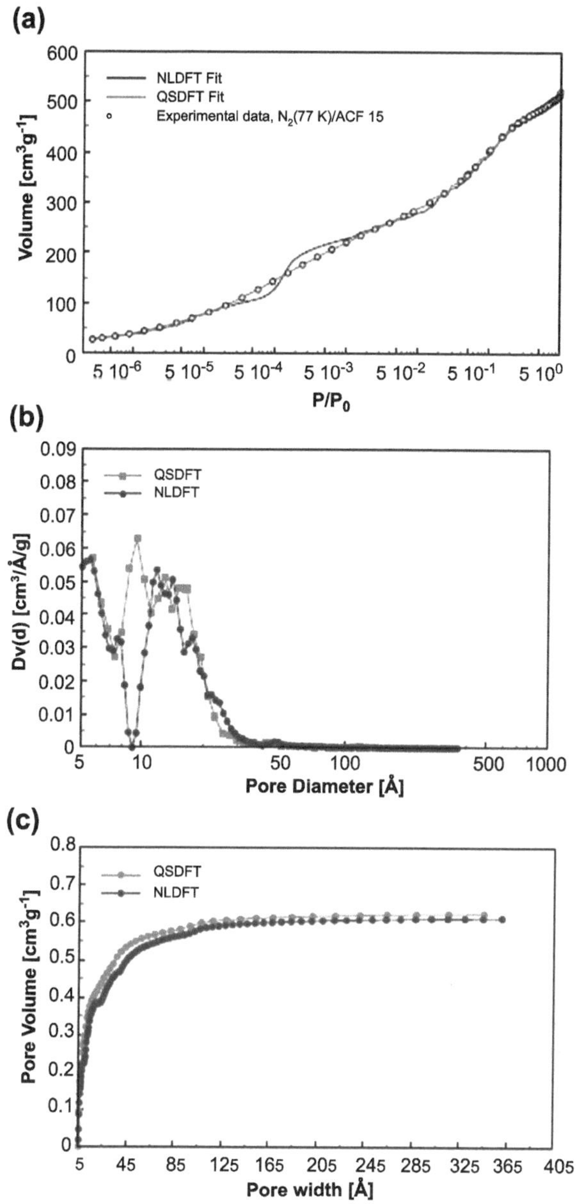

FIGURE 4.9 Comparison of the NLDFT and QSDFT methods for nitrogen (77 K) adsorption on activated carbon fiber ACF-15. (a) Experimental isotherm plotted on a semi-logarithmic scale with the NLDFT and QSDFT theoretical isotherms, (b) NLDFT and QSDFT pore-size distributions calculated from the nitrogen isotherm, and (c) QSDFT and NLDFT cumulative pore volumes. *(Reprinted from [15], with Permission from Elsevier)*

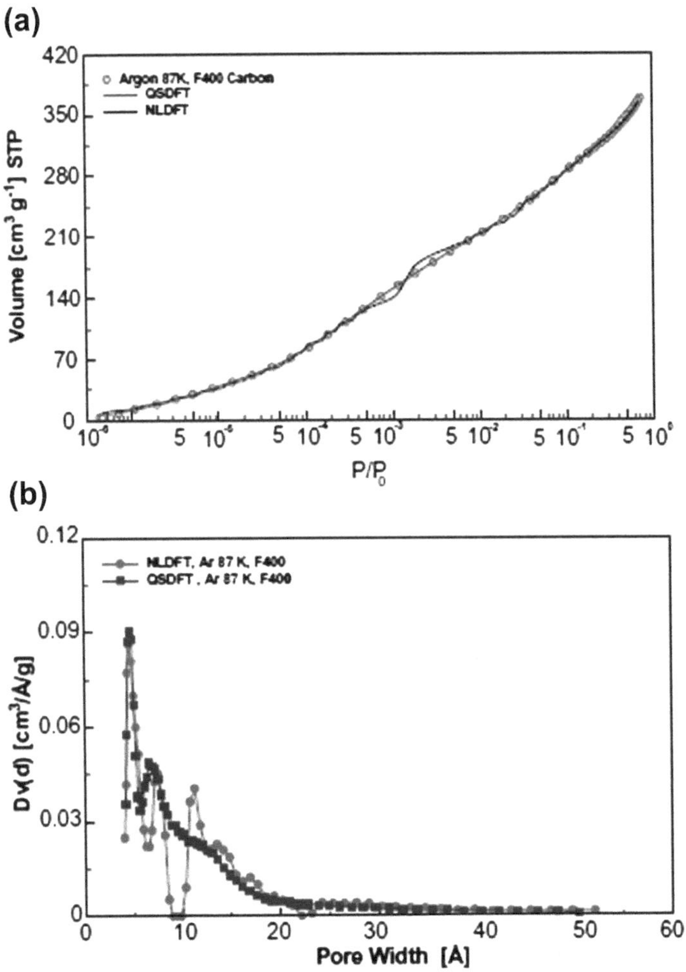

FIGURE 4.10 Pore-size distributions calculated for granular-activated carbon Calgon F400. (a) Experimental argon (87 K) isotherm plotted on a semi-logarithmic scale together with the NLDFT and QSDFT theoretical isotherms and (b) NLDFT and QSDFT pore-size distributions calculated from the argon isotherm. *(Reprinted from [15], with Permission from Elsevier)*

adsorption and desorption data (Fig. 4.11b). An excellent fit of the QSDFT model to both the adsorption branch and desorption branch (Fig. 4.11a) of the experimental isotherm is obtained by applying (i) the adsorption branch kernel,

FIGURE 4.11 Nitrogen (77 K) adsorption on a CMK-3 sample. (a) Experimental isotherm with fit-from-QSDFT cylindrical adsorption and cylindrical equilibrium kernels, (b) pore-size distributions from QSDFT cylindrical adsorption and cylindrical equilibrium kernels derived from adsorption and desorption branches of the experimental isotherm, and (c) pore-size distributions calculated from the nitrogen isotherm using the BJH and QSDFT methods.

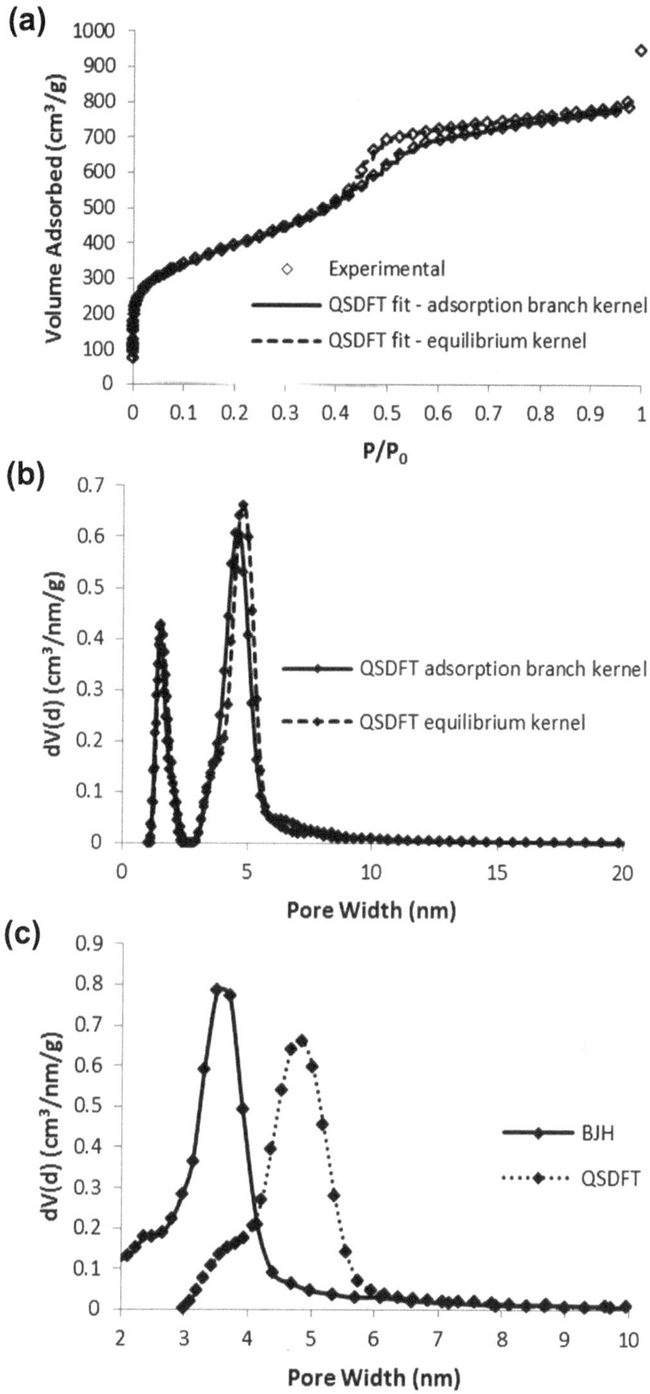
(a)
Volume Adsorbed (cm³/g)
1000
900
800
700
600
500
400
300
200
100
0
0 0.1 0.2 0.3 0.4 0.5 0.6 0.7 0.8 0.9 1
P/P0
Experimental
QSDFT fit - adsorption branch kernel
QSDFT fit - equilibrium kernel
(b)
dV(d) (cm³/nm/g)
0.7
0.6
0.5
0.4
0.3
0.2
0.1
0
0 5 10 15 20
Pore Width (nm)
QSDFT adsorption branch kernel
QSDFT equilibrium kernel
(c)
dV(d) (cm³/nm/g)
0.9
0.8
0.7
0.6
0.5
0.4
0.3
0.2
0.1
0
2 3 4 5 6 7 8 9 10
Pore Width (nm)
BJH
QSDFT

which takes correctly into account the delay in condensation due to metastable pore fluid in the region of hysteresis, to the adsorption data and (ii) the equilibrium QSDFT kernel on the desorption data. Overlapping pore-size distributions were observed, indicating that, for this particular sample, the desorption branch essentially represents evaporation under equilibrium conditions, i.e. hysteresis is here essentially caused by metastable pore fluid due to the existence of metastable adsorption films associated with the pore condensation. The obtained QSDFT mean mesopore diameter (4.8 nm) agrees well with the pore size derived from a geometrical model of CMK-3 (assuming hexagonally arranged carbon rods according to [112]) coupled with X-ray diffraction data (5 nm) obtained on this sample confirming that the QSDFT method accurately calculates the pore size from the experimental data [16].

Figure 4.11c compares the QSDFT mesopore-size distribution of this CMK-3 sample with the prediction of the BJH method. The BJH pore-size distribution curve is shifted significantly (ca. 25 %) to lower pore size, similar to the results obtained for silica mesoporous molecular sieves [11,12] and confirming that macroscopic, thermodynamic methods not only fail to provide a pore-size distribution over the complete micro- and mesopore range but also have a tendency to underestimate the pore size even in the range where they can be applied. This is further demonstrated in Fig. 4.12 which shows the micropore-size distributions calculated using the HK and QSDFT methods from a nitrogen adsorption isotherm obtained on a typical activated carbon sample (Fig. 4.12a). The HK pore-size distribution is shifted to smaller pore widths as compared to the QSDFT pore-size distribution, which is expected (see Fig. 4.6 which compares pore-filling pressures for nitrogen (77 K) in slit-pore carbons as a function of pore width as predicted by DFT, HK, and modified Kelvin (MK) equations).

Furthermore, novel QSDFT hybrid methods assuming cylindrical pore geometry for micro- and narrow mesopores, but spherical pore geometry for pores of width larger than 5 nm were applied to study the pore-size distribution of novel three-dimensionally ordered mesoporous (3DOm) carbons [16]. Here, good agreement with pore-size results from independent methods has also been found.

The combination of various adsorptives (argon and nitrogen) coupled with a proper QSDFT method allows us to obtain information about pore morphology and pore topology of complex micro–mesoporous carbons, as it has recently been demonstrated for the characterization of lignocellulosic-derived activated carbons [32]. Figure 4.13a shows argon (87 K) and nitrogen (77 K) isotherms on such samples with a type H3/H4 hysteresis loop indicating clearly the presence of complex meso- and microporosity. As discussed earlier, the use of different adsorptives, such as nitrogen and argon, allows one to identify and separate the underlying adsorption and hysteresis mechanisms. By comparing the pore-size distribution curves obtained from the argon and nitrogen desorption branches, one can differentiate whether pore blocking/percolation or cavitation-induced evaporation occurs, which provides

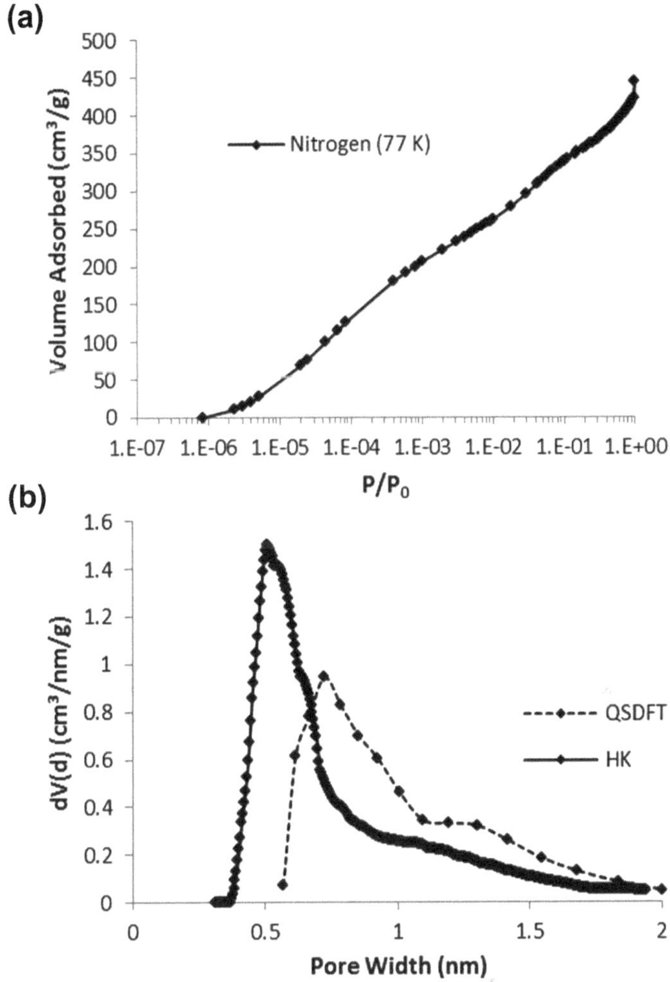

FIGURE 4.12 (a) Nitrogen (77 K) sorption isotherm for a microporous-activated carbon and (b) pore-size distributions calculated from the nitrogen isotherm using the HK and QSDFT methods.

additional information about pore network properties. It is interesting to note that the argon and nitrogen pore-size distribution curves from the desorption branches agree well for the mesopores with diameters up to 4 nm, but then deviate significantly for the pore-size range between ca. 4 and 6 nm (Fig. 4.13b). In line with our discussion of adsorption mechanism, this indicates that the steps down in the desorption branches around $P/P_0 = 0.4$ are associated with cavitation-induced evaporation from larger mesopores, which are only accessible by pore entrances with widths smaller than ca. 5–6 nm. In this case, no further quantitative information about the size of the necks can be obtained from the desorption pore-size distributions, and the argon and nitrogen

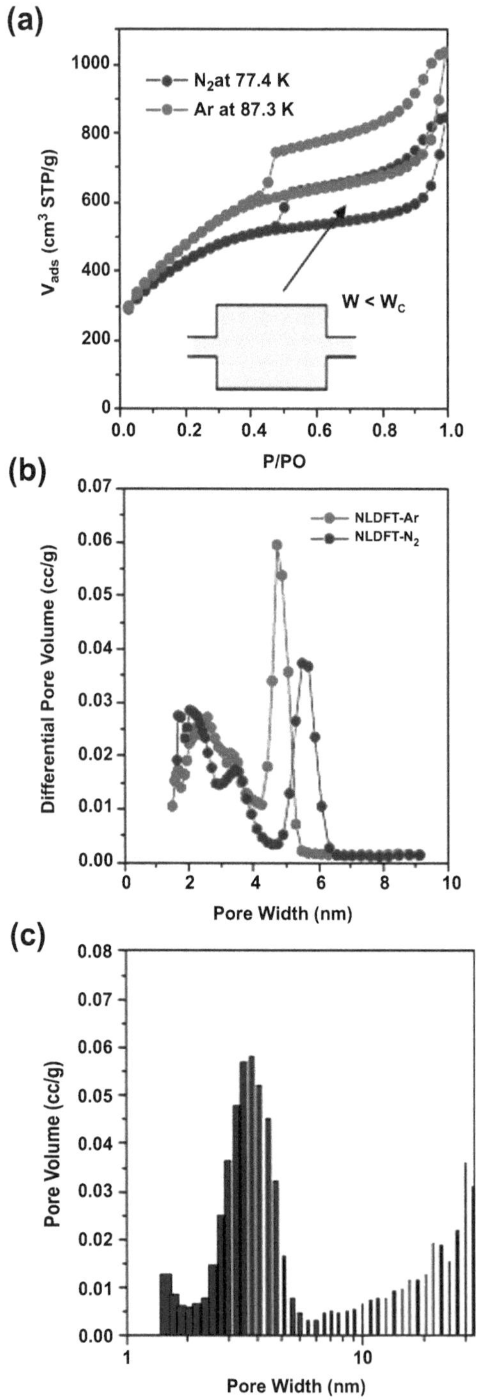

FIGURE 4.13 (a) Nitrogen (77 K) and argon (87 K) adsorption–desorption isotherms for a lignocellulosic-derived carbon. Inset shows a schematic representation of the large mesoporous cavities connected to the external surface through narrower mesopore necks, (b) pore-size distributions calculated from the desorption branches of the nitrogen and argon isotherms using NLDFT (assuming a cylindrical pore model), and (c) pore-size distribution calculated from the adsorption branch using a QSDFT metastable adsorption branch model (assuming a slit/cylindrical pore model). *(Reprinted from [32], with Permission from Elsevier)*

pore-size distribution peaks in this region reflect an artificial pore-size distribution. An accurate pore-size distribution of this material could only obtained from the adsorption branch by applying a QSDFT model to the nitrogen data which takes into account delay of condensation due to metastable pore fluid in cylindrical carbon mesopores. This distribution is shown in Fig. 4.13c in the form of a histogram, clearly indicating that this sample consists of a bimodal

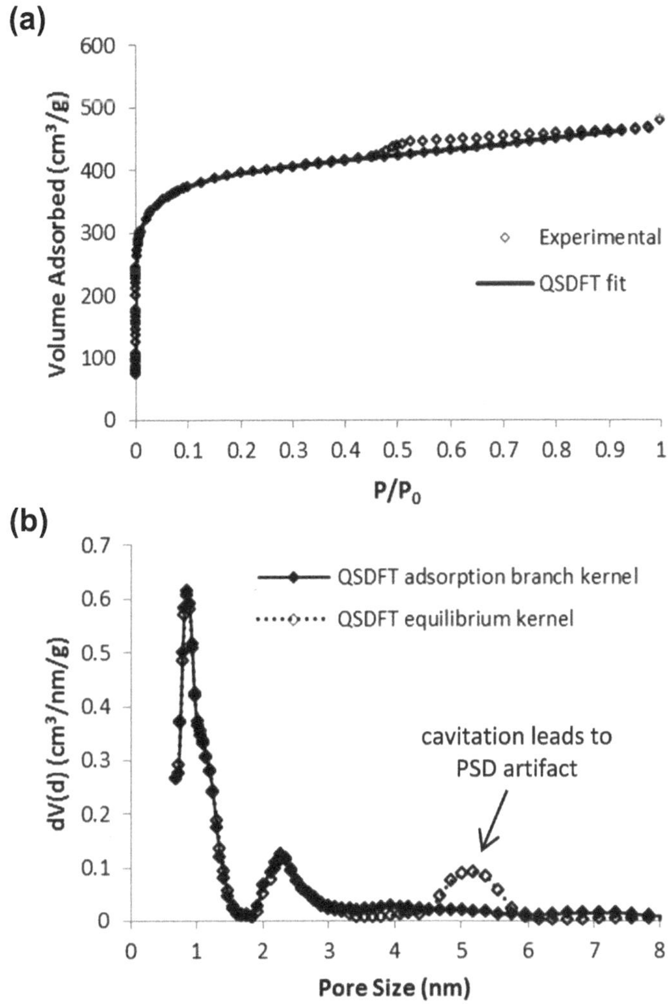

FIGURE 4.14 (a) Carbide-derived carbon nitrogen (77 K) adsorption isotherm with QSDFT fit and (b) pore-size distribution calculated from the desorption branch using a QSDFT equilibrium kernel illustrating the artifact due to cavitation at around 5 nm and pore-size distribution calculated from the adsorption branch using a QSDFT metastable adsorption branch kernel.

mesopore-size distribution, whereby some of the larger mesopores are only accessible by significantly smaller entrances.

Another interesting example is given in Fig. 4.14 which shows the nitrogen (77 K) adsorption isotherm of a typical carbide-derived carbon [113,114]. The isotherm reveals a type H4 hysteresis loop indicating the presence of micro- and mesopores. The steep stepdown in the isotherm around a relative pressure of 0.4 indicates the presence of cavitation in this sample. In this case, an accurate pore-size distribution must be calculated from the adsorption branch of the isotherm. Hence, a hybrid QSDFT method which assumes slit-pore geometry for the micropores and a cylindrical pore model to describe correctly the adsorption/condensation mechanism in the mesopores, i.e. takes into account the delay in condensation due to metastable adsorption fluids, was applied. Figure 4.14a demonstrates that this model fits the experimental adsorption isotherm and the resulting pore-size distribution is shown in Fig. 4.14b. In addition, the pore-size distribution was obtained from the desorption branch by applying a slit/cylinder QSDFT equilibrium transition model. The obtained pore-size distribution agrees with the pore-size distribution derived from the adsorption branch with the exception of the artificial peak at 5 nm which is due to cavitation-induced evaporation, indicating that some larger pores are only accessible through necks/entrances which are smaller than a critical width (around 5–6 nm for N_2/77 K adsorption).

4.6. CONCLUSIONS

During recent years, major progress has been achieved in the understanding of adsorption and phase behavior in ordered micro- and mesoporous materials, leading to major advances in structural characterization by physical adsorption, including the development of advanced theoretical procedures based on statistical mechanics known as density functional theory (NLDFT or QSDFT) and molecular simulation (GCMC and MD). Methods such as these describe the configuration of adsorbed molecules on a molecular level, and their application has contributed to a better quantitative description of adsorption hysteresis. In certain cases, it is now possible to obtain reliable information from both the adsorption and desorption branches of the hysteresis loop, which is crucial for pore-size characterization of carbons consisting of an interconnected micro–mesoporous network. It has been demonstrated that the application of DFT-based methods provides a much more accurate and comprehensive pore-size analysis compared to macroscopic thermodynamic methods (e.g. Kelvin equation, BJH, and Horvath–Kawazoe), i.e. pore-size and pore-volume information can be obtained over the complete micro- and mesopore range. However, some classical methods, such as the Dubinin–Raduskevich-based approaches, allow one to obtain reliable micropore volumes (in agreement with DFT results). Furthermore, the recently developed QSDFT approach quantitatively accounts for surface heterogeneity effects, providing a reliable pore-size assessment of disordered micro- and mesoporous carbons. Coupling such advanced theoretical approaches with a combination of adsorption experiments with various adsorptives (carbon dioxide, argon, and nitrogen) not only leads to a complete micro- and mesopore-size distribution, but also suggests useful information about pore morphology and topology. More work is needed to further advance the assessment of surface chemistry of nanoporous carbons by physical adsorption. It is also of interest to investigate and quantify the effects of adsorption-induced carbon pore deformation and the resulting challenges for physical adsorption characterization [115,116].

ACKNOWLEDGMENT

We would like to thank Prof. Stefan Kaskel for providing a carbide-derived carbon sample, Prof. Michael Froeba for CMK-3 mesoporous carbon, and Prof. Catherine Morlay for CMC2 activated carbon.

REFERENCES

[1] T.J. Bandosz, M.J. Biggs, K.E. Gubbins, Y. Hattori, T. Iiyama, K. Kaneko, J. Pikunic, K. Thomson, Molecular models of porous carbons, Chem. Phys. Carbon 28 (2003) 41–228.
[2] T.J. Mays, Active carbon fibers, in: Carbon Materials for Advanced Technologies, Elsevier, 1999.
[3] G.M. Roy, Activated Carbon Applications in the Food and Pharmaceutical Industries, CRC Press, 1995.
[4] H. Marsh, F. Rodríguez-Reinoso, Activated Carbon, Elsevier, 2006.
[5] F. Rodríguez-Reinoso, The role of carbon materials in heterogeneous catalysis, Carbon 36 (1998) 159–175.
[6] J. Rouquerol, D. Avnir, C.W. Fairbridge, D.H. Everett, J.H. Haynes, N. Pernicone, J.D.F. Ramsay, K.S.W. Sing, K.K. Unger, Recommendations for the characterization of porous solids, Pure Appl. Chem. 66 (1994) 1739–1758.
[7] S.J. Gregg, K.S.W. Sing, Adsorption, Surface Area and Porosity, Academic Press, 1982.
[8] F. Rouquerol, J. Rouquerol, K. Sing, Adsorption by Powders and Porous Solids, Elsevier, 1999.
[9] S. Lowell, J.E. Shields, M.A. Thomas, M. Thommes, Characterization of Porous Solids and Powders: Surface Area, Pore Size and Density, Springer, 2004.
[10] A.V. Neimark, K.S.W. Sing, M. Thommes, Surface area and porosity, in: Handbook of Heterogeneous Catalysis, Wiley, 2008.
[11] M. Thommes, Physical adsorption characterization of nanoporous materials, Chem. Ing. Tech. 82 (2010) 1059–1073.
[12] M. Thommes, Physical adsorption characterization of ordered and amorphous mesoporous materials, in: Nanoporous Materials Science and Engineering, World Scientific, 2004.
[13] K.S.W. Sing, D.H. Everett, R.A.W. Haul, L. Moscou, R.A. Pierotti, J. Rouquérol, T. Siemieniewska, Reporting physisorption data for gas/solid systems with special reference to the determination of surface area and porosity, Pure Appl. Chem. 57 (1985) 603–619.

[14] P.I. Ravikovitch, A.V. Neimark, Density functional theory model of adsorption on amorphous and microporous silica materials, Langmuir 22 (2006) 11171–11179.

[15] A.V. Neimark, Y. Lin, P.I. Ravikovitch, M. Thommes, Quenched solid density functional theory and pore size analysis of micro-mesoporous carbons, Carbon 47 (2009) 1617–1628.

[16] G.Y. Gor, M. Thommes, K.A. Cychosz, A.V. Neimark, Quenched solid density functional theory method for characterization of mesoporous carbons by nitrogen adsorption, Carbon 50 (2012) 1583–1590.

[17] M. Thommes, Textural characterization of zeolites and ordered mesoporous materials by physical adsorption, Stud. Surf. Sci. Catal. 168 (2007) 495–523.

[18] D. Lässig, J. Lincke, J. Moellmer, C. Reichenbach, A. Moeller, R. Gläser, G. Kalies, K.A. Cychosz, M. Thommes, R. Staudt, H. Krautscheid, A microporous copper metal-organic framework with high H_2 and CO_2 adsorption capacity at ambient pressure, Angew. Chem. Int. Ed. 50 (2011) 10344–10348.

[19] J. Silvestre-Albero, A. Silvestre-Albero, F. Rodríguez-Reinoso, M. Thommes, Physical characterization of activated carbons with narrow microporosity by nitrogen (77.4 K), carbon dioxide (273 K) and argon (87.3 K) adsorption in combination with immersion calorimetry, Carbon (2012). doi: 10.1016/j.carbon.2011.09.005.

[20] M. Thommes, R. Köhn, M. Fröba, Sorption and pore condensation behavior of nitrogen, argon, and krypton in mesoporous MCM-48 silica materials, J. Phys. Chem. B 104 (2000) 7932–7943.

[21] J. Garrido, A. Linares-Solano, J.M. Martín-Martínez, M. Molina-Sabio, F. Rodríguez-Reinoso, R. Torregrosa, Use of N_2 vs. CO_2 in the characterization of activated carbons, Langmuir 3 (1987) 76–81.

[22] D. Cazorla-Amorós, J. Alcañiz-Monge, A. Linares-Solano, Characterization of activated carbon fibers by CO_2 adsorption, Langmuir 12 (1996) 2820–2824.

[23] J. García-Martínez, D. Cazorla-Amorós, A. Linares-Solano, Further evidences of the usefulness of CO_2 adsorption to characterise microporous solids, Stud. Surf. Sci. Catal. 128 (2000) 485–494.

[24] P.I. Ravikovitch, A. Vishnyakov, R. Russo, A.V. Neimark, Unified approach to pore size characterization of microporous carbonaceous materials from N_2, Ar, and CO_2 adsorption isotherms, Langmuir 16 (2000) 2311–2320.

[25] A. Vishnyakov, P.I. Ravikovitch, A.V. Neimark, Molecular level models for CO_2 sorption in nanopores, Langmuir 15 (1999) 8736–8742.

[26] R.V.R.A. Rios, J. Silvestre-Albero, A. Sepúlveda-Escribano, M. Molina-Sabio, F. Rodríguez-Reinoso, Kinetic restrictions in the characterization of narrow microporosity in carbon materials, J. Phys. Chem. C 111 (2007) 3803–3805.

[27] G. Reichenauer, Micropore adsorption dynamics in synthetic hard carbons, Adsorption 11 (2005) 467–471.

[28] S. Furmaniak, A.P. Terzyk, P.A. Gauden, P.J.F. Harris, P. Kowalczyk, The influence of carbon surface oxygen groups on dubinin-astakhov equation parameters calculated from CO_2 adsorption isotherm, J. Phys. Condens. Matter 22 (2010) 085003.

[29] Y. Zhu, S. Murali, M.D. Stoller, K.J. Ganesh, W. Cai, P.J. Ferreira, A. Pirkle, R.M. Wallace, K.A. Cychosz, M. Thommes, D. Su, E.A. Stach, R.S. Ruoff, Carbon-based supercapacitors produced by activation of graphene, Science 332 (2011) 1537–1541.

[30] J. Jagiello, M. Thommes, Comparison of DFT characterization methods based on N_2, Ar, CO_2, and H_2 adsorption applied to carbons with various pore size distributions, Carbon 42 (2004) 1227–1232.

[31] J. Jagiello, W. Betz, characterization of pore structure of carbon molecular sieves using DFT analysis of Ar and H_2 adsorption data, Microporous Mesoporous Mater. 108 (2008) 117–122.

[32] A. Silvestre-Albero, M. Gonçalves, T. Itoh, K. Kaneko, M. Endo, M. Thommes, F. Rodríguez-Reinoso, J. Silvestre-Albero, Well-defined mesoporosity on lignocellulosic-derived activated carbons, Carbon 50 (2012) 66–72.

[33] J.K. Brennan, T.J. Bandosz, K.T. Thomson, K.E. Gubbins, Water in porous carbons, Colloids Surf. A 187–188 (2001) 539–568.

[34] D.D. Do, H.D. Do, A model for water adsorption in activated carbon, Carbon 38 (2000) 767–773.

[35] D.D. Do, H.D. Do, Modeling of adsorption on nongraphitized carbon surface: GCMC simulation studies and comparison with experimental data, J. Phys. Chem. B 110 (2006) 17531–17538.

[36] K. Kaneko, Y. Hanzawa, T. Iilyama, T. Kanda, T. Suzuki, Cluster-mediated water adsorption on carbon nanopores, Adsorption 5 (1999) 7–13.

[37] T. Kimura, H. Kanoh, T. Kanda, T. Ohkubo, Y. Hattori, Y. Higaonna, R. Denoyel, K. Kaneko, Cluster-associated filling of water in hydrophobic carbon micropores, J. Phys. Chem. B 108 (2004) 14043–14048.

[38] P. Lodewyckx, E.F. Vansant, Water isotherms of activated carbons with small amounts of surface oxygen, Carbon 37 (1999) 1647–1649.

[39] P. Lodewyckx, E. Raymundo-Piñero, H. Wullens, M. Vaclavikova, F. Beguin, Water isotherms of structurally identical carbons with different amounts of surface oxygen groups, in: Proceedings of the International Carbon Conference, 2008. Nagano, Japan.

[40] J.-C. Liu, P.A. Monson, Does water condense in carbon pores? Langmuir 21 (2005) 10219–10225.

[41] J.-C. Liu, P.A. Monson, Monte Carlo simulation study of water adsorption in activated carbon, Ind. Eng. Chem. Res. 45 (2006) 5649–5656.

[42] A.M. Slasli, M. Jorge, F. Stoeckli, N.A. Seaton, Modelling of water adsorption by activated carbons: effects of microporous structure and oxygen content, Carbon 42 (2004) 1947–1952.

[43] A.M. Slasli, M. Jorge, F. Stoeckli, N.A. Seaton, Water adsorption by activated carbons in relation to their microporous structure, Carbon 41 (2003) 479–486.

[44] F. Stoeckli, A. Lavanchy, The adsorption of water by active carbons, in relation to their chemical and structural properties, Carbon 38 (2000) 475–477.

[45] A. Striolo, K.E. Gubbins, M.S. Gruszkiewicz, D.R. Cole, J.M. Simonson, A.A. Chialvo, P.T. Cummings, T.D. Burchell, K.L. More, Effect of temperature on the adsorption of water in porous carbons, Langmuir 21 (2005) 9457–9467.

[46] T. Ohba, H. Kanoh, K. Kaneko, Cluster-growth-induced water adsorption in hydrophobic carbon nanopores, J. Phys. Chem. B 108 (2004) 14964–14969.

[47] P.D. Sullivan, B.R. Stone, Z. Hashisho, M.J. Rood, Water adsorption with hysteresis effect onto microporous activated carbon fabrics, Adsorption 13 (2007) 173–189.

[48] M. Thommes, C. Morlay, R. Ahmad, J.P. Joly, Assessing surface chemistry and pore structure of active carbons by a combination of physisorption (H_2O, Ar, N_2, CO_2), XPS and TPD-MS, Adsorption 17 (2011) 653–661.

[49] H. Kuwabara, T. Suzuki, K. Kaneko, Ultramicropores in microporous carbon fibres evidenced by helium adsorption at 4.2 K, J. Chem. Soc. Faraday Trans. 87 (1991) 1915–1916.

[50] K. Kaneko, N. Setoyama, T. Suzuki, H. Kuwabara, Ultramicroporosity of porous solids by He adsorption, in: Proc. IVth Int. Conf. on Fundamentals of Adsorption, 1992. Kyoto.

[51] M. Polanyi, Über die Adsorption vom Standpunkt des Dritten Wärmesatzes, Verh. Dtsch. Phys. Ges. 16 (1914) 1012–1016.
[52] M.M. Dubinin, D.P. Timofeev, Zh. Fiz. Khim. 22 (1948) 133.
[53] L.D. Gelb, K.E. Gubbins, R. Radhakrishnan, M. Sliwinska-Bartkowiak, Phase separation in confined systems, Rep. Prog. Phys. 62 (1999) 1573–1659.
[54] G. Mason, The effect of pore space connectivity on the hysteresis of capillary condensation in adsorption–desorption isotherms, J. Colloid Interface Sci. 88 (1982) 36–46.
[55] G.C. Wall, R.J.C. Brown, The determination of pore-size distributions from sorption isotherms and mercury penetration in interconnected pores: the application of percolation theory, J. Colloid Interface Sci. 82 (1981) 141–149.
[56] A.V. Neimark, Percolation theory of capillary hysteresis phenomena and its application for characterization of porous solids, Stud. Surf. Sci. Catal. 62 (1991) 67–74.
[57] M. Parlar, Y.C. Yortsos, Percolation theory of vapor adsorption–desorption processes in porous materials, J. Colloid Interface Sci. 124 (1988) 162–176.
[58] H. Liu, L. Zhang, N.A. Seaton, Sorption hysteresis as a probe of pore structure, Langmuir 9 (1993) 2576–2582.
[59] H. Liu, N.A. Seaton, Determination of the connectivity of porous solids from nitrogen sorption measurements – III. Solids containing large mesopores, Chem. Eng. Sci. 49 (1994) 1869–1878.
[60] F. Rojas, I. Kornhauser, C. Felipe, J.M. Esparza, S. Cordero, A. Domínguez, J.L. Riccardo, Capillary condensation in heterogeneous mesoporous networks consisting of variable connectivity and pore-size correlation, Phys. Chem. Chem. Phys. 4 (2002) 2346–2355.
[61] H.-J. Woo, L. Sarkisov, P.A. Monson, Mean-field theory of fluid adsorption in a porous glass, Langmuir 17 (2001) 7472–7475.
[62] L. Sarkisov, P.A. Monson, Modeling of adsorption and desorption in pores of simple geometry using molecular dynamics, Langmuir 17 (2001) 7600–7604.
[63] P.I. Ravikovitch, A.V. Neimark, Density functional theory of adsorption in spherical cavities and pore size characterization of templated nanoporous silicas with cubic and three-dimensional hexagonal structures, Langmuir 18 (2002) 1550–1560.
[64] P.I. Ravikovitch, A.V. Neimark, Experimental confirmation of different mechanisms of evaporation from ink-bottle type pores: equilibrium, pore blocking, and cavitation, Langmuir 18 (2002) 9830–9837.
[65] A. Vishnyakov, A.V. Neimark, Monte Carlo simulation test of pore blocking effects, Langmuir 19 (2003) 3240–3247.
[66] B. Libby, P.A. Monson, Adsorption/desorption hysteresis in inkbottle pores: a density functional theory and Monte Carlo simulation study, Langmuir 20 (2004) 4289–4294.
[67] M. Thommes, B. Smarsly, M. Groenewolt, P.I. Ravikovitch, A.V. Neimark, Adsorption hysteresis of nitrogen and argon in pore networks and characterization of novel micro- and mesoporous silicas, Langmuir 22 (2006) 756–764.
[68] A. Schreiber, S. Reinhardt, G.H. Findenegg, The lower closure point of the adsorption hysteresis loop of fluids in mesoporous silica materials, Stud. Surf. Sci. Catal. 144 (2002) 177–184.
[69] C.J. Rasmussen, A. Vishnyakov, M. Thommes, B.M. Smarsly, F. Kleitz, A.V. Neimark, Cavitation in metastable liquid nitrogen confined to nanoscale pores, Langmuir 26 (2010) 10147–10157.
[70] J. Rouquerol, P. Llewellyn, F. Rouquerol, Is the BET equation applicable to microporous adsorbents? Stud. Surf. Sci. Catal. 160 (2007) 49–56.

[71] T. Keii, T. Takagi, S. Kanetaka, A new plotting of the BET method, Anal. Chem. 33 (1961) 1965–1965.
[72] P. Schneider, Adsorption isotherms of microporous–mesoporous solids revisited, Appl. Catal. A 129 (1995) 157–165.
[73] T.A. Centeno, F. Stoeckli, The assessment of surface areas in porous carbons by two model-independent techniques, the DR equation and DFT, Carbon 48 (2010) 2478–2486.
[74] K.E. Gubbins, Theory and simulation of adsorption in micropores, in: Physical Adsorption: Experiment, Theory and Applications, Kluwer Academic Publishers, 1997.
[75] C. Lastoskie, K.E. Gubbins, N. Quirke, Pore size distribution analysis of microporous carbons: a density functional theory approach, J. Phys. Chem. 97 (1993) 4786–4796.
[76] K.E. Gubbins, Y.-C. Liu, J.D. Moore, J.C. Palmer, The role of molecular modeling in confined systems: impact and prospects, Phys. Chem. Chem. Phys. 13 (2011) 58–85.
[77] M.J. Biggs, A. Buts, Virtual porous carbons: what they are and what they can be used for, Mol. Sim. 32 (2006) 579–593.
[78] K.T. Thomson, K.E. Gubbins, Modeling structural morphology of microporous carbons by reverse Monte Carlo, Langmuir 16 (2000) 5761–5773.
[79] S.K. Jain, K.E. Gubbins, R.J.-M. Pellenq, J.P. Pikunic, Molecular modeling and adsorption properties of porous carbons, Carbon 44 (2006) 2445–2451.
[80] T.X. Nguyen, N. Cohaut, J.-S. Bae, S.K. Bhatia, New method for atomistic modeling of the microstructure of activated carbons using hybrid reverse Monte Carlo simulation, Langmuir 24 (2008) 7912–7922.
[81] N.A. Seaton, J.P.R.B. Walton, N. Quirke, A new analysis method for the determination of the pore size distribution of porous carbons from nitrogen adsorption measurements, Carbon 27 (1989) 853–861.
[82] J.P. Olivier, W.B. Conklin, M.V. Szombathely, Determination of pore size distribution from density functional theory: a comparison of nitrogen and argon results, Stud. Surf. Sci. Catal. 87 (1994) 81–89.
[83] A.V. Neimark, The method of indeterminate lagrange multipliers in nonlocal density functional theory, Langmuir 11 (1995) 4183–4184.
[84] L. Gurvich, J. Phys. Chem. Soc. Russ. 47 (1915) 805–827.
[85] M.M. Dubinin, L.V. Radushkevitch, Dokl. Akak. Nauk. SSSR 55 (1947) 331.
[86] M.M. Dubinin, V.A. Astakhov, Description of adsorption equilibria of vapors on zeolites over wide ranges of temperature and pressure, Adv. Chem. Series 102 (1971) 69–85.
[87] H.F. Stoeckli, A generalization of the dubinin-radushkevich equation for the filling of heterogeneous micropore systems, J. Colloid Interface Sci. 59 (1977) 184–185.
[88] M.M. Dubinin, H.F. Stoeckli, Homogeneous and heterogeneous micropore structures in carbonaceous adsorbents, J. Colloid Interface Sci. 75 (1980) 34–42.
[89] F. Stoeckli, E. Daguerre, A. Guillot, Correspondence, Carbon 37 (1999) 2075–2077.
[90] F. Stoeckli, M.V. López-Ramón, D. Hugi-Cleary, A. Guillot, Micropore sizes in activated carbons determined from the Dubinin–Radushkevich equation, Carbon 39 (2001) 1115–1116.
[91] C. Scherdel, G. Reichenauer, M. Wiener, Relationship between pore volumes and surface areas derived from the evaluation of N_2-sorption data by DR-, BET- and t-plot, Microporous Mesoporous Mat. 132 (2010) 572–575.
[92] B.C. Lippens, J.H. de Boer, Studies on pore systems in catalysts: V. The *t* method, J. Catal. 4 (1965) 319–323.
[93] D. Atkinson, P.J.M. Carrott, Y. Grillet, J. Rouquerol, K.S.W. Sing, in: Proc. 2nd Int. Conf. on Fundamentals of Adsorption, Engineering Foundation, 1987.

[94] P.J.M. Carrott, K.S.W. Sing, Gas chromatographic study of microporous carbons, J. Chromatogr. A 406 (1987) 139–144.
[95] K. Kaneko, C. Ishii, M. Ruike, H. Kuwabara, Origin of superhigh surface area and microcrystalline graphitic structures of activated carbons, Carbon 30 (1992) 1075–1088.
[96] D.H. Everett, J.C. Powl, Adsorption in slit-like and cylindrical micropores in the Henry's law region. A model for the microporosity of carbons, J. Chem. Soc., Faraday Trans. 1 (72) (1976) 619–636.
[97] G. Horvath, K. Kawazoe, Method for the calculation of effective pore size distribution in molecular sieve carbon, J. Chem. Eng. Japan 16 (1983) 470–475.
[98] A. Saito, H.C. Foley, Curvature and parametric sensitivity in models for adsorption in micropores, AIChE J. 37 (1991) 429–436.
[99] C.M. Lastoskie, A modified Horvath–Kawazoe method for micropore size analysis, Stud. Surf. Sci. Catal. 128 (2000) 475–484.
[100] P. Tarazona, Free-energy density functional for hard spheres, Phys. Rev. A 31 (1985) 2672–2679.
[101] P. Tarazona, R. Evans, A simple density functional theory for inhomogeneous liquids, Mol. Phys. 52 (1984) 847–857.
[102] J.P. Olivier, Improving the models used for calculating the size distribution of micropore volume of activated carbons from adsorption data, Carbon 36 (1998) 1469–1472.
[103] D.A. Soares Maia, J.C.A. de Oliveira, J.P. Toso, K. Sapag, R.H. López, D.C.S. Azevedo, C. L. Cavalcante, G. Zgrablich, Characterization of the PSD of activated carbons from peach stones for separation of combustion gas mixtures, Adsorption 17 (2011) 853–861.
[104] V.A. Bakaev, Rumpled graphite basal plane as a model heterogeneous carbon surface, J. Chem. Phys. 102 (1995) 1398–1404.
[105] A.R. Turner, N. Quirke, A grand canonical Monte Carlo study of adsorption on graphitic surfaces with defects, Carbon 36 (1998) 1439–1446.
[106] S.M.P. Lucena, C.A.S. Paiva, P.F.G. Silvino, D.C.S. Azevedo, C.L. Cavalcante, The effect of heterogeneity in the randomly etched graphite model for carbon pore size characterization, Carbon 48 (2010) 2554–2565.
[107] P.I. Ravikovitch, J. Jagiello, D. Tolles, A.V. Neimark, Extended Abstracts of Carbon Conference, Lexington, KY, 2001.
[108] S.K. Bhatia, Density functional theory analysis of the influence of pore wall heterogeneity on adsorption in carbons, Langmuir 18 (2002) 6845–6856.
[109] J. Jagiello, J.P. Olivier, A simple two-dimensional NLDFT model of gas adsorption in finite carbon pores. Application to pore structure analysis, J. Phys. Chem. C 113 (2009) 19382–19385.
[110] P.I. Ravikovitch, A. Vishnyakov, A.V. Neimark, Density functional theories and molecular simulations of adsorption and phase transitions in nanopores, Phys. Rev. E 64 (2001) 011602.
[111] S. Jun, S.H. Joo, R. Ryoo, M. Kruk, M. Jaroniec, Z. Liu, T. Ohsuna, O. Terasaki, Synthesis of new, nanoporous carbon with hexagonally ordered mesostructure, J. Am. Chem. Soc. 122 (2000) 10712–10713.
[112] S.H. Joo, R. Ryoo, M. Kruk, M. Jaroniec, Evidence for general nature of pore interconnectivity in 2-dimensional hexagonal mesoporous silicas prepared using block copolymer templates, J. Phys. Chem. B 106 (2002) 4640–4646.
[113] Y. Gogotsi, A. Nikitin, H. Ye, W. Zhou, J.E. Fischer, B. Yi, H.C. Foley, M.W. Barsoum, Nanoporous carbide-derived carbon with tunable pore size, Nat. Mater. 2 (2003) 591–594.

[114] E. Kockrick, C. Schrage, L. Borchardt, N. Klein, M. Rose, I. Senkovska, S. Kaskel, Ordered mesoporous carbide derived carbons for high pressure gas storage, Carbon 48 (2010) 1707–1717.

[115] E.A. Ustinov, D.D. Do, Effect of adsorption deformation on thermodynamic characteristics of a fluid in slit pores at sub-critical conditions, Carbon 44 (2006) 2652–2663.

[116] P. Kowalczyk, A. Ciach, A.V. Neimark, Adsorption-induced deformation of microporous carbons: pore size distribution effect, Langmuir 24 (2008) 6603–6608.

[117] F. Rouquerol, J. Rouquerol, B. Imelik, Validite de la loi B.E.T. dans le cas de l'adsorption d'azote, d'argon et de butane sur des adsorbants poreux, Bull. Soc. Chim. France (1964) 635–639.

[118] M. Kruk, M. Jaroniec, A. Sayari, Adsorption study of surface and structural properties of MCM-41 materials of different pore sizes, J. Phys. Chem. B 101 (1997) 583–589.

[119] M. Jaroniec, L.A. Solovyov, Improvement of the Kruk-Jaroniec-Sayari method for pore size analysis of ordered silicas with cylindrical mesopores, Langmuir 22 (2006) 6757–6760.

Chapter 5

Water Adsorption by Carbons. Hydrophobicity and Hydrophilicity

Ajna Tóth and Krisztina László
Department of Physical Chemistry and Materials Science, Budapest University of Technology and Economics, Budapest, Hungary

Chapter Outline

5.1. Introduction 147
5.2. Water Adsorption Isotherms 151
5.3. Surface Chemistry 155
5.3.1. Oxygen Functionalities 157
5.3.2. Nitrogen-Containing Functional Groups 159
5.3.3. Other Heteroatoms 161
5.4. Kinetics of Water Adsorption 164
5.5. Coadsorption of Water in Multicomponent Vapor Systems 164
5.6. Concluding Remarks 166
References 166

5.1. INTRODUCTION

The versatility of activated carbons makes them the most frequently employed adsorbents. Since ancient times the high affinity of carbon for a wide diversity of chemical species has established it as a multipurpose adsorbent in both gas and liquid, most often aqueous, phases. In the former of these two modes, they are used extensively not only in daily life as domestic or office air filters but also on an industrial scale for gas purification. Being general adsorbents they are the prime choice for dealing with industrial accidents or terrorist attacks. Under real operating conditions, water molecules from the surrounding atmosphere tend to infiltrate and preempt adsorption sites reserved for other target molecules or to influence the kinetics of the adsorption, thereby compromising the role of the carbon. By contrast, in aqueous-phase

Novel Carbon Adsorbents. DOI: 10.1016/B978-0-08-097744-7.00005-3
Copyright © 2012 Elsevier Ltd. All rights reserved.

applications the wettability of carbon by water is of decisive interest on account of its excellent transport properties.

Porous carbons exist in different forms (activated carbon, activated carbon fibers, carbon aerogels, monoliths, etc.), but their surface properties can be defined ultimately as a system of more or less organized hydrophobic pore walls of different shapes, diameter, and connectivities, decorated to a greater or lesser degree with heteroatoms at their edges or within the wall.

The outstanding sorption performance of active carbons stems from a unique combination of geometrical and chemical characteristics. Carbon adsorbents with tailor-made surface area and pore hierarchy are commercially available, e.g. [1].

Porous carbon, by virtue of its high surface area and complex porosity, is considered as a general adsorbent, as the interaction of carbon surfaces with most organic molecules is dominated by dispersion forces, i.e., it is nonspecific. The sorption performance, however, is strongly influenced by the chemical nature of the molecules to be adsorbed. From small-angle X-ray scattering (SAXS) a pore-filling factor p can, for example, be deduced, such that $p=0$ when the pores are empty and $p=1$ when they are full. In this expression, q defines the spatial resolution of the experiment, and the corresponding distance scale is $w=2\pi/q$. Figures 5.1a and b demonstrate how the spatial distribution of the interacting fluid depends on both the polarity and the surface chemistry of the porous carbon samples (Table 5.1). In the micro- and mesoporous range, for water vapor at relative pressure $P/P_0=0.5$ (P is the actual pressure and P_0 is the saturation pressure at the temperature of the observation), the value of $p(q) \approx 0.9$ in the oxidized carbon (APETB), but is below 0.6 in the other two samples, while hexane vapor adsorbs strongly in all samples [2]. For n-hexane vapor, the dispersion interaction ensures complete filling of the narrow pores, while larger pores are unpopulated. The value of transfer wave vector q at which $p(q)$ vanishes for n-hexane coincides with the critical size of this molecule. For water vapor at $P/P_0=0.5$, cluster formation and the degree of filling were found to depend on the extent of oxidation. In the highly oxidized sample, a cooperative mechanism enhances the filling of even larger pores [3]. While the pores corresponding to the applied relative pressure are completely filled by n-hexane at $P/P_0=0.4$, water only partially fills the pores, even in the most oxidized carbon.

The most outstanding, and also the most practically relevant, molecule from the point of view of adsorption is water. The interaction of carbon materials with water is a two-sided phenomenon. Hydrophilicity is disadvantageous when water competes with the adsorptive that is to be removed, but it is desirable in several situations, e.g., in electrochemical applications. The state-of-the-art knowledge on water vapor adsorption on porous carbons was also overviewed in two excellent comprehensive publications in the last decade [5,6].

In addition to sorption-related applications, surface hydrophobicity is a genuine challenge when carbon materials such as fibers or nanotubes are embedded in a polymer matrix. Applications of such carbon materials are

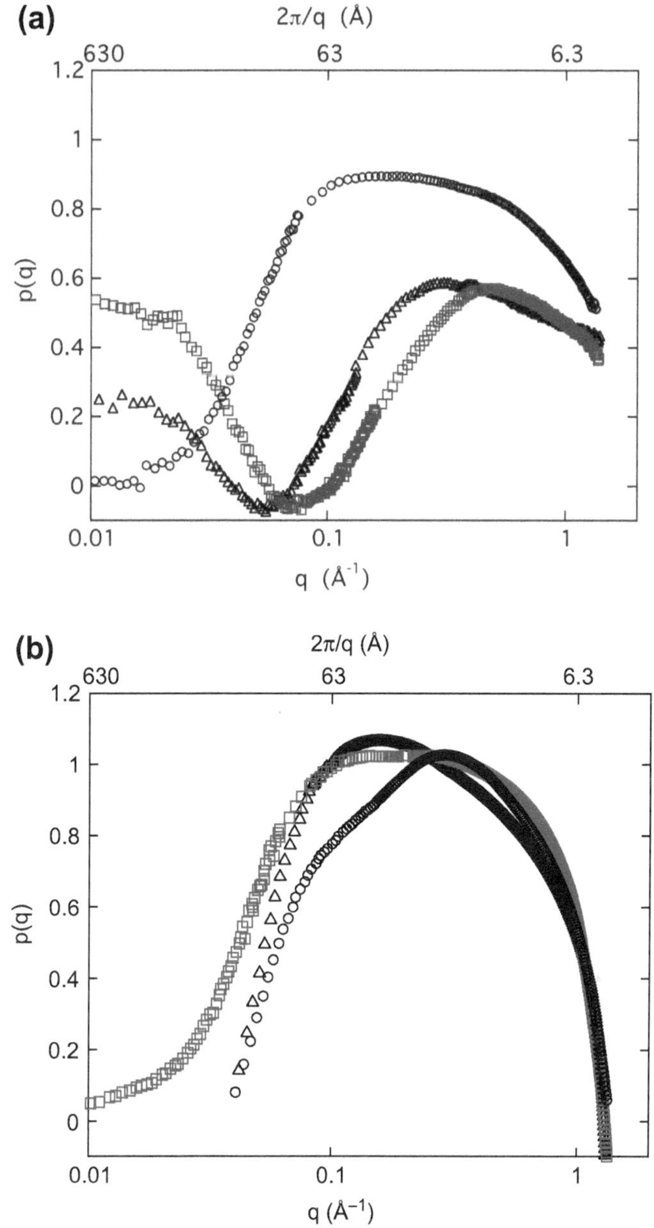

FIGURE 5.1 Comparison of the pore-filling factor $p(q)$ of equilibrated (a) water vapor ($P/P_0 = 0.5$) and (b) n-hexane vapor ($P/P_0 = 0.4$) in ash-free microporous carbons Δ: APETW, □: APETA, ○: APETB (Table 5.1). In the formalism of scattering, q is the spatial resolution of the measurement ($q = (4\pi/\lambda)\sin(\theta/2)$, with λ the wavelength of the X-rays and θ the scattering angle). Upper scale represents the nominal equivalent distance $w = 2\pi/q$. In the figures only 20% of the data are plotted. *(From [2])*

TABLE 5.1 Surface Characteristics of Polymer-based Ash-free Microporous Carbons in Fig. 5.1

		APETW	APETA	APETB
surface area*, m^2/g	S_{BET}	1156	1114	304
	S_{DR}	1352	1293	337
	S_x (air)[a]	1930	1810	920
	O/C % from XPS	6.4	9.9	26.6

** from BET and DR models and SAXS measurements [4], respectively.*

limited by their poor compatibility with certain liquids, polymers, etc., especially with polar compounds. The interaction between hydrophobic carbon materials and either hydrophobic or hydrophilic media is weak, as only long-range secondary interactions may occur. These interaction forces are easily overcome, thus restricting the applicability of such carbons in composite materials. While this problem appears to be solved for carbon fibers, full exploitation of the outstanding mechanical potential of multiwall carbon nanotubes (MWCNTs) (e.g., high modulus and high tensile strength coupled with low weight) is still limited by their poor dispersibility and wettability. As these materials have become available in high-quality form and on an 'industrial' scale, it is now a pressing problem for carbon scientists to render these materials hydrophilic without loss of their strength, which is greater than that of stainless steel.

The spectacular rise of the nanocarbon industry carries with it the ecological concern of the wettability of carbon nanoparticles. Although the as-produced carbon nanoparticles such as carbon nanotubes (CNTs) have poor wettability and dispersibility, functionalized (e.g., O- and N-doped) carbon nanoparticles exhibit pH-dependent properties that influence their colloidal stability, which ultimately determines their fate and bioavailability in the environment [7].

Modeling of the hydrophobic/hydrophilic behavior of carbon nanoparticles, especially carbon nanotubes, has been widely employed. In spite of a sustained experimental research effort in the field, however, the water vapor adsorption behavior of CNTs remains poorly known. Their unique physical and chemical features (size, shape, surface area, sorption capacity, catalytic activity, etc.) guarantee that the comparison with porous carbons is not trivial. Even more systematic experimental work is needed to make their adsorption behavior understood. The large variety of impurities further complicates the picture. Owing to both space limitations and the complexity of the hydrophobic and hydrophilic properties of such carbon nanoparticles, particularly carbon nanotubes, the adsorption of water vapor in these systems is not discussed here.

5.2. WATER ADSORPTION ISOTHERMS

In spite of their intrinsically hydrophobic character, nanoporous carbon materials can adsorb a large amount of water. The limited affinity of water toward untreated porous carbon surfaces, however, is clearly indicated by the type V vapor adsorption isotherms of carbon surfaces [8]. At low relative pressure uptake is negligible, and substantial sorption in untreated carbons starts only at higher relative pressures, generally around 0.5. The carbon–water interaction is particularly sensitive to structural changes in the morphology of the adsorbent, since the sorption interaction is influenced both by the pore structure and by the surface chemistry [6,9]. The position of the upturn point is defined by the overall feature of the carbon surface (Fig. 5.2). In practice, however, the distinction and the separation of these two factors are very difficult. Simulation studies on a realistic carbon structure also shows that the presence of water vapor dramatically affects the connectivity of the pores [10].

Completely reversible water uptake isotherms are very rare. Water vapor adsorption in hydrophobic nanopores exhibits a unique hysteresis that is not explained by capillary condensation theory [12]. The raw material and the conditions of the activation process, e.g., burn-off, temperature, choice of the activating agent, etc., which determine three-dimensional factors such as pore shape and pore connectivity, all influence the position, the extension, and the width of the hysteresis loop. Low burn-off activated carbons, typically having narrow (ultra)micropores, often yield a water adsorption isotherm with no explicit hysteresis loop. By contrast, carbons with wider pores do exhibit

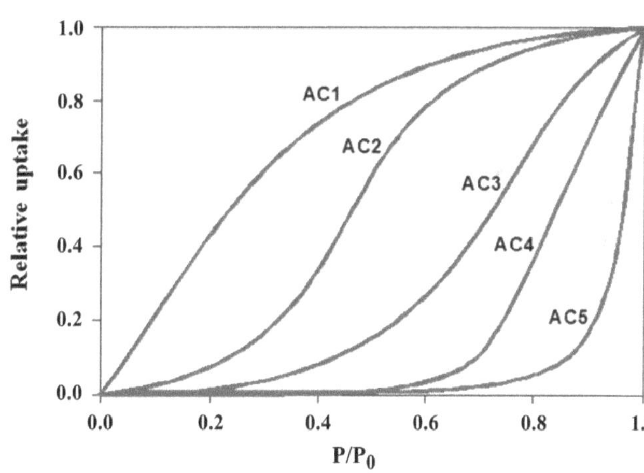

FIGURE 5.2 Water vapor adsorption isotherms on porous carbons of different origins. AC1: polyacrylonitrile-based ultramicroporous carbon, AC2: peat-based carbon, AC3: coal-based activated carbon, AC4: furfurol-based carbon, AC5: mesoporous carbon from molybdenum carbide. *(Adapted from [11])*

hysteresis. The hysteresis indicates that during the adsorption or/and desorption process the water is in metastable state. SAXS studies, aided by molecular simulation, on practically hydrophobic activated carbon fibers showed that the adsorbed water molecules assemble into different molecular structures during adsorption and desorption [13]. The transformation between these various structures is kinetically forbidden for micropore widths larger than about 1.1 nm. This kinetically forbidden transformation gives rise to metastable structures during adsorption, and is believed to be responsible for the existence of the hysteresis loop (Fig. 5.3) [14]. The low density (0.81–0.86 g cm^{-3}) of the water adsorbed in the hydrophobic carbon micropores suggests that its structure is close to that of solid water [15]. The finding of Nakamura et al. that the width of the hysteresis loop depends on the equilibration time is also consistent with the metastability of the adsorption branch [16].

Adsorption isotherms can be measured in both static and dynamic conditions. Nowadays, automatic instruments using either the gravimetric or volumetric principle are available on the market. The breakthrough technique, traditionally employed in filter tests, can also be used. The isotherms are generally measured at well-controlled ambient temperature in the range 293–303 K.

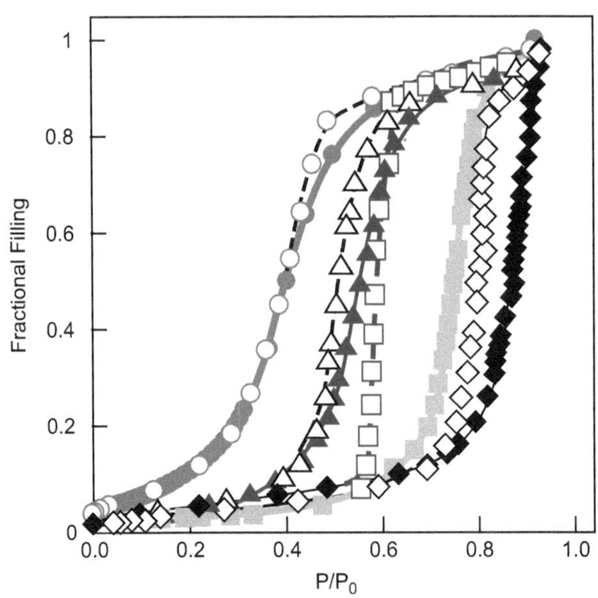

FIGURE 5.3 Adsorption isotherms of H_2O at 303 K for activated carbon fibers with average pore width $w = 0.7$ (circles), 0.9 (triangles), 1.1 (squares), and 1.3 nm (diamonds), respectively. Solid symbols: adsorption; open symbols: desorption. *(Adapted with permission from Ref. [14]. Copyright 2011 American Chemical Society)*

Water vapor isotherms can also be obtained from wide-angle X-ray scattering (WAXS) measurements, by comparing the amplitude of the signals from the adsorbed water and from bulk water in the transfer momentum region $q > 1\ \text{Å}^{-1}$. Such measurements, which detect the total amount of adsorbed molecules, rather than their spatial distribution as in SAXS, have shown good agreement with gravimetric isotherms on microporous polymer-based carbons (Fig. 5.4).

The critical parameters of these measurements are the temperature of both the sample preparation and the adsorption, and the equilibration time. The sample preparation temperature must be optimized to remove all adsorbed impurities (including water) but must conserve the surface chemistry and the texture of the porous carbon. During the measurement the temperature must also be carefully controlled to prevent 'conventional' condensation close to $P/P_0 = 1$.

Owing to the generally weak interaction between the carbon surface and the water molecules, the adsorption of water, particularly at low relative water pressure, may be extremely slow. Ultramicroporosity can delay intrusion of the water molecules into the narrowest pores [18]. Adsorbed water molecules anchored to functional groups near the entrance of the pores may also limit access to narrow pores.

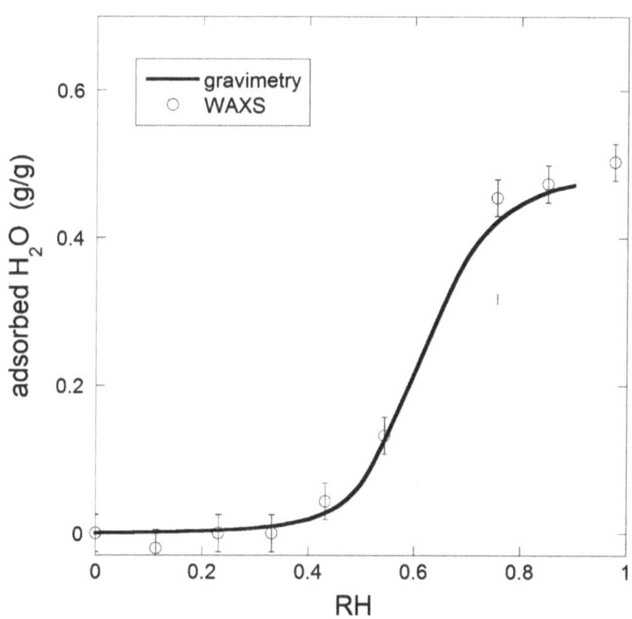

FIGURE 5.4 Comparison of the measurements of the total adsorbed water in a microporous carbon measured by WAXS (open symbols) and by gravimetry (continuous line). The equilibration took 8 weeks. *(From [17])*

The generally accepted mechanism of water adsorption was first proposed by Dubinin and Serpinsky [19]. It involves the following steps: (1) adsorption of water on surface functional groups, (2) adsorption of water around previously adsorbed water molecules and subsequent cluster formation, (3) pore filling occurs at a relative pressure of about $P/P_0 = 0.5$, and (4) a plateau is reached at high pressure when all pores are filled. Dubinin also found that that the formation of a water monolayer on carbon black is complete only at values of P/P_0 as high as 0.6 [20].

Polar surface groups act as primary active centers, which interact with water molecules by hydrogen bonding. When surface oxygen is present, at low relative pressures and submonolayer surface coverages, water adsorption is significantly influenced by specific interactions [21,22]. Where two or more sites are positioned at a separation suitable for both water–site and water–water bonding, the formation of water clusters is facilitated by cooperative adsorption [3, 6, 23]. Secondary adsorption then occurs at the primary adsorbed water molecules and a structured adsorbed layer forms at the adsorbent surface. The total adsorption is then the sum of adsorption at all available sites, both primary and secondary. The initial stages of water adsorption are thus analogous to nucleation. The precise shape of the resulting isotherm depends on the surface concentration of primary sites, which also controls the value of the limiting water adsorption, and, together with the distribution of these sites, the density of the adsorbed water [24,25].

High-sensitivity calorimetric measurements of the differential heat of adsorption (ΔH_{ads}) on high surface area porous carbons have confirmed that water adsorbs *via* three mechanisms: i) chemical adsorption, $\Delta H_{ads} > 50$ kJ/mol, ii) condensation, $\Delta H_{ads} \approx 42$ kJ/mol, and iii) physical adsorption, $\Delta H_{ads} < 42$ kJ/mol. The absolute and relative amounts of water adsorption associated with each mechanism are a function of surface chemistry [26].

Several semiempirical isotherm equations have been developed with fitting parameters that predict the concentration of primary active sites, the molecular size of water clusters, and the water adsorption capacity on the surface functional groups as well as in the micropores, and that can also estimate equilibrium rate constants [27–32]. Originally, each of these models was applied to a specific kind of carbon adsorbent.

One of the simplest water adsorption isotherm models, based on the Polányi potential theory, is the Dubinin–Astakhov equation [8]. However, it has been reported that this equation is not able to fit the initial section of the isotherm, particularly in the case of high-oxygen-content carbons [33]. The most commonly employed phenomenological description of water vapor isotherms is the Dubinin–Serpinsky (DS) equation, given by [34]

$$\frac{P}{P_0} = \frac{n}{k_1(n_0 + n)(1 - k_2 n)} \tag{5.1}$$

where n is the number of adsorbed water molecules at relative pressure P/P_0, n_0 is the number of primary active sites, k_1 is the ratio of the rate constant for adsorption and desorption, while k_2 is a parameter affecting the maximum adsorbable amount of water. According to this model, all the primary active centers have the same adsorption energy. The water molecules adsorb on these sites in a 1:1 stoichiometric ratio and become secondary sites for further molecules. It was recently calculated by Stoeckli and others that the stoichiometry, and thus the interaction energy per site, indeed depends on the quality of the oxygen functionality, resulting in an average molar contribution of 6.2–6.8 kJ/mol for each species [35,36]. This equation generally provides a good description of water vapor adsorption isotherms in the relative pressure region 0.4–0.8. At lower relative pressures, however, the D'Arcy–Watt isotherm (Eqn 5.2) [37] may give better agreement with experimental data. This equation has three terms, representing respectively water that is strongly bound, weakly bound, and sorbed in multimolecular water clusters:

$$n = \frac{KK'\frac{P}{P_0}}{1+K\frac{P}{P_0}} + C\frac{p}{p_0} + \frac{kk'\frac{P}{P_0}}{1-k\frac{P}{P_0}} \tag{5.2}$$

Here P/P_0 is the relative vapor pressure of water, and K' and k' are related to the number of strong and secondary sites, respectively. C is a function of the number and the strength of weak binding sites (i.e., the Henry constant), K is a function of the primary site–water interaction energy, and k is a function of the water–water interaction. The coefficients of the D'Arcy–Watt equation have thermodynamic significance and can be used to calculate the total strength of water binding and the number of binding sites for each of the three regions of water binding.

The Do equation is based on the growth of the water cluster on the functional groups and the penetration of these clusters into the micropores as pentamers of size 0.6 nm [38]. A detailed discussion of the mathematical representations of water vapor adsorption on carbon surfaces has been published by Furmaniak et al. [39].

5.3. SURFACE CHEMISTRY

It is well established that carbon as such is a hydrophobic material. The contact angle of water on a graphite surface is about 90°. In microstructured "bucky" paper membranes, however, a contact angle of 113° has been measured [40]. Water exhibits a relatively small dispersion interaction (35–40 mJ m^{-2}) on oxygen-free carbon surfaces, which, depending on the exact surface packing model used, yields an interaction energy of approximately 2 kJ/mol [41]. This feature of carbon may be advantageous if adsorption of water is not desired, because adsorption of water reduces the pore volume available to the target molecules. The poor wettability of carbon by water may, however, limit several practical applications of carbon materials.

The concentration of polar sites, which are in most cases covalently bound oxygen functionalities and possibly some electron donor–acceptor sites (Fig. 5.5), determines the hydrophilicity of the carbon structure. The accelerated formation of such sites when the carbon is exposed to air and water vapor (e.g., during storage) is responsible for “aging” of activated carbons. Such functional groups, however, are always present on carbon surfaces, since chemisorption of oxygen on the high surface area carbon from the ambient environment is in practice a spontaneous process. Heat treatment of the carbon in an inert or reductive medium may reduce the water uptake through thermal desorption and reduction, respectively, of the oxygen-containing primary adsorption sites.

To strengthen the weak interaction between the surrounding medium and the carbon material calls for chemical approaches. Indeed, introduction of heteroatoms appears to offer a possible route for overcoming these obstacles. The introduction of heteroatoms into the carbon skeleton renders the carbon surface more attractive to polar molecules [42]. Affinity between the carbon surface and the fluid phase plays a major role in adsorption processes, as the generally nonspecific interaction affects the distribution of the adsorbed molecules within the pores [43].

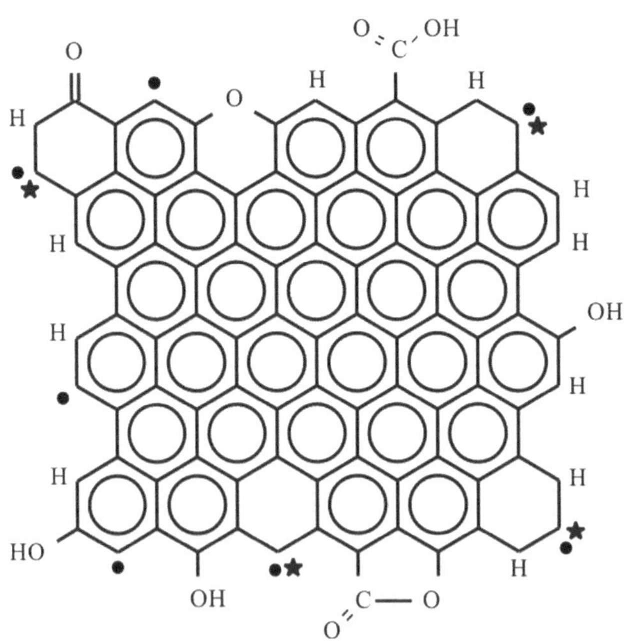

FIGURE 5.5 Chemical heterogeneities along the edge of a C, O, and H-containing activated carbon. ● and ●* represent an unpaired σ electron and in-plane σ pair, respectively (* is a localized π electron). *(From [48])*

The species, abundance, and the distribution of the noncarbon elements and their surface functional groups fundamentally depend on the raw material and the treatment applied during the manufacture of the carbon. Hydrophobic/hydrophilic properties are altered during activation. The hydrophobicity of carbon cryogels increases with increasing pyrolysis temperature [44]. Oxidation is performed most often either in gas phase (air or oxygen) or by wet oxidation (H_2O_2, HNO_3, H_2SO_4, $(NH_4)_2S_2O_3$, etc.). Oxidation may change the textural properties and deteriorate the surface area as well as the microporosity of the starting carbon e.g. [45,46].

The challenge is that the outcome of the various chemical treatments also depends strongly on the type of carbon material. There is an emerging need for controlled and/or uniform functionalization of carbon surfaces, which could be of particular interest for CNTs, which are high-cost materials. Selective hydroxylation can be achieved, e.g., by means of the Fenton reaction [47].

5.3.1. Oxygen Functionalities

The most common heteroatom in the carbon materials is oxygen. This tends to bond along the edges of the graphene layers in the form of various functional groups that are well known in organic chemistry (Fig. 5.5). These groups may be of acidic or basic character.

The water adsorption properties of carbon materials reflect their surface chemistry and polarity and, since oxide species are intrinsic to the surface chemistry, they are of great interest in carbon surface science, hence have been much studied. It is well established that, in general terms, the water adsorption properties of carbon surfaces vary proportionally with the concentration of polar functional groups [49]. Information on surface composition can be derived by several direct and indirect methods, e.g., immersion or/and titration calorimetry, temperature-programmed desorption (TPD), acid/base titration methods, X-ray photoelectron spectroscopy XPS, and FTIR [50–54]. The determination of the nuclear magnetic resonance spin–spin relaxation time of the adsorbed water as a function of P/P_0 and/or the adsorbed amount of water provides a means of analyzing changes in the state of the water during the adsorption process. In the absence of primary active sites, the relaxation time of the protons in the micropores was found to be 21 ms [55]. A critical assessment of some of these techniques was published in the recent past [56]. In the case of oxygen functionalities, however, a comparison of the results from these techniques (Figs 5.6 and 5.7) shows that the thermodynamic properties of carbon surfaces can be confidently predicted from any of these techniques alone [41].

The strongest adsorption centers are carboxylic groups [57]. The derived H-bond energy between water and surface groups lies in the range 10–14 kJ/mol, which is consistent with the mechanism of physisorption and cluster formation [26]. Molecular dynamic simulations on water adsorption on COOH and OH decorated graphite surfaces show that at low water coverage the adsorption

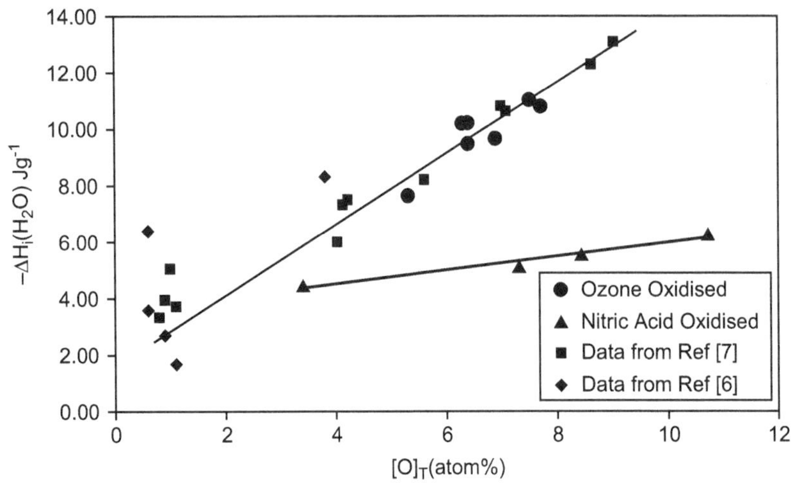

FIGURE 5.6 Correlation between the enthalpy of immersion in water $\Delta H_{i(H_2O)}$ and total surface oxygen $[O]_T$ measured by XPS for ozone and nitric acid-oxidized carbon blacks. *(From [41])*

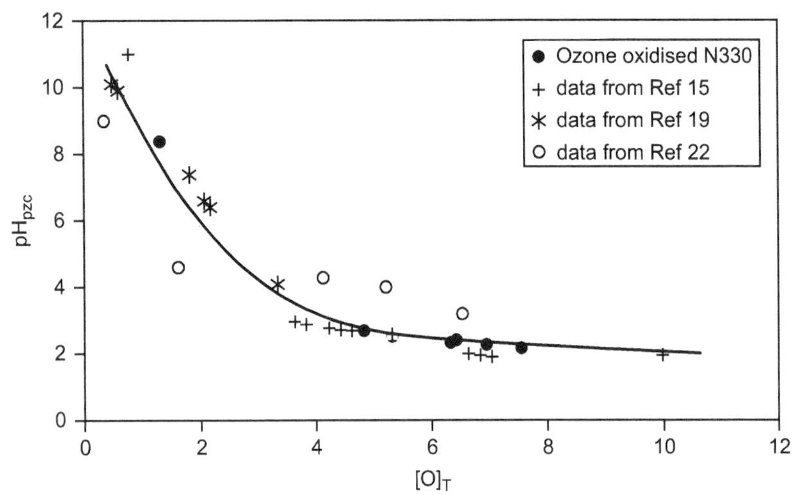

FIGURE 5.7 Correlation between pH_{pzc} and total oxygen concentration $[O]_T$ for ozone-oxidized carbon black (from XPS, expressed in atom %) and active carbons (from TPD and titration, expressed in mmol g^{-1}; 1 at.% = 1 mmol g^{-1} = 10^{17} atoms per m^2 of carbon surface). *(From [41])*

process is driven mainly by the water–water and the water–COOH site interactions. When the density of COOH sites is quite low, the predominance of the water–water interactions leads to clustering of the water molecules above the surface, with the resulting water aggregate being preferentially attached to the COOH sites. When the density of COOH sites is sufficiently high, and for small water coverages, the water molecules prefer to be adsorbed on as large as possible a number of COOH sites rather than forming a single aggregate on the surface. In contrast, the water–OH site interactions appear quite weak, and the water molecules tend to disregard the presence of these sites when the surface also contains COOH sites [58].

As noted above, chemistry–water adsorption performance relationships are now well developed for microporous carbons and carbon blacks, but there are very few reports, if any, of experimentally derived water adsorption characteristics for carbon nanotubes. There is an urgent need, however, to integrate these materials into the general framework of carbon adsorption and to establish behavioral links with other types of carbon surfaces/structures. Reliable experimental information is also required to determine the compatibility of the behavior of CNTs with general water adsorption models that were originally developed for activated carbon and related materials [19].

5.3.2. Nitrogen-Containing Functional Groups

Nitrogen is the next most extensively investigated heteroatom after oxygen (Fig. 5.8). Nitrogen atoms in the carbon structure behave like polar

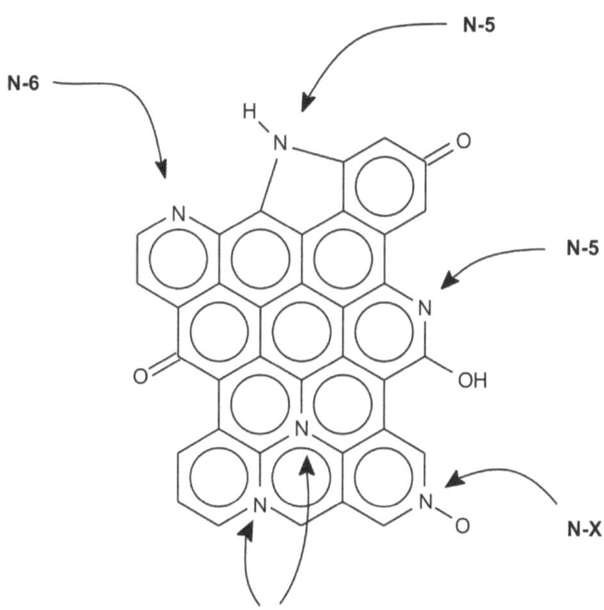

FIGURE 5.8 Nitrogen-containing functional groups in activated carbon. N-6: pyridinic N, N-5: pyrrolic/pyridine, N-Q: quaternary N and N-X: N-oxides. *(From [60])*

sites: N-containing carbons exhibit higher surface polarity than carbons without nitrogen, increasing either the negative (pyridinic nitrogen) or the positive charge (N atoms substituting C atom in the graphitic layer) of the surface [59]. Nitrogen induces basicity in the structure.

The influence of N-containing groups on the adsorption performance of porous carbon is far less explored than that of oxygen groups. It is generally recognized, however, that N-containing porous carbon has stronger affinity for water than the N-free carbon with similar pore structure and surface oxygen content. Figure 5.9 compares the water vapor uptake of three polymer-based microporous carbons with surface compositions listed in Table 5.2 [61]. Comparison of the isotherms leads us to conclude that increasing the surface concentration of the functional groups increases the water uptake at low relative humidity. Moreover, although APETOX and APAN contain the same atomic % of heteroatoms, for APAN a strong enhancement in water uptake is observed. The presence and the in-plane distribution of the nitrogen functionalities may be responsible for this effect.

The enhancing role of nitrogen in water vapor uptake has also been recently confirmed in long-range ordered microporous carbons with narrow pore-size distribution, where the N was introduced by acetonitrile CVD (N/C atomic ratio ca. 0.06) [62]. The H_2O adsorption–desorption isotherm of these carbons is of

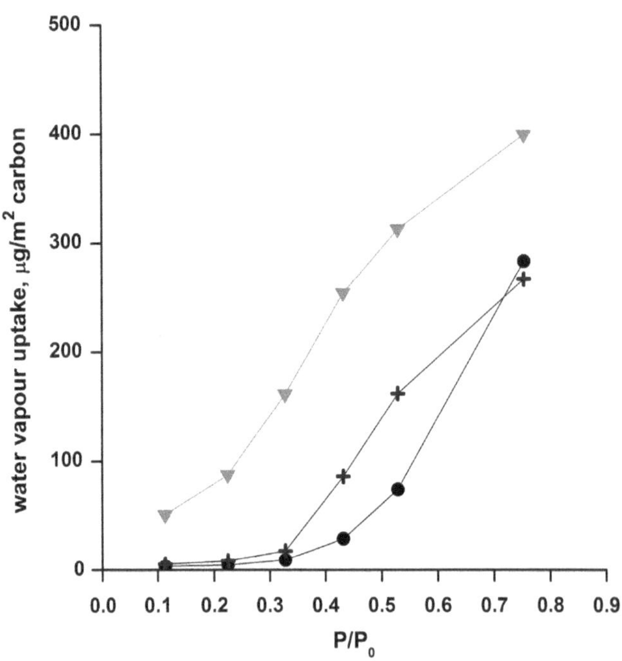

FIGURE 5.9 Equilibrium adsorption of water vapor at 20 °C on polymer-based microporous carbons with closely similar micropore size distributions. Black circles: APET; blue crosses: APETOX; green triangles: APAN. Description of the samples is given in Table 5.2. *(Adapted from [61])*

TABLE 5.2 Characteristics of the Polymer-based Carbons in Fig. 5.9

Sample		Surface Area	Total Carbon	Total Oxygen	Total Nitrogen
Label	Description	m^2/g		at. %	
APET	Made from polyethyleneterephthalate	1440	95.7	4.3	–
APETOX	APET oxidized with HNO_3	1510	90.0	10.0	–
APAN	Made from polyacrylonitrile	356	89.3	5.4	5.3*

** 32% pyridine-like, 34% pyrrolic and/or pyridon N, 24% quaternary N and 10% N-oxide.*

type V, which exhibits a steep uptake around $P/P_0 = 0.5$ and pronounced hysteresis. N-doping results in a shift of the uptake pressure to lower values than for N-free samples. [63,64].

Figure 5.10 shows the water adsorption isotherms of a mesoporous melamine–formaldehyde carbon aerogel (MFCAG, Table 5.3) at 298 K. The steep rise in water adsorption at higher relative pressures can be attributed to capillary condensation of water in the mesopores, rather than to micropore filling as in the case of the standard type V isotherms of activated carbons [65]. The nitrogen atoms are believed to be responsible for the enhanced uptake. Like an oxygen functional group, pyridinic nitrogen acts as a primary adsorption center and affects mainly the filling of the micropores. The effect of quaternary groups as primary adsorption centers is less significant. Their role in capillary condensation, however, was found to be remarkable, since quaternary nitrogen provides a relatively polarized surface with high water wettability [60].

No particular enhancement in the adsorption properties caused by nitrogen doping was observed, however, when low N-content (0.6 atomic %) MWCNTs were exposed to water vapor [66].

Much remains to be discovered about how nitrogen atoms, especially the different N-species, influence the hydrophilicity of porous carbons. These effects are much less well documented than those of O-functionalities. Nevertheless, N-containing carbons show great promise in the expanding field of biomedical applications of carbon materials.

5.3.3. Other Heteroatoms

Chemical treatment, in addition to impregnation or doping techniques, is often used to enhance the performance of porous carbons. Incorporation of sulfur

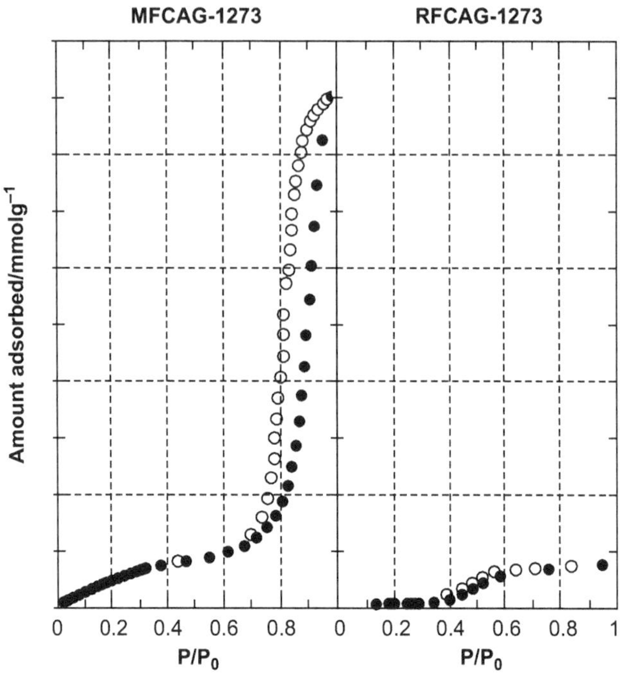

FIGURE 5.10 Adsorption isotherms of water (298 K) on N-containing (MFCAG) and N-free (RFCAG) carbon aerogels heat-treated at 1273 K. *(Adapted from [65])*

TABLE 5.3 Characteristics of Carbon Aerogels in Fig. 5.10

Sample	C	H	N	Diff.	O*	N/C**
			wt%			Atomic ratio
MFCAG-1273	79.40	1.24	14.66	4.71	1.48	0.158
RFCAG-1273	95.99	3.14	0	2.96	1.61	0

** Oxygen content remaining after evacuation at 383 K, determined from TPD measurements.*
*** N/C atomic ratio calculated from the carbon and nitrogen content from elemental analysis.*

atoms into the aromatic rings of the carbon matrix has been achieved by H_2S treatment of oxidized carbons in inert atmosphere. The temperature of the treatment is reported to be capable of tuning the O/S ratio (Fig. 5.11) [67]. Adsorption may occur through sulfur–sulfur bridges, while sulfur and sulfur–oxygen groups present in larger pores enhance the adsorption *via* specific acid–base and polar interactions.

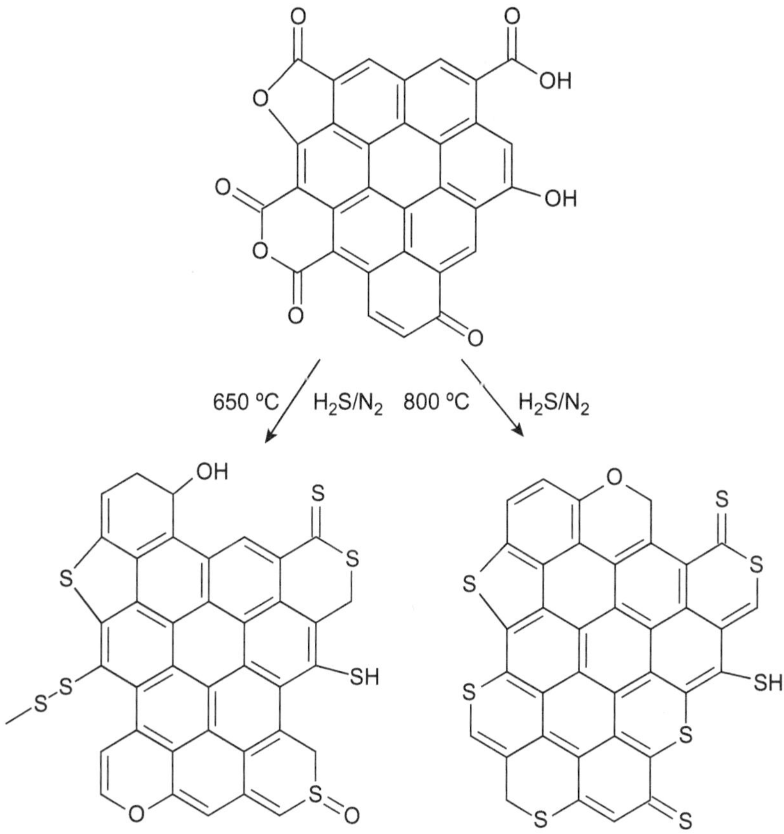

FIGURE 5.11 Introduction of S heteroatoms by high-temperature H_2S treatment. *(From [67])*

To overcome potential failures in applications, such as from sensitivity to erosion and susceptibility to oxidation, recent investigations have turned to boron, silicon, and phosphorus [68–75]. A detailed discussion of the influence of such modifications, as well as of metal impregnation, on the hydrophilic/hydrophobic behavior of porous carbons lies beyond the scope of this chapter and is therefore not undertaken here.

Hydrophobicity of the carbon surface can be enhanced by fluorination. Plasma-enhanced chemical vapor deposition (PECVD) with perfluorohexane (PFH) reduces the water uptake without reducing the adsorption capacity with respect to organic vapors [76]. The superhydrophobic polymer layer reduced the heat of immersion with water to $-12\ J\ g^{-1}$ and yielded a contact angle of 120°. Below $P/P_0 = 0.3$ no water adsorption was detected, and water uptake close to the saturation was reduced by 40%. In addition to their scientific interest [77], such materials show strong promise in military applications and in nuclear separation.

5.4. KINETICS OF WATER ADSORPTION

Few papers have been devoted to this topic, perhaps because the process is slow. First-order kinetics is most frequently used, based on the linear driving force model, e.g. [18]. This model was used by Fletcher et al. in their investigations into the effect of functional groups on the rate of adsorption [78]. Relationships were observed between water adsorption kinetics at low surface coverage and the type and concentration of oxygen surface functional groups. When carboxylic groups were present in significant amounts, a two-component kinetic model with two exponentials was used, in which the slow kinetic component was associated with adsorption on these groups. In the opposite case, carbons in which carboxylic groups exist only at low concentration, the results could be satisfactorily explained by a stretched exponential kinetic model. In the range $P/P_0 < 0.2$, the adsorption rate depended linearly on pressure, with carbons having the lowest concentration of functional groups exhibiting the highest rates. These authors proposed a site-to-site hopping model between groups for the mechanism of diffusion. In the primary mechanism, the strong interactions between water and carboxylic acid groups hinder diffusion and accordingly reduce the rate constants, with the result that the water molecules form a structured surface layer with a reduced degree of freedom. In the secondary mechanism, the water adsorbs on the already adsorbed water molecules *via* H-bonds, i.e., the adsorbed phase is more fluid. By comparing the rate of adsorption and desorption in the hydrophobic pores, Nakamura et al. found that their distributions are different. The broad distribution of the adsorption rate is in agreement with cluster formation [16]. Clusters enter the micropores when their dispersion potential becomes sufficiently high [38]. Do and Do suggested that this occurs when the five water molecules are assembled. The mobility of the previously adsorbed water species as well as the site at which where assembly occurs remain to be elucidated. The value of the heat of adsorption (ca. –40 kJ/mol at 298 K) suggests that this is the energy barrier that must be overcome. Bradley et al. recently proposed that water cluster formation takes place in the free vapor phase at a critical relative pressure, most probably in the wider micropores. Such spaces contain more functional groups and the cooperative mechanism enhances the concentration of water molecules. These clusters can then adsorb in the ultramicropores by the dispersion interaction [76] (Fig. 5.12).

5.5. COADSORPTION OF WATER IN MULTICOMPONENT VAPOR SYSTEMS

Coadsorption of water with other molecules, such as volatile organic compounds, either consecutively or simultaneously, has been the subject of several studies. Various techniques have been employed for this purpose, mainly based on breakthrough or consecutive equilibrium techniques [79–83]. Such adsorption methods yield overall adsorbed amounts. The isotope

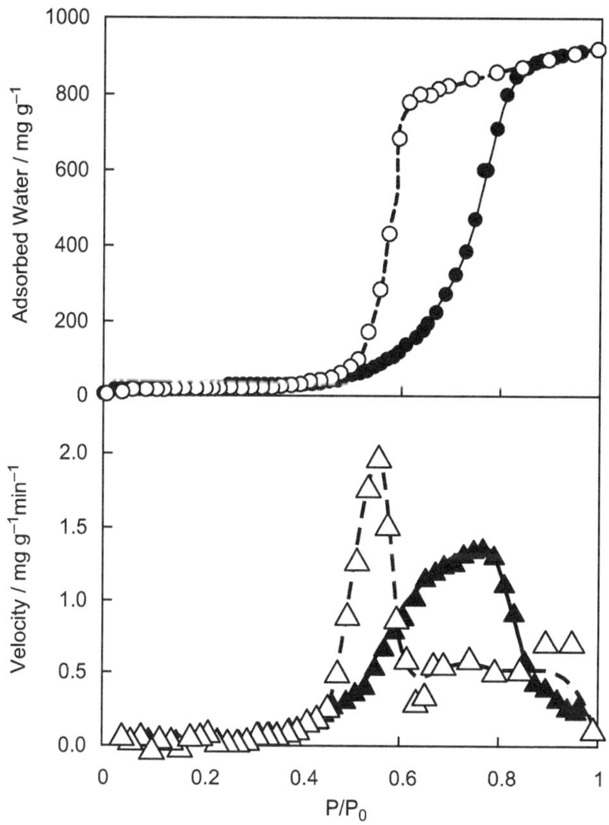

FIGURE 5.12 Water adsorption isotherm and the rate of water adsorption on a pitch-based carbon fiber at 303 K. Full symbols: adsorption, open symbols: desorption. *(From [16])*

exchange technique applied by Rynders et al. was able to simultaneously measure multicomponent gas adsorption kinetics and equilibria and self-diffusivities of the components in a single, isothermal experiment without disturbing the adsorbed phase [84].

To study the spatial organization of adsorbed molecules, X-ray microtomography offers interesting possibilities [85], but this technique presents a challenge for molecules composed of light atoms, such as water, and the resolution is not very high. SAXS or small-angle neutron scattering (SANS) observations also yield complementary direct information, not only on the location of the adsorption sites, but also on the surface area, S_X, both of the dry carbon [4] and that of the liquid in contact with the carbon [2]. The effects of adsorption of a single species of molecule can be monitored by SAXS alone. In the presence of more than one species, however, SANS combined with contrast variation can be used together with selective isotopic substitution, e.g., by replacing hydrogen by deuterium. Measurements of the SANS response of an oxidized high surface area carbon have shown that in the simultaneous presence of toluene vapor and water vapor at $P/P_0 = 0.87$, both species of molecule coexist in the pores, in which the major constituent is toluene, while the volume fraction of water is only 12% [86].

5.6. CONCLUDING REMARKS

The intention of this outline is to show that surface hydrophobicity and hydrophilicity play an important role in the adsorption of water by carbons. A variety of experimental approaches has shown the importance of polar groups, which act as nucleation centers at the different stages of cluster formation and pore filling, whether in low-resolution experiments such as gas adsorption techniques, in higher-resolution scattering observations such as SAXS or SANS, or again in local methods such as proton magnetic resonance.

It cannot be overemphasized, however, that the conclusions discussed here, drawn from these experimental and theoretical considerations for porous carbons, are not necessarily valid for CNTs. The extension of such generalizations calls for extreme caution. Similarly, results from CNT simulations are not directly transposable to porous carbon materials. On the other hand, experimental and theoretical results on both 'conventional' porous carbons and porous carbon nanoparticles may improve understanding of these still open questions. The same holds for the synergism between the experimental and computational efforts.

It should also be mentioned that several outstanding questions in water adsorption remain incompletely understood, or even open. Among these figure the kinetics of adsorption at the molecular level and the effect of swelling of the carbon matrix in the presence of adsorbed molecules. There is ample room for concerted experimental and theoretical work on the hydrophobic effect or the enthalpy/entropy balance of the water adsorption.

ACKNOWLEDGMENTS

The work was performed with the support of the FP7 framework program FRESP – Advanced first response respiratory protection (grant number 218138). This chapter is related to the scientific program "Development of quality-oriented and harmonized R + D + I strategy and functional model at BME" supported by the New Széchenyi Plan (Project ID: TÁMOP-4.2.1/B-09/1/KMR-2010-0002). We express our thanks to E. Geissler and B. Berke for their invaluable assistance.

REFERENCES

[1] H. Marsh, F. Rodríguez-Reinoso, Activated Carbon, Elsevier, 2006.

[2] K. László, E. Geissler, Surface chemistry and contrast-modified SAXS in polymer-based activated carbons, Carbon 44 (2006) 2437–2444.

[3] E.A. Muller, L.F. Rull, L.F. Vega, K.E. Gubbins, Adsorption of water on activated carbons: a molecular simulation study, J. Phys. Chem. 100 (1996) 1189–1196.
[4] K. László, O. Czakkel, K. Josepovits, C. Rochas, E. Geissler, Influence of surface chemistry on the SAXS response of polymer-based activated carbons, Langmuir 21 (2005) 8443–8451.
[5] J.K. Brennan, T.J. Bandosz, K.T. Thompson, K.E. Gubbins, Water in porous carbons, Colloids Surf. A Physicochem. Eng. Asp. 187–188 (2001) 539–568.
[6] D. Mowla, D.D. Do, K. Kaneko, Adsorption of water vapor on activated carbon: a brief review, in: L.R. Radovic (Ed.), Chemistry and Physics of Carbon, vol. 28, Marcel Dekker, 2003, pp. 229–262.
[7] S.J. Klaine, P.J.J. Alvarez, G.E. Batley, T.F. Fernandes, R.D. Handy, D. Lyon, S. Mahendra, M.J. McLaughlin, J.R. Lead, Nanomaterials in the environment: behavior, fate, bioavailability, and effects, Environ. Toxicol. Chem. 27 (9) (2008) 1825–1851.
[8] F. Rouquerol, J. Rouquerol, K. Sing, Adsorption by Powders and Porous Solids, Academic Press, 1999.
[9] A.M. Slasli, M. Jorge, F. Stoeckli, N.A. Seaton, Water adsorption by activated carbons in relation to their microporous structure, Carbon 41 (2003) 479–486.
[10] J.K. Brennan, K.T. Thomson, K.E. Gubbins, Adsorption of water in activated carbons: effects of pore blocking and connectivity, Langmuir 18 (2002) 5438–5447.
[11] R.S. Vartapetyan, A.M. Voloshchuk, Adsorption mechanism of water molecules on carbon adsorbents, Uspekhi khimii 64 (11) (1995) 1055–1072 (in Russian).
[12] Y. Hanzawa, K. Kaneko, Lack of predominant adsorption of water vapour on carbon mesopores, Langmuir 13 (1997) 5802–5804.
[13] T. Ohba, K. Kaneko, Cluster-associated filling of water molecules in slit-shaped graphitic nanopores, Mol. Phys. 105 (2007) 139–145.
[14] T. Ohba, K. Kaneko, Kinetically forbidden transformations of water molecular assemblies in hydrophobic micropores, Langmuir 27 (2011) 7609–7613.
[15] T. Iiyama, M. Ruike, K. Kaneko, Structural mechanism of water adsorption in hydrophobic micropores from in situ small angle X-ray scattering, Chem. Phys. Lett. 331 (2000) 359–364.
[16] M. Nakamura, T. Ohba, P. Branton, H. Kanoh, K. Kaneko, Equilibrium-time and pore width dependent hysteresis of water adsorption isotherm on hydrophobic microporous carbons, Carbon 48 (2010) 305–312.
[17] E. Geissler, K. László, Applications of small angle X-ray scattering techniques for characterizing high surface area carbons. Chapter 5, in: S. Mikhalovsky, A. Khajibaev (Eds.), Biodefence NATO Science for Peace and Security Series A Chemistry and Biology, 2011, pp. 41–49.
[18] P. Lodewyckx, The effect of water uptake in ultramicropores on the adsorption of water vapour in activated carbon, Carbon 48 (2010) 2549–2553.
[19] M.M. Dubinin, V.V. Serpinsky, Isotherm equation for water-vapor adsorption by microporous carbonaceous adsorbents, Carbon 5 (1981) 402–403.
[20] M.M. Dubinin, Water vapour adsorption and the microporous structures of carbonaceous adsorbents, Carbon 18 (1980) 355–364.
[21] A. Andreu, H.F. Stoeckli, R.H. Bradley, Specific and non-specific interactions on non-porous carbon black surfaces, Carbon 45 (2007) 1854–1864.
[22] M.V. Lopez-Ramon, H.F. Stoeckli, C. Moreno-Castilla, F. Carrasco-Marin, Specific and nonspecific interactions between methanol and ethanol and active carbons, Langmuir 16 (2000) 5967–5972.
[23] E.A. Muller, K.E. Gubbins, Molecular simulation study of hydrophilic and hydrophobic behavior of activated carbon surfaces, Carbon 36 (1998) 1433–1438.
[24] R.Sh. Vartapetyan, A.M. Voloshchuk, E.B. Shumilina, The critical size of clusters of water-molecules on a carbon surface, Russ. Chem. Bull. 42 (1993) 46–48.

[25] R.H. Bradley, A. Andreu, H.F. Stoeckli, Correlation of water adsorption isotherms with oxygen groups on carbon black surfaces, in: International Conference on Carbon. July 2007, Seattle, USA, 2007, CD ROM of Extended Abstract.

[26] J. Phillips, D. Kelly, L. Radovic, F. Xie, Microcalorimetric study of the influence of surface chemistry on the adsorption of water by high surface area carbons, J. Phys. Chem. B 104 (34) (2000) 8170–8176.

[27] A.M. Slasli, M. Jorge, F. Stoeckli, N.A. Seaton, Modelling of water adsorption by activated carbons: effects of microporous structure and oxygen content, Carbon 42 (2004) 1947–1952.

[28] O. Talu, F. Meunier, Adsorption of associating molecules in micropores and application to water on carbon, AIChE J. 42 (1996) 809–819.

[29] S.W. Rutherford, Modeling water adsorption in carbon micropores: Study of water in carbon molecular sieves, Langmuir 22 (2006) 702–708.

[30] S. Lagorsse, M.C. Campo, F.D. Magalhaes, A. Mendes, Water adsorption on carbon molecular sieve membranes: experimental data and isotherm model, Carbon 43 (2005) 2769–2779.

[31] T. Zimny, G. Finqueneisel, L. Cossarutto, J.V. Weber, Water vapor adsorption on activated carbon preadsorbed with naphthalene, J. Colloid Interface Sci. 285 (2005) 56–60.

[32] G. Marban, A.B. Fuertes, Co-adsorption of n-butane/water vapour mixtures on activated carbon fibre-based monoliths, Carbon 42 (2004) 71–81.

[33] P. Lodewyckx, E.F. Vansant, Water isotherms of activated carbons with small amounts of surface oxygen, Carbon 37 (1999) 1647–1649.

[34] M.M. Dubinin, E.D. Zaverina, V.V. Serpinsky, The sorption of water vapour by active carbon, J. Chem. Soc. (1955) 1760–1766.

[35] F. Stoeckli, L. Currit, A. Laederach, T.A. Centeno, Water-adsorption in carbons described by the Dubinin-Astakhov and Dubinin-Serpinski equations, J. Chem. Soc. Faraday Trans. 90 (1994) 3689–3691.

[36] F. Carrasco-Marin, A. Mueden, T.A. Centeno, F. Stoeckli, C. Moreno-Castilla, Water adsorption on activated carbons with different degrees of oxidation, J. Chem. Soc. Faraday Trans. 93 (1997) 2211–2215.

[37] R.L. D'Arcy, I.C. Watt, Analysis of sorption isotherms of nonhomogeneous sorbents, J. Chem. Soc. Faraday Trans. 66 (1970) 1236–1245.

[38] D.D. Do, H.D. Do, A model for water adsorption in activated carbon, Carbon 38 (2000) 767–773.

[39] S. Furmaniak, P.A. Gauden, A.P. Terzyk, G. Rychlicki, Water adsorption on carbons - Critical review of the most popular analytical approaches, Adv. Colloid Interface Sci. 137 (2008) 82–143.

[40] L.F. Dumée, K. Sears, J. Schütz, N. Finn, C. Huynh, S. Hawkins, M. Duke, S. Gray, Characterization and evaluation of carbon nanotube Bucky-Paper membranes for direct contact membrane distillation, J. Memb. Sci. 351 (2010) 36–43.

[41] R.H. Bradley, R.A. Daley, F. La Goff, Polar and dispersion interactions at carbon surfaces - further development of the XPS based model, Carbon 40 (3) (2002) 1173–1179.

[42] C.A. Leon y Leon, L.R. Radovic, Interfacial chemistry and electrochemistry of carbon surfaces, in: P.A. Thrower (Ed.), Chemistry and Physics of Carbon, vol. 24, Marcel Dekker, 1994, pp. 213–310.

[43] L.R. Radovic, C. Moreno-Castilla, J. Rivera-Utrilla, Carbon materials as adsorbents in aqueous solutions, in: L.R. Radovic (Ed.), Chemistry and Physics of Carbon, vol. 27, Marcel Dekker, 2001, pp. 227–405.

[44] T. Yamamoto, T. Sugimoto, T. Suzuki, S.R. Mukai, H. Tamon, Preparation and characterization of carbon cryogel microspheres, Carbon 40 (2002) 1345–1351.

[45] K. László, K. Marthi, C. Rochas, F. Ehrburger-Dolle, F. Livet, E. Geissler, Morphological investigation of chemically treated PET based activated carbons, Langmuir 20 (2004) 1321–1328.
[46] Z. Wu, P.A. Webley, D. Zhao, Comprehensive study of pore evolution, mesostructural stability, and simultaneous surface functionalization of ordered mesoporous carbon (FDU-15) by wet oxidation as a promising adsorbent, Langmuir 26 (2010) 10277–10286.
[47] R.H. Bradley, K. Cassity, R. Andrews, M. Meier, S. Osbeck, A. Andreu, C. Johnston, A. Crossley, Surface Studies of Hydroxylated Multi-Wall Carbon Nanotubes, Appl. Surf. Sci., 258 (2012) 4835–4843.
[48] L.R. Radovic, Surface chemistry of activated carbon materials: state of the art and implications for adsorption, in: J.A. Schwarz, C.I. Contescu (Eds.), Surfaces of nanoparticles and porous materials, Marcel Dekker, 1999, pp. 529–565.
[49] P.L. Walker, J. Janov, Hydrophilic oxygen complexes on activated graphon, J. Colloid. Interface Sci. 28 (1968) 449–458.
[50] J.L. Figueiredo, M.F.R. Pereira, M.M.A. Freitas, J.J.M. Órfão, Modification of the surface chemistry of activated carbons, Carbon 37 (1999) 1379–1389.
[51] K. László, E. Tombácz, K. Josepovits, Effect of activation on the surface chemistry of carbons from polymer precursors, Carbon 39 (2001) 1217–1228.
[52] H.F. Gorgulho, J.P. Mesquita, F. Gonçalves, M.F.R. Pereira, J.L. Figueiredo, Characterization of the surface chemistry of carbon materials by potentiometric titrations and temperature-programmed desorption, Carbon 46 (2008) 1544–1555.
[53] M.V. López-Ramón, F. Stoeckli, C. Moreno-Castilla, F. Carrasco-Marín, Specific and non-specific interactions of water molecules with carbon surfaces from immersion calorimetry, Carbon 38 (2000) 825–829.
[54] F. Stoeckli, A. Lavanchy, The adsorption of water by active carbons, in relation to their chemical and structural properties, Carbon 38 (2000) 475–494.
[55] G. Sh. Gogelashvili, R. Sh. Vartapetyan, D.V. Ladychuk, YuB. Grunin, E.V. Khozina, Specific features of the adsorption and nuclear magnetic relaxation of water molecules in active carbons. 1. Relaxation between the spin-spin relaxation of adsorbed water molecules and structural parameters of microporous active carbons, Colloid J. 65 (5) (2003) 545–551.
[56] H.P. Boehm, Surface oxides on carbon and their analysis: a critical assessment, Carbon 40 (2002) 145–149.
[57] C.L. McCallum, T.J. Bandosz, S.C. McGrother, E.A. Muller, K.E. Gubbins, A molecular model for adsorption of water on activated carbon: comparison of simulation and experiment, Langmuir 15 (1999) 533–544.
[58] S. Picaud, B. Collignon, P.N.M. Hoang, J.C. Rayez, Adsorption of water molecules on partially oxidized graphite surfaces: a molecular dynamics study of the competition between OH and COOH sites, Phys. Chem. Chem. Phys. 10 (2008) 6998–7009.
[59] J. Lahaye, G. Nange, A. Bagreev, V. Strelko, Porous structure and surface chemistry of nitrogen containing carbons from polymers, Carbon 37 (1999) 585–590.
[60] F. Kapteijn, J.A. Moulijn, S. Matzner, H.P. Boehm, The development of nitrogen functionality in model chars during gasification in CO_2 and O_2, Carbon 37 (1999) 1143–1150.
[61] K. László, Characterization and adsorption properties of polymer based microporous carbons with different surface chemistry, Micropor. Mesopor. Mat. 80 (2005) 205–211.
[62] P.-X. Hou, H. Orikasa, T. Yamazaki, K. Matsuoka, A. Tomita, N. Setoyama, Y. Fukushima, T. Kyotani, Synthesis of nitrogen-containing microporous carbon with a highly ordered structure and effect of nitrogen doping on H_2O adsorption, Chem. Mater. 17 (2005) 5187–5193.

[63] C.M. Yang, K. Kaneko, Adsorption properties of nitrogen-alloyed activated carbon fiber, Carbon 39 (2001) 1075–1082.

[64] L. Cossarutto, T. Zimny, J. Kaczmarczyk, T. Siemieniewska, J. Bimer, J.V. Weber, Transport and sorption of water vapour in activated carbons, Carbon 39 (2001) 2339–2346.

[65] T. Matsuoka, H. Hatori, M. Kodama, J. Yamashita, N. Miyajima, Capillary condensation of water in the mesopores of nitrogen-enriched carbon aerogels, Carbon 42 (2004) 2346–2349.

[66] A. Tóth, K.V. Voitko, O. Bakalinska, G.P. Prykhod'ko, I. Bertóti, V.M. Gun'ko, K. László, Influence of surface functionalization on the morphology and surface hydrophilicity of MWCNT as seen by vapour adsorption, Carbon 50 (2) (2012) 577–585.

[67] M. Seredych, M. Khine, T.J. Bandosz, Enhancement in dibenzothiophene reactive adsorption from liquid fuel via incorporation of sulfur heteroatoms into the nanoporous carbon matrix, Chem. Sus. Chem. 4 (2011) 139–147.

[68] A. Schenk, B. Winter, C. Lutterloh, J. Biener, U.A. Schubert, J. Kueppers, The origin of reduced chemical erosion of graphite based materials induced by boron doping, J. Nucl. Mater. 220–222 (1995) 767–770.

[69] Y.J. Lee, H.J. Joo, L.R. Radovic, Preferential distribution and oxidation inhibiting/catalytic effects of boron in carbon fiber reinforced carbon (CFRC) composites, Carbon 41 (2003) 2591–2600.

[70] L.E. Jones, P.A. Thrower, The effect of boron on carbon-fiber microstructure and reactivity, J. Chim. Phys. 84 (1987) 1431–1438.

[71] L.R. Radovic, M. Karra, K. Skokova, P.A. Thrower, The role of substitutional boron in carbon oxidation, Carbon 36 (1998) 1841–1854.

[72] S.G. Oh, N.M. Rodriguez, In-situ electron microscopy studies of the inhibition of graphite oxidation by phosphorous, J. Mater. Res. 8 (1993) 2879–2888.

[73] Y.J. Lee, L.R. Radovic, Oxidation inhibition effects of phosphorus and boron in different carbon fabrics, Carbon 41 (2003) 1987–1997.

[74] H.B. Park, Y.M. Lee, Pyrolytic carbon-silica membrane: a promising membrane material for improved gas separation, J. Memb. Sci. 213 (2003) 263–272.

[75] J.F. Yang, G.J. Zhang, N. Kondo, T. Ohji, Synthesis and properties of porous Si_3N_4/SiC nanocomposites by carbothermal reaction between Si_3N_4 and carbon, Acta Mater. 50 (2002) 4831–4840.

[76] R.H. Bradley, M.W. Smith, A. Andreu, M. Falco, Surface studies of novel hydrophobic active carbons, Appl. Surf. Sci. 257 (2011) 2912–2919.

[77] G. Li, K. Kaneko, S. Ozeki, F. Okino, R. Ishikawa, M. Kanda, H. Touhara, Water rejective nature of fluorinated microporous carbon fibers, Langmuir 11 (1995) 716–717.

[78] A.J. Fletcher, Y. Uygur, K.M. Thomas, Role of surface functional groups in the Adsorption kinetics of water vapor on microporous activated carbons, J. Phys. Chem. C 111 (23) (2007) 8349–8359.

[79] E.N. Rudisill, J.J. Hacskaylo, M.D. LeVan, Coadsorption of hydrocarbons and water on BPL activated carbon, Ind. Eng. Chem. Res. 31 (1992) 1122–1130.

[80] N. Qi, M.D. LeVan, Coadsorption of organic compounds and water vapor on BPL activated carbon. 5. Methyl ethyl ketone, methyl isobutyl ketone, toluene, and modeling, Ind. Eng. Chem. Res. 44 (2005) 3733–3741.

[81] F. Delage, P. Pré, P. Le Cloirec, Effects of moisture on warming of activated carbon bed during VOC adsorption, J. Environ. Eng. 125 (1999) 1160–1167.

[82] A.W. Heinen, J.A. Peters, H. van Bekkum, Competitive adsorption of water and toluene on modified activated carbon supports, Appl. Catal. A: Gen. 194–195 (2000) 193–202.

[83] P.J. Reucroft, P.B. Rao, G.B. Freeman, Binary vapour adsorption by activated carbon, Carbon 21 (3) (1983) 171–176.

[84] R.M. Rynders, M.B. Rao, S. Sircar, Isotope exchange technique for measurement of gas adsorption equilibria and kinetics, AIChE J. 43 (10) (1997) 2456–2470.

[85] A. Léonard, H. Wullens, S. Blacher, P. Marchot, D. Toye, M. Crine, P. Lodewyckx, In situ observation of wall effects in activated carbon filters by X-ray microtomography, Sep. Purif. Technol. 64 (2008) 127–130.

[86] K. László, O. Czakkel, B. Demé, E. Geissler, Simultaneous adsorption of toluene and water vapour on a high surface area carbon, Carbon, submitted.

Chapter 6

The Basicity of Carbons

Miguel A. Montes-Morán
Instituto Nacional del Carbón, INCAR-CSIC, Oviedo, Spain

Dimas Suárez
Departamento de Química Física y Analítica, Universidad de Oviedo, Oviedo, Spain

J. Angel Menéndez and Enrique Fuente
Instituto Nacional del Carbón, INCAR-CSIC, Oviedo, Spain

Chapter Outline

6.1. Introduction 173
6.2. The 6 Ws of Carbon Basicity 174
6.2.1. What? 174
6.2.2. Which? 174
6.2.3. Where? 175
6.2.4. How? 175
6.2.5. When? 177
6.2.6. Why? 179
6.3. Oxygen-containing Basic Groups 180
6.4. Heteroatom (Other than Oxygen) Functional Groups 184
6.4.1. Nitrogen 184
6.4.2. Phosphorus and Sulfur 188
6.5. Basic Sites on Carbon Atoms of the Basal Plane and Graphene Edges 189
6.6. Inorganic (or Mineral) Matter 193
6.7. Conclusions (and Some Open Questions) 196
References 198

6.1. INTRODUCTION

Acidity, defined as the quality or state of being acid, is possibly the chemical concept that has been most successfully adopted in our daily language. Basicity, on the other hand, is much less known and its use seems more restricted to scientific and technical disciplines. Both concepts are a consequence of the natural inclination of humankind for classifying things. It was in the mid-XVII century when Robert Boyle grouped a series of inorganic

Novel Carbon Adsorbents. DOI: 10.1016/B978-0-08-097744-7.00006-5
Copyright © 2012 Elsevier Ltd. All rights reserved.

compounds under a common label – acids. Bases were also identified as substances that showed different, mostly opposite, properties to those of acids [1]. Acids and bases tend to react (aka neutralization), and the driving force of such reactions soon became a fundamental type of interactions in chemistry and life sciences. Different theories were proposed to understand and predict acid–base interactions. Theories of Arrhenius, Brønsted–Lowry, and Lewis led consecutively to different definitions of acids and bases to cover as much a broad spectrum of chemical and biological substances as possible. Acid–base characteristics have been also identified in solids since they are relevant to both their practical and industrial applications [2,3]. This is indeed the case of carbon materials whose basicity is covered in this chapter. However, assessing the basicity of carbons at either the molecular or the phenomenological level can be a challenging problem. To facilitate the readership to comprehend all the issues at stake, the main aspects of carbon basicity are introduced next by answering a checklist of six questions.

6.2. THE 6 Ws OF CARBON BASICITY

6.2.1. What?

In this chapter, a generalized definition of what it is meant by basicity is adopted, involving the transference of protons between two different entities. Thus, basicity is defined here as the ability of carbon materials to act as proton sinks when faced with a suitable proton source. In this way, basicity embraces not only the concepts developed by Brønsted and Lowry but also those of the Lewis theory, as far as protons will always behave as electron acceptors.

6.2.2. Which?

A tentative definition of "carbon material" that was proposed more than thirty years ago and seems still valid for (or accepted by) the carbon community comprises a wide diversity of materials [4]. We intend to concentrate on the basicity of sp^2 carbons. The amount of materials that must be covered after such restriction is nevertheless considerable, from activated carbons to graphenes, from natural graphites to glassy carbons. This comment is stressed here to point out that, generally speaking, "carbon material" is not a solid with a definite chemical formulation. However, dealing with the basicity of carbons rather than just, for example, the basicity of carbon fibers involves the assumption that there are common factors that can be identified as responsible for the basicity of most, if not all, carbon materials. However, it should be clear that every carbon material presents its own peculiarities that may affect the way they interact with protons, both phenomenologically and in terms of practical outcomes.

6.2.3. Where?

With the previous definitions in hand, the next to elucidate is what does specifically make carbons to behave as proton sinks. A possible answer to that question entails the identification of basic sites on carbons, which denotes the existence of particular locations on the materials where protons will eventually gather. Traditionally, basic sites are assumed to occur on the solid surface. In other words, basicity is related to the surface of carbons and it is commonly referred to as "carbon surface basicity". This description assumes that protons probing the basicity of carbon materials will only access the outermost layers of the solid, which seems reasonable inasmuch as the structural integrity of the solid surface stands the conditions employed for testing basicity (see below). Also notice that, for some bidimensional carbon materials (graphenes), it would make little sense to discriminate between the basicity of the (bulk) material and that of its surface.

6.2.4. How?

The basicity of a given carbon material will be, in principle, determined by the amount and strength of the basic sites present on its surface. Understanding the basicity of carbons has been demonstrated far from being a simple task. There are three main reasons for this: (i) basic sites on carbons coexist with acid sites that can act as proton sources, i.e. carbon surfaces are amphoteric; (ii) the intrinsic complexity, chemically speaking, of the carbon surfaces that makes quite an elusive task the direct identification of possible basic sites; and (iii) difficulties in discriminating between acids that are only physically adsorbed on carbon surfaces and those that are chemically neutralized by carbon surfaces. Thus, most experimental studies on the basicity of carbons are limited to a rough description of the overall acid–base interactions between the carbon material and a liquid or gaseous proton source. Since adsorption deals with the interaction of adsorbates with solid surfaces, the experimental methodologies developed for testing carbon basicity may be considered as adsorption studies on carbon materials involving a particular probe (proton source). This description allows us to stress a critical subject during the evaluation of carbon basicity by current experimental procedures: carbon basicity not only depends, as quoted above, on the amount and strength of basic sites but also is strongly conditioned by the accessibility of those sites. In that sense, all diffusion-related features that are overlooked when dealing with homogeneous solutions could actually determine the output of carbon basicity measurements. This is particularly, but not exclusively, important when characterizing the basicity of porous materials.

So far, the most extended methodologies for testing the basicity of carbons involve proton sources in aqueous solutions, from relatively simple pH determinations to potentiometric titrations [5–8]. All these methodologies are adapted from traditional titration techniques developed for homogeneous aqueous solutions. Since carbon materials do not, in principle, dissolve in water, those homogeneous solutions are replaced by solid suspensions, entangling a series of implications that need to be considered. First, as just mentioned, diffusion-related parameters such as solid particle size, amount of solid tested per volume of water solution, temperature, ionic strength, and ionic size affect this type of measurements. Second, knowledge of water–carbon interaction is required to understand carbon basicity. This interaction, which is described in terms of hydrophilicity or hydrophobicity, must also play a crucial role when measuring the basicity of carbons in aqueous suspensions. One may question, for example, if an aqueous-based methodology is adequate to test hydrophobic basic sites (i.e. pure C basal planes) on carbon surfaces [9].

Carbon basicity measurements based on the adsorption of gases or vapors as proton sources are less commonly reported. This is important since the ability of a carbon surface to act as a proton sink might be different in (dry) gas-phase and liquid-phase applications. Volumetric or gravimetric analyses of the adsorption of acidic gases or vapors are a suitable technique for studying the basicity of carbons, although studies seem restricted to highly porous materials [10,11]. Other approaches based on the adsorption of different gas or vapor probes have been alternatively used to provide information of the basicity of carbons, including solids of low specific area. Gas–solid chromatography (or, as recently renamed, inverse gas chromatography) has been demonstrated to be a powerful tool to monitor changes in the (Lewis) basicity of different carbons materials [12–15]. The sensitivity of the technique makes possible to work at a very low adsorbate concentration, virtually in the Henry's law regime. After adoption of donor–acceptor concepts to describe the energetics of the gas–solid interaction, elution of selected molecules with well-known physicochemical properties may eventually lead to the parametrization of the carbon surface basicity (donor numbers).

Adsorption energetics, whether in gas or liquid phase, has been also used to account for the basicity of carbons. Calorimetric techniques allow one to detect changes in the energy of interaction (i.e. heat of adsorption) between an acidic probe and the surface of carbons [16–18]. If the textural properties of a series of materials are assumed to be very similar, then those changes can be related to the basic character of the solid surface.

Adsorption of selected probes can be also followed by different spectroscopic techniques. The approach is relatively simple and is based either on the detection of molecular properties of the probe adsorbed on the carbon surface or on the changes observed on the properties of those surfaces after adsorption. FTIR, solid-NMR, XPS, and ESR are some examples of useful techniques for that purpose. In principle, it should be possible to quantify the amount of basic sites on carbons. This is, however, a rather optimistic scenario for most carbon materials due to both matrix effects and the complexity of such matrix (i.e. the carbon material). Semiquantitative analysis of basic sites can be eventually performed after designing very specific surface reactions, in which only the basic sites, or even a certain type of basic site, are prone to react with a selected molecule [19,20]. Normally in such cases, previous adoption of models to describe basic sites on carbons is required (see below).

Back into aqueous solutions, basic carbons acting as proton sinks will exhibit an excess of positive charge on their surface. Measuring those charges would then render an indication of the basicity of the materials. Conventional methodologies comprise the estimation of the electrokinetic potential (ζ-potential) and mass titrations [21]. Comparison of the results obtained with these two surface-charge measuring techniques, i.e. the so-called isoelectric point and the point of zero charge, respectively, constitutes a good example of the relevance of the location of the basic sites when testing carbon materials. Thus, for porous carbons, electrokinetic measurements, unlike mass titration, are strongly determined by basic groups on the outermost surface of the material. In other words, basic sites located on the inner pores are not "accessible" when probed with this particular technique. As a consequence, electrokinetic measurements and mass titration of the same (porous) carbon may provide information regarding the basicity of the surface that may differ qualitatively (acid vs. basic character) [21,22].

Electrochemical measurements can also be a tool for determining the basic character of a carbon material, although more knowledge is required to establish direct relationships between the electrochemical response of carbons and their basic character [23,24]. When a powdered activated carbon electrode is immersed for the first time into an electrolyte solution, it exhibits large electrochemical capacities, about 100 F g^{-1} and more, due to the large pore surface and electrolyte ion adsorption. Such double-layer capacity corresponds roughly to 1 mmol g^{-1} of adsorbed ionic species, thus suggesting that a significant fraction of the acid/base species in aqueous solution could interact with the carbon surface groups to build up an electrical double layer rather than giving a proton exchange reaction with those groups. In other words, the acid/base buffer capacity of some carbon materials is probably entangled with its electrochemical capacitance and, consequently, both the physical adsorption effects and the intrinsic basicity of carbon functionalities would play an important role in the electrochemical response of the carbon electrode.

6.2.5. When?

All these techniques just discussed above are able to quantify the ability of a series of carbons to gather protons on their surfaces, thus ranking them in terms of their basic strength. Yet the reason for this ability, i.e. the nature of the basic sites on carbons, remains unclear [21,25–27]. Looking for evidence of those basic sites has been the main purpose of a number of investigations from almost a hundred years ago. The approach at those times was based on a detailed observation of certain physicochemical reactions promoted by carbon materials that also behaved as proton sinks [28]. Most of these carbons were obtained after high-temperature (>1000 K) treatment of different precursors (or low-temperature synthetized carbons) under nonoxidant atmosphere, a procedure that constitutes since then an everyday recipe for the preparation of basic carbons. The study of the interaction of these high-temperature-treated carbon surfaces with molecular oxygen led to the first models accounting for the basicity of carbons, which were based on oxygen functional groups [29–31]. Further work was carried out to understand the basicity of carbons with very low oxygen contents [28].

Landmarks on the Basicity of Carbons

- Burstein and Frumkin (1929): "Electrochemical model" [28]
- Garten and Weiss (1957): "Chromene-type surface oxides" [29]
- Boehm and Voll (1971): "Pyrone-type surface oxides" [30]
- León y León, Solar, Calemma and Radovic (1992): "Basal planes" [27]

The use of specific chemical surface characterization techniques in more recent years has shed some light on the identification of basic sites on carbons. As mentioned above, the chemical complexity of most carbon surfaces makes it quite difficult for a given technique to univocally identify specific moieties as being responsible for the basic character of the solids. Consequently, most of the surface characterization methodologies, which are nowadays popular among carbon scientists, need to make assumptions related to the types of basic sites they are testing. Techniques such as TPD [32], FTIR [33], or XPS [34], for example, are commonly used to detect and even quantify the basicity of carbons, assuming that mainly oxygen groups are responsible for the basicity of carbon materials, which may not be always true. On the other hand, the nature of carbon surfaces always compels to combine several surface characterization techniques to obtain as much a complete description as possible. Unfortunately, results obtained when describing the basicity of carbons by different techniques are still sometimes difficult to explain [35].

Nevertheless, there seems to be a compromise within the scientific community in assuming that basicity of carbons should be understood as a combination of different contributors, or basic site families (Scheme 6.1):

a) oxygen functional groups,

b) heteroatom (other than oxygen) functional groups,

c) π-electrons and edge-electron pairs, and

d) inorganic matter.

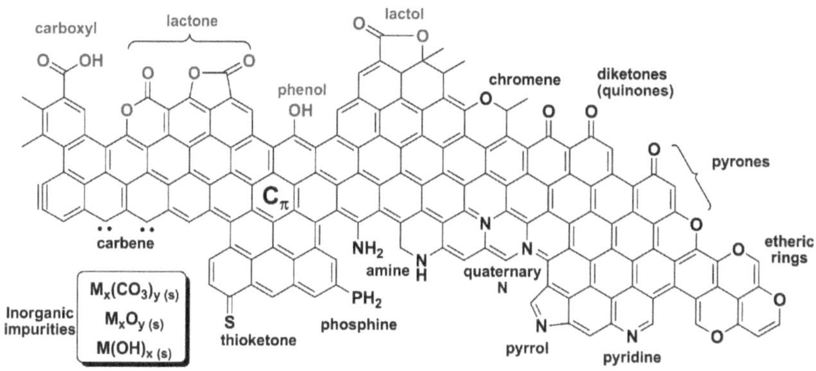

SCHEME 6.1 An overview of basic sites on carbons; some acidic sites are also included in light grey.

The description of these families in terms of occurrence and basic strength constitutes the main core of this chapter. On the other hand, basic sites that may arise from the impregnation of the carbon surfaces with chemicals, or by anchoring/grafting of molecules, are not considered here; otherwise, the compilation of sites would be limitless.

6.2.6. Why?

There are several applications of carbons for which solid surfaces exhibiting a basic character is a prerequisite. A comprehensive list of such applications is beyond the scope of this chapter. The reader may find representative examples in most of the references quoted in the References section. Nevertheless, we present now some examples intending to illustrate why, for certain applications, basic carbons should be the best choice. In general, basic carbons tend to adsorb acidic molecules; so basic carbons are preferable to neutral or acidic ones in applications involving the adsorption of acidic compounds. Many examples of this obvious, but sometimes ignored, behavior can be found in gas-phase and aqueous-solution applications.

Thus, for example, basic carbons are required for abatement of sour gas (i.e. H_2S) emissions [36], SO_2 control [37], and CO_2 capture [38]. Another interesting feature of basic carbons used in gas-phase applications is that, in general, they are more resistant to aging than acidic ones. However, this is due to the fact that they have lower oxygen content than acidic carbons so they are more hydrophobic, which contributes to a lower adsorption of undesirable moisture that eventually may block some of the porosity [39].

On the other hand, adsorption in aqueous phase is often governed by the electrostatic properties of both the adsorbent and adsorbate. Thus, basic carbons, which are positively charged in aqueous solution with pH above their point of zero charge, are preferable for adsorbing anions or negatively charged species. Several examples of this important feature of basic carbons can be found in [40]. Interestingly, for most of these applications it seems that what is relevant is how basic the carbon is (i.e. its point of zero charge), no matter the origin of its basicity (i.e. the type of basic site). However, the distribution of these basic sites between the inner/outer carbon surfaces can be very important [41].

Less abundant but not least important are those applications in which the basic properties of carbons confer them catalytic properties, for example, the selective catalytic reduction (SCR) of nitrogen oxides with ammonia over activated carbon, which requires the occurrence of basic nitrogen sites on the surface of the material [42]. In these cases, it is very likely that the nature of the basic centers may play an important role in the catalytic activity of the carbon [43]. However, the degree of knowledge achieved in this matter is so far quite limited. Furthermore, the lack of understanding of carbon basicity leaves sometimes relevant experimental results unexplained. The basicity of carbons is thus important not only for already-known processes but also for searching and/or controlling the performance of carbon materials in new applications.

6.3. OXYGEN-CONTAINING BASIC GROUPS

Oxygen is the second most frequent element present on the surface of carbon materials and, therefore, characterization of the structure and activity of oxygen functionalities attached to carbon sheets has deserved a great deal of attention from carbon science. Of particular interest has been the Boehm titration method [5,21], which estimates the total number of the oxygenated surface groups and classifies them into two families attending to their acidic or basic character in aqueous solution, either as direct proton donors/acceptors or as immediate precursors of acid/base groups formed upon hydrolysis processes (Scheme 6.1).

Concerning the acid functional groups (i.e. carboxyl groups, lactones, and phenol and lactol groups), there is a general agreement about their qualitative and quantitative imprints on various experimental characterization techniques as well as their influence on the adsorption and catalysis exerted by activated carbons. Unfortunately, this has not been the case for those basic oxygen-containing functionalities that have been proposed in the literature such as chromene structures, diketone, or quinone groups, and pyrone-like groups (see Scheme 6.1).

Difficulties in discriminating among the hypothetical oxygen-containing basic sites on carbon materials arise from different factors. To begin with, basic carbons in which oxygen is the only heteroatom are relatively scarce. Moreover, the total amount of oxygen in these materials is rather low (2–4 %) given that their preparation from various precursor materials generally involves high-temperature thermal treatments under reductive conditions in order to promote the desorption of oxygenated acidic groups that are thermally more labile. The remaining oxygen-containing groups can be expected to be a combination of sp^2 and sp^3 O atoms bound to the carbon edges that would constitute the basic functional groups. However, their molecular differentiation in terms of functionalities through characterization analyses (e.g. XPS and TPD) has remained largely elusive. Another source of confusion has been the uncertainty about the actual strength as bases of these groups, because the small organic compounds that were initially considered as reference compounds have quite low pK_a values (e.g. the monocyclic γ-pyrones are well-known organic compounds which exhibit a rather poor basicity with a pK_a value of ~−1), which cannot explain the pK_a distribution observed in oxygen-containing basic carbons by means of continuous equilibrium potentiometric titration [44].

Further complications in unraveling the relationship between carbon basicity and oxygen result from the commonly observed amphoteric character of carbon surfaces, containing both acidic and basic surface groups [45], the appearance of hysteresis in the titration curves of some basic materials when performed upward and downward [46], and, especially, the triggering of electrochemical processes by proton binding [44]. Even other authors have proposed that oxygen-containing groups on carbon surfaces are primarily reducing agents rather than Brønsted bases so that they would first transfer electrons to oxygen molecules, thus yielding a positively charged carbon skeleton and hydroxide anions, which would be formed as a final product of O_2 reduction [46].

To fill the gap of fundamental knowledge about the basicity of oxygen-containing groups, some intrinsic properties of these groups have been characterized at the molecular level by means of high-level quantum chemical calculations on large heterocyclic compounds containing up to nine carbon rings [27,47–49]. According to these studies, basic carbon surfaces containing a *mixture* of sp^2 and sp^3 O atoms located at their edges would be thermodynamically more favorable than surfaces containing either all-sp^2 or all-sp^3 O atoms. More particularly, bicyclic pyrones would be especially stable structures, the adjacent basal plane contributing further to stabilize them [47]. This hypothesis is further reinforced by the presence of IR absorption bands at $1700\,cm^{-1}$ and $1580\,cm^{-1}$ detected on basic carbon materials, which, according to theoretical calculations [50], can be assigned to polyaromatic systems in which both cyclic ketone and cyclic ether rings coexist on the carbon surface.

Concerning the intrinsic basicity of polycyclic pyrones in aqueous solution, quantum chemical pK_a calculations have convincingly shown that protonation on the carbonylic O atom would be greatly favored by electronic conjugation throughout the carbon skeleton (Scheme 6.2) [48,49]. In fact, the computed pK_a's for the most likely pyrone groups, ~5–6 and ~7–8 for bicyclic and tricyclic structures, respectively, explain well the continuous pK_a distribution observed in a basic activated carbon [44]. Furthermore, pyrones in which the ketone and etheric rings would be separated by 2–3 carbon rings

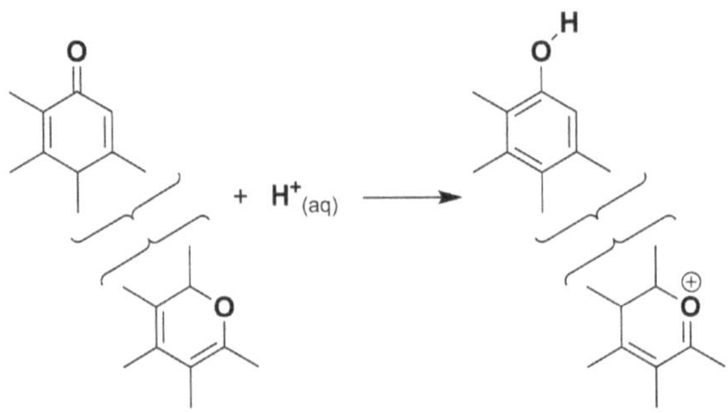

SCHEME 6.2 Pyrone-like model acting as proton sink.

could result in strong basic sites (pK_a ~10–13). Besides these theoretical results suggesting that polycyclic pyrones should be positively considered as organic bases, it may also be interesting to note that protonated polycyclic pyrone-like compounds can be found in natural products such as vitisin-B, which is found in red wines, and that can be isolated as stable organic cations in the form of solid salts [51].

Perhaps the main conclusions drawn from the theoretical analyses are: (i) the term "pyrone groups on carbon surfaces" should be interpreted in terms of a *global* distribution of peripheral etheric and carbonyl rings rather than *specific* functional groups and (ii) these groups exhibit a clear structure–basicity relationship. This picture somehow supersedes the original proposals of weakly basic organic functionalities (ketones, etheric rings, and chromenes) accounting for carbon basicity. In addition, it has been shown theoretically that pyrone-like model compounds may have standard electrochemical potentials (E^0) for protonation → reduction processes (see Scheme 6.3) that lie in the range observed in some basic carbons ($E^0 = 0$–1 V) [47]. This suggests that pyrone-like groups may be responsible of the redox activity of basic carbon surfaces instead of hydroquinone groups, which are generally the most favored candidates [52]. However, it must be emphasized that many molecular details of the hypothetical redox processes involving pyrone groups are completely unknown and, therefore, whether or not polycyclic pyrone groups can explain

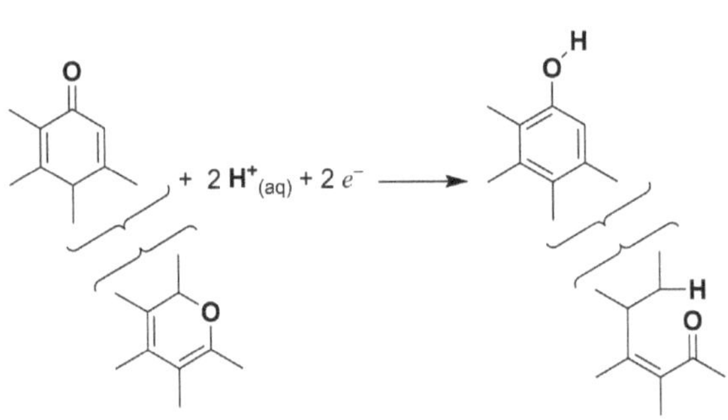

SCHEME 6.3 Potential redox activity of pyrone-like structures.

the redox activity and other physicochemical features associated with proton uptake is still an unsettled question.

In spite of its limitations, pyrone-like structures constitute an attractive model that can help to both rationalize different properties exhibited by basic carbons and promote further experimental and theoretical research in order to improve our understanding of its structural, energetic, and chemical implications. As a matter of fact, the recent literature contains numerous examples of the characterization and/or application of all kinds of carbon materials in which their oxygen-containing basic groups can play an important role. For example, the surface chemistry of activated carbons usually requires the presence of basic sites in order to catalyze various reactions (dehydrogenation of alcohols, oxidation of NO, SO_2, and H_2S; catalytic ozonation, catalytic wet air oxidation, etc.) [53]. As a clear correlation has been found in some processes between catalytic activity and the concentration of "carbonyl/quinone" surface groups, increasing our mechanistic knowledge of these important reactions will also require determining the actual structure and catalytic role of those basic oxygen-containing groups. Similarly, the increase of the adsorption capacity of basic activated carbons (containing ~4% O) functionalized with organic π-compounds toward metal cations and chromate ions has been broadly related to the presence of the carbon basic sites, but the influence of the oxygen-containing groups has not been determined yet [54,55]. On the other hand, it has been claimed that basic oxygen groups (ascribed to "semiquinones") on the surface of Pt-doped activated carbons can increase significantly the reversible hydrogen storage capacity, but again no clear explanation for this effect is available yet and, therefore, further study of the role of the basic sites in hydrogen spillover is needed [56].

These selected examples and others strongly suggest that carbon science may benefit from the "pyrone model" in order to formulate and test molecular hypothesis about the activity of basic activated carbons. This is indeed the case of recent studies on the electrochemical behavior of carbons as the electrode materials in supercapacitors for which perfect correlations have been found between the total number of basic groups and the gravimetric capacitance [57]. Interestingly, the electrochemically active oxygen functionality has been identified as "pyrone oxygen" (defined as carbonylic and etheric oxygen in close vicinity) basing on the results of XPS O 1s data, although further characterization of the electrochemical properties of pyrone compounds (e.g. by cyclic voltammetry) was deemed to be required [57,58].

In addition to conventional activated carbons, characterization of the basic oxygen-containing functionalities in carbon nanomaterials has deserved increasing attention during the last few years. However, in some cases, it turns out that researchers are apparently unaware of current paradigms on carbon basicity. Thus, it has been reported that the adsorption against lead ions of graphite nanosheets (GNSs) is enhanced by heat treatment, which, in turn, increases the basicity of the GNSs [59]. This effect was interpreted merely in terms of the Lewis carbon basicity of the basal plane even though the GNSs retain a considerable amount of carbonyl and etheric oxygen after heat treatment. Another interesting case of nanocarbon with a basic character is diamond nanoparticles obtained by annealing aggregated nanodiamond powder in hydrogen gas. During this process, a large fraction of the initial nanodiamond oxygen groups are reduced and the relatively large (>100 nm) aggregates are broken down into much smaller nanoclusters (4 nm) that can be dispersed in aqueous solution *via* high-power ultrasound and high-speed centrifugation. The resulting stable dispersion has a high positive zeta potential (i.e. a basic character) and the nanostructures still possess oxygen groups as suggested by FTIR absorption in the 1500–1700 cm^{-1} region. However, the generation of a positive zeta potential was considered an unusual result that was ascribed to basal plane basicity as "protonation of conventional oxygen functional groups is rather unlikely to occur" [60].

Finally, it may be interesting to comment on the reduction of graphite oxide to yield reduced graphite oxides (rGOs), which are considered as a valuable precursor in the search for large-scale production processes of GNSs. However, the majority of reduction techniques report the presence of a significant fraction of residual oxygen in rGO, which represents a fundamental limitation that must be overcome in order to achieve properties that resemble those of mechanically exfoliated graphene. Most interestingly, molecular dynamics simulations have revealed new insight into the structure of progressively reduced graphene oxide during thermal annealing by showing that highly stable "carbonyl and ether groups" are formed (even at high-temperature annealing) that hinder its complete reduction to GNS [61]. Although the total oxygen content of rGOs is larger (~18%) than that of basic activated carbons, these theoretical and experimental results on the structure of rGOs are in consonance with the previous proposals about the thermodynamic stability of "pyrone groups" on thermally activated carbon surfaces.

6.4. HETEROATOM (OTHER THAN OXYGEN) FUNCTIONAL GROUPS

6.4.1. Nitrogen

The basic nature of nitrogen-containing organic compounds determines important acid–base interactions that are in the core of both organic and life chemistry. Nitrogen is thus expected to be present in carbon materials prepared from a wide variety of precursors, from biomass to polymers [62–64]. In addition, specific treatments have been formulated to increase the amount of nitrogen and/or to produce specific nitrogen compounds on the surface of carbons [62].

A systematic description of possible nitrogen basic sites on carbon surfaces is quite challenging. Some of the most relevant are included in Scheme 6.1. In principle, aliphatic amines, if present, would be basic sites of moderate strength ($pK_a < 11$). Significant quantities of these compounds are only expected after grafting treatments in which side amino groups are anchored onto the carbon surfaces through chemical reaction with previous existing functional groups (i.e. oxygen-based) [65]. This would then exclude aliphatic amines from the carbon basic sites of concern in the present chapter, for reasons mentioned above. Nonetheless, they are named here in realm of the wide interest and application of amine-grafted carbons.

Cyclic amines and imines are however expected to occur on carbon surfaces submitted to NH_3 or HCN treatments at high temperature (above 1000 K). Under such conditions, aromatic amines (i.e. anilines) are also prone to be formed. Anilines are known to be moieties of lower basic strengths than aliphatic amines, in spite of certain special molecular arrangements [66,67]. According to Stöhr et al. [68], gaseous ammonia decomposes to radical species such as NH_2, NH, and atomic H at high temperatures. The high reactivity of those radicals would eventually lead to the formation of NH_2 (i.e. anilines) or NH (i.e. cyclic amines and imines) functional groups attached to the carbon rings. On the other hand, the occurrence of these basic nitrogen groups on carbons treated at high temperatures may be questioned. Indeed, the stability of these particular groups with temperature is rather limited. The amination or ammoxidation of carbons at temperatures above 500 K lead to the formation of amides, imides, lactams (pyridones), or nitriles, groups of a very weak basic character, which are further transformed into pyridine- or pyrrole-like functional groups at higher temperatures (Scheme 6.4.) [69,70]. In addition, detection of amino groups on these ammonia-treated carbons is hampered by the limitations of the surface characterization techniques. For example, the literature values reported for the XPS binding energies of amine, amide, and imide nitrogen functional groups on carbons are within an interval of less than 0.4 eV [68,70–73].

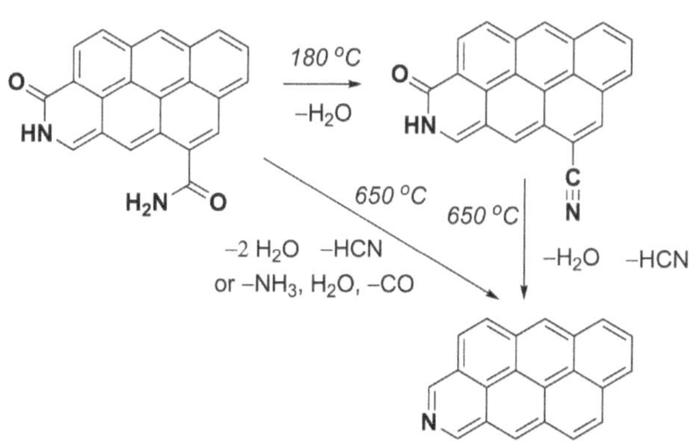

SCHEME 6.4 Thermal stability of nitrogen surface groups on carbons, after [69,70].

The insertion of the N atoms in the C (sp^2) ring structure of a carbon material confers stability to the resulting nitrogen functional groups to withstand higher temperatures than oxygen functional groups [62,74]. A plausible route for such insertion, which ends with the formation of pyridine- and pyrrole-like structures, has been inferred from different experimental studies (Scheme 6.4), including high-temperature treatments of carbon precursors (or carbon materials) impregnated with different nitrogen promoters (urea, for example) [75]. Whereas pyrrole is a very weak base, pyridine-like compounds exhibit a mild basic character. Theoretical calculations have shown that pK_a values of pyridine-like compounds are not significantly affected as the number of the fused rings of cluster models increases [27]. Experimental results of selected compounds also confirm this trend, thus suggesting that pyridine-like groups on carbons would exhibit pK_a values in the range of 4–6 units. It was then suggested that nitrogen functional groups could hardly account for the basic sites with p$K_a > 8$ that are commonly detected during potentiometric titrations of nitrogen-doped basic carbons.

There are some experimental evidences confirming that carbons treated under gaseous ammonia at high temperatures show higher HCl uptakes in aqueous solution than their corresponding counterparts treated under inert atmosphere [62,74]. A more detailed study shows that nitrogen-enriched activated carbons prepared from physical mixtures of sucrose and organic nitrogen additives and treated up to 1373 K under inert atmosphere have always a higher concentration of basic sites than the sucrose-only carbon treated under similar conditions [76]. Differences between the concentration of basic sites of the N-carbons and the reference, sucrose-only carbon are however less noticeable when expressed per unit area, i.e. in $\mu mol\, m^{-2}$ rather than in $\mu mol\, g^{-1}$. In addition, there seems to be no direct correlation between the final N content of the modified carbons and the concentration of their basic sites [76]. Recent studies carried out by the Prof. Bandosz group with carbons prepared from direct carbonization of commercial polymers reveal little pH differences if the carbons were previously impregnated with urea [77]. Interestingly, pK_a distributions derived from proton binding curves show that, in spite of the total amount ($mmol\, g^{-1}$) of surface groups being higher on the N-doped carbons, the concentration of the strongest basic groups detected ($10 < pK_a < 11$) was considerably higher on the heat-treated, nondoped carbon. In other words, the ability of N-enriched carbons of higher HCl uptakes than the heat-treated counterparts might be a consequence of more (or more accessible) basic sites per unit area, rather than stronger, acting as proton sinks.

An additional consideration in this subject should be paid to the mutual "interactions" between nitrogen and oxygen surface functional groups and how this could eventually affect the basicity of those groups. The pyrone-like model proposed by Voll and Boehm to explain the basicity of heat-treated carbons is sustained on experimental evidences describing the formation of basic oxygen surface functional groups after exposure of the heat-treated carbons to molecular oxygen (in air or water) [31]. Thus, the basicity of the oxygen surface groups on heat-treated carbons could be different when nitrogen atoms have been incorporated to the carbon layers. Theoretical calculations have relativized the extent of such effect: the basic strength of simple pyrone-like models does not seem to change significantly when N atoms are inserted in the fused rings (Scheme 6.5) [27].

As illustrated in the previous discussion, and notwithstanding the considerable amount of work devoted to N-doped carbons, the nature of the N-based basic sites on carbon materials still has loose ends. A final thought is devoted here to the role that the so-called quaternary nitrogen sites on carbons may play in their ability to gather protons [58,72,78,79]. According to spectroscopic evidences, the occurrence of such species is very common on N-enriched carbon surfaces and may eventually account for up to 90 % of the surface nitrogen [78,79]. Thus, deconvolution procedures of the N1s XP spectra of such carbons assign the band centered at ca. 401.5 eV to the binding energy of N atoms inserted in the valleys of zigzag edges or in the skeleton of the sp^2 C layers (Scheme 6.1). In both situations, the nitrogen atoms incorporate the two p-electrons to the π-cloud of the graphene layers, in a similar way of a pyrrole-like structure, thus altering the electronic mobility of the solid. Semiempirical quantum calculations predict an important increase of the electron-donor (i.e. Lewis basic) character of carbon cluster models containing quaternary nitrogen [80]. Moreover, a model for the formation of quaternary nitrogen based on pyridone-like structures has been recently proposed [74]. Pyridones, which involve the presence of both N and O heteroatoms on carbons, are plausible nitrogen groups that have been considered in studies dealing with the evolution of nitrogen in chars during gasification under different atmospheres [72]. After the insertion of the quaternary nitrogen in the valley position of a zigzag edge, this model would assume the prevalence of a *zwitterionic* structure that resembles a pyrone-like moiety in which the sp^3 oxygen is replaced by a positively charged N atom [74]. The oxygen atom would take on a negative charge, thus exhibiting a moderate basic character with an estimated pK_a of 8–11, similar to that of phenolic compounds (Scheme 6.6) [35]. As a consequence, both nitrogen and nitrogen/oxygen functional groups would explain the pK_a distribution of nitrogen-doped carbons discussed above, i.e. eventually

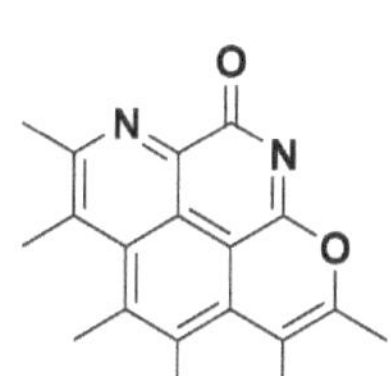

SCHEME 6.5 A possible model of nitrogen and oxygen functionalities coexisting on a carbon surface.

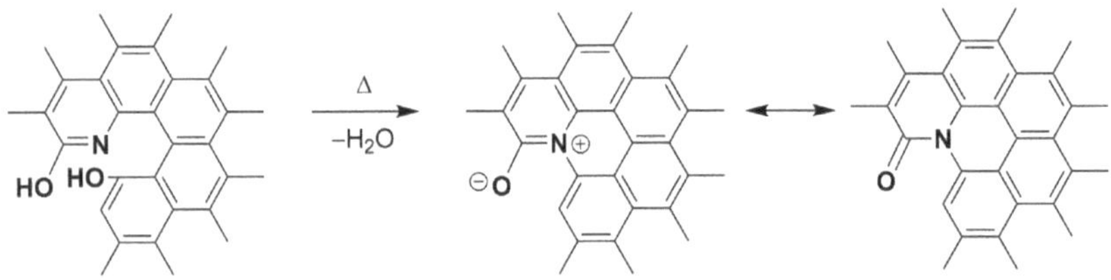

SCHEME 6.6 Suggested pathway for the formation of quaternary N structures on carbons, after [74].

there would be no need for assuming the presence of oxygen-only basic groups on the surface of such carbon materials.

6.4.2. Phosphorus and Sulfur

The semiempirical calculations of Strelko et al. also predicted a significant increase of the electron-donor capacity of carbons if phosphorus, rather than N atoms, were inserted accordingly into the carbon clusters [80]. P has become a relatively common heteroatom present on carbons since the growing interest on the activation of lignocellulosic precursors with phosphoric acid [81]. So far, however, most studies have described the incorporation of phosphorus into the carbon layers through the formation of phosphates, phosphonates, and phosphate esters [82,83], which are nonbasic moieties. Other possibilities of P insertion, i.e. in a similar fashion to N in pyridines, would eventually lead to phosphinine-type functional groups, which might have much lower basic character than the nitrogen counterparts. So, despite the possibility of "quaternary phosphorus" formation on carbons, only phosphine-type compounds could be P-basic sites on the surface of carbon materials. Phosphines, like amines, have been only reported to occur after the design of specific anchoring routes to attach homogeneous catalysts on different carbons [84,85].

Sulfur is the last heteroatom that would be considered in this section as a possible contributor to the basicity of carbons. As in the case of nitrogen, sulfur may be present in certain carbon precursors or incorporated to a carbon material by specific treatments, mainly with sulfurous gases [86–88]. Sulfur functional groups on carbons have been also proposed from different evidences pointing to the formation of C=S groups that would be rather stable with temperature [89,90]. Combining XPS and FTIR analyses, Chang suggested the possibility of having carbon–sulfur functional groups on S-doped carbons that would mimic carbon–oxygen groups already discussed, i.e. thiocarbonyls and thiolactones [87]. Sulfur–oxygen functional groups (sulfone and sulfoxide) have been also proposed to be present on carbons treated with sulfur agents [91]. In general, S-carbons are neutral or slightly acidic. A recent work shows that the amount of total groups with $pK_a > 8$ is much lower on an S-doped carbon than that of the heat-treated (under nitrogen) counterpart [92]. Still, more work is required to clarify the role of carbon–sulfur moieties in acid–base interactions. For example, the possible role of thiocarbonyl groups as proton sinks should be revisited: previous studies indicate that, in general, thiocarbonyl derivatives are more basic than their carbonyl counterparts, due to differences in electronegativity and polarizability of S and O atoms [93].

6.5. BASIC SITES ON CARBON ATOMS OF THE BASAL PLANE AND GRAPHENE EDGES

Could "pure" carbons (i.e. those carbons composed exclusively by the carbon element) act as proton sinks? An affirmative answer to this question would mean that these carbons, exempt of surface groups, should have basic centers allocated either in the carbon atoms of the basal planes or in those at the graphene edges. However, an attempt to give an answer leads us to another intriguing question: do "pure" carbons actually exist, or necessarily unsaturated carbon atoms at the graphene edges need to be saturated with, at least, a heteroatom or hydrogen? The answer to this question is not straightforward. Thus, as it will be discussed later, unsaturated carbon edge sites have been postulated to exist (at least in theory) at room temperature, i.e. "pure" carbons are theoretically possible. However, in practice, carbons are so reactive that upon exposure to air they tend to react with oxygen and/or moisture, giving rise to oxygen and/or hydrogen-containing surface groups that saturate, if not all, at least some of those edge carbon atoms. The presence of the "heteroatom" surface groups should have an effect on the electron density of the graphene sheet, localizing or delocalizing this density depending on their electron affinity. Hence, the characteristics (i.e. strength as bases) of the "mostly C" basic sites would be inevitably altered by the presence of heteroatoms. Assuming these difficulties in the study of the degree of basicity that can be attributed exclusively to the basal planes or unsaturated carbon edges, what follows is a nearly chronological account of a series of articles that attempts to give some insight into this controversial issue.

Using several experimental techniques and two series of "highly pure" carbons (along with a revision of the pertinent literature), León y León et al., in their influential paper published in 1992, gave experimental evidence that protons can be adsorbed on basal planes due to electrostatic interaction with the π-electron system of the graphene structures [28]. These authors postulated the existence of C_π sites in which electron donor–acceptor (EDA) interactions might take place. These sites would behave as Lewis basic sites (i.e. electron pair donors) of an unclear nature that, for convenience, are normally expressed with the following equation:

$$C_\pi + 2H_2O \rightarrow C_\pi - H_3O^+ + OH^-$$

According to this hypothesis, "the C_π sites are probably located at π-electron-rich regions within the basal planes of carbon crystallites, away from the crystallite edges" [28]. In principle, delocalized π-electrons of the basal planes should be abundant enough to represent an important percentage of the basic sites of a given carbon. However, in the case of porous carbons, the authors comment that "the EDA interaction of H_3O^+ ions with the basal planes of porous microcrystalline carbons is limited by electrical, electronic, and geometric effects and, consequently, only a small fraction of each basal plane can hold EDA complexed H_3O^+ ions" [28].

In spite of some scarce studies that supported the key role of oxygen groups in understanding the basicity of carbons [94,95], the paper of León y León et al. soon opened a controversy that is still far from being settled. Thus, flow calorimetric measurements on different carbons ranging from highly oxidized acidic materials to basic carbons with low oxygen content gave an estimation of the heats of neutralization of "generic" acid and basic sites [96]. Although only two basic carbons were considered, the authors tentatively extrapolated their thermochemical measurements to obtain the heat of neutralization of the basic sites "at 0% oxygen content", yielding a value of –38 kJ mol^{-1} (–42 kJ mol^{-1} for the acid sites). The study concluded that basicity has little to do with the surface oxygen content and that, attending to the heat of neutralization, the "generic" basic site should be regarded more as a Lewis type rather than a Brønsted type. A different study, based on XPS measurements, seemed to confirm that the basicity of a series of carbon blacks is mainly associated with a Lewis basicity type of interaction, hence related to the EDA concept of the basal plane [97].

The ability of fine graphite particles for uptake of very small (but measureable) quantities of acid was then reported by Boehm [25,21]. pH determination on aqueous suspensions of highly pure ($C > 99\%$), small particle graphite renders almost neutral values [27]. These are two examples that would suggest that, even in the case that these basic sites (basal planes) were present in large concentration, this would only satisfy one of the requirements to be the main contributors to the carbon basicity: it would be also necessary for these groups to have a relatively high basic strength.

Montes-Morán et al. thus made use of ab initio calculations to analyze the H_3O^+–π interaction in cluster models [98]. Three main conclusions were drawn in this study: (i) The pillared disposal of graphene layers of the basic structural units (BSU) [99] increases the basic strength of the C_π sites with respect to a single graphene sheet. (ii) A decrease in the interplanar distance between graphene layers reinforces the H_3O^+–π interaction energy, reaching a maximum at 0.33 nm for a pyrene–benzene–H_3O^+ model. (ii) The effect of increasing the area of the basal plane (from coronene–H_3O^+ to $C_{54}H_{18}$–H_3O^+ cluster models) is moderate. However, the basic strength in the gas phase of the basal planes seems to be much lower than that of the pyrone-like groups as suggested also by computational calculations (see Section 6.3). For example, the energy for the acid–base reaction of H_3O^+ with pyrone-type structures in the gas phase is about 30–40 kcal mol^{-1} higher (in absolute value) than the estimated binding energy of H_3O^+ on basal planes [27]. This suggests that, in terms of their intrinsic basic strength, C_π sites would have a moderate contribution compared to other "heteroatom"-based basic groups.

This conclusion would put into question the prevalence of Lewis-basic type of interactions in carbons with low oxygen content [28,96,97]. Indeed, the heat of neutralization reported by Barton et al. (–38 kJ mol^{-1}) [96] is not too far from that of the standard enthalpy of neutralization (–57.3 kJ/mol for the $H^+ + OH^- \rightarrow H_2O$ reaction). Note also that this figure is in consonance with expectation for the chemical reaction of weak Brønsted bases like the postulated bicyclic pyrone groups (see Section 6.3). Similarly, although the study of Darmstad and Roy [97] adds substantial support to the basicity of the basal planes and disfavors the role of the O-containing functional groups, it cannot be regarded as conclusive due to the approximations involved in the data analyses (e.g. the same surface density of C atoms is assumed for structurally heterogeneous materials; all the surface O atoms are apparently assigned to basic groups). Besides, although their conclusion was supported by the lack of correlation between the surface concentration of O as measured by XPS and the total amount of basic sites as determined by the depletion technique (a portion of an HCl/NaCl solution is titrated after equilibration in contact with the carbon black), no data about either the point of zero charge or the global basic/acid character of the carbons are provided. Finally, the experimental evidences of the protonation of basal planes of [28] have been recently questioned [21] on the basis of the oxygen content selected to calculate the HCl/O ratio for a carbon black with low oxygen content (sample M1 in [28]). Although the new numbers reported in [21] are possibly incorrect (hence the conclusions of [28] would prevail), an important observation should remain, i.e. the HCl/O ratio calculations in the paper of León y León et al. [28] are nonetheless strongly determined by the accuracy of measuring very low O contents on carbons.

The controversy thus remains open, although we endorse the position of Boehm: "Most likely, basicity of oxygen-containing groups exists in addition to the π basicity of basal planes" [21]. Furthermore, more aspects should be considered, as far as a "separate" view of the basicity of the basal plane vs. "heteroatomic" basic sites would always tell us only a part of the story. The strength as a base of the C_π sites is expected to strongly depend on both the amount and type of surface groups that may (or may not) be present at the edges of the graphene layers. Thus, it seems reasonable to think that when electron-withdrawing surface groups (like oxygen-containing surface groups) decorates the edges of the basal planes, the basicity of the C_π sites would decrease since these groups would tend to decrease the π-electron density of the graphene layers. Conversely, the presence of electron-donating functional groups would increase the π-electron density and so the strength as bases of the C_π sites [100]. In other words, the strength as bases of the C_π sites is affected by the presence of surface groups and a more sophisticated model than the previous described in [98] should take into account these effects (Fig. 6.1).

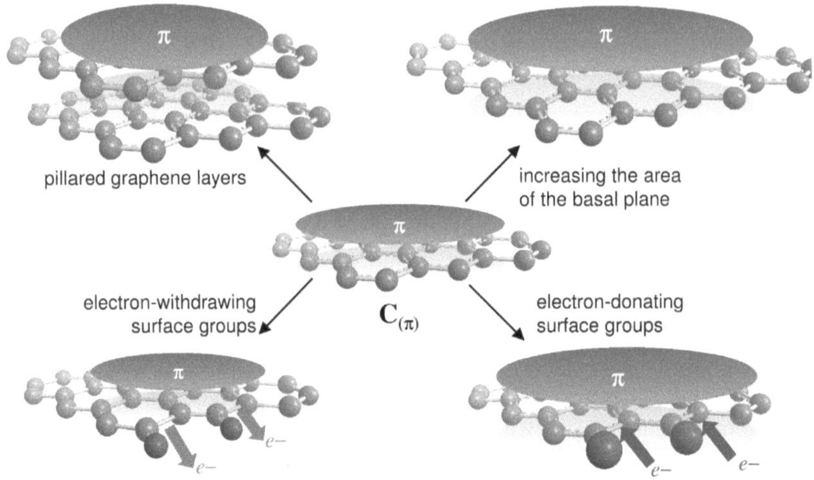

FIGURE 6.1 Schematic illustration of the effects of different contributions on the π-electron density of the basal plane. The cooperative interactions between the basal plane and surface functional groups are also represented.

Other potential source of carbon basicity, different from "heteroatoms" surface groups, could be allocated in certain carbon atoms of the graphene edges. Yet in 1960 Coulson [101] proposed (primarily based on intuition) that "[I]f the third valence is not used, the atom may … pair with a neighboring edge atom to form a partially triple bond … or revert to a divalent state based on s^2p^2." Using the terminology of modern organic chemistry, the free armchair sites correspond to o-benzyne (or carbyne)-type moieties, while the free zigzag sites are carbene type. Later on, the theoretical work of Nakada et al. [102] recognizes that "Finite graphite systems having a zigzag edge exhibit a special edge state. The corresponding energy bands are almost flat at the Fermi level and thereby give a sharp peak in the density of states. The charge density in the edge state is strongly localized on the zigzag edge sites […] the nanometer size effect is crucial for determining the relative importance of the edge state." At the same time, evidences of oxygen-free carbon (edge) sites, which remained unoxidized after air exposure (at room temperature) of activated carbons previously subjected to inert-gas high-temperature treatments, were presented [103,104].

The relevance, stability, and potential for magnetism of these sites were recognized by Radovic and Bockrath [105] more than fifty years after Coulson's proposal, suggesting for the chemical nature of free edge sites those sketched in Scheme 6.7. The authors reached the conclusion that these sites not only are stable at room temperature, but can also account for the origin of ferromagnetism in carbons.

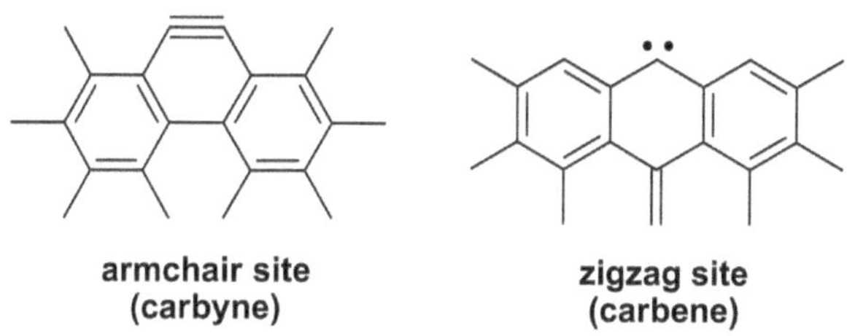

SCHEME 6.7 Carbyne and carbene sites on carbon materials.

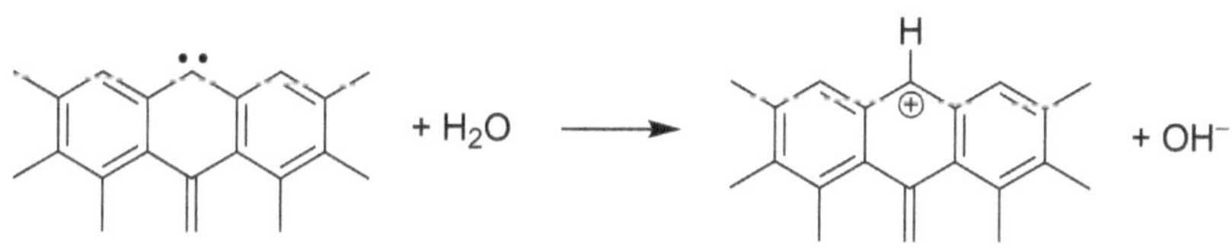

SCHEME 6.8 Carbene site acting as an hypothetical proton sink, after [105].

Moreover, in this paper, another relevant (and so far quite ignored) feature regarding the origin of carbon basicity is pointed out: the potential contribution of the carbine sites to the carbon basicity according to the mechanism depicted in Scheme 6.8.

Whether these sites are abundant or not on the surface of different carbon materials as well as their relative basic strength are questions that require further investigation. Nevertheless, regarding their abundance, the occurrence of these unsaturated valence sites on graphene layers might be of capital importance in nanocarbons such as nanotubes or graphenes, where the concentration of edge sites at the periphery of either flat or rolled graphene sheets would be relatively high in comparison to the size of the basal plane. Moreover, if porous carbons can be regarded as made up of micrographene structures, they may have many edge sites on the periphery of their micrographene constituents [102].

6.6. INORGANIC (OR MINERAL) MATTER

The definition of carbon material stated in the Introduction (section 6.1.) is wide enough to include sp^2 carbons with significant quantities of inorganic impurities in their composition. This is the case for most carbons for which, according to literature, the amount of inorganic (or mineral) matter ranges from less than 1 w% to 15–20 w% [81,106]. As it has been stressed several times through the chapter, only inorganic matter remaining after the preparation of carbons will be here of interest, i.e. the basicity of carbons impregnated with basic salts or other basic inorganic compounds is not considered.

In that sense, the presence of inorganic species should be due to the contributions of both the precursor and the method of preparation of a given carbon. Attending to the type of precursor, two families of raw materials are used as precursors for the production of a vast amount of commercial carbons, namely coal-derived and biomass-derived carbons [81,106]. In spite of some heavy metals found in coal ashes, the ashes from the two raw materials contain almost the same elements, namely Na, K, Ca, Mg, Si, Al, Fe, P, and S.

The use of specific techniques such as low-temperature ashing combined with XRD and FTIR characterization allowed some researchers, in the nineties of the last century, to obtain additional information regarding the mineralogical species that were present in coal and biomass [107–110]. More recently, Vassiliev et al. have performed an exhaustive analysis of the mineral matter of both coal and biomass [111–114]. Conclusions of interest (from the point of view of the study of the basicity of carbons) are the occurrence of silicates, phosphates, sulfates, carbonates, and oxides/hydroxides in both types of precursors. In the case of biomass solids, chlorides, nitrates, and organic (i.e. oxalate) salts were also identified [113,114].

The conditions used for the preparation of a specific carbon from these precursors will determine the evolution of the inorganic species. Specifically, knowledge about the transformation of the different compounds with increasing temperature under nonoxidizing atmospheres is crucial. A detailed description of such transformations would take too long to be included here. To simplify, let us assume that the inorganic species that would eventually remain in the carbons obtained can be classified in the following two groups. On the one hand, oxides/hydroxides and carbonates of alkaline (Na and K) and alkaline-earth (Ca and Mg) elements, which are soluble in water and will undergo hydrolysis reactions; these species, if some additional conditions are fulfilled (see below), should end being responsible for the contribution of the inorganic matter to the basicity of the resulting carbons. On the other hand, the rest of species that would be insoluble in water or that, being soluble, will not undergo hydrolysis.

The basicity of the inorganic (or mineral) matter on carbon has a special casuistic that has not been so far contemplated in the Chapter. Inorganic compounds on carbons are not "functional groups" attached to the carbon clusters but remain as "independent" grains. As for the rest of sites, these grains must be accessible to proton sources in order to contribute to the carbon basicity. In aqueous solutions, the inorganic particles having contact with water molecules will undergo hydrolysis reactions that will progressively dissolve them. This behavior opens up the possibility of the ionic species moving into the water phase. Basicity is thus transferred from the carbon surface to the water solution, i.e. we might consider a homogeneous, rather than a heterogeneous, basicity when dealing with inorganic matter sites. The same ionic species may also intervene in other homogeneous reactions in aqueous media.

TABLE 6.1 Solubility of Alkaline and Alkaline-earth Hydroxide and Carbonates. Data Collected from [116]

Solubility g/100 ml H_2O (25 °C)		Solubility Product Constant (K_s) (25 °C)	
NaOH	109.2	$Ca(OH)_2$	5.47×10^{-6}
KOH	110.0	$Mg(OH)_2$	1.2×10^{-11}
Na_2CO_3	23.8	$CaCO_3$	4.80×10^{-9}
K_2CO_3	112.3	$MgCO_3$	2.6×10^{-5}

Hence, the basicity of the inorganic matter on carbons in aqueous solution will depend not only on the particular basic strength of a given inorganic compound and its accessibility, but also on its relative solubility. Table 6.1 shows the solubility of different alkaline and alkaline-earth (hydro)oxides and carbonates. Na and K compounds are much more soluble in water than Ca and Mg counterparts. Nonetheless, even low soluble Ca and Mg oxides or carbonates may render moderately basic solutions ($9 < pH < 11$) [115]. These pH values are well above the pK_a values of most of the basic sites of carbons discussed until now. In other words, it is quite likely that, if present in significant quantities, (basic) inorganic matter will be the strongest contributor to the basicity of carbons, especially in aqueous solutions.

The previous assertion lies on, overall, empirical data of pH measurements of different carbon suspensions in aqueous solutions. We do believe that only carbons containing significant amounts of (basic) inorganic compounds bring about water suspensions with $pH > 10$ (Table 6.2). For these basic carbons,

TABLE 6.2 Ash Content and Point of Zero Charge of Several Commercial Activated Carbons

Company	Raw Material	Activating Agent	Reference	ASH (% weight)	pH
Norit	Peat	Steam	RO-3515	6	10.6
Norit	Peat	Steam	RO-3515 (washed)	4.7	9.8
Norit	Peat	Steam	PK1-3	8.6	11.7
Chemviron Carbon	Bituminous coal	Steam	Filtrasorb 400	6.9	9.7

dissolving in water significant amounts of the inorganic matter tends to reduce the basicity of the resulting solid (Table 6.2). Moreover, "dry methodologies" of preparing carbon materials from coal or biomass precursors would lead to basic carbons. This is the case for most, if not all, commercial activated carbons prepared by physical (steam) activation. A final consideration to the inorganic matter is related to the physical integrity of carbons containing significant amounts of soluble grains. If the grains are incrusted in small size pores, dissolving them may provoke the collapse of the pores to eventually cause the disintegration of the carbon structure.

6.7. CONCLUSIONS (AND SOME OPEN QUESTIONS)

There are still reasons to state that the nature of basic sites on the surface of carbons is less clear than that of the acidic sites, as pointed out by Boehm. By keeping in mind all the difficulties and problems arising in the assessment of carbon basicity that are discussed in Section 6.2, researchers concerned with any of the various manifestations of carbon basicity would have to carefully balance all the factors involved in their particular systems in order to find out useful structure–activity relationships. In addition, all the key facts outlined for the various functionalities and species potentially contributing to carbon basicity should also be considered. In some cases, the entanglement among the various factors affecting carbon basicity could largely hamper the unambiguous distinction of their individual contributions and/or the magnitude of cooperative effects.

In this chapter, four main contributors have been identified as responsible for the basicity of carbons, namely surface oxygen groups (specifically, pyrone-type groups), surface nitrogen (and, possibly, some phosphorus and sulfur) functional groups, basal planes and edges of graphene layers, and some inorganic species. Their occurrence and relative strength as bases have been discussed. In addition, the concept of site accessibility has been remarked as a necessary condition for a site to contribute effectively to the basicity of carbons.

Overall, the impact of N-groups on carbon basicity is quite well understood in terms of the influence of different nitrogen compounds in the preparation of these materials, the experimental characterization of the N-containing groups, the well-known basic strength of the N-functionalities, and so on. However, the contribution of the so-called quaternary nitrogen to carbon basicity is worth considering.

The abundance and qualitative importance of mineral matter for assessing carbon basicity should not be overlooked. Actually, we reckon that mineral matter is the main contributor to the basicity of carbons showing points of zero charge above 10 pH units.

A major question that remains unsettled is the contribution of the basal planes to carbon basicity. There are strong evidences that the carbon basal planes can absorb substantial amounts of strong acids through donor–acceptor or cation–π interactions, thus acting as effective Lewis bases. However, the actual strength of these interactions and their dependence with microporous structure are not clear. Similarly, the precise influence of unsaturated carbon edges (if they exist) on carbon basicity remains largely unknown.

However, we do consider that ascribing the basic behavior of carbons mainly to Lewis-type interactions is not an adequate description. Carbon materials that have a low oxygen-content, having no mineral matter or significnant concentrations of any other heteroatom, tend to show basic properties in aqueous suspensions. The basicity of these materials most likely reflects the complex interplay between the Lewis basicity of the basal plane (resulting in the physical adsorption of strong acids such as HCl) and the Brønsted basicity exhibited by the combination of etheric and ketone O-groups (pyrone-like) decorating the edges of the carbon sheets.

The actual relationship of potential redox processes involving quinone groups and/or pyrones and carbon basicity is still an open question. For example, it is not clear whether or not O-containing groups reacting with acids can be oxidized irreversibly and what consequences for carbon basicity can be expected. This opens up a new perspective in understanding the fate of basic sites as the carbons become acidic (after oxidation treatments).

From the analysis of the recent literature, it seems that the current picture of the basicity of the novel forms of nanocarbon materials is somewhat obscured by the intrinsic difficulties characteristic of conventional carbon materials. Therefore, in order to understand the basicity, electrochemistry, and other properties of these materials containing O- and/or N-groups, new structure–activity relationships will have to be established at the molecular level during the forthcoming years.

Overall, fundamental research is still needed. Both experimental and theoretical modeling would be required to answer specific questions. From knowledge to be gained, many practical applications could benefit as suggested by the current examples in the literature.

Some Open Questions

- How basic can a carbon material be?
- Are pores on carbons basic sites?
- Are aqueous-based methodologies adequate for testing hydrophobic sites?
- What is the contribution of quaternary nitrogen to the basicity of N-doped carbons?
- What is the extent of cooperative effects between two or more types of basic sites?
- Do basic sites on carbons endure oxidizing treatments?
- Do basic sites on carbons endure reactions with acids?

REFERENCES

[1] W.B. Jensen, The origin of the term "base", J. Chem. Educ. 83 (2006) 1130.

[2] G. Busca, Bases and basic materials in chemical and environmental processes. Liquid versus solid basicity, Chem. Rev. 110 (2010) 2217–2249.

[3] G. Busca, Bases and basic materials in chemical and environmental processes: a review of commercial processes, Ind. Eng. Chem. Res. 48 (2009) 6486–6511.

[4] K.-H. Köchling, B. McEnaney, F. Rozploch, E. Fitzer, International committee for characterization and terminology of carbon "first publication of 30 tentative definitions", Carbon 20 (1982) 445–449.

[5] H.-P. Boehm, E. Diehl, W. Heck, R. Sappok, Surface oxides of carbon, Angew. Chem. Int. Ed. Engl. 3 (1964) 669–677.

[6] S. Neffe, Evaluation of the pH-metric method for the determination of acidic groups on the surface of oxidized carbons, Carbon 25 (1987) 441–443.

[7] J. Jagiełło, T.J. Bandosz, J.A. Schwarz, Carbon surface characterization in terms of its acidity constant distribution, Carbon 32 (1994) 1026–1028.

[8] A.M. Puziy, O.I. Poddubnaya, J.A. Ritter, A.D. Ebner, C.E. Holland, Elucidation of the ion binding mechanism in heterogeneous carbon-composite adsorbents, Carbon 39 (2001) 2313–2324.

[9] A. Bismarck, C. Wuertz, J. Springer, Basic surface oxides on carbon fibers, Carbon 37 (1999) 1019–1027.

[10] C.L. Mangun, K.R. Benak, J. Economy, K.L. Foster, Surface chemistry, pore sizes and adsorption properties of activated carbon fibers and precursors treated with ammonia, Carbon 39 (2001) 1809–1820.

[11] S.-J. Park, B.-J. Kim, Influence of oxygen plasma treatment on hydrogen chloride removal of activated carbon fibers, J. Colloid Interface Sci. 275 (2004) 590–595.

[12] R. Menzel, A. Lee, A. Bismarck, M.S.P. Shaffer, Inverse gas chromatography of as-received and modified carbon nanotubes, Langmuir 25 (2009) 8340–8348.

[13] E. Díaz, S. Ordóñez, A. Vega, Adsorption of volatile organic compounds onto carbon nanotubes, carbon nanofibers, and high-surface-area graphites, J. Colloid Interface Sci. 305 (2007) 7–16.

[14] M.A. Montes-Morán, A. Martínez-Alonso, J.M.D. Tascón, Adsorption of polar probe molecules on plasma-oxidised high strength carbon fibres, Fuel Process. Technol. 77–78 (2002) 359–364.

[15] E. Papirer, E. Brendle, F. Ozil, H. Balard, Comparison of the surface properties of graphite, carbon black and fullerene samples, measured by inverse gas chromatography, Carbon 37 (1999) 1265–1274.

[16] M.V. López-Ramón, F. Stoeckli, C. Moreno-Castilla, F. Carrasco-Marín, On the characterization of acidic and basic surface sites on carbons by various techniques, Carbon 37 (1999) 1215–1221.

[17] F. Stoeckli, C. Moreno-Castilla, F. Carrasco-Marín, M.V. López-Ramón, Distribution of surface oxygen complexes on activated carbons from immersion calorimetry, titration and temperature-programmed desorption techniques, Carbon 39 (2001) 2235–2237.

[18] F. Xie, J. Phillips, I.F. Silva, M.C. Palma, J.A. Menéndez, Microcalorimetric study of acid sites on ammonia- and acid-pretreated activated carbon, Carbon 38 (2000) 691–700.

[19] M.R. Alexander, F.R. Jones, Effect of electrolytic oxidation upon the surface chemistry of type A carbon fibres-Part II, analysis of derivatised surface functionalities by XPS, and TOF SIMS, Carbon 33 (1995) 569–580.

[20] L.A. Langley, D.E. Villanueva, D.H. Fairbrother, Quantification of surface oxides on carbonaceous materials, Chem. Mater. 18 (2006) 169–178.

[21] H.-P. Boehm, Surface chemical characterization of carbons from adsorption studies, in: Adsorption by Carbons, Elsevier, 2008.

[22] J.A. Menéndez, M.J. Illán-Gómez, C.A. León y León, L.R. Radovic, On the difference between the isoelectric point and the point of zero charge of carbons, Carbon 33 (1995) 1655–1657.

[23] S. Biniak, A. Swiatkowski, M. Pakula, Electrochemical studies of phenomena at active carbon–electrolyte solution interfaces, in: , Chemistry and Physics of Carbon, vol. 27, Marcel-Dekker, 2001.

[24] M. Pakuła, S. Biniak, A. Świątkowski, S. Neffe, Influence of progressive surface oxidation of nitrogen-containing carbon on its electrochemical behaviour in phosphate buffer solutions, Carbon 40 (2002) 1873–1881.

[25] H.P. Boehm, Some aspects of the surface chemistry of carbon blacks and other carbons, Carbon 32 (1994) 759–769.

[26] H.P. Boehm, Surface oxides on carbon and their analysis: a critical assessment, Carbon 40 (2002) 145–149.

[27] M.A. Montes-Morán, D. Suárez, J.A. Menéndez, E. Fuente, On the nature of basic sites on carbon surfaces: an overview, Carbon 42 (2004) 1219–1225.

[28] C.A. León y León, J.M. Solar, V. Calemma, L.R. Radovic, Evidence for the protonation of basal plane sites on carbon, Carbon 30 (1992) 797–811.

[29] R. Burstein, A. Frumkin, On the behavior of outgassed activated carbon towards electrolytes, Z. Phys. Chem. A 141 (1929) 219–220 (in German).

[30] V.A. Garten, D.E. Weiss, A new interpretation of the acidic and basic structures of activated carbon in relation to its behavior as a catalyst and adsorbent. II. The chromene–carbonium ion couple in carbon, Aust. J. Chem. 10 (1957) 309–328.

[31] H.P. Boehm, M. Voll, Basic surface oxides on carbon-I. Adsorption of acids, Carbon 8 (1970) 227–240 (in German).

[32] J.L. Figueiredo, M.F.R. Pereira, M.M.A. Freitas, J.J.M. Órfão, Modification of the surface chemistry of activated carbons, Carbon 37 (1999) 1378–1389.

[33] J. Zawadzki, Infrared spectroscopy in surface chemistry of carbons, in: Chemistry and Physics of Carbon, vol. 21, Marcel Dekker, 1989.

[34] A. Proctor, P.M.A. Sherwood, X-ray photoelectron spectroscopic studies of carbon fibre surfaces-II: The effect of electrochemical treatment, Carbon 21 (1983) 53–59.

[35] H.F. Gorgulho, J.P. Mesquita, F. Gonçalves, M.F.R. Pereira, J.L. Figueiredo, Characterization of the surface chemistry of carbon materials by potentiometric titrations and temperature-programmed desorption, Carbon 46 (2008) 1544–1555.

[36] T.J. Bandosz, On the adsorption/oxidation of hydrogen sulfide on activated carbons at ambient temperatures, J. Colloid Interface Sci. 246 (2002) 1–20.

[37] P. Davini, Desulphurization properties of active carbons obtained from petroleum pitch pyrolysis, Carbon 37 (1999) 1363–1371.

[38] M.G. Plaza, C. Pevida, A. Arenillas, F. Rubiera, J.J. Pis, CO_2 capture by adsorption with nitrogen enriched carbons, Fuel 86 (2007) 2204–2212.

[39] E.A. Müller, L.F. Rull, L.F. Vega, K.E. Gubbins, Adsorption of water on activated carbons: a molecular simulation study, J. Phys. Chem. 100 (1996) 1189–1196.

[40] L.R. Radovic, C. Moreno-Castilla, J. Rivera-Utrilla, Carbon materials as adsorbents in aqueous solutions, in: Chemistry and Physics of Carbon, vol. 27, Marcel-Dekker, 2001.

[41] G. de la Puente, J.A. Menéndez, On the contribution of oxygen-containing surface groups in carbons and their influence on the preparation of carbon supported molybdenum catalysts, Solid State Ionics 112 (1998) 103–111.
[42] G.S. Szymański, T. Grzybek, H. Papp, Influence of nitrogen surface functionalities on the catalytic activity of activated carbon in low temperature SCR of NOx with NH_3, Catal. Today 90 (2004) 51–59.
[43] L.R. Radovic, F. Rodríguez-Reinoso, Carbon materials in catalysis, in: Chemistry and Physics of Carbon, vol. 25, Marcel Dekker, 1997.
[44] A. Contescu, M. Vass, C. Contescu, K. Putyera, J.A. Schwarz, Acid buffering capacity of basic carbons revealed by their continuous pK distribution, Carbon 36 (1998) 247–258.
[45] K. László, K. Josepovits, E. Tombácz, Analysis of active sites on synthetic carbon surfaces by various methods, Anal. Sci. 17 (2001) i1741–i1744.
[46] V.V. Strelko, N.T. Kartel, I.N. Dukhno, V.S. Kuts, R.B. Clarkson, B.M. Odintsov, Mechanism of reductive oxygen adsorption on active carbons with various surface chemistry, Surf. Sci. 548 (2004) 281–290.
[47] E. Fuente, J.A. Menéndez, D. Suárez, M.A. Montes-Morán, Basic surface oxides on carbon materials: a global view, Langmuir 19 (2003) 3505–3511.
[48] D. Suárez, J.A. Menéndez, E. Fuente, M.A. Montes-Morán, Contribution of pyrone-type structures to carbon basicity: an ab initio study, Langmuir 15 (1999) 3897–3904.
[49] J.A. Menéndez, D. Suarez, E. Fuente, M.A. Montes-Morán, Contribution of pyrone-type structures to carbon basicity: theoretical evaluation of the pK_a of model compounds, Carbon 37 (1999) 1002–1006.
[50] E. Fuente, J.A. Menéndez, M.A. Díez, D. Suárez, M.A. Montes-Morán, Infrared spectroscopy of carbon materials: a quantum chemical study of model compounds, J. Phys. Chem. B 107 (2003) 6350–6359.
[51] M.J. Cejudo-Bastante, I. Hermosín-Gutiérrez, M.S. Pérez-Coello, Micro-oxygenation and oak chip treatments of red wines: effects on colour-related phenolics, volatile composition and sensory characteristics. Part II: Merlot wines, Food Chem. 124 (2011) 738–748.
[52] B.D. Epstein, E. Dalle-Molle, J.S. Mattson, Electrochemical investigations of surface functional groups on isotropic pyrolytic carbon, Carbon 9 (1971) 609–615.
[53] J.L. Figueiredo, M.F.-R. Pereira, The role of surface chemistry in catalysis with carbons, Catalysis Today 150 (2010) 2–7.
[54] M. Gutiérrez-Valero, M.L. Godino-Salido, P. Arranz-Mascaró, R. López-Garzón, R. Cuesta, J. García-Martín, Adsorption of designed pyrimidine derivative ligands on an activated carbon for the removal of Cu(II) ions from aqueous solution, Langmuir 23 (2007) 5995–6003.
[55] J. García-Martín, R. López-Garzón, M.L. Godino-Salido, M. Gutiérrez-Valero, P. Arranz-Mascaró, R. Cuesta, F. Carrasco-Marín, Ligand adsorption on an activated carbon for the removal of chromate ions from aqueous solutions, Langmuir 21 (2005) 6908–6914.
[56] Q. Li, A.D. Lueking, Effect of surface oxygen groups and water on hydrogen spillover in Pt-doped activated carbon, J Phys. Chem. C 115 (2011) 4273–4282.
[57] M. Seredych, D. Hulicova-Jurcakova, G.Q. Lu, T.J. Bandosz, Surface functional groups of carbons and the effects of their chemical character, density and accessibility to ions on electrochemical performance, Carbon 46 (2008) 1475–1488.
[58] D. Hulicova-Jurcakova, M. Seredych, G.Q. Lu, T.J. Bandosz, Combined effect of nitrogen- and oxygen-containing functional groups of microporous activated carbon on its electrochemical performance in supercapacitors, Adv. Funct. Mater. 19 (2009) 438–447.

[59] Z.-H. Huang, X. Zheng, W. Lv, M. Wang, Q.-H. Yang, F. Kang, Adsorption of lead(II) ions from aqueous solution on low-temperature exfoliated graphene nanosheets, Langmuir 27 (2011) 7558–7562.
[60] O.A. Williams, J. Hees, C. Dieker, W. Jäger, L. Kirste, C.E. Nebel, Size-dependent reactivity of diamond nanoparticles, ACS Nano 8 (2010) 4824–4830.
[61] A. Bagri, C. Mattevi, M. Acik, Y.J. Chabal, M. Chhowalla, V.B. Shenoy, Structural evolution during the reduction of chemically derived graphene oxide, Nat. Chem. 2 (2010) 581–587.
[62] H.-P. Boehm, Catalytic properties of nitrogen containing carbons, in: Carbon Materials for Catalysis, Wiley, 2009.
[63] J.P. Boudou, Ph. Parent, F. Suárez-García, S. Villar-Rodil, A. Martínez-Alonso, J.M.D. Tascón, Nitrogen in aramid-based activated carbon fibers by TPD, XPS and XANES, Carbon 44 (2006) 2452–2462.
[64] R.H. Bradley, X. Ling, I. Sutherland, G. Beamson, XPS determination of surface nitrogen species on PAN carbon fibres, Carbon 32 (1994) 185–186.
[65] L. Zhao, Z. Bacsik, N. Hedin, W. Wei, Y. Sun, M. Antonietti, M.-M. Titirici, Carbon dioxide capture on amine-rich carbonaceous materials derived from glucose, ChemSusChem 3 (2010) 840–845.
[66] R.W. Alder, P.S. Bowman, W.R.S. Steele, D.R. Winterman, The remarkable basicity of 1,8-bis(dimethylamino)naphthalene, Chem. Commun. (London) 13 (1968) 723–724.
[67] D. Suárez, J.A. Menéndez, E. Fuente, M.A. Montes-Morán, Pyrone-like structures as novel oxygen-based organic superbases, Angew. Chem. Int. Ed. 39 (2000) 1320–1323.
[68] B. Stöhr, H.P. Boehm, R. Schlögl, Enhancement of the catalytic activity of activated carbons in oxidation reactions by thermal treatment with ammonia or hydrogen cyanide and observation of a superoxide species as a possible intermediate, Carbon 29 (1991) 707–720.
[69] R.J.J. Jansen, H. van Bekkum, Amination and ammoxidation of activated carbons, Carbon 32 (1994) 1507–1516.
[70] R.J.J. Jansen, H. van Bekkum, XPS of nitrogen-containing functional groups on activated carbon, Carbon 33 (1995) 1021–1027.
[71] J.R. Pels, F. Kapteijn, J.A. Moulijn, Q. Zhu, K.M. Thomas, Evolution of nitrogen functionalities in carbonaceous materials during pyrolysis, Carbon 33 (1995) 1641–1653.
[72] F. Kapteijn, J.A. Moulijn, S. Matzner, H.-P. Boehm, Development of nitrogen functionality in model chars during gasification in CO_2 and O_2, Carbon 37 (1999) 1143–1150.
[73] S. Biniak, G. Szymánski, J. Siedlewski, A. Światkowski, The characterization of activated carbons with oxygen and nitrogen surface groups, Carbon 35 (1997) 1799–1810.
[74] R. Arrigo, M. Hävecker, S. Wrabetz, R. Blume, M. Lerch, J. McGregor, E.P.J. Parrott, J.A. Zeitler, L.F. Gladden, A. Knop-Gericke, R. Schlögl, D.S. Su, Tuning the acid/base properties of nanocarbons by functionalization via amination, J. Am. Chem. Soc. 132 (2010) 9616–9630.
[75] J. Rivera-Utrilla, M. Sánchez-Polo, Ozonation of naphthalenesulphonic acid in the aqueous phase in the presence of basic activated carbons, Langmuir 20 (2004) 9217–9222.
[76] D. Mang, H.P. Boehm, K. Stanczyk, H. Marsh, Inhibiting effect of incorporated nitrogen on the oxidation of microcrystalline carbons, Carbon 30 (1992) 391–398.
[77] M. Seredych, D. Hulicova-Jurcakova, T.J. Bandosz, Effect of the incorporation of nitrogen to a carbon matrix on the selectivity and capacity for adsorption of dibenzothiophenes from model diesel fuel, Langmuir 26 (2010) 227–233.
[78] C.-M. Yang, M. El-Merraoui, H. Seki, K. Kaneko, Characterization of nitrogen-alloyed activated carbon fiber, Langmuir 17 (2001) 675–680.

[79] C.-M. Yang, K. Kaneko, Adsorption properties of nitrogen-alloyed activated carbon fiber, Carbon 39 (2001) 1075–1082.

[80] V.V. Strelko, V.S. Kuts, P.A. Thrower, On the mechanism of possible influence of heteroatoms of nitrogen, boron and phosphorus in a carbon matrix on the catalytic activity of carbons in electron transfer reactions, Carbon 38 (2000) 1499–1524.

[81] H. Marsh, F. Rodriguez-Reinoso, Activated Carbon, Elsevier, 2006.

[82] M. Jagtoyen, F. Derbyshire, Activated carbons from yellow poplar and white oak by H_3PO_4 activation, Carbon 36 (1998) 1085–1097.

[83] A.M. Puziy, O.I. Poddubnaya, R.P. Socha, J. Gurgul, M. Wisniewski, XPS and NMR studies of phosphoric acid activated carbons, Carbon 46 (2008) 2113–2123.

[84] J.A. Díaz-Auñón, M.C. Román-Martínez, C. Salinas-Martínez de Lecea, H. Alper, Strategies for the heterogenization of rhodium complexes on activated carbon, Stud. Surf. Sci. Catal. 143 (2002) 295–304.

[85] A.B. Dongil, B. Bachiller-Baeza, A. Guerrero-Ruiz, I. Rodríguez-Ramos, Chemoselective hydrogenation of cinnamaldehyde: A comparison of the immobilization of Ru–phosphine complex on graphite oxide and on graphitic surfaces, J. Catal. 282 (2011) 299–309.

[86] E. Papirer, V.T. Nguyen, J.-B. Donnet, Introduction of sulfur groups onto the surface of carbon blacks, Carbon 16 (1978) 141–144.

[87] C.H. Chang, Preparation and characterization of carbon-sulfur surface compounds, Carbon 19 (1981) 175–186.

[88] C. Petit, K. Kante, T.J. Bandosz, The role of sulfur-containing groups in ammonia retention on activated carbons, Carbon 48 (2010) 654–667.

[89] R.B. Puri, R.S. Hazra, Carbon–sulphur surface complexes on charcoal, Carbon 9 (1971) 123–134.

[90] H.E. Blayden, J.W. Patrick, Solid complexes of carbon and sulphur-I. Sulphurised polymer carbons, Carbon 5 (1967) 533–544.

[91] T.J. Bandosz, C.O. Ania, Surface chemistry of activated carbons and its characterization, in: Activated Carbon Surfaces in Environmental Remediation, Elsevier, 2006.

[92] M. Seredych, T.J. Bandosz, Investigation of the enhancing effects of sulfur and/or oxygen functional groups of nanoporous carbons on adsorption of dibenzothiophenes, Carbon 49 (2011) 1216–1224.

[93] J.-L.M. Abboud, O. Mo, J.L.G. De Paz, M. Yanez, M. Esseffar, W. Bouab, M. El-Mouhtadi, R. Mokhlisse, E. Ballesteros, M. Herreros, H. Homan, C. Lopez-Mardomingo, R. Notario, Thiocarbonyl versus carbonyl compounds: A comparison of intrinsic reactivities, J. Am. Chem. Soc. 115 (1993) 12468–12476.

[94] E. Papirer, S. Li, J.-B. Donnet, Contribution to the study of basic surface groups on carbons, Carbon 25 (1987) 243–247.

[95] E. Papirer, J. Dentzer, S. Li, J.-B. Donnet, Surface groups on nitric acid oxidized carbon black samples determined by chemical and thermodesorption analyses, Carbon 29 (1991) 69–72.

[96] S.S. Barton, M.J.B. Evans, E. Halliop, J.A.F. MacDonald, Acidic and basic sites on the surface of porous carbon, Carbon 35 (1997) 1361–1366.

[97] H. Darmstadt, C. Roy, Surface spectroscopic study of basic sites on carbon blacks, Carbon 41 (2003) 2662–2665.

[98] M.A. Montes-Morán, J.A. Menéndez, E. Fuente, D. Suárez, Contribution of the basal planes to carbon basicity: An ab initio study of the H_3O^+-π interaction in cluster models, J. Phys. Chem. B 102 (1998) 5595–5601.

[99] K. Oshida, T. Nakazawa, K. Osawa, M. Endo, Nano structure of low crystallinity carbon materials analyzed by using high energy X-ray diffraction, Cent. Eur. J. Phys. 7 (2009) 232–236.
[100] L.R. Radovic, I.F. Silva, J.I. Ume, J.A. Menéndez, C.A. León y León, A.W. Scaroni, An experimental and theoretical study of the adsorption of aromatics possessing electron-withdrawing and electron-donating functional groups by chemically modified activated carbons, Carbon 9 (1997) 1339–1348.
[101] C.A. Coulson, The electronic structure of the boundary atoms of a graphite layer, in: Proceedings of the Fourth Conference on Carbon, Pergamon Press, 1960.
[102] K. Nakada, M. Fujita, G. Dresselhaus, M.S. Dresselhaus, Edge state in graphene ribbons: nanometer size effect and edge shape dependence, Phys. Rev. B 54 (1996) 17954–17961.
[103] J.A. Menéndez, J. Phillips, B. Xia, L.R. Radovic, On the modification and characterization of chemical surface properties of activated carbon: in the search of carbons with stable basic properties, Langmuir 12 (1996) 4404–4410.
[104] J.A. Menéndez, B. Xia, J. Phillips, L.R. Radovic, On the modification and characterization of chemical surface properties of activated carbon: microcalorimetric, electrochemical, and thermal desorption probes, Langmuir 13 (1997) 3414–3421.
[105] L.R. Radovic, B. Bockrath, On the chemical nature of graphene edges: origin of stability and potential for magnetism in carbon materials, J. Am. Chem. Soc. 127 (2005) 5917–5927.
[106] I.S. Roskill, The Economics of Activated Carbon, eighth ed., Roskill Information Services, Ltd, 2008.
[107] M.R. Martínez-Tarazona, J.M. Palacios, A. Martínez-Alonso, J.M.D. Tascón, The characterization of organomineral components of low-rank coals, Fuel Process. Technol. 25 (1990) 81–87.
[108] M.R. Martínez-Tarazona, A. Martínez-Alonso, J.M.D. Tascón, Interactions between carboxyl groups and inorganic elements in Spanish brown coals, Fuel 69 (1990) 362–367.
[109] M.R. Martínez-Tarazona, D.A. Spears, J.M. Palacios, A. Martínez-Alonso, J.M.D. Tascón, Mineral matter in coals of different rank from the Asturian Central basin, Fuel 71 (1992) 367–372.
[110] S. Pusz, A. Krzton, J.L. Komraus, M.R. Martínez-Tarazona, A. Martínez-Alonso, J.M.D. Tascón, Interactions between organic matter and minerals in two bituminous coals of different rank, Int. J. Coal Geol. 33 (1997) 369–386.
[111] S.V. Vassilev, C.G. Vassileva, A new approach for the combined chemical and mineral classification of the inorganic matter in coal. 1. Chemical and mineral classifications systems, Fuel 88 (2009) 235–245.
[112] S.V. Vassilev, C.G. Vassileva, D. Baxter, L.K. Andersen, A new approach for the combined chemical and mineral classification of the inorganic matter in coal. 2. Potential applications of the classification systems, Fuel 88 (2009) 246–254.
[113] S.V. Vassilev, D. Baxter, L.K. Andersen, C.G. Vassileva, An overview of the chemical composition of biomass, Fuel 89 (2010) 913–933.
[114] S.V. Vassilev, D. Baxter, L.K. Andersen, C.G. Vassileva, T.J. Morgan, An overview of the organic and inorganic phase composition of biomass. Fuel 94 (2012) 1–33.
[115] W. Stumm, J.J. Morgan, Aquatic Chemistry, Wiley Interscience, 1996.
[116] W.M. Haynes, CRC Handbook of Chemistry and Physics, ninety first ed., CRC Press, 2010.

Chapter 7

Adsorption by Carbon Gels

Alain Celzard, Vanessa Fierro and Gisèle Amaral-Labat
Institut Jean Lamour - UMR CNRS 7198, Université de Lorraine, ENSTIB, Épinal, France

Chapter Outline

7.1. Introduction to Carbon Gels 207
7.1.1. Preparation of Carbon Gels 208
7.1.2. Structure of Carbon Gels 210
7.2. Adsorption for Porosity Characterization 212
7.3. Adsorption in the Gas Phase 219
7.3.1. Adsorption of Permanent Gases, Separation, and Storage 220
7.3.2. Adsorption of Condensable Vapors and Detection 222
7.4. Adsorption in the Liquid Phase 225
7.4.1. Adsorption of Organic Compounds 225
7.4.2. Adsorption of Halides 229
7.4.3. Adsorption of Metals 230
7.4.4. Adsorption in the Framework of Energy Applications 233
7.5. Conclusions 237
References 239

7.1. INTRODUCTION TO CARBON GELS

Carbon gels are attractive, versatile, materials which have been suggested for various applications such as adsorbents, porous electrodes for electric double-layer supercapacitors or secondary batteries, filling material for HPLC columns, catalyst supports, and heat insulators at high temperature [1–6]. These solids indeed present high pore volumes based on narrow pores, and especially possess a suitable mesoporosity allowing a fast transfer of molecules throughout the pore network. They are also electrically conducting and present

Novel Carbon Adsorbents. DOI: 10.1016/B978-0-08-097744-7.00007-7
Copyright © 2012 Elsevier Ltd. All rights reserved.

a high chemical resistance. Carbon gels are prepared by pyrolysis in inert atmosphere of porous organic precursors, usually highly porous dried gels based on thermoset resins.

7.1.1. Preparation of Carbon Gels

Organic gels are usually formed by the addition and/or condensation polymerization of low-molecular-weight, multifunctional, oligomers in aqueous or alcoholic solutions, leading to aquagels or alcogels, respectively. A colloidal suspension of solid particles, the "sol", is first formed. The particles grow and finally coalesce at the gel point, marking the sol–gel transition at which the viscosity of the medium diverges [7]. The incipient wet gel thus formed behaves as a giant molecule whose size is that of the container in which it was prepared. The gel then strengthens progressively as the residual, nonreacted, oligomers make bonds with the developing network. This phenomenon called aging or curing, allows reaching favorable conditions for drying the gel with the lowest number of cracks. Wet gels are thus semisolid systems comprising two phases, solid and fluid, embedded in each other in such a way that the pores of the solid, filled with solvent, are of colloidal dimensions [8]. In other words, gels comprise a continuous solid skeleton, consisting of chains of rounded particles arranged in pearl necklace structure immersed in a continuous liquid phase. Continuity of the solid phase ensures the flexibility of gels. A schematic view of the sol–gel transition is given in Fig. 7.1(a). It is clear that, according to such structure, high pore volumes may be obtained when the liquid is removed from the wet gel.

Removing the liquid leads to materials whose porosity depends on composition and initial dilution of the solid phase, and also on the way the wet gels have been dried. The solvent saturating the porosity may indeed be evacuated according to three methods: subcritical drying, supercritical drying, and freeze drying.

The first one (path #1 in Fig. 7.1(b)) is the most simple and leads to xerogels. However, capillary forces, exerted at the meniscus between solvent and vapor, induce a strong shrinkage and may destroy the tenuous structure of the solid phase. Hence, most of times, xerogels present a rather low porosity and a number of cracks. The latter may be avoided, nevertheless, through a cautious convective drying [9]. The other two drying methods are very interesting for getting highly porous dry gels maintaining the initial nanostructure with a very low shrinkage.

On the one hand, supercritical evacuation (path #2 in Fig. 7.1(b)) consists of heating the wet gel under pressure, far beyond the critical point of the solvent, so that liquid–gas menisci and hence capillary stresses vanish. The temperature is then kept constant while the supercritical solvent is evacuated and exchanged by inert gas at the same pressure, leading to aerogels. For doing so, the liquid contained in the porosity of the wet gel is habitually exchanged with liquid CO_2, a solvent having low critical temperature of 31.1 °C and moderate critical

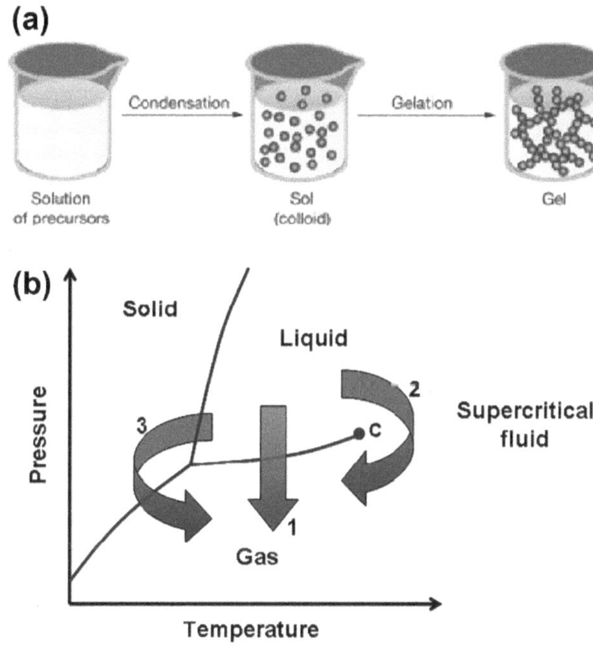

FIGURE 7.1 (a) Principles of a sol–gel reaction. *(Reproduced from https://www.llnl.gov/str/May05/Satcher.html with Permission.)* (b) Different ways of drying: "the resultant wet gel."

pressure of 7.38 MPa. However, organic solvents such as acetone are also possible. It should be observed that, in many papers, the term "aerogel" is improperly used for designating materials having a low density but having been dried in subcritical conditions, i.e. xerogels.

On the other hand, cryogels are obtained if the solvent has been previously frozen and sublimated at low pressure (path #3 in Fig. 7.1(b)). Both supercritical and freeze drying induce very low shrinkage, so that low-density dry organic gels are recovered. However, the pore structure is not the same, that of cryogels being generally coarser [10]. Freeze-drying aquagels, in which water ice crystals act as templates, may even produce very wide pores, sometimes called "megalopores", leading to microhoneycomb structures [11,12]. The phenomenon can be minimized, though never completely avoided, through the use of solvents presenting much lower volume changes upon freezing, such as *tert*-butanol (2-methyl-2-propanol).

Despite their high cost, resins based on resorcinol–formaldehyde (RF) were the most studied precursors of carbon gels. RF resins are indeed easily gelled in water, are easy to handle, and lead to nice, monolithic, large, and reproducible elastic aquagels. Figure 7.2 shows the reaction of resorcinol with formaldehyde in water. Each resorcinol molecule reacts with two formaldehyde molecules, and the successive condensations lead to RF polymer chains. These chains are

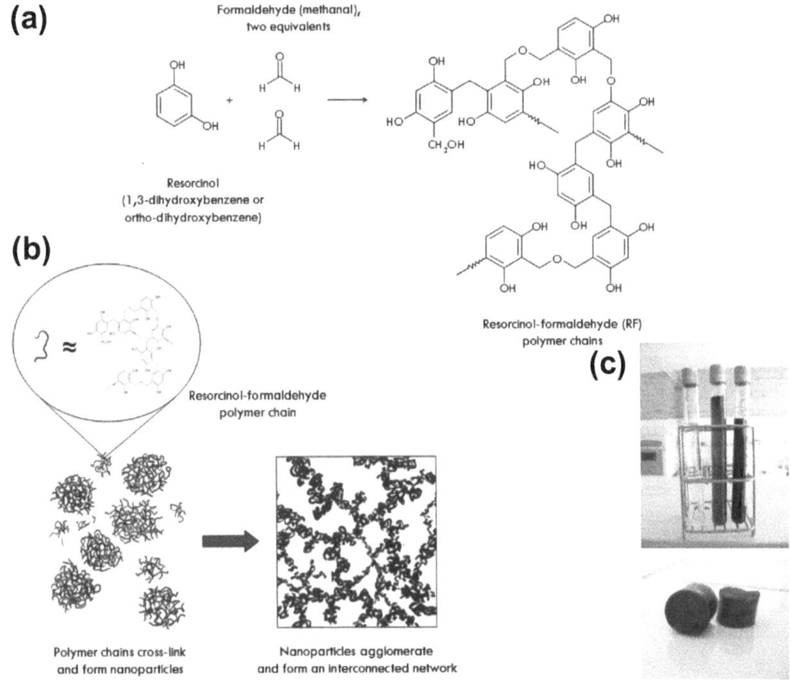

FIGURE 7.2 Formation of a resorcinol–formaldehyde (RF) gel in water. (a) Chemical structure, (b) Development of the porous structure. *(Reproduced from http://www.aerogel.org/?p=71 with Permission.)* (c) Typical RF Aquagels.

progressively cross-linked, so that highly porous, roughly spherical, particles appear. Finally, the aggregation of such particles gives rise to a solid gel having a typical nodular structure. Figure 7.2 also shows sealed tubes in which gelation occurred, as well as the RF aquagels, having their usual red–brown color.

7.1.2. Structure of Carbon Gels

Due to the versatility of the sol–gel process, not only monolithic organic gels are produced, but also beads, powders, or thin films can be prepared. After drying, their controlled pyrolysis may lead to carbon gels presenting a similar pore structure. For that purpose, the final temperature should not be too high, typically lower than 1000 °C, otherwise pore collapse might occur [13]. Given the structure presented in Fig. 7.2(b), carbon gels are expected to be materials having a hierarchical pore texture, such that micropores are essentially located within the nodules, whereas mesopores and macropores correspond to the spaces between the nodules and between the chains of nodules, respectively. Such a structure allows an easy penetration of gaseous or liquid molecular species and their easy access to the smallest pores. Moreover, the porosity of carbon gels can be significantly developed by activation: physically with steam or carbon dioxide, or chemically with either strong acids or molten metallic salts such as alkaline hydroxides. Activated carbon gels are thus a new family of porous materials with enhanced pore volumes that can be easily derived from organic and/or pyrolyzed gels.

Figure 7.3(a) shows the typical nodular structure of carbon gels as revealed by scanning electron microscopy. Yet, other pore structures like that presented in Fig. 7.3(b) have also been reported. This kind of texture might occur from a reaction-induced phase separation mechanism, such as spinodal decomposition [14]. However, homogeneous gelation most of the times leads to agglomerated nodules, so the same structure as that of RF gels is expected. The nodular structure has indeed been observed for most other carbon gels, whatever the composition of their organic precursors, provided they are based on thermoset resins. Some materials, such as "carbon nanotube aerogels" [15,16] or "graphene aerogels" [17,18], are indeed not representative of the materials treated here. They are called "aerogels" only because they are obtained from the supercritical or freeze drying of material formerly suspended in solution, and therefore presenting very low bulk densities. They will not be considered extensively here, because they behave like pure nanotubes or pure graphene from the point of view of their adsorption properties.

The pore structure of carbon gels is particularly interesting because it can be easily tailored. As explained above, different drying modes lead to different total pore volumes and different pore-size distributions. Besides, for a given drying mode, the degree of dilution of the solid phase has a dramatic impact on the resultant bulk density, and hence on the pore-size distribution as well. Finally, the pH of the initial solution also has a significant influence on the pore texture, higher pHs generally leading to smaller nodules [11,19]. For example,

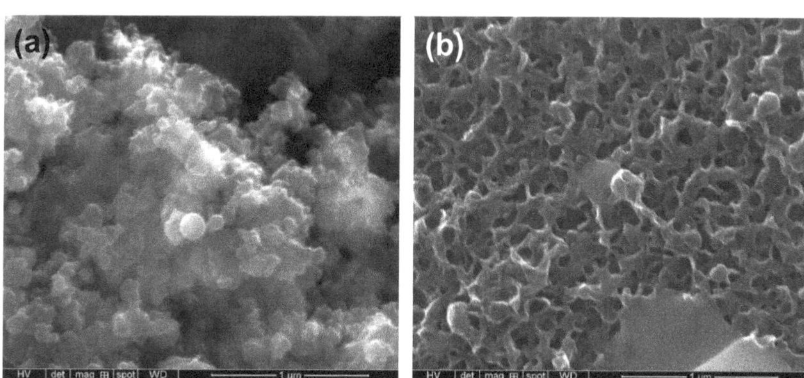

FIGURE 7.3 Scanning electron microscope (SEM) images of two kinds of carbon gels at the same magnification: (a) typical nodular structure (here in a tannin–formaldehyde carbon aerogel), (b) alternative possible structure (here in a tannin–resorcinol–formaldehyde carbon cryogel).

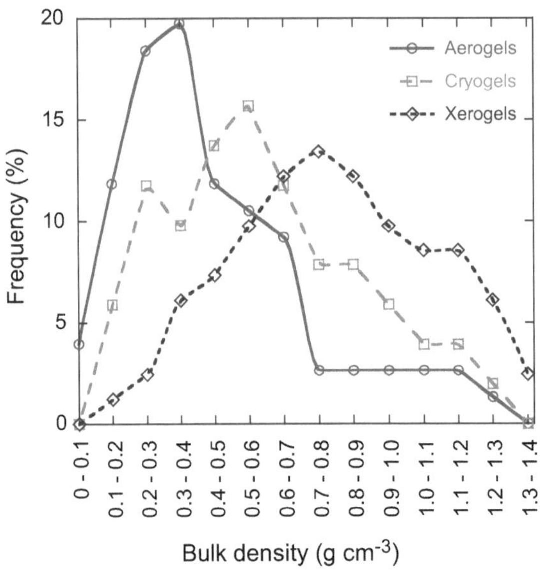

FIGURE 7.4 Distributions of bulk density of carbon gels (after the data reported in about one hundred scientific papers).

in the case of carbon gels made from phenolic resins, high pH values lead to almost nonporous materials, whereas high micro- and mesopore volumes are obtained around pH 5. Interestingly, the effects of pH and dilution may be varied separately, leading to a wide range of porous materials.

Bulk densities of carbon gels that have been reported in about one hundred papers examined by the present authors are given in Fig. 7.4. The peaks of the distributions of densities are around 0.3–0.4, 0.5–0.6, and 0.7–0.8 g cm^{-3} for carbon aerogels, cryogels, and xerogels, respectively. Full widths at half maximum of such distributions correspond to density ranges of 0.1–0.6, 0.4–1.2, and 0.15–0.85 g cm^{-3} for carbon aerogels, xerogels and cryogels, respectively. Only a few values outside these ranges have been reported.

The skeletal density of carbon gels, determined from helium pycnometry of thoroughly crushed carbon gel samples, is always found to be close to 2.0 ± 0.1 g cm^{-3}. The corresponding porosity of carbon gels thus ranges from 30 to 97.5%. Given that the aforementioned ranges of density largely overlap, "dense" aerogels and lightweight xerogels can be obtained, depending on both the precursor and the preparation conditions. All these different characteristics have consequences on the porous structures of carbon gels, and hence on their adsorption properties and related potential applications.

7.2. ADSORPTION FOR POROSITY CHARACTERIZATION

Pyrolysis of organic gels induces isotropic shrinkage. Given that the precursor does not melt, the initial pore structure is globally maintained, but some changes occur: macropore and mesopore volumes generally decrease and increase, respectively. These effects are due to the shrinkage of the material. At the same time, micropores are produced and surface area increases accordingly, due to the evolution of gases during carbonization [20]. These phenomena are observed as long as the pyrolysis temperature is not too high, typically lower than 1000 °C. At higher temperatures, micropores quickly disappear, whereas the volume of mesopores decreases more slowly [21,22].

The narrow porosity in carbon gels is most of the times evidenced through nitrogen adsorption–desorption experiments at 77 K, whereas wide pores are generally investigated by mercury intrusion. Nitrogen adsorption can be carried out with gels before and after pyrolysis, so that the origin of the resultant pore texture of carbon gels can be clearly understood. For instance, in the case of resorcinol–formaldehyde gels prepared with a constant dilution ratio $D=5.7$ ($D=$ molar ratio $W/(R+C+F)$, where W, R, C, and F stand for water, resorcinol, "catalyst" NaOH, and formaldehyde respectively), the porosity of dry, organic xerogels changes dramatically with the initial pH of the solution. The materials are purely macroporous at $pH=2.00$–4.00 (type II isotherm), micro–macroporous at $pH=5.00$–5.45 (combination of types I and II isotherms), micro–mesoporous at $pH=5.80$–6.50 (combination of types I and IV isotherms), purely microporous at $pH=7.35$ (type I isotherms), and finally nonporous at all at $pH\geq 8.00$ [23]. The corresponding carbon xerogels are thus micro–macroporous, then micro–mesoporous, and finally nonporous when the pH of the initial solution increased from 2 to 6.5. On average, increasing the pH decreased the pore size. Such behaviors are clearly observed in Fig. 7.5.

Nitrogen adsorption is the standard way of characterizing carbon gels, whose microporosity is generally not so narrow for justifying the use of CO_2 as a probe molecule. CO_2 adsorption has indeed been used in a few cases only, and no significant difference between results obtained from N_2 or CO_2 was observed. However, using CO_2 instead of N_2 is really meaningful in the case of activated carbon gels [24].

The way the nitrogen adsorption–desorption isotherms are next treated for deriving pore texture parameters such as micropore and mesopore volumes as well as pore-size distributions depends on the authors. Almost all researchers use to apply the well-known Brunauer–Emmett–Teller (BET) method [25] for determining the total surface area. In a very few cases only, the surface area was determined by the α_S method [26], small-angle X-ray scattering (SAXS), or small-angle neutron scattering (SANS). When the surface areas were measured by both BET and SAXS methods and compared, a good agreement has been found between the two data sets [27–29].

The total pore volume of carbon gels can be calculated from their bulk density (see Fig. 7.4) and from the skeletal density of the constitutive carbon. Now, the pore volume that can be measured by nitrogen adsorption at saturation ($p/p_0 \rightarrow 1$) is generally significantly lower than the total pore volume. It can thus be concluded that macropores are present, and the latter are indeed particularly well evidenced by mercury porosimetry. Because of macropores, and given that

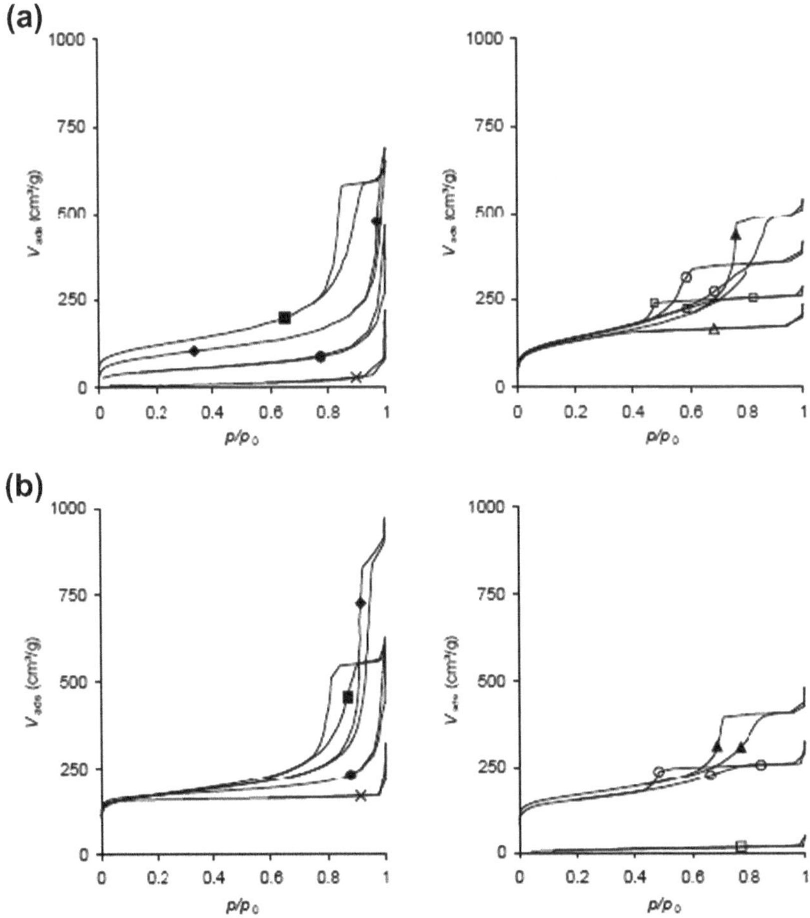

FIGURE 7.5 Nitrogen adsorption–desorption isotherms of RF xerogels obtained at a constant dilution ratio of 5.7 (see text) but different pH values: (a) after drying, (b) after pyrolysis: (×) pH = 2.00, superimposed to pH = 4.00; (●) pH = 5.00; (♦) pH = 5.45; (■) pH = 5.80; (▲) pH = 6.00; (○) pH = 6.25; (□) pH = 6.50; (Δ) pH = 7.35. The isotherm at pH = 7.35 after pyrolysis is superimposed to that at pH = 6.50. Isotherms at pH = 8.00 are not shown (nonporous materials). *(Reproduced from [23] with Permission)*

micropores are also most of the times present, the corresponding nitrogen adsorption–desorption isotherms may be combinations of type I and II isotherms. However, macropores are hardly detected by nitrogen adsorption, so the corresponding isotherms rather belong to type I in this case. Most carbon gels also comprising large mesopore volumes, combinations of types I and IV are also very common. Typical examples of nitrogen adsorption–desorption isotherms of carbon gels presenting different pore textures are presented in Fig. 7.6.

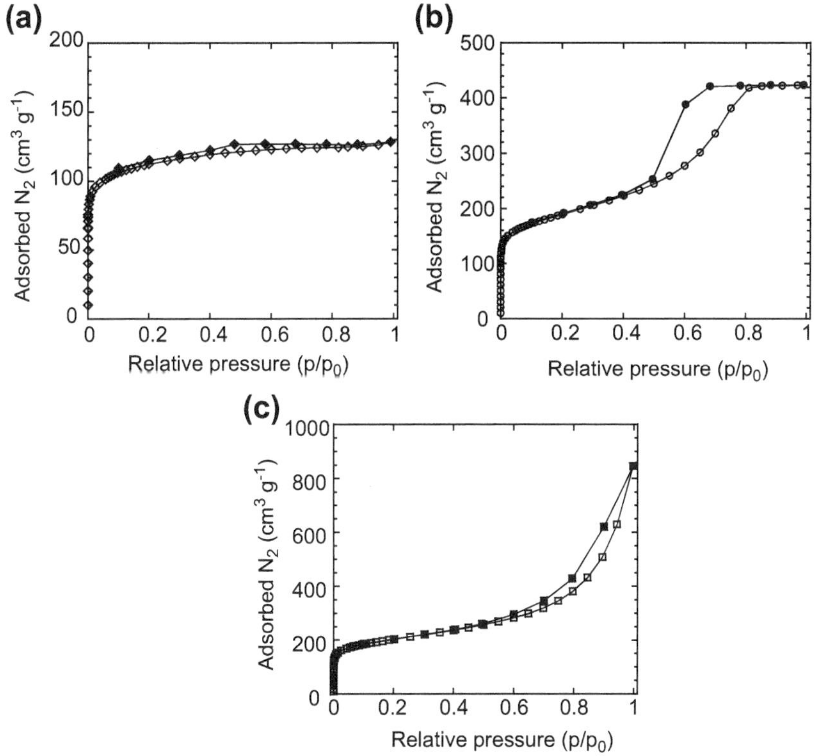

FIGURE 7.6 Nitrogen adsorption–desorption isotherms of carbon gels having different pore textures: (a) Type I (carbon cryogel from tannin–formaldehyde prepared at pH 5.3), (b) combination of types I and IV with an H2 hysteresis loop (carbon cryogel from resorcinol–formaldehyde (RF) at a R/water mass ratio of 0.5, R/F molar ratio of 0.5, and R/potassium carbonate molar ratio of 100), (c) combination of types I and IV with a H3 hysteresis loop (same as (b) but the R/water mass ratio was 0.1). *(After [31] and [32])*

Most of the times, hysteresis is observed and is either H2 or H3 type. In some cases, the loop is so narrow with nearly parallel adsorption and desorption branches that it is difficult to decide whether the loop is H2 or H1 type. It may be useful to recall that type H1 hysteresis is observed when pores are uniform in size and in shape. H2 type corresponds to inkbottle-shaped pores, i.e. whose pore body is wider than the pore mouth, whereas type H3 hysteresis is characteristic of a very wide distribution of pore sizes [30]. Figure 7.6(a), with a very narrow hysteresis, is typical of an almost purely microporous carbon gel. Figure 7.6(b) is a combination of type I and IV isotherms with a H2 hysteresis loop, characterizing the presence of micropores and inkbottle mesopores. Finally, Fig. 7.6(c) is a combination of type I and IV isotherms with an H3 hysteresis loop, indicating the presence of micropores and wide mesopores with a broad size distribution that continues into the macropore domain. The hysteresis loop is indeed very narrow, the adsorption and desorption branches being almost vertical and nearly parallel above 0.8 relative pressure, confirming the presence of a significant outer surface.

From such isotherms, the typical surface areas are within the ranges presented in Fig. 7.7, whether they were determined from BET analysis, SAXS, or α_S method. The peaks of the distributions of surface areas are around 600–700 $m^2 g^{-1}$ for both carbon aerogels and cryogels, against 500–600 $m^2 g^{-1}$ for carbon xerogels. Full widths at half maximum of such distributions correspond to surface area ranges of 500–800, 350–780, and 350–700 $m^2 g^{-1}$ for carbon aerogels, xerogels, and cryogels, respectively. Exceptionally high values have been reported for some gels, as high as those of activated carbons, i.e. much higher than 1000 $m^2 g^{-1}$ and sometimes above 2500 $m^2 g^{-1}$. These findings, not shown in Fig. 7.7, were explained by the presence of NaOH [33] or KOH [34] used as gelation catalysts, and which behaved as activating agents during the pyrolysis of the organic gels. On the other hand, surface areas close to zero were also observed in the case of poorly porous carbon xerogels.

Activated carbon gels are just briefly considered here because they behave as ordinary activated carbons. Surface areas as high as 2200, 2500, and 2800 $m^2 g^{-1}$ have indeed been reported for carbon cryogels activated with H_3PO_4 [32], xerogels activated with KOH [24], and carbon aerogels activated with KOH [35], respectively. It is interesting to note that, through chemical activation of organic gels with H_3PO_4, the volume of micropores is

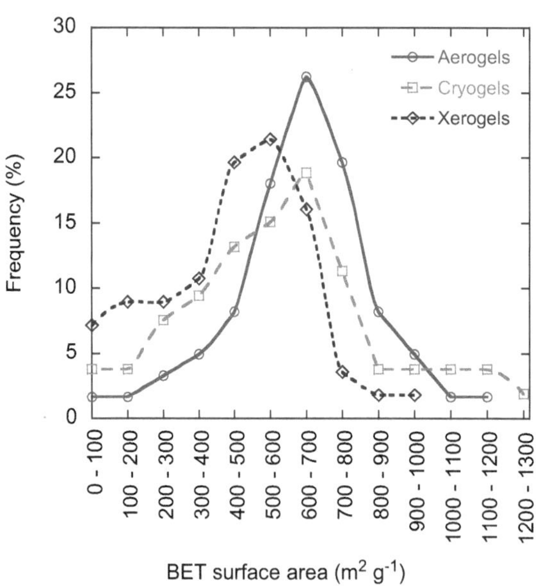

FIGURE 7.7 Distributions of BET surface areas of carbon gels (after the data reported in about one hundred scientific papers).

dramatically increased but, in contrast, lower or much higher mesopore volumes can be obtained, depending on the amount of acid [32,36]. It is even possible to produce bimodal activated carbons combining micropores and macropores in the near absence of mesoporosity by chemical activation of organic gels [32]. High adsorption/desorption and diffusion kinetics are thus expected in such kind of porous materials. Additional references to physically activated carbon gels may be found in [37].

In the literature, determination of pore volumes and pore-size distributions is carried out through the application of various methods, and all possible combinations of models have been reported. Most micropore volumes have been calculated by application of the t-plot method [38], followed by Dubinin–Radushkevich (DR) equation [39] and occasionally α_S method. The t-plot method also allows the determination of both micropore and external surface areas. Micropore-size distributions were not often calculated, because the interest of carbon gels usually lies in their mesoporosity. When required, density functional theory (DFT) [40] and exceptionally Horwath–Kawazoe (HK) method [41] have been used.

Mesopore volumes were very often derived from the method of Dollimore and Heal (DH) applied to the desorption branch of the isotherms [42]. However, the most frequent method used for determining the mesopore-size distribution of carbon gels is by far that of Barret, Joyner, and Halenda (BJH) [43] applied to the desorption branch of the isotherms, immediately followed by the DH method. Both DH and BJH methods also allowed determining the mesopore-size distribution, the mesopore surface area and the average mesopore diameter. Mesopore volumes were also sometimes derived from the Broeckoff–de Boer method [44] or even by application of the α_S model.

In a number of cases, mesopore volumes were deduced from the difference between the volume of nitrogen adsorbed at saturation, V_{sat}, and the micropore volume calculated according to one of the aforementioned DR, t-plot, or α_S models. Such a method may be inaccurate in the absence of well-defined plateau when the relative pressure is close to one. This is the reason why V_{sat}, sometimes called Gurvitch volume [45], has been measured at $p/p_0 = 0.90$, 0.95, 0.975, 0.99, or 0.995, depending on the authors. In other papers, the pore volumes were calculated by combination of several different techniques. For example, the total pore volume was either derived from both bulk and skeletal densities of the carbon gel, as already discussed in this section, or calculated as the sum of volumes measured by adsorption on the one hand and by mercury porosimetry on the other. The macropore volume could thus be deduced from the knowledge of the other kinds of pore volumes.

As far as mercury intrusion techniques are used as supplementary tool of porosity characterization, extreme care must be taken to the possible shrinkage of the samples induced by the isostatic pressure. Compression of the gel may indeed occur below a critical pressure, above which intrusion of mercury really begins. Such behavior may lead to considerable errors in pore volume measurements and related pore-size distributions. Fortunately, it can be detected by the characteristic shape of the intrusion curves, and by the weighing of samples submitted to mercury porosimetry experiments carried out at various pressures. These aspects are out of the scope of the present paper and have been discussed elsewhere [46,47].

It is difficult to decide, among all the suggested methods, which are the ones most suitable for investigating the porosity of carbon gels. This is probably the reason why most of them have been already applied and why no clear consensus has emerged. However, the results that can be obtained by application of the most usual models are not so different, as demonstrated later. BJH, DH, and DFT methods have indeed been applied in Fig. 7.8 for determining the mesopore-size distributions for the most usual nitrogen adsorption–desorption isotherms characterizing carbon gels, i.e. combinations of types I and IV having H2 or H3 hysteresis loops as those presented in Fig. 7.6(b) and (c), respectively. Micropore and mesopore volumes have been calculated as well, based on DR, t-plot, and α_S models on the one hand, and based on BJH or DH models and Gurvitch volume at $p/p_0 = 0.95$ on the other (see Table 7.1).

The agreement between BJH and DH mesopore-size distributions is excellent, supporting the frequent use of these methods for characterizing carbon gels. The DFT model leads to broader distributions shifted toward higher pore diameters. These findings are systematically observed, whether the pores were assumed to be cylindrical or slit-shaped. In the present examples, the micropore volume measured by the DR method is significantly higher than that derived from the t-plot or the α_S method. However, in other works, such difference is not so high, even if the trend is the same,

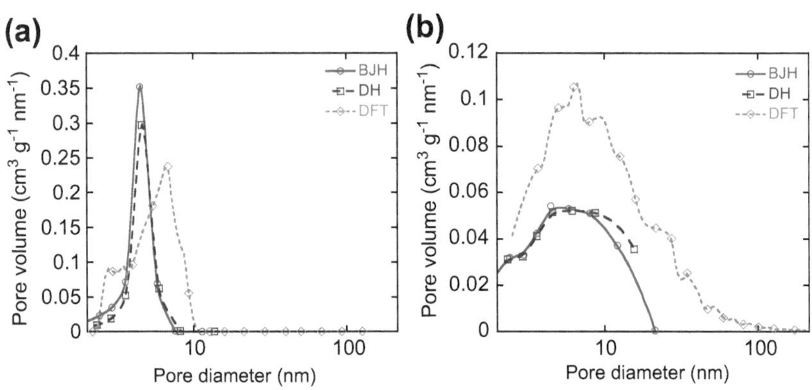

FIGURE 7.8 Pore-size distributions (range of pore width 2–200 nm) calculated by application of Barret–Joyner–Halenda (BJH), Dollimore–Heal (DH), and density functional theory (DFT) models, assuming cylindrical pores in carbon cryogels whose isotherms are given: (a) in Figure 7.6(b), (b) in Fig. 7.6(c).

TABLE 7.1 Pore Texture Parameters Derived from Fig. 7.6(a) and (b) by Application of Different Models

		Micropore Volume ($cm^3\ g^{-1}$)			Mesopore Volume ($cm^3\ g^{-1}$)		
Material	N_2 Isotherm	DR	t-plot	α_S	BJH	DH	$V_{0.95} - V_{\text{t-plot}}$
Carbon cryogel of Fig. 7.6(b)	Type I-IV with H2 loop	0.26	0.12	0.10	0.62	0.52	0.53
Carbon cryogel of Fig. 7.6(a)	Type I-IV with H3 loop	0.28	0.13	0.13	1.25	0.78	0.84

i.e. DR volumes are always slightly higher than micropore volumes measured by other methods [48].

From these results and from the critical analysis of the literature dealing with carbon gels, the present authors recommend the investigation of micro- and mesoporosity through the use of t-plot and BJH or DH methods, respectively.

According to most published works, average pore diameters, whether calculated from DH or BJH methods, have progressively narrower distributions from carbon aerogels to cryogels and xerogels: 5–75, 10–30, and 5–15 nm, respectively, are typical values. The mean pore size also decreases in the same way: 20–25, 10–20, and 5–10 nm for carbon aerogels to cryogels and xerogels, respectively. However, all these values just correspond to trends, and it should be recalled that dramatic changes of pore textures are always possible through the use of different preparation conditions such as pH, nature of catalyst, temperature of gelation, and aging. Therefore, separating the effects of synthesis parameters from those induced by the drying mode is very difficult. Additionally, doping carbon gels with metals is also able to modify the pore structure at different scales and therefore change the shape of their isotherms. A number of different materials may thus be prepared, whose main adsorption properties are described in the next section.

7.3. ADSORPTION IN THE GAS PHASE

Carbon gels, either as such or modified by hetero-element doping or by activation, have been investigated for the adsorption of permanent gases or condensable vapors. The aim of these works was to more thoroughly characterize the carbon gels, or to suggest potential applications such as gas storage and VOC removal or detection.

7.3.1. Adsorption of Permanent Gases, Separation, and Storage

Adsorption of permanent gases such as N_2, O_2, CO_2, C_2H_6, and C_4H_{10} has been studied on carbon cryogel and xerogel microspheres at room temperature by Yamamoto et al. [22,49]. The aim was to test carbon gels as materials for gas separation through pressure- or temperature-swing adsorption. Carbon gels are indeed expected to be among the most suitable adsorbents for that purpose, due to their easily tunable pore structure. The main results may be summarized as follows.

In [49], all tested cryogels and xerogels had rather similar micropore volumes and therefore presented nearly identical nitrogen and oxygen adsorption isotherms at 298 K (see Fig. 7.9(a)). However, the adsorption rate at a given equilibrium pressure of each gas was much higher in carbon gels whose pores were much wider than the kinetic diameters of N_2 and O_2: 0.36 and 0.35 nm, respectively [49]. Nitrogen molecules being larger than oxygen ones, their diffusion was slower in ultramicropores, so that differences of adsorption rates were observed only in carbon gels whose ultramicropore width was within the range 0.33–0.40 nm, i.e. close to the molecular size of N_2 and O_2 molecules (see Fig. 7.9(b)). Different sizes of pores at the surface of the gels could indeed be evidenced in such materials using the molecular probe method [22], i.e. adsorbing at 298 K molecules having different minimum molecular dimensions: CO_2 (0.33 nm), C_2H_6 (0.40 nm), and C_4H_{10} (isobutane: 0.50 nm).

From Fig. 7.9(b), it is clear that separation of air is thus possible on materials having narrow pores. Moreover, the strong dependence on temperature of the adsorbed amounts of N_2 and O_2 (not shown) supports the applicability of carbon gels to temperature-swing adsorption.

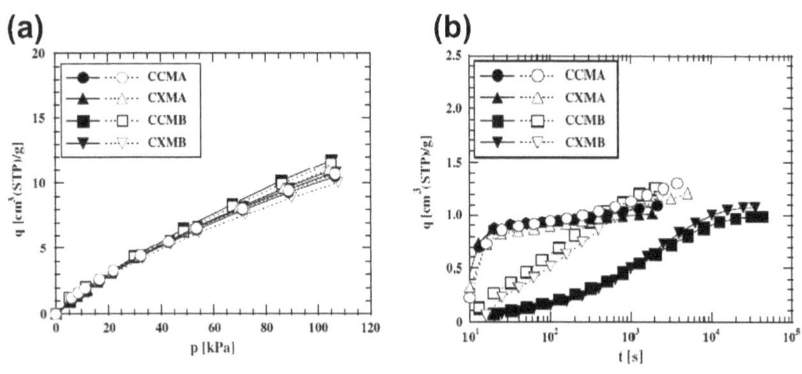

FIGURE 7.9 Adsorption at 298 K of nitrogen (closed symbols) and oxygen (open symbols) on carbon cryogel (CCMA and CCMB) and carbon xerogel (CXMA and CXMB) microspheres. The pores of CCMA and CXMA were wider than those of CCMB and CXMB, but all micropore volumes were rather similar: (a) adsorption isotherms at equilibrium. (b) Transient changes of adsorbed amounts, the equilibrium pressure of each gas being 5.8 kPa. *(Reprinted from [49], with permission from Elsevier)*

Carbon gels have been suggested as adsorbents for hydrogen storage. However, it is well-known that hydrogen requires extremely narrow pores for being suitably retained in the carbonaceous structure. This is the reason why raw carbon gels were not investigated as such for that purpose, but were either activated or doped with metal nanoparticles [50]. After activation, the adsorbed amount at 77 K and 3.5 MPa followed Chahine' rule [51], i.e. the hydrogen storage capacity increased by 1 wt. % every 500 $m^2\ g^{-1}$ of BET surface area. For example, excess adsorption isotherms and correlations between excess hydrogen uptake and micropore volume on the one hand, and BET surface area on the other, are given in Fig. 7.10 for raw carbon aerogel and carbon aerogels activated with CO_2. The highest excess uptake reported so far with this kind of material is 5.2 wt. % at 77 K and 3.5 MPa [35].

Doping carbon gels with nitrogen [1] or with cobalt nanoparticles [52,53] was shown to increase the isosteric heat of adsorption and therefore enhance the hydrogen storage capacity. H_2 adsorption at 77 K on carbon xerogels (whether as such or after ammoxidation) was controlled by textural properties, without a clear effect of nitrogen doping. In contrast, the isotherms near room temperature presented higher initial slopes in nitrogen-doped xerogels, indicating a higher binding energy [1]. A much higher heat of adsorption was also measured when carbon aerogels were doped with Co nanoparticles [53]. Doping decreased the surface area and hence the excess hydrogen storage capacity. However, expressed as adsorption capacity per unit surface area, Co-doped aerogel presented higher performances than those of pure carbon: 1.3–1.8 wt. % H_2 per 500 $m^2\ g^{-1}$ [52].

Carbon gels have also been suggested as porous hosts for storing ammonia borane which can be decomposed in order to produce hydrogen [54,55]. Such an indirect way of storing hydrogen does not involve adsorption and is consequently not detailed here.

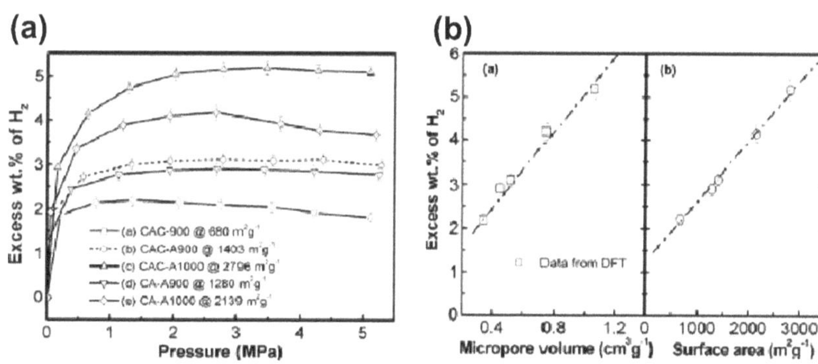

FIGURE 7.10 (a) Excess adsorption isotherms of hydrogen at 77 K on raw carbon aerogel (CAC-900) and carbon aerogels activated with CO_2 at 900 or 1000 °C during 1 h (CAC-A and CA-A). (b) Correlations between excess uptakes and pore texture parameters. *(Reprinted from [35], with permission from Elsevier)*

Another energy vector of interest is methane. Very strong adsorption of methane in carbon aerogel at room temperature has been reported on the basis of studies of molecular mobility by neutron scattering experiments [56]. The diffusivity of methane in a carbon aerogel characterized by an average pore diameter of 4.8 nm, as measured by the BJH method, was found to be two orders of magnitude lower than that of bulk methane, irrespective of the pressure (see Fig. 7.11(a)). This finding suggests a very strong confinement of methane in the pores. The highest diffusion coefficient (8.55×10^{-10} m^2 s^{-1}) was found at a pressure of 2.75 MPa. This value corresponds to that of liquid methane below 100 K, and is much lower than what can be measured in other porous adsorbents such as zeolites, microporous silica, or metal organic frameworks.

From neutron scattering studies, it could be deduced that narrow pores become filled at low pressure with adsorbed methane having liquid-like density and diffusivity. As the pressure increases up to 2.75 MPa, pores with increasingly wider diameters are filled with liquid-like methane (see Fig. 7.11(b)). Above 2.75 MPa, intermolecular collisions become important, resulting in a lower molecular mobility, just like in bulk methane. Interestingly, the diffusivity of methane in carbon aerogel bearing adsorbed moisture is slightly higher than in the dry gel. This finding was explained by the hydrophobicity of methane enhancing its density fluctuations when in contact with water [56].

7.3.2. Adsorption of Condensable Vapors and Detection

Studies of steam adsorption onto carbon gels are scarce, despite their interest as soon as adsorption of water may compete with that of other molecules. Water

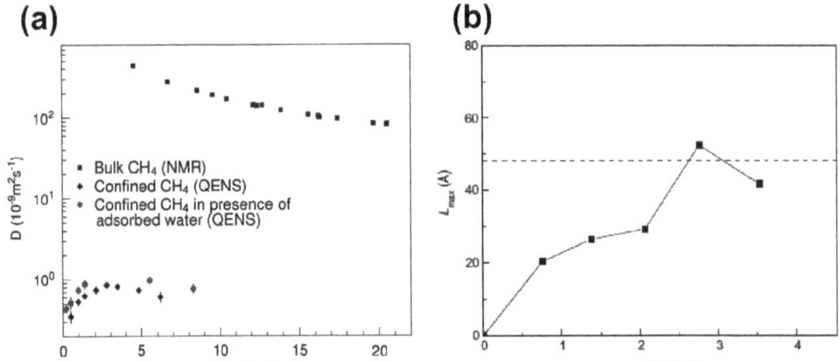

FIGURE 7.11 (a) Self-diffusivity of methane at 298 K in dry and wet carbon aerogel (characterized by an average pore diameter of 4.8 nm according to BJH method), measured by quasi-elastic neutron scattering (QENS) and compared to that of bulk methane. (b) Maximum pore width completely filled with liquid-like methane as a function of pressure. The dashed line corresponds to the average pore width of the carbon aerogel. *(Reprinted from [56], with permission from Elsevier)*

vapor adsorption isotherms are of type V, as shown in Fig. 7.12, indicating that the interactions between water molecules are much stronger than between water and carbon. Carbon cryogels were thus found to be rather hydrophobic, especially at low relative pressure for which the water uptake was very low and controlled by the temperature of pyrolysis of the organic gel precursors. At p/p_0 around 0.5, condensation in mesopores occurred, except in one sample in which the surface pores were too narrow. Hydrophobicity of carbon gels could thus be controlled by their temperature of carbonization [22].

Carbon gels have been studied as adsorbents of other condensable vapors, especially VOCs, with different aims in view: abatement, separation, or detection. Carbon aerogels indeed present both microporosity for adsorption and mesoporosity for efficient transport of VOCs toward the micropores. Additionally, micro- and mesoporosity may be independently controlled through suitably chosen synthesis parameters.

Abatement of benzene, toluene, and o- and m-xylene was investigated with carbon aerogels under dynamic conditions in either dry or wet air [57]. Breakthrough curves showed very steep increases in the VOC concentration at the outlet of the column after the breakthrough point, see Fig. 7.13(a). From such curves, the adsorbed amounts were determined and seen to increase in the order benzene < toluene < xylenes, i.e. in the order of increasing relative pressure. As expected, the adsorbed amount per unit weight of adsorbent increased with the degree of activation of the gels, but the adsorbed amount per unit volume of adsorbent bed stopped increasing above a given burn-off, due to the corresponding decrease of bulk density of the carbon monoliths. When air was wet, the VOC capacity was a few % lower than in dry air because of

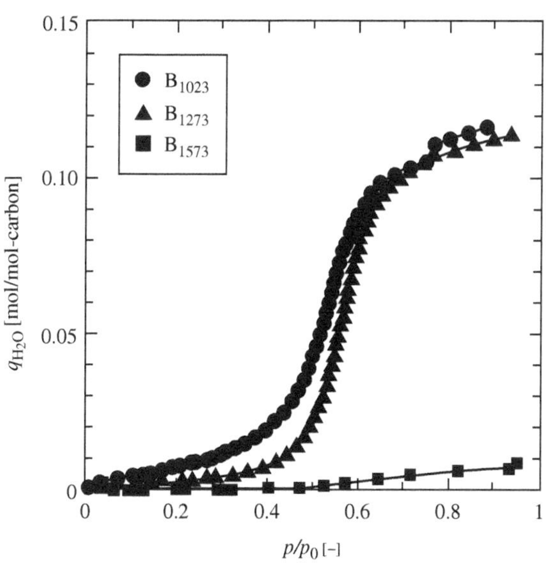

FIGURE 7.12 Adsorption isotherms of steam at 298 K onto carbon cryogel microspheres prepared at pyrolysis temperatures ranging from 1023 to 1573 K. *(Reprinted from [22], with permission from Elsevier)*

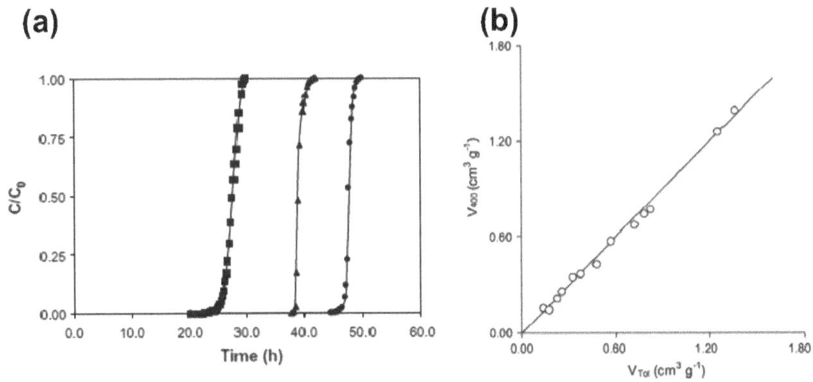

FIGURE 7.13 (a) Breakthrough curves of toluene (741 ppmv) from dry air (25 °C, 60 $cm^3 min^{-1}$) for raw carbon aerogel (■) and carbon aerogel activated with CO_2 at 16% (▲) and 27% (●) of burn-off. *(Reprinted from [57], with permission from Elsevier)* (b) Relationship between amount of toluene desorbed at 400 °C (V_{400}) and amount of toluene adsorbed at saturation at 25 °C (V_{Tol}). *(Reprinted from [58], with permission from Elsevier)*

competition between water and hydrocarbon molecules for adsorption sites on the carbon surface.

In another study, the adsorption of toluene was shown to be completely reversible at 400 °C in carbon aerogels, whether they were previously activated with steam or not (see Fig. 7.13(b)) [58]. The adsorption capacities at 25 °C, up to 1.18 g g^{-1}, were higher than in all other carbonaceous adsorbents reported so far. Assuming that adsorbed toluene is liquid, such exceptional performances were explained by the fact that not only the whole microporosity but also part of the mesoporosity contributed to the adsorption of toluene. The adsorption capacity was not affected after several adsorption (25 °C)–desorption (400 °C) cycles [58].

The excellent adsorption performances of carbon gels toward VOCs combined with their intrinsic electrical conduction properties were applied for preparing devices for the detection of organic vapors. For example, conducting composites made of carbon aerogel dispersed into polystyrene were shown to be sensitive to vapors of n-hexane, ethyl acetate, THF, chloroform, and acetone [59]. When air saturated with such chemicals was put into contact with these composites, their electrical resistance increased dramatically. The process was totally reversible as soon as solvent vapor was replaced by dry air. Materials made from carbon aerogels were more sensitive than those based on carbon black because of their higher dynamic adsorption capacity [59]. Ultrasensitive (in the ppm range) and reversible chemical sensors based on changes of electrical resistance were also made from multiwalled carbon nanotubes aerogels [60].

Finally, adsorption in the gas phase is related to another application of major importance: catalysis. This area is too specific and too broad for being considered here. However, the following features are worth citing. Most carbon gels have been used as catalysts after doping with various metals and subsequent activation. The presence of metal changed, sometimes significantly, both the pore texture and the surface area of the support, and also the degree of organization of the carbon at the nanometer scale. Consequently, the catalytic activity was controlled not only by the nature, the oxidation state, and the dispersion of the metal, but also by the availability of the active sites which depends on the porosity. Reactions of isomerization, decomposition, combustion, organic synthesis, carbon nanotube and nanofilament growth, hydrogenation, hydrodechlorination, and reduction have been reported in a review [20] as well as in several more recent works, e.g. [5,61–63].

7.4. ADSORPTION IN THE LIQUID PHASE

As demonstrated above, carbon gels are, most of the times, highly mesoporous adsorbents. Their average pore size makes them ideal materials for adsorbing dissolved and solvated compounds. Moreover, diffusion of the latter may be quite fast in mesopores, unlike what happens in the microporosity. Even when such liquid species are small enough for entering micropores, the mesopore network remains necessary for getting high adsorption–desorption kinetics. This is the reason why porous electrodes for supercapacitors have been suggested as a potential use of carbon gels from the beginning of their discovery. In the following, adsorption of organic compounds from water or from other solvents, halides, and metal ions or complexes will be considered. Finally, energy applications such as supercapacitors and fuel cells will be discussed in relation to the adsorption phenomenon.

7.4.1. Adsorption of Organic Compounds

Carbon cryogel microspheres possessing controlled pore structure were tested as adsorbents for organic molecules in water, e.g. phenol and dyes such as Black 5 and Red E [64]. The adsorption of phenol, the smallest molecule of the study, was found to be controlled by surface area. Carbon cryogel thus adsorbed less phenol than a commercial activated carbon having higher BET surface area and correspondingly higher micropore volume. However, the adsorption constants of the Langmuir equation that could fit the isotherms showed that the affinity of cryogel toward phenol was higher than that of activated carbon.

In contrast, adsorption of large molecules such as dyes was found to be controlled by mesopores – the higher the mesopore size, the higher the adsorbed amount. The isotherms were best fitted by the Freundlich model, and differences of adsorption capacities were explained in terms of molecular size, structure, and polarity. For each carbon gel, less Red E (asymmetric and bearing a carboxylic acid) was systematically adsorbed than Black 5 (symmetric and flat), see Fig. 7.14 [64].

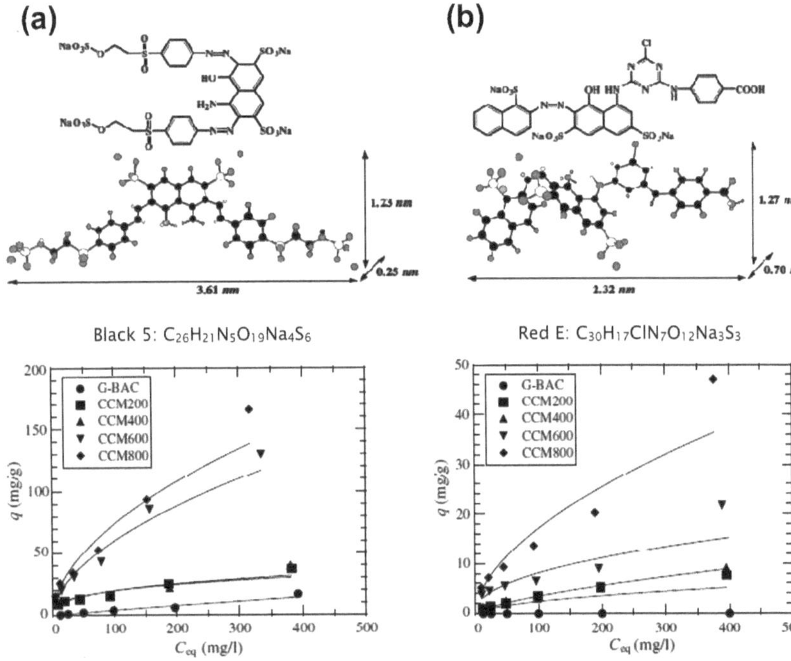

FIGURE 7.14 Adsorption of dyes from water on carbon cryogel microspheres (CCM) having different pore textures (the mesopore volume and the mesopore size increase from 200 to 800 at constant micropore volume), and on an almost purely microporous commercial activated carbon (G-BAC): (a) Black 5, (b) Red E. *(Reprinted from [64], with permission from Elsevier)*

A recent work confirmed the applicability of carbon xerogels to the removal of cationic dyes from water, such as methylene blue and Rhodamine B. These compounds were shown to be available to 1/3 to 4/5 of the total surface area, depending on the dye and on the synthesis parameters. This is a good result, considering that the investigated gels were dried subcritically before pyrolysis and were not activated subsequently. Thus, the adsorption capacities of such carbon xerogels were comparable to those of conventional activated carbons, making them good candidates for remediation processes [65].

Persistent organic pollutants such as polycyclic aromatic hydrocarbons (PAHs) could be successfully removed from water using the solid-phase extraction (SPE) technique. For that purpose, composite cryogels based on multiwalled carbon nanotubes and polyvinyl alcohol were used [66]. Benzo[a] anthracene, benzo[b]fluoranthene and benzo[a]pyrene could be concentrated so efficiently that their detection at levels at low as 5, 8, and 5 ng L^{-1} in water, respectively, was possible. Comparison with a commercial SPE sorbent showed that the latter did not show significantly higher performances. On the contrary, the cryogel could operate at higher loading flow rate without sorbent clogging, required a shorter time for completion without loss of extraction efficiency, and was cheaper than the commercial sorbent [66].

Removal of aqueous organic pollutants can also be done by adsorption–catalytic process (ACP), which is based on a concentration by adsorption on a porous material followed by photocatalytic degradation or a catalytic oxidation. A detailed discussion of such processes using metal-doped carbon gels has been given in [67]. For example, carbon cryogel beads have been tested and compared according to two ways: as such for adsorbing phenol on the one hand, and after doping with metal oxide for destroying adsorbed phenol in the presence of ozone on the other [68]. Adsorption isotherms of aqueous phenol on doped and nondoped carbon cryogel were very similar and both Langmuir types, indicating that phenol molecules were preferably adsorbed in micropores. Metal (Mn, Fe, Co, and Ni) oxides loaded at levels ranging from 0.5 to 5 wt. % accelerated the adsorption rate of phenol. Additionally, due to their catalytic activity for degrading phenol in the presence of ozone, doped cryogels significantly enhanced the removal efficiency of aqueous phenol. All other things being equal, the performances of the ACP process increased with the average mesopore width, in favor of higher diffusion rate of both phenol and reaction products [68].

A few applications of carbon gels in the area of life sciences have been reported, especially as adsorbents of antibiotics [69] or as supports for enzyme immobilization with applications as biosensors [70]. Removal of antibiotic ciprofloxacin was faster on carbon xerogel than on granular activated carbon and multiwalled carbon nanotubes, due to its much larger pore size: more than 30 nm against 3–4 nm for the other adsorbents. Carbon xerogels had a surface area two times lower than that of activated carbon and bore acidic surface groups, detrimental to adsorption of ciprofloxacin. Therefore, the saturation capacity of carbon xerogel, derived from its Langmuir-type adsorption isotherm, was two times lower than that of activated carbon, around 110 mg g^{-1}. However, the adsorption capacity of carbon xerogel normalized by the BET surface area was the same as that of activated carbon, also bearing the same kinds of moieties [69].

Peroxidase enzyme was immobilized at the surface of pure carbon xerogel and nitrogen-doped carbon xerogels. The biosensor was immersed in hydrogen peroxide solution, and the oxidation of phenol catalyzed by the enzyme attached to the carbon surface was investigated. The enzymatic activity was found to increase in the order: graphite powder (used as reference) < nitrogen-free carbon xerogel < nitrogen-doped carbon xerogel. At constant surface area, carbon xerogels having higher nitrogen contents led to higher activity, allowing a higher sensitivity of the biosensor and a better analytical response to phenol concentration [70].

Water is not the only solvent from which organic compounds could be successfully removed by carbon gels. Sulfur-containing chemicals are poisons for the noble-metal catalysts used in reforming, water gas shift reactors, and fuel cells, and should thus be removed from most liquid fuels. Organosulfur molecules are generally aromatic compounds that survived the usual HDS process, such as dibenzothiophene (DBT).

Adsorption of DBT from hexadecane, considered to be a model of diesel fuel, was found to be faster and higher in carbon aerogels having higher mesopore volumes and broader mesopores, at constant micropore volume (see Fig. 7.15) [71]. Saturation capacity of the best gel estimated by application of Langmuir's model, 15.1 mg of sulfur per g of dry carbon aerogel, was found to be only slightly lower than that of Cu–Y zeolite, one of the sorbents having the best performances for desulfurization of commercial diesel. The presence of naphthalene, a molecule having similar structure and electronic properties as those of DBT, decreased the adsorption of DBT by a few % only, suggesting a good selectivity of carbon aerogel toward sulfur compounds [71]. Adsorption of heavier molecules such as 4,6-dimethyldibenzothiophene (4,6-DMDBT) in the presence of DBT was also found to be possible, although 4,6-DMDBT was adsorbed more slowly and at lower levels, due to steric effects. Dynamic adsorption of both compounds was shown to be controlled by inner-particle diffusion [72].

Separation of toluene, benzene, and o-xylene was demonstrated feasible on the basis of carbon cryogels used as packing material for HPLC columns [22]. Figure 7.16 thus shows that such organic compounds could be separated within a few minutes, thanks to carbon cryogel microspheres having a high mesopore volume (0.90 $cm^3\ g^{-1}$) and a comparatively low amount of micropores (micropore volume: 0.15 $cm^3\ g^{-1}$, BET surface area: 525 $m^2\ g^{-1}$). Improved separation performances are expected through the use of monodisperse carbon cryogel microspheres having lower micropore volumes [22].

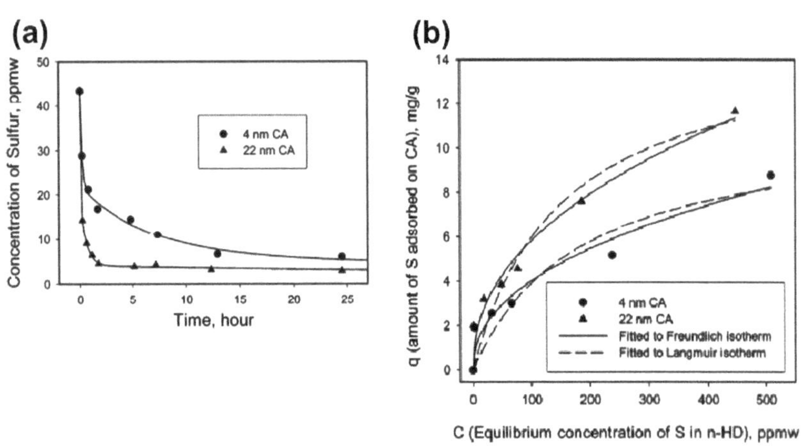

FIGURE 7.15 Adsorption of dibenzothiophene from n-hexadecane (n-HD), using carbon aerogels (CA) having average pore diameters of 4 and 22 nm: (a) Adsorption kinetics, (b) Adsorption isotherms at equilibrium. *(Reprinted with permission from [71]. Copyright (2012) American Chemical Society.)*

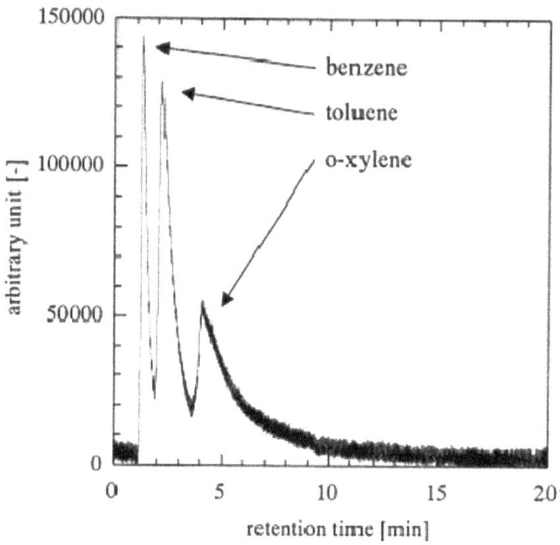

FIGURE 7.16 Chromatogram of a mixture of benzene, toluene, and o-xylene dissolved in acetonitrile flowing at 1.5 $cm^3\ min^{-1}$ on carbon cryogel microspheres (see text, diameter 10–20 μm) packed in an HPLC stainless steel column (25 cm × 2.1 cm I.D.) at 318 K. *(Reprinted from [22], with permission from Elsevier)*

7.4.2. Adsorption of Halides

The removal of halides is of special interest given that they may form very toxic compounds after applying oxidative treatments such as those used for producing drinking water. Additionally, such by-products have very low taste and odor thresholds. Removal of halides was investigated by adsorption on Ag-doped carbon aerogels [73,74]. The adsorption kinetics was independent of the anion but, after equilibrium, the adsorption capacities were very different and increased in the order: $I^- < Br^- < Cl^-$, except at low halide concentration (see Fig. 7.17(a)). These results were obtained for a solution of pH close to 7, whereas the adsorbent had a pH of the point of zero charge (pH_{PZC}) of 4.5 only. The carbon surface was thus negatively charged, not favoring the adsorption process. However, silver nanoparticles considerably improved the adsorption capacity, compared to that of Ag-free adsorbent. Higher loading levels led to improved adsorption despite the corresponding decrease of surface area, suggesting a specific Ag–halide interaction at the surface, see Fig. 7.17(b) [73].

Adsorption of bromide and iodide was lowered by more than 30% in the presence of chloride or organic matter in water, due to competition for adsorption sites and pore blocking, respectively. Nevertheless, silver-doped carbon aerogels were much more efficient for adsorbing bromide (up to 3 $\mu mol\ g^{-1}$) than conventional activated carbon for water treatment (less than 0.5 $\mu mol\ g^{-1}$), despite the fact that activated carbon had a positively charged surface, i.e. more favorable for the adsorption of anionic species [74].

Performances of silver-doped carbon aerogels designed for the same purpose could be significantly improved by activation with CO_2. Such a treatment both developed the micropore volume and shifted the average mesopore

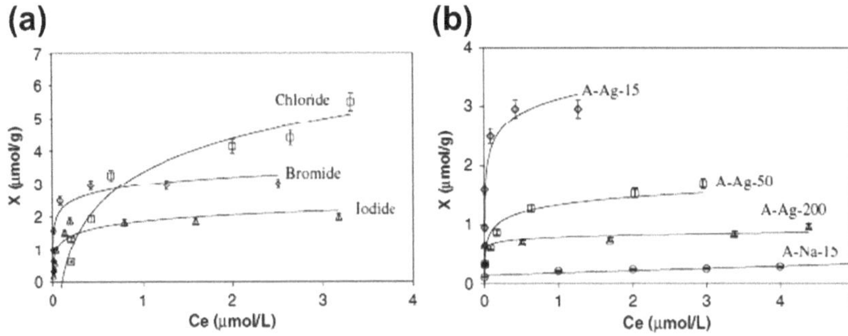

FIGURE 7.17 (a) Adsorption isotherms of halides at 25 °C from water at pH 7 onto a carbon aerogel containing 10 wt. % Ag. (b) Adsorption isotherms of bromide (same conditions) onto carbon aerogels doped with different amounts of Ag: 0% (A-Na-15); 4% (A-Ag-200); 6% (A-Ag-50) and 10% (A-Ag-15). *(Reprinted from [73], with permission from Elsevier)*

diameter toward higher values. It also increased the pH_{PZC} above 9, making the surface positively charged for the adsorption of halides at pH 7. The adsorption capacity of Br^- and I^- was therefore doubled with reference to nonactivated Ag-doped aerogels. Dynamic adsorption capacities were found to be much lower than those measured in batch, suggesting that diffusion toward part of the Ag sites was difficult. After use, these materials could be regenerated with aqueous ammonia with little change in the column characteristics. Ag-doped carbon aerogels are thus interesting adsorbents for water treatment, considering that Ag is also highly bactericidal [74].

Removal of chloride from water is also possible by electrosorption. Carbon aerogel has been compared to other adsorbents: activated carbon (granular and cloth) and multiwalled carbon nanotubes. The adsorbed amount of ions was found to increase with surface area and micropore volume, so that carbon aerogel presented much lower performances than activated carbons. However, normalizing the results with respect to BET surface area, it was concluded that all tested carbon surfaces were equivalent for the electrosorption of NaCl, and that both mesopores and micropores were effective in the process. Difference of pore textures only led to different kinetics, adsorption being the fastest on carbon aerogel and nanotubes [75].

7.4.3. Adsorption of Metals

Industrial wastewaters may be contaminated by heavy metals. The feasibility of using carbon aerogels for adsorbing them has been investigated in the case of mercury, lead, nickel, and cadmium [76–79]. Initial metal ion concentrations ranged from 5 to 70 mg L^{-1}, and different pH values and temperatures were tried. Adsorption of metals indeed depends both on the nature of the adsorbent (surface functions, pH_{PZC}, and pore texture) and on the speciation of the metals.

At any pH, the removal of metal ions was found to be higher in the presence of carbon aerogel than by precipitation in the absence of carbon, even at high pH [76]. Increasing the pH is in favor of higher adsorption because metal species M^{2+} hydrolyze into $M(OH)^+$ and $M(OH)_2$, and hence bear a lower electrostatic charge. Less repulsion and less competition with H^+ for adsorption thus exist at higher pH. Additionally, an adsorption mechanism by ion exchange also occurs between metal ions and H^+ ions at the carbon surface, such as

$$2 > C = O - H^+ + M^{2+} \rightarrow (> C = O)_2M^{2+} + 2H^+$$

It is then expected that the exchange capacity of carbon gels increases with the amount of oxygenated groups, whose dissociation increases with the concentration of metal ions.

Adsorption kinetics revealed that, in the aforementioned concentration conditions, equilibrium is attained within 2 h. The curves presented two different slopes, see Fig. 7.18(a), corresponding to surface layer diffusion and to pore diffusion at low and high agitation times, respectively. The pseudo-second-order equation was found to be the most suitable for fitting the experimental data. At equilibrium, the adsorption isotherms (see Fig. 7.18(b)) were best described by Langmuir equation: saturation capacities followed the order Pb(II) $\geq$ Hg(II) (around 35 mg g^{-1}) > Ni(II) (around 3 mg g^{-1}) [76].

Additional experiments revealed the endothermic character of the adsorption of metal ions onto carbon aerogel surface. Therefore, the metal uptake increased with temperature, as shown in Fig. 7.19(a) in the case of Hg(II). The latter was almost completely removed from water at pH 5–7 and 70 °C, provided the amount of carbon aerogel was enough, see Fig. 7.19(b) [77]. However, such high performances decreased in the presence of other metal ions, due to competition and surface charge screening by the additional ions. The saturation adsorption capacity of Pb(II) for the same carbon aerogel as in Figs 7.18 and 7.19 thus decreased from 35 mg g^{-1}, when lead was alone, to 25 mg g^{-1} in the presence of Hg(II), and to 8 mg g^{-1} in the presence of both Hg(II) and Cd(II) [78].

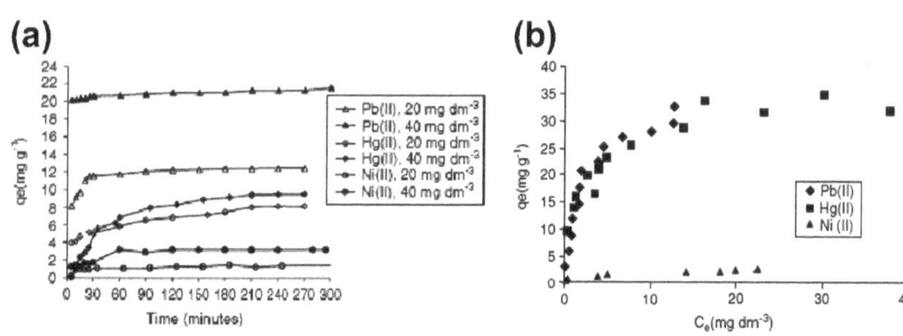

FIGURE 7.18 Adsorption of lead, mercury, and nickel from water at pH 4.5 and 37 °C on carbon aerogel (BET = 700 $m^2 g^{-1}$, pH_{PZC} = 4.87, V_{micro} = 0.10 $cm^3 g^{-1}$, V_{meso} = 1.36 $cm^3 g^{-1}$): (a) kinetics, (b) isotherms at equilibrium. *(Reproduced from [76] with Permission from John Wiley and Sons)*

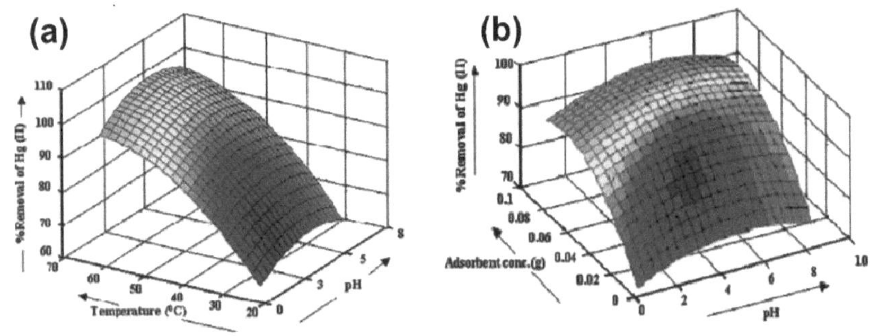

FIGURE 7.19 Removal of mercury from water by adsorption onto carbon aerogel (same characteristics as in Fig. 7.18) as a function of: (a) temperature and pH, (b) amount of carbon gel and pH. *(Reprinted from [77], with permission from Elsevier)*

Adsorption from the liquid phase is also a process by which carbon gels may be doped with metals. When metal complexes are dissolved in water, their adsorption onto a carbon surface critically depends on its pH_{PZC}. For instance, at pH lower than pH_{PZC}, the functional groups are protonated so that the surface is positively charged. Conversely, surfaces are deprotonated and hence negatively charged at pH higher than pH_{PZC}. Therefore, species such as $[PtCl_6]^{2-}$ and $[Pt(NH_3)_4]^{2+}$ are adsorbed through coulombic interactions when their aqueous solutions have a pH below and above pH_{PZC}, respectively. High amounts of metals may be adsorbed this way called the "strong electrostatic adsorption" method, with the advantage of retaining a high dispersion of the complexes during subsequent drying and reduction steps.

Figure 7.20(a) shows, for example, the amount of $[PtCl_6]^{2-}$ and $[Pt(NH_3)_4]^{2+}$ adsorbed at the surface of carbon xerogels before and after oxidation with nitric acid, having pH_{PZC} of 9.4 and 2.4, respectively [61]. Such curves allow determining the optimal pH for adsorption. Oxidation significantly decreased the mesopore volume, whereas the micropore volume remained almost constant. Figure 7.20(b) shows that the weight loading of metal complex increased linearly with the mesopore surface area [61]. The Pt loading thus ranged from 8 to 10 wt.% and from 1 to 5 wt.% for nonoxidized and oxidized carbons, respectively. A high mesopore surface was thus in favor of adsorption.

Supercritical fluid-aided deposition is another way of incorporating metallic nanoparticles into many possible porous substrates. For that purpose, an organometallic complex is dissolved in a fluid such as supercritical carbon dioxide ($scCO_2$), adsorbed on the support, and reduced to its metal. The gas-like diffusivity and viscosity of $scCO_2$ indeed affords fast diffusion and permeation, its liquid-like viscosity permits the dissolution of a number of complexes, and its low surface tension allows a good wetting of narrow pores usually not available to liquid solvents. Deposition of highly dispersed platinum and ruthenium nanoparticles onto carbon aerogels has been reported through the use of $scCO_2$ [80–83].

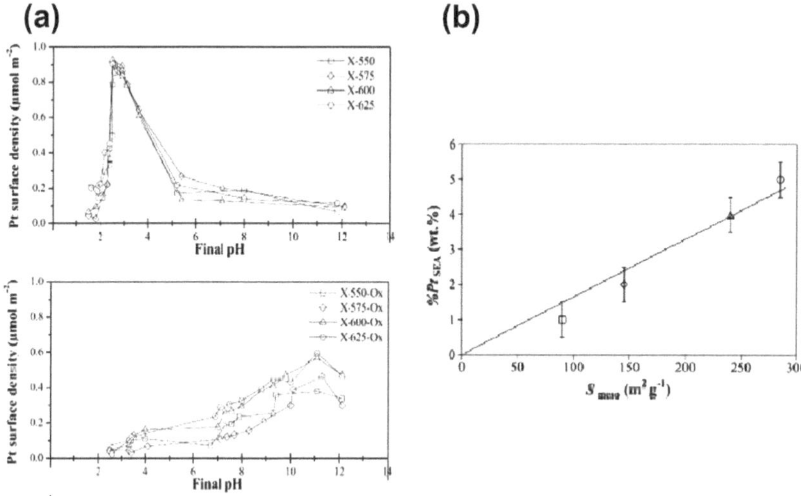

FIGURE 7.20 (a) Uptakes of chloroplatinic acid (H_2PtCl_6) on nonoxidized carbon xerogels having different mesopore volumes (top), and of tetra-ammine platinum chloride ($Pt(NH_3)_4Cl_2$) on oxidized carbon xerogels (bottom). (b) Platinum weight loading versus the mesopore surface area of oxidized carbon xerogels. *(Reprinted from [61], with permission from Elsevier)*

For example, high uptakes of dimethyl(1,5-cyclooctadiene) platinum(II) dissolved in $scCO_2$ were observed for two carbon aerogels having BET surface areas around 700 $m^2\ g^{-1}$ but having average pore sizes of 4 and 22 nm [80,81]. Adsorption isotherms were best described by Langmuir's equation, and the highest uptake was observed for the carbon having the highest mesopore volume and the highest average pore size at constant micropore volume. High Pt loadings ranging from 10 to 40 wt. % combining nanoparticle sizes as low as 1 nm could be obtained [80,81].

The same qualitative results were obtained with different organometallic Ru complexes [82,83]. The adsorption kinetics was extremely fast, except when the pore size was close to that of the complex. The isotherms were fully reversible, and the uptake was the highest for the most mesoporous carbon aerogels. The surface area occupied by the complexes, calculated from the geometry of the molecule and from the adsorbed amount, was very close to the mesopore surface area, suggesting that a monolayer covers the accessible surface. Such large molecules probably do not enter micropores, explaining why the mesoporosity is so important [82,83].

7.4.4. Adsorption in the Framework of Energy Applications

Carbon gels have been suggested for long as porous electrodes for electrochemical energy storage. Electrochemical capacitors may store energy either on the basis of purely electrostatic attraction between ions and the charged surface of carbon or additionally involve faradaic reactions similar to those occurring in accumulators. These devices are then called electrical double-layer capacitors (EDLCs) and supercapacitors (SCs), respectively. Their capacitances are related to the chemical nature of the surface (presence/absence of electroactive species) and to the extent of the surface area available to the electrolyte.

Besides having a high capacitance, additional requirements for making competitive electrodes are high electrical conductivity, good wettability of the carbon surface, long cycle life, good charge/discharge efficiency, high specific peak power, low self-discharge, and adequate operating voltage [84]. Most of these features may be controlled by pore structure and surface modification of carbon gels.

Under application of an electrical polarization, solvated ions diffuse in mesopores for being finally adsorbed in micropores. A higher capacitance is thus a priori expected from a porous material having a higher surface area. However, the surface area measured by gas adsorption may be significantly higher than the one electrochemically available for the ions of the liquid electrolyte. Therefore, most of the times, no correlation between capacitance and BET surface area or micropore volume was found, except in rare occasions and only for carbon gels prepared from the same raw materials in similar conditions [2,31,85]. However, even for carbons made from identical precursors, and for which a constant specific capacitance was expected, no proportionality between surface area and capacitance was observed in many cases, due to different accessibility of micropores to the ions of the electrolyte [86].

The majority of published works agree that micropores determine the high surface area on which ions are adsorbed, whereas mesopores are essential for ion transfer [84]. Hence, the higher accessibility of ions to the electrochemically active surface, the higher the capacitance of the EDLC. However, contradictory results describing the effect of pore size on capacitance have been reported in the literature. Some authors have claimed that carbon aerogels having pore diameters within the range 3–13 nm are those showing the highest capacitance values [87], and that mesopores broader than 20 nm are unnecessarily wide [88]. Other works have suggested that pores larger than only 0.5 nm could be accessible electrochemically to aqueous electrolyte, but the electrical double layer could not be fully developed in micropores [89,90]. In these conditions, the existence of micropores should reduce significantly the overall capacitance.

However, some papers reported that the presence of supermicropores (0.7–2 nm) advantageously improves the capacitance [91,92]. Some observations also suggest that ions can diffuse in supermicropores and are adsorbed as nonsolvated in ultramicropores (<0.7 nm). The latter might thus contribute significantly to the value of capacitance in aqueous medium [93].

This controversy comes from the very different results that carbon materials may generate. The authors of the present chapter indeed observed that, for carbon cryogels prepared in rather similar conditions (except initial pH) but made from phenol–formaldehyde or from tannin–formaldehyde resins, the capacitance was controlled by mesopores or by micropores, respectively [31,86]. It seems very difficult to foresee, only from the pore structure estimated by adsorption methods, what will be the electrochemical behavior of a given carbon gel.

Therefore, very different double-layer capacitances, calculated from the ratio of the specific capacitance to the BET surface area, from 4 to 32 μF cm^{-2}, have been reported for carbon gels. For comparison, the specific capacity generally accepted for a clean graphite surface is 20 μF cm^{-2} [94,95], whereas an average value of 25 μF cm^{-2} has been reported for other carbon materials [84].

Nonfunctionalized carbon gels investigated by cyclic voltammetry at low scan rate (1–2 mV s^{-1}) present curves like those of Fig. 7.21(a) at 2 mV s^{-1}, i.e. have a nearly rectangular, box-like shape. Moreover, the cycles are symmetrical and reversible. Such behavior is typical of an ideal capacitor. When the scan rate increases, pore availability becomes critical. The cycles become narrower and more rounded, see Fig. 7.21(a), implying the decrease of the specific capacitance. Such a decrease is related to the difficulty for the electrolyte to diffuse in the small pores of the carbon gel, leading to lower available electroactive surface area. The internal resistance is thus significantly increased, leading to a major drop of the capacitance and to an expected slower charging process (see Fig. 7.21(b)).

Pores wider than 1.1 nm can be electrochemically accessed in less than 0.1 s, whereas pores having widths of 0.6 nm require about 5 s for being reached [96]. In relation to this, it has been shown recently that, for carbon cryogels being mainly micro- and macroporous (i.e. having a very low mesopore volume), ultramicropores mainly account for the capacitance within the

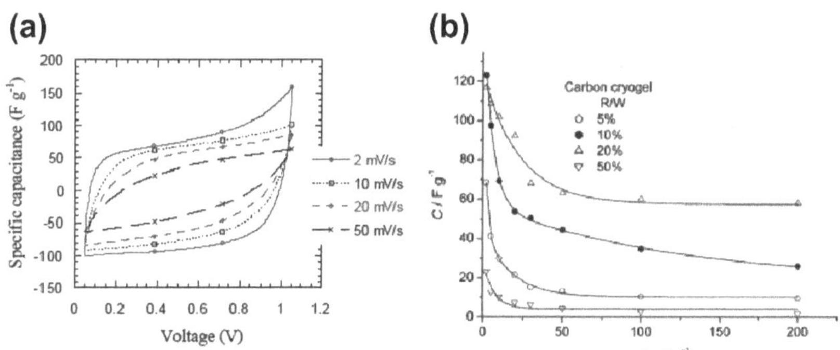

FIGURE 7.21 (a) Cyclic voltammograms obtained at different scan rates for porous electrodes made from tannin-formaldehyde carbon cryogels. (b) Decrease of specific capacitance with scan rate for carbon cryogels made from resorcinol-formaldehyde at different resorcinol/water (R/W) weight ratios. *(Reprinted from [31] and [94], with permission from Elsevier)*

range of scan rates 2–20 mV s^{-1}, whereas supermicropores control the capacitance above 50 mV s^{-1} [31].

Typical capacitances reported for carbon gels were in the range 30–150 F g^{-1}, without correlation with BET surface area as explained above. Extreme values of 15–200 F g^{-1} have been sometimes found. Activated carbon gels presented improved capacitances, up to 300 F g^{-1}, whether the activation was physical (in air at 450 °C) [97] or chemical (KOH at 700 °C) [34]. Performances in terms of specific capacitances were most of the times independent from the type of carbon gel: xero-, cryo-, or aerogel.

Improvement of capacitance was achieved by modification of carbon gel surfaces. Pseudocapacitance effect produces a deviation from the rectangular shape of voltammograms, with reversible redox peaks connected with pseudo-faradaic reactions. These phenomena indeed occurred by oxidation of the carbon surface (see Fig. 7.22(a)) [97], or by doping with hetero-elements such as B and N [85] or with electroactive oxides such as MnO_2 [98,99] or RuO_2 [100,101] (see Fig. 7.22(b) for example).

Reversible energy storage of around 250 mAh g^{-1} in carbon aerogels was also observed in aqueous electrolyte (KOH 6 mol L^{-1}) from the contribution of two mechanisms: electrical double-layer (around 130 F g^{-1}) and hydrogen adsorption [102]. Hydrogen may indeed be formed at low potentials by the reduction of water into OH^- and nascent H. The latter first adsorbs on the carbon surface and recombines into H_2, which can next diffuse in the bulk of the carbon gel and finally be adsorbed in the pores [103].

Such electrochemically induced adsorption improved the energy storage capacity of carbon gels. After discharge, the measured current showed that some hydrogen had not been oxidized and hence remained adsorbed in the

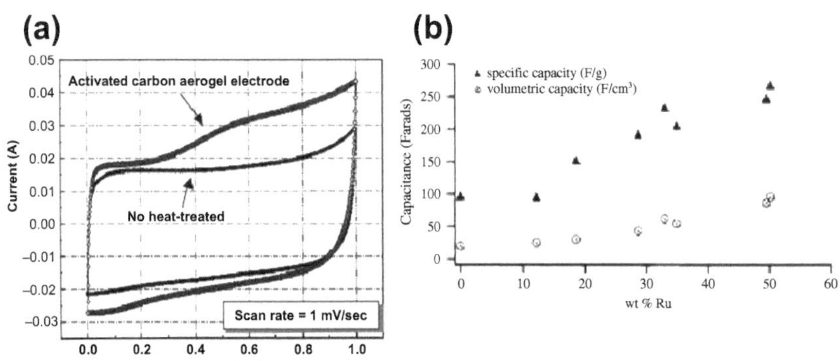

FIGURE 7.22 Improvement of specific capacitance of electrochemical capacitor electrodes based on carbon gels: (a) Effect of activation (heat-treatment of 1 h in air at 450 °C) on the voltammograms, *(Reprinted from [97], with permission from Elsevier)* (b) Effect of the content of Ru on the specific capacitance. *(Reprinted with permission from [100]. Copyright (2012) American Chemical Society.)*

porosity. This amount was low, however, less than 1 wt. %, and decreased with the organization of the carbon, and hence with the temperature of pyrolysis [102]. 1000 °C seemed to be a good compromise for both retaining structural disorder and getting high electrical conductivity. The optimal pore size for the adsorption of hydrogen is close to 0.7 nm [93], so that the nanotexture (ultramicroporosity) governs the performance, whereas other pores only control the kinetics. The performances of carbon gels for electrochemical hydrogen storage in aqueous medium may thus be estimated by CO_2 adsorption, which is a much better indicator than N_2 for that purpose [102].

Metal-doped carbon gels have been tested as catalysts in direct methanol (DMFCs) and polymer electrolyte membrane fuel cells (PEMFCs). In such applications, high electrical conductivity, well-connected broad porosity, chemical purity, and possibility of having evenly distributed nanoparticles of noble metals without agglomeration are key parameters that can be met by carbon gels. Thus, for DMFCs, highly mesoporous carbon cryo- and xerogels doped with Pt–Ru presented more than double specific catalytic activity than that of Pt–Ru supported on Vulcan carbon black [104]. In [105], the performances of commercial catalysts were attained with much less Pt–Ru loading, using a carbon aerogel as support. These favorable features were explained on the basis of high dispersion of metal on carbon gels and wide mesopores, allowing high diffusion of fuel and excellent contact between metal, fuel and liquid electrolyte or proton conductor.

Similar conclusions were obtained in the case of PEMFCs based on carbon aerogels, i.e. high fractions of surface Pt were observed on such supports, leading to better electrochemical activity than commercial Pt-doped carbon blacks [106,107]. Just like for DMFCs, the highest performances of PEMFCS, up to 0.8 mW cm^{-2}, were reached with carbon aerogels having the broadest mesopores [108].

Finally, carbon aerogels [109], cryogels [110], and xerogels [3] have been successfully used in the preparation of composite electrodes for Li-ion batteries. The relevant mechanisms do not involve adsorption and are thus not detailed here.

7.5. CONCLUSIONS

Carbon gels are rather new materials, first described around two decades ago, which received considerable attention due to their special characteristics. They can indeed be prepared in the form of monoliths, beads, powders or films, and their pore structure may be tailored through the choice of precursors and the control of synthesis parameters. A very broad range of carbon materials can thus be prepared, from almost nonporous to highly micro-, meso-, and even macroporous, having various possible surface areas.

Carbon gels may be derived from a number of thermoset resins or nonfusible polymers, and even from carbon nanotubes. All are derived from solutions or suspensions in a solvent that has been subsequently removed by evaporation, supercritical extraction, or lyophilization, leading to xero-, aero-, and cryogels, respectively. Each kind of material has its own porous structure. As a result, a carbon gel may be described as an interconnected and open network made of primary particles having typical dimensions of a few tens to a few hundreds nanometers. When the gels are made from resins, the porous structure is also particularly sensitive to gelation parameters such as nature of catalyst, pH, temperature, and aging step. Interestingly, the volumes of micro- and mesopores can be controlled independently, the former being located inside the primary particles, whereas the latter correspond to interparticle spaces.

Pyrolysis of organic gels generally retains the initial structure, but changes occur induced by evolution of gases and shrinkage. The possible presence of transition metals may favor the partial graphitization of the carbon and decrease the structural disorder at the nanometer scale. The surface of carbon gels may also be modified chemically, thus allowing different values of pH of the point of zero charge, and can be dramatically developed by activation. Hence, functionalized and activated carbon gels are additional materials having potentially new adsorption properties.

Even if any kind of porous structure is virtually possible with carbon gels, most of them exhibit high mesopore volumes, associated with variable microporosity, depending on the cases. Hence, they are best characterized by nitrogen adsorption experiments using t-plot and Dollimore–Heal (DH) or Barrett–Joyner–Halenda (BJH) methods for microporosity and mesoporosity, respectively. By application of DH or BJH models, average and peak pore widths can be determined, on the basis of which the adsorption behaviors of different carbon gels can be easily explained.

The way carbon gels are prepared and their resulting well-developed, accessible, surface area make them easy to dope with metals. The latter indeed remain in a highly dispersed state and are readily accessible to fluids throughout the interconnected network of mesopores. This is the reason why many metal-doped carbon gels were proved to be very efficient catalysts for a number of reactions in the gas phase. Other works have emphasized the interest of carbon gels, doped or not, for air separation, gas storage, and abatement or detection of VOCs in air.

However, due to their significant mesoporosity, the major applications of carbon gels are related to adsorption and catalysis in liquid phase. Wide pores indeed allow fast kinetics and/or removal of large molecules such as dyes or organosulfur compounds from water and liquid fuels, respectively. Very good results were also obtained in adsorption of charged species, such as halides and heavy metals from water, and in separation of a series of homologous compounds. Besides, high-power densities were measured when Pt-doped carbon aerogels were used as catalysts in DM and PEM fuel cells.

Finally, adsorption of ions on charged and highly conducting carbon gel surfaces for storing electrochemical energy is one of the most promising applications. Whether they were used as such or activated, or doped or functionalized for allowing additional pseudo-faradaic reactions, high electrochemical capacitances and high cyclability could be observed for porous electrodes based on carbon gels. Moreover, such devices are presently commercialized.

Given the great potential of carbon gels, it is most likely that new or improved functionalities will be suggested in the near future.

ACKNOWLEDGMENTS

The authors gratefully acknowledge the financial support of the CPER 2007-2013 "Structuration du Pôle de Compétitivité Fibres Grand'Est" (Competitiveness Fibre Cluster), through local (Conseil Général des Vosges), regional (Région Lorraine), national (DRRT and FNADT) and European (FEDER) funds.

REFERENCES

[1] K.Y. Kang, B.I. Lee, J.S. Lee, Hydrogen adsorption on nitrogen-doped carbon xerogels, Carbon 47 (2009) 1171–1180.

[2] A. Halama, B. Szubzda, G. Pasciak, Carbon aerogels as electrode materials for electrical double layer supercapacitors – synthesis and properties, Electrochim. Acta 55 (2010) 7501–7505.

[3] X.Y. Yuan, Y.J. Chao, Z.F. Ma, X. Deng, Preparation and characterization of carbon xerogel (CX) and CX-SiO composite as anode material for lithium-ion battery, Electrochem. Commun. 9 (2007) 2591–2595.

[4] T. Yamamoto, T. Ohmori, Y.H. Kim, Preparation and characterization of monodisperse carbon cryogel microspheres, Microporous Mesoporous Mater. 112 (2008) 211–218.

[5] R.A. Catalao, F.J. Maldonado-Hodar, A. Fernandes, C. Henriques, M.F. Ribeiro, Reduction of NO with metal-doped carbon aerogels, Appl. Catal. B Env. 88 (2009) 135–141.

[6] L.W. Hrubesh, R.W. Pekala, Thermal properties of organic and inorganic aerogels, J. Mater. Res. 9 (1994) 731–738.

[7] A. Celzard, J.F. Marêché, Applications of the sol–gel process suing well-tested recipes, J. Chem. Educ. 79 (2002) 854–859.

[8] C.J. Brinker, G.W. Scherer, Sol – gel science. The physics and chemistry of sol–gel processing, Academic Press, 1990.

[9] A. Léonard, N. Job, S. Blacher, J.P. Pirard, M. Crine, W. Jomaa, Suitability of convective air drying for the production of porous resorcinol-formaldehyde and carbon xerogels, Carbon 43 (2005) 1808–1811.

[10] H. Tamon, H. Ishizaka, T. Yamamoto, T. Suzuki, Preparation of mesoporous carbon by freeze drying, Carbon 37 (1999) 2049–2055.

[11] N. Job, A. Théry, R. Pirard, J. Marien, L. Kocon, J.N. Rouzaud, F. Beguin, J.P. Pirard, Carbon aerogels, cryogels and xerogels: influence of the drying method on the textural properties of porous carbon materials, Carbon 43 (2005) 2481–2494.

[12] H. Nishihara, S.R. Mukai, H. Tamon, Preparation of resorcinol–formaldehyde carbon cryogel microhoneycombs, Carbon 42 (2004) 889–901.

[13] U. Fischer, R. Saliger, V. Bock, R. Petricevic, J. Fricke, Carbon aerogels as electrode material in supercapacitors, J. Porous Mater. 4 (1997) 281–285.
[14] R.W. Pekala, D.W. Schaefer, Structure of organic aerogels. 1. Morphology and scaling, Macromolecules 26 (1993) 5487–5493.
[15] A.G. Yodh, Nanomaterials: carbon nanotube aerogel, Adv. Compos. Bull. (2007) 3.
[16] M.A. Worsley, S.O. Kucheyev, J.H. Satcher, A.V. Hamza, T.F. Baumann, Mechanically robust and electrically conductive carbon nanotube foams, Appl. Phys. Lett. 94 (2009) 073115.
[17] X. Zhang, Z. Sui, B. Xu, S. Yue, Y. Luo, W. Zhan, B. Liu, Mechanically strong and highly conductive graphene aerogel and its use as electrodes for electrochemical power sources, J. Mater. Chem. 21 (2011) 6494–6497.
[18] M.A. Worsley, P.J. Pauzauskie, T.Y. Olson, J. Biener, J.H. Satcher Jr., T.F. Baumann, Synthesis of graphene aerogel with high electrical conductivity, J. Am. Chem. Soc. 132 (2010) 14067–14069.
[19] A. Szczurek, G. Amaral-Labat, V. Fierro, A. Pizzi, E. Masson, A. Celzard, The use of tannin to prepare carbon gels. Part I: carbon aerogels, Carbon 49 (2011) 2773–2784.
[20] C. Moreno-Castilla, F.J. Maldonado-Hódar, Carbon aerogels for catálisis applications: an overview, Carbon 43 (2005) 455–465.
[21] M. Inagaki, Pores in carbon materials – importance of their control, New Carbon Mater. 24 (2009) 193–232.
[22] T. Yamamoto, T. Sugimoto, T. Suzuki, S.R. Mukai, H. Tamon, Preparation and characterization of carbon cryogel microspheres, Carbon 40 (2002) 1345–1351.
[23] N. Job, (2005). Matériaux carbonés poreux de texture contrôlée synthétisés par procédé sol–gel et leur utilisation en catalyse hétérogène. PhD thesis, Liège University (Belgium).
[24] L. Zubizarreta, A. Arenillas, J.P. Pirard, J.J. Pis, N. Job, Tailoring the textural properties of activated carbon xerogels by chemical activation with KOH, Microporous Mesoporous Mater. 115 (2008) 480–490.
[25] S. Brunauer, P.H. Emmett, E. Teller, Adsorption of gases in multimolecular layers, J. Am. Chem. Soc. 60 (1938) 309–319.
[26] F. Rouquerol, J. Rouquerol, K. Sing, Adsorption by powders and porous solids, Academic Press, 1999.
[27] V. Bock, A. Emmerling, J. Fricke, Influence of monomer and catalyst concentration on RF and carbon aerogel structure, J. Non Cryst. Solids 225 (1998) 69–73.
[28] R. Saliger, V. Bock, R. Petricevic, T. Tillotson, S. Geis, J. Fricke, Carbon aerogels from dilute catalysis of resorcinol with formaldehyde, J. Non Cryst. Solids 221 (1997) 144–150.
[29] O. Czakkel, E. Geissler, I.M. Szilagyi, E. Szekely, K. Laszlo, Cu-doped resorcinol-formaldehyde (RF) polymer and carbon aerogels, J. Colloid Interface Sci. 337 (2009) 513–522.
[30] K.S.W. Sing, D.H. Everett, R.A.W. Haul, L. Moscou, R.A. Pierotti, J. Rouquerol, T. Siemienewska, Reporting physisorption data for gas–solid systems, Pure Appl. Chem. 57 (1985) 603–619.
[31] G. Amaral-Labat, A. Szczurek, V. Fierro, N. Stein, C. Boulanger, A. Pizzi, A. Celzard. Pore structure and electrochemical performances of tannin-based carbon cryogels. Biomass Bioenerg. 39 (2012) 274–282.
[32] A. Szczurek, G. Amaral-Labat, V. Fierro, A. Pizzi, A. Celzard, Bimodal activated carbons derived from resorcinol-formaldehyde cryogels, Sci. Technol. Adv. Mater. 12 (2011) 035001.
[33] Y. Zhu, H. Hu, W. Li, X. Zhang, Cresol-formaldehyde based carbon aerogel as electrode material for electrochemical capacitor, J. Power Sources 162 (2006) 738–742.
[34] Y. Zhu, H. Hu, W. Li, X. Zhang, Resorcinol-formaldehyde based porous carbon as an electrode material for supercapacitors, Carbon 45 (2007) 160–165.

[35] H.Y. Tian, C.E. Buckley, S.B. Wang, M.F. Zhou, Enhanced hydrogen storage capacity in carbon aerogels treated with KOH, Carbon 47 (2009) 2128–2130.

[36] F.L. Conceiçao, P.J.M. Carrott, M.M.L. Ribeiro Carroot, New carbon materials with high porosity in the 1-7 nm range obtained by chemical activation with phosphoric acid of resorcinol-formaldehyde aerogels, Carbon 47 (2009) 1874–1877.

[37] S.A. Al-Muhtaseb, J.A. Ritter, Preparation and properties of resorcinol–formaldehyde organic and carbon gels, Adv. Mater. 15 (2003) 101–114.

[38] B.C. Lippens, J.H. de Boer, Pore system in catalysts. V. The t-plot method, J. Catal. 4 (1965) 319–323.

[39] M.M. Dubinin, Fundamentals of the theory of adsorption in micropores of carbon adsorbents – characteristics of their adsorption properties and microporous structures, Carbon 27 (1989) 457–467.

[40] P. Tarazona, Solid–fluid transition and interfaces with density functional approaches, Surf. Sci. 331–333 (1995) 989–994.

[41] G. Horvath, K. Kawazoe, Method for the calculation of effective pore size distribution in molecular sieve carbon, J. Chem. Eng. Jpn 16 (1983) 470–475.

[42] D. Dollimore, G.R. Heal, An improved method for the calculation of pore-size distribution from adsorption data, J. Appl. Chem. 14 (1964) 109–114.

[43] E.P. Barret, L.G. Joyner, P.P. Halenda, The determination of pore volume and area distribution in porous substances. I. Computations from nitrogen isotherms, J. Am. Chem. Soc. 73 (1951) 373–380.

[44] A.J. Lecloux, Texture of catalysts, in: J.R. Anderson, M. Boudart (Eds.), Catalysis: Science and Technology, Springer, 1981, pp. 171–230.

[45] S.J. Gregg, K.S.W. Sing, Adsorption, Surface Area and Porosity, Academic Press, 1982.

[46] R. Pirard, A. Rigacci, J.C. Marechal, D. Quenard, B. Chevalier, P. Achard, J.P. Pirard, Characterization of hyperporous polyurethane-based gels by non-intrusive mercury porosimetry, Polymer 44 (2003) 4881–4887.

[47] N. Job, R. Pirard, J.P. Pirard, C. Alié, Non intrusive mercury porosimetry: pyrolysis of resorcinol–formaldehyde xerogels, Part. Part. Syst. Char. 23 (2006) 72–81.

[48] C. Scherdel, G. Reichenauer, Carbon xerogels synthesized phenol–formaldehyde gels, Microporous Mesoporous Mater. 126 (2009) 133–142.

[49] T. Yamamoto, A. Endo, T. Ohmori, M. Nakaiwa, Porous properties of carbon gel microspheres as adsorbents for gas separation, Carbon 42 (2004) 1671–1676.

[50] H. Kabbour, T.F. Baumann, J.H. Satcher, A. Saulnier, C.C. Ahn, Toward new candidates for hydrogen storage: high-surface-area carbon aerogels, Chem. Mater. 18 (2006) 6085–6087.

[51] R. Chahine, P. Benard, in: P. Kittel (Ed.), Advance in Cryogenic Engineering, vol. 43, Plenum Press, 1998, p. 1257.

[52] H.Y. Tian, C.E. Buckley, D.A. Sheppard, M. Paskevicius, N. Hanna, A synthesis method for cobalt doped carbon aerogels with high surface area and their hydrogen storage properties, Int. J. Hydrogen Energy 35 (2010) 13232–13246.

[53] H.Y. Tian, C.E. Buckley, M. Paskevicius, D.A. Sheppard, S.B. Wang, C.J. Webb, E.M. Gray, Nanoscale cobalt doped carbon aerogel: microstructure and isosteric heat of hydrogen adsorption, Int. J. Hydrogen Energy 36 (2011) 10855–10860.

[54] S. Sepehri, A. Feaver, W.K.J. Shaw, C.J. Howard, Q. Zhang, T. Autrey, G. Cao, Spectroscopic studies of dehydrogenation of ammonia borane in carbon cryogel, J. Phys. Chem. B 111 (2007) 14285–14289.

[55] A. Feaver, S. Sepehri, P. Shamberger, A. Stowe, T. Autrey, G. Cao, Coherent carbon cryogel – ammonia borane nanocomposite for H_2 storage, J. Phys. Chem. B 111 (2007) 7469–7472.

[56] S.M. Chathoth, E. Mamontov, Y.B. Melnichenko, M. Zamponi, Diffusion and adsorption of methane confined in nano-porous carbon aerogel: a combined quasi-elastic and small-angle neutron scattering study, Microporous Mesoporous Mater. 132 (2010) 148–153.
[57] F. Carrasco-Marín, D. Fairén-Jiménez, C. Moreno-Castilla, Carbon aerogels from gallic acid-resorcinol mixtures as adsorbents of benzene, toluene and xylenes from dry and wet air under dynamic conditions, Carbon 47 (2009) 463–469.
[58] F.J. Maldonado-Hodar, C. Moreno-Castilla, F. Carrasco-Marín, A.F. Pérez-Cadenas, Reversible toluene adsorption on monolithic carbon aerogels, J. Hazard. Mater. 148 (2007) 548–552.
[59] B. Zhang, X. Dong, W. Song, D. Wu, R. Fu, B. Zhao, M. Zhang, Electrical response and adsorption performance of novel composites from polystyrene filled with carbon aerogel in organic vapors, Sens. Actuators B Chem. 132 (2008) 60–66.
[60] J. Zou, J. Liu, A.J. Karakoti, A. Kumar, D. Joung, Q. Li, S.I. Khondaker, S. Seal, L. Zhai, Ultralight multiwalled carbon nanotube aerogel, ACS Nano 4 (2010) 7293–7302.
[61] S. Lambert, N. Job, L. D'Souza, M.F. Ribeiro Pereira, R. Pirard, B. Heinrichs, J.L. Figueiredo, J.P. Pirard, J.R. Regalbuto, Synthesis of very highly dispersed platinum catalysts supported on carbon xerogels by the strong electrostatic adsorption method, J. Catal. 261 (2009) 23–33.
[62] N. Job, B. Heinrichs, F. Ferauche, F. Noville, J. Marien, J.P. Pirard, Hydrodechlorination of 1,2-dichloromethane on Pd–Ag catalysts supported on tailored texture carbon xerogels, Catal Today 102–103 (2005) 234–241.
[63] F.J. Maldonado-Hodar, Removing aromatic and oxygenated VOCs from polluted air stream using Pt-carbon aerogels: assessment of their performance as adsorbent and combustion catalysts. J. Hazard. Mater. 194 (2011) 216–222.
[64] S.-I. Kim, T. Yamamoto, A. Endo, T. Ohmori, M. Nakaiwa, Adsorption of phenol and reactive dyes from aqueous solution on carbon cryogel microspheres with controlled porous structure, Microporous Mesoporous Mater. 96 (2006) 191–196.
[65] B.S. Girgis, A.A. Attia, N.A. Fathy, Potential of nano-carbon xerogels in the remediation of dye-contaminated water discharges, Desalination 265 (2011) 169–176.
[66] P. Kueseng, C. Thammakhet, P. Thavarungkul, P. Kanatharana, Mutiwalled carbon nanotubes/cryogel composite, a new sorbent for determination of trace polycyclic aromatic hydrocarbons, Microchem. J. 96 (2010) 317–323.
[67] M. Sanchez-Polo, J. Rivera-Utrilla, J. Méndez-Díaz, J. López-Peñaver, Metal-doped carbon aerogels. New materials for water treatments, Ind. Eng. Chem. Res. 47 (2008) 6001–6005.
[68] T. Yamamoto, S.-I. Kim, J. Chaichanawong, E. Apiluck, T. Ohmori, Removal of aqueous pollutants by adsorption-catalytic process using mesoporous carbon beads loaded with metal oxides, Appl. Catal. B Environ. 88 (2009) 455–461.
[69] S.A.C. Carabineiro, T. Thavorn-Amornsri, M.F.R. Pereira, J.L. Figueiredo Comparison between activated carbon, carbon xerogel and carbon nanotubes for the adsorption of the antibiotic ciprofloxacin. Catal. Today, in press.
[70] H.F. Gorgulho, F. Gonçalves, M.F.R. Pereira, J.L. Figueiredo, Synthesis and characterization of nitrogen-doped carbon xerogels. Proc Intern Conf Carbon '09. Biarritz, (2009), France, 14–19 June.
[71] S. Haji, C. Erkey, Removal of dibenzothiophene from model diesel by adsorption on carbon aerogels for fuel cell applications, Ind. Eng. Chem. Res. 42 (2003) 6933–6937.
[72] D. Jayne, Y. Zhang, S. Haji, C. Erkey, Dynamics of removal of organosulphur compounds from diesel by adsorption on carnbon aerogels for fuel cell applications, Int. J. Hydrogen Energy 30 (2005) 1287–1293.

[73] M. Sanchez-Polo, J. Rivera-Utrilla, E. Salhi, U. von Gunten, Ag-doped carbon aerogels for removing halide ions in water treatment, Water Res. 41 (2007) 1031–1037.

[74] M. Sanchez-Polo, J. Rivera-Utrilla, E. Salhi, U. von Gunten, Removal of bromide and iodide anions from drinking water by silver-activated carbon aerogels, J. Colloid Interface Sci. 300 (2006) 437–441.

[75] I. Villar, D.J. Suarez-De la Calle, Z. González, M. Granda, C. Blanco, R. Menéndez, R. Santamaría, Carbon materials as electrodes for electrosorption of NaCl in aqueous solutions, Adsorption 17 (2010) 467–471.

[76] J. Goel, K. Kadirvelu, C. Rajagopal, V.K. Garg, Investigation of adsorption of lead, mercury and nickel from aqueous solutions onto carbon aerogel, J. Chem. Technol. Biotechnol. 80 (2005) 469–476.

[77] J. Goel, K. Kadirvelu, C. Rajagopal, V.K. Garg, Removal of mercury(II) from aqueous solution by adsorption on carbon aerogel: response surface methodological approach, Carbon 43 (2005) 197–200.

[78] J. Goel, K. Kadirvelu, C. Rajagopal, V.K. Garg, Removal of lead(II) from aqueous solution by adsorption on carbon aerogel using a response surface methodological approach, Ind. Eng. Chem. Res. 44 (2005) 1987–1994.

[79] J. Goel, K. Kadirvelu, C. Rajagopal, V.K. Garg, Cadmium(II) uptake from aqueous solution by adsorption on carbon aerogel using a response surface methodological approach, Ind. Eng. Chem. Res. 45 (2006) 6531–6537.

[80] C.D. Saquing, T.T. Cheng, M. Aindow, C. Erkey, Preparation of platinum/carbon aerogel nanocomposites using a supercritical deposition method, J. Phys. Chem. B 108 (2004) 7716–7722.

[81] C.D. Saquing, D. Kang, M. Aindow, C. Erkey, Investigation of the supercritical deposition of platinum nanoparticles into carbon aerogels, Microporous Mesoporous Mater. 80 (2005) 11–23.

[82] Y. Zhang, D. Kang, M. Aindow, C. Erkey, Preparation and characterization of ruthenium/ carbon aerogel nanocomposites via a supercritical fluid route, J. Phys. Chem. B 109 (2005) 2617–2624.

[83] Y. Zhang, B. Cangul, Y. Garrabos, C. Erkey, Thermodynamics and kinetics of adsorption of bis(2,2,6,6-tetramethyl-3,5-heptanedionato) (1,5-cyclooctadiene) ruthenium (II) on carbon aerogel from supercritical CO_2 solution, J. Supercrit. Fluids 44 (2008) 71–77.

[84] E. Frackowiak, F. Béguin, Carbon materials for electrochemical storage of energy in capacitors, Carbon 39 (2001) 937–950.

[85] S. Sepehri, B.B. Garcia, Q. Zhang, G. Cao, Enhanced electrochemical and structural properties of carbon cryogels by surface chemistry alteration with boron and nitrogen, Carbon 47 (2009) 1436–1443.

[86] A. Szczurek, K. Jurewicz, G. Amaral-Labat, V. Fierro, A. Pizzi, A. Celzard, Structure and electrochemical capacitance of carbon cryogels derived from phenol–formaldehyde resins, Carbon 48 (2010) 3874–3883.

[87] S. Escribano, S. Berthon, J.L. Ginoux, P. Achard, Caracterization of carbon aerogels, in: Extended Abstracts, Eurocarbon'98, 1998, pp. 841–842.

[88] Y.Z. Wei, B. Fang, S. Iwasa, M. Kumagai, A novel electrode material for electric double-layer capacitors, J. Power Sources 141 (2005) 386–391.

[89] K.L. Yang, T.Y. Ying, S. Yiacoumi, C. Tsouris, E.S. Vittoratos, Electrosorption of ions from aqueous solutions by carbon aerogel: an electrical double-layer model, Langmuir 17 (2001) 1961–1969.

[90] T.Y. Ying, K.L. Yang, S. Yiacoumi, C. Tsouris, Electrosorption of ions from aqueous solutions by nanostructured carbon aerogel, J. Colloid Interface Sci. 250 (2002) 18–27.

[91] K. Jurewicz, C. Vix-Guterl, E. Frackowiak, S. Saadallah, M. Reda, J. Parmentier, J. Patarin, F. Béguin, Capacitance properties of ordered porous carbon materials prepared by a templating procedure, J. Phys. Chem. Solids 65 (2004) 287–293.
[92] P. Simon, Y. Gogotsi, Review: materials for electrochemical capacitors, Nat Mater. 7 (2008) 845–854.
[93] C. Vix-Guterl, E. Frackowiak, K. Jurewicz, M. Friebe, J. Parmentier, F. Béguin, Electrochemical energy storage in ordered porous carbon materials, Carbon 43 (2005) 1293–1302.
[94] B. Babic, B. Kaluderovic, L. Vracar, N. Krstajic, Characterization of carbon cryogel synthesized by sol–gel polycondensation and freeze-drying, Carbon 42 (2004) 2617–2624.
[95] R. Saliger, U. Fischer, C. Herta, J. Fricke, High surface area carbon aerogels for supercapacitors, J. Non Cryst. Solids 225 (1998) 81–85.
[96] D. Qu, H. Shi, Studies of activated carbons used in double-layers capacitors, J. Power Sources 74 (1998) 99–107.
[97] S.W. Hwang, S.H. Hyun, Capacitance control of carbon aerogels, J. Non Cryst. Solids 347 (2004) 238–245.
[98] G. Lv, D. Wu, R. Fu, Preparation and characterizations of MnO_2-dispersed carbon aerogel as supercapacitor electrode material, J. Non Cryst. Solids 355 (2009) 2461–2465.
[99] Y.J. Lee, J.C. Jung, S. Park, J.G. Seo, S.H. Baeck, J.R. Yoon, J. Yi, I.K. Song, Effect of preparation method on electrochemical property of Mn-doped carbon aerogel for supercapacitor, Current Appl. Phys. 11 (2011) 1–5.
[100] J.M. Miller, B. Dunn, Morphology and electrochemistry of ruthenium/carbon aerogel nanostructures, Langmuir 15 (1999) 799–806.
[101] J.M. Miller, B. Dunn, T.D. Tran, R.W. Pekala, Deposition of ruthenium nanoparticles on carbon aerogels for high energy density supercapacitor electrodes, J. Electrochem. Soc. 144 (1997) L309–L311.
[102] A. Théry, Relations entre l'organisation multi échelles des aérogels de carbone et leurs propriétés de stockage électrochimique de l'hydrogène. PhD thesis, Orléans University, France, 2004.
[103] K. Jurewicz, E. Frackowiak, F. Béguin, Towards the mechanism of electrochemical hydrogen storage in nanostructured carbon materials, Appl. Phys. A Mater. 78 (2004) 981–987.
[104] C. Arbizzani, S. Beninati, E. Manferrari, F. Soavi, M. Mastragostino, Cryo- and xerogel carbon supported PtRu for DMFC anodes, J. Power Sources 172 (2007) 578–586.
[105] H. Du, B. Li, F. Kang, R. Fu, Y. Zeng, Carbon aerogel supported Pt-Ru catalysts for using as the anode of direct methanol fuel cells, Carbon 45 (2007) 429–435.
[106] H.J. Kim, W.I. Kim, T.J. Park, H.S. Park, D.J. Suh, Highly dispersed platinum-carbon aerogel catalyst for polymer electrolyte membrane fuel cells, Carbon 46 (2008) 1393–1400.
[107] E. Guilminot, F. Fischer, M. Chatenet, A. Rigacci, S. Berthon-fabry, P. Achard, E. Chainet, Use of cellulose-based carbon aerogels as catalyst support for PEM fuel cell electrodes: electrochemical characterization, J. Power Sources 166 (2007) 104–111.
[108] A. Smirnova, X. Dong, H. Hara, A. Vasiliev, N. Sammes, Novel carbon aerogel-supported catalysts for PEM fuel cell application, Int. J. Hydrogen Energy 30 (2005) 149–158.
[109] G.X. Wang, J.H. Ahn, J. Yao, S. Bewlay, H.K. Liu, Nanostructured Si–C composite anodes for lithium-ion batteries, Electrochem. Commun. 6 (2004) 689–692.
[110] T. Hasegawa, S.R. Mukai, Y. Shirato, H. Tamon, Preparation of carbon gel microspheres containing silicon powder for lithium ion battery anodes, Carbon 42 (2004) 2573–2579.

Chapter 8

Adsorption by Phosphorus-Containing Carbons

Alexander M. Puziy
Institute for Sorption and Problems of Endoecology, National Academy of Sciences of Ukraine, Kyiv, Ukraine

Juan M.D. Tascón
Instituto Nacional del Carbón, INCAR-CSIC, Oviedo, Spain

Chapter Outline

8.1. Introduction 245
8.2. Methods of Phosphorus Introduction 246
8.3. Chemical State of Phosphorus Heteroatoms 246
8.4. Adsorption by Phosphorus-Containing Carbons 247
8.4.1. Adsorption of Water (Hydrophilicity) 247
8.4.2. Adsorption of Ammonia 248
8.4.3. Adsorption of Protons 250
8.4.4. Adsorption of Metal Ions 254
8.4.5. Adsorption of Organics 262
8.5. Conclusions 263
References 263

8.1. INTRODUCTION

Adsorption in general and adsorption by carbons in particular are present in many natural physical, biological, and chemical systems and are widely used in industrial applications [1]. Activated carbon is the most commonly employed general-purpose adsorbent which has been used to great advantage for almost four thousand years. Even now new applications for carbon adsorbents are being found and their potential is far from being exhausted. It is well known that the properties of carbon adsorbents are determined by both their porous structure and their surface chemistry. The surface chemistry of activated carbon is a result of the presence of heteroatoms such as oxygen, nitrogen, or phosphorus originating from the precursor or activation agent.

Novel Carbon Adsorbents. DOI: 10.1016/B978-0-08-097744-7.00008-9
Copyright © 2012 Elsevier Ltd. All rights reserved.

The most widely investigated heteroatoms are oxygen and nitrogen while studies of phosphorus-containing carbons are rare [2,3]. Only the oxidation inhibition of carbon materials by phosphorus compounds has been studied to any great depth due to its practical importance for the aerospace industry [4,5]. However, other properties of phosphorus-containing carbons related to the phosphorus compounds present in the structure have not yet received sufficient attention. The absence of systematic studies on phosphorus-containing carbons is surprising considering that widely used phosphoric acid activation leads to the incorporation of phosphorus in activated carbon. In this chapter, adsorption properties specific to phosphorus-containing carbons are considered.

8.2. METHODS OF PHOSPHORUS INTRODUCTION

Phosphorus-containing carbons may be obtained by the carbonization of a carbonaceous precursor physically mixed with a phosphorus compound [6–20] or covalently bonded with a phosphorus compound [21–26] or, alternatively, by modifying a carbon material with a phosphorus compound [27–33]. Another method of preparing phosphorus-containing carbons is through the co-deposition of carbon and phosphorus from vapor phase [34–39].

The carbonization of a carbonaceous precursor impregnated with phosphoric acid (phosphoric acid activation) is widely used for the production of highly porous carbon adsorbents [40–42]. This process not only enhances the porosity of carbons but also leads to the incorporation of phosphorus compounds into the carbon backbone – a fact that is often overlooked.

8.3. CHEMICAL STATE OF PHOSPHORUS HETEROATOMS

To understand the properties of phosphorus-containing carbons, the chemical structure of the phosphorus present in the carbon must be known. Several hypothetical structures have been proposed [4,13–16,22–24,28,30,32,33,43–45]. They show a tetra-coordinated phosphorus atom surrounded only by oxygen atoms (a phosphate-like structure) or by oxygen and carbon atoms (a phosphonate structure). The difference between the two models is in the way the phosphorus compound is connected to the graphene layers – via a C–O–P or a C–P linkage. The presence of elemental phosphorus has also been reported in carbons obtained at high temperatures (>1000 °C) [22]. In addition, the occurrence of phosphorus in a substitutional position in carbons obtained by CVD from phosphorus and carbon sources has been observed [34–39].

A systematic study using FTIR, XPS, and ^{31}P-NMR methods has increased our understanding of the chemical structure of phosphorus compounds in phosphorus-containing carbons obtained from the heat treatment of a phosphorylated polymer [21] or polymeric [6,7,46,47] and lignocellulosic [8,9,47] precursors impregnated with H_3PO_4. It has been shown that independently of the preparation method phosphorus exists in the form of condensed phosphates bound to the carbon matrix by C–O–P linkage. Carbons obtained at 500–700 °C may contain phosphonates (C–P linkage) while the reduction of pentavalent to zerovalent (elemental) phosphorus occurs at 1000 °C [47]. With the increase in carbonization temperature up to 800 °C, the phosphorus and oxygen content in carbon increases, but decreases at higher temperatures due to the destruction of the C–O–P linkage and the volatilization of phosphorus compounds accompanied by the reduction of phosphorus [6–9]. Along with phosphorus heteroatoms, P-doped carbons contain a significant amount of oxygen, implying the presence of not only phosphate-like but also P-free oxygen-containing surface groups. With the increase in heat treatment temperature up to 800 °C, the atomic ratio O/P in carbon decreases to a value of 4 which is characteristic of phosphates.

Thus, the most abundant chemical form of phosphorus in phosphorus-containing carbons is the phosphate-like structure. This is probably because the phosphorus atom has great difficulty in breaking all the P–O bonds present in the precursor and so cannot be substitutionally incorporated into the carbon backbone. Another reason may be the theoretically demonstrated ability of phosphorus atoms in a substitutional position to chemisorb oxygen, which leads to the formation of P–O bonds [48]. This explanation is supported by the formation of phosphate structures in phosphorus-containing carbon obtained from oxygen-free sources (C_{60} and P) [39].

The presence of condensed phosphates endows phosphorus-containing carbons with a number of properties. Unlike heteroatom-free carbons, phosphorus-containing carbons have a hydrophilic and acidic surface, and are able to adsorb cations from aqueous solutions and to enhance acid-catalyzed reactions [49–53]. These properties make them similar to oxidized carbons that have oxygen-containing surface groups, such as carboxylic, lactone, keto-enol, and phenolic groups [54].

8.4. ADSORPTION BY PHOSPHORUS-CONTAINING CARBONS

8.4.1. Adsorption of Water (Hydrophilicity)

Kuzin et al. showed that the PCl_3 treatment of carbon from novolac phenol–formaldehyde resin significantly increases the uptake of water vapor while the adsorption of benzene remains essentially unchanged [55]. The most pronounced increase was observed at low relative pressures – the number of primary adsorption centers rose from 0.11 to 4.71 mmol/g, which is comparable to 6.15 mmol/g for oxidized carbon. At a relative pressure 0.5, a phosphorus-containing carbon was observed to adsorb a 13-times higher amount of water than the parent carbon.

Siedlewski et al. studied the hydrophilicity of phosphorus-containing carbon from polyfurfuryl alcohol [31]. Thermogravimetric investigation showed that carbon modified with PCl_3 at 500–800 °C could adsorb five times

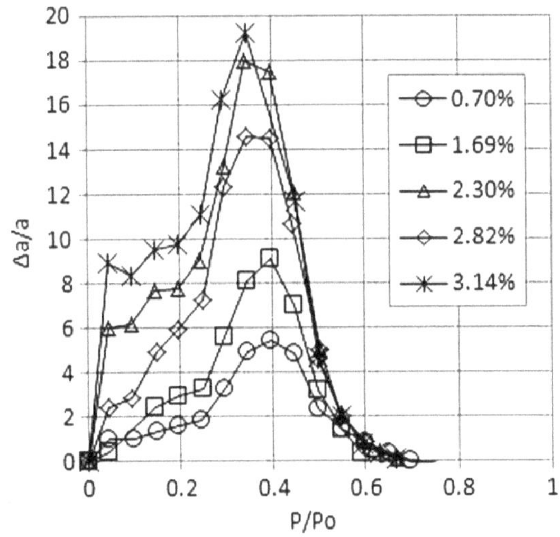

FIGURE 8.1 Difference between water uptake by PCl_3-modified and -non-modified carbon. Phosphorus content shown on the graph in percentages. *(Adapted from Ref. [56])*

more water (estimated from the mass loss in the temperature range of 20–200 °C accompanied by a highly endothermic effect) than nonmodified carbon. Further investigation [56] showed that the shape of water adsorption isotherms for PCl_3-modified carbons is the same as for oxidized carbon. The difference between water uptake by modified and nonmodified carbons revealed a pronounced increase in adsorption at low relative pressures with two maxima at 0.05 and 0.35–0.4 (Fig. 8.1). The magnitude of water adsorption increase correlated with the phosphorus content (Fig. 8.2).

McKee et al. reported an increase in the hydrophilicity of graphite after impregnation with organo-phosphorus esters followed by heat treatment [33]. The phosphates were more effective in promoting water adsorption than the phosphites, probably because three water molecules could be hydrogen-bonded to the phosphate group ($-OPO_3$) compared to only two to the phosphite group ($-OPO_2$) on the graphite surface.

8.4.2. Adsorption of Ammonia

Kuzin et al. [55] found that treatment of phenol–formaldehyde-based carbon with PCl_3 at 800 °C increases the adsorption of ammonia by 2–3 times (Fig. 8.3). The adsorption of ammonia on carbon containing 3.5% of phosphorus was almost the same as on oxidized carbon. Since the porous structure of carbons remained intact, the increase was ascribed exclusively to the enhanced acidic properties of the modified carbon.

Puziy et al. studied the acid properties of phosphorus-containing carbons by the quasi-equilibrium desorption of ammonia [49]. The total amount of surface groups obtained from ammonia desorption was very close to that determined by

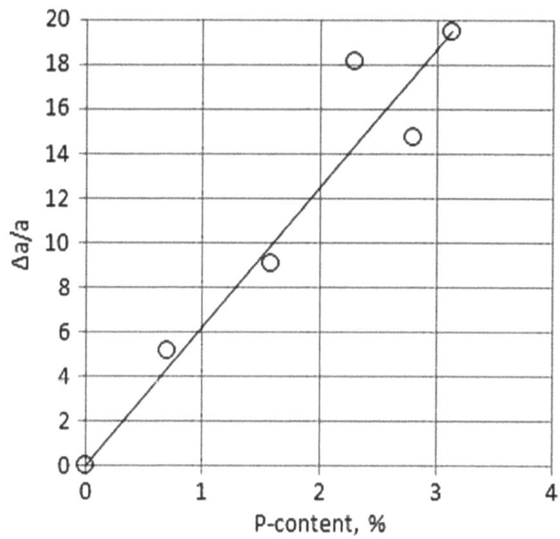

FIGURE 8.2 Increase in water adsorption at a relative pressure of 0.35–0.4 as a function of phosphorus content of PCl_3-modified carbons. *(Adapted from Ref. [56])*

potentiometric titration. Calculated adsorption energy distribution showed a decrease in the total amount of surface groups from 5.9 to 2.3 mmol/g with the increase in heat-treatment temperature of carbon to 600 °C. This decrease was accompanied by the formation of an increasing number of different types of surface groups from two to five through the formation of new groups with higher energies.

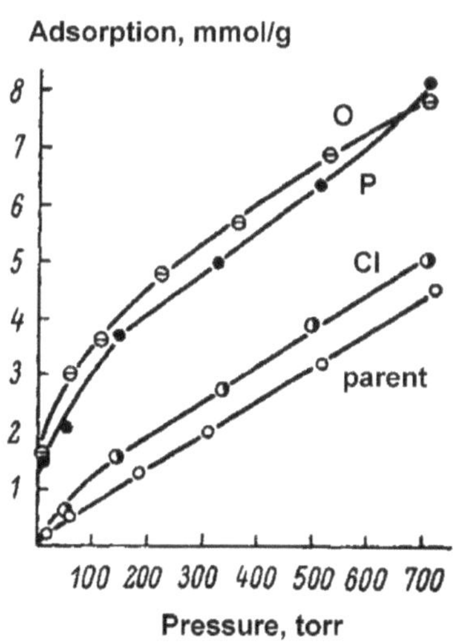

FIGURE 8.3 Ammonia adsorption isotherms at 20 °C on phenol–formaldehyde-based carbon as well as on PCl_3-treated, oxidized, and chlorinated carbons. *(Adapted from Ref. [55])*

8.4.3. Adsorption of Protons

Kuzin and Koemets [29] observed a linear relationship between phosphorus content and NaOH uptake for phenol–formaldehyde-based carbons treated with PCl_3 at 800 °C, at a ratio of one phosphorus atom per one NaOH molecule (Fig. 8.4). The authors used the same PCl_3 treatment and attained different phosphorus contents by varying the carbon burn-off. As a result, the structure of the phosphorus-containing surface groups in all three carbons remained most probably unchanged.

Lubliner et al. [30] reported an increase in NaOH uptake followed by a leveling-off with increasing phosphorus content in cellulose-derived carbon fibers treated with PCl_3 (Fig. 8.5). After comparing the NaOH uptakes with the results of the chemical analysis of P, the authors found that in the early stages of phosphorylation carbon fibers contained less phosphorus than one might be led to expect from the NaOH uptake for a monoprotic derivative of

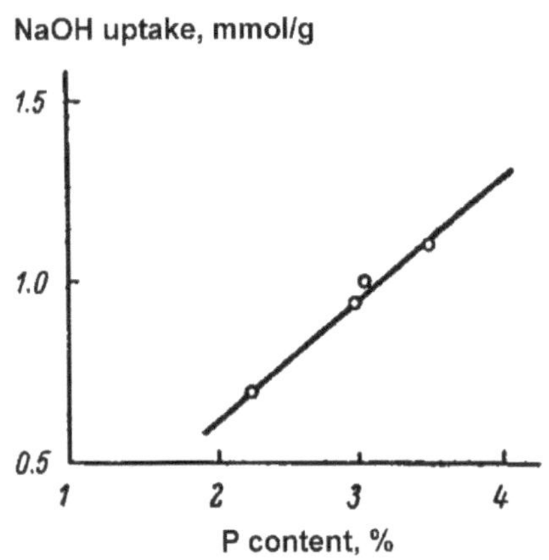

FIGURE 8.4 Relationship between phosphorus content and NaOH uptake for phenol–formaldehyde-based carbons with different burn-offs treated with PCl_3 at 800 °C. *(Adapted from Ref. [29])*

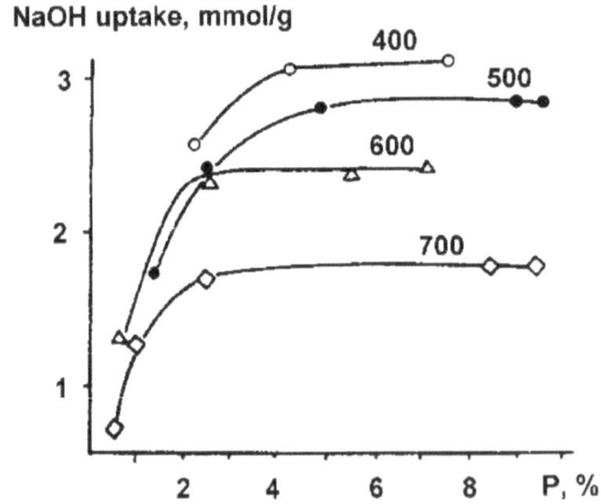

FIGURE 8.5 Relationship between phosphorus content and NaOH uptake for PCl_3-treated cellulose-based carbon fibers obtained at temperatures shown in the graph. *(Adapted from Ref. [30])*

phosphinic acid R–P(O)H(OH). They attributed this to the additional presence of carboxylic groups. For carbon fibers with a high phosphorus content, NaOH uptake was less than expected due to the partial condensation of phosphinic acid derivatives.

Siedlewski et al. [31] observed a linear increase in NaOH uptake with increasing phosphorus content for polyfurfuryl alcohol-derived carbons modified with PCl_3 (Fig. 8.6). The increase in the NaOH uptake was accompanied by a decrease (to the point of disappearance) in HCl uptake. The authors pointed out that the cation-exchange capacity of carbon containing 3.14% phosphorus is of the same order as for carbons oxidized with nitric acid, and suggested that phosphorus-containing carbons might be used as cation exchangers.

Puziy et al. [6–9] observed that with the increase in carbonization temperature NaOH uptake rose to 2.2–2.5 mmol/g at 700–800 °C but decreased at higher temperatures (Fig. 8.7). The cation-exchange capacity of phosphorus-containing carbons was scarcely affected by the impregnation ratio in the range of 0.6–0.9. However, the tendency was for it to decrease in the impregnation ratio range of 1.0–1.1.

Phosphorus-containing carbon obtained from a nitrogen-containing copolymer (polyimide) had a cation-exchange capacity of up to 3.1–3.7 mmol/g [57], which was slightly higher than that of carbons obtained from nitrogen-free precursors [6–9]. The reason is due to the enhanced interaction of phosphoric acid with the basic nitrogen-containing groups of the polymer precursor. This conclusion is supported by the increasing cation-exchange capacity of carbon with the increase in the nitrogen-containing component in the copolymer [10].

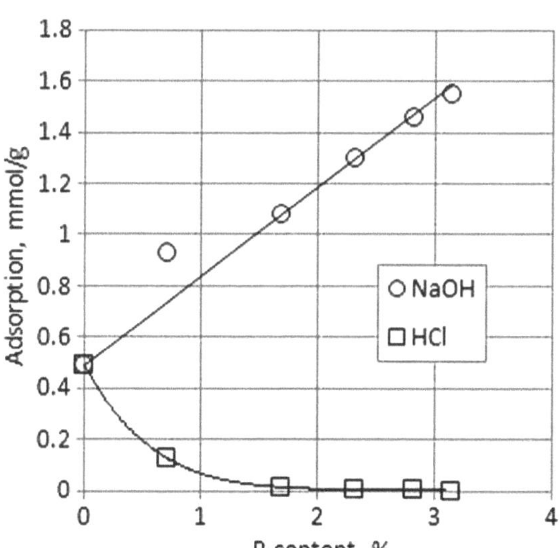

FIGURE 8.6 Dependence of NaOH and HCl adsorption on the phosphorus content of PCl_3-modified carbons from polyfurfuryl alcohol. *(Adapted from Ref. [31])*

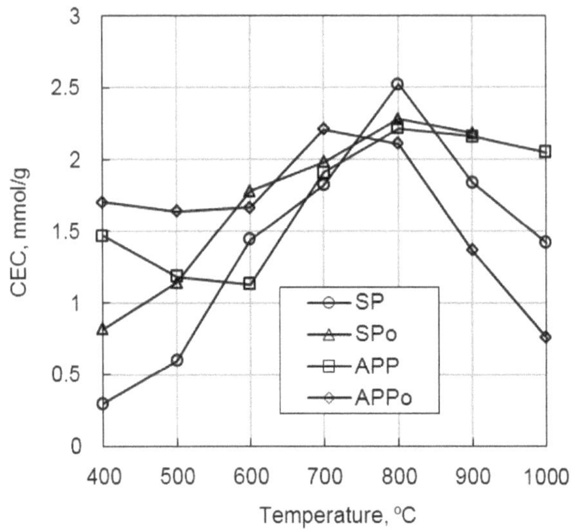

FIGURE 8.7 Temperature dependence of the cation-exchange capacity of phosphorus-containing carbons from styrene–divinylbenzene (SP, SPo) and from fruit stones (APP, APPo) obtained in an argon or air atmosphere (index "o"). *(Data compiled from Refs. [6–9])*

The cation-exchange capacity of phosphorus-containing carbons was shown to be chemically stable in both acid and basic media. The boiling of phosphorus-containing carbons in 0.5 M H_2SO_4 or 1 M NaOH solutions practically did not alter their cation-exchange capacity [6–9]. The treatment of phosphorus-containing carbon fibers with 2 M HCl or HNO_3 for one day did not cause any change in cation-exchange capacity either [30].

Guo and Rockstraw reported that with the increase in the impregnation ratio (from 0.5 to 2.5) or final activation temperature (from 170 to 500 °C), the phosphorus contents increased in carbons from cellulose and Kraft lignin [58]. Whereas the concentration of surface groups reached a maximum at 250 °C, it decreased at higher temperatures up to 400 °C and leveled off at 400–500 °C. The authors concluded that the acidic surface groups must consist of both the temperature-sensitive and the temperature-insensitive types. Temperature-sensitive groups are unstable at higher temperature, decompose at 250 °C, and disappear almost completely at 400 °C. The same dependence of acid surface group concentration on temperature was observed for carbons from pecan shell [59].

Using potentiometric titration, Bandosz et al. [60] found that for phosphorus-containing carbons from wood the surface heterogeneity and the total number of surface groups increased with increasing activation temperature from 300 to 500 °C. On the basis of the values of the dissociation constant, the surface groups were assigned to carboxylic acids (p*K* 3–8), phenols (p*K* 8–10), phosphonic acids ($–PO_3H_2$; p*K* 7–9), phosphonous acids ($–PO_2H_2$; p*K* 3–4.5), and phosphines ($-PR_3$, pK 3–6.5; 8–9). It was demonstrated that phosphorus-containing chemical groups play a dominant role in the acidity of the carbon obtained at 500 °C.

Puziy et al. used potentiometric titration to study the acid–base properties of phosphorus-containing carbons [6–11,21,27,61–64]. Figure 8.8 shows proton-binding isotherms for phosphorus-containing carbon from the polymer precursor SP800, additionally oxidized phosphorus-containing carbon SP800-Ox, and oxidized carbon from the same precursor SCS-Ox. Although all carbons have nearly the same point of zero charge (about 1.5), phosphorus-containing carbon SP800 shows a higher cation-exchange capacity at pH 3–10 than oxidized carbon SCS-Ox. The additional oxidation of phosphorus-containing carbon significantly increases its cation-exchange capacity (see Fig. 8.8, carbon PS800 Ox).

The calculated proton affinity distributions reflect four types of acid surface groups for phosphorus-containing carbon SP800 (Fig. 8.9, Table 8.1). On the basis of the similarity of the dissociation constants of the surface groups and functional groups in simple compounds, the most acidic surface group with pK 2.3 can be ascribed to strong acidic carboxylic or phosphate groups. For the purpose of reference, the pK_1 of orthophoshoric acid is 2.148 and carboxylic groups of benzenehexacarboxylic (mellitic) acid have a pK from 1.4 to 7.0 [65]. The surface group with pK 4.7 can be ascribed to weak carboxylic acid. For reference purposes, benzoic acid has a pK of 4.204. The surface group with a pK of 6.8 could be ascribed either to the second dissociation constant of the

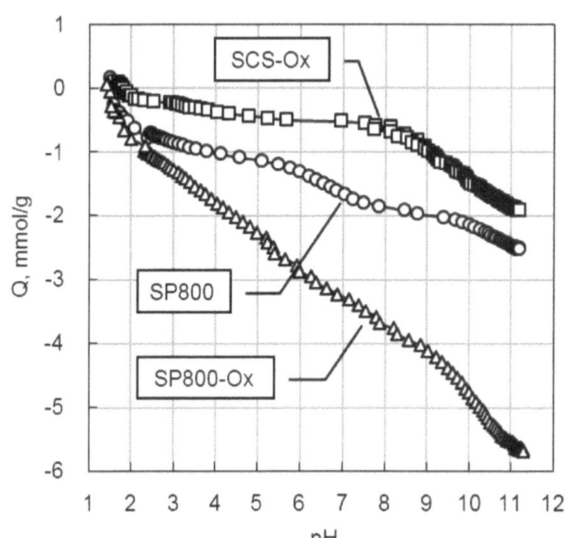

FIGURE 8.8 Proton-binding isotherms for phosphorus-containing carbon from the polymer precursor SP800, the additionally oxidized phosphorus-containing carbon SP800-Ox, and oxidized carbon from the same precursor SCS-Ox. *(Adapted from Refs. [63,64])*

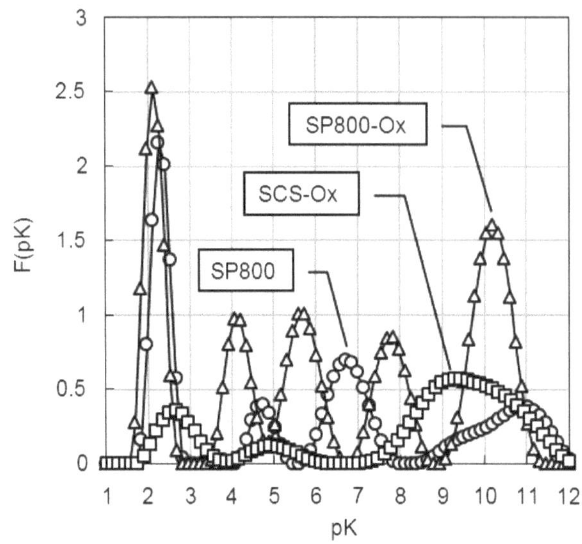

FIGURE 8.9 Proton affinity distributions on phosphorus-containing carbon from the polymer precursor SP800, the additionally oxidized phosphorus-containing carbon SP800-Ox, and the oxidized carbon from the same precursor SCS-Ox. *(Adapted from Refs. [63,64])*

phosphate groups (pK_2 7.199) or to hydroxyl groups formed by keto-enol tautomerism [66,67]:

For reference, diacetylacetone has a pK of 7.42. Surface groups with a pK of 10.5 can be ascribed to phenolic groups. Upon oxidation of the phosphorus-containing carbon SP800, the total amount of surface groups increased. In addition, there is a new type of surface group with a pK of 7.8 that can be ascribed to enol or lactone groups.

Phosphorus-containing nanostructured carbon with a high content of acidic surface groups (4.1–4.7 mmol/g) was recently obtained by the template method using sucrose and silica gel [11]. About half of the total amount of surface groups was strongly acidic and has been ascribed to phosphate groups (Fig. 8.10).

8.4.4. Adsorption of Metal Ions

8.4.4.1. Ion Exchange

The presence of acidic surface groups confers on phosphorus-containing carbons the ability to exchange surface protons with ions from aqueous solutions. This property of phosphorus-containing carbons was patented long ago [68], although no explanation of the underlying phenomenon was given.

TABLE 8.1 Acid-base characteristics of phosphorus-containing carbon from the polymer precursor SP800, the additionally oxidized phosphorus-containing carbon SP800-Ox and the oxidized carbon from the same precursor SCS-Ox. Adapted from Refs. [63,64]

			Strong acid carboxylic or phosphate		Weak acid carboxylic		Phosphate or enol		Enol or lactone		Phenolic	
Carbon	PZC	Q_{tot}	Q_1	pK_1	Q_2	pK_2	Q_3	pK_3	Q_4	pK_4	Q_5	pK_5
SP800	1.55	3.05	1.22	2.30	0.28	4.74	0.80	6.75	-	-	0.75	10.52
SP800-Ox	1.47	5.86	1.46	2.14	0.75	4.17	0.97	5.69	0 83	7.77	1.85	10.18
SCS-Ox	1.82	2.13	0.38	2.71	0.15	4.98	-	-	-	-	1.60	9.68

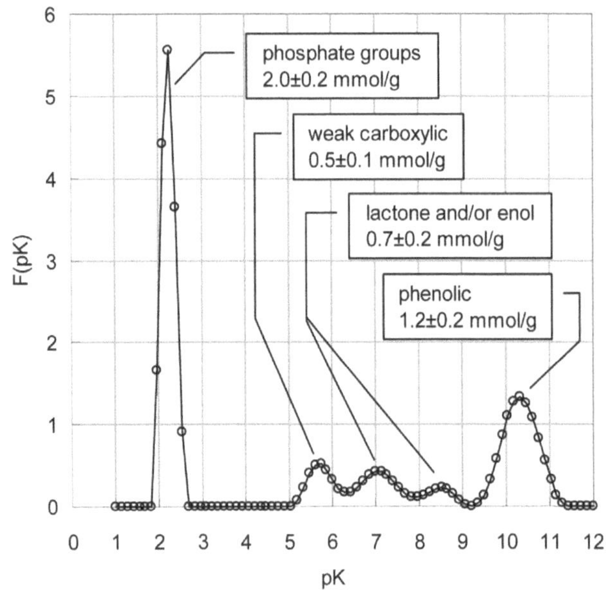

FIGURE 8.10 Proton affinity distributions for nanostructured functionalized carbon. *(Reprinted from Ref. [11] with permission from Elsevier)*

Amelin and Kariakin [69] investigated the cation exchange of alkali and alkali earth metal ions in the wood-based commercial carbon BAU treated with H_3PO_4 at 500 °C. The carbon modified in this way had an exchange capacity 2.1 mmol/g. The affinity sequence for metal ions was $Ba^{2+} > Ca^{2+} > Mg^{2+} > Li^{+} > K^{+} > Na^{+}$.

Toles et al. [70] studied copper adsorption on carbons obtained from almond and pecan shells prepared by chemical activation with H_3PO_4 and physical activation with CO_2. The copper uptake was similar or higher in the case of CO_2-activated carbons from both precursors despite the much lower BET surface area than that of their acid-activated counterparts. Oxidation in air at 300 °C significantly increased the adsorption of copper on acid-activated carbons while the increase in Cu uptake in the case of CO_2-activated oxidized carbons was only moderate. Copper uptake by phosphoric acid-activated carbons is proportional to the number of groups that can be titrated with bases such as sodium ethoxide, sodium hydroxide, sodium carbonate, and sodium bicarbonate [71,72]. A linear relationship between the concentration of surface groups and copper uptake suggested that most of the copper adsorption was due to surface sites containing acidic hydrogen (e.g. carboxylic, phenolic, and other alcohols).

Strelko Jr. et al. reported a higher affinity of phosphorus-containing carbon (obtained from phosphorylated phenol–formaldehyde resin) to lead than to copper [23]. They attributed the ability to adsorb heavy metal ions to surface acidic groups containing oxygen and phosphorus. Breakthrough experiments showed that the affinity of phosphorus-containing carbon to metal ions was Pb > Cu > Cd ≥ Ni (Fig. 8.11). Because the solubility of metal phosphates follows the same sequence, the authors [23] ascribed the adsorption to the formation of insoluble surface metal-phosphate species.

Other investigators also reported an ion-exchange mechanism for the adsorption of mercury [73], lead [74], copper [75], or chromium (VI) [76]. The adsorption of metal ions may include the formation of surface complexes with the participation of the functional groups on the surface of carbon. An oxidizing atmosphere during carbonization promotes the incorporation of phosphorus into the resulting carbon and stimulates the formation of acid surface groups [75].

Puziy et al. studied heavy metal ion adsorption on phosphorus-containing carbons [6–9,63,77]. They showed that phosphorus-containing carbons have a higher adsorption capacity than oxidized carbon prepared from the same precursor (Fig. 8.12). This enhanced capacity is especially pronounced in acid

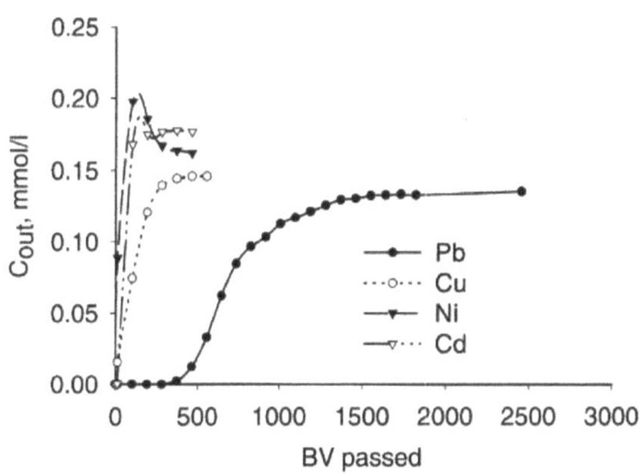

FIGURE 8.11 Breakthrough of heavy metal ions on phosphorus-containing carbon. Feed: Pb^{2+} 0.143mmol/l, Cu^{2+} 0.158 mmol/l, Ni^{2+} 0.163 mmol/l, Cd^{2+} 0.148 mmol/g. Flow rate 15 BV/h, pH 4.8. *(Reprinted from Ref. [23] with permission from Elsevier)*

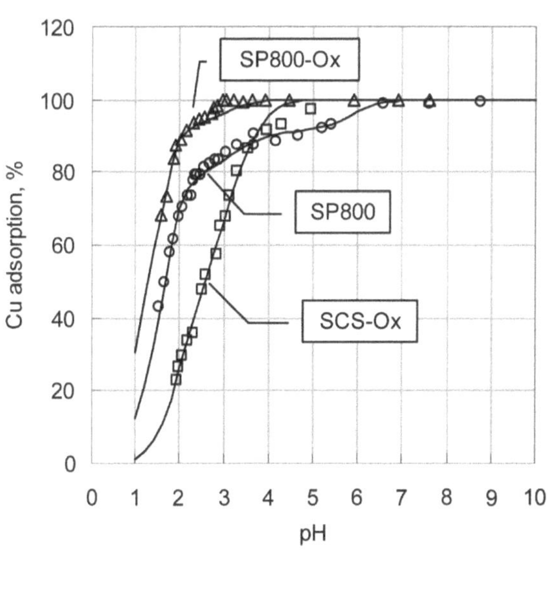

FIGURE 8.12 Adsorption of Cu(II) by the phosphorus-containing carbon SP800, the additionally oxidized phosphorus-containing carbon SP800-Ox, and the oxidized carbon SCS-Ox from a 0.001 M solution. *(Adapted from Refs. [63,64])*

media. Significant progress in understanding the mechanism of metal ion binding by phosphorus-containing carbons was made by using a two-stage approach [6–9,63,77,78]. The method is based on two sets of experimental data (potentiometric titration and metal ion adsorption) and is characterized by a stage-by-stage calculation of the parameters that characterize the surface groups (concentration and dissociation constants) followed by an evaluation of their interaction with metal ions (constants of surface complex formation). This method revealed the composition of the adsorbed Cu species and the crucial role of phosphate groups in metal ion binding (Fig. 8.13).

Column adsorption revealed the superior adsorption performance of phosphorus-containing carbon over oxidized carbon and even over ion exchangers (Fig. 8.14). The breakthrough volume for phosphorus-containing carbon was 4 times higher than that of oxidized carbon and 1.4–2.2 times higher than that of ion exchangers, whereas the saturated adsorption capacity was 4.1 times higher than for oxidized carbon and 1.3–2.3 times higher than for the ion exchangers.

8.4.4.2. Reduction–Oxidation Adsorption

Siedlewski et al. [31] reported a pronounced decrease in the ability of PCl_3-modified carbons to reduce Fe^{3+} (estimated as Fe^{3+} uptake from 1 M HCl) with increasing phosphorus content in these carbons (Fig. 8.15).

Chen and Zeng reported [79] that activated carbon fibers obtained by the phosphoric acid activation of sisal fiber at 850 °C showed an enhanced reduction–adsorption capacity to $Ag(NH_3)_2^+$ compared to carbon fibers obtained by other methods ($ZnCl_2$ activation, steam activation followed by oxidation with H_2O_2, $KMnO_4$, or HNO_3). Although the authors did not distinguish between adsorption and reduction, the enhanced amount of silver

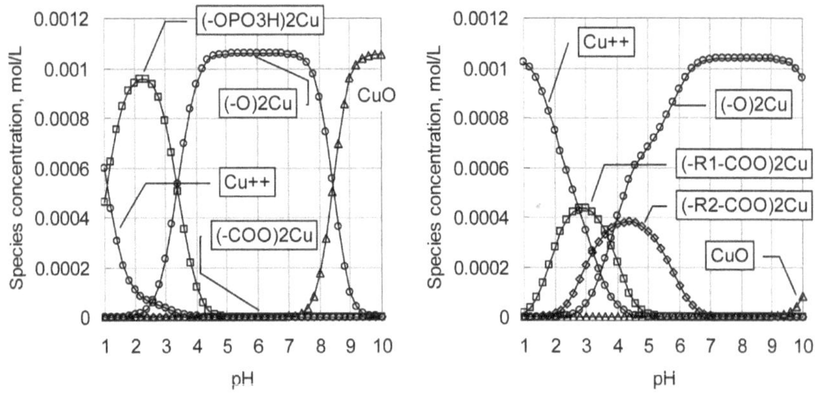

FIGURE 8.13 Cu species distribution on the phosphorus-containing carbon SP800-Ox (a) and the oxidized carbon SCS-Ox (b). *(Adapted from Refs. [63,64])*

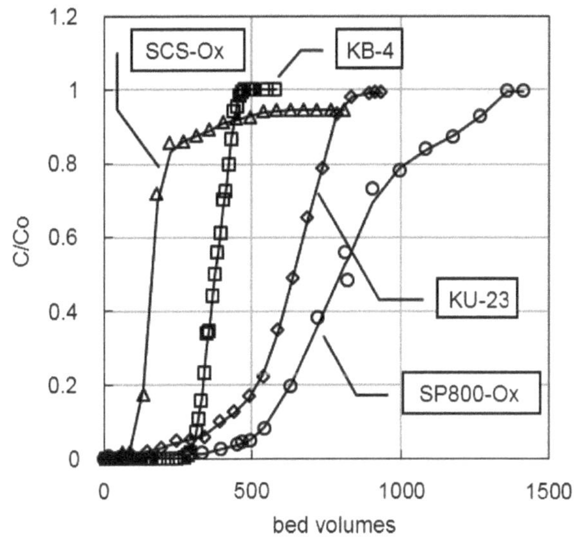

FIGURE 8.14 The breakthrough adsorption of copper on phosphorus-containing carbon SP800-Ox, the oxidized carbon SCS-Ox, the weak carboxylic acid (KB-4), and strong sulfonic acid (KU-23) ion exchangers. *(Adapted from Ref. [64])*

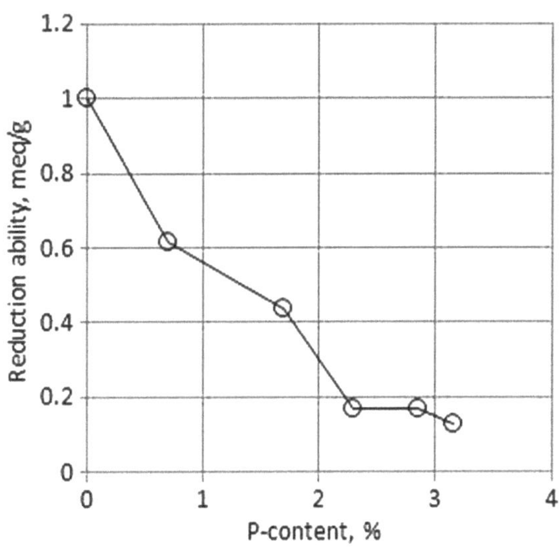

FIGURE 8.15 Dependence of reduction ability for Fe^{3+} on phosphorus content of PCl_3-modified carbons from polyfurfuryl alcohol. *(Adapted from Ref. [31])*

particles on the surface of the H_3PO_4 activated carbon fiber was evidenced by electron microscopy. The authors suggested that H_3PO_4 activated carbon fibers could be used for the recovery of noble metals as well as for the preparation of catalysts.

Siedlewski et al. [31] observed an increase in the oxidative properties for potassium iodide in N,N-dimethylformamide with the increasing phosphorus content of PCl_3-modified carbons from polyfurfuryl alcohol (Fig. 8.16). The oxidation ability of P-containing carbons increased with the increase in phosphorus content to 2.5%. Any further increase in P-content caused a small decrease in oxidation ability.

González-Serrano et al. [80] investigated the adsorption of chromium (VI) on lignin-based phosphoric acid-activated carbon. The adsorption of Cr(VI) increased with both the activation temperature and the impregnation ratio. The differences in adsorption capacities cannot be explained solely by the micropore volumes. Adsorption at pH 3 was about two times higher than at pH 7. The authors assumed that at pH 3 the reduction of Cr(VI) to Cr(III) could be taking place on the carbon surface, since reduction is thermodynamically favored by the hydroquinone-type surface groups.

8.4.4.3. Electroadsorption of Ions

Schönfelder et al. [12] reported an almost linear increase in the reversible lithium intercalation capacity of carbon from epoxy resin with the rise in the phosphorus-doping level (Table 8.2). The introduction of phosphorus into the carbon increased the global charge/discharge coulombic efficiencies to around 80% during the first cycle (Table 8.2). With further cycles, efficiencies stabilized at around 98%. However, the total capacities decreased. A doping level with 5–10% of phosphorus was the most effective in terms of improving capacity and efficiency. Lower dopant levels reduced these characteristics, while higher dopant levels (13%) favored a saturation point where it became difficult to accommodate more phosphorus into the carbon structure.

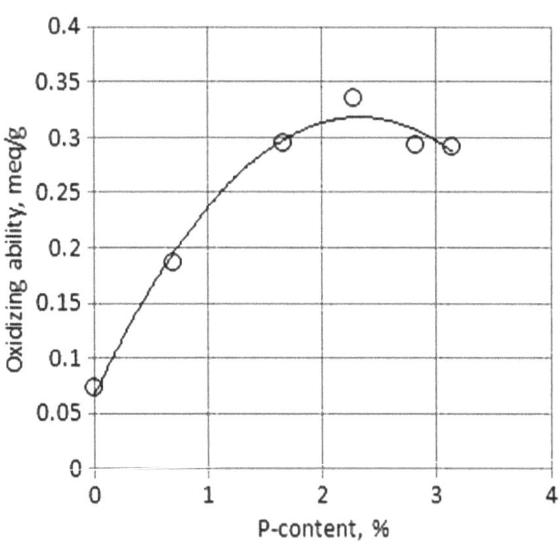

FIGURE 8.16 Dependence of oxidative ability to KI on phosphorus content of PCl_3-modified carbons from polyfurfuryl alcohol. *(Adapted from Ref. [31])*

TABLE 8.2 Phosphorus Content in Carbon from Epoxy Resin and Reversible Capacity Together with Coulombic Efficiency

Heat Treatment Temperature, °C	P-content of Carbon, wt.%	Reversible Capacity, mAh/g	First Cycle Coulombic Efficiency, %
900	0	368	68
900	2.84	415	68
900	5.28	532	77
900	9.89	550	80
900	13.9	540	70
1000	0	380	70
1000	2.9	424	66
1000	5.17	467	83
1000	9.61	515	80
1000	13.3	530	78
1100	5.2	419	80
1100	9.47	454	81
1100	10.26	464	78
1200	4.99	361	83
1200	9.38	432	81
1200	8.97	417	75

Reprinted from Ref. [12] with permission from Elsevier

Wu et al. [13] reported an increase in the reversible lithium capacity of carbon obtained by carbonizing a mixture of PAN and H_3PO_4 while increasing the amount of added acid. The improvement in charging behavior with the addition of H_3PO_4 was attributed to an increase in the cyclization process, the regularity of the carbon structure, the interlayer distance d_{002}, and possibly also to an increase in the number of micropores.

Wang et al. [14] showed the important role of phosphorus heteroatoms for the electrosorption of Li ions on P-doped disordered carbon. They showed an almost linear increase in the capacity of a Li-ion rechargeable battery with increasing phosphorus content in phosphorus-doped disordered carbon (Fig. 8.17). The increase in capacity was attributed to additional Li species in the proximity of phosphorus species.

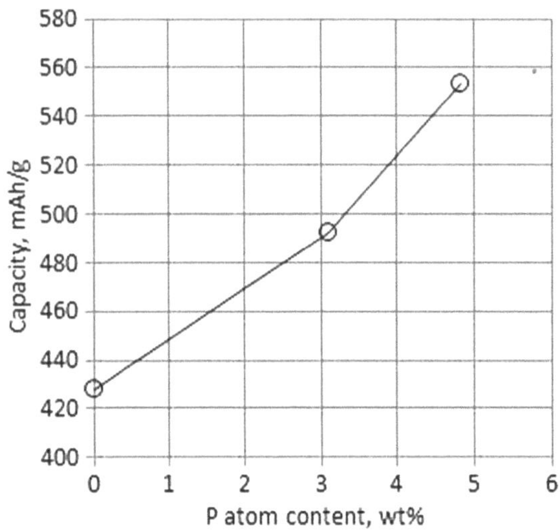

FIGURE 8.17 The relationship between the P atom content and the capacity of carbon. *(Adapted from Ref. [14])*

Xiang et al. [24] observed a decrease in reversible electrochemical lithium insertion capacity upon the P-doping of carbon at 500 °C. On the other hand, the addition of phosphoric acid to the precursor (phenolic resin) increased the reversible capacity of carbon obtained at 900 °C. A cyclic discharge/charge analysis of the following 20 cycles showed that the reversible capacity of carbons prepared at 900 °C deteriorated only slightly and the Faradic efficiency was close to 100% from the second cycle.

Hulicova-Jurcakova et al. [81] showed the superior performance of phosphorus-containing carbons used as electrodes in a supercapacitor with aqueous (1 M H_2SO_4) electrolyte. Phosphorus-containing carbons displayed a higher capacity (galvanostatic measurements) up to 192–212 F/g and a higher chemical stability than phosphorus-free carbon. The chemical stability of phosphorus-containing carbon enabled (i) the potential window to be extended from 1.3 to 1.5 V and (ii) a stable performance to be attained over 15000 charge–discharge cycles at a loading rate of 5 A/g without any significant loss of capacity. The combined effect of a high capacity and an extended potential window led to a higher energy density, which reached 16.3 W h/kg at a power density of 33 W/kg — 3 times higher than that of a Norit carbon. Statistical analysis showed that the key factors determining the capacity of carbon are the phosphorus content and surface area of pores in the size range of 0.65–0.83 nm as measured by carbon dioxide adsorption.

8.4.5. Adsorption of Organics

Altenor et al. [82] investigated phenol and methylene blue (MB) adsorption on phosphorus-containing carbons from vetiver roots. It was found that the mechanism of organics adsorption is complex, as it involves not only the porous structure but also the surface chemistry. The presence of micropores favors phenol adsorption, while developed mesoporosity favors MB adsorption. Basic groups favor phenol adsorption, whereas acidic groups such as carboxylic groups favor MB adsorption due to electrostatic interactions between MB molecules and deprotonated surface groups.

Seredych et al. [83] investigated the adsorptive removal of dibenzothiophene (DBT) and 4,6-dimethyldibenzothiophene (DMDBT) from model diesel fuel on polymer-derived carbons obtained by the carbonization of poly(4-styrenesulfonic acid-co-maleic acid) sodium salt followed by heat treatment in the presence of phosphoric acid. The results indicated that the presence of phosphorus, especially in the form of pyrophosphates and P_2O_5, increases the capacity and selectivity of carbons for the removal of dibenzothiophenes. It also affects the adsorption mechanism. Phosphorus suppresses the oxidation reactions of DBT and DMDBT. Owing to the possible presence of bulky phosphorus groups in pores of 1–3 nm thiophenic molecules are strongly adsorbed there via dispersive forces. An acidic environment also enhances adsorption via acid–base interactions. The physical adsorption mechanism and the stability of the surface make these carbons attractive candidates for thermal regeneration.

8.5. CONCLUSIONS

A specific feature of phosphorus-containing carbons is the presence of polyphosphates, which along with their porous structure determines their adsorption properties. Polyphosphates impart hydrophilic and acidic properties to carbon adsorbents. Thus, phosphorus-containing carbons are promising candidates as adsorbents for basic substances from both gas and liquid phases. They have a great potential for water treatment and engine fuel conditioning. Their enhanced electrochemical properties make phosphorus-containing carbons promising electrode materials for lithium batteries and electrochemical double-layer capacitors.

ACKNOWLEDGMENTS

Financial support from the Spanish CSIC is gratefully acknowledged.

REFERENCES

[1] E.J. Bottani, J.M.D. Tascón, (Eds.), Adsorption by Carbons, Elsevier Science, 2008.
[2] F.G.R. Gimblett, J.J. Freeman, K.S.W. Sing, Element-containing carbon fibres: recent USSR/Eastern European research in fibre technology, J. Mater. Sci. 24 (1989) 3799–3812.
[3] T.J. Bandosz, C.O. Ania, Surface chemistry of activated carbons and its characterization, in: T.J. Bandosz (Ed.), Activated Carbon Surfaces in Environmental Remediation, Academic Press, 2006, pp. 159–229.
[4] D.W. McKee, Oxidation protection of carbon materials, in: P.A. Thrower (Ed.), Chemistry and Physics of Carbon, vol. 23, Marcel Dekker, Inc, 1991, pp. 173–232.

[5] D.W. McKee, Inhibition of carbon gasification, in: J. Lahaye, P. Ehrburger (Eds.), Fundamental Issues in Control of Carbon Gasification Reactivity, Kluwer Academic Publishers, 1991, pp. 483–514.
[6] A.M. Puziy, O.I. Poddubnaya, A. Martínez-Alonso, F. Suárez-García, J.M.D. Tascón, Synthetic carbons activated with phosphoric acid. I. Surface chemistry and ion binding properties, Carbon 40 (2002) 1493–1505.
[7] A.M. Puziy, O.I. Poddubnaya, A. Martínez-Alonso, F. Suárez-García, J.M.D. Tascón, Synthetic carbons activated with phosphoric acid. III. Carbons prepared in air, Carbon 41 (2003) 1181–1191.
[8] A.M. Puziy, O.I. Poddubnaya, A. Martínez-Alonso, F. Suárez-García, J.M.D. Tascón, Surface chemistry of phosphorus-containing carbons of lignocellulosic origin, Carbon 43 (2005) 2857–2868.
[9] A.M. Puziy, O.I. Poddubnaya, A. Martínez-Alonso, A. Castro-Muñiz, F. Suárez-García, J.M.D. Tascón, Oxygen and phosphorus enriched carbons from lignocellulosic material, Carbon 45 (2007) 1941–1950.
[10] A.M. Puziy, O.I. Poddubnaya, B. Gawdzik, M. Sobiesiak, M.M. Tsyba, Phosphoric acid activation-Functionalization and porosity modification, Appl. Surf. Sci. 253 (2007) 5736–5740.
[11] A.M. Puziy, O.I. Poddubnaya, C.A. Reinish, M.M. Tsyba, L.I. Mikhalovska, S.V. Mikhalovsky, One-pot preparation of functionalized nanostructured carbons, Carbon 49 (2011) 599–604.
[12] H.H. Schönfelder, K. Kitoh, H. Nemoto, Nanostructure criteria for lithium intercalation in non-doped and phosphorus-doped hard carbons, J. Power Sources 68 (1997) 258–262.
[13] Y. Wu, S. Fang, Y. Jiang, Carbon anodes for a lithium secondary battery based on polyacrylonitrile, J. Power Sources 75 (1998) 201–206.
[14] S. Wang, H. Matsui, Y. Matsumura, ^{7}Li-NMR study of lithium charged phosphorus-doped disordered carbon, Synth. Met. 103 (1999) 2521–2522.
[15] R. Fu, L. Liu, W. Huang, P. Sun, Studies on the structure of activated carbon fibers activated by phosphoric acid, J. Appl. Polym. Sci. 87 (2003) 2253–2261.
[16] M. Wiśniewski, A. Pacholczyk, A.P. Terzyk, G. Rychlicki, New phosphorus-containing spherical carbon adsorbents as promising materials in drug adsorption and release, J. Colloid Interface Sci. 354 (2011) 891–894.
[17] F. Suárez-García, S. Villar-Rodil, C.G. Blanco, A. Martínez-Alonso, J.M.D. Tascón, Effect of phosphoric acid on chemical transformations during Nomex pyrolysis, Chem. Mater. 16 (2004) 2639–2647.
[18] F. Suárez-García, A. Martínez-Alonso, J.M.D. Tascón, Nomex polyaramid as a precursor for activated carbon fibres by phosphoric acid activation. Temperature and time effects, Microporous Mesoporous Mater. 75 (2004) 73–80.
[19] F. Suárez-García, A. Martínez-Alonso, J.M.D. Tascón, Activated carbon fibers from Nomex by chemical activation with phosphoric acid, Carbon 42 (2004) 1419–1426.
[20] A. Castro-Muñiz, F. Suárez-García, A. Martínez-Alonso, J.M.D. Tascón, Activated carbon fibers with a high content of surface functional groups by phosphoric acid activation of PPTA, J. Colloid Interface Sci. 361 (2011) 307–315.
[21] A.M. Puziy, O.I. Poddubnaya, The properties of synthetic carbon derived from nitrogen- and phosphorus-containing polymer, Carbon 36 (1998) 45–50.
[22] R. Imamura, K. Matsui, S. Takeda, J. Ozaki, A. Oya, A new role for phosphorus in graphitization of phenolic resin, Carbon 37 (1999) 261–267.
[23] V. Strelko Jr., M. Streat, O.P. Kozynchenko, Preparation, characterisation and sorptive properties of polymer based phosphorus-containing carbon, React. Funct. Polym. 41 (1999) 245–253.

[24] H.-Q. Xiang, S.-B. Fang, Y.-Y. Jiang, Lithium insertion in carbons prepared from phosphorus-containing polymers, J. Power Sources 94 (2001) 85–91.
[25] I.S. Skorynina, S.S. Gusev, N.K. Vorobyova, I.N. Ermolenko, Study of pyrolysis of cellulose modified with condensed phosphates, Vestsi Akad. Navuk Belarusk. SSR 3 (1970) 29–34.
[26] I.N. Ermolenko, I.I. Vygovskiy, I.P. Lubliner, Investigation of structure and properties of carbon fibers containing phosphorus and metal, Vestsi Akad. Navuk Belarusk. SSR Ser. Khim. Navuk 4 (1974) 78–81.
[27] A.M. Puziy, O.I. Poddubnaya, B. Gawdzik, M. Sobiesiak, M.M. Tsyba, Functionalization of carbon and silica gel by phosphoric acid, Adsorpt. Sci. Technol. 25 (2007) 531–542.
[28] S. Labruquère, R. Pailler, R. Naslain, B. Desbat, Oxidation inhibition of carbon fibre preforms and C/C composites by H_3PO_4, J. Eur. Ceram. Soc. 18 (1998) 1953–1960.
[29] I.A. Kuzin, L.A. Koemets, Preparation of phosphorylated active carbon, Zh. Prikl. Khim. 43 (1970) 695–698.
[30] I.P. Lubliner, I.N. Ermolenko, A.E. Kofman, D.K. Kuk, Investigation of phosphorylation of carbon fibers, Vestsi Akad. Navuk Belarusk. SSR 3 (1973) 80–85.
[31] J. Siedlewski, W. Śmigiel, A. Arcimowicz, Badanie własności wegli aktywowanych modyfikowanych trójchlorkiem fosforu, Koks Smola Gaz 11 (1975) 300–302.
[32] X. Wu, L.R. Radovic, Inhibition of catalytic oxidation of carbon/carbon composites by phosphorus, Carbon 44 (2006) 141–151.
[33] D.W. McKee, C.L. Spiro, E.J. Lamby, The inhibition of graphite oxidation by phosphorus additives, Carbon 22 (1984) 285–290.
[34] S. Marinkovic, C. Suznjevic, A. Tukovic, I. Dezarov, D. Cerovic, Preparation and properties of phosphorus-containing pyrocarbon—I. Chemical-vapour-codeposition of carbon and phosphorus, Carbon 11 (1973) 217–220.
[35] S. Marinkovic, C. Suznejevic, A. Tukovic, I. Dezarov, D. Cerovic, Preparation and properties of phosphorus-containing pyrocarbon — part II. Effect of heat-treatment on phosphorus-containing pyrocarbon, Carbon 12 (1974) 57–62.
[36] S. Marinkovic, Substitutional solid solubility in carbon and graphite, in: P.A. Thrower (Ed.), Chemistry and Physics of Carbon, vol. 19, Marcel Dekker, Inc, 1984, pp. 1–64.
[37] M. Rusop, T. Soga, T. Jimbo, Defect studies and photoelectrical properties of phosphorus doped amorphous carbon films, Diamond Relat. Mater. 13 (2004) 2197–2202.
[38] M. Rusop, T. Soga, T. Jimbo, Structural, bonding and physical characteristics of phosphorus-doped hydrogenated amorphous carbon films grown by plasma-enhanced chemical vapor deposition, Thin Solid Films 482 (2005) 280–286.
[39] N. Feldtner, W. Brockner, P. Scharff, M.M. Dadras, Novel carbon materials obtained by reactions of C_{60} fullerene with phosphorus at high temperature, J. Non-Cryst. Solids 333 (2004) 301–306.
[40] M. Smíšek, S. Černý, M. Brychta, M. Flašar, M. Kůstka, H. Stuchlík, V. Valter, Aktivní uhlí, SNTL (1964).
[41] H. Kienle, E. Bäder, Aktivkohle und ihre Industrielle Anwendung, Ferdinand Enke Verlag, 1980.
[42] H. Marsh, F. Rodríguez-Reinoso, Activated Carbon, Elsevier Ltd, 2006.
[43] D.W. McKee, Effect of adsorbed phosphorus oxychloride on the oxidation behavior of graphite, Carbon 10 (1972) 491–497.
[44] Y.-J. Lee, L.R. Radovic, Oxidation inhibition effects of phosphorus and boron in different carbon fabrics, Carbon 41 (2003) 1987–1997.
[45] S.G. Oh, N.M. Rodriguez, In situ electron microscopy studies of the inhibition of graphite oxidation by phosphorus, J. Mater. Res. 8 (1993) 2879–2888.

[46] A.M. Puziy, O.I. Poddubnaya, A.M. Ziatdinov, On the chemical structure of phosphorus compounds in phosphoric acid-activated carbon, Appl. Surf. Sci. 252 (2006) 8036–8038.

[47] A.M. Puziy, O.I. Poddubnaya, R.P. Socha, J. Gurgul, M. Wiśniewski, XPS and NMR studies of phosphoric acid activated carbons, Carbon 46 (2008) 2113–2123.

[48] J. Dai, J. Yuan, Modulating the electronic and magnetic structures of P-doped graphene by molecule doping, J. Phys. Condens. Matter 22 (2010) 225501.

[49] A.M. Puziy, O.I. Poddubnaya, Y.N. Kochkin, N.V. Vlasenko, M.M. Tsyba, Acid properties of phosphoric acid activated carbons and their catalytic behavior in ethyl-tert-butyl ether synthesis, Carbon 48 (2010) 706–713.

[50] J. Bedia, J.M. Rosas, J. Márquez, J. Rodríguez-Mirasol, T. Cordero, Preparation and characterization of carbon based acid catalysts for the dehydration of 2-propanol, Carbon 47 (2009) 286–294.

[51] J. Bedia, R. Ruiz-Rosas, J. Rodríguez-Mirasol, T. Cordero, A kinetic study of 2-propanol dehydration on carbon acid catalysts, J. Catal. 271 (2010) 33–42.

[52] J. Bedia, J.M. Rosas, D. Vera, J. Rodríguez-Mirasol, T. Cordero, Isopropanol decomposition on carbon based acid and basic catalysts, Catal. Today 158 (2010) 89–96.

[53] J. Bedia, R. Barrionuevo, J. Rodríguez-Mirasol, T. Cordero, Ethanol dehydration to ethylene on acid carbon catalysts, Appl. Catal. B 103 (2011) 302–310.

[54] H.P. Boehm, Some aspects of the surface chemistry of carbon blacks and other carbons, Carbon 32 (1994) 759–769.

[55] I.A. Kuzin, A.I. Loskutov, V.F. Palfitov, L.A. Koemets, Investigation of the effect of active carbon surface chemistry on adsorption of water, carbon dioxide and ammonia, Zh. Prikl. Khim. 45 (1972) 760–765.

[56] J. Siedlewski, W. Śmigiel, A. Arcimowicz, Badanie właściwości sorpcyjnych i katalitycznych wegli aktywowanych modyfikowanych trójchlorkiem fosforowym, Przem. Chem. 58 (1979) 677–678.

[57] M. Sobiesiak, B. Gawdzik, A.M. Puziy, O.I. Poddubnaya, Phosphoric acid and steam as activation agents for carbonized porous polymer surfaces, Adsorpt. Sci. Technol. 24 (2006) 167–175.

[58] Y. Guo, D.A. Rockstraw, Physical and chemical properties of carbons synthesized from xylan, cellulose, and Kraft lignin by H_3PO_4 activation, Carbon 44 (2006) 1464–1475.

[59] Y. Guo, D.A. Rockstraw, Physicochemical properties of carbons prepared from pecan shell by phosphoric acid activation, Bioresour. Technol. 98 (2007) 1513–1521.

[60] H. Benaddi, T.J. Bandosz, J. Jagiello, J.A. Schwarz, J.N. Rouzaud, D. Legras, F. Béguin, Surface functionality and porosity of activated carbons obtained from chemical activation of wood, Carbon 38 (2000) 669–674.

[61] A.M. Puziy, O.I. Poddubnaya, Characterization of surface heterogeneity of carbon-composite adsorbents, Mater. Sci. Forum 308-311 (1999) 908–916.

[62] D.J. Malik, V. Strelko Jr., M. Streat, A.M. Puziy, Characterisation of novel modified active carbons and marine algal biomass for the selective adsorption of lead, Water Res. 36 (2002) 1527–1538.

[63] A.M. Puziy, O.I. Poddubnaya, V.N. Zaitsev, A bibasic method for investigation of the mechanism of heavy metal ion adsorption on heterogeneous adsorbents, Rep. Natl. Acad. Sci. Ukraine 11 (2010) 119–125.

[64] A.M. Puzii, Methods of production, structure, and physicochemical characteristics of phosphorylated carbon adsorbents, Theor. Exp. Chem. 47 (2011) 277–291.

[65] J.A. Dean, Lange's Handbook of Chemistry, fifteenth ed. McGraw-Hill, Inc, 1999.

[66] J.D. Roberts, M.C. Caserio, Basic Principles of Organic Chemistry, second ed., W.A. Benjamin, Inc, 1977.

[67] M.B. Smith, J. March, March's Advanced Organic Chemistry: Reactions, Mechanisms, and Structure, sixth ed., John Wiley & Sons, Inc, 2007.
[68] Verfahren zum Entfernen von Kationen aus Wasser oder wässerigen Lösungen. Patent DE733410, 1943.
[69] A.N. Amelin, Y.V. Kariakin, Investigation metal ion exchange constants on carbons activated with phosphoric acid, Zh. Prikl. Khim. 45 (1971) 2357–2359.
[70] C.A. Toles, W.E. Marshall, M.M. Johns, Granular activated carbons from nutshells for the uptake of metals and organic compounds, Carbon 35 (1997) 1407–1414.
[71] C.A. Toles, W.E. Marshall, M.M. Johns, Surface functional groups on acid-activated nutshell carbons, Carbon 37 (1999) 1207–1214.
[72] C.A. Toles, W.E. Marshall, M.M. Johns, Phosphoric acid activation of nutshells for metals and organic remediation: process optimization, J. Chem. Technol. Biotechnol. 72 (1998) 255–263.
[73] T. Budinova, E. Ekinci, F. Yardim, A. Grimm, E. Björnbom, V. Minkova, M. Goranova, Characterization and application of activated carbon produced by H_3PO_4 and water vapor activation, Fuel Process. Technol. 87 (2006) 899–905.
[74] K. Li, Z. Zheng, Y. Li, Characterization and lead adsorption properties of activated carbons prepared from cotton stalk by one-step H_3PO_4 activation, J. Hazard. Mater. 181 (2010) 440–447.
[75] K.T. Klasson, L.H. Wartelle, J.E. Rodgers III, I.M. Lima, Copper(II) adsorption by activated carbons from pecan shells: effect of oxygen level during activation, Ind. Crops Prod. 30 (2009) 72–77.
[76] P. Álvarez, C. Blanco, M. Granda, The adsorption of chromium (VI) from industrial wastewater by acid and base-activated lignocellulosic residues, J. Hazard. Mater. 144 (2007) 400–405.
[77] A.M. Puziy, O.I. Poddubnaya, V.N. Zaitsev, O.P. Konoplitska, Modeling of heavy metal ion binding by phosphoric acid activated carbon, Appl. Surf. Sci. 221 (2004) 421–429.
[78] A.M. Puziy, O.I. Poddubnaya, J.A. Ritter, A.D. Ebner, C.E. Holland, Elucidation of the ion binding mechanism in heterogeneous carbon-composite adsorbents, Carbon 39 (2001) 2313–2324.
[79] S. Chen, H. Zeng, Improvement of the reduction capacity of activated carbon fiber, Carbon 41 (2003) 1265–1271.
[80] E. Gonzalez-Serrano, T. Cordero, J. Rodríguez-Mirasol, L. Cotoruelo, J.J. Rodriguez, Removal of water pollutants with activated carbons prepared from H_3PO_4 activation of lignin from kraft black liquors, Water Res. 38 (2004) 3043–3050.
[81] D. Hulicova-Jurcakova, A.M. Puziy, O.I. Poddubnaya, F. Suárez-García, J.M.D. Tascón, G.Q. Lu, Highly stable performance of supercapacitors from phosphorus-enriched carbons, J. Am. Chem. Soc. 131 (2009) 5026–5027.
[82] S. Altenor, B. Carene, E. Emmanuel, J. Lambert, J.-J. Ehrhardt, S. Gaspard, Adsorption studies of methylene blue and phenol onto vetiver roots activated carbon prepared by chemical activation, J. Hazard. Mater. 165 (2009) 1029–1039.
[83] M. Seredych, C.T. Wu, P. Brender, C.O. Ania, C. Vix-Guterl, T.J. Bandosz, Role of phosphorus in carbon matrix in desulfurization of diesel fuel using adsorption process, Fuel 92 (2012) 318–326.

Chapter 9

Porous Carbide-Derived Carbons

Nadejda Popovska-Leipertz
Lehrstuhl für Chemische Reaktionstechnik, Universität Erlangen-Nürnberg, Erlangen, Germany

Chapter Outline

9.1. Introduction 269
9.2. Processing of Porous Carbide-Derived Carbons (Cdcs) using Fibrous Carbide Templates 271
9.2.1. Processing of Fibrous Carbide Templates 271
9.2.2. Etching of the Carbide Templates with Chlorine 271
9.3. Characterization of the Porous CDCs 277
9.3.1. Degree of Structural Order of CDC 277
9.3.2. Oxidation Behavior 279
9.3.3. Pore Structure of CDC 280
9.3.4. Mechanical Properties of CDC 283
9.4. Cdcs with Enhanced Porosity by Post-Synthesis Treatment in Carbon Dioxide 284
9.4.1. Post-Synthesis Treatment of CDC in CO_2 286
9.5. Concluding Remarks 290
References 292

9.1. INTRODUCTION

Porous solids are of great technological importance due to their ability to interact with solids, liquids, and gases not only on the surface, but through their bulk. Carbons are multipurpose materials, which can be synthesized in variety of forms and microstructures using gaseous, liquid, or solid precursors. Although large pores can be produced and well controlled in a variety of materials, nanoporous carbon because of its high specific surface area with a tunable, very narrow pore-size distribution, has received much attention

Novel Carbon Adsorbents. DOI: 10.1016/B978-0-08-097744-7.00009-0
Copyright © 2012 Elsevier Ltd. All rights reserved.

recently duc to many potential energy-related applications such as batteries, supercapacitors, fuel cells, and gas storages.

Carbide-derived carbon (CDC) represents a new class of these nanoporous carbons, produced by selective thermo-chemical etching of inorganic carbon-containing materials, such as metal carbides transforming the carbide structure into carbon. The structure of the CDC obtained by different methods, such as halogenation, hydrothermal treatment, or vacuum decomposition, can be designed by the initial carbide structure and shape, with the opportunity for further structural modifications by controlling the process parameters such as temperature, pressure, etching agent type, and concentration [1]. CDC ranges from extremely disordered to highly ordered structures. A comprehensive review about CDCs has been published recently [2]. It provides detailed information about how a variety of carbon structures can be produced from different carbides and how these structures can be controlled, and describes CDC properties beneficial for a number of applications.

Chlorination is the most economic and scalable method for CDC synthesis using a large number of binary and ternary carbides. Among them, SiC is the most investigated precursor for this process. Titanium carbide (TiC) is also a very promising candidate for producing CDC because of the small and uniform carbon–carbon distance in the lattice, leading to a highly porous carbon with narrow pore-size distribution [1]. Typically, the structure and the shape of the carbide precursor are maintained after chlorination. Therefore, the CDC synthesis is referred to as a conformal transformation process [2].

Most investigations into the field of CDC are based on chlorination of metal carbides in the form of powder. Biomorphic carbide ceramics are a special form of template that can be obtained from carbonized paper by the chemical vapor infiltration and reaction (CVI-R) technique [5–7]. These biomorphic fibrous SiC and TiC templates undergo conformal transformation into porous carbon by chlorination. The process can be controlled in order to obtain full conversion to pure carbon or to perform it as partial etching producing metal carbide coated with porous carbon. The formation of carbon on the surface is uniform and does not change the shape of the initial templates allowing preparation of structures with complex geometry and high mechanical stability. The idea is to use these structures as adsorbent or catalyst support because of their easier separation from the reaction mixture. Tribology is also a promising application field. The processing of porous CDC using biomorphic fibrous carbide templates was investigated in detail in a dissertation published last year [9].

This chapter summarizes the results on studying the conversion of biomorphic fibrous carbides such as SiC and TiC into porous carbon (CDC). The kinetics of the chlorination process as well as the influence of the process parameters, such as temperature, composition of the reactive gas, and addition of transition metal catalyst on the microstructure of the resulting CDC, are discussed. Since the pore volume of CDC is limited by the metal content in the carbide precursor used, additional porosity has been developed by post-synthesis treatment of CDC in carbon dioxide.

9.2. PROCESSING OF POROUS CARBIDE-DERIVED CARBONS (CDCs) USING FIBROUS CARBIDE TEMPLATES

The idea of the biomorphic ceramics is to use natural structures, such as wood or cellulose fibers, as template for processing highly porous materials after conversion into ceramic. Flat paper sheets have the advantage to be formed previously to the conversion into complex three-dimensional shapes, obtaining in this way ceramic structures with high surface to volume ratios for different gas fluid dynamics [4].

9.2.1. Processing of Fibrous Carbide Templates

Biomorphic SiC and TiC ceramics were produced from paper preforms by the chemical vapor infiltration and reaction (CVI-R) technique in a three-step process described elsewhere [5–7]. The ceramization route involves first carbonization of the fibrous paper structures, where, due to pyrolysis reactions, the organic molecules of the paper fibers are converted into a carbon lattice, the so-called bio-carbon template (C_b). In a second step, the C_b templates were infiltrated by chemical vapor infiltration with appropriate gaseous precursors, whereby a ceramic layer was deposited around the carbon fibers. The third step in the ceramization process is a thermal treatment implying the chemical reaction. Solid–solid or gas–solid reactions can be performed either simultaneously during the infiltration or as a separate high-temperature processing step, depending on the deposited layer. During the thermal treatment, the deposited layer has to react with the carbon from the template forming the desired ceramic layer.

The precursor system used to form SiC and TiC layers on carbon fibers of the carbon templates was CH_3SiCl_3/H_2 and $TiCl_4/CH_4/H_2$, respectively, which decomposes according to Eqns (9.1) and (9.2):

$$C_b + CH_3SiCl_3 \xrightarrow{H_2} C_b - SiC + HCl \tag{9.1}$$

$$C_b + TiCl_4 + CH_4 \xrightarrow{H_2} C_b - TiC + TiC_b + HCl \tag{9.2}$$

An overall flow chart of processing of fibrous carbide templates by the CVI-R technique is shown in Fig. 9.1.

9.2.2. Etching of the Carbide Templates with Chlorine

The biomorphic SiC and TiC ceramics were covered with porous carbon by selective etching of the metal in chlorine or chlorine/hydrogen atmosphere in a range of temperatures between 400 and 1200 °C, according to Eqn (9.3). The processing schema is presented in Fig. 9.2:

$$MeC + Cl_2 \rightarrow MeCl_4 + C \tag{9.3}$$

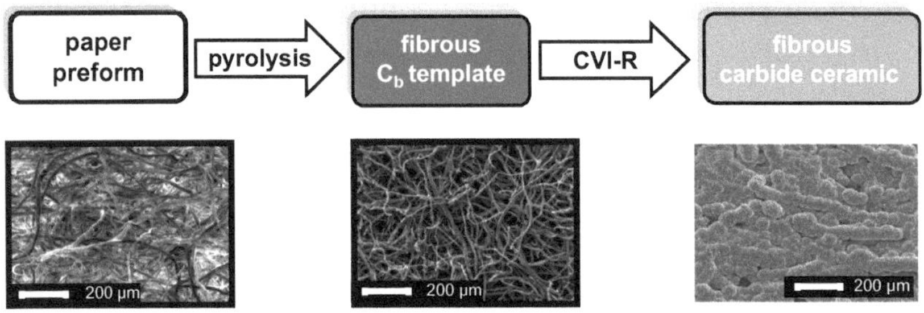

FIGURE 9.1 Processing of fibrous carbide templates by the CVI-R technique.

After chlorination, the furnace was cooled down to room temperature under helium/hydrogen flow to remove the residual chlorine and the reaction by-products from the pores of the samples. A detailed description of the chlorination apparatus and the applied procedure used in this study can be found in [3] and [8].

CDCs derived from paper preforms show high shape stability and very good mechanical performances. The cross-section of the sample as shown in Fig. 9.2 consists of a core of the initial carbon template (C_b), surrounded by a not-converted carbide precursor, followed by a layer of highly porous carbon (CDC). Depending on the degree of conversion porous CDC–metal carbide composites with different compositions for many applications can be obtained.

The surface morphology of biomorphic fibrous carbides before and after chlorination was investigated by scanning electron microscopy (SEM) and is presented exemplarily for TiC in Fig. 9.3. The original fiber structure of the carbon biotemplate was maintained over all process steps. The EDX analysis shows that, after the etching process, no metal could be detected on the fiber surface.

As mentioned previously [1], the type of carbide precursor and the process parameters, such as temperature, etching agent, and addition of catalyst, affect

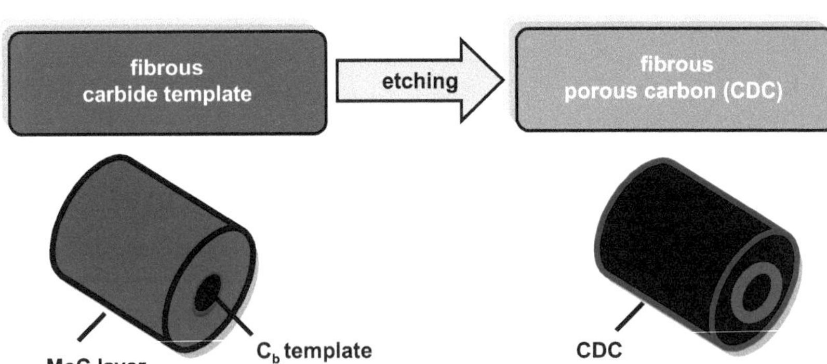

FIGURE 9.2 Processing of CDC by chlorination of fibrous carbide templates.

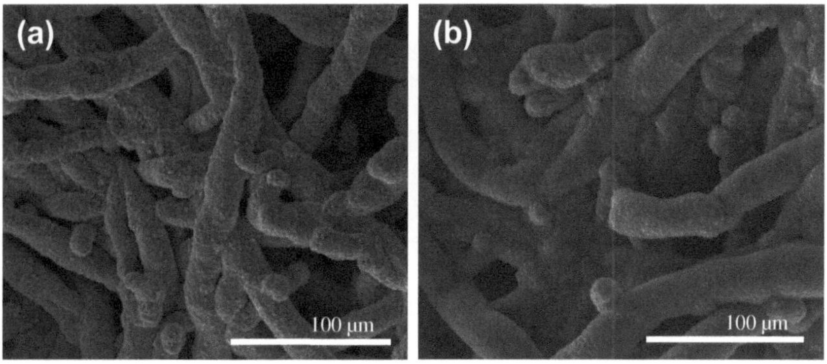

FIGURE 9.3 SEM images of fibrous TiC ceramic (a) and TiC-CDC composite (b) [8].

both the chlorination kinetics and the microstructure of the resulting carbon, as will be discussed in the next sections.

9.2.2.1. Effect of the Process Conditions on the Etching Rate

Kinetic investigations showed that temperatures below 650 °C were not sufficient to remove Si from SiC, because of its lower reactivity compared to TiC. In the case of TiC-CDC, a deviation from the linear run of the kinetic curves was observed at a reaction time of 1 h, as can be seen in Fig. 9.4. The assumption

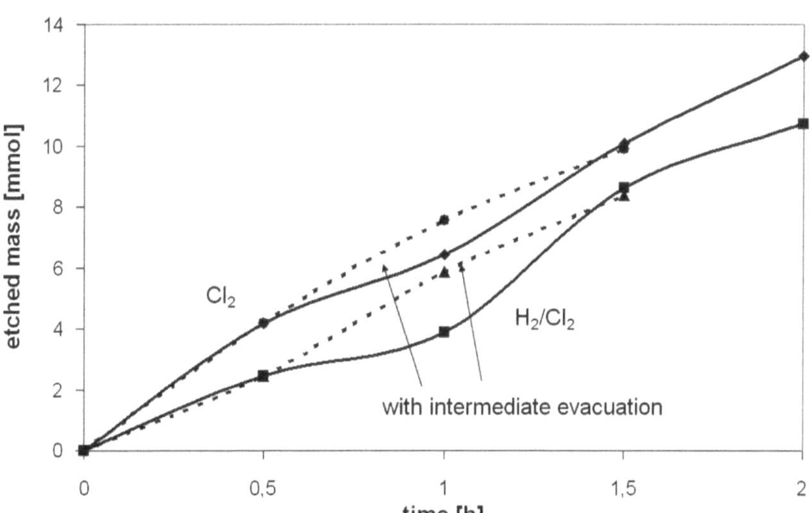

FIGURE 9.4 Kinetic curves of the etching of biomorphic TiC with chlorine and hydrogen/chlorine mixture ($H_2/Cl_2 = 0.25$) at 800 °C [3].

that gaseous by-products occupied the pores leading to decrease in the etching rate was confirmed by an experiment with intermediate evacuation. Such an effect was not observed with SiC as a precursor because of its lower reactivity evolving less gaseous by-products [3].

Based on these results, it can be concluded that the conversion of the carbide to carbon can well be controlled varying the etching time. After chlorination of 20 min, about 20% of the biomorphic TiC was converted into carbon. An almost complete conversion of 90% has been achieved after a reaction time of 90 min [9,31].

The effect of the chlorine concentration was investigated up to 0.6 mol m^{-3} [9]. In this range, the etching rate and thus the carbide conversion to carbon increases linearly indicating a first order reaction regarding chlorine (Eqn (9.3)).

In order to compare the etching behavior of TiC and SiC in chlorine or hydrogen/chlorine atmosphere, experiments were performed at 800 °C, because of the lower reactivity of SiC, as mentioned above. Generally, it is well known that addition of hydrogen to chlorine leads to a reduced reaction rate [2], but almost no details about the effect of the hydrogen/chlorine ratio on the etching process are given in the literature. In order to clarify this, a test series with different hydrogen/chlorine ratios was conducted at 800 °C for both biomorphic TiC and SiC precursors [3]. An almost linear decrease in the etching rate with increasing hydrogen/chlorine ratio was detected, and it dropped to zero at a value in the range of 0.65–0.72 (Fig. 9.5). The HCl formed during the reaction seems not to be effective enough to remove Si, as described in [10]. Chen et al. varied the H_2/Cl_2 ratio in the range 0–1.86 and reported that the reaction became slower as H_2 concentration was increased and ceased at $H_2/Cl_2 > 1$. No extraction of Si could be reached at a ratio of 1 and 1.86. It can be concluded that the reaction rate decelerates with increasing amount of hydrogen and stops at a critical value of hydrogen/chlorine ratio R_{crit}, which depends mainly on the carbide precursor used. This ratio for biomorphic carbides has been calculated evaluating the curves in Fig. 9.5, according to Eqns (9.4) and (9.5):

$$\text{Etching rate} = -A(\text{ratio } H_2/Cl_2) + B \qquad (9.4)$$

$$(\text{ratio } H_2/Cl_2)_{crit} = B/A \qquad (9.5)$$

where: A = rate of decrease of the etching rate with increasing H_2/Cl_2 ratio, mmol/h

B = etching rate without addition of hydrogen, mmol/h

R_{crit} = ratio H_2/Cl_2 where no etching occurs

The calculated values for A, B, and R_{crit} are given in Table 9.1

The etching rate of 6.3 mmol/h for TiC in pure chlorine is higher than that for SiC, which was found to be 4.6 mmol/h, but compared to SiC etching it

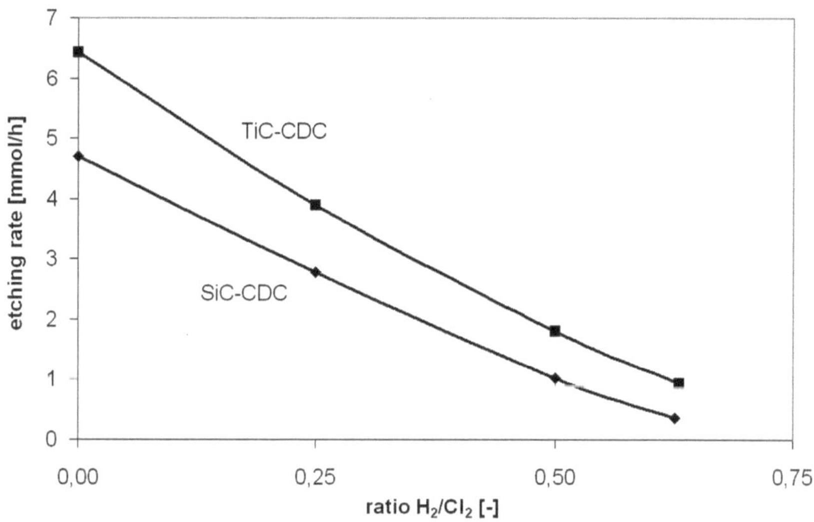

FIGURE 9.5 Influence of the hydrogen/chlorine ratio in the reaction gas on the etching rate of TiC and SiC fibrous templates [3].

TABLE 9.1 Calculated Values for *A*, *B*, and R_{crit} from Fig. 9.5 and Eqn (9.5)

	TiC-CDC	SiC-CDC
A, mmol/h	8.75	7.02
B, mmol/h	6.28	4.63
R_{crit}	0.72	0.66

decreases more rapidly to zero at a critical hydrogen/chlorine ratio, which was calculated to be 0.72 for TiC and 0.66 for SiC.

Another way to create new carbon structures is catalyst-assisted chlorination of metal carbides. Cobalt(II), nickel(II), and ferric(III) chloride were used as catalysts present during chlorination and were found to affect the carbon microstructure [11,14]. It was shown that increasing the concentration of the catalyst leads to a decrease in the amount of micropores in the resulting carbon. A tendency to form barrel-like multiwall nanoparticles was also found by chlorination in the presence of catalytically active metals. The chlorination of TiC [11] and SiC [13] with addition of a catalyst led to formation of better ordered carbon structures at lower temperatures, as indicated by TEM and Raman microscopy. The results on investigating the effect of transition metal catalysts in form of salts or oxides on the chlorination process of biomorphic TiC and SiC ceramics and on the microstructure of the resulting CDC have been published in Ref. [12]. The investigated catalysts include salts of Fe(III), Ru(III), as well as bimetallic Ru(III)/Fe(III).

The presence of a catalyst enhances the etching of titanium by chlorine resulting in a 12–30% higher etching rate depending on the type of the catalyst. The experiments with biomorphic SiC were carried out in a temperature range of 800–1200 °C because silicon cannot be extracted with etching rates comparable to titanium at lower temperatures [3]. The highest etching rate without a catalyst was found at 1000 °C, whereas in the case of a catalyst present during chlorination of SiC almost no temperature dependence of the etching rate could be observed, as can be seen in Fig. 9.6. However, similar to the results with TiC the use of Ru(III) or a bimetallic Ru(III)/Fe(III) catalyst increases the etching rate. The initial value of 4.7 mmol/h at 800 °C without a catalyst was increased to 7.1 and 7.5 mmol/h with addition of Ru(III) or a bimetallic catalyst, respectively. In the case of SiC, the influence of the catalyst on the increasing etching rate was with 14–58% more strongly pronounced C compared to TiC. All these results have been achieved using a very low catalyst concentration of only 0.5 wt.%.

The chlorination of biomorphic carbides in the presence of a catalyst affects the microstructure of the resulting CDC significantly, as will be discussed in Section 9.3.

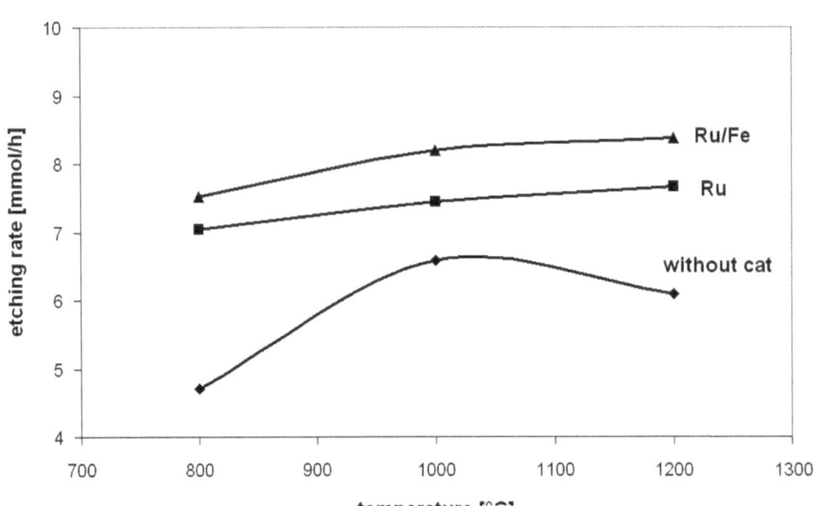

FIGURE 9.6 Effect of transition metal catalyst (0.5 wt.%) on the etching rate of a fibrous SiC template at different temperatures [12].

9.3. CHARACTERIZATION OF THE POROUS CDCs

9.3.1. Degree of Structural Order of CDC

Raman spectroscopy is a suitable method to determine the degree of structural order of carbon materials. The two characteristic peaks in Raman spectra around 1330–1342 cm^{-1} (D-band) and 1585–1600 cm^{-1} (G-band) confirmed the conversion of metal carbides to carbon after chlorination without and with addition of Ru(III) as a catalyst. Only Raman spectra for the CDC samples chlorinated in the presence of Ru(III) catalyst were observed because of the strongest visible influence.

To estimate the structural order and graphite crystallinity of different CDCs, the area ratio between the D and the G-band (A_D/A_G) was determined and is presented in Table 9.2.

A much higher intensity of the D-band compared to the G-band is evident in the whole range of chlorination temperatures indicating a disordered structure of CDC independent of the carbide precursor and the presence of a catalyst. However, an increase in reaction temperature leads to a decrease in the FWHM of both bands suggesting an increase in structural carbon ordering. This conclusion is more strongly pronounced for chlorination in the presence of the Ru(III) catalyst.

Further evidence for the development of ordered carbon structures is indicated by the appearance and the shape of the second order Raman peaks at 2700 cm^{-1} (G′) and 2940 cm^{-1} (D″) [15]. The G′-band, which suggests an enhanced carbon order, is pronounced for TiC-CDC obtained only at elevated temperature such as 1200 °C (Fig. 9.7a). A catalyst added to the precursor carbide seems to enhance the graphitization expressed by a clearly visible G′-band even at temperatures as low as 1000 °C (Fig. 9.7b). Similar results were obtained for chlorination of TiC in the presence of other catalysts, such as Fe (III).

TABLE 9.2 Area Ratio of D- and G-bands from the Raman Spectra of CDC as a Function of the Carbide Precursor, Chlorination Temperature, and Catalyst Presence

	TiC-CDC			SiC-CDC		
T,°C/cat	-	Ru	Ru/Fe	-	Ru	Ru/Fe
800	4.7	3.4	3.4	4.7	4.9	3.9
1000	3.9	2.3	2.0	4.2	4.5	3.6
1200	1.5	1.8	1.6	3.3	1.8	1.8

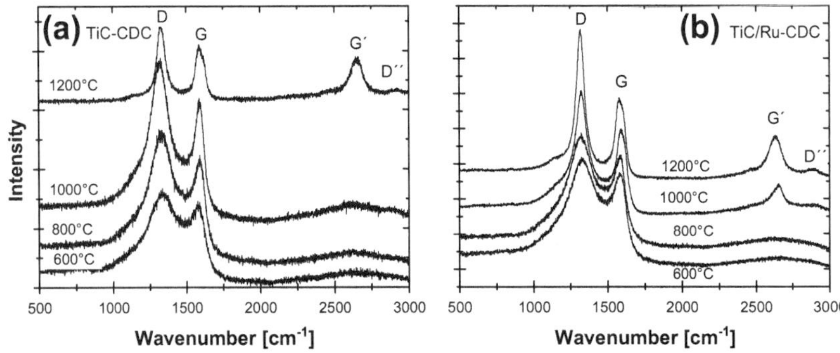

FIGURE 9.7 Raman spectra of CDC from fibrous TiC chlorinated without (a) and in the presence of Ru(III) catalyst (b) [12].

The Raman spectra of SiC-CDC confirm the conversion of SiC to carbon in the temperature range from 800 to 1200 °C. Disordered carbon is again produced in all experiments. Also in the case of SiC, an increase in reaction temperature leads to a decrease in the FWHM of the D-band suggesting a slightly improved carbon ordering. However, a second order G′-band was not clearly detected indicating that starting graphitization of SiC-CDC is not as strongly pronounced as for chlorination of TiC-CDC at temperatures as high as 1000 and 1200 °C.

It can be concluded that addition of catalyst leads to an improved ordered carbon structure mainly in the temperature range 600–1000 °C. At temperatures as high as 1200 °C, no catalyst is necessary to enhance the microstructural order of the resulting CDC.

The same tendency of an improved ordered microstructure of CDC with increasing reaction temperature can be also seen in the presence of a bimetallic catalyst. Over the whole temperature range, the $FWHM_D$ values are below those for samples chlorinated in the presence of either Ru(III), or Fe (II), or Fe (III) catalyst [12]. Therefore, it can be concluded that the highest degree of crystallinity can be achieved by chlorination of TiC in the presence of a bimetallic catalyst at 1200 °C; however, a considerable amount of disordered structure still remains in the final material.

It can be summarized that a well-ordered carbon structure cannot be obtained by chlorination of SiC but by chlorination of TiC, which can in general be graphitized at lower reaction temperatures. For TiC-CDC, a sequence to higher ordered microstructure has been estimated as follows: no catalyst < Fe < Ru < Ru/Fe catalyst, indicating a synergetic effect of the bimetallic catalyst.

The dependence of the degree of ordering of CDC on the etching temperature and the presence of a catalyst during the chlorination was additionally examined by HRTEM [12]. A reaction temperature of 800 °C results in a predominantly amorphous carbon structure, whereas at 1200 °C large areas of carbon with improved order can be found in TiC/Ru-CDC. The micrographs of

TiC/RuFe-CDC chlorinated at 1200 °C confirm the change from amorphous to higher ordered structure, depending on the reaction temperature and catalyst type. The dark octagon in the micrographs was assigned to the catalyst $RuCl_3$, which was confirmed by analytical electron microscopy using EDS in the TEM. Obviously, in the surrounding area close to the catalyst nanoparticles an improved carbon ordering occurs.

The mechanism for the formation of improved ordered carbon structures in the vicinity of the catalyst nanoparticles was not investigated in detail. It is well known that transition metals, such as Fe, Ni, or Co, can interact with carbonaceous materials [16]. Carbon species may diffuse and dissolve into the metal particles. This leads to supersaturation resulting in participation of carbon from the solid solution as graphitic carbon surrounding the catalyst nanoparticle. A similar mechanism was proposed also for the catalytic growth of carbon nanotubes (CNTs) by chemical vapor deposition [17].

9.3.2. Oxidation Behavior

The oxidation behavior of CDC obtained from different carbide precursors and chlorination temperatures was investigated by thermal gravimetric analysis (TGA) [3]. A small piece of the carbon–ceramic composite was heated with 5 °C/min to 750 °C and kept for two hours. The mass loss during heating was plotted versus the temperature in order to determine the burn-off temperature of the CDC. To evaluate exactly the burn-off temperature, the first deviation of the TG curves was determined. For each sample, the temperature at which oxidation was started and the temperature at maximum burn-off rate were analyzed; Table 9.3.

A clear dependence of burn-off temperature on carbide precursor and chlorination temperature can be seen and is due to the different structural ordering of the resulting CDC. The better ordered carbon oxidizes at higher temperatures, whereas amorphous carbon is much easier to be oxidized.

TABLE 9.3 Effect of the Carbide Precursor and the Chlorination Temperature on the Oxidation Behavior of the CDC [3]

	TiC-CDC		SiC-CDC	
$T_{etching}$, °C	$T_{ox\ start}$, °C	$T_{ox\ max}$, °C	$T_{ox\ start}$, °C	$T_{ox\ max}$, °C
600	420	525	444	561
800	480	600	439	588
1000	500	605	448	583
1200	540	650	498	620

For SiC-CDC, no structural changes occur at 600–1000 °C, which was also confirmed by the TG experiments. According to the results of the Raman analysis, a small increase in ordering at 1200 °C could be seen, and this was again supported by TG analysis. Therefore, the oxidation behavior of CDC obtained at different chlorination temperatures and carbide precursor correlates with its microstructure, as determined by Raman spectroscopy.

9.3.3. Pore Structure of CDC

The porous structure of CDC prepared from biomorphic TiC and SiC without and in presence of different catalysts was characterized by low temperature nitrogen adsorption experiments. Adsorption isotherms for samples chlorinated in the presence of a catalyst show a similar behavior, as described in previous work for TiC-CDC [8]. The specific surface area (SSA) was calculated according to the BET theory [17]. The temperature dependence of SSA for TiC-CDC obtained without and in the presence of catalyst is presented in Fig. 9.8.

All curves in Fig. 9.8 show a similar run with maximum SSA at 800 °C. This tendency of showing maximum SSA at intermediate reaction temperatures was already described in our previous works [3,8] and also in the literature about chlorination of boron carbide [19]. CDC produced at low temperature has an amorphous structure and contain probably pores that are too small to be accessible to nitrogen molecules [1]. With increasing reaction temperature, better ordered domains, such as multiwalled graphitic structures, start forming

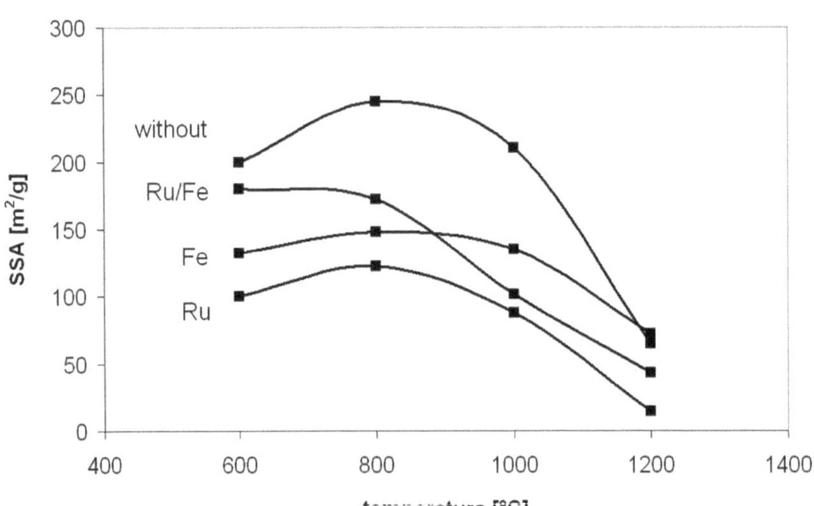

FIGURE 9.8 Specific surface area (SSA) of TiC-CDC obtained at different chlorination temperatures without and in the presence of catalyst [12].

within the CDC structure. The small areas in between these walls are generally not accessible to N_2, and therefore the amount of improved organized multi-walled graphite ribbons could decrease the SSA.

Despite the higher etching rate in the presence of a catalyst, the curves with catalysts are shifted to lower SSA values, which may be caused by an improved structural ordering of the resulting CDC. According to the reports in the literature [20], addition of metal chlorides to titanium carbide does not affect the carbon structure obtained after chlorination at temperatures below 700 °C. However, we have found that a higher amount of well-ordered carbon structures appears to be formed, which causes a strong decline in SSA, especially at lower temperatures. The strongest influence of a catalyst on the SSA was found in the case of Ru(III). For Fe(III) and the bimetallic Ru(III)/Fe(III) catalysts, a difference in SSA is mainly found for temperatures below 1200 °C.

In the case of SiC-CDC, the decrease in SSA in the presence of catalyst is not as strongly pronounced as with TiC. Additionally, the decline of SSA at chlorination temperatures above 1000 °C is very weak, which leads to the assumption that carbon ordering of SiC-CDC is not as developed at higher temperatures as in the case of TiC-CDC. It can be concluded that the effect of a catalyst on SSA is not as strong as in the case of a TiC precursor. This assumption was confirmed by the results of the Raman spectroscopy [12].

Fig. 9.9 shows the isotherms measured by nitrogen sorption experiments of samples chlorinated at 800 °C without and in the presence of a bimetallic catalyst.

Isotherms of type I according to the Brunauer classification were found [18], indicating that CDC obtained at 800 °C is microporous. Additionally, some mesoporosity is also present because of the weak slope of the isotherm for pressure >0.1, which does not show a real horizontal plateau. This is associated with the filling of mesopores. The sample obtained in the presence of a catalyst exhibits a hysteresis loop because of capillary condensation and mesoporosity. As described in previous publications [3,8], the pore size of CDC is influenced by the etching temperature. At temperatures up to 1000 °C, microporous CDC was formed, but in the case of TiC loaded with a bimetallic Ru(III)/Fe(III) catalyst before chlorination a hysteresis can already be seen at a reaction temperature of 800 °C. The hysteresis loop is more pronounced at an increased etching temperature, which suggests a higher amount of mesopores. Its amount increases just slightly with increasing chlorination temperature above 800 °C, which is negligibly lower than without the addition of a catalyst. At 1200 °C, the material is still mainly microporous but also some mesopores can be found.

By additional evaluation of N_2 adsorption data, the micropore volume fraction of each sample could be estimated and presented in Fig. 9.10.

The CDC obtained at 800 °C is mainly microporous with a micropore fraction of about 80% independent of the carbide precursor. Increased

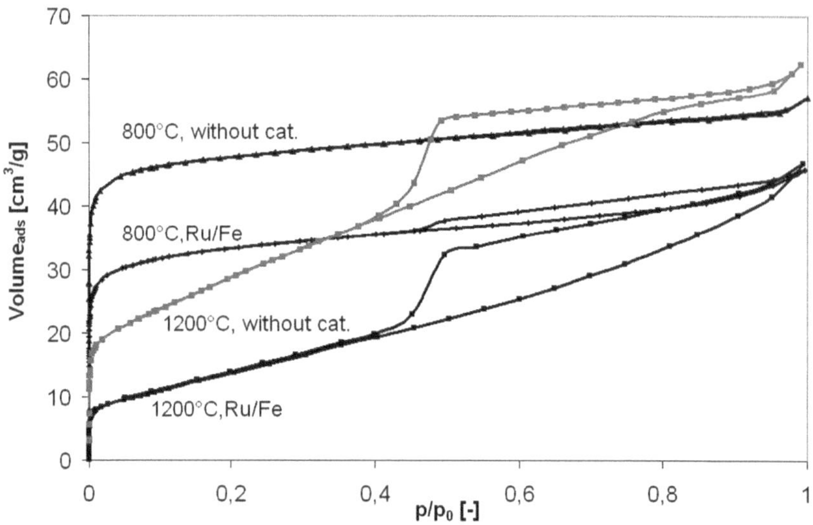

FIGURE 9.9 Nitrogen adsorption isotherms of TiC-CDC obtained at 800 and 1200 °C without and in presence of a bimetallic catalyst [12].

reaction temperature and presence of catalyst promotes the formation of mesopores by reducing micropore volume. The pore size of micropores is below 1 nm, as shown in previous publications [3,8]. Higher temperature favors the formation of mesopores, maintaining the average size of the

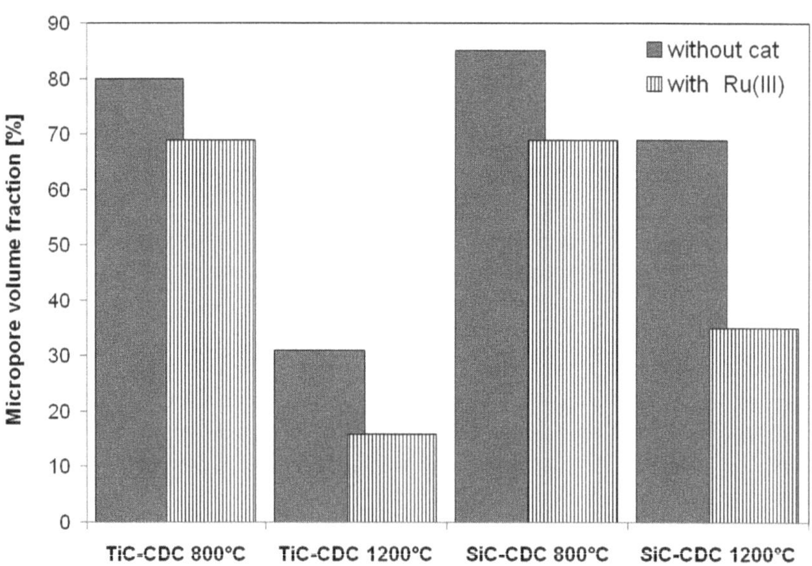

FIGURE 9.10 Micropore volume fraction of TiC-CDC and SiC-CDC obtained at different temperatures without and in the presence of a Ru catalyst [12].

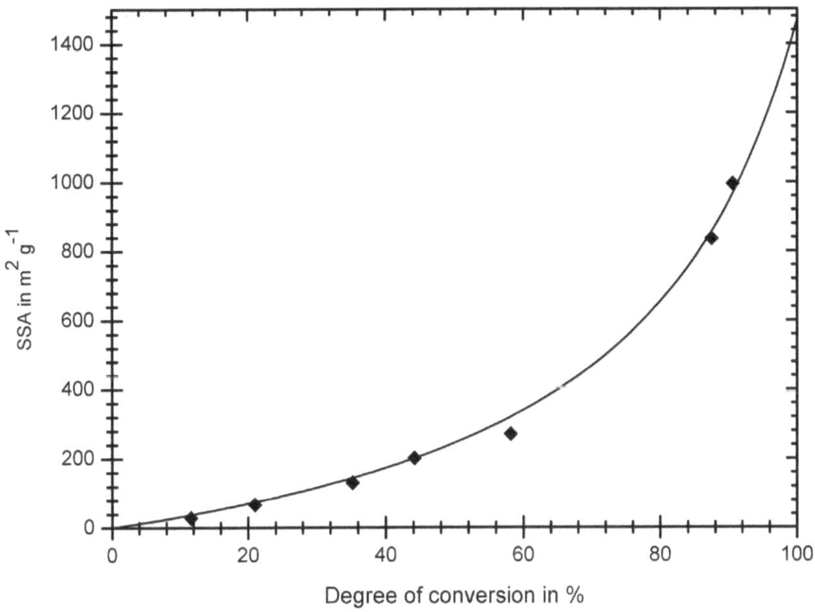

FIGURE 9.11 Specific surface area (SSA) of TiC-CDC as a function of the degree of conversion [31].

micropores. The produced mesopores are in the range of 3–5 nm. The amount of micropores diminishes at higher reaction temperatures, as shown in Fig. 9.10.

It was found [31] that the specific surface area of CDC increases non-linearly with the degree of conversion of the carbide to carbon (Fig. 9.11) and is created mainly in the last 10% conversion.

This behavior can be explained well under the assumption that the specific surface area is only supplied by the microporous carbon and not by the unconverted carbide template. Thus, a high degree of conversion is necessary in order to obtain CDC with high specific surface area.

9.3.4. Mechanical Properties of CDC

The mechanical stability of the CDCs was estimated by measuring their bending strength [8,9]. The initial biomorphic fibrous ceramics show a bending strength of 25–36 MPa at a porosity of 60%. The bending strength decreases almost linearly with increased conversion to carbon, as can be seen in Fig. 9.12.

However, a bending strength of 10–12 MPa was measured even after complete conversion of the carbides to carbon. The biomorphic CDC–carbide composites retain the high mechanical stability of the initial carbide

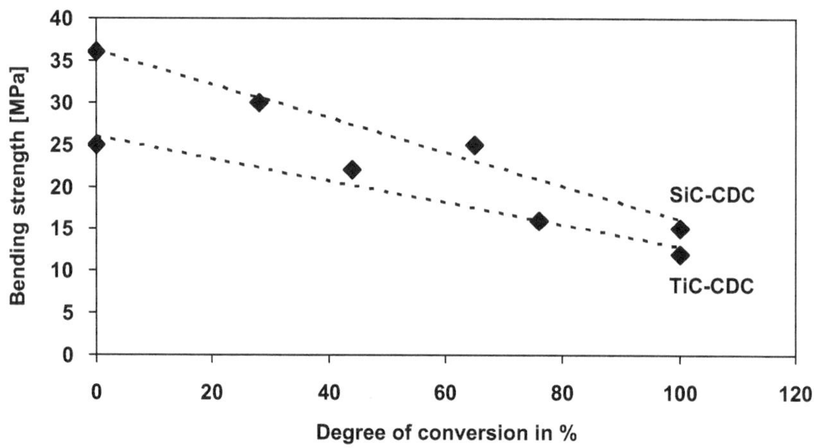

FIGURE 9.12 Bending strength of TiC-CDC and SiC-CDC as a function of conversion to carbon at a chlorination temperature of 800 °C.

precursor. The CDC layer is adherent, and the composites show high shape stability.

9.4. CDCs WITH ENHANCED POROSITY BY POST-SYNTHESIS TREATMENT IN CARBON DIOXIDE

CDC materials show excellent properties but concerning the values for SSA and pore volume, they are not able to compete against activated carbons. The reason is that the pore volume of the CDC is limited by the metal content in the carbide precursor used. The partial oxidation of carbon is an easy and effective way to enhance its porosity. Chemical activation with KOH, NaOH, $ZnCl_2$ [21], or various other methods such as thermal activation in CO_2 [22], or steam [23] are known.

The activation of CDC is a relatively new research field. Several investigations into the activation of CDC have been published. Portet et al. [24] reported the activation of CDC by KOH to increase the volume of small pores <2 nm, which led to significant increase in the electrical double layer performance of the material. The thermal activation of CDC was investigated by oxidation in air and carbon dioxide. The reaction with O_2 leads to a very rapid enlargement of the micropore volume and conversion into mesopores due to its high reactivity. Since it is difficult to control the porosity during O_2 treatment, CO_2 was used as a mild oxidant and subsequently to tune successfully the pore structure. Short activation times lead to an increase in the micropore volume, whereas high temperatures and longer treatment times favor the formation of mesopores. Optimum activation conditions for TiC-CDC obtained after chlorination at 600 °C were a relatively low temperature of about 875 °C and longer activation time (>8 h) resulting in a low mass loss. CO_2 activation at optimal conditions resulted in an SSA_{BET} of >3000 $m^2 g^{-1}$ and a pore volume of 1.3–1.4 $cm^3 g^{-1}$, leading to increased gravimetric H_2 uptake of 4.6 wt% [25].

Different activation procedures of CDC show influence on their pore structure and hence on the methane uptake. High activation temperatures and longer treatment times result in a widening of micropores and in the formation of mesopores. The activation of CDC with CO_2 and KOH follows a large increase in SSA_{BET} and pore volume. A methane uptake of 18.5 wt% at 25 °C and 60 bar was obtained with an activated CDC sample [26]. Post-synthesis treatment is proven as an efficient process for hydrogen storage in CDC materials. Since activation of CDC in CO_2 yields high micropore volume, a hydrogen uptake of 4.7 wt% was obtained with CO_2 post-treatment of TiC chlorinated at 600 °C [27].

The general objective of this work was to study the possibility of developing additional porosity by post-synthesis treatment of CDC produced from biomorphic carbide–ceramic structures. Since the porosity is dependent on the experimental conditions used in the activation procedure, the influence of the activation temperature as well as the treatment time and the flow rate of CO_2 on the resulting pore structure were investigated [28].

The pore structure of the activated CDC samples was characterized by gas sorption measurements using a Quadrasorb (N_2 at −196 °C), NOVA 3000e (CO_2 at 0 °C) from Quantachrome Instruments. The specific surface area, SSA_{BET}, was calculated using Brunauer–Emmett–Teller (BET) analysis from both N_2 and CO_2 sorption. The pore-size distribution was determined from the isotherms according to the non-local density functional theory model (NLDFT), which assumed cylindrical/slit pores. The pore-size distribution of pores below 15 Å was calculated from CO_2 sorption. The micropore volume V_{mic} was estimated by the NLDFT method. The mesopore volume, V_{meso}, was calculated by subtracting the micropore volume from the total pore volume, V_{tot}.

For the post-synthesis treatment, CDC samples derived from biomorphic TiC and SiC ceramics were used. The metal was completely removed from the carbide by reaction with chlorine so that the samples consist of pure carbon comprising C-biotemplate (C_b) and CDC. The CDC/C_b samples before the activation process exhibit a CDC amount of 65–70 wt.-%.

Temperatures in the range of 800–900 °C are reported in the literature for the activation of various carbon materials with CO_2 [23]. The activation of CDC was performed usually at temperatures in the range 875–990 °C [25,26]. First, appropriate treatment temperature for the activation of CDC derived from biomorphic carbide ceramic structures was determined using thermogravimetric analysis (TGA). Since the treatment temperature of 800 °C causes only a very small mass loss, the temperature increase to 850 °C led to significant enhancment of the gasification reaction. The temperature is generally chosen to be the lowest possible one, since Osswald et al. [25] suggested as optimum activation conditions for CDC low temperatures and longer activation times.

Based on the TGA results, a temperature of 850 °C was selected for the post-synthesis treatment of biomorphic carbide structures.

9.4.1. Post-Synthesis Treatment of CDC in CO_2

For the activation experiments, CDCs were used that were obtained after chlorination of biomorphic TiC and SiC ceramics at 800 °C. Previous investigations revealed that the highest SSA_{BET} and pore volume were achieved after chlorination at this temperature so that these CDCs are a good starting material for a further increase in the porosity [3,8].

TiC- and SiC-CDCs were treated in carbon dioxide at 850 °C for different times up to 9 h. The mass loss during the gasification reaction shows a linear dependence on the treatment time for both CDC samples, independent of the carbide precursor. The rate of gasification was a little higher for SiC-CDC, which showed a higher mass loss than TiC-CDC at comparable activation time. This is due to the slightly higher value for SSA_{BET} of SiC-CDC before the activation treatment, compared to TiC-CDC. Therefore, a higher surface area was available for the reaction with CO_2 and higher mass loss could be obtained. The maximum activation time of 9 h resulted in a mass loss of about 55 wt% and 50 wt% for activation of SiC-CDC and TiC-CDC, respectively. This range of mass loss is typical for the production of activated carbon materials [28].

Figure 9.13 represent the nitrogen sorption isotherms for SiC-CDC before and after post-synthesis treatment in CO_2 [29].

The samples were treated for different times (0–9 h) in CO_2 flow. The isotherms for the untreated CDC samples as well as 3 h activated samples are clearly of type I, according the IUPAC classification, indicating the microporous character of the original CDC as well as for the short time activated CDC sample. It can be concluded that the original microporous structure can be maintained after the gasification reaction. However, the post-treatment significantly increases the adsorption capacity of the material by creating additional porosity. In Fig. 9.13, the isotherms of samples that are treated for longer times of 6–9 h in CO_2 display a transition of type I to type IV isotherm. An activation of 6 h suggests a widening of micropores without any remarkable mesopores in the material. This changed after an activation of 9 h, especially in the case of SiC-CDC, proven by the different shape of the isotherm and the detected hysteresis indicating a mesoporosity. Activation times up to 6 h led to a micropore widening, as suggested by the lower sharpness of the knee of the isotherm with longer treatment times. At activation time up to 9 h the amount of micropores remain almost constant but additional mesopores were generated. The total amount of adsorbed nitrogen was increased with treatment time in the whole range studied. The widening of micropores was clearly described in the literature [22,23].

An obvious increase in adsorption capacity can also be noted for the activation of TiC-CDC. Again, as in the case of activated SiC-CDC, the isotherm of non-activated TiC-CDC was of type I whereas the activation during 9 h leads to

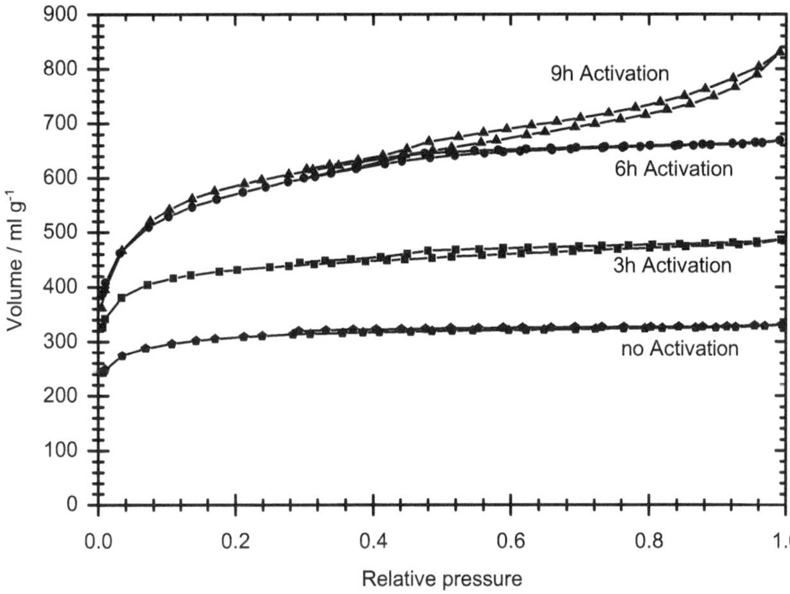

FIGURE 9.13 Nitrogen sorption isotherms for SiC-CDC before and after post-synthesis treatment in CO_2 [28].

type IV isotherm, suggesting the original microporous character of CDC and its following altering during gasification reaction to pore widening and formation of mesopores. However, comparing the isotherms for both carbide precursors it can be estimated that the activation of SiC-CDC shows higher adsorption capacity of N_2 for activation times of 6 and 9 h. Less mass loss during activation leads to a reduced adsorption capacity in the case of TiC-CDC.

The CO_2 sorption experiments provide additional information about the micropore structure [29]. For non-activated SiC-CDC, pores of a size of 3.5 Å were obtained. Additionally, the main size range of the pores was between 4 and 7 Å and some pores were found in the range 8–9 Å. The curve for the sample that was activated for 3 h shows the same behavior until a pore size of 5 Å. At bigger pore sizes, the activation treatment for 3 h clearly enhances the cumulative pore volume. An activation time of 9 h led to a further increase in the cumulative pore volume, but the range of small pore size was significantly decreased. A moderate activation time of 3 h creates additional pores with a similar size as the non-activated CDC, except the smallest pore size of 4–5 Å. The activation time of 9 h results in a even decreased pore volume compared to the non-activated CDC, if just the pore size <12 Å is considered. The total volume of pores that can be detected by CO_2 sorption measurement was higher than the non-activated CDC. Hence, this fact points to the widening of the micropores with longer activation time.

The differential pore-size distribution obtained from N_2 sorption analysis for TiC-CDC highlights this effect also for the other carbide precursor. Figure 9.14 shows the formation of higher pore volume after the activation reaction. It can also be observed that the formed pores after activation of 9 h are not only of a microporous nature but a considerable amount of mesopores was formed. This confirms the results found after activation of SiC-CDC at longer times that resulted as well in the formation of mesopores.

Additionally to the cumulative pore volume of pores <15 Å, the N_2 sorption was evaluated according to the total pore volume. In Fig. 9.15, the total pore volume of activated SiC-CDC samples (0 – 9 h activation time) is plotted.

Additionally, the cumulative pore volumes of pores that are <20 Å, <15 Å and <7 Å were added. In this way, the widening of pores can be clearly observed. The total pore volume shows a progressive increase with the activation time. It should be noted that the total pore volume consists almost completely of micropores <20 Å, more precisely of pores <15 Å. This fact is true for activation times up to 6 h. After an activation treatment of 9 h, the total pore volume was still increased but the micropore volume stagnates suggesting the formation of larger pores at these conditions.

Summarizing, it can be concluded that the total pore volume and especially the micropore volume of CDC can be increased by treatment in CO_2 at moderate activation times up to 6 h. Longer activation times lead to the

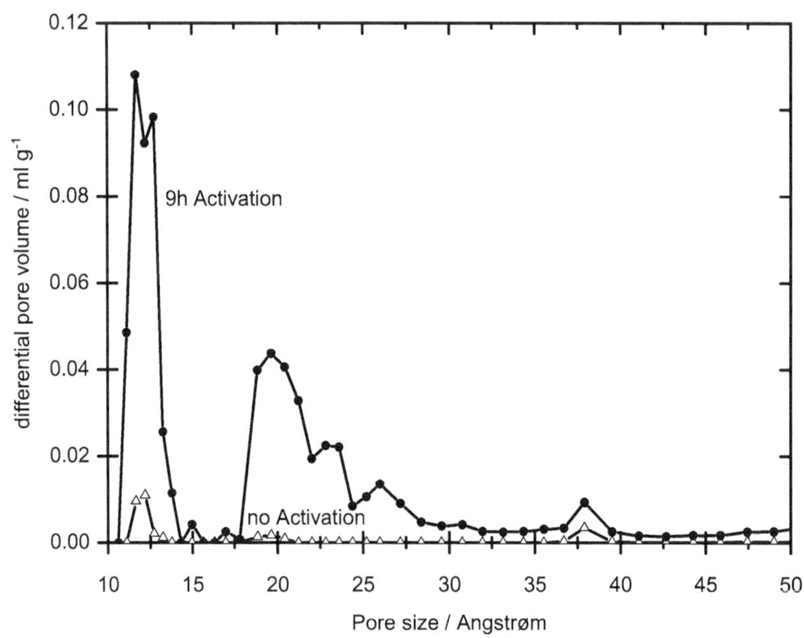

FIGURE 9.14 Differential pore-size distribution of non-activated and activated (9 h) TiC-CDC [29].

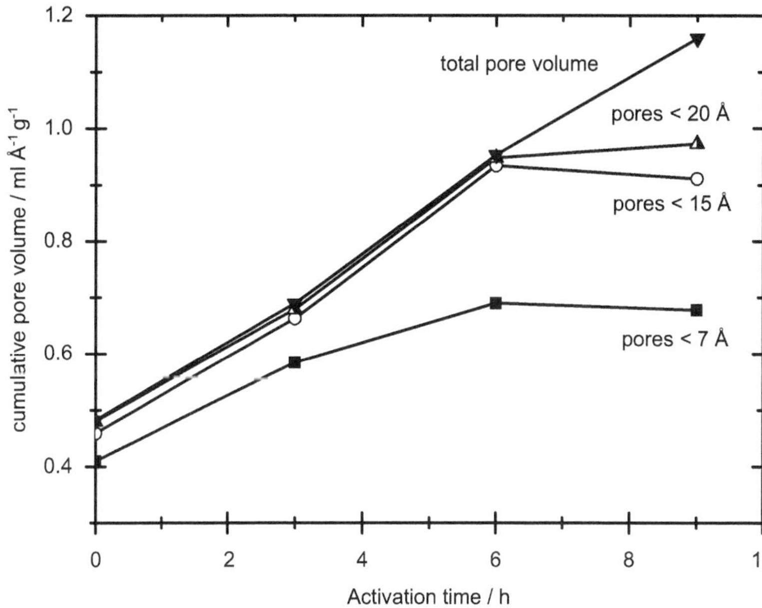

FIGURE 9.15 Cumulative pore volume of SiC-CDC as a function of activation time [29].

formation of larger micropores and additional mesopores. The generation of micro- and mesopores during the gasification reaction was described by Rodriguez-Reinoso suggesting the two simultaneous actions such as the development of micropores in the first steps of gasification and further widening of the micropores after that. The second effect was observed predominantly for burn-off in the range of 40–60 wt% [30].

In Table 9.4, all data about the pore structure of SiC-CDC and TiC-CDC before and after post-synthesis treatment in CO_2 are summarized. The SSA_{BET} as well as the SSA_{DFT} are increased with increasing activation time. An activation time of 9 h leads to a doubling of the SSA value. Therefore, activation of CDCs in CO_2 is an effective and easy way to enhance their porosity. Similar results concerning values for SSA and pore volume were found for the activation of TiC-CDC obtained from TiC powder [27].

The morphology of the CDC samples before and after post-synthesis treatment in carbon dioxide is shown in the SEM images in Fig. 9.16.

The original fiber structure was maintained over all process steps, but a difference in the fiber size can be observed. After the activation process, the fibers became smaller in diameter, which is probably attributed to the activation since the CDC process is known to be conformal in size. The activation leads to a mass loss of about 50 wt%. Additionally, there are cracks visible on the fiber surface after the CO_2 treatment.

TABLE 9.4 Pore Analysis of CDC Before and After Post-Synthesis Treatment in CO_2

CDC	t_{act} (h)	Burn-off (%)	SSA_{BET} $m^2\,g^{-1}$	V_{tot} $cm^3\,g^{-1}$	$V_{tot\ DFT}$ $cm^3\,g^{-1}$	$V_{mic\ DFT}$ $cm^3\,g^{-1}$	V_{mic}/V_{tot} DFT (%)
SiC-CDC	–	–	1182	0.51	0.48	0.48	100
	3	23	1621	0.75	0.69	0.68	98.6
	6	42	2087	1.03	0.95	0.95	100
	9	56	2196	1.28	1.16	0.97	83.6
TiC-CDC	–	–	977	0.41	0.38	0.37	97.4
	6	36	1811	0.80	0.75	0.68	90.7
	9	50	2240	1.11	1.05	0.77	73.3

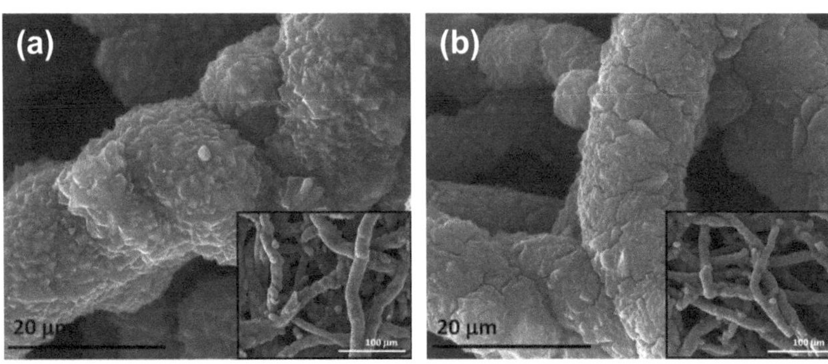

FIGURE 9.16 SEM images of TiC-CDC before (a) and after post-synthesis treatment in CO_2 (b) ($T=850$ °C, $t=9$ h) [29].

Summarizing the results presented above, it can be concluded that an enhancement of the porosity of CDC is possible by post-synthesis thermal treatment in CO_2. The microporous character of CDC can be maintained at temperatures as high as 850 °C if activation time does not exceed 6 h. Longer reaction times lead to a change in the pore structure at the expense of micropores. Depending on the subsequent application of activated CDC, the treatment time has to be adjusted to achieve the desired pore structure.

9.5. CONCLUDING REMARKS

Biomorphic carbide ceramics – TiC and SiC – derived from paper preforms by the CVI-R technique were converted into porous carbon by selective etching of the metal in chlorine or hydrogen/chlorine mixture in a temperature range 400–1200 °C. The pore structure and degree of structural order of CDCs can be controlled varying the carbide precursor and reaction temperature and adding a catalytic active metal such as Fe, Ru, or bimetallic Fe/Ru in the form of salts during the chlorination process. Partial conversion of the carbides results in CDC–carbide composites with controllable composition and microstructure.

Kinetic investigations showed that temperatures below 650 °C were not sufficient to remove Si from SiC, because of its lower reactivity compared to TiC. In the case of TiC-CDC, a deviation from the linear run of the kinetic curves was observed at a reaction time of 1 h. The assumption that gaseous by-products occupied the pores leading to decrease in the etching rate was confirmed by an experiment with intermediate evacuation. Such an effect was not observed with SiC as a precursor because of its lower reactivity evolving less gaseous by-products. The conversion of the biomorphic carbide to carbon can well be controlled varying the etching time. After a chlorination time of 20 min, about 20% of the biomorphic TiC was converted into carbon. An almost complete conversion of 90% could be achieved after about 90 min of chlorination.

Addition of hydrogen to the reactive gas inhibits the chlorination reaction. A linear decrease in the etching rate with increasing hydrogen/chlorine ratio was observed for both carbide precursors used. A critical ratio, where no appreciable etching takes place, was estimated to be 0.72 for TiC-CDC and 0.66 for SiC-CDC.

The presence of a transition metal catalyst in form of salts during the chlorination enhances the etching rate, but no significant effect of the catalyst type and the carbide precursor could be found. Additionally, the presence of a catalyst with a concentration as low as 0.5 wt.% improved the degree of structural order of the resulting CDC at reaction temperatures as high as 600 °C for TiC-CDC but 1000 °C for the SiC precursor. The effect of the catalyst type on increased carbon ordering was found to be as follows: no catalyst $<$ Fe $<$ Ru $<$ Ru/Fe catalyst, indicating a synergetic effect of the bimetal catalyst.

It was found that the specific surface area of CDC increases non-linearly with the degree of conversion of the carbide to carbon and is created mainly in the last 10% conversion. Thus, a high degree of conversion is necessary in order to obtain CDC with high SSA.

The pore volume in the CDC is limited by the metal content in the carbide precursor used. Enhanced specific surface area up to 2300 $m^2\,g^{-1}$ and a pore volume of about 1.3 $cm^3\,g^{-1}$ have been achieved by the high temperature post-synthesis treatment of biomorphic CDC in carbon dioxide. The pore structure was clearly influenced by the activation time using a relatively low activation temperature of 850 °C. Shorter reaction times up to 3 h maintain the original microporous character of CDC leading at the same time to an increase in the total pore volume and the micropore volume but keeping the pore size the same. A longer time of about 6 h resulted in a widening of the micropores maintaining also the microporous character of the material. A further increase in the activation time to 9 h led to the formation of mesopores. Therefore, adjusting the time of thermal treatment is a useful tool to control the pore structure of CDC. The microporous character of the CDC samples can be maintained by using moderate activation times.

The unique properties of porous CDC, such as high specific surface area and tunable pore size with a narrow pore-size distribution, make it an ideal material for sorbents, supercapacitor electrodes, or gas storage. CDC derived from biomorphic fibrous carbide structures has additionally the advantage of shape stability and very good mechanical performances, which make it suitable for advanced engineering applications, such as tribology or catalyst support, because of the easy separation from the reaction mixture.

ACKNOWLEDGMENTS

The author thanks the German Research Foundation (DFG) and Federal Ministry for Education and Research (BMBF) for the financial support in a frame of different grants. She likes to thank all students and co-workers who have contributed to this work, especially Dr. Daniela Almeida Streitwieser, Dr. Hanadi Ghanem, Dr. Martina Schmierler (born Kormann), Prof. Dr. Bastian Etzold, and Mr. Helmut Gerhard.

REFERENCES

[1] Y. Gogotsi, Carbon Nanomaterials, CRC Taylor & Francis, Boca Raton, 2006.
[2] V. Presser, M. Heon, Y. Gogotsi, Carbide-derived carbons – from porous networks to nanotubes and graphene, Adv. Funct. Mater. 21 (5) (2011) 810–833.
[3] M. Kormann, H. Gerhard, N. Popovska, Comparative study of carbide-derived carbons obtained from biomorphic TiC and SiC structures, Carbon 47 (2009) 242–250.
[4] T. Ota, M. Imaeda, H. Takase, M. Kobayashi, N. Kinoshita, T. Hirashita, H. Miyazaki, Y. Hikichi, Porous titania ceramic prepared by mimicking silicified wood, J. Am. Ceram. Soc. 83 (6) (2000) 1521–1523.
[5] N. Popovska, M. Kormann, Carbide-derived carbon: a novel approach for creating porous carbon materials with tuneable microstructure, JOM 62 (6) (2010) 44–49.
[6] D. Almeida Streitwieser, N. Popovska, H. Gerhard, G. Emig, Application of the chemical vapor infiltration and reaction technique (CVI-R) for the preparation of high porous biomorphic SiC ceramics derived from paper, J. Europ. Ceram. Soc. 25 (6) (2005) 817–828.
[7] N. Popovska, D. Almeida Streitwieser, Xu Chen, H. Gerhard, Paper derived biomorphic porous titanium carbide and titanium oxide ceramics produced by chemical vapor infiltration and reaction (CVI-R), J. Europ. Ceram. Soc. 25 (6) (2005) 829–836.
[8] M. Kormann, H. Ghanem, H. Gerhard, N. Popovska, Processing of carbide-derived carbon (CDC) using biomorphic porous titanium carbide ceramics, J. Europ. Ceram. Soc. 28 (2008) 1297–1303.
[9] M. Schmierler, Herstellung poroeserKohlenstoffe aus biomorphen Karbidkeramiken, Dissertation, Shaker Verlag Aachen, 2010.

[10] X. Chen, D.R. Cantrell, K. Kohlhaas, S. Stankovich, J.A. Ibers, M. Jaroniec, et al., Carbide-derived carbon and novel core-shell nanowires, Chem. Mater. 18 (2006) 753–758.

[11] J. Leis, A. Perkson, M. Arulepp, P. Nigu, G. Svensson, Catalytic effects of metals of the iron subgroup on the chlorination of titanium carbide to form nanostructural carbon, Carbon 40 (2002) 1559–1564.

[12] M. Kormann, H. Gerhard, C. Zollfrank, H. Scheel, N. Popovska, Effect of transition metal catalysts on the microstructure of carbide-derived carbon, Carbon 47 (2009) 2344–2351.

[13] M. Käärik, M. Arulepp, M. Karelson, J. Leis, The effect of graphitization catalyst on the structure and porosity of SiC derived carbons,, Carbon 46 (2008) 1579–1587.

[14] A. Perkson, J. Leis, M. Arulepp, M. Käärik, S. Urbonaite, G. Svensson, Barrel-like carbon nanoparticles from carbide by catalyst assisted chlorination, Carbon 41 (2003) 1729–1735.

[15] Y.J. Lee, The second order Raman spectroscopy in carbon crystallinity, J. Nucl. Mater. 325 (2004) 174–179.

[16] W. Lian, H. Song, X. Chen, L. Li, J. Huo, M. Zhao, G. Wang, The transformation of acetylene black into onion-like hollow carbon nanoparticles at 1000°C using an iron catalyst, Carbon 46 (2008) 525–530.

[17] A.C. Dupuis, The catalyst in the CCVD of carbon nanotubes – a review, Prog. Mater. Sci. 50 (2005) 929–961.

[18] P.A. Webb, C. Orr, Analytical Methods in Fine Particle Technology, Micromeritics Instrument Corporation, 1997.

[19] R.K. Dash, A. Nikitin, Y. Gogotsi, Microporous carbon derived from boron carbide, Microporous Mesoporous Mater. 72 (2004) 203–208.

[20] J. Leis, A. Perkson, M. Arulepp, P. Nigu, G. Svensson, Catalytic effects of metals of the iron subgroup on the chlorination of titanium carbide to form nanostructural carbon, Carbon 40 (2002) 1559–1564.

[21] D.C.S. Azevedo, J.C.S. Araújo, M. Bastos-Neto, A.E.B. Torres, E.F. Jaguaribe, C.L. Cavalcante, Microporous activated carbon prepared from coconut shells using chemical activation with zinc chloride, Microporous Mesoporous Mater. 100 (2007) 361–364.

[22] E. Vilaplana-Ortego, J.A. Maciá-Agulló, J. Alcañiz-Monge, D. Cazorla-Amorós, A. Linares-Solano, Comparative study of the micropore development on physical activation of carbon fibers from coal tar and petroleum pitches, Microporous Mesoporous Mater. 112 (2008) 125–132.

[23] S. Román, J.F. Gonzáles, C.M. Gonzáles-García, F. Zamora, Control of pore development during carbon dioxide and steam activation of olive stones, Fuel Process. Technol. 89 (2008) 715–720.

[24] C. Portet, M.A. Lillo-Ródenas, A. Linares-Solano, Y. Gogotsi, Capacitance of KOH activated carbide-derived carbons, Phys. Chem. Chem. Phys. 11 (2009) 4943–4945.

[25] S. Osswald, C. Portet, Y. Gogotsi, G. Laudisio, J.P. Singer, J.E. Fischer, V.V. Sokolov, J.A. Kukushkina, E.A. Kravchik, Porosity control in nanoporous carbide-derived carbon by oxidation in air and carbon dioxide, J. Solid State Chem. 182 (2009) 1733–1741.

[26] S.-H. Yeon, S. Osswald, Y. Gogotsi, J.P. Singer, J.M. Simmons, J.E. Fischer, M.A. Lillo-Ródenas, A. Linares-Solano, Enhanced methane storage of chemically and physically activated carbide-derived carbon, J. Power Sources 191 (2009) 560–567.

[27] Y. Gogotsi, C. Portet, S. Osswald, J.M. Simmons, T. Yildirim, G. Laudisio, G.E. Fischer, Importance of pore size in high pressure hydrogen storage by porous carbons, Int. J. Hydrogen Energy 34 (2009) 6314–6319.

[28] G. San Miguel, G.D. Fowler, C.J. Sollars, A study of the characteristics of activated carbons produced by steam and carbon dioxide activation of waste tyre rubber, Carbon 41 (2003) 1009–1016.
[29] M. Kormann, N. Popovska, Processing of carbide-derived carbons with enhanced porosity by activation with carbon dioxide, Microporous and Mesoporous Mater. 130 (2010) 167–173.
[30] F. Rodríguez-Reinoso, M. Molina Sabio, M.T. González, The use of steam and CO_2 as activating agents in the preparation of activated carbons, Carbon 33 (1995) 15–23.
[31] Ph. Becker, F. Glenk, M. Kormann, N. Popovska, B. Etzold, Chlorination of titanium carbide for the processing of nanoporous carbon: a kinetic study, Chem. Eng. J. 159 (2010) 236–241.

Chapter 10

Zeolite-Templated Carbon – Its Unique Characteristics and Applications

Hirotomo Nishihara and Takashi Kyotani
Institute for Multidisciplinary Research for Advanced Materials, Tohoku University, Katahira, Sendai, Japan

Chapter Outline

10.1. Introduction **295**
10.2. Synthesis Method **297**
10.2.1. Zeolite 297
10.2.2. Carbon Introduction 299
10.3. Structure of ZTC **301**
10.3.1. Bulk Shape 301
10.3.2. Ordered Structure 301
10.3.3. Textural Properties 304
10.4. Molecular Structure of ZTC **305**
10.4.1. Information Obtained from Various Kinds of Analysis 305
10.4.2. Construction of Possible Molecular Model 307
10.4.3. Origin of the Extremely Large Surface Area of ZTC 311
10.4.4. Comparison with Ordered Mesoporous Carbons 313
10.5. Application for Hydrogen Storage **314**
10.6. Application for Electrical Double-Layer Capacitors **316**
10.7. General Conclusion and Perspectives **317**
References **318**

10.1. INTRODUCTION

Zeolites have an open structure, which contains regular three-dimensional nanochannels. The shape and the volume of these channels are strictly

Novel Carbon Adsorbents. DOI: 10.1016/B978-0-08-097744-7.00010-7
Copyright © 2012 Elsevier Ltd. All rights reserved.

controlled by the framework topology. These channels are the basis of the action of zeolites as molecular-sieve adsorbents or catalysts. In addition, the well-defined nanospace in zeolites can be utilized for synthesizing new materials. The filling of the zeolite nanospace with carbon followed by the extraction of the carbon from the zeolite framework should give rise to the formation of a carbon with a porous structure reflecting the pore structure of the original zeolite template, as shown in the scheme in Fig. 10.1. This is called the "template carbonization method". Base on this idea, about 20 years ago, we started a research to synthesize a unique carbon structure using zeolite Y as a template and polyfurfuryl alcohol as a carbon precursor [1]. Furfuryl alcohol was introduced in the nanochannels of zeolite Y, polymerized and then carbonized. The carbon was liberated from the zeolite network by washing with HF and HCl solutions. The carbon thus obtained exactly reflected the morphology of the zeolite crystalline structure. Some carbon samples had BET surface areas about 1140 $m^2\ g^{-1}$, but, regrettably, it turned out that there was no ordering in pore structure. We therefore examined other combinations of precursors and templates under a wide range of experimental conditions. As a result, in 1997, we obtained carbon with much higher surface area, 2260 $m^2\ g^{-1}$, with a micropore volume as high as 1.1 $cm^3\ g^{-1}$ [2]. However, a substantial volume of mesopores was also present, indicating that the control of porosity was not perfect.

Six years before our publication in 1997 [2], Enzel et al. already made a similar research [3]. They pyrolyzed polyacrylonitrile in the channels of zeolite Y and mordenite, and then liberated the resulting carbons from the zeolites using HF solution to prepare conductive materials by the template method. The carbon obtained had much lower electronic conductivity than the carbon prepared from bulk polyacrylonitrile, suggesting that the spatial limitation in the zeolite channels prevents the formation of the graphitized structure with high conductivity. In 1992, Cordero et al. prepared a carbon/zeolite

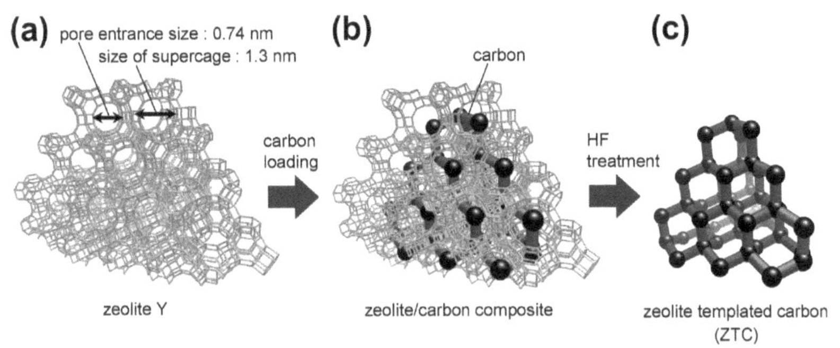

FIGURE 10.1 Synthesis scheme of the zeolite-templated carbon (ZTC). (a) Zeolite Y template. A Si–O–Si bond is expressed by a line. (b) Carbon/zeolite composite. (c) ZTC liberated from the zeolite template. *(Reprinted with permission from Ref [23]. Copyright 2009 Elsevier)*

composite by the carbon deposition from propylene pyrolysis over zeolite cracking catalyst, followed by the dissolution of the zeolite with HF and HCl solutions to isolate the pyrolytic carbon, and compared the high-temperature oxidation rate of the resulting carbon with other pyrolytic carbons [4]. Six years later, this group reported more detailed characterization results for the resulting carbons [5], which were very similar to those obtained in our study [2]. Johnson et al. reported the carbon preparation using different types of zeolite materials (Y, β, L) [6]. Meyers et al. also used different zeolites (Y, β, ZSM-5) for synthesizing new carbons [7]. These studies showed that all zeolite templates except the L type can endow the resulting carbons with porosity to a greater or lesser extent. However, none of these porous carbons reflected the ordered channel structures of the original zeolite templates. In other words, no one had succeeded in preparing an ordered porous carbon as a negative replica of the zeolite template, until we achieved success in 2000 [8].

The ordered porous carbon we finally prepared is named as zeolite-templated carbon (ZTC). Since the size of the zeolite nanochannels is too small (<1.3 nm) to allow the formation of usual carbon stacking structure, ZTC has an extraordinary unique structure, which has never been seen in any carbonaceous materials ever reported. Due to such a unique structure, ZTC is characterized with an extremely large surface area and micropore volume up to 4000 $m^2\ g^{-1}$ and 1.8 $cm^3\ g^{-1}$, respectively, and thereby adsorbs a large amount of molecules or ions. Such properties make ZTC very attractive adsorbents, especially as high-performance energy storage media. In this chapter, we first explain the synthesis method in great detail, where we place special emphasis on how the regular three-dimensional structure of zeolite can be replicated in the resulting carbon structure. Second, we introduce the morphology and structure of ZTC, as well as its unique pore structure. Third, we propose a possible molecular structure of ZTC and emphasize that it is a completely new nanocarbon form. Lastly, we demonstrate the excellent performance of ZTC in the application for hydrogen storage and electric double-layer capacitors.

10.2. SYNTHESIS METHOD

To synthesize ZTC, carbon must be uniformly introduced into narrow zeolite nanochannels. The key factors for successful synthesis are: (i) choice of zeolite template and (ii) carbon introduction, which are explained in the following sections.

10.2.1. Zeolite

Zeolite is a general name given to a series of crystalline aluminosilicates. Thus far, more than 190 different zeolite crystals have been discovered; however, ZTC can be synthesized only from the specific zeolites having appropriate pore structure, Si/Al ratio, and cation type.

10.2.1.1. Pore Structure

Each type of zeolite has its own pore structure. For ZTC synthesis, the pore entrance size should be as large as possible, from the following two reasons:

1. To avoid pore blocking during carbon introduction.
2. To form robust carbon framework which can retain the ordered network structure after the template removal.

The pore entrance of zeolite is usually a ring of O–Si–O chain, and the pore entrance size depends on the number of O atoms included in the ring. For example, the pore entrances of zeolite A and MFI are oxygen 8- and 10-membered rings, respectively, and they are too small to allow enough carbon introduction inside the zeolite micropores [9]. On the other hand, zeolites with 12-membered ring, such as zeolite Y (X), zeolite β, zeolite L, and mordenite, enable enough carbon filling [9].

Not only pore entrance size but also pore connection topology is significantly important. To obtain porous carbon replica with ordered framework like the structure shown in Fig. 10.1c, zeolite should have three-dimensionally connected micropores. Thus far, only zeolite Y (X) [8–10], zeolite β [9,11], and EMT [12] have been successfully used as hard templates for obtaining porous ZTC with high structure regularity. On the other hand, zeolites with one-dimensional micropores, such as zeolite L and mordenite, give one-dimensional carbon frameworks which do not retain structure regularity of the zeolite templates [9].

10.2.1.2. Si/Al Ratio

Zeolites usually have strong acidity, which plays a significant role for the uniform carbon deposition by chemical vapor deposition (CVD). This is one of the key processes to obtain ZTC with high surface area (see Section 10.2.2 for details). It is well known that the acid sites in zeolites catalyze the formation of polycyclic aromatic compounds called "coke". In the CVD process, due to the catalysis, it is possible to convert the hydrocarbon gas below its decomposition temperature into a carbonaceous structure inside the zeolite micropores, and a significant deposition of pyrolyzed carbon outside the zeolite particles can be avoided. Zeolites with poor acidity such as silicalite-1 therefore cannot be used as hard templates.

The origin of the strong acidity is Al-substituted sites in the zeolite framework. Accordingly, smaller Si/Al ratio gives larger acid site amount. The preferable Si/Al ratio for ZTC synthesis is below 20. It should be noted that zeolites containing too large amount of Al generally have poor heat stability. For example, zeolite A (Si/Al $=1$) and zeolite X (Si/Al <2) lose their crystal structures around 900 °C even if they contain carbon inside their micropores.

10.2.1.3. Cation Type

Each Al site in the zeolite framework has an exchangeable cation, and the cation type dominates the strength of the acid catalysis. Proton (H^+) gives a very strong acidity, thereby enhancing carbon deposition. In this case, zeolite micropores are easily blocked and carbon is deposited not only inside zeolite micropores but also outside. Transition metal ions, such as Ni^{2+} or Co^{2+}, give further high catalytic activity, and in some cases, even carbon nanotubes are generated [13,14]. The moderate catalytic activity is provided by Na^{+} cation, producing ZTC with high structure regularity as well as high surface area [8].

10.2.2. Carbon Introduction

There are three methods for the carbon introduction into the zeolite micropores.

1. Polymer carbonization which consists of the introduction of organic monomers, followed by polymerization and carbonization [2,6,15–18].
2. CVD method using small organic molecules as carbon sources [2,5,11,19,20].
3. Tow-step method which consists of the polymer carbonization followed by further CVD [8,21].

10.2.2.1. Polymer Carbonization

In the polymer carbonization, organic monomers are first introduced inside the nanochannels of zeolite and then they are polymerized. If the polymerization is proceeded by acid catalyst like the case of poly(furfuryl alcohol), zeolite itself can catalyze the polymerization. If this is not the case like poly(acrylonitrile), γ-ray radiation can be used for the polymerization in the zeolite [2]. This method allows uniform carbon introduction and is useful especially for zeolites with large particle sizes [16]. However, a large fraction of the organic polymer formed inside zeolite micropores is generally decomposed during carbonization step, and the resulting carbon framework tends not to be robust enough to retain high structure regularity after the removal of the zeolite template. Therefore, it is difficult to obtain ZTC with very high surface area by the polymer carbonization.

10.2.2.2. CVD Method

CVD method allows stuffing larger amount of carbon into zeolite micropores than the case of the polymer carbonization, and therefore the resulting carbon framework can be strong enough. In this method, a carbon source, e.g. hydrocarbon gas, is attached to zeolite below the decomposition temperature of the carbon source gas itself, and the zeolite catalyzes carbon deposition. Considering such a mechanism, the carbon source molecules should be as small as possible so that the source molecules can pass through the zeolite micropores even after narrowing by carbon deposition. Figure 10.2 illustrates the pore entrance (oxygen 12-membered ring) of zeolite Y before (Fig. 10.2a) and after (Fig. 10.2b) carbon deposition, together with several hydrocarbons

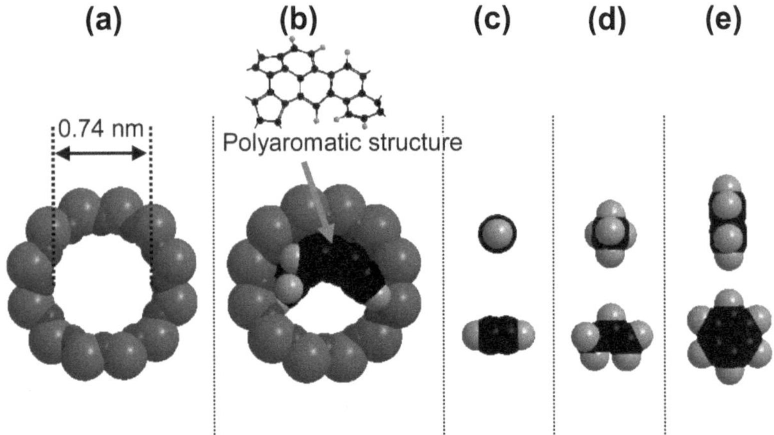

FIGURE 10.2 The pore entrance structure of zeolite Y (a) before and (b) after carbon deposition, and molecular images of (c) acetylene, (d) propylene, and (e) benzene (views from different angles). Polyaromatic structure shown in (b) is a fragment of the molecular model of ZTC shown later (Fig. 10.11c).

(Fig. 10.2c–e). Note that polyaromatic structure shown in Figure 10.2b is a fragment of the molecular model of ZTC shown later (Fig. 10.11c), i.e. this is a part of Fig. 10.11c. Acetylene (Fig. 10.2c) and propylene (Fig. 10.2d) can pass through the narrowed micropore (Fig. 10.2b), whereas it seems to be difficult for benzene (Fig. 10.2e). Acetylene is the smallest hydrocarbon and very well suitable as a CVD source for producing high-quality ZTC easily [19]. Not only hydrocarbons but also nitrogen-containing compounds, such as acetonitrile, can be used as a CVD source, and in this case, nitrogen-doped ZTC is obtained [11,22].

The operating conditions of CVD, such as temperature, period, and concentration of the CVD source gas, significantly affect the quality of the resulting ZTC.

As for the CVD temperature, higher temperature generally enhances carbon deposition. However, if the CVD temperature is too high, the CVD source molecule is decomposed not only in zeolite nanochannels but also in a vapor phase, and a large amount of carbon is deposited also outside zeolite particles. The self-decomposition temperature of acetylene and propylene are about 650 and 750 °C, respectively. In order to avoid carbon deposition outside the zeolite particles, the CVD temperature should be lower than the self-decomposition temperature.

The longer CVD time also increases the amount of carbon deposition. Below the self-decomposition temperatures, however, carbon deposition can be saturated after a certain period. For example, in the case of acetylene CVD on NaY zeolite at 600 °C, carbon deposition amount at first increases with increasing the CVD time, and the amount reaches ca. 28 wt% after 4 h. This is the maximum and no further carbon is deposited anymore.

The concentration of the CVD source gas is also a key factor for the carbon deposition. Under a fixed CVD temperature and time, larger concentration enhances carbon deposition.

Since the carbon deposited by CVD does not yet have a robust network structure at this moment, if zeolite is removed soon after the CVD, the ordered structure is likely to be lost. To stabilize the carbon framework, a high-temperature treatment is essential. The temperature of this treatment should be as high as possible. The upper limit of the temperature depends on the heat stability of zeolite: most of zeolites are not stable above 1000 °C.

10.2.2.3. Two-step Method

The two-step method takes over the advantages of each of two methods, i.e. uniform carbon deposition by the polymer carbonization and robust carbon framework by the CVD method. It is therefore possible to obtain ZTC with high structure regularity having very high surface area. Our group have actually demonstrated the synthesis of ZTC with an extremely high surface area of 4100 m^2 g^{-1} [21].

10.3. STRUCTURE OF ZTC

ZTC has a very unique structure which is completely different from conventional porous carbons, such as activated carbons, and this gives rise to its extraordinary adsorption properties. In this chapter, its basic structure found by electron microscopy, X-ray diffraction (XRD), and N_2 physisorption is explained.

10.3.1. Bulk Shape

ZTC is obtained as a negative replica of zeolite template. As a result, the bulk shape of the zeolite template is also replicated by ZTC. Figure 10.3 shows scanning microscopy (SEM) images of two zeolites with different particle sizes (Fig. 10.3a: 200 nm, Fig. 10.3b: 2 μm) and the resulting ZTCs from them (Fig. 10.3c and d). It is clearly seen that the bulk shapes of the zeolite templates are well transferred to the corresponding ZTCs. Thus, the bulk shape of ZTC can be controlled by changing the morphology of the zeolite template.

10.3.2. Ordered Structure

Figure 10.4 shows transmission electron microscopy (TEM) images of ZTC synthesized from zeolite Y. A low magnification image (Fig. 10.4a) shows that the shape of ZTC is polygonal particles which are derived from zeolite template, as was found from Fig. 10.3a. Besides bulk shape, microscopic

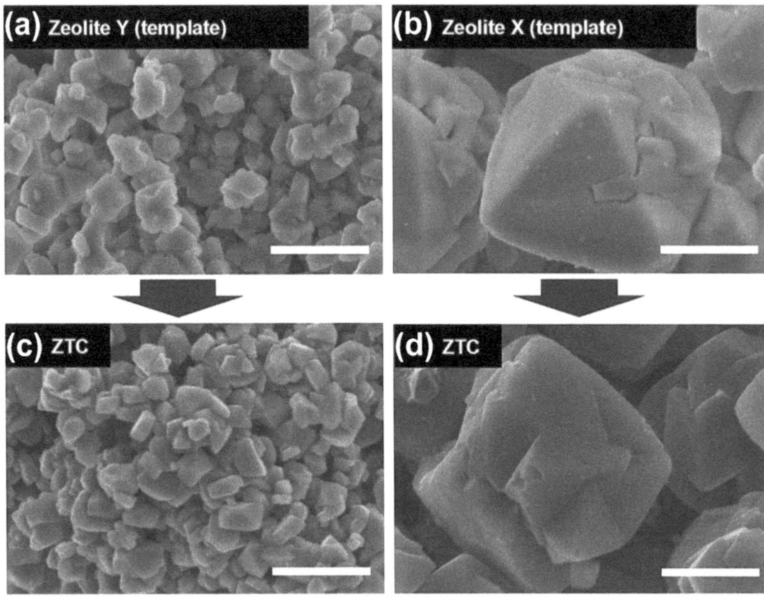

FIGURE 10.3 SEM images of (a) zeolite Y (particle size is ca. 200 nm), (b) zeolite X (particle size is ca. 2 μm), (c) ZTC from zeolite Y and (d) ZTC from zeolite X. Scale bar is 1 μm.

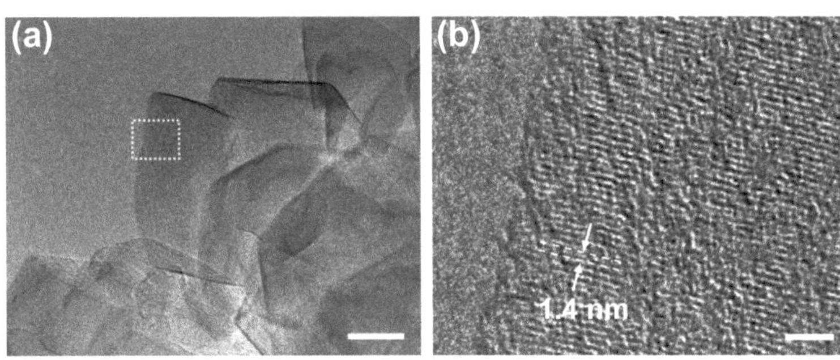

FIGURE 10.4 TEM images of ZTC obtained from zeolite Y. (b) shows an enlarged image of a part surrounded with a white square in (a). Scale bars in (a) and (b) are 100 and 10 nm, respectively.

structure of zeolite template is replicated by ZTC. A high magnification image (Fig. 10.4b) clearly shows zeolite-like fringes with a periodicity of 1.4 nm, corresponding to zeolite (111) plane.

Figure 10.5 shows XRD patterns of zeolite Y and the resulting ZTC. Generally, disordered carbons show broad peaks around $2\theta = 26°$ and $44°$ corresponding to carbon (002) and (10) peaks, respectively. However, ZTC with high structure regularity usually does not show these peaks, indicating that

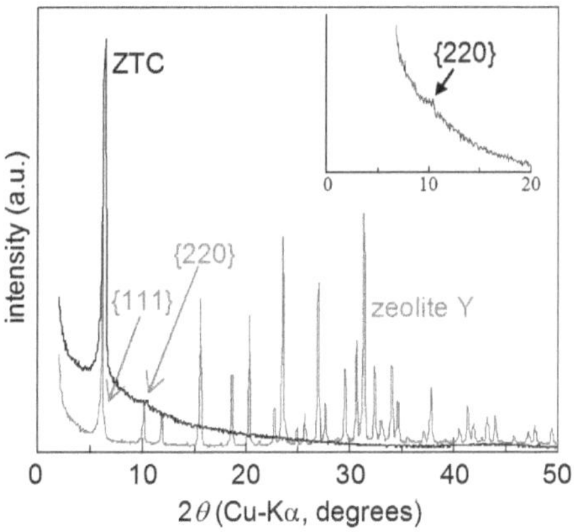

FIGURE 10.5 XRD patterns of zeolite Y and ZTC obtained as a replica. *(Reprinted with permission from Ref [23]. Copyright 2009 Elsevier)*

this carbon contains neither graphene-stacking structure nor large and flat graphene plane. Instead, ZTC has a sharp peak around $2\theta = 6.4°$ and also has a weak peak at $2\theta = 10.4°$. These peaks correspond to (111) and (220) planes derived from the zeolite template [23]. If the synthesis of ZTC is not successful, its XRD pattern shows broad peaks of carbon (002) and (10). There are two possible reasons for the appearance of these peaks. One possibility is the formation of disordered carbon shell outside zeolite micropores as shown in Fig. 10.6. In this case, the surface area of ZTC becomes lower, because such outside carbon is usually nonporous. The other possibility is a collapse of carbon framework. If carbon introduction into zeolite micropores is not enough, or if heat treatment for structure stabilization is not enough, carbon framework is partly destroyed upon the zeolite removal and graphene sheets formed inside the zeolite nanochannels are aggregated and/or stacked. Also in this case, the surface area of ZTC decreases significantly.

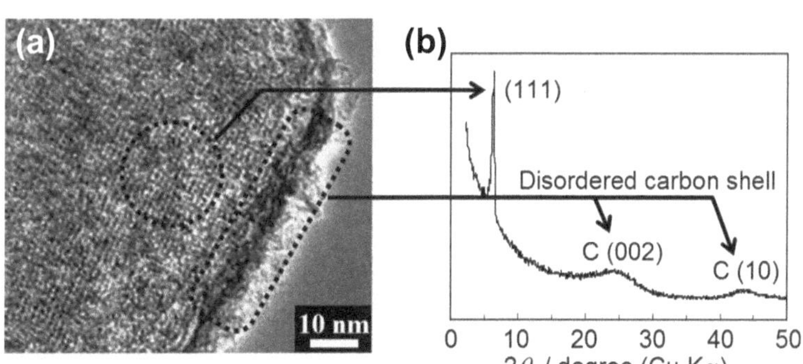

FIGURE 10.6 (a) A TEM image and (b) XRD patterns of ZTC containing a disordered carbon shell formed outside zeolite micropores.

10.3.3. Textural Properties

10.3.3.1. Analysis by N_2 Physisorption

Figure 10.7a shows N_2 isotherms of ZTC synthesized from zeolite Y and commercial activated carbons: MSC30 (Kansai Coke and Chemicals Co., Ltd.) and A20 (Unitika, Ltd.). ZTC has a typical type I isotherm by IUPAC definition, indicating that it is a microporous carbon. On the other hand, the isotherms of A20 and MSC30 are slightly deviated from the typical type I isotherm, with regard to adsorption increase around $p/p_0 = 0.1 \sim 0.3$, especially in MSC30. The adsorption in this pressure range corresponds to the presence of small mesoporous with the size of 2–4 nm [24–26]. Figure 10.7b shows pore-size distributions calculated by the density functional theory (DFT) method. As was indicated from their isotherms, MSC30 and A20 have broad distributions in the range of 1–3.5 nm. On the other hand, ZTC shows a sharp peak at 1.2 nm.

The BET surface areas and pore volumes calculated from the N_2 isotherms are summarized in Table 10.1. MSC30 is a highly activated carbon produced by KOH chemical activation, and has a very high BET surface area of $2690\ m^2\ g^{-1}$, which is almost the upper limit for conventional activated carbons. However, ZTC has a further higher BET surface area of $4030\ m^2\ g^{-1}$. Another remarkable feature of ZTC is its very large micropore volume ($1.69\ cm^3\ g^{-1}$) without mesopores. In the case of a conventional activation method, not only micropores but also mesopores are inevitably formed, especially by deep activation for obtaining high surface area. MSC30 thus has a large total pore volume of $1.73\ cm^3\ g^{-1}$, but the fraction of micropores is no more than 65% due to the presence of significant mesopores. In the case of ZTC, 95% of its total pore volume is of micropore volume (Table 10.1). Having very high surface area without the presence of mesopores is important especially for energy-storage applications such as hydrogen storage and electric

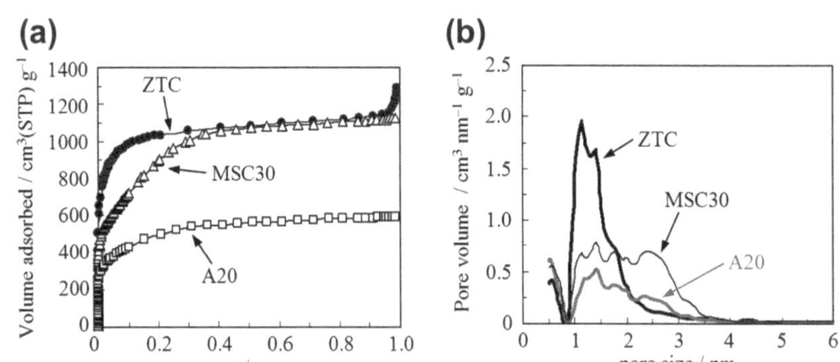

FIGURE 10.7 (a) N_2 adsorption–desorption isotherms and (b) pore-size distributions of ZTC synthesized from zeolite Y and commercially activated carbons (MSC30 and A20).

TABLE 10.1 BET Surface Areas and Pore Volumes of ZTC and Activated Carbons

Sample	S_{BET}[a] $m^2\ g^{-1}$	V_{micro}[b] $cm^3\ g^{-1}$	V_{meso}[c] $cm^3\ g^{-1}$	V_{total}[d] $cm^3\ g^{-1}$
ZTC	4030	1.69	0.09	1.78
MSC30	2690	1.12	0.61	1.73
A20	1650	0.64	0.27	0.91

[a]*BET surface area (p/p_0 = 0.01–0.05).*
[b]*Micropore volume calculated by the Dubinin–Radushkevich equation.*
[c]$V_{total} - V_{micro}$.
[d]*Total pore volume estimated from the adsorption of N_2 at p/p_0 = 0.96.*

double-layer capacitors, because mesopores usually lower the density of the porous materials, resulting in a lower energy density per unit volume.

10.4. MOLECULAR STRUCTURE OF ZTC

Since we first succeeded in the synthesis of ZTC in 2000 [8], many researchers have been working on the development of several types of ZTCs, using several types of zeolite templates and carbon sources. However, its molecular structure had been completely unclear for a long period, because lattice fringes of graphene sheets can never be observed even with high-resolution TEM (Fig. 10.4b). Then, we made a great effort on analyzing ZTC structure with a variety of methods, and finally unveiled that its framework is comprised of cross-linked buckybowl-like nanographenes, which is a completely new nanocarbon form of sp^2 carbon network. In this section, the process of determining the molecular structure is presented.

10.4.1. Information Obtained from Various Kinds of Analysis

It has been known that ZTC is a negative replica of zeolite template from the results of SEM, TEM, and XRD, as shown in Figs 10.3–10.5. TEM and XRD results clearly indicate that ZTC has an ordered network structure which is replicated along with the pore network structure of the zeolite, and the ordered network has the same symmetry structure ($Fd\overline{3}m$) as C–C network in a diamond crystal (Fig. 10.1c). As was described in the Section 10.3.2, the XRD pattern of ZTC does not show carbon (002) nor (10) peaks, indicating that this carbon contains neither graphene-stacking structure nor large and flat graphene plane. This is reasonable considering the size of the zeolite pores. Figure 10.8 compares the sizes of the zeolite pore and several molecules, benzene, coronene, and fullerene (C_{60}). The spherical zeolite pore (supercage) is only 1.3 nm,

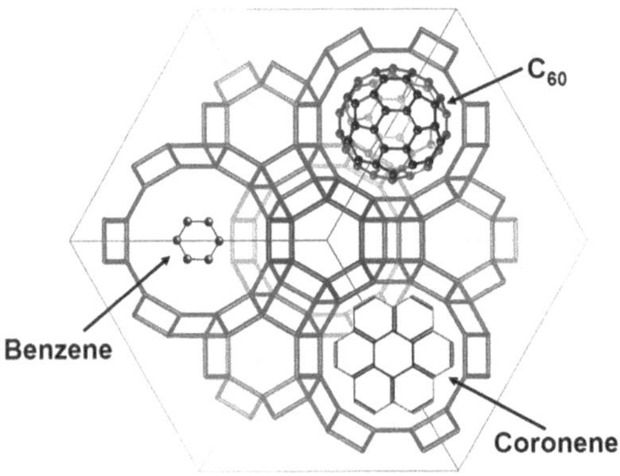

FIGURE 10.8 Comparison of a supercage of zeolite Y and several molecules. A Si–O–Si bond is expressed by a line. Dodecagons surrounding benzene, coronene, or C_{60} correspond to the supercage. This image is from the (111) direction of zeolite Y.

and the size barely allows coronene or C_{60} to be included. Accordingly, a large graphite crystal never forms in such a small nanospace due to the structure hindrance. The ZTC framework must be comprised of very small carbonaceous fragments whose sizes are less than 1.3 nm.

ZTC has several characteristics which are typically seen in sp^2-carbon materials, such as black color, hydrophobicity, and electrical conductivity. The results of ^{13}C NMR [27] and electron energy-loss near-edge fine structure spectra [23] indeed show the presence of sp^2 carbons in ZTC, and sp^3 carbons are not detected. ZTC is therefore mainly comprised of sp^2 carbons.

In order to obtain further information on the sp^2-carbon framework, we measured Raman spectrum of ZTC and compared it with those of the reference carbons (Fig. 10.9). ZTC exhibits some typical Raman features for carbon

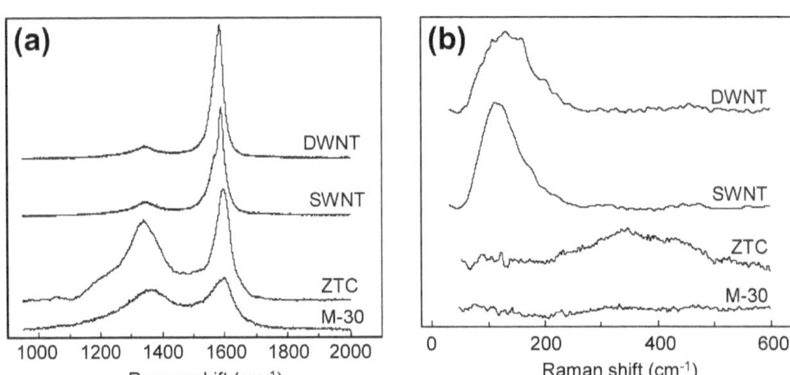

FIGURE 10.9 Raman spectra of ZTC and some reference carbons [double- and single-walled carbon nanotubes (DWCNT and SWCNT) and activated carbon M-30] recorded with 488 nm laser line. (a) High-frequency Raman spectra and (b) low-frequency Raman spectra. *(Reprinted with permission from Ref [23]. Copyright 2009 Elsevier)*

materials: the tangential G band around 1598 cm^{-1} and the defect-induced D band around 1345 cm^{-1}. Such intense G band directly indicates the significant presence of graphene sheets in ZTC, and D band indicates that the graphene has defects such as edge site. Additionally, the XRD pattern and the TEM observation did not show any evidence for sheet stacking of the nanographenes (Figs 10.4 and 10.5). These findings indicate that this carbon is made up by the assembly of single nanographene sheets, but not by the assembly of stacked graphene sheets, as is common for carbon materials. In addition to the intense G and D bands, we observed a broad Raman feature below 500 cm^{-1}, i.e. in a frequency range similar to that commonly found for the radial breathing mode (RBM) vibrations of SWCNTs or DWCNTs [28,29]. Figure 10.9b shows the low-frequency Raman spectrum of ZTC, together with those of some reference carbon materials, i.e. SWCNTs, DWCNTs and activated carbon M-30 [23]. We see from Fig 10.9b that ZTC has a broad Raman feature between 230 and 480 cm^{-1}. In marked contrast, the M-30 sample does not have any features in this frequency range. Such Raman features as observed for ZTC have also been reported for other carbons that consist of, or at least contain, curved graphitic sheets of a very small diameter of curvature (below 2–3 nm), including some types of bent graphitic layers [30,31], carbon onions [32]. The band at 230–480 cm^{-1} observed in the Raman spectrum of ZTC can therefore also be considered as an indication of the presence of curved graphene sheets with a very small diameter of curvature in this sample. The presence of such curved graphene sheets in ZTC is reasonable, because the sheets must be forced to accommodate themselves into the more or less spherical cavities (~1.3 nm in diameter) of the zeolite Y supercages and the ~0.74 nm wide necks interconnecting adjacent cavities.

10.4.2. Construction of Possible Molecular Model

Based on the experimental evidences described in the precious section, we attempted to construct a reasonable molecular model for ZTC that could fit into the three-dimensionally ordered pores of zeolite Y [23]. This model must be consistent with the structural information obtained by the different characterization techniques (Raman, XRD, TEM, ^{13}C NMR, and electron energy-loss near-edge fine structure spectra). In addition, we imposed the condition that the model structure must be energetically stable upon the geometry optimization calculation using the semi-empirical Hamiltonian PM5 in MOPAC. To adapt a graphene sheet to the zigzagged channels of the zeolite, the sheet has to be artificially curved, thereby consisting of not only hexagons but also pentagons and/or heptagons. There are very few possibilities to arrange these polygons in such strictly regulated nanospace. Moreover, the graphene sheet should display a tetrahedral network structure as shown in Fig. 10.1c. One possibility would be a negatively curved graphene network that is known as Schwarzite [33,34]. Figure 10.10(a and b) shows two of such possible carbon models where the

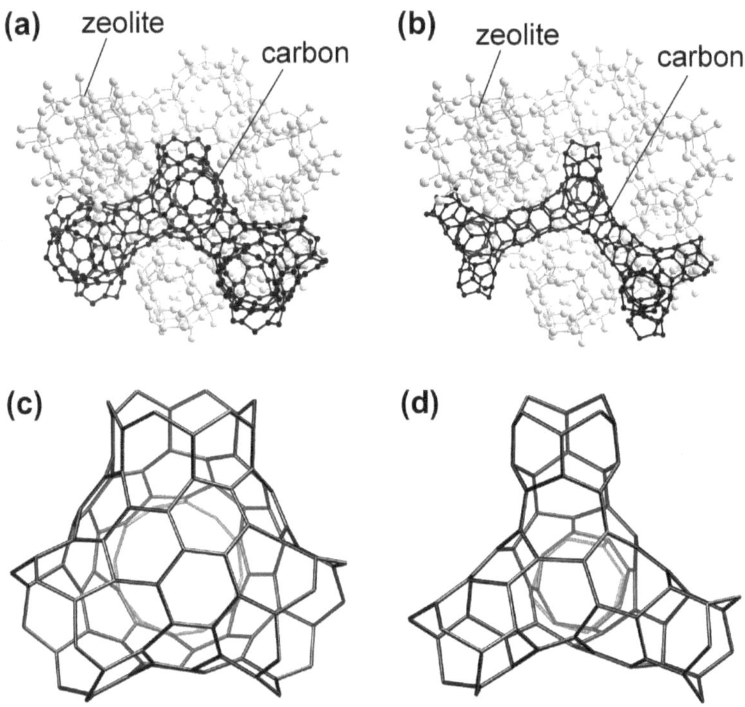

FIGURE 10.10 Two hypothetical molecular structures of ZTC. The images (a) and (b) show the two ZTC models fitted into the framework of zeolite Y. (a) and (b) are comprised of building units (c) C_{88} and (d) C_{68}, respectively, both of which have twelve heptagons. Note that one supercage of zeolite Y contains one building unit for both models. *(Reprinted with permission from Ref [23]. Copyright 2009 Elsevier)*

carbon atoms (black spheres) are interlinked with each other by sp^2-bonds to form a single and extremely curved graphene sheet with the three-dimensionally regular ordering. Each model is made up of building units, each of which fits into one supercage of zeolite Y. The units of models shown in Fig. 10.10a and b contain 88 and 68 carbon atoms (Fig. 10.10c and d), which respectively correspond to 0.67 and 0.52 g g^{-1} of zeolite at the stage of the zeolite/carbon composites. A similar molecular model was also proposed by Roussel et al. with a statistical mechanics approach [35]. However, also these models have much larger amount of carbon against zeolite template compared to the amount experimentally obtained (0.29 g g^{-1} of zeolite). If 50 carbon atoms were removed from the building unit of the model shown in Fig. 10.10a, its carbon content would become similar to the actual value. In the case of the model shown in Fig. 10.10b, 30 atoms must be eliminated. The removal of so many carbon atoms from each unit, however, inevitably leads to the collapse of the ordered structures of both models. Carbon frameworks composed of tubular graphenes are therefore different from the actual structure.

We then made every effort to construct other possible models, but always failed because the actual number of carbon atoms in zeolite is too small to construct a robust model. As mentioned before, in Fig. 10.10a and b, each building unit is tetrahedrally connected to four other units as shown in Fig 10.1c. However, we realized that, even if each building unit was not connected to all four other ones, but only to three others (Fig. 10.11a), the ordered structure could still be kept robust enough. Using this unsaturated structure, we again tried to construct an idealized molecular model for ZTC. Since the building unit shown in Fig. 10.11a (a black ball with three sticks) has a threefold axis, an idealized molecular model as the building unit should also have a threefold rotational symmetry. We finally found that the carbon model

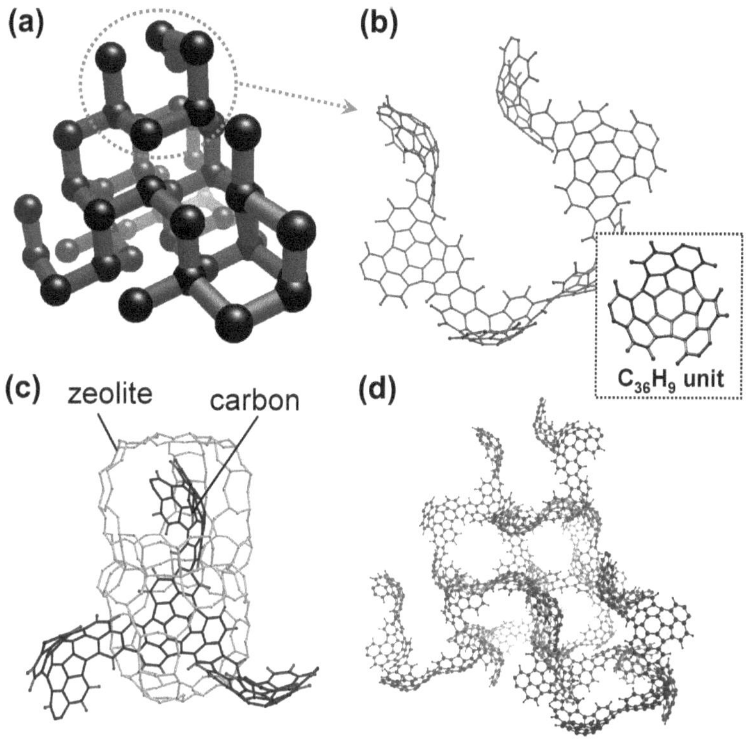

FIGURE 10.11 Idealized molecular model containing only carbon and hydrogen atoms. (a) A ball and stick model for the structure of the ZTC framework. Black sphere and gray stick correspond to a basic buckybowl unit and the connection between two units, respectively. (b) An idealized molecular model constructed with six buckybowl units of $C_{36}H_9$ (inset). All the graphene edges are passivated with hydrogen atoms. The structure in (b) corresponds to the part indicated by a dotted circle in the network structure in (a). (c) The molecular model constructed with four buckybowl units with zeolite framework. (d) A large-sized molecular model that is constructed with the buckybowl units following the regularity of the zeolite Y template. The model (d) has the same network structure as the model (a). *(Reprinted with permission from Ref [23]. Copyright 2009 Elsevier)*

proposed in Fig. 10.11b met all the requirements. In this model, a basic building unit corresponding to a black sphere, shown in the inset of Fig. 10.11b, is a small curved graphene ($C_{36}H_9$, containing 3 pentagons), which is interlinked with three other units as indicated by the ordered framework structure in Fig. 10.11a. Note that the unit is quite similar to fullerene fragments, so-called buckybowls. The framework structure of this model looks unstable, but surprisingly the structure was energetically stable when the whole model shown in Fig. 10.11b was subjected to the geometry optimization calculation using the molecular orbital theory. Figure 10.11c shows the ZTC model embedded in the zeolite framework. It is clearly found that the present molecular model can fit the pore system of the zeolite. The amount of carbon in the model was calculated to be 0.28 g g^{-1} of zeolite, which is very similar to the experimental value (0.29). In Fig. 10.11d, we present a large-sized idealized molecular model prepared from the buckybowls. Single and curved graphenes make up the three-dimensionally ordered network structure. We must emphasize that the steric hindrance imposed by the zeolite channels and the necessary number of carbon atoms are so strict that using sp^2-carbon atoms it was essentially impossible to make up any different network structures other than the present one.

ZTC actually contains a significant amount of oxygen as high as 8.9 wt% [23], which should be included in the idealized model of ZTC. Thus, we carried out the detailed analysis for oxygen functional groups. From the joint analysis using temperature programmed desorption (TPD) and FT-IR, we determined the quantities of oxygen functional groups, and constructed a modified molecular model in which the edge sites of the buckybowl units are occupied by appropriate types and amounts of functional groups, as shown in Fig. 10.12 [23]. Note that the whole structure (with the functional groups) shown in Fig. 10.12 was found to be energetically stable by the geometry optimization calculation using the molecular orbital theory.

In order to confirm the reliability of the proposed molecular model (Fig. 10.12), we compared numerical data of the model with these obtained by experiments, as summarized in Table 10.2 [23]. All of the values obtained from the model and experiments agree with each other very well.

Thus, we have proposed a possible molecular structure of ZTC, which possesses all the characteristics deduced from the experimental analyses of this material. This ordered microporous carbon is comprised of the assembly of single, nonstacked nanometer-sized graphene fragments, these nanographenes being curved like buckybowls due to the steric hindrance of the zeolite nano-channels in which they are formed. This three-dimensionally ordered and highly porous buckybowl-based structure is unique among any sp^2-bonded carbon materials produced so far. Although the actual carbon material will differ a little from this ideal model, the former can be thought of as a distorted version of the latter (*via* addition/removal of carbon atoms from, or twisting of, the buckybowls, etc), which still retains main features of the idealized model. The present model can therefore be considered as a reasonable starting point for

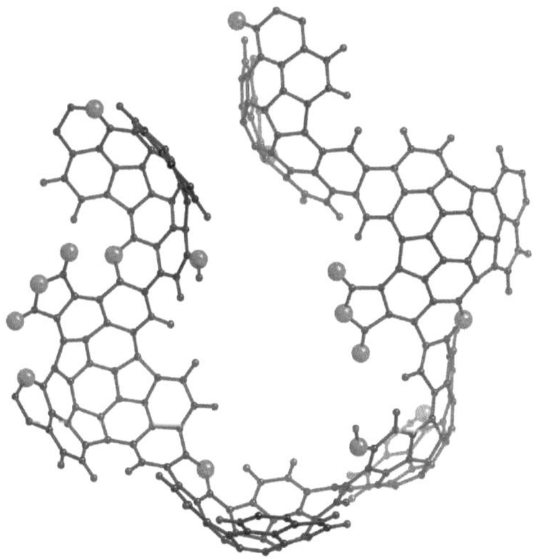

FIGURE 10.12 An idealized model of ZTC containing functional groups (7 esters, 2 phenolic hydroxyl groups, 2 acid anhydride and 1 carboxyl group), adjusted to reflect the results of FT-IR and TPD. Large spheres indicate O atoms. *(Reprinted with permission from Ref [23]. Copyright 2009 Elsevier)*

future refinements of the structure of this quite novel carbon material and also for the theoretical modeling and understanding of ever-found and to-be-found interesting properties of ZTC, such as its high H_2 adsorption capacity and high performance as an electrode material for electric double-layer capacitors, as explained in the later sections.

10.4.3. Origin of the Extremely Large Surface Area of ZTC

ZTC shows an extremely high BET surface area of 4030 $m^2\ g^{-1}$ (Table 10.1). It should be noted that this value is estimated by the BET method, which usually

TABLE 10.2 Comparison of Numerical Data Obtained From Experiments and Molecular Model Shown in Figure 10.12

	Experiment [23]	Model
Carbon amount in carbon/ zeolite composite (wt%)	22	23
Graphene curvature (nm)	0.5–1.1	0.9–1.3
Pore diameter (nm)	1.0–1.5	1.0–1.7
Micropore volume ($cm^3\ g^{-1}$)	1.8	1.71
Surface area ($m^2\ g^{-1}$)[a]	3730	3432

[a]*Surface area calculated by the SPE method [36].*

causes an overestimation especially in the calculation for microporous carbons [36]. For ZTC, its real surface area can be actually estimated to be 3432 m^2 g^{-1}, from its molecular structure shown in Fig. 10.12 [23]. Thus, the geometrical surface area of ZTC is over the theoretical surface area of both sides of a single graphene, 2627 m^2 g^{-1}. The reason for this extraordinary high surface area is the contribution of the edge planes. Figure 10.13 illustrates geometrical surface areas of graphene (Fig. 10.13a), polycyclic aromatic compounds (Fig. 10.13b and c), and benzene (Fig. 10.13d). It is well known that the theoretical surface area of both sides of an infinity large graphene is 2627 m^2 g^{-1} (Fig. 10.13a), where the contribution of edge planes (Fig. 10.13b) is not considered. However, the contribution of the edge plane becomes significant when the size of graphene is less than several nanometers, and therefore, the geometrical surface area could be larger than 2627 m^2 g^{-1} in nanographenes (Fig. 10.13b–d). Accordingly, ZTC which is composed of the nanographene network (Fig. 10.12) has the extraordinary high surface area due to a large amount of edge planes.

Recently, “graphene” or “graphene-based materials” have attracted significant attention and many researchers are trying to use them for various kinds of applications including for adsorbents, considering its large theoretical surface area of 2627 m^2 g^{-1}. It should, however, be noted that most of graphene surface is usually hidden and not fully available for adsorption. For example, the actual BET surface area of the reported “graphene-based material” was as low as 705 m^2 g^{-1} [37], indicating that more than 2/3 surface of the graphenes is not exposed to the outside. This is due to the aggregation of graphenes by strong van der Waals interaction. In order to expose the entire surface, graphene has to be formed into a self-standing three-dimensional network, like a pillared graphene which has been proposed by Froudakis et al. in their theoretical

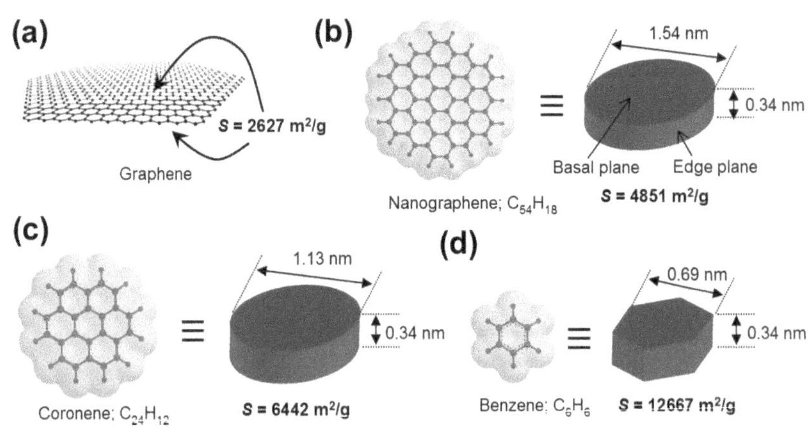

FIGURE 10.13 Molecular models and the corresponding geometrical surface areas of (a) graphene, (b) nanographene ($C_{54}H_{18}$), (c) coronene, and (d) benzene.

work [38]. In this sense, ZTC does have an ideal three-dimensional structure (Fig. 10.11d) where the entire surface of the nanographene components is fully exposed to outside and can contribute to adsorption.

10.4.4. Comparison with Ordered Mesoporous Carbons

The first report for the synthesis of ZTC was actually done by our group in 1997 [2]. In this report, we used zeolite as a template and synthesized microporous carbons basically according to the same synthesis scheme as Fig. 10.1. Unfortunately, at that time, we did not confirm the replication of zeolite-ordered structure by the obtained carbons. We have then confirmed that ZTC has an ordered structure derived from zeolite template in 2000 [8].

In 1999, two Korean groups separately developed the method to obtain ordered mesoporous carbons (OMCs), according to the similar synthesis scheme to Fig. 10.1 but using mesoporous silica as hard templates (Fig. 10.14) [39,40]. Mesopores silica has much larger pores (>2 nm) than zeolite (<1.3 nm), and therefore, carbon introduction can be more easily achieved than the case of zeolite. From its easy synthesis procedure, OMC attracted a great deal of attention, and rapidly spread worldwide.

Because of their similar synthesis procedure, sometimes ZTC and OMC are not clearly distinguished from each other, rather they are regarded as similar materials having uniform nanopores. However, they are completely different carbon materials based on their molecular-level structures, as zeolites are different from mesoporous silicas. Figure 10.15 illustrates molecular-level structures of OMC [39–41] and ZTC together with their specific features. OMC has a uniform structure derived from the mesoporous silica template, but the framework itself is a disordered carbon which contains graphene-stacking structure and random micropores, like activated carbons. Thus, the microscopic structure of OMC is intrinsically different from that of ZTC, i.e. single nanographene network. The different framework structures make their pore structure also very much different. OMC has a uniform mesopores and also has random micropores inside the framework. Accordingly, OMC has a bimodal pore-size distribution both in micro- and mesopore regions. On the other hand,

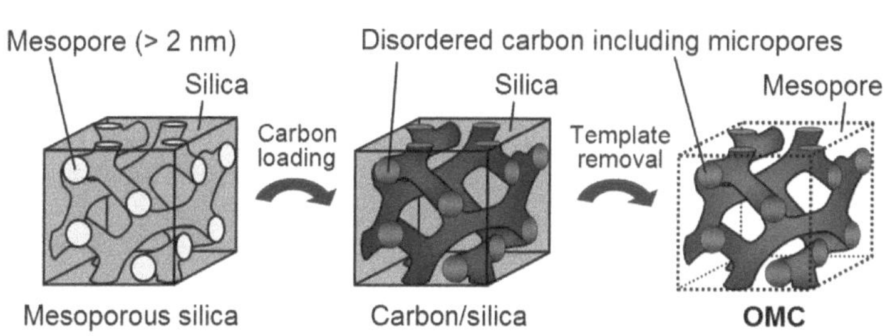

FIGURE 10.14 Synthesis scheme of the ordered mesoporous carbon (OMC).

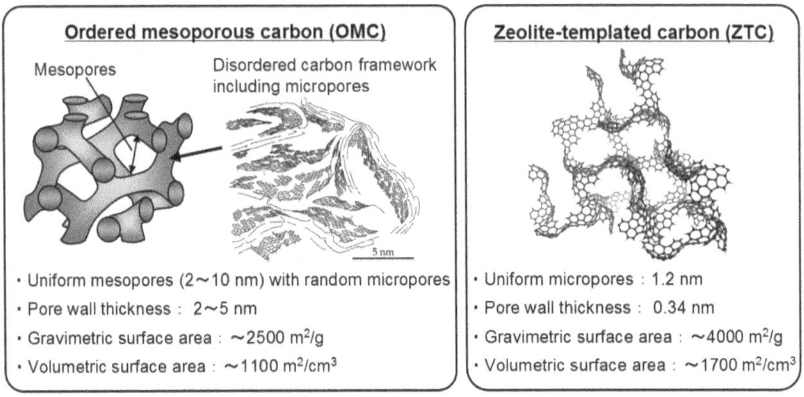

FIGURE 10.15 Illustrations and specific features of ordered mesoporous carbon (OMC) and zeolite-templated carbon (ZTC).

ZTC is comprised of single nanographenes forming unimodal micropores with the size of 1.2 nm, and there are no other extrapores, such as ultramicropores, mesopores, and macropores. Due to the presence of micropores, OMC has a high gravimetric surface area of 2500 $m^2\ g^{-1}$, whereas its volumetric surface area is no more than 1100 $m^2\ cm^{-3}$, because the presence of large mesopores makes the density of carbon significantly low. Note that the volumetric surface area is calculated by (gravimetric surface area) $\times$ (density of carbon particle). On the other hand, ZTC has not only high gravimetric surface area (ca. 4000 $m^2\ g^{-1}$) but also very high volumetric surface area (1700 $m^2\ cm^{-3}$), due to the absence of larger pores. The volumetric surface area is very important to achieve high energy density per unit volume.

10.5. APPLICATION FOR HYDROGEN STORAGE

A fuel cell vehicle is a next-generation automobile that uses an energy source other than crude oil derivatives. However, there are several problems that should be solved before practical use. One of them is how to store hydrogen; that is, the development of a safe and efficient hydrogen-storage system is required [42]. There are several methods to store hydrogen: cryogenic liquid, compressed gas, metal hydrides, and adsorption. Each of them has advantages and disadvantages. Among these methods, adsorption of hydrogen in porous solids, especially carbon materials, is an attractive choice. They enjoy a long lifecycle, high durability, and hydrogen release at low temperature but lack sufficient storage capacity at room temperature. On the basis of physisorption, large surface area [43,44], large micropore volume [45–47], and a suitable pore size [45,46,48] are highly desirable for carbon materials to store enough hydrogen at room temperature. Highly microporous carbons are therefore the most promising, rather than carbon nanotubes or carbon nanofibers. In the conventional chemical activation, larger surface area can be obtained under severe activation conditions, but at the same time, such activation significantly increases the amount of unnecessary mesopores. It is thus very difficult to achieve very high surface area and appropriate pore size simultaneously by conventional activation methods.

ZTC can overcome this problem. Several groups have reported the hydrogen-storage performance of several types of ZTCs at −196 °C [11,20,35,49,50]. It is noteworthy that Mokaya and co-workers [11] showed that ZTC exhibits the highest hydrogen uptake capacity among porous materials, including porous carbons, carbon nanotubes, zeolites, and metal–organic frameworks. In addition, Lachawiec and Yang [51] have reported noticeable large hydrogen uptake on ZTC at room temperature (up to 10 MPa). Our group has reported that ZTC exhibits a superior storage capacity especially at ultrahigh pressure range at room temperature [52]. Figure 10.16 shows hydrogen adsorption isotherms of ZTC and a high surface area activated carbon (MSC30, see Table 10.1) up to 34 MPa. In MSC30, its isotherm apparently levels off above 20 MPa, and it almost reaches the ceiling at 30 MPa. However, ZTC does not level off up to 34 MPa. Cazorla-Amorós and co-workers [45] have measured high-pressure hydrogen isotherms of various kinds of activated carbons at room temperature up to 70 MPa. They reported that most of the activated carbons reached a plateau above 20 MPa, but some showed almost no leveling off even at 70 MPa. They explained that there is an optimal pore size that can enhance the adsorption. The optimal pore size changes with the storage pressure. Kowalczyk et al. [48] have theoretically predicted that, in the case of carbon slit pores at 30 °C, 0.65 nm is optimum below 30 MPa and the optimal pore size is increased to 2.13 nm at 50 MPa. Thanks to the zeolite template, ZTC contains only such suitable-sized micropores (about 1.2 nm) for the enhancement effect. ZTC thus can exhibit one of the highest storage amounts (2.2 wt % at 30 °C and 34 MPa) among any pure carbon materials.

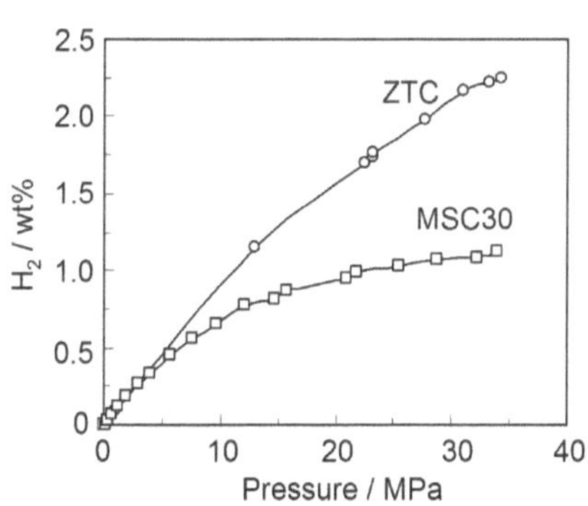

FIGURE 10.16 Hydrogen adsorption isotherms of ZTC and MSC30 up to 34 MPa, measured at 30 °C. *(Reprinted with permission from Ref [52]. Copyright 2009 American Chemical Society)*

10.6. APPLICATION FOR ELECTRICAL DOUBLE-LAYER CAPACITORS

Electric double-layer capacitor (EDLC) is one of the key energy-storage devices for the development of clean energy technology including next-generation electric vehicles [53]. EDLC is charged in principle through the physisorption of electrolyte ions onto nanopore surfaces of a carbon electrode. Before the ions are adsorbed, they need to migrate from a bulk solution to the electrode surface through the narrow nanopores. Thus, the reduction of the ion-transfer resistance in the nanopores is one of the key issues to realize a high-power density EDLC [54–56]. Thus far, great efforts have been made to produce nanostructured electrode materials that possess efficient paths for ion diffusion. The successful examples are single-walled carbon nanotubes (SWCNTs) [57–59], templated mesoporous carbons (TMCs) [54,55], and hierarchical porous carbons (HPCs) [60,61]. These high-rate materials are characterized with large diffusion paths, the sizes of which are much larger than the size of solvated ions (~0.5 to 1.4 nm) [62], allowing the ions to move very rapidly during the charge/discharge processes. Consequently, these materials can retain their capacitance even at an ultrahigh current, i.e. exhibiting a high-rate performance, which is essential to construct a high-power density EDLC. However, paths that are too large make the electrode density significantly low and thus seriously decrease the volumetric capacitance, eventually resulting in the low volumetric energy density of the final EDLC device [60]. To avoid this problem, the size of the diffusion path should be as small as possible, but the fast ion transport has to be retained.

We have demonstrated that it is possible to meet the two seemingly contradictory requirements with ZTC, which can be clearly distinguished from TMCs with narrow pore-size distributions in the mesopore range (>2 nm) and therefore with the low volumetric capacitance [56,63]. Table 10.3 summarizes gravimetric (S_g) and volumetric (S_v) surface areas for several high-rate materials reported so far. In principle, S_g and S_v have directly to do with the gravimetric and the volumetric capacities, respectively. It is clear that ZTC has a great potential for higher capacity, especially for the volumetric one, compared with the others.

Figure 10.17 shows the change of the volumetric capacitances with the current density in 1 M Et_4NBF_4/propylene carbonate (PC) electrolyte solution at 25 °C. Two types of ZTCs are compared with activated carbons. ZTC-S has a higher surface area of 3690 $m^2\ g^{-1}$, while ZTC-L has a lower surface area of 2910 $m^2\ g^{-1}$ but has a better electric conductivity. Note that a larger current density corresponds to a faster discharge rate. In the low-current region below 2 $A\ g^{-1}$, all the ZTCs and the activated carbons show larger volumetric capacities than those of the high-rate materials (SWCNT [58] shows 22 $F\ cm^{-3}$ at 10 $A\ g^{-1}$, and HPC [60] shows 39 $F\ cm^{-3}$ at 4 $A\ g^{-1}$) due to their high volumetric surface areas. However, A20, a microporous activated carbon (Fig. 10.7b), cannot retain the capacitance, which decreases rapidly with increasing the current. MSC30, a carbon with some mesopores (Fig. 10.7b), retains 40% of its initial capacitance even at an ultrahigh current density of

TABLE 10.3 Gravimetric (S_g) and Volumetric (S_v) Surface Areas of the High-Rate Materials Reported So Far [63]

Carbon	S_g [$m^2\ g^{-1}$]	S_v [$m^2\ cm^{-3}$][a]	Reference
SWCNT	~1000	~600	[58]
TMC	~1700	~900	[55]
HPC	~2060	<510[b]	[60,61]
ZTC	~3040	~1460	[56]

[a] *$S_v = S_g\,\rho$, where ρ is the particle density of carbon. ρ was calculated by $\rho = \{V_{total} + 1/\rho_{carbon}\}^{-1}$, where V_{total} is the total pore volume estimated from N_2 isotherm (77 K) and ρ_{carbon} is the true density of carbon (2 g cm^{-3}).*

[b] *ρ of HPC was calculated without considering the total macropore volume because of a lack of the data in the reference. Therefore S_v of HPC could actually be much less than 510 $m^2\ cm^{-3}$.*

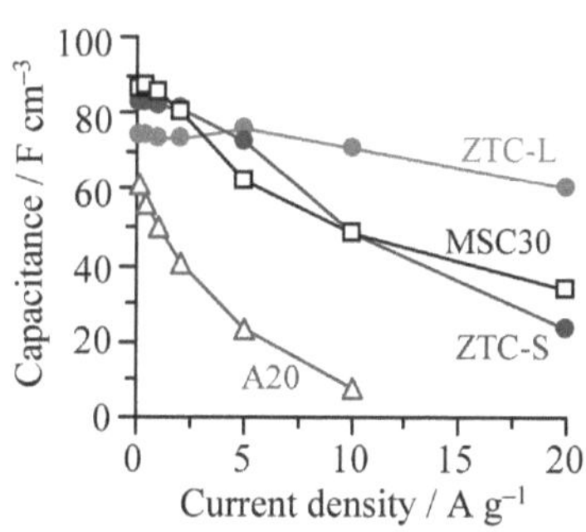

FIGURE 10.17 Volumetric capacitance (C_v) versus current density for two types of ZTCs (ZTC-S and ZTC-L) and activated carbons (MSC30 and A20). C_v was calculated by $C_v = C_s\,F$, where C_s is a gravimetric capacitance obtained by galvanostatic charge–discharge cycling, and F is the particle density (see Table 10.3). *(Reprinted with permission from Ref [63]. Copyright 2011 American Chemical Society)*

20 A g^{-1}. These results are in accord with the general tendency in the references [55,56,60,61], i.e. larger nanopores can accelerate ion-transfer, and as a result, a better-rate performance is achieved. On the other hand, ZTCs can well retain the capacitance in the ultrahigh current region, although their nanopore sizes (~1.2 nm) are less than those of MSC30. This high-rate performance can be ascribed to their 3D and mutually connected nanopore arrangement.

10.7. GENERAL CONCLUSION AND PERSPECTIVES

ZTC has a 3D ordered framework composed of buckybowl-like nanographenes, thereby making it extraordinary sp^2-based carbon material. In addition, ZTC has a huge adsorption capability together with a uniform micropore size (ca. 1.2 nm), and such characteristics make this material a very attractive adsorbent, especially for energy storage such as hydrogen storage and EDLC as described in this chapter. With its extremely unique structure and properties, ZTC can be a next-generation 3D graphene-based architecture which follows fullerene (0D), carbon nanotube (1D), and graphene (2D).

Since the discovery of ZTC [1,2,8], many different research groups have dealt with this very unique carbon material, and the works related to ZTC are now becoming to be a new research field. As described in the Section 10.2, various types of ZTCs with different structures have been synthesized so far. However, there is still great possibility for the synthesis of further new ZTCs. One possible method is the use of new zeolite templates which meet the requirements described in the Section 10.2.1, for producing new types of ZTCs with different framework regularities from the conventional ZTCs synthesized with FAU, BEA, and EMT. There are actually many possible candidates for a hard template, considering the numerous types of zeolite crystals (>190). Even by using the same zeolite template, various different types of ZTCs are possibly produced by changing the carbon source, carbon introduction amount into zeolite, and heat-treatment conditions. In addition, postsynthesis modification is expected to be further developed because ZTC has a large amount of edge sites which are fully exposed and highly reactive. ZTC-based composite materials are also of interest. There have been several reports for Pt-loaded ZTC with regard to its promising applications in fuel cells [10,64] or hydrogen storage [52,65]. Not only Pt but also other kinds of metals, metal oxides, and organic compounds could be combined with ZTC to form new composite materials with superior performances.

It should be emphasized that ZTC is attractive not only as high-performance sorbents and/or energy storage media but also as model porous materials for understanding the basic phenomena related to micropores, as we have demonstrated in the research on EDLC [56]. We believe that a family of ZTC materials will be further developed hereafter, which could open the door for new nanocarbon technology involving batteries, gas storage, catalysts, and any other fields.

ACKNOWLEDGMENTS

This work was partially supported by KAKENHI (Grant-in-Aid for Scientific Research) on the Priority Area 'New Materials Science Using Regulated Nano Spaces-Strategy in Ubiquitous Elements' from the Ministry of Education, Culture, Sports, Science and Technology of Japan (T.K.) for the Sections 10.3 and 10.4; by NEDO (T.K.) for the Section 10.5; by Grant-in-Aid for Scientific Research on the Innovative Areas: "Fusion Materials" (Area no. 2206) from the Ministry of Education, Culture, Sports, Science and Technology (H.N.) for the Sections 10.3 and 10.4; and by Japan Society of the Promotion of Science, Grant-in-Aid for young scientists (A), 23685041 (H.N.) for the Sections 10.2 and 10.6.

REFERENCES

[1] T. Kyotani, T. Nagai, A. Tomita, Preparation of New Carbon from Zeolite as Template. Extended Abstracts of Carbon '92, 1992. pp. 437–439.

[2] T. Kyotani, T. Nagai, S. Inoue, A. Tomita, Formation of new type of porous carbon by carbonization in zeolite nanochannels, Chem. Mater. 9 (1997) 609–615.

[3] P. Enzel, T. Bein, Poly(acrylonitrile) chains in zeolite channels – polymerization and pyrolysis, Chem. Mater. 4 (1992) 819–824.
[4] T. Cordero, P.A. Thrower, L.R. Radovic, On the oxidation resistance of carbon–carbon composites obtained by chemical vapor infiltration of different carbon cloths, Carbon 30 (1992) 365–374.
[5] J. Rodriguez-Mirasol, T. Cordero, L.R. Radovic, J.J. Rodriguez, Structural and textural properties of pyrolytic carbon formed within a microporous zeolite template, Chem. Mater. 10 (1998) 550–558.
[6] S.A. Johnson, E.S. Brigham, P.J. Ollivier, T.E. Mallouk, Effect of micropore topology on the structure and properties of zeolite polymer replicas, Chem. Mater. 9 (1997) 2448–2458.
[7] C.J. Meyers, S.D. Shah, S.C. Patel, R.M. Sneeringer, C.A. Bessel, N.R. Dollahon, R.A. Leising, E.S. Takeuchi, Templated synthesis of carbon materials from zeolites (Y, beta, and ZSM-5) and a montmorillonite clay (K10): physical and electrochemical characterization, J. Phys. Chem. B 105 (2001) 2143–2152.
[8] Z.X. Ma, T. Kyotani, A. Tomita, Preparation of a high surface area microporous carbon having the structural regularity of Y zeolite, Chem. Commun. (2000) 2365–2366.
[9] T. Kyotani, Z.X. Ma, A. Tomita, Template synthesis of novel porous carbons using various types of zeolites, Carbon 41 (2003) 1451–1459.
[10] F.B. Su, H.J. Zeng, Y.J. Yu, L. Lv, J.Y. Lee, X.S. Zhao, Template synthesis of microporous carbon for direct methanol fuel cell application, Carbon 43 (2005) 2368–2373.
[11] Z.X. Yang, Y.D. Xia, R. Mokaya, Enhanced hydrogen storage capacity of high surface area zeolite-like carbon materials, J. Am. Chem. Soc. 129 (2007) 1673–1679.
[12] F.O.M. Gaslain, J. Parmentier, V.P. Valtchev, J. Patarin, First zeolite carbon replica with a well resolved X-ray diffraction pattern, Chem. Commun. (2006) 991–993.
[13] K. Hernadi, A. Fonseca, J.B. Nagy, D. Bernaerts, A. Fudala, A.A. Lucas, Catalytic synthesis of carbon nanotubes using zeolite support, Zeolites 17 (1996) 416–423.
[14] K. Mukhopadhyay, A. Koshio, T. Sugai, N. Tanaka, H. Shinohara, Z. Konya, J.B. Nagy, Bulk production of quasi-aligned carbon nanotube bundles by the catalytic chemical vapour deposition (CCVD) method, Chem. Phys. Lett. 303 (1999) 117–124.
[15] A. Garsuch, O. Klepel, Synthesis of ordered carbon replicas by using Y-zeolite as template in a batch reactor, Carbon 43 (2005) 2330–2337.
[16] A. Garsuch, O. Klepel, R.R. Sattler, C. Berger, R. Glaser, J. Weitkamp, Synthesis of a carbon replica of zeolite Y with large crystallite size, Carbon 44 (2006) 593–596.
[17] F.B. Su, L. Lv, T.M. Hui, X.S. Zhao, Phenol adsorption on zeolite-templated carbons with different structural and surface properties, Carbon 43 (2005) 1156–1164.
[18] F.B. Su, X.S. Zhao, L. Lv, Z.C. Zhou, Synthesis and characterization of microporous carbons templated by ammonium-form zeolite Y, Carbon 42 (2004) 2821–2831.
[19] P.X. Hou, T. Yamazaki, H. Orikasa, T. Kyotani, An easy method for the synthesis of ordered microporous carbons by the template technique, Carbon 43 (2005) 2624–2627.
[20] Z.X. Yang, Y.D. Xia, X.Z. Sun, R. Mokaya, Preparation and hydrogen storage properties of zeolite-templated carbon materials nanocast via chemical vapor deposition: effect of the zeolite template and nitrogen doping, J. Phys. Chem. B 110 (2006) 18424–18431.
[21] K. Matsuoka, Y. Yamagishi, T. Yamazaki, N. Setoyama, A. Tomita, T. Kyotani, Extremely high microporosity and sharp pore size distribution of a large surface area carbon prepared in the nanochannels of zeolite Y, Carbon 43 (2005) 876–879.
[22] P.X. Hou, H. Orikasa, T. Yamazaki, K. Matsuoka, A. Tomita, N. Setoyama, Y. Fukushima, T. Kyotani, Synthesis of nitrogen-containing microporous carbon with a highly ordered

structure and effect of nitrogen doping on H_2O adsorption, Chem. Mater. 17 (2005) 5187–5193.
[23] H. Nishihara, Q.H. Yang, P.X. Hou, M. Unno, S. Yamauchi, R. Saito, J.I. Paredes, A. Martinez-Alonso, J.M.D. Tascon, Y. Sato, M. Terauchi, T. Kyotani, A possible buckybowl-like structure of zeolite templated carbon, Carbon 47 (2009) 1220–1230.
[24] M. Kruk, M. Jaroniec, R. Ryoo, S.H. Joo, Characterization of ordered mesoporous carbons synthesized using MCM-48 silicas as templates, J. Phys. Chem. B 104 (2000) 7960–7968.
[25] J.S. Lee, S.H. Joo, R. Ryoo, Synthesis of mesoporous silicas of controlled pore wall thickness and their replication to ordered nanoporous carbons with various pore diameters, J. Am. Chem. Soc. 124 (2002) 1156–1157.
[26] C. Vix-Guterl, S. Saadallah, L. Vidal, M. Reda, J. Parmentier, J. Patarin, Template synthesis of a new type of ordered carbon structure from pitch, J. Mater. Chem. 13 (2003) 2535–2539.
[27] Z.X. Ma, T. Kyotani, Z. Liu, O. Terasaki, A. Tomita, Very high surface area microporous carbon with a three-dimensional nano-array structure: synthesis and its molecular structure, Chem. Mater. 13 (2001) 4413–4415.
[28] M.S. Dresselhaus, G. Dresselhaus, R. Saito, A. Jorio, Raman spectroscopy of carbon nanotubes, Phys. Rep. 409 (2005) 47–99.
[29] M.S. Dresselhaus, P.C. Eklund, Phonons in carbon nanotubes, Adv. Phys. 49 (2000) 705–814.
[30] Y. Gogotsi, J.A. Libera, N. Kalashnikov, M. Yoshimura, Graphite polyhedral crystals, Science 290 (2000) 317–320.
[31] J.J. Wang, M.Y. Zhu, R.A. Outlaw, X. Zhao, D.M. Manos, B.C. Holloway, V.P. Mammana, Free-standing subnanometer graphite sheets, Appl. Phys. Lett. 85 (2004) 1265–1267.
[32] D. Roy, M. Chhowalla, H. Wang, N. Sano, I. Alexandrou, T.W. Clyne, G.A.J. Amaratunga, Characterisation of carbon nano-onions using Raman spectroscopy, Chem. Phys. Lett. 373 (2003) 52–56.
[33] T. Lenosky, X. Gonze, M. Teter, V. Elser, Energetics of negatively curved graphitic carbon, Nature 355 (1992) 333–335.
[34] V. Rosato, M. Celino, G. Benedek, S. Gaito, Thermodynamic behavior of the carbon schwarzite fcc $(C_{36})_2$, Phys. Rev. B 60 (1999) 16928–16933.
[35] T. Roussel, A. Didion, R.J.M. Pellenq, R. Gadiou, C. Bichara, C. Vix-Guterl, Experimental and atomistic simulation study of the structural and adsorption properties of Faujasite zeolite-templated nanostructured carbon materials, J. Phys. Chem. C 111 (2007) 15863–15876.
[36] K. Kaneko, C. Ishii, M. Ruike, H. Kuwabara, Origin of superhigh surface-area and micro-crystalline graphitic structures of activated carbons, Carbon 30 (1992) 1075–1088.
[37] H.S. Zhou, S.M. Zhu, M. Hibino, I. Honma, M. Ichihara, Lithium storage in ordered mesoporous carbon (CMK-3) with high reversible specific energy capacity and good cycling performance, Adv. Mater. 15 (2003) 2107–2111.
[38] K.S. Xia, Q.M. Gao, C.D. Wu, S.Q. Song, M.L. Ruan, Activation, characterization and hydrogen storage properties of the mesoporous carbon CMK-3, Carbon 45 (2007) 1989–1996.
[39] R. Ryoo, S.H. Joo, S. Jun, Synthesis of highly ordered carbon molecular sieves via template-mediated structural transformation, J. Phys. Chem. B 103 (1999) 7743–7746.
[40] J. Lee, S. Yoon, T. Hyeon, S.M. Oh, K.B. Kim, Synthesis of a new mesoporous carbon and its application to electrochemical double-layer capacitors, Chem. Commun. (1999) 2177–2178.
[41] M. Sevilla, S. Alvarez, T.A. Centeno, A.B. Fuertes, F. Stoeckli, Performance of templated mesoporous carbons in supercapacitors, Electrochim. Acta 52 (2007) 3207–3215.
[42] L. Schlapbach, A. Zuttel, Hydrogen-storage materials for mobile applications, Nature 414 (2001) 353–358.

[43] M.K. Haas, J.M. Zielinski, G. Dantsin, C.G. Coe, G.P. Pez, A.C. Cooper, Tailoring single walled carbon nanotubes for hydrogen storage, J. Mater. Res. 20 (2005) 3214–3223.

[44] Y. Kojima, Y. Kawai, A. Koiwai, N. Suzuki, T. Haga, T. Hioki, K. Tange, Hydrogen adsorption and desorption by carbon materials, J. Alloys Compd. 421 (2006) 204–208.

[45] M.A. de la Casa-Lillo, F. Lamari-Darkrim, D. Cazorla-Amoros, A. Linares-Solano, Hydrogen storage in activated carbons and activated carbon fibers, J. Phys. Chem. B 106 (2002) 10930–10934.

[46] M. Jorda-Beneyto, F. Suarez-Garcia, D. Lozano-Castello, D. Cazorla-Amoros, A. Linares-Solano, Hydrogen storage on chemically activated carbons and carbon nanomaterials at high pressures, Carbon 45 (2007) 293–303.

[47] H. Takagi, H. Hatori, Y. Soneda, N. Yoshizawa, Y. Yamada, Adsorptive hydrogen storage in carbon and porous materials, Mater. Sci. Eng. B 108 (2004) 143–147.

[48] P. Kowalczyk, H. Tanaka, R. Holyst, K. Kaneko, T. Ohmori, J. Miyamoto, J. Phys. Chem. B 109 (2005) 17174–17183.

[49] L. Chen, R.K. Singh, P. Webley, Synthesis, characterization and hydrogen storage properties of microporous carbons templated by cation exchanged forms of zeolite Y with propylene and butylene as carbon precursors, Microporous Mesoporous Mater. 102 (2007) 159–170.

[50] A. Pacula, R. Mokaya, Synthesis and high hydrogen storage capacity of zeolite-like carbons nanocast using as-synthesized zeolite templates, J. Phys. Chem. C 112 (2008) 2764–2769.

[51] A.J. Lachawiec, R.T. Yang, Isotope tracer study of hydrogen spillover on carbon-based adsorbents for hydrogen storage, Langmuir 24 (2008) 6159–6165.

[52] H. Nishihara, P.X. Hou, L.X. Li, M. Ito, M. Uchiyama, T. Kaburagi, A. Ikura, J. Katamura, T. Kawarada, K. Mizuuchi, T. Kyotani, High-pressure hydrogen storage in zeolite-templated carbon, J. Phys. Chem. C 113 (2009) 3189–3196.

[53] R. Kotz, M. Carlen, Principles and applications of electrochemical capacitors, Electrochim. Acta 45 (2000) 2483–2498.

[54] K. Jurewicz, C. Vix-Guterl, E. Frackowiak, S. Saadallah, A. Reda, J. Parmentier, J. Patarin, F. Beguin, Capacitance properties of ordered porous carbon materials prepared by a templating procedure, J. Phys. Chem. Solids 65 (2004) 287–293.

[55] W. Xing, S.Z. Qiao, R.G. Ding, F. Li, G.Q. Lu, Z.F. Yan, H.M. Cheng, Superior electric double layer capacitors using ordered mesoporous carbons, Carbon 44 (2006) 216–224.

[56] H. Nishihara, H. Itoi, T. Kogure, P.X. Hou, H. Touhara, F. Okino, T. Kyotani, Investigation of the ion storage/transfer behavior in an electrical double-layer capacitor by using ordered microporous carbons as model materials, Chem. Eur. J. 15 (2009) 5355–5363.

[57] K.H. An, W.S. Kim, Y.S. Park, J.M. Moon, D.J. Bae, S.C. Lim, Y.S. Lee, Y.H. Lee, Electrochemical properties of high-power supercapacitors using single-walled carbon nanotube electrodes, Adv. Funct. Mater. 11 (2001) 387–392.

[58] D.N. Futaba, K. Hata, T. Yamada, T. Hiraoka, Y. Hayamizu, Y. Kakudate, O. Tanaike, H. Hatori, M. Yumura, S. Iijima, Shape-engineerable and highly densely packed single-walled carbon nanotubes and their application as super-capacitor electrodes, Nat. Mater. 5 (2006) 987–994.

[59] C.M. Niu, E.K. Sichel, R. Hoch, D. Moy, H. Tennent, High power electrochemical capacitors based on carbon nanotube electrodes, Appl. Phys. Lett. 70 (1997) 1480–1482.

[60] S.W. Woo, K. Dokko, H. Nakano, K. Kanamura, Preparation of three dimensionally ordered macroporous carbon with mesoporous walls for electric double-layer capacitors, J. Mater. Chem. 18 (2008) 1674–1680.

[61] H. Yamada, H. Nakamura, F. Nakahara, I. Moriguchi, T. Kudo, Electrochemical study of high electrochemical double layer capacitance of ordered porous carbons with both meso/macropores and micropores, J. Phys. Chem. C 111 (2007) 227–233.

[62] C.M. Yang, Y.J. Kim, M. Endo, H. Kanoh, M. Yudasaka, S. Iijima, K. Kaneko, Nanowindow-regulated specific capacitance of supercapacitor electrodes of single-wall carbon nanohorns, J. Am. Chem. Soc. 129 (2007) 20–21.

[63] H. Itoi, H. Nishihara, T. Kogure, T. Kyotani, Three-Dimensionally arrayed and mutually connected 1.2 nm-nanopores for high-performance electric double layer capacitor, J. Am. Chem. Soc. 133 (2011) 1165–1167.

[64] E.N. Coker, W.A. Steen, J.T. Miller, A.J. Kropf, J.E. Miller, The preparation and characterization of novel Pt/C electrocatalysts with controlled porosity and cluster size, J. Mater. Chem. 17 (2007) 3330–3340.

[65] Z. Wang, F.H. Yang, R.T. Yang, Enhanced hydrogen spillover on carbon surfaces modified by oxygen plasma. J. Phys. Chem. C, 114 (2010) 1601–1609.

Chapter 11

Adsorption by Soft-Templated Carbons

Joanna Górka
Oak Ridge National Laboratory, Chemical Sciences Division, Oak Ridge, TN, USA

Sheng Dai
Oak Ridge National Laboratory, Chemical Sciences Division, Oak Ridge, USA, and Department of Chemistry, University of Tennessee, Knoxville, USA

Chapter Outline

11.1. Introduction 323
11.2. Overview of Methods Used to Synthesize Ordered Mesoporous Carbons 324
11.2.1. Hard-Templating Synthesis of Mesoporous Carbons 324
11.2.2. Soft-Templating Synthesis of Mesoporous Carbons 325
11.3. Tailoring Porosity of Carbon Adsorbents 327
11.4. Surface Modification of Mesoporous Carbon Materials 333
11.4.1. Functionalization During Synthesis 333
11.4.2. Postsynthesis Modification 338
11.5. Macroscopic Morphologies of Soft-Templated Carbons 340
11.6. Summary and Outlook 342
References 344

11.1. INTRODUCTION

Nanoporous carbons include a broad spectrum of structures ranging from highly microporous active carbons to carbon nanotubes and ordered mesoporous carbons. The properties of each group of nanoporous carbons can be controlled by the choice of carbon precursor and the method of carbonization and activation, which directly affect their pore structure and surface chemistry.

Novel Carbon Adsorbents. DOI: 10.1016/B978-0-08-097744-7.00011-9
Copyright © 2012 Elsevier Ltd. All rights reserved.

Both pore structure and surface chemistry are crucial for a number of applications, including adsorption, separation, catalysis, and energy conversion and storage.

This chapter explores porous carbon adsorbents, with the main focus on mesoporous carbons obtained by a soft-templating approach. Since this method was elaborated not long ago, there is a need to take a closer look at the strategy for preparing well-ordered mesoporous carbons by self-assembly of phenolic resins in the presence of triblock copolymers. The methods established to monitor the adsorption and structural properties of carbon are discussed, including procedures that exclusively address tuned mesoporosity as well as ones that lead to greatly enhanced microporosity and specific surface area. All attempts undertaken so far toward surface modification of polymer-templated carbons can be divided into two categories: i) direct functionalization during synthesis by using different carbon precursors or by adding inorganic salts to introduce metal/metal oxide nanoparticles to the carbon matrix and ii) post-synthesis functionalization. To provide a complete view of the versatility of carbons obtained by this method, the macroscopic morphologies possible with these carbons are addressed.

11.2. OVERVIEW OF METHODS USED TO SYNTHESIZE ORDERED MESOPOROUS CARBONS

11.2.1. Hard-Templating Synthesis of Mesoporous Carbons

The porous carbons (e.g., activated carbon, carbon black, and activated carbon fiber) are very efficient adsorbents in water and air-purification applications owing to their large specific surface areas, pore volumes, chemical inertness, and good mechanical stability. However, the micropores exclusively present in porous carbons largely restrict mass transfer, which is essential in wastewater treatment applications. Taking this into account, the so-called hard-templating method (Fig. 11.1) [1], which for the first time afforded ordered mesoporous

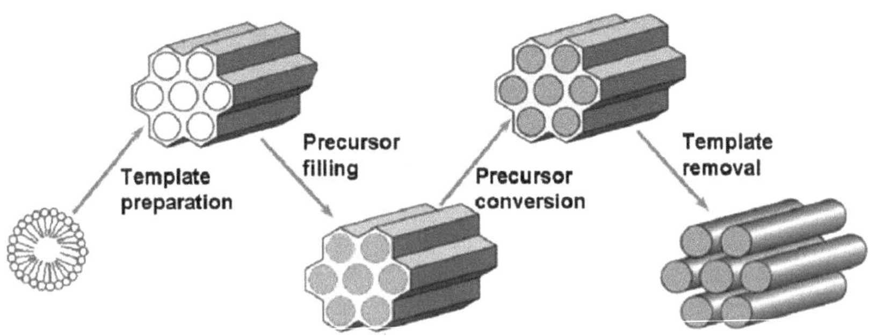

FIGURE 11.1 Schematic illustration of hard-templating method. *(Reprinted with permission from Ref. 10. Copyright 2008 American Chemical Society)*

carbons (OMCs), fundamentally changed porous carbon materials research [2–4]. During the first step of the synthesis, the pores of an ordered mesoporous silica (OMS) are used as a template and are filled with a suitable carbon precursor. After carbonization and silica template removal, the final carbon material is a reverse replica of the template used and possesses characteristics such as high surface area, large pore volumes, and ordered mesopores. All these parameters can easily be tuned by using different silica mesostructures as templates.

In addition to OMSs, colloidal silica particles [5–7] and colloidal crystals [8,9] can be used as templates. In particular, a broad variety of commercially available silica colloids provide exciting opportunities to tune the size of the mesopores in carbon. The preparation procedure involves mixing colloidal silica particles with carbon precursor, followed by carbonization and silica dissolution. The resulting carbon exhibits a disordered but interconnected and uniform network of mesopores whose diameters resemble the size of the silica nanoparticles used [5–7]. In contrast, the colloidal crystal templating method leads to ordered carbons [8,9].

The simplicity of the hard-templating approach is the main reason for its popularity; however, this method has considerable disadvantages that need to be taken into account for large-scale production. The need to prepare a siliceous template and dissolve it increases the number of preparation steps and the total cost of synthesis. Moreover, silica scaffold etching is usually performed using sodium hydroxide or hydrofluoric acid, both of which are unsafe to handle and potentially dangerous from an environmental perspective, especially considering the large quantities of material required to obtain the final carbon product.

11.2.2. Soft-Templating Synthesis of Mesoporous Carbons

More recently, ordered and disordered carbons prepared by the self-assembly of organic carbon precursors and triblock copolymers (soft templates) have been reported [10,11]. This new strategy takes advantage of the favorable organic–organic interactions between some thermosetting polymers and thermally decomposable surfactants to form ordered polymer–polymer nanocomposites. During the initial thermal treatment of the thermally decomposable surfactant, the thermally unstable copolymer (a soft template) is removed while the thermosetting polymer is carbonized by heating at higher temperatures to form the carbon pore walls. In comparison with hard-templated carbons, the development of carbons obtained from polymeric precursors represents a major improvement in materials science since fewer steps are required and the block copolymers used are biodegradable and can be recovered easily.

In the very first report of the soft-templating synthesis of ordered mesoporous carbon films, polystyrene-block-poly(4-vinylpyridine) was used as a structure-directing agent [12]. Later, two independent groups propounded the use of poly(ethylene oxide)–poly(propylene oxide)–poly(ethylene oxide) (PEO–PPO–PEO) triblock copolymers as structure-forming agents (which were widely used in the preparation of ordered mesoporous silicas) for the preparation of OMCs [13,14]. In addition to thin films, these two groups also elaborated the recipes for the preparation of carbons in the form of fibers and monoliths. Zhao's method [14] is based on the prepolymerization of phenol and formaldehyde in the presence of sodium hydroxide as catalyst to obtain a low-molecular-weight copolymer called *resol* to use as a carbon precursor (Fig. 11.2). Subsequently, the as-obtained resol after neutralization interacts with block copolymer to assemble into polymeric mesostructures. The complementary procedure proposed by Dai et al. [13] involves phloroglucinol and formaldehyde as the carbon-yielding component (Fig. 11.3). Under acidic conditions, the phloroglucinol–formaldehyde self-assembles in the presence of block copolymer to give a polymeric composite that, after thermal treatment, is transformed into the final carbon. This truly one-pot synthesis has the great advantage of avoiding the prepolymerization and pH adjustment traditionally required for the synthesis in alkaline media.

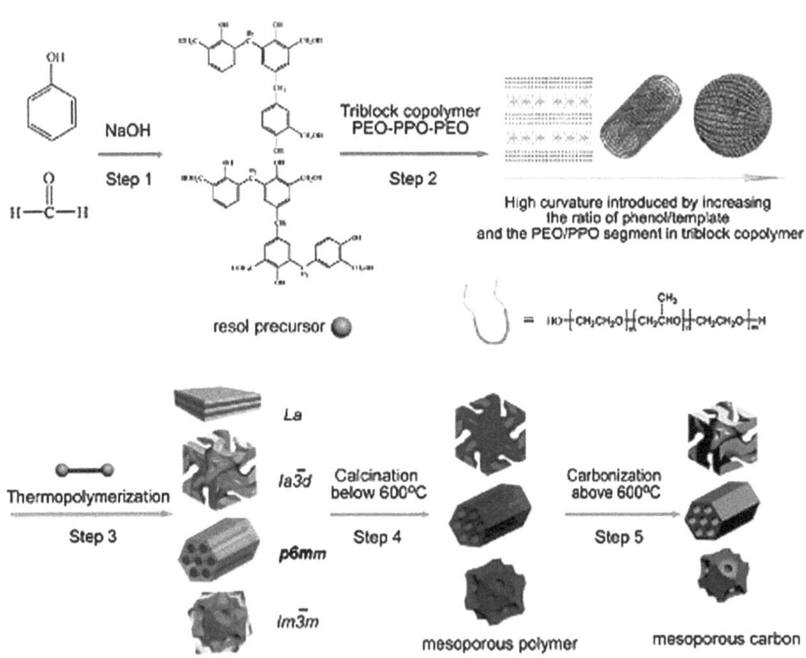

FIGURE 11.2 Scheme for the soft-templated synthesis of ordered mesoporous carbons under basic conditions. *(Reprinted with permission from Ref. 14. Copyright 2006 American Chemical Society)*

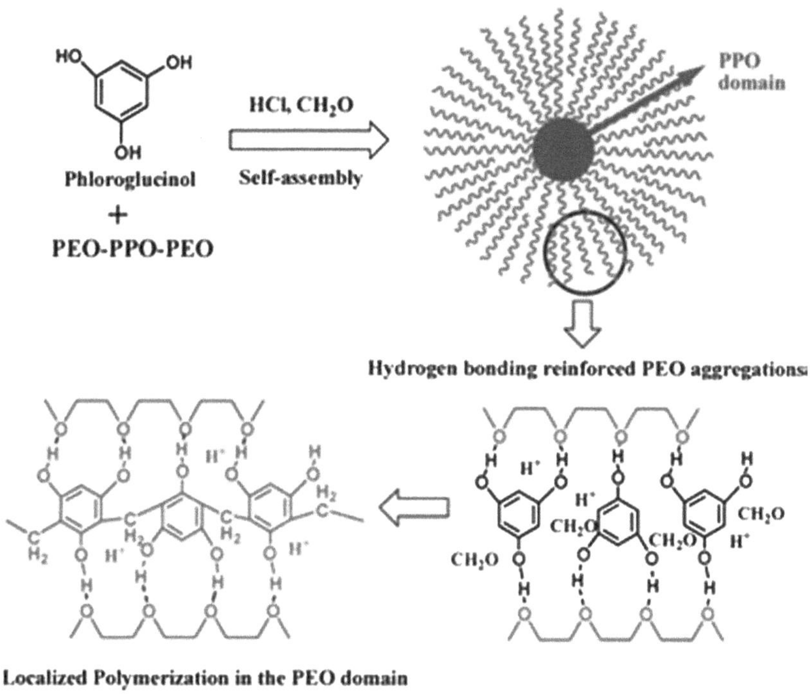

FIGURE 11.3 Schematic illustration of the polymerization of phloroglucinol and formaldehyde under acidic conditions. *(Reprinted with permission from Ref. 13. Copyright 2006 American Chemical Society)*

11.3. TAILORING POROSITY OF CARBON ADSORBENTS

As mentioned earlier, one of the most important requirements for a good adsorbent is enhanced textural properties, which include a large and accessible surface area for good contact between adsorbate and adsorbent, the presence of mesopores to facilitate higher mass transfer, and a large pore volume.

Until recently, most ordered mesoporous carbons (OMCs) were obtained by using ordered mesoporous silicas (OMSs) [2,15] or colloidal silica crystals [8,16,17] as hard templates. In the former case, the resulting carbons are inverse replicas of the OMS templates with pore widths determined by the silica wall thickness, which usually does not exceed 6 nm. However, the use of colloidal silica crystals or colloidal silica yields carbons with spherical pores of tailored sizes in the mesopore range from ~5–6 nm to 50 nm and the macropore range (above 50 nm), thanks to the availability of silica colloids in this range. Additionally, the broad variety of commercially available silica colloids reduces the total cost of the synthesis. Thus, colloidal templating seems to be a very promising method for the preparation of ordered macroporous carbons [8,16,17] and polymers with large pores [18]. Initially, the colloid-templated carbons possessed a high specific surface area and a large mesopore volume [5] but, at the same time, a quite broad pore-size distribution (PSD) due to aggregation of silica colloids. To prevent such aggregation, a stabilizing surfactant was added [6]. Although the strategy of using colloidal silica as a structure-directing agent is straightforward, it is still a challenge to obtain uniform and spherical mesopores that accurately resemble the size of the silica colloids used [19]. This refers especially to carbons obtained by the carbonization of polymeric precursors formed in the presence of colloidal silica, where templating can occur on the colloidal aggregates instead of on single silica colloids, which results in a broad pore-size distribution [5]. Initially, this difficulty was circumvented by using mesophase pitch particles as the carbon precursor [20–22]. The interpenetration of silica colloids into the mesophase pitch was controlled effectively by adjusting the heating, stabilization, and carbonization conditions to ensure that carbon precursor completely covered the surface of the silica colloids, which is the main requirement for obtaining spherical mesopores that reflect the size of single colloids.

A complete filling of the template pores produces true inverse carbon replicas [17,19]. Theoretical and experimental studies showed that the complete filling of pores of a colloidal crystal template with carbon precursor yielded carbons with a pore volume of ~1.36 cm^3/g [23]. On the other hand, the formation of a thin carbon film on the template surface produced carbons with much larger pore volumes. For instance, it was shown that the formation of phenolic resin films just on the surface of colloidal silica afforded carbons with pore volumes as high as 9 cm^3/g [23,24].

The use of colloidal silica facilitates the formation of carbon monoliths. Monolithic colloidal silica templates can easily be fabricated by evaporating the solvent from a colloidal silica solution and compressing the dried silica into desired forms [25]. In the next step, the voids between colloids of the silica monolith are filled with carbon precursor, which creates a carbon monolith with uniform spherical mesopores after carbonization and silica dissolution.

There are some applications that rely solely on the well-developed porosity of the material. One of them is the adsorption of large molecules. A pioneering work in this area [26] demonstrated vitamin E adsorption over two different hard-templated mesoporous carbons (CMK-1 and CMK-3), and the results were compared to those obtained for entirely microporous (activated) carbons. It was found that vitamin E adsorption depends not only on the solvent used but also on the textural characteristics of the adsorbent. The highest vitamin E adsorption capacity was observed for CMK-3 carbon, which possesses the highest total and mesopore volume. The experiments with increasing concentration of vitamin E showed almost the same adsorption from diluted solutions and quite pronounced differences from more concentrated solutions for all adsorbents. This led to the conclusion that not BET surface area but the presence of mesopores is a key factor for enhanced adsorption of large molecules such as vitamin E.

Higher hydrophobicity of the carbon surface in comparison to silica improves adsorbate–adsorbent interactions, which was demonstrated by Vinu et al. [27] for amino acids. Adsorption of L-histidine onto two different carbon supports (mesoporous CMK-3 and activated carbon) and mesoporous silica (SBA-15) was investigated for solutions with different pH values. The results clearly show that the amount of amino acid adsorbed onto CMK-3 carbon is much higher than the other adsorbents tested. The maximum loading of L-histidine was observed at a pH corresponding to the isoelectric point of L-histidine, and is probably caused by enhanced hydrophobic interactions between histidine and the carbon surface. This translates to close packing of histidine molecules inside the mesopores. In addition, the same concept was utilized for the selective adsorption and separation of tea components (catechin and tannic acid) on mesoporous carbons possessing spherical pore geometry, the so-called carbon nanocages [28].

In the case of the soft-templating method, one of the simplest ways to tune mesoporosity and pore geometry is to use copolymers with a different ratio of PEO/PPO block such as Pluronic F127, F108, P123, etc. Studies showed the possibility of obtaining a whole sequence of mesostructures from lamellar to bicontinuous cubic, hexagonal, and body-centered cubic by increasing the PEO-to-PPO ratio. Moreover, the use of reverse amphiphilic triblock copolymer PPO_{53}–PEO_{136}–PPO_{53} [29] led to the formation of mesoporous carbons with a face-centered cubic structure and a bimodal pore-size distribution. This is important for the fact that the diffusion rate is strongly related to the pore geometry, which is another way to control and design the properties of the material. Deng et al. [30] reported thick-walled mesoporous carbons prepared by utilizing a weakly hydrophobic poly(ethylene oxide)-b-poly(methyl methacrylate) (PEO-b-PMMA) diblock copolymer. In this case, the ester groups from PMMA block interact more strongly with hydrophilic phenolic carbon precursor, facilitating the formation of thick carbon walls.

Another method is analogous to those used for tuning the pore size of ordered mesoporous silicas and involves small organic molecules as pore expanders. Mesoporous carbons with ultra-large pores were synthesized by utilizing diblock copolymer PEO-b-PS as a template, homopolystyrene (h-PS) as a pore expander, and resol as a carbon precursor (Fig. 11.4) [31]. The addition of h-PS up to 20 wt% relative to PEO-b-PS afforded highly uniform mesopores in the range from 26 nm to 40 nm, while higher loadings caused multimodal pore-size distributions.

Recently, Fulvio and coworkers [32] obtained interesting results when the typical soft-templated synthesis of mesoporous carbons was performed with the use of carbon black (CB) and carbon onions (COs). The gradual increase in the content of CB and CO led to mesopore enlargement caused by the inability of the hydrophilic parts of the triblock copolymer to interpenetrate the carbon particles as it is observed in the pure phenolic resin. This simple approach, the so-called "brick-and-mortar" self-assembly, extended the mesopore regime of

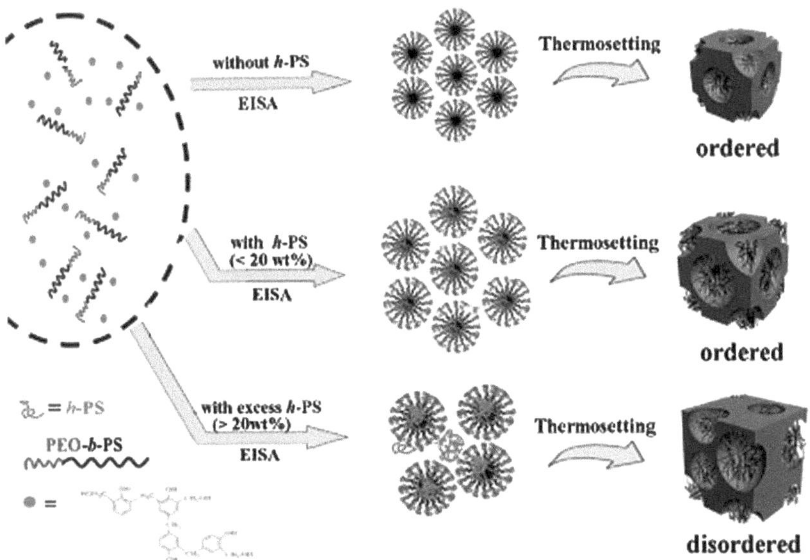

FIGURE 11.4 Formation of ultra-large-pore mesoporous carbon with the use of homopolymer h-PS as the pore expander. *(Reprinted with permission from Ref. 31. Copyright 2008 American Chemical Society)*

soft-templated carbons from typically obtained 5–8 nm to 10–20 nm without using any organic additives or blends of surfactants, which might be quite expensive.

Commonly, both hard- and soft-templating strategies most often produce carbons with ordered and uniform mesopores. However, numerous applications, especially gas storage, removal of pollutants of different sizes, heterogeneous catalysis, and energy storage and conversion [33–36], require carbons with structures that contain both micropores and mesopores [3,4,37–40]. Therefore, the creation of additional microporosity in mesoporous carbons without diminishing their mesoporosity is a very important task. In general, different strategies can be used for the development of microporosity in mesoporous carbons. One of these is the postsynthesis activation of soft- and hard-templated carbons with CO_2 or KOH in order to generate microporosity in the mesopore walls.

So far, mesoporous carbons obtained by inverse replication of ordered mesoporous silicas such as SBA-15 were activated with CO_2 [41–44] and KOH [44,45]. This activation resulted in a substantial increase of microporosity and a simultaneous reduction of mesoporosity [45]. The same was recently reported for mesoporous carbons fabricated by using colloidal silica as a hard template [25]. Also, the soft-templated mesoporous carbons were activated by KOH [46–49] and CO_2 [50,51]. In this case, the pH conditions for polymerization of carbon precursors and self-assembly seem to strongly influence not only the average mesopore size (4 nm vs. 8 nm in basic and acidic media, respectively) but also the degree of cross-linking affecting the pore wall thickness and its stability [14,52], which is especially important in postsynthesis activation. Carbons synthesized under acidic conditions possess exceptional resistance to a wide range of activating agents and temperatures, even graphitization [47,51–53], compared to those obtained by using a basic synthesis route [48]. This was quite a unique finding, bearing in mind that, for instance, KOH activation is considered to be a harsh activation method compared to CO_2 or water vapor activation. Each of the activating agents employed reacts with the carbon framework in a different way, which translates to different carbon burn-off and porosity development; however, the fully preserved mesoporosity is a common feature for this kind of carbon (Fig. 11.5). Another example that proves the exceptional stability of carbons prepared under acidic conditions involved NH_3 activation at high temperatures [54,55]. This method, except for creating micropores, which normally takes place during the activation process, also induced the replacement reaction between oxygen-bearing species in the carbon matrix and NH_3, resulting in nitrogen-doped carbons.

Another alternative for creating microporosity in mesoporous carbons involves the incorporation of TEOS during soft-templating synthesis [56–58]. These works show that a large amount of TEOS can be added during the formation of phenolic resin in the presence of block copolymer template. The

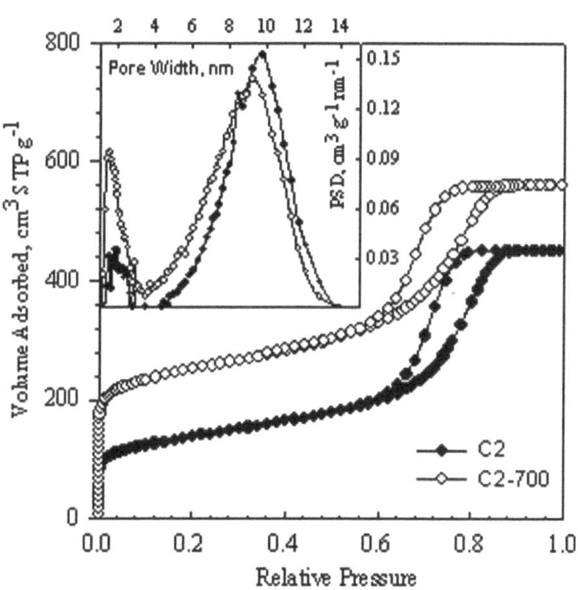

FIGURE 11.5 Nitrogen adsorption isotherms and the corresponding pore-size distributions (PSDs) for the mesoporous phenolic resin-based carbon prepared at 850 °C before (filled circles) and after (open circles) KOH activation at 700 °C. *(Reprinted with permission from Ref. 49. Copyright 2008 Elsevier)*

resulting phenolic resin–TEOS–block copolymer composites were converted into mesoporous silica–carbon materials by controlled heating and carbonization under flowing inert gas. Dissolution of homogeneously distributed silica in these composite materials resulted in the creation of micropores in the mesopore walls.

In addition, carbons with hierarchical porosity and distinct pore geometry in the system (spherical vs. cylindrical pores) can easily be prepared through the use of silica colloids [59]. The first implementation of this idea was presented by Deng et al. [59], who used silica colloidal crystals to prepare macro–mesoporous carbons. The resulting porous carbons have highly ordered macropores resembling the size of the colloids used (230–430 nm) and

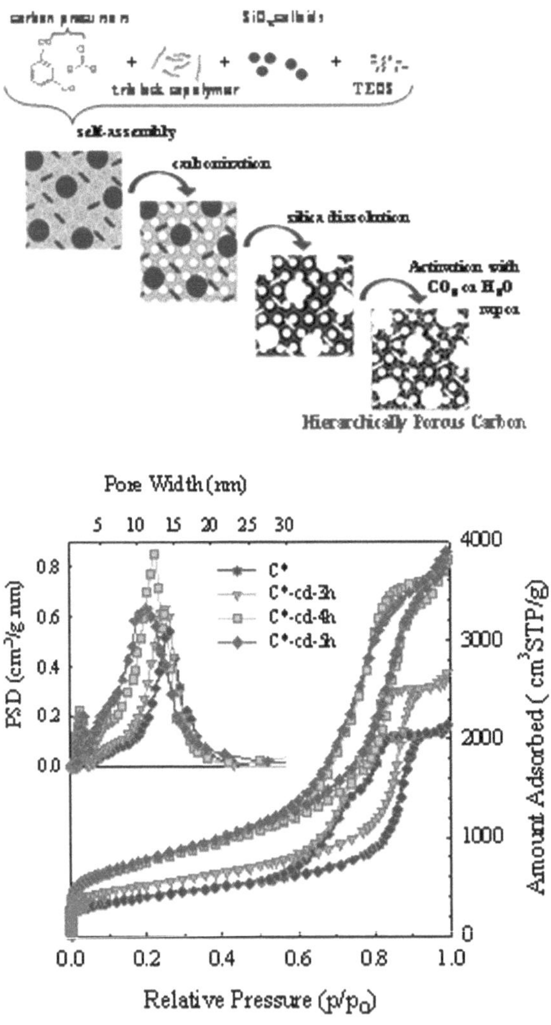

FIGURE 11.6 Schematic illustration of a combined soft- and hard-templating synthesis of mesoporous carbons followed by post-synthesis activation with CO_2 or H_2O and nitrogen adsorption isotherms and the corresponding pore-size distributions for the carbons prepared with 20 nm silica colloids and activated with CO_2. *(Reprinted with permission from Ref. 60. Copyright 2011 Elsevier)*

interconnected through 30–65 nm windows, and smaller mesopores of ~11 nm originating from soft templating. Górka and Jaroniec [60] reported the preparation of hierarchical phenolic resin-based carbons by a combination of soft and hard templating followed by postsynthesis activation (Fig. 11.6). Triblock copolymers and silica nanoparticles used as soft and hard templates, respectively, afforded two distinct pore geometries, while the addition of TEOS during the synthesis and postsynthesis activation with CO_2 or water vapor tailored the microporosity of the final carbons. It was shown that the combination of all the aforementioned techniques led to an impressively high surface area of ~2800 m^2/g and a total pore volume of ~6.00 cm^3/g with almost untouched mesopore structure. Recently, Mayes et al. [61] presented a completely different approach aiming at hierarchical carbon preparation. In this work, glyoxal instead of formaldehyde was used as a cross-linker in the synthesis of polymer-templated carbons. It appeared that phloroglucinol reacts with glyoxal in the presence of triblock copolymer in the same fashion as with formaldehyde, forming a well-ordered network of mesopores with diameters of ~7.5 nm and, on top of that, creating additional macroporosity up to 200 nm. Both the glyoxal-based and the formaldehyde-based carbons were tested as capacitive deionization electrode materials for the removal of salts from brackish water. The results clearly show a threefold increase in ion uptake kinetics for glyoxal–phloroglucinol carbon, which is probably owing to its hierarchical porosity.

11.4. SURFACE MODIFICATION OF MESOPOROUS CARBON MATERIALS

11.4.1. Functionalization During Synthesis

In general, the inertness of the carbon surface is considered very beneficial especially when it comes to applications requiring good mechanical stability at wide ranges of temperature, pressure, and pH. On the other hand, the low reactivity of carbon makes postsynthesis functionalization quite challenging. In some cases, however, a carefully selected carbon precursor can bring the desired properties without dealing with troublesome postsynthesis modification. This is especially true for the soft-templating method, in which a set of specific requirements for the carbon precursor must be applied successfully. In regard to the carbon precursor, the carbon-yielding component must be able to self-assemble with the pore-forming component to form a highly cross-linked polymeric matrix that can sustain the decomposition of the pore-forming element and retain the nanostructure.

Most reports employ phenol, resorcinol, phloroglucinol, or even mixtures of these [62] as carbon precursors mostly, but not only, because of the effectiveness of hydrogen bond formation between the soft template and carbon precursors. Other important factors that favor phenolic resins over any other carbon precursor are their ability for thermosetting, high carbon yield after carbonization, and good miscibility with copolymers used as templates. Tanaka et al. [63] reported the use of triethyl orthoacetate (EOA) together with resorcinol–formaldehyde as coprecursors for carbon synthesis. The resulting carbon films exhibited an enhanced periodicity proved by both XRD and TEM measurements. Another group reported the preparation of ordered and disordered carbons from resorcinol–furfural oligomers [64] and showed that the degree of polymerization and molecular architecture determines the self-assembly and the quality of the final product. Recently, other polymeric precursors such as melamine [65], urea–formaldehyde [66], and even novolac resins [67] were considered as alternatives for the originally used phenolic resins. The advantage of using carbon precursors containing heteroatoms is the possibility of tailoring the surface chemistry of the resulting products according to the specific application. For instance, the nitrogen-containing ordered mesoporous polymers obtained from urea–phenol–formaldehyde resin [66] are claimed to be more efficient adsorbents toward heavy metal ions due to the greatly improved hydrophilic properties of the polymer surface and higher affinity toward metal ions. The aforementioned features can be utilized in catalysis, proton-conducting membranes, adsorbents for CO_2 capture, or in "reactive templates" for the preparation of nanostructured metal nitrides. It has been shown that the highest nitrogen content (~3 wt%) was obtained for a sample prepared with a 1:1 molar ratio of urea to phenol. The amount of nitrogen can easily be augmented by using a nitrogen-rich precursor such as hexamethoxymethylmelamine (HMMM) [65]. HMMM, typically used as an additive to a melamine/formaldehyde system to prevent the self-condensation of the latter, was used in that work as a single carbon precursor, which in the presence of Pluronic F127 under acidic conditions condensed into a nitrogen-rich mesoporous framework. The resulting polymers exhibited much larger mesopores than those obtained from a urea–phenol–formaldehyde system (7.8 nm vs. 3.5 nm, respectively). In spite of differences in mesoporosity, the nitrogen-containing precursors altered the strength of the hydrogen bond formed between the carbon precursor and the soft template, which governs the self-assembly. Another popular type of phenolic resin, novolac, possesses an abundance of hydroxyl groups that could be utilized successfully in the soft-templating synthesis of carbons; however, its linear structure and thermoplastic properties make it unstable at the high temperatures required for carbonization. To overcome this problem, a new method was proposed recently that involves the addition of a curing agent that can promote the cross-linking of the linear novolac resins, leading to the formation of stable, thermosetting polymeric frameworks [67].

Adsorption, catalysis, separations, and energy-related applications often require mesoporous carbons with incorporated inorganic species or carbon–inorganic nanocomposites. Inorganic species can be incorporated into mesoporous carbons during their synthesis or by postsynthesis infiltration [26,68].

The latter strategy, although often used, has several shortcomings, including limited loading of inorganic species, difficulties in controlling their size and distribution, and possible pore blocking; however, the aforementioned limitations do not appear when inorganic species are incorporated during carbon synthesis. Therefore, several attempts have been made to incorporate inorganic species during hard-templating [25] and soft-templating [58,69–72] synthesis of mesoporous carbons.

Colloidal templating is well suited for the incorporation of inorganic nanoparticles into carbons because these particles can be easily coassembled with silica colloids to form a hard template [25]. In contrast to the postsynthesis modification of carbons, which does not allow for easy control of the size and amount of the nanoparticles formed, colloidal templating permits incorporation of a large amount of inorganic nanoparticles of desired size and shape without affecting the pore size of the resulting carbon composites; the pore size is determined by the size of the silica colloids only. A shortcoming of this approach is its limitation to nanoparticles, which can survive the removal of a siliceous template.

In comparison to hard-templating synthesis, the soft-templating synthesis is better suited for the incorporation of inorganic nanoparticles as well as for their formation from suitable inorganic precursors. It was shown previously that a direct addition of metal salts into a synthesis mixture represents a convenient way to form well-dispersed nanoparticles in a carbon matrix, but, at the same time, it is difficult to control the size of these particles as well as the formation of carbon mesostructure, which often becomes microporous [73,74]. Thus, application of the latter type of carbons may be limited due to poor mass transfer in the micropores. In spite of the aforementioned shortcomings, the recently presented [69,70] idea of using block copolymers as soft templates for the carbon structure formation in the presence of metal salts that undergo transformations into nanoparticles is a very promising approach for the design of nanostructured carbon-based hybrid materials.

In some of the first reports, it was shown [57,58] that tetraethyl orthosilicate (TEOS) can easily participate in the self-assembly of organic species used for the synthesis of mesoporous carbons, leading to the uniform distribution of all species in the silica–carbon composites. The addition of other inorganic components (e.g., iridium, titanium, and nickel components) during self-assembly synthesis of carbons resulted in the formation of ordered mesostructures with inorganic nanoparticles [69–72,75]. The titania-containing mesostructures were synthesized with the use of prehydrolyzed $TiCl_4$ as an inorganic precursor [69,76,77]. In contrast to silica, titania easily nucleates, leading to TiO_2 nanocrystals being embedded in the pore walls of ordered mesoporous carbons. The uniform distribution of titania nanocrystals proved the formation of an interpenetrating hybrid framework during the coassembly process. The same group recently modified this approach to obtain crystalline C–TiO_2 composites [76]. It was shown that the use of an acid–base pair ($TiCl_4$

and $Ti(OC_4H_7)_4$) as titania precursors instead of a single source can greatly improve the titania content in the sample, even up to 87 wt%. Moreover, the mesostructured composites still exhibit good adsorption properties, such as a relatively high specific surface area of ~200 m^2/g and a relatively large pore volume.

OMCs with magnetic properties are of great interest because of their possible applications (e.g., magnetic storage media) [78]. Since the discovery of soft-templating synthesis of mesoporous carbons, this easy and cost-effective approach seems to be a good alternative to the quite laborious hard-templating method. In a recent report [79,80], the one-pot synthesis of γ-Fe_2O_3-containing mesoporous carbon obtained by the coassembly of block copolymer with resol and ferric citrate was demonstrated (Fig. 11.7) [79]. It was shown that the samples with low γ-Fe_2O_3 content (such as 9.0 wt%) possessed an ordered two-dimensional hexagonal (p6 mm) structure, uniform mesopores (~4.0 nm), high surface areas (up to 590 m^2/g), and pore volumes (up to 0.48 cm^3/g). All maghemite/carbon nanocomposites exhibited excellent magnetic properties, where the saturation magnetization strength was tuned easily by increasing the γ-Fe_2O_3 loading. However, γ-Fe_2O_3 also lowered the values of the BET surface area and the pore volume.

The ferromagnetic properties of Ni-OMC composites, especially their self-protection against acid leaching, make them attractive materials for magnetic

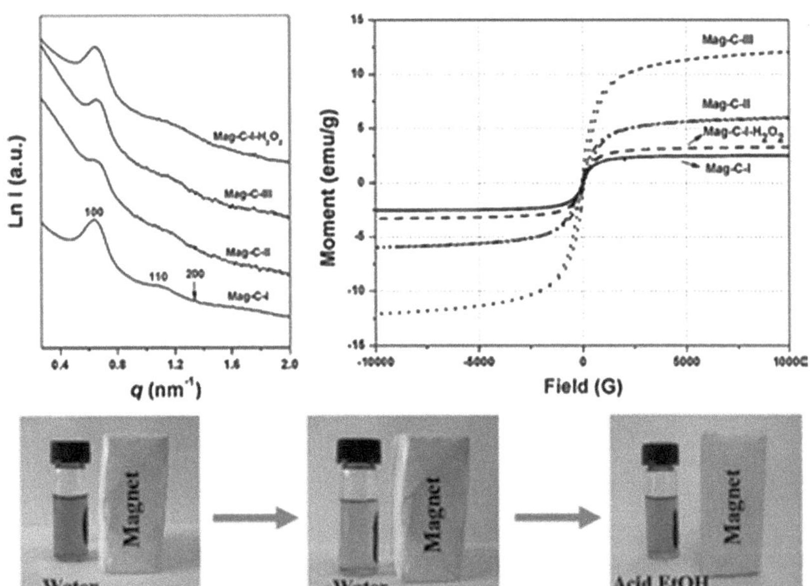

FIGURE 11.7 SAXS pattern, magnetization curves of mesostructured iron-containing carbons, and optical photos of the adsorption and separation of fuchsin base onto iron-containing carbons. *(Reprinted with permission from Ref. 79. Copyright 2009 Royal Society of Chemistry)*

separations [71]. Also, for the first time, metal-containing carbons with a cubic structure of Im3 m symmetry have been reported. The different synthesis conditions required for the formation of cages instead of channels also affect the growth of nickel nanocrystals, which showed nearly the same size (~20 nm) regardless of the amount of nickel salt used in the synthesis. In contrast, work reported by Wang and Dai [81] shows that the average size of nickel particles increased with the metal loading. There is some evidence that nanoparticles are formed in the carbon matrix and on the outer surface of hexagonally ordered carbons as well. Besides that, the resulting nickel–carbon composites exhibited good structural properties such as large and uniform mesopores (~7 nm), good total pore volume (0.46–0.68 cm^3/g), and high BET surface area (up to 660 m^2/g). The idea of using metal salt for the generation of nanoparticles in OMCs and tetraethyl orthosilicate (TEOS) as a mesostructure-reinforcing agent was reported by Zhou et al. [72]. Depending on the $NiCl_2$ concentration, the Ni–C composite obtained after silica dissolution exhibited high specific surface area (1220 m^2/g).

The fact that the addition of foreign species into a mixture of block copolymer and carbon precursor does not disintegrate the self-assembly process provides a great opportunity for the one-pot synthesis of organic–inorganic composites. Previously, methods such as surface functionalization [82,83], filling the pores of mesoporous silicas with organic moieties [84,85], or direct synthesis of periodic organosilicas (PMOs) [86,87] were used to achieve ordered mesoporous organic–silica nanocomposites but with moderate success. The risk of nonuniform distribution or pore blockage and the necessity of using rather expensive organosilanes largely limited the preparation of that type of material. The soft-templating approach helps to overcome the aforementioned drawbacks. Górka et al. [88] demonstrated the preparation of soft-templated mesoporous polymeric composites and carbons in the presence of organosilanes. It has been shown that the addition of a large heterocyclic organosilane molecule such as tris(3-trimethoxysilylpropyl)isocyanurate (ICS) does not interfere with a mesophase formation, leading to uniform incorporation of ICS into the polymer and carbon framework. Based on the TGA and IR data obtained, they determined and confirmed the presence of isocyanurate rings in the final polymer products. The thermal treatment at higher temperatures resulted in mesoporous carbons doped with nitrogen and silicon atoms possessing high surface area, pore volume, and mesopores in the 8–10 nm range.

Recently, another group reported an interesting approach to the synthesis of ordered mesoporous poly(furfuryl alcohol) (PFA)–silica nanocomposites [89]. The use of furfuryl alcohol (FA) as a carbon precursor requires the use of a special alkoxysilane as a linker between PFA/TEOS and a soft template in order to improve the assembly of PFA/TEOS in hydrophilic domains of block copolymers. The resulting composites possessed high organic content even up to 60 wt%, although the structural and adsorption properties slowly deteriorated with the increasing amount of PFA in the samples. In addition, the hydrophilic/hydrophobic nature of PFA–silica nanocomposite was evaluated on the basis of water and toluene adsorption. Since PFA oligomers are hydrophobic, their larger loadings enhance affinity toward toluene molecules.

11.4.2. Postsynthesis Modification

Although all examples of nanocomposites discussed above were obtained through one-pot synthesis, there are a few interesting postsynthesis modification methods involving soft-templated carbons. The simple bridging of inorganic groups to organic frameworks was demonstrated to obtain SO_3H-functionalized mesoporous polymers [90]. When carefully controlled, this facile sulfonation method allows the generation of a substantial number of Brőnsted acid sites without losing the structural ordering of the polymer framework. The strong oxidation conditions, with the exception of creating carboxylic groups that allow for the grafting of many other functional groups on the surface, may easily cause structure deterioration. The latter, along with pore evolution and simultaneous surface functionalization, was studied by Wu et al. [91] by wet oxidation of OMCs with acidic $(NH_4)_2S_2O_8$ (APS), HNO_3, and H_2O_2 solutions. It was found that OMC periodicity can still be maintained after treatment carried out in 1 M acidic APS for 3 days at 120 °C and/or in 15 M HNO_3 for 12 h at 70 °C (Fig. 11.8). The same conditions were applied for the wet oxidation of mesoporous carbon obtained by the hard-templating method to reveal the much lower oxidation stability of the hard-templated carbons. In comparison to untreated carbon, the specific surface area and microporosity decrease severely, with simultaneous increase in the density of carboxylic groups introduced onto the carbon surface. As-prepared materials were tested for immobilization/adsorption of heavy metal ions, organic molecules, and drugs. The carboxylic groups proved to trap effectively Fe, Cu, Cd, and Pb ions, giving considerably higher retrieved immobilization capacities than those found in the literature for adsorbents such as clays, activated carbons, or mesoporous silicas. Moreover, surface-oxidized carbons appeared to be effective adsorbents for the organic dye fuchsin and ibuprofen, proving again high capacities and excellent binding capabilities.

All aforementioned examples show that the acid treatment performed to introduce surface groups usually leads to more or less pronounced structure deterioration. In contrast, Mayes et al. [92] demonstrated that a simple stirring of OMCs with phosphoric acid overnight at room temperature produces phosphorylated mesoporous carbons with unchanged structural and adsorption properties of the starting carbon material. Additionally, the phosphorus content evaluated with XPS showed 1% phosphorus in the modified carbon, which is comparable to results obtained for carbons synthesized with the addition of phosphoric acid [93].

Even though polymer-templated carbons proved to be much more mechanically stable than their hard-templated analogs, there is still a need to explore

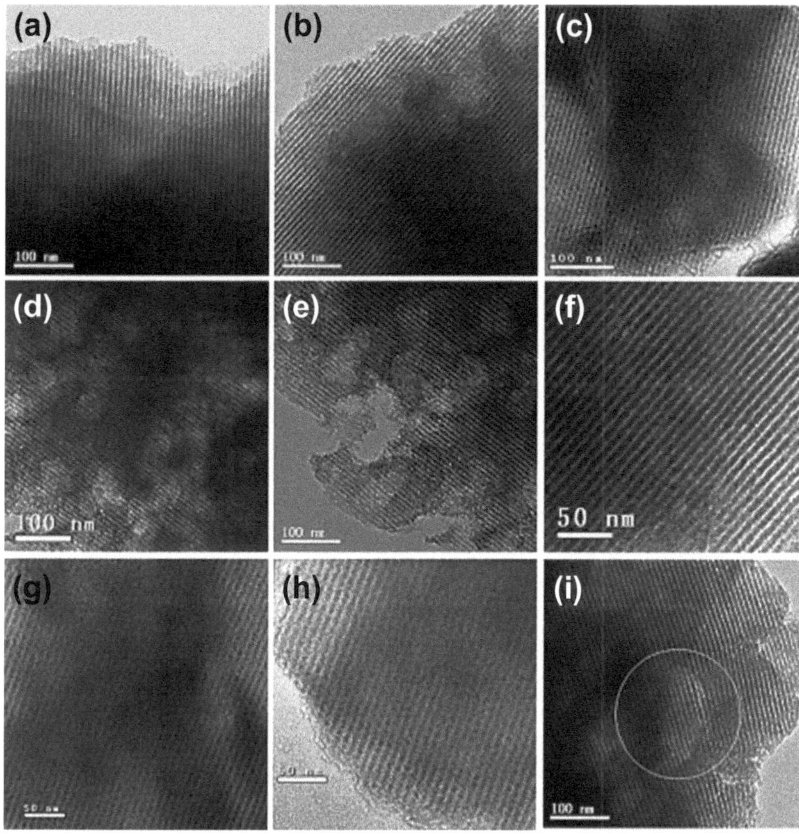

FIGURE 11.8 TEM images of (a) the pristine mesoporous carbon FDU-15 material and the products after the treatments in a 1.0 M acidic APS solution at (b) 30, (c) 60, (d) 90, and (e) 120 °C for 6 h and at 60 °C for (f) 0.5, (g) 3.0, (h) 12, and (i) 24 h. *(Reprinted with permission from Ref. 91. Copyright 2010 American Chemical Society)*

other modification routes, including the infiltration of mesoporous carbons with suitable precursors or surface modification. In the case of surface modification, two methods have been used thus far. The first method involves chloromethylation of the aromatic moieties of the mesoporous polymers and subsequent amination with methylamine, dimethylamine, or ethylinediamine [94]. This two-step functionalization method allows the incorporation of amino or diamino groups into the polymer framework to obtain a stable base catalyst with controllable basicity and mesostructure. The second method [95], Prato's reaction (Fig. 11.9), utilizes the 1,3 dipolar cycloaddition of azomethine ylides, which was successfully applied for the organic modification of carbon nanotubes, nanohorns, and fullerenes. The thermal condensation of α-amino acid with aldehyde generates azomethine ylides, which after the cycloaddition reaction form pyrrolidines attached to the surface of ordered mesoporous

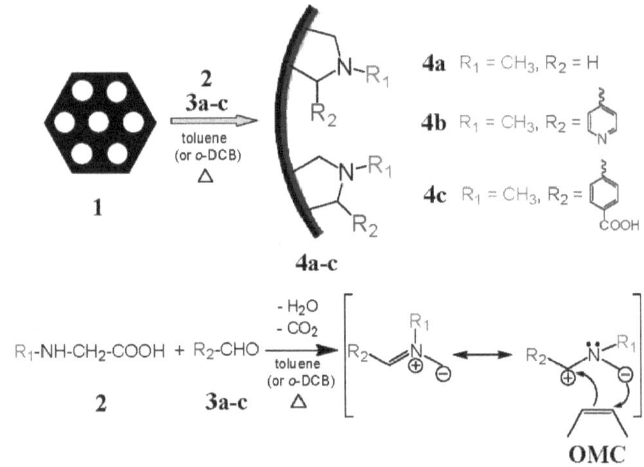

FIGURE 11.9 Functionalization of ordered mesoporous carbons with azomethine ylides. *(Reprinted with permission from Ref. 95. Copyright 2008 American Chemical Society)*

carbons. The ylides addition reduced the pore volume and the surface area; however, the mesoporosity remained untouched. The same work also shows that amorphous carbon has a preferential reactivity for ylide addition, in contrast to graphitic carbon. Prato's reaction also offers the possibility to integrate two different functional groups simultaneously, which could be a very simple method of preparing carbon-based cooperative catalysts.

Another example of the postsynthesis modification of carbons was based on the simple solution infiltration method [96–98]. It was demonstrated by Ryoo et al. [98] for CMK-3 mesoporous carbons that the controlled impregnation of pores with organic monomers results in a polymer–carbon composite that exhibits the surface properties of the polymer and the stability of the carbon. The synthesis conditions were adjusted so that the monomers would fill the micropores and form a thin film on the carbon surface, leaving a volume of mesopores accessible for potential guest molecules. A similar approach was applied to the preparation of novel cathode material for Li/S batteries [96], where the micropores of the soft-templated carbon were filled with elemental sulfur to obtain a hierarchically structured sulfur–carbon nanocomposite.

11.5. MACROSCOPIC MORPHOLOGIES OF SOFT-TEMPLATED CARBONS

Many practical applications demand not only a well-defined porous structure but also specific macroscopic morphologies. Most materials can be shaped and processed after synthesis, but in the case of soft-templated carbons, mesostructure formation and morphology growth occur simultaneously; so synthesis conditions need to be planned ahead to address both issues. It was demonstrated that the polymer solution obtained after phase separation can easily be used for the preparation of carbons in the form of monoliths, fibers, and films [12,13,99] utilizing such popular processing techniques as spin coating or fiber extrusion. Regarding the preparation of monolithic carbons, the most commonly used evaporation-induced self-assembly (EISA) method [12–15,63,99] affords carbons with different shapes and sizes depending on the mold used; however, the extensive material shrinkage and the formation of microcracks lower the quality of these materials considerably. To avoid these problems, a series of extended low-temperature treatments is performed prior to carbonization. This procedure, called a low-temperature autoclaving (LTA) route [100], led to crack-free mesoporous carbons with macroscopic dimensions and highly ordered p6 mm mesostructure. The carbon films are usually prepared by spin coating or dip coating [13] on a substrate with the polymeric solution containing a carbon precursor and a template. Both methods result in well-ordered carbon films, where the film thickness can be controlled to some extent by adjusting the density of the polymeric solution spread on the substrate. A more effective way to substantially reduce film thickness without losing good material quality is the stepwise assembly approach [12]. In the first part, the solution containing block copolymer and resorcinol is deposited on a substrate, which then is exposed to formaldehyde gas for *in situ* polymerization and cross-linking. A similar strategy was recently demonstrated for the preparation of 15 nm thin films from benzyl alcohol [101]. Instead of the usual monoliths, fibers, and films, other interesting carbon morphologies have been synthesized such as membranes [14], spheres [102], rods [103], and single crystals [105]. These can be achieved either by

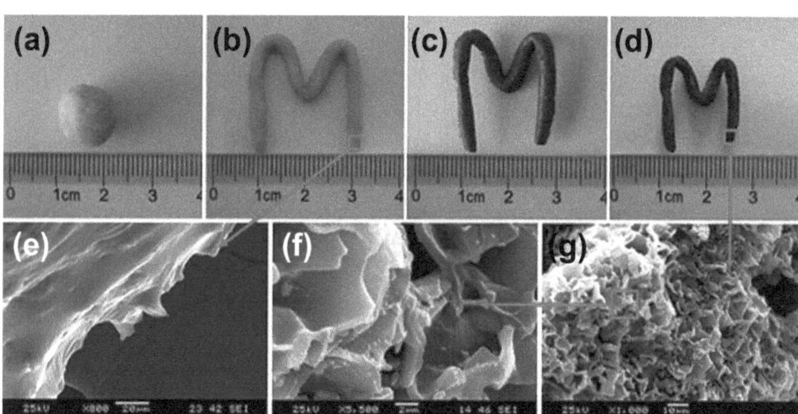

FIGURE 11.10 Photographs (a, b, c and d) of mesostructured plasticine, mesoporous polymer and mesoporous carbon: (a) as-made MC-29 plasticine with a sphere shape; (b) as-made MC-29 with an "M" shape; (c) mesoporous polymer MC-29-350 with an "M" shape; and (d) mesoporous carbon MC-29-600 with an "M" shape. SEM images (e, f, and g) of the as-made MC-29 membrane; and (e) MC-29-600 after calcination at 600 °C under nitrogen with a high (f) and low (g) magnification. *(Reprinted with permission from Ref. 106. Copyright 2009 Elsevier)*

changing the solvent system from water–ethanol [13,15,63] to an entirely aqueous [103–105] or ethanol solution [15] or by varying the PEO/PPO ratio and the concentration of triblock copolymer. Interestingly, the addition of reverse triblock copolymer 25R4 [106,107] as an intercalating soft matter resulted in "plasticine," easily shaped organic–organic hybrid mesostructures (Fig. 11.10) that retain their shape and mesoporosity after carbonization.

11.6. SUMMARY AND OUTLOOK

Each passing year brings a larger number of publications in the area of soft-templated carbons, indicating a growing interest in this simpler, greener, and less-expensive synthesis of ordered mesoporous carbons. The existing literature in this relatively new research area shows an enormous potential in the development of soft-templated OMCs and OMC-based composites for advanced applications ranging from adsorption and catalysis to environmental processes, energy storage, and conversion. A comparison of hard- and soft-templating synthesis showed that the latter is more feasible for the preparation of ordered mesoporous carbons and carbon-based materials. The reduced number of preparation steps and the greener synthesis route, along with more freedom in the design of the structural properties of the final material, make soft-templating synthesis very appealing to both industry and academia.

A vast amount of work was devoted to elaborating the synthesis procedure to better understand the phenomena governing the carbon's mesophase formation from phenolic resins and triblock copolymers and to gain some control of the mesoscopic architectures and surface properties of the latter. Toward this goal, different block copolymers were used to tailor pore diameters and their symmetry as well as pore wall thickness, also bearing in mind the mechanical properties of the carbon. Another method for tuning mesoporosity and enhancing total pore volume employs small organic molecules used as expanders. All these efforts aim at the development of effective adsorbents for wastewater treatment requiring fast molecular diffusion and uniform mesopores to adsorb/immobilize bulky pollutants such as dyes, organics, and biomolecules, etc. [108].

Another important area of soft-templated carbon research is devoted to the elaboration of efficient ways to improve the BET surface area and micropore volume. The general synthesis procedure results in mostly mesoporous carbons having microporosity below 15% of the total pore volume, which is quite unexpected, especially compared to traditional carbon adsorbents (activated carbons). The postsynthesis activation and/or the addition of small organo-silane (TEOS) gave excellent results in creating additional microporosity. It was shown that regardless of the activating medium (KOH, CO_2, or water vapor), the polymer-templated carbons easily sustain harsh activation conditions without losing mesoporous structure. This is one of the most important features distinguishing soft-templated carbons from their hard-templated analogs. It was shown that the specific surface area can easily reach 1500 m^2/g or even be doubled if the postsynthesis activation is combined with the addition of TEOS as an alternative microporogen. High-surface-area carbons can possibly be used for hydrogen storage applications and can also provide the possibility to introduce active sides for almost any guest. So far, large numbers of reports demonstrate the feasibility of using inorganic salts during soft-templating synthesis to obtain carbons with incorporated metal/metal oxide nanoparticles. Except for the obvious applications in catalysis, that kind of hybrid porous material can also be promising for CO_2 capture, as was shown in the case of carbon-supporting calcium oxide nanoparticles [109]. This is also a good method for manipulating the acidity/basicity of the carbon surface.

Another method for carbon framework modification during synthesis relies on the use of different carbon precursors already possessing heteroatoms or substituting the part of typically used carbon precursors with components containing functional groups of choice. While the former has the advantage of forming a mesostructure with uniformly distributed heteroatoms, the latter, if the thermal treatment is performed at low temperatures, results in mesoporous polymers decorated with organic groups. The further carbonization of these polymers yields carbons doped with heteroatoms.

Regarding postsynthesis functionalization, simple acid treatment appears to be a very effective way to introduce carboxylic, sulfonic, and phosphoric groups onto the carbon surface; however, if not controlled well, severe structural damage may occur. Also, some more advanced surface modification reactions were reported, including amination and those based on diazonium chemistry. However, to date only a few works deal with this subject.

Even though much has been done in the area of polymer-templated carbons, some challenges remain before these carbon materials can be implemented practically. One of them is to improve the performance of carbon materials in terms of higher selectivity and molecular recognition. To fulfill this requirement, stronger emphasis should be placed on the development or improvement of surface modification techniques that would allow introducing multifunctional groups to the surface in a controlled and even manner without pore blockage. Some of the methods, which still have not been used for the modification of soft-templated carbons and already proved their potential for other nanomaterials, involve the use of ultrasounds, microwaves, and radical reactions [110]. In addition, some other carbon precursors should be explored, bearing in mind that the ability to form ordered and stable mesostructures could also donate some intrinsic microporosity.

ACKNOWLEDGMENTS

The authors would like to thank the financial support from the DOE Office of Nuclear Energy.

REFERENCES

[1] R. Ryoo, S.H. Joo, S. Jun, Synthesis of highly ordered carbon molecular sieves via template-mediated structural transformation, J. Phys. Chem. B 103 (1999) 7743–7746.

[2] R. Ryoo, S.H. Joo, M. Kruk, M. Jaroniec, Ordered mesoporous carbons, Adv. Mater. 13 (2001) 677–681.

[3] L. Lee, S. Han, T. Hyeon, Synthesis of new nanoporous carbon materials using nanostructured silica materials as templates, J. Mater. Chem. 14 (2004) 478–486.

[4] J. Lee, J. Kim, T. Hyeon, Recent progress in the synthesis of porous carbon materials, Adv. Mater. 18 (2006) 2073–2094.

[5] S. Han, T. Hyeon, Novel silica-sol mediated synthesis of high surface area porous carbons, Carbon 37 (1999) 1645–1647.

[6] S. Han, T. Hyeon, Simple silica-particle template synthesis of mesoporous carbons, Chem. Commun. (1999) 1955–1956.

[7] S. Han, K. Sohn, T. Hyeon, Fabrication of new nanoporous carbons through silica templates and their application to the adsorption of bulky dyes, Chem. Mater. 12 (2000) 3337–3341.

[8] A.A. Zakhidov, R.H. Baughaman, Z. Iqbal, C. Cui, I. Khayrullin, S.O. Dantas, J. Marti, V.G. Ralchenko, Carbon structures with three-dimensional periodicity at optical wavelengths, Science 282 (1998) 897–901.

[9] J.-S. Yu, S.B. Yoon, G.S. Chai, Ordered uniform porous carbon by carbonization of sugars, Carbon 39 (2001) 1442–1446.

[10] Y. Wan, Y. Shi, D.Y. Zhao, Supramolecular aggregates as templates: ordered mesoporous polymers and carbons, Chem. Mater. 20 (2008) 932–945.

[11] C.D. Liang, Z.J. Li, S. Dai, Mesoporous carbon materials: synthesis and modification, Angew. Chem. Int. Ed. 47 (2008) 3696–3717.

[12] C. Liang, K.L. Hong, G.A. Guiochon, J.W. Mays, S. Dai, Synthesis of a large-scale highly ordered porous carbon film by self-assembly of block copolymers, Angew. Chem. Int. Ed. 43 (2004) 5785–5789.

[13] C. Liang, S. Dai, Synthesis of mesoporous carbon materials via enhanced hydrogen-bonding interaction, J. Am. Chem. Soc. 128 (2006) 5316–5317.

[14] Y. Meng, D. Gu, F. Zhang, Y. Shi, L. Cheng, D. Feng, Z. Wu, Z. Chen, Y. Wan, A. Stein, D.Y. Zhao, A family of highly ordered mesoporous polymer resin and carbon structures from organic–organic self-assembly, Chem. Mater. 18 (2006) 4447–4464.

[15] S. Jun, S.H. Joo, R. Ryoo, M. Kruk, M. Jaroniec, Z. Liu, T. Ohsuna, O. Terasaki, Synthesis of new, nanoporous carbon with hexagonally ordered mesostructure, J. Am. Chem. Soc. 122 (2000) 10712–10713.

[16] A. Stein, F. Li, N.R. Denny, Morphological control in colloidal crystal templating of inverse opals, hierarchical structures, and shaped particles, Chem. Mater. 20 (2008) 649–666.

[17] S.B. Yoon, G.S. Chai, S.K. Kang, J.S. Yu, K.P. Gierszal, M. Jaroniec, Graphitized pitch-based carbons with ordered nanopores synthesized by using colloidal crystals as templates, J. Am. Chem. Soc. 127 (2005) 4188–4189.

[18] S.A. Johnson, P.J. Ollivier, T.E. Mallouk, Ordered mesoporous polymers of tunable pore size from colloidal silica templates, Science 283 (1999) 963–965.

[19] K.P. Gierszal, S.B. Yoon, J.S. Yu, M. Jaroniec, Adsorption and structural properties of mesoporous carbons obtained from mesophase pitch and phenol-formaldehyde carbon precursors using porous templates prepared from colloidal silica, J. Mater. Chem. 16 (2006) 2819–2823.

[20] Z. Li, M. Jaroniec, Colloidal imprinting: a novel approach to the synthesis of mesoporous carbons, J. Am. Chem. Soc. 123 (2001) 9208–9209.
[21] Z. Li, M. Jaroniec, Synthesis and adsorption properties of colloid-imprinted carbons with surface and volume mesoporosity, Chem. Mater. 6 (2003) 1327–1333.
[22] Z. Li, M. Jaroniec, Colloid-imprinted carbons as stationary phases for reversed-phase liquid chromatography, Anal. Chem. 76 (2004) 5479–5485.
[23] P.K. Gierszal, M. Jaroniec, Carbons with extremely large volume of uniform mesopores synthesized by carbonization of phenolic resin film formed on colloidal silica template, J. Am. Chem. Soc. 128 (2006) 10026–10027.
[24] K.P. Gierszal, M. Jaroniec, Large pore volume carbons with uniform mesopores and macropores: synthesis, characterization, and relations between adsorption parameters of silica templates and their inverse carbon replicas, J. Phys. Chem. C 111 (2007) 9742–9748.
[25] M. Jaroniec, J. Choma, J. Górka, A. Zawislak, Colloidal silica templating synthesis of carbonaceous monoliths assuring formation of uniform spherical mesopores and incorporation of inorganic nanoparticles, Chem. Mater. 20 (2008) 1069–1075.
[26] M. Hartmann, A. Vinu, G. Chandrasekar, Adsorption of vitamin E on mesoporous carbon molecular sieves, Chem. Mater 17 (2005) 829–833.
[27] A. Vinu, K.Z. Hossain, G. Satish Kumar, K. Ariga, Adsorption of L-histidine over mesoporous carbon molecular sieves, Carbon 44 (2006) 530–536.
[28] K. Ariga, A. Vinu, M. Miyahara, J.P. Hill, T. Mori, One-pot separation of tea components through selective adsorption on pore-engineered nanocarbon, carbon nanocage, J. Am. Chem. Soc. 129 (2007) 11022–11023.
[29] Y. Huang, H. Cai, T. Yu, F. Zhang, F. Zhang, Y. Meng, D. Gu, Y. Wan, X. Sun, B. Tu, D.Y. Zhao, Formation of mesoporous carbon with a face-centered-cubic Fdm structure and bimodal architectural pores from the reverse amphiphilic triblock copolymer PPO-PEO-PPO, Angew. Chem. Int. Ed. 46 (2007) 1089–1093.
[30] Y. Deng, C. Liu, D. Gu, T. Yu, B. Tu, D.Y. Zhao, Thick wall mesoporous carbons with a large pore structure templated from a weakly hydrophobic PEO–PMMA diblock copolymer, J. Mater. Chem. 18 (2008) 91–97.
[31] Y.H. Deng, J. Liu, C. Liu, D. Gu, Z.K. Sun, J. Wei, J.Y. Zhang, L.J. Zhang, B. Tu, D.Y. Zhao, Ultra-large-pore Mesoporous carbons templated from poly(ethylene oxide)-b-polystyrene diblock copolymer by adding polystyrene homopolymer as a pore expander, Chem. Mater. 20 (2008) 7281–7286.
[32] P.F. Fulvio, R.T. Mayes, X. Wang, S.M. Mahurin, J.C. Bauer, V. Presser, J. McDonough, Y. Gogotsi, S. Dai, "Brick-and-mortar self-assembly approach to graphitic mesoporous carbon nanocomposites, Adv. Fun. Mater. 21 (2011) 2208–2215.
[33] D. Saha, S.G. Deng, Hydrogen adsorption on ordered mesoporous carbons doped with Pd, Pt, Ni, and Ru, Langmuir 25 (2009) 12550–12560.
[34] D. Saha, S.G. Deng, Enhanced hydrogen adsorption in ordered mesoporous carbon through clathrate formation, Int. J. Hydrogen Energy 34 (2009) 8583–8588.
[35] Z.P. Sun, X.G. Zhang, Y.Y. Liang, H. Tong, R.L. Xue, S.D. Yang, H.L. Li, Ordered mesoporous carbons (OMCs) as supports of electrocatalysts for direct methanol fuel cells (DMFCs): effect of the pore characteristics of OMCs on DMFCs, J. Electroanal. Chem. 633 (2009) 1–6.
[36] H.Q. Li, R.L. Liu, D.Y. Zhao, Y.Y. Xia, Electrochemical properties of an ordered mesoporous carbon prepared by direct tri-constituent co-assembly, Carbon 45 (2007) 2628–2635.
[37] H.F. Yang, D.Y. Zhao, Synthesis of replica mesostructures by the nanocasting strategy, J. Mater. Chem. 15 (2005) 1217–1231.

[38] A. Vinu, T. Mori, K. Ariga, New families of mesoporous materials, Sci. Tech. Adv. Mater. 7 (2006) 753–771.

[39] A.H. Lu, F. Schuth, Nanocasting: a versatile strategy for creating nanostructured porous materials, Adv. Mater. 18 (2006) 1793–1805.

[40] J. Lee, J. Kim, T. Hyeon, Synthesis of new nanostructured carbon materials using silica nanostructured templates by Korean research groups, Int. J. Nanotech. 3 (2006) 253–279.

[41] K. Xia, Q. Gao, C. Wu, S. Song, M. Ruan, Activation, characterization and hydrogen storage properties of the mesoporous carbon CMK-3, Carbon 45 (2007) 1989–1996.

[42] K.S. Xia, Q.M. Gao, S.Q. Song, C.D. Wu, J.H. Jiang, J. Hu, L. Gao, CO_2 activation of ordered porous carbon CMK-1 for hydrogen storage, Int. J. Hydrogen Energy 33 (2008) 116–123.

[43] K.S. Xia, Q.M. Gao, J.H. Jiang, J. Hu, Hierarchical porous carbons with controlled micropores and mesopores for supercapacitor electrode materials, Carbon 46 (2008) 1718–1726.

[44] M. Armandi, B. Bonelli, E.I. Karaindrou, C.O. Arean, E. Garrone, Post-synthesis modifications of SBA-15 carbon replicas: improving hydrogen storage by increasing microporous volume, Catal. Today 138 (2008) 244–248.

[45] M. Choi, R. Ryoo, Mesoporous carbons with KOH activated framework and their hydrogen adsorption, J. Mater. Chem. 17 (2007) 4204–4209.

[46] W. Xing, C.C. Huang, S.P. Zhuo, X. Yuan, G.Q. Wang, D. Hulicova-Jurcakova, Z.F. Yan, G.Q. Lu, Hierarchical porous carbons with high performance for supercapacitor electrodes, Carbon 47 (2009) 1715–1722.

[47] W. Xing, S.P. Zhuo, X.L. Gao, Preparation of hierarchical porous carbon by post activation, Mater. Lett. 63 (2009) 1311–1313.

[48] Y. Yan, J. Wei, F.Q. Zhang, Y. Meng, B. Tu, D.Y. Zhao, The pore structure evolution and stability of mesoporous carbon FDU-15 under CO_2, O_2 or water vapor atmospheres, Micropor. Mesopor. Mater. 113 (2008) 305–314.

[49] J. Górka, A. Zawislak, J. Choma, M. Jaroniec, KOH activation of mesoporous carbons obtained by soft-templating synthesis, Carbon 46 (2008) 1159–1161.

[50] J. Jin, S. Tanaka, Y. Egashira, N. Nishiyama, KOH activation of ordered mesoporous carbons prepared by a soft-templating method and their enhanced electrochemical properties, Carbon 48 (2010) 1985–1989.

[51] D. Wu, Y. Liang, Q.X. Yang, C. Zou, Z.H. Li, G.F. Lv, X.H. Zeng, R. Fu, Preparation of activated ordered mesoporous carbons with a channel structure, Langmuir 24 (2008) 2967–2969.

[52] Q.X. Wang, C.D. Liang, S. Dai, Facile synthesis of ordered mesoporous carbons with high thermal stability by self-assembly of resorcinol–formaldehyde and block copolymers under highly acidic conditions, Langmuir 24 (2008) 7500–7505.

[53] X. Wang, J.S. Lee, C. Tsouris, D.W. DePaoli, S. Dai, Preparation of activated mesoporous carbons for electrosorption of ions from aqueous solutions, J. Mater. Chem. 20 (2010) 4602–4608.

[54] X.Q. Wang, J.S. Lee, Q. Zhu, J. Liu, Y. Wang, S. Dai, Ammonia-treated ordered mesoporous carbons as catalytic materials for oxygen reduction reaction, Chem. Mater. 22 (2010) 2178–2180.

[55] Y. Shao, X. Wang, M. Engelhard, C. Wang, S. Dai, J. Liu, Z. Yang, Y. Lin, Nitrogen-doped mesoporous carbon for energy storage in vanadium redox flow batteries, J. Power Sour. 195 (2010) 4375–4379.

[56] R. Liu, Y. Shi, Y. Wan, Y. Meng, F. Zhang, D. Gu, Z. Chen, B. Tu, D.Y. Zhao, Triconstituent co-assembly to ordered mesostructured polymer–silica and carbon–silica nanocomposites and large-pore mesoporous carbons with high surface areas, J. Am. Chem. Soc. 128 (2006) 11652–11662.
[57] Y. Deng, T. Yu, Y. Wan, Y. Shi, Y. Meng, D. Gu, L. Zhang, Y. Huang, C. Liu, X. Wu, D.Y. Zhao, Ordered mesoporous silicas and carbons with large accessible pores templated from amphiphilic diblock copolymer poly(ethylene oxide)-b-polystyrene, J. Am. Chem. Soc. 129 (2007) 1690–1697.
[58] P.H. Lin, C.Y. Chang-Chien, C.Y. Tang, C.Y. Lin, Synthesis of p6mm hexagonal mesoporous carbons and silicas using Pluronic F127-PF resin polymer blends, Micropor. Mesopor. Mater. 93 (2006) 344–348.
[59] Y.H. Deng, C. Liu, T. Yu, F. Liu, F.Q. Zhang, Y. Wan, L.J. Zhang, C.C. Wang, B. Tu, P.A. Webley, H.T. Wang, D.Y. Zhao, Facile synthesis of hierarchically porous carbons from dual colloidal crystal/block copolymer template approach, Chem. Mater. 19 (2006) 3271–3277.
[60] J. Górka, M. Jaroniec, Hierarchically porous phenolic resin-based carbons obtained by block copolymer-colloidal silica templating and post-synthesis activation with carbon dioxide and water vapor, Carbon 49 (2011) 154–160.
[61] R.M. Mayes, C. Tsouris, J.O. Kiggins Jr.J.O. Kiggins, S.M. Mahurin, D.W. DePaoli, D. Dai, Hierarchical ordered mesoporous carbon from phloroglucinol-glyoxal and its application in capacitive deionization of brackish water, J. Mater.Chem. 20 (2010) 8674–8678.
[62] S. Tanaka, Y. Katayama, M.P. Tad, H.W. Hillhouse, Y. Miyake, Fabrication of continuous mesoporous carbon films with face-centered orthorhombic symmetry through a soft templating pathway, J. Mater. Chem. 17 (2007) 3639–3645.
[63] S. Tanaka, N. Nishiyama, Y. Egashira, K. Ueyama, Synthesis of ordered mesoporous carbons with channel structure from an organic–organic nanocomposite, Chem. Commun. (2005) 2125–2127.
[64] D.H. Long, W.M. Oiao, L. Zhan, X.Y. Liang, L.C. Ling, Effect of template and precursor chemistry on pore architectures of triblock copolymer-templated mesoporous carbons, Micropor. Mesopor. Mater. 121 (2009) 58–66.
[65] K. Kailasam, Y.S. Jun, P. Katekomol, J.D. Eppig, W.H. Hong, A. Thomas, Mesoporous melamine resins by soft templating of block-co-polymer mesophases, Chem. Mater. 22 (2010) 428–434.
[66] J. Yang, Y. Zhai, Y. Deng, D. Gu, Q. Li, Q. Wu, Y. Huang, B. Tu, D.Y. Zhao, Direct triblock-copolymer-templating synthesis of ordered nitrogen-containing mesoporous polymers, J. Coll. Inter. Sci. 342 (2010) 579–585.
[67] Y. Huang, J.P. Yang, H.Q. Cai, Y.P. Zhai, D. Feng, Y.H. Deng, B. Tu, D.Y. Zhao, A curing agent method to synthesize ordered mesoporous carbons from linear novolac phenolic resin polymers, J. Mater. Chem. 19 (2009) 6536–6541.
[68] A. Stein, Z. Wang, M.A. Fierke, Functionalization of porous carbon materials with designed pore architecture, Adv. Mater. 21 (2008) 265–293.
[69] R. Liu, Y. Ren, Y. Shi, F. Zhang, L. Zhang, B. Tu, D.Y. Zhao, Controlled synthesis of ordered mesoporous C–TiO2 nanocomposites with crystalline titania frameworks from organic–inorganic–amphiphilic coassembly, Chem. Mater. 20 (2008) 1140–1146.
[70] P. Gao, A. Wang, T. Zhang, Synthesis of highly ordered Ir-containing mesoporous carbon materials by organic–organic self-assembly, Chem. Mater. 20 (2008) 1881–1888.
[71] J. Yao, L. Li, H. Song, C. Liu, X. Chen, Synthesis of magnetically separable ordered mesoporous carbons from F127/[Ni(H2O)6](NO3)2/resorcinol-formaldehyde composites, Carbon 47 (2009) 436–444.

[72] J. Zhou, J. He, Y. Guo, X. Chen, T. Wang, D. Sun, D. Wang, Z. Di, $NiCl_2$ assisted synthesis of ordered mesoporous carbon and a new strategy for a binary catalyst, J. Mater. Chem. 18 (2008) 5776–5781.

[73] C. Moreno-Castilla, F.J. Maldonado-Hódar, J. Rivera-Utrilla, E. Rodriguez-Castellon, Group 6 metal oxide-carbon aerogels. Their synthesis, characterization and catalytic activity in the skeletal isomerization of 1-butene, Appl. Catal. A 183 (1999) 345–356.

[74] M.L. Rojas-Cervantes, M. Alonso, J. Diaz-Teran, A.J. Lopez-Peinado, R.M. Martin-Aranda, V. Gomez-Serrano, Basic metal–carbons catalysts prepared by sol–gel method, Carbon 42 (2004) 1575–1582.

[75] H. Wang, A.Q. Wang, X.D. Wang, T. Zang, One-pot synthesized MoC imbedded in ordered mesoporous carbon as a catalyst for N_2H_4 decomposition, Chem. Commun. (2008) 2565–2567.

[76] X. Qian, Y. Wan, Y. Wen, N. Jia, X. Li, D.Y. Zhao, Synthesis of ordered mesoporous crystalline carbon-anatase composites with high titania contents, J. Coll. Inter. Sci. 328 (2008) 367–373.

[77] C.H. Huang, D. Gu, D.Y. Zhao, R.A. Doong, Direct synthesis of controllable microstructures of thermally stable and ordered mesoporous crystalline titanium oxides and carbide/carbon composites, Chem. Mater. 22 (2010) 1760–1767.

[78] T. Hyeon, Chemical synthesis of magnetic nanoparticles, Chem. Commun. (2003) 927–934.

[79] Y. Zhai, Y. Dou, X. Liu, B. Tu, D.Y. Zhao, One-pot synthesis of magnetically separable ordered mesoporous carbon, J. Mater. Chem. 19 (2009) 3292–3300.

[80] J.S. Li, J. Gu, H.J. Li, Y. Liang, Y.X. Hao, X.Y. Sun, L.J. Wang, Synthesis of highly ordered Fe-containing mesoporous carbon materials using soft templating routes, Micropor. Mesopor. Mater. 128 (2010) 144–149.

[81] X. Wang, S. Dai, A simple method to ordered mesoporous carbons containing nickel nanoparticles, Adsorption 15 (2009) 138–144.

[82] Y. Yang, Y.F. Lu, M.C. Lu, J. Huang, R. Haddad, G. Xomeritakis, N.G. Liu, A.P. Malanoski, D. Sturmayr, H.Y. Fan, D.Y. Sasaki, R.A. Assink, J.A. Shelnutt, F. van Swol, G.P. Lopez, A.R. Burns, C.J. Brinker, Functional nanocomposites prepared by self-assembly and polymerization of diacetylene surfactants and silicic acid, J. Am. Chem. Soc. 125 (2003) 1269–1277.

[83] A. Shimojima, Z. Liu, T. Oshuna, O. Terasaki, K. Kuroda, Self-assembly of designed oligomeric siloxanes with alkyl chains into silica-based hybrid mesostructures, J. Am. Chem. Soc. 127 (2005) 14108–14116.

[84] S.B. Yoon, J.Y. Kim, J.S. Yu, A direct template synthesis of nanoporous carbons with high mechanical stability using as-synthesized MCM-48 hosts, Chem. Commun. (2002) 1536–1537.

[85] M. Choi, F. Kleitz, D.N. Liu, H.Y. Lee, W.S. Ahn, R. Ryoo, Controlled polymerization in mesoporous silica toward the design of organic−inorganic composite nanoporous materials, J. Am. Chem. Soc. 127 (2005) 1924–1932.

[86] T. Asefa, M.J. MacLachan, N. Coombs, G.A. Ozin, Periodic mesoporous organosilicas with organic groups inside the channel walls, Nature 402 (1999) 867–871.

[87] S. Inagaki, S. Guan, Y. Fukushima, Y. Oshuna, O. Terasaki, Novel mesoporous materials with a uniform distribution of organic groups and inorganic oxide in their frameworks, J. Am. Chem. Soc. 121 (1999) 9611–9614.

[88] J. Górka, M. Jaroniec, Tailoring adsorption and framework properties of mesoporous polymeris composites and carbons by addition of organosilanes during soft-templating synthesis, J. Phys. Chem. 114 (2010) 6298–6303.

[89] Y. Zhai, B. Tu, D.Y. Zhao, Organosilane-assisted synthesis of ordered mesoporous poly(furfuryl alcohol) composites, J. Mater. Chem. 19 (2009) 131–140.
[90] R. Xing, N. Liu, Y. Liu, H. Wu, Y. Jiang, L. Chen, M. He, P. Wu, Novel solid acid catalysts: sulfonic acid group-functionalized mesostructured polymers, Adv. Func. Mater. 17 (2007) 2455–2461.
[91] Z. Wu, P.A. Webley, D.Y. Zhao, Comprehensive study of pore evolution, mesostructural stability, and simultenous surface functionalization of ordered mesoporous carbon (FDU-15) by wet oxidation as a promising adsorbent, Langmuir 26 (2010) 10277–10286.
[92] R.M. Mayes, P.F. Fulvio, Z. Ma, S. Dai, Phosphorylated mesoporous carbon as a solid acid catalyst, Phys. Chem. Chem. Phys. 13 (2011) 2492–2494.
[93] X. Zhao, A. Wang, J. Yan, G. Sun, L. Sun, T. Zhang, Synthesis and electrochemical performance of heteroatom-incorporated ordered mesoporous carbons, Chem. Mater. 22 (2010) 5463–5473.
[94] R. Xing, H. Wu, X. Li, Z. Zhao, Y. Liu, L. Chen, P. Wu, Mesopolymer solid base catalysts with variable basicity: preparation and catalytic properties, J. Mater. Chem. 19 (2009) 4004–4011.
[95] X.Q. Wang, D.E. Jiang, S. Dai, Surface modification of ordered mesoporous carbons via 1,3-dipolar cycloaddition of azomethine ylides, Chem. Mater. 20 (2008) 4800–4802.
[96] C.D. Liang, N.J. Dudney, J.Y. Howe, Hierarchically structured sulfur/carbon nanocomposite material for high-energy lithium battery, Chem. Mater. 21 (2009) 4724–4730.
[97] Q. Zhu, S.H. Zhou, X.Q. Wang, S. Dai, Controlled synthesis of mesoporous carbon modified by tungsten carbides as an improved electrocatalyst support for the oxygen reduction reaction, J. Power Sour. 193 (2009) 495–500.
[98] M. Choi, R. Ryoo, Ordered nanoporous polymer-carbon composites, Nature Mater 2 (2003) 473–476.
[99] Y. Huang, H.Q. Cai, D. Feng, D. Gu, Y.H. Deng, B. Tu, D.Y. Zhao, One-step hydrothermal synthesis of ordered mesostructured carbonaceous monoliths with hierarchical porosities, Chem. Commun. (2008) 2641–2643.
[100] L. Liu, F.Y. Wang, G.S. Shao, Z.Y. Yuan, A low-temperature autoclaving route to synthesize monolithic carbon materials with an ordered mesostructures, Carbon 48 (2010) 2089–2099.
[101] J. Jin, N. Nishiyama, Y. Egashira, K. Ueyama, Vapor phase synthesis of ultrathin carbon films with a mesoporous monolayer by a soft-templating method, Chem. Commun. (2009) 1371–1373.
[102] Y. Yan, F.Q. Zhang, Y. Meng, B. Tu, D.Y. Zhao, One-step synthesis of ordered mesoporous carbonaceous spheres by an aerosol-assisted self-assembly, Chem. Commun. (2007) 2867–2869.
[103] F.Q. Zhang, Y. Meng, D. Gu, Y. Yan, Z.X. Chen, B. Tu, D.Y. Zhao, An aqueous cooperative assembly route to synthesize ordered mesoporous carbons with controlled structures and morphology, Chem. Mater. 18 (2006) 5279–5288.
[104] F.Q. Zhang, D. Gu, T. Yu, F. Zhang, S.H. Xie, L.J. Zhang, Y.H. Deng, Y. Wan, B. Tu, D.Y. Zhao, Mesoporous carbon single-crystals from organic–organic self-assembly, J. Am. Chem. Soc. 129 (2007) 7746–7747.
[105] D. Gu, H. Bongard, Y. Meng, K. Miyasaka, O. Terasaki, F.Q. Zhang, Y.H. Deng, Z.X. Wu, D. Feng, Y. Fang, B. Tu, F. Schuth, D.Y. Zhao, Growth of single-crystal mesoporous carbons with Im3m symmetry, Chem. Mater. 22 (2010) 4828–4833.
[106] X.F. Qian, Z. Wang, Y. Wan, From polymeric "plasticine" to shape-controlled mesoporous carbon, J. Coll. Inter. Sci. 335 (2009) 222–227.

[107] X.F. Qian, H. Li, Y. Wan, Structure design of mesoporous carbons by blending PEO–PPO–PEO-type and PPO–PEO–PPO-type amphiphilic block copolymers in organic–organic self-assembly, Micropor. Mesopor. Mater. 141 (2011) 26–37.
[108] X. Zhuang, Y. Wan, C. Feng, Y. Shen, D.Y. Zhao, Highly efficient adsorption of bulky dye molecules in wastewater on ordered mesoporous carbons, Chem. Mater. 21 (2009) 706–716.
[109] L. She, J. Li, Y. Wan, X. Yao, B. Tu, D.Y. Zhao, Synthesis of ordered mesoporous MgO/carbon composites by a one-pot assembly of amphiphilic triblock copolymers, J. Mater. Chem. 21 (2011) 795–800.
[110] H. Xu, K.S. Suslick, Sonochemical preparation of functionalized graphenes, J. Am. Chem. Soc. 133 (2011) 9148–9151.

Chapter 12

Hydrothermal Carbons: Synthesis, Characterization, and Applications

Maria-Magdalena Titirici
Max Planck Institute of Colloids and Interfaces, Potsdam, Germany

Chapter Outline

12.1. Introduction 352
12.2. Formation Mechanism and Final Chemical Structure of HTC Materials 352
12.2.1. Carbohydrate-Based HTCs 353
12.2.2. Lignocellulosic-Biomass-Derived HTCs 356
12.3. Porous HTC Materials 357
12.3.1. Chemical Activation 358
12.3.2. Nanocasting, Soft Templating, and the Use of Natural Gelators 362
12.4. Functionalization of HTC Materials 362
12.4.1. Postmodification 363
12.4.2. "*In situ*" Functionalization 364
12.5. HTC Nanocomposites 367
12.5.1. Coating of Preformed Nanostructures 367
12.5.2. Loading Inorganic Nanostructures onto HTC Materials 370
12.5.3. *In situ* HTC Nanocomposites 370
12.6. Applications of HTC 373
12.6.1. Agricultural Applications: Biochar 374
12.6.2. Adsorption 376
12.6.3. Heterogeneous Catalysis 380
12.6.4. Magnetic HTCs 383
12.6.5. Photocatalysis 385
12.6.6. Energy Storage 386
12.6.7. Biological Applications 394
12.7. Conclusions 395
References 395

Novel Carbon Adsorbents. DOI: 10.1016/B978-0-08-097744-7.00012-0
Copyright © 2012 Elsevier Ltd. All rights reserved.

12.1. INTRODUCTION

Starting with the discovery of fullerenes [1] and carbon nanotubes, [2] the interest in the science related to valuable carbon materials raised significantly. This is motivated by the large number of applications in carbon fixation [3], catalyst supports, [4,5] adsorbents, [6,7] gas storage, [8] electrodes, [9] carbon fuel cells [10], and cell biology [11]. Many synthetic methods, such as pyrolysis, high-voltage-arc discharge, and laser ablation, have been reported for the preparation of amorphous or crystalline nanostructured carbon materials with different sizes, shapes, and chemical compositions [12–14].

Considering the energy shortage, environmental crisis, and the ever-increasing customer demands, the discovery of facile, low-cost, environmentally friendly, and nontoxic routes to produce novel valuable carbon materials is a key issue in today's society. So far, excluding activated carbons, only little research has been done on the synthesis and characterization of the structure of carbon materials based on natural resources. This is surprising, given that carbon material synthesis has been done since the beginning of civilization on the basis of biomass (through coal formation) with the petrochemical age being only a late deviation. Therefore, an approach toward advanced carbon synthesis based on renewable resources, mimicking the natural process of coal formation, would be a significant operation, as the final products represent a significant proportion of modern materials.

This is where the highly unconventional technique of hydrothermal carbonization, denoted from here on as HTC, comes into play. In HTC, carbohydrates and raw lignocellulosic biomasses are converted into valuable carbon materials using water as a carbonization medium at mild temperatures (<200 °C) under self-generated pressures [15]. Even though this methodology was developed almost 100 years ago, its full potential, as a synthetic route for carbon materials having important applications in several fields such as catalysis, energy storage, CO_2 sequestration, water purification, and agriculture, has been revealed only in the past few years [16]. This chapter offers an overview on the recently developed HTC technology including formation mechanism, nanostructuring, and functionalization as well as its applications.

12.2. FORMATION MECHANISM AND FINAL CHEMICAL STRUCTURE OF HTC MATERIALS

Before going into the formation mechanism of HTC and the characterization of its final chemical structure, some important advantages of this unconventional carbonization technique should be summarized:

1. Carbonization temperatures are low (180–250 °C).
2. Carbonization takes place in water under self-generated pressures, thus avoiding precursor drying costs [17].

3. Spherical micro-sized particles are generally obtained [18].
4. Controlled porosity can be easily introduced using nanocasting procedures, [19] natural templates [20,21], or activation procedures/thermal treatments [22].
5. Carbonaceous materials can be combined with other components, e.g. inorganic nanoparticles to form composites with special chemico-physical properties [23].
6. The resulting carbon particles have many polar oxygenated groups residing at the surface, which can be used for functionalization [24].
7. The surface chemistry and electronic properties can be easily controlled *via* additional thermal treatment, while the morphology and porosity remain unaffected.
8. Is carbon negative, meaning that it can bind CO_2 from the plant precursor [18].

12.2.1. Carbohydrate-Based HTCs

The elaboration of a general reaction mechanism for the transformation of carbohydrates as well as raw biomass into HTC is crucial since only a complete understanding of the chemical transformations occurring during the HTC process will allow control over morphology, porosity, and functionality. Due to the complex composition of lignocellulosic biomass, initially simple carbohydrates have been employed as model systems in order to study their transformation mechanism into HTC.

The decomposition of sugars under subcritical aqueous conditions takes place *via* dehydration and retro-aldol condensation [25]. At lower temperatures ($T < 300\,°C$), dehydration is favored [26]. HMF and furfural are well-known dehydration products of sugars [27]. Furfural is generated from pentoses, such as xylose, whereas hydroxymethylfurfural (HMF) is built from hexoses, such as fructose and glucose. Glucose is the most abundant sugar unit in biomass and the major product of acidic hydrolysis of lignocellulosic biomass [26]. Therefore, glucose is the most used precursor to produce HTC so far. Under acidic hydrolytic conditions, the hydrothermal process allows effective conversion of glucose to HMF supported by the Lobry de Bruyn–Alberta van Ekstein rearrangement (BER) which implied that glucose isomerizes to fructose and then dehydrates to HMF [28]. Since HMF is one of the most promising platform chemicals on the way to substitute petrochemicals by renewable compounds, a great deal of research was carried out on the effective production of the reactive furan derivative [29]. The main challenges are to find the appropriate dehydration catalysts and its fast extraction from the aqueous medium. With increasing the residence time in aqueous medium, HMF successively decomposes into levulinic acid (LA) and formic acid (FA) on the one hand and into polymeric carbonaceous material on the other hand [25]. While most researchers are trying to avoid such resinification or hydrothermal carbonization of HMF, Titirici and others rediscovered these processes for the production of green and valuable carbon and carbon hybrid materials [3, 30–33]. Depending on the applied conditions i.e. the grade of carbonization, the products are called humines, [34] hydrothermal carbons [18], or hydrochars [35]. As mentioned above, carbonaceous materials prepared at 140–220 °C are called hydrothermal carbons (HTC).

Figure 12.1 shows typical HTC morphologies obtained upon hydrothermal carbonization of various mono- and disaccharides and their decomposition products. It can be observed that the morphologies are very similar for hexose-derived carbons such as glucose and fructose as well as for their decomposition product HMF (Fig. 12.1a,b,c) consisting of spherically shaped particles aggregated together. Hydrothermal treatment of pentoses (e.g. xylose, Fig. 12.1d) as well as of its decomposition product furfural (Fig. 12.1e) leads to separate carbon spheres with diameters between 100 and 500 nm. Sucrose (Fig. 12.1f) is a disaccharide formed out of glucose and fructose bond between C1 on the glucosyl subunit and C2 on the fructosyl unit. The resulting HTC material is composed of spherical particles whose size varies between 700 nm and 2 μm. Upon HTC conditions, sucrose is hydrolyzed into glucose and fructose. Fructose starts forming HTC *via* HMF route while glucose isomerizes first to fructose and then forms HTC *via* the same HMF route. Due to differences in the kinetic of nucleation and growth, the polydispersity is higher in the case of disaccharides as well as in the case of easily hydrolyzable polysaccharides such as starch or amylopectin [18].

Two main structural models are available so far in literature regarding the carbohydrate-derived HTCs: The first one, proposed by Sevilla et al., is based on FT-IR, Raman, and XPS spectroscopy and considers the carbonaceous scaffold as a polyaromatic lignin-type network decorated with chemical functions [36]. The second one, proposed by Baccile and Titirici, is based on

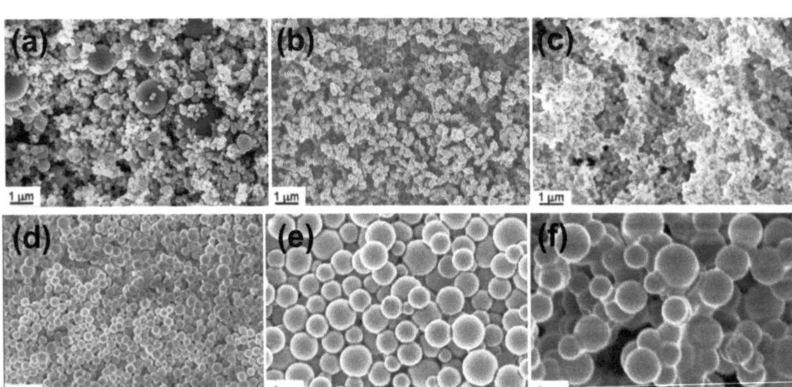

FIGURE 12.1 SEM micrographs of HTC materials obtained at 180 °C from: (a) glucose; (b) fructose; (c) hydroxymethylfurfural (HMF); (d) xylose; (e) furfural; (f) sucurose.

advanced solid-state ^{13}C NMR experiments and highlights the importance of furans rather than aromatic rings in the carbonaceous scaffold [37]. For these specific materials, solid-state NMR provides a better spectral resolution with respect to FT-IR or XPS, which are otherwise important for surface characterization of HTC. However, the possibility of using selective pulse sequences in ^{13}C NMR leads to the identification of separate chemical species without ambiguity. In addition, the poor spectral resolution of FT-IR and XPS does not allow a reliable quantification of the carbon species while ^{13}C NMR should be able to provide much better results from this point of view. Based on the ^{13}C NMR studies complemented by GC–MS, a formation mechanism and a final structure for the monosaccharide derived HTC at 180 °C were proposed. Glucose was selected as a good model carbon source and ^{13}C-enriched samples were prepared. ^{13}C NMR was used to investigate the chemical composition and local structure of the hydrothermally treated materials. Several standard MAS NMR techniques (singe pulse (SP), cross-polarization (CP) and inversion recovery cross-polarization (IRCP)), which are mainly employed in the study of carbon structures, have been used to identify the amount and type of sp^2 and sp^3 carbon sites. In addition, bond filtering techniques based on *J*-coupling (insensitive nuclei-enhanced by polarization transfer (INEPT)) provided a clear-cut distinction between protonated and quaternary carbons. Finally, the connectivity between carbon species was identified using a 2D ^{13}C–^{13}C correlation experiment based on double-quantum excitation of through-space dipolar-coupled nuclei. It was found that about 60% of the carbon atoms belong to a cross-linked furan-based structure. Furan moieties are directly linked either *via* the R-carbon or *via* sp^2- or sp^3-type carbon groups, where cross-linking can occur. Additional cross-linking sites are located at the -carbons of the furan ring (Fig. 12.2).

HTC was proposed to form upon the dehydration of hexoses (glucose and fructose) to HMF and of pentoses (xylose) to furfural. Once these compounds are formed, they then undergo a very complex chemical cascade involving a simultaneous combination of ring-opening reactions to create diketones that can further undergo aldol-type condensations with the furan ring, while Diels–Alder reactions may lead to more aromatic features, with concurrent polycondensation reactions also occurring. In summary, a set of dehydrations, polymerization (condensation and addition), and aromatization reactions will lead to the final chemical structure of HTC, which is composed of condensed furan rings bridged by aliphatic regions with terminal hydroxyl and carbonyl functional groups (Fig. 12.2). Upon the "polymerization" of HMF or furfural, nucleation takes place followed by growth of the particles upon further incorporation of HMF-derived monomers, leading to spherically shaped particles (Fig. 12.1).

Further experiments demonstrated that all sugars in their hexose form, regardless of their complexity, dehydrate into HMF, which finally condenses to a carbon-like material having morphological similarities and the same chemical

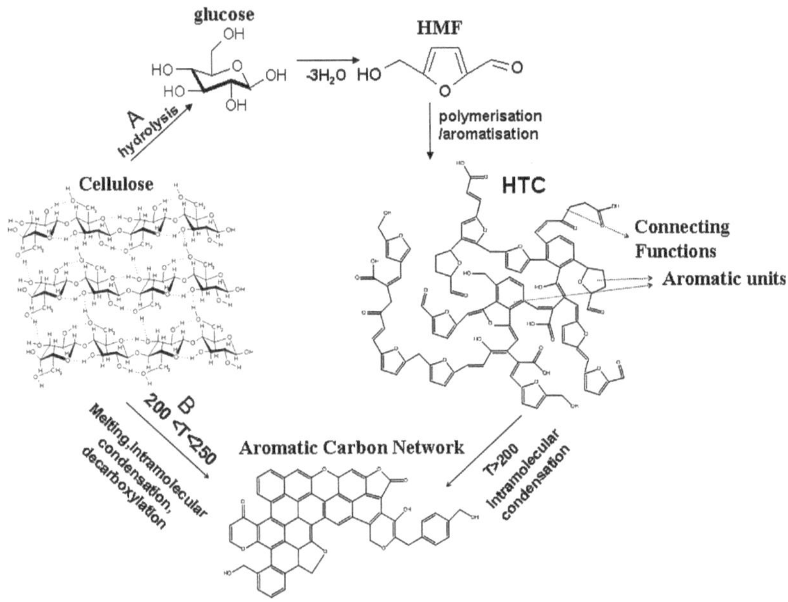

FIGURE 12.2 Conversion of cellulose into HTC: (a) via HMF resulting in a furan–rich aromatic network; (b) direct aromatization.

and structural composition. By contrast, xylose dehydrates into furfural, which in turn reacts to provide very similar carbon materials as obtained from pure furfural. ^{13}C solid-state NMR shows that the local structures of these two families of carbons, from hexoses and pentoses, are slightly different, with more aromatic features in the case of pentose due to the lack of one additional carbon bridged to a hydroxyl group. The use of easily hydrolyzable polycarbohydrates does not harm the outcome of the hydrothermal carbonization reaction, and remarkable similarities between the products of homologous series do occur, with respect to both morphology and local structural connectivity [18].

12.2.2. Lignocellulosic-Biomass-Derived HTCs

Lignocellulosic biomasses represent the most abundant renewable resources on Earth and are very complex systems composed of three main components: cellulose, hemicellulose, and lignin. The relative proportion of cellulose varies according to the biomass origin from 20% to 45% for agricultural wastes [38]. Chemically, cellulose is a linear and syndiotatic polymer of β-D-glucose [39]. The anhydroglucose units (AGU) are bonded to each other through a glycosidic linkage at positions 1 and 4 and the number of repeating AGU defines the degree of polymerization (DP) of cellulose on which its solubility depends. A relatively dense network of intermolecular H bonds is established in polymers with more than 30 AGU. Consequently, these macromolecules have high resistance against chemical and biological transformations and are insoluble in most of the common solvents. Hemicellulose, the second biomass component, contains many different sugar monomers, has a random structure, is amorphous, and easy hydrolyzable. Unlike carbohydrate-based cellulose and hemicellulose, lignin is a complex, cross-linked, 3D biopolymer with phenyl propane units and hydrophobic properties. Lignin is concentrated between the outer layers of the biomass fibers leading to structural rigidity and holding the polysaccharide fibers together. The decomposition and degradation of lignin and hemicelluloses are much easier than for cellulose [40,41]. Based on this different reactivity, it is feasible to separate cellulose and hemicelluloses from lignin, especially using alkaline treatments [42].

Very recently, Titirici et al. embarked on a project to examine the HTC of microcrystalline cellulose as well as of real lignocellulosic biomass (i.e. rye straw) and compare the chemical structure of cellulose and rye straw HTC materials to the ones derived from glucose [43]. ^{13}C solid-state NMR was used once more for the characterization and comparison of HTC materials derived from glucose, cellulose, and rye straw. In the case of glucose, changing the processing temperature allows control of both the particle diameter and particle-size distribution. The final carbon chemical structure can be switched from a carbonaceous polyfuran rich in oxygen-containing functional groups to a carbon network of extensive aromatic domains (Fig. 12.2). On the other hand, in the case of cellulose and generally of lignocellulosic biomass, a fundamental difference in the HTC mechanism is observed. The polyfuranic intermediate, which is characteristic for glucose-derived HTC materials at either low processing temperatures or short reaction times, cannot be observed in cellulose-derived HTC carbons. In contrast, the biomass-derived HTC materials show a well-developed aromatic nature, formed already during the early stages of HTC reaction. Thus, under mild hydrothermal operating conditions (180–250 °C) cellulose reacts according to a reaction scheme that can be more associated with the classical pyrolysis process, even if the exact chemical paths are not clear yet (Fig. 12.2, Route B). This study also showed how lignin is only mildly affected by HTC process and how its presence influences the final material structuring. Therefore, converting cellulose and lignocellulosic biomasses into HTC with the same chemical structure and morphology like glucose remains a very challenging task in HTC.

12.3. POROUS HTC MATERIALS

One of the main disadvantages of the HTC materials is that they present limited porosity and surface area. For enhanced performance, such textural nanoscale porosity is highly desirable. The easiest possibility to increase the surface area of HTC by two orders of magnitude is thermal treatment under an inert atmosphere, where small organic molecules are removed, thus reaching a comparable specific surface to conventional charcoal and ultramicropores (<0.5 nm). N_2 adsorption is not a suitable method to characterize the pore properties of HTC carbons since they only contain ultramicropores. The N_2 is too large to penetrate the HTC pores and nonreversible type I adsorption isotherms are obtained. Figure 12.3a,b shows the CO_2 adsorption isotherms and pore-size distribution of glucose-derived HTC made at 180 °C and further calcined at higher temperatures under an inert atmosphere (i.e. N_2 gas). Interestingly, independent of the carbohydrate precursor the adsorption isotherms, pore volume, and pore size of the HTC materials before and after thermal treatment are very similar (Fig. 12.3c,d).

Besides microporosity, the thermal treatment also removes the polar oxygenated groups from the surface increasing the electronic conductivity of the resulting materials, beneficial for electrochemical applications.

12.3.1. Chemical Activation

One of the most commonly used methodologies to introduce (micro)porosity into carbons is activation, especially chemical activation [44].

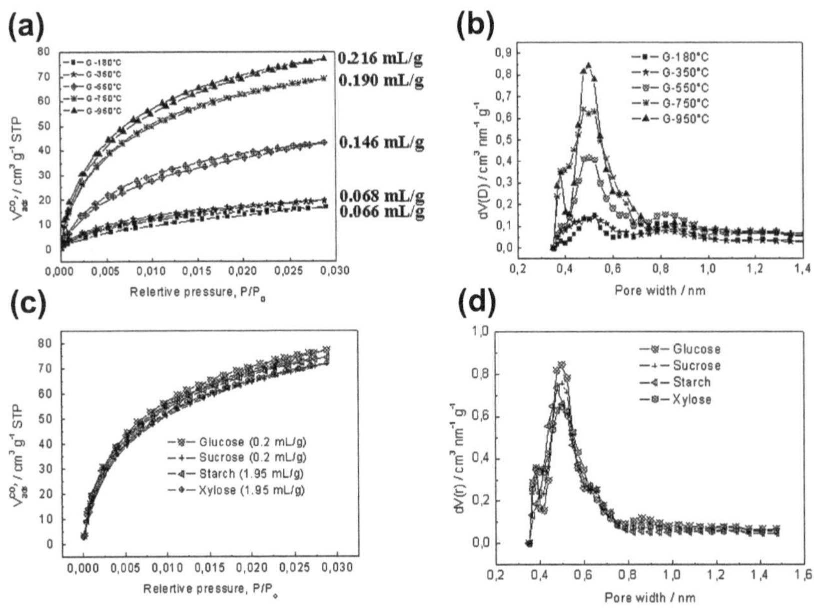

FIGURE 12.3 (a) CO_2 adsorption isotherms for glucose-derived HTC followed by thermal treatment at various temperatures; (b) Pore size distributions of the glucose-HTC materials calcined at different temperatures determined using CO_2 adsorption; (c) CO_2 adsorption isotherms for HTC materials from glucose, sucrose, starch and xylose after additional thermal treatment at 950 °C; (d) Pore size distributions obtained using CO_2 adsorption of the HTC materials from glucose, sucrose, starch and xylose after additional thermal treatment at 950 °C.

The relationship between the chemical structure of HTC carbons and their KOH chemical activation was investigated [45]. For the purpose of this investigation, HTC carbons derived from glucose at different HTC temperatures (sample names *x*–*y*, where *x* is the initial of the carbon precursor and *y* the HTC temperature) and then activated at the same processing conditions (KOH ratio = 3, activation temperature 750 °C).

The first effect of chemical activation on HTC carbons is a complete loss of initial spherical morphology and the formation of bulk-like carbon structures with rough surfaces due to the presence of micropores.

All the N_2 isotherms of the ACs produced from glucose-derived HTC carbons are characterized by a type I profile (Fig. 12.4a). A more attentive analysis reveals that G-240 °C has the largest N_2 uptake arising from a higher porosity development than in the cases of G-180 °C and G-280 °C. Furthermore, interestingly, G-280 °C N_2 isotherm shows a much sharper knee at low relative pressure than G-180 °C and G-240 °C. This suggests that the pore-size distribution of this former sample is much narrower than for the two latter ones.

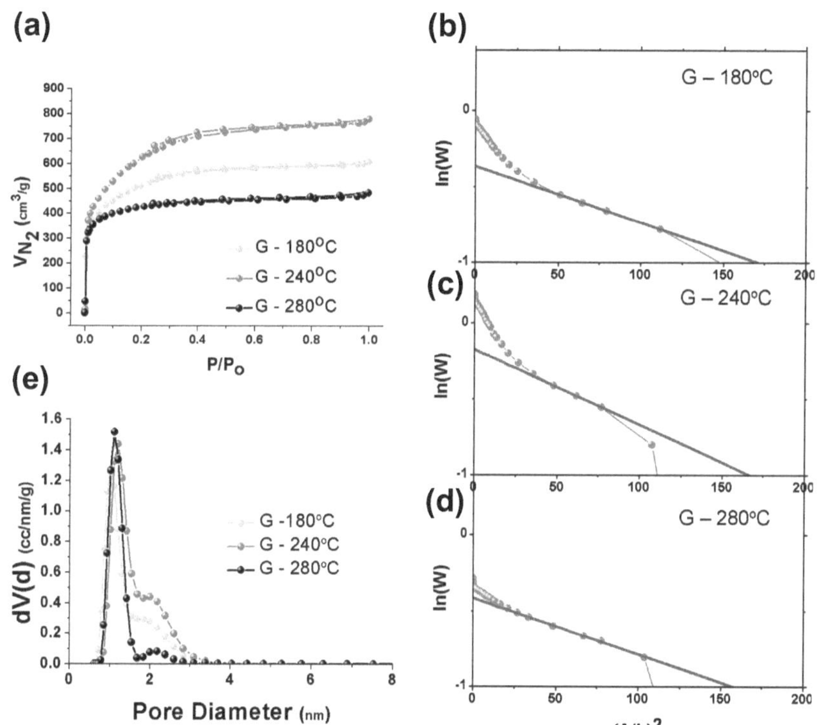

FIGURE 12.4 (a) N_2 adsorption isotherms, (b), (c), (d) application of DR equation to N_2 adsorption isotherms and (e) N_2 adsorption QSDFT pore size distribution of ACs (G-180 °C, G-240 °C, G-280 °C).

TABLE 12.1 Calculated Surface Area (S) and Pore Volume (V) Using Different Models (subscript) on Either N_2 or CO_2 Isotherms for G-180 °C, G-240 °C, and G-280 °C

Sample	$S_{N2\text{-}BET}$ (m^2/g)	$V_{N2\text{-}Ex}$ (cm^3/g)[a]	$V_{N2\text{-}DR}$ (cm^3/g)	$V_{CO2\text{-}NLDFT}$ (cm^3/g)	$V_{CO2\text{-}DR}$ (cm^3/g)
G-180	1859	0.94	0.69	0.34	0.31
G-240	2264	1.21	0.84	0.40	0.35
G-280	1610	0.73	0.66	0.41	0.41

[a]*Calculated from the experimental isotherm*

Therefore, it can be deduced that G-180 °C and G-240 °C have a higher tendency to develop larger pores during chemical activation than G-280 °C.

Apart from the total micropore volume (Table 12.1), applying the Dubinin–Radushkevich (DR) equation to the N_2 isotherms (Fig. 12.4b–d) can also provide useful information about the AC's PSD. At low $(A/b)^2$ values the deviation of the experimental results curve (dotted line) from the linear fit is an estimate of the PSD of the sample. Larger deviations correspond to samples with a greater pore fraction out of the micropore range (i.e. in the mesopore range). According to this criterion, G-240 °C (Fig. 12.4c) is characterized by the widest PSD and largest fraction of mesopores. G-180 °C (Fig. 12.4b) shows a similar trend, although the fraction of mesopores is certainly smaller since the deviation from the linear fit is less pronounced. Contrarily to these first two samples, G-280 °C (Fig. 12.4c) shows only a negligible deviation from the linear fit, suggesting a minor presence of mesopores in this latter sample. The findings obtained from the DR analysis are in good agreement with the QSDFT-PSD (Fig. 12.4e). The amount of mesopores present in the G-280 °C sample, as already mentioned, is negligible. G-180 °C and G-240 °C are instead characterized by a large fraction of mesopores, with the latter sample exhibiting the highest amount.

In order to assess the pore volume and the PSD of the narrow microporosity region present in G-180 °C, G-240 °C and G-280 °C, CO_2 adsorption were used, since N_2 sorption is not as effective in probing this pore range. It was evident from the measured CO_2 isotherms that G-280 °C has the largest gas uptake followed by G-240 °C and then G-180 °C. This trend differs from the one observed in N_2 sorption and it suggests that G-280 °C generates prevalently micropores (predominantly narrow micropores) with a narrow PSD. On the other hand, G-180 °C and especially G-240 °C have the tendency to generate ACs with a broader PSD. This was also confirmed by the NLDFT PSDs derived from the CO_2 isotherms. Even in the lower micropore range (pore width

<1.0 nm), the PSD of G-280 °C is shifted toward smaller pore widths than in the case of G-180 °C and G-240 °C.

Using the previously discussed models, the key parameters defining the porosity of the ACs (i.e. surface area and pore volume) could be calculated (Table 12.1).

Combining the different pore volume values (i.e. $V_{N2\text{-}QSDFT}$, $V_{N2\text{-}DR}$, $V_{CO2\text{-}NLDFT}$, and $V_{CO2\text{-}DR}$), it is possible to estimate the fraction of each porosity range (i.e. narrow micropores, supermicropores, and mesopores) contributing to the total pore volume. $V_{CO2\text{-}NLDFT}$ and $V_{CO2\text{-}DR}$ can be considered as an approximate estimate of narrow micropore volume ($V_{CO2\text{-}DR}$ values are used for the analysis). $V_{N2\text{-}DR}$ is an estimate of the overall microporous volume. $V_{N2\text{-}EX}$ (calculated directly from the experimental isotherm) corresponds to the total pore volume.

The porosity analysis overview for the chemical activation of cellulose and rye-straw-derived HTC carbons shows very similar trends to the ones observed for glucose. The highest porosity development is observed for the sample synthesized at 240 °C as indicated by its highest surface area and total pore volume values. The samples made at 200 °C and 240 °C show very similar pore volume distribution across the three different porosity ranges. On the other hand, the samples synthesized at 280 °C are characterized by the lowest surface area and total pore indicating a lower extent of activation.

Overall, this analysis highlights that the HTC temperature extensively affects the porosity of the derived ACs. HTC carbons synthesized at higher temperatures (e.g. 280 °C) generate ACs with a lower porosity development and narrower PSDs while the ones produced at 180–240 °C, upon KOH activation, develop a much larger porosity characterized by a greater mesopores fraction. These trends can be explained by taking into account the dependence of the chemical nature of HTC carbons upon the synthesis temperature: higher temperatures generate HTC carbons with a higher degree of aromatization resulting in enhanced chemical stability and structural order. As observed for the hydroxide activation of several coals, both features are detrimental to the reactivity of the carbon substrate leading to a reduced porosity development.

Additionally, we have also investigated the effect of KOH ratio on the developed porous properties of HTCs. Low KOH ratio values (2:1) produce a lower extent of activation and favor the formation of microporous carbons with a narrow PSD. Using higher amounts of KOH leads to higher porosity development, as indicated by larger surface area and total pore volume values. However, their increase is partially the result of pore widening, since the pore-volume distribution is clearly shifted more toward the supermicropore and mesopore ranges. These observations are in agreement with the trends generally observed during KOH activation of petroleum-derived carbonaceous substrates [46] confirming the similar reactivity of HTC carbons during chemical activation and their resultant high suitability as AC precursors.

12.3.2. Nanocasting, Soft Templating, and the Use of Natural Gelators

Except for the very well known process of chemical activation, [46] in the literature there are a multitude of techniques for introducing porosity into carbon materials [47]. In hard templating, a preformed porous solid is infiltrated with the carbon precursor and the carbonization occurs inside the pores, followed by original scaffold removal, yielding a carbon porous replica. This technique was intensively used for the replication of ordered silica materials (e.g. SBA-15 [48]) into the corresponding ordered porous materials CMK-3 and CMK-5 [49,50]. More recently, a soft-templating approach in carbon was developed by Zhao et al. [51] and Dai et al. [52] who demonstrated a direct route to ordered carbons *via* organic–organic block copolymer self-assembly (e.g. Pluronic® F127) and suitable aromatic carbon precursor (e.g. phloroglucinol or resorcinol).

One of the main limitations regarding the chemistry of these reported carbons is the relatively chemically condensed pore walls/surfaces, which inhibits facile postchemical modification. Therefore, it would be advantageous from a sustainable and economic footing if renewable precursors could be employed in a simple, scalable (potentially carbon neutral) process (i.e. HTC) in the direct synthesis of such ordered porous carbonaceous materials. This has been demonstrated by combining the HTC of glucose with hard templates for the production of mesoporous spheres [53], ordered carbon materials (HTC-CMK-3, -5,) [54], tubular carbon (Fig. 12.1d) [24], and hollow carbon spheres [55] (Fig. 12.5).

However, the hard templating using silica is tedious because of problematic infiltration procedures as well as template removal which normally involve harsh and nonenvironmentally friendly conditions. Soft templating was also successfully implemented in HTC. The use of Pluronic® amphiphilic block copolymers as structure-directing agents and fructose as a carbon source enabled the production of HTC materials with well-defined ordered micro- or mesopores (Fig. 12.5) [19].

For example, a porous nitrogen-doped carbon was originally prepared using the crustacean shells of shrimps and lobsters as natural templates Here, the $CaCO_3$ nanoparticles served as removable hard templates, while chitin as a (nitrogen-doped) carbon source [20]. Another unique procedure to produce porous nitrogen-doped carbons was based on the natural interactions between carbohydrates (glucose) and gelating proteins. Depending on the applied conditions, either nitrogen-doped carbon nanoparticles [56] or monolithic hierarchically porous nitrogen-doped carbons [21] were obtained in the absence of sacrificial templates.

12.4. FUNCTIONALIZATION OF HTC MATERIALS

Another very important prerequisite for a successful performance of HTC in various applications, as it will be outlined in the last section of this chapter, is

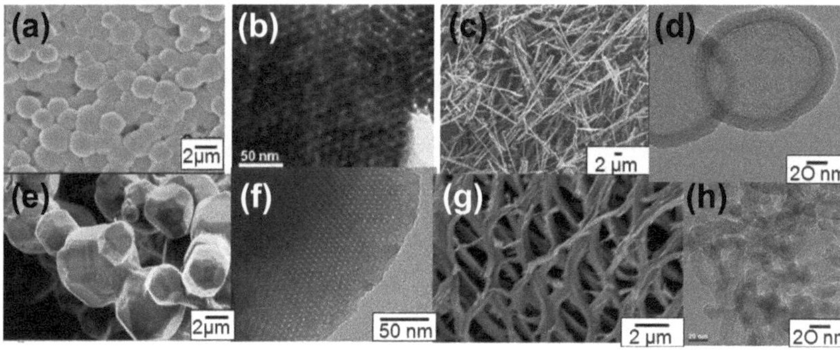

FIGURE 12.5 Porous HTCs: (a) Mesoporous HTC microspheres obtained by replication of silica using glucose *(Taken with permission from M.-M. Titirici, A. Thomas, M. Antonietti, Advanced Functional Materials 2007, 17, 1010)*. (b) Ordered mesoporous HTC obtained by replication of SBA-15 using furfural *(Taken with permission from M.-M. Titirici, A. Thomas, M. Antonietti, Journal of Materials Chemistry 2007, 17, 3412)*. (c) HTC tubular carbons obtained using a macroporous alumina membrane as template and furfural as a carbon source *(Taken with permission from S. Kubo, I. Tan, R. J. White, M. Antonietti, M.-M. Titirici, Chemistry of Materials 2010, 22, 6590)*. (d) HTC hollow spheres obtained using latex nanoparticles as sacrificial templates and glucose as a carbon source *(Taken with permission from R. J. White, K. Tauer, M. Antonietti, M.-M. Titirici, J. Amer. Chem. Soc. 2011, 132, 17360)*. (e; f) Micro-mesoporous ordered HTC particles obtained in the presence of amphiphilic block copolymers as soft templates and fructose as a carbon source *(Taken with permission from S. Kubo, R. J. White, N. Yoshizawa, M. Antonietti, M.-M. Titirici, Chemistry of Materials, 23, 2011, 4882)*. (g) nitrogen doped porous carbon obtained using shirmp/lobster food wastes after hydrothermal carbonization and removal of $CaCO_3$ (h) porous nitrogen doped carbon aerogel obtained using the natural interactions between carbohydrates and gelating proteins (i.e. ovalbumin) *(Taken with permission from R. J. White, N. Yoshizawa, M. Antonietti, M.-M. Titirici, Green Chem. 2011, 13, 2428.)*

the material functionality. Two possibilities of functionalization of the HTC materials exist: *in situ* in one-step process or postmodification.

12.4.1. Postmodification

It was explained earlier that HTCs contain polar oxygenated groups such as –OH, –C=O, and –COOH. The flexibility of HTC materials is that such groups either can be used for further functionalization or can be removed by heat treatment depending on the desired application.

The hydroxyl or carboxylic groups located on the surface of HTC materials could be chemically transformed into amino groups employing reagents such as 3-chloropropylamine [54] or tris(2-aminoethyl)amine [57]. Such amino-modified materials were further used to graft various organic polymers such as a thermoresponsive polymer (N-isopropylacrlyamide) [24]. The resulting thermoresponsive carbon materials, where the surface can be changed from hydrophilic (below its lower critical solution temperature LCST = 32 °C) to hydrophobic (above 32 °C), represent a very powerful class of materials and they could be, for example, used for temperature-driven chromatographic separations or adsorption as well as platforms for drug delivery, especially in combination with hollow carbon structures.

The possibility of straightforward functionalization in a similar fashion to silica surface, using either small organic molecules or organic polymers, can open up new perspectives for the application of such types of materials as stationary phases in chromatographic separations.

Besides amines, acidic groups can be postintroduced on the surface of such carbons. Sevilla et al. demonstrated the successful functionalization of HTC materials with sulfonic groups and their use as acid catalysts for the production of biofuels [58]. Shoujian Li and co-workers reported that a large number of oxygen-containing functional groups (OFGs) could be additionally created on the surface of hydrothermal carbon (HTC) by simply heating it at low temperatures (300 °C) in air [59].

Furthermore, the character of the surface and the type of functional groups can be tuned by simple postcalcination under an inert atmosphere. By heat treatment at 350 °C the OH groups disappear and the surface contains only –C=O and COOH groups. Additional treatment at 500 °C removes some of the carbonyl groups while most of the carboxylic groups are maintained [24]. Further treatment at higher temperatures removes most of the oxygenated groups and converts the carbon structure into a turbostatic-like disordered carbon structure with aromatic character and hydrophobic properties. This is one of the main advantages of HTC, the tunability of its surface properties with simple temperature treatments, which is otherwise difficult to be achieved with classical carbons.

12.4.2. "*In situ*" Functionalization

HTC materials can also be functionalized in a one-step process. Similarly to sol–gel technology this implies the use of precursors or coreactants containing the desired functionality.

Carboxylic or imidazole-containing HTC materials have been produced using the hydrothermal carbonization in the presence of small amounts of vinylic monomers (e.g. acrylic acid or vinyl imidazole) and used for either water purification [60] or heterogeneous base catalysis [61], as it will be described in the applications subchapter.

In this approach, organic monomers can be partially replaced by the controlled dehydration products of carbohydrates, which can be then copolymerized or can undergo cycloaddition reaction with functional comonomers to give a new type of hybrid between carbon and polymer latex. These latexes consist mainly of renewable raw materials, having a low cost base and expanding heterophase polymerization processes to carbonaceous materials.

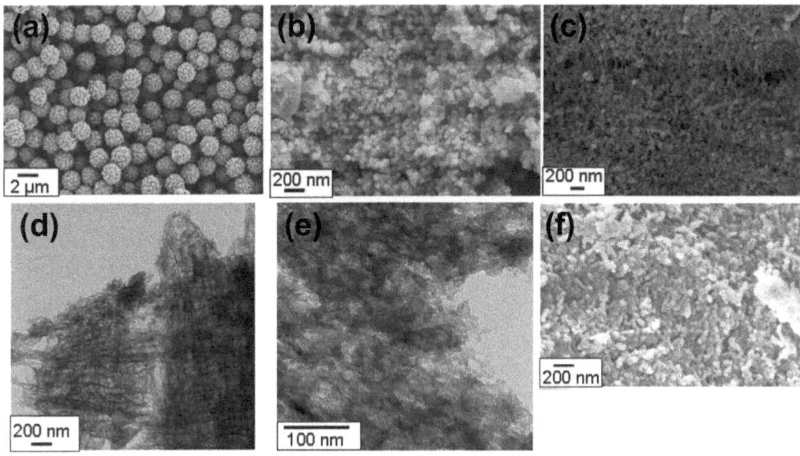

FIGURE 12.6 (a) SEM micrograph of the carboxylic rich HTC using glucose and acrylic acid; (b) SEM of HTC obtained using glucosamine as a precursor; (c) SEM of HTC obtained using chitosan as a precursor; (d) TEM of nitrogen doped carbon using shrimp waste as precursor *(Taken with permission from R. J. White, M. Antonietti, M.-M. Titirici, Journal of Materials Chemistry 2009, 19, 8645)*; (e) TEM of nitrogen doped carbon using lobster waste as precursor; (f) SEM of the carbon nanoparticles obtained using glucose in the presence of ovalbumine *(Taken with permission from N. Baccile, M. Antonietti, M.-M. Titirici, Chemsuschem **2010**, 3, 246)*

This synthetic strategy can be extended to various compositions of different water-soluble organic monomers yielding materials that combine the surface properties of the polymers with the structural, mechanical, thermal, and electric properties of the carbon framework. Interestingly, the morphology of such materials changes from smooth spheres to raspberry-like macroporous particles (Fig. 12.6a). As discussed in the first subchapter related to transformation mechanism, in the early stages of HTC small nuclei are formed. Acrylic acid reacts *via* cycloaddition with the decomposition monomers of glucose and stabilizes these small initial particles which assemble later into micrometer spheres formed out of small interconnected particles.

Another example of *in situ* functionalization is represented by production of nitrogen-doped carbon materials. They can be produced either using amino-containing carbohydrates, such as chitosan or glucosamine [62], or by the addition of nitrogen-containing molecules such as proteins [21,63] or amino acids [64].

Titirici et al. have demonstrated that it is possible to synthesize nitrogen-containing carbonaceous materials by hydrothermal treatment of amino-containing carbohydrates at 180 °C. (Fig. 12.6b,c) Besides being green, cheap, and sustainable, this process has the advantage to introduce nitrogen dopants into the carbonaceous scaffold, which are known to be very favorable for catalysis, adsorption, or energy storage devices. Temperature treatments showed that all materials retain the nitrogen content almost unchanged up to 750 °C. This is due to the fact that upon hydrothermal treatment glucose decomposes into HMF and ammonia. They undergo a chemical reaction, *via* Maillard-type chemistry followed by cycloaddition reactions, forming an extended nitrogen-containing aromatic network containing pyrrole-type nitrogen functionalities. In contrast to the hydrothermal carbons obtained from pure carbohydrates, these types of N-containing materials have an increased level of aromatic character already present at 180 °C. This is one of the most important benefits of the nitrogen-containing monomers. A similar situation occurs upon employing pure carbohydrates hydrothermally treated in the presence of amino acids. Detailed investigations on reaction mechanism were performed and the final chemical structure of the nitrogen-doped hydrothermal carbons using ^{13}C and ^{15}N solid-state NMR studies and ^{15}N-glycine and ^{13}C-glucose as precursors was elucidated [64]. The combination of one-dimensional (CP, IRCP) and two-dimensional (SQ-DQ ^{13}C, double CP ^{13}C (^{1}H ^{15}N)) experiments helped drawing the carbon-to-carbon and carbon-to nitrogen chemical arrangements. We were able to propose several possible structures that include the presence of pyrrole motifs connected to furans.

Prawn and lobster shells, an abundant marine chitin (poly-β(1/4)-N-acetyl-D-glucosamine) biomass source and high-volume food waste, have been also successfully converted into nitrogen-doped carbons in a simple three-step procedure to yield a highly textured, high surface area, and high pore volume material. The preformed natural inorganic ($CaCO_3$)/organic (chitin) shell composite provides the precursor structure, which is simply converted *via* hydrothermal carbonization to yield nitrogen-rich carbon material templated by the mineral scaffold. The inorganic component is removed by washing with acetic acid, leading to enhanced mesoporous properties. This approach renders materials with interesting nanoscale morphology and an accessible interconnected pore structure [20] (Fig. 12.6d,e)).

Another interesting methodology for *in situ* functionalization of HTC is represented by hydrothermal treatment of glucose in the presence of ovalbumin. Here, the protein plays a double role: on one hand, it reacts with glucose according to the mechanism previously discussed and secondly it gels and thus stabilizes the small HTC nuclei formed in the incipient states of the process. Depending on the reaction time and drying methodology, either small carbon nanoparticles [56] (Fig. 12.6f) or porous monolithic gels [21] (Fig. 12.5e) can be obtained. The intelligent application of a secondary biomass precursor (e.g. protein) as a nitrogen donor and highly effective surface stabilizing agent represents a novel, template-free approach to the facile synthesis of a new generation of tunable porous carbons. Materials are of high specific surface area composed of mesopores, combined with an excellent and macroporous transport architecture – structuring not easily accessible by other synthetic approaches. Such pore structuring potentially circumvents the "*plumbing*" problems associated with more regular mesoporous solids prepared by templating. The carbonization temperature provides a control vector over surface chemistry/carbon aromatization/nitrogen condensation and by implication surface polarity.

12.5. HTC NANOCOMPOSITES

Controlled synthesis of carbonaceous nanocomposites has become a hot research area and achieved many important results, because of their combined and improved properties with highly potential values in many fields. As in the case of functionalization, two main methodologies can be distinguished for the production of hydrothermal carbon nanocomposites: postmodification and *in situ* synthesis. The postmodification method implies coating of preformed nanostructures or decorating of inorganic nanostructures on the carbonaceous materials. Furthermore, carbonaceous materials can act as sacrificial templates or support shells for the synthesis of hollow and complex nanostructures, as well as the free-standing membranes.

12.5.1. Coating of Preformed Nanostructures

One important advantage of the HTC process is that it can be rationally used to coat carbonaceous shells on the surface of preformed nanostructures such as inorganic, organic, and polymer nanostructures. The coating process is greatly dependent on the surface properties, e.g. surface charge and hydrophilic/hydrophobic character, of such nanostructures. These carbonaceous shells not only act as a protective layer on the surface of such nanostructures but also bring up unique chemical or physical properties. The final nanocomposites have shown the important applications in many fields, such as catalysis, fuel cells, drug delivery, and bioimaging [65].

A particularly good example is the synthesis of uniform Te@carbonaceous composite nanocables *via* the HTC process using preformed Te nanowires with diameters of several nanometers and length up to several micrometers and glucose as starting reagent (Fig. 12.7a–b) [66]. Te nanowires could not only restrain the usual homogeneous nucleation of carbonaceous spheres, but also promote the heterogeneous deposition of carbonaceous shell on the backbone of Te nanowires. The diameter of the final nanocables could be easily adjusted by changing the hydrothermal carbonization time, or the ratio of the tellurium and glucose.

Another example of HTC composites prepared using this methodology are the silica/carbon-encapsulated core–shell spheres that could be synthesized by the HTC process with silica spheres as the template and glucose as reagent [67]. The as-synthesized SiO_2@C nanocomposites could be further used as template for preparation of SiO_2@C@SiO_2 composites, SiO_2@SiO_2 materials with vacant region between shells, noble-metal nanoparticle-loaded

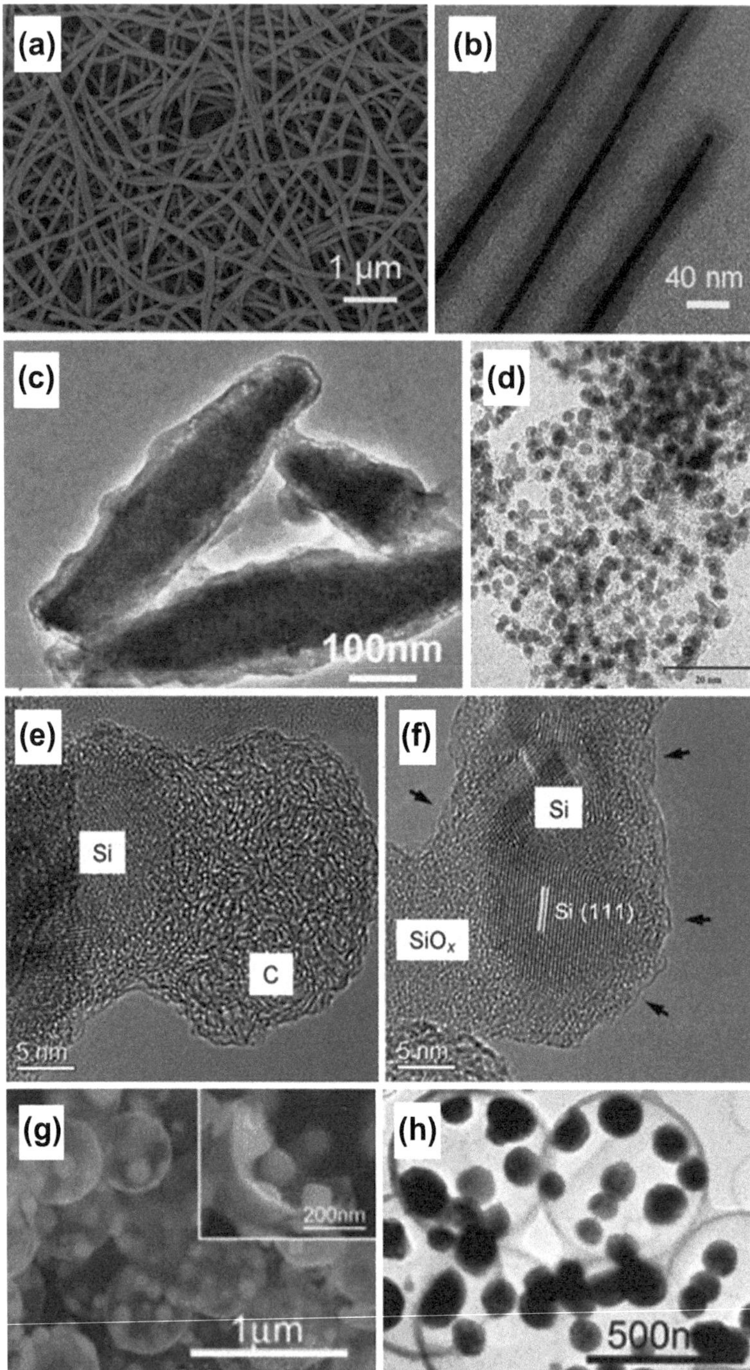
(a)
1 µm
(b)
40 nm
(c)
100nm
(d)
(e)
Si
C
5 nm
(f)
Si
Si (111)
SiOx
5 nm
(g)
200nm
1µm
(h)
500n

SiO_2@C composites, and hollow carbonaceous capsules through different synthetic processes. The carbonaceous shell could protect the nanostructures, after additional heat treatment and graphitization and could greatly improve the physical properties of nanocomposites. For example, carbon-coated Fe_3O_4 nanospindles could be synthesized by the HTC process and the subsequent heat treatment (Fig. 12.7c) [68]. These composites could be used as a superior anode material for lithium-ion batteries with high coulombic efficiency in the first cycle, high reversible capacity, enhanced cycling performance, and high rate capability compared with bare hematite spindles and commercial magnetite particles. The improved performance mainly attributed to the uniform and continuous carbon-coating layers increase the electronic conductivity of electrodes, forming and stabilizing uniform and thin solid electrolyte interphase film on the surface. Similarly, carbon-coated Si nanocomposites through the HTC process and heat treatment have displayed significant improved lithium-storage performance because of the generation of a passivated layer of SiO_x and carbon (Fig. 12.7e–f) [69]. A carbonaceous shell has been used as a powerful tool for the smart design synthesis process, forming more complex nanocomposites. For instance, the synthesis of the coaxial SnO_2@carbon hollow nanospheres by removing silica from the original SiO_2@SnO_2@carbon structures has been well designed [70]. The as-synthesized nanocomposites were high-energy anode materials for the lithium-ion battery with exceptional cycle life for at least several hundred cycles and rate capability. Further, similar complex nanostructures of tin-nanoparticles encapsulated in elastic hollow carbon spheres (TNHCs) have been successfully synthesized (Fig. 12.7g–h) [71]. SiO_2 spheres were used as the core for the deposition of SnO_2 shell *via* the hydrolysis of Na_2SnO_3. After etching the core of SiO_2 by sodium hydroxide solution, SnO_2 hollow spheres were coated with the carbonaceous shell *via* the HTC process. Finally, after heat treatment at 700 °C for 4 h under N_2 atmosphere, the complex TNHCs hybrid materials were synthesized. The hybrid material was a promising anode material for high-performance

FIGURE 12.7 (a-b) SEM and TEM images of Te@carbonaceous composite nanocables. *(Reproduced with permission from H. S. Qian, S. H. Yu, L. B. Luo, J. Y. Gong, L. F. Fei, X. M. Liu, Chem. Mater. 2006, 18, 2102)*; (c) TEM images of carbon-coated Fe_3O_4 nanospindles. *(Reproduced with permission from W. M. Zhang, X. L. Wu, J. S. Hu, Y. G. Guo and L. J. Wan, Adv. Funct. Mater., 2008, 18, 3941.)*; (d) TEM images of Carbon-decorated FePt nanoparticles. *(Reproduced with permission from N. Caiulo, C. H. Yu, K. M. K. Yu, C. C. H. Lo, W. Oduro, B. Thiebaut, P. Bishop and S. C. Tsang, Adv. Funct. Mater., 2007, 17, 1392)*; (e-f) TEM images of carbon-coated Si nanocomposites. *(Reproduced with permission from Y.-S. Hu, R. Demir-Cakan, M.M. Titirici, J. O. Muller, R. Schlogl, M. Antonietti, J. Maier, Angew. Chem. Int. Ed. 2008, 47, 1645)*; (g-h) SEM and TEM images of tin-nanoparticles encapsulated in elastic hollow carbon spheres. *(Reproduced with permission from W. M. Zhang, J. S. Hu, Y. G. Guo, S. F. Zheng, L. S. Zhong, W. G. Song, L. J. Wan, Adv. Mater. 2008, 20, 1160.)*

lithium-ion batteries with high specific capacity, high volume capacity, and good cycle performance.

12.5.2. Loading Inorganic Nanostructures onto HTC Materials

The important advantage of HTC materials is, as previously mentioned, their rich surface functional groups, such as hydroxyl, aldehyde, and carboxyl groups. These surface functional groups have shown remarkable reactivity toward various inorganic precursors For example, based on the redox reaction, they could *in situ* reduce and stabilize noble-metal ions forming very fine noble-metal nanostructures, such as Ag, Au, Pt, and Pd [72–75]. The obtained noble-metal nanostructures have uniform sizes and can be easily modified with biomolecules or drugs [11,76]. Titirici and Palkovitz reported the loading of Pt nanoparticles onto nitrogen-doped carbon materials [77]. The resulting composites were used as efficient catalysts for the oxidation of methane to methanol with superior performance to the standard dichlorobipyrimidyl platinum (II) complex reported by Periana and co-workers [78–81].

The so-called "golden fleece" nanocomposites consist of uniform carbonaceous nanofibers loaded with noble-metal nanoparticles with different diameters and distributions depending on the noble metal (Fig. 12.8) [73]. Carbonaceous nanofibers produced by the HTC process have rich surface functional groups that can *in situ* reduce noble-metal ions into noble-metal nanoparticles, such as Pd, Pt, and Au. These nanocomposites have shown high permeation and stability, high surface area, and chemical reactivity, which are the ideal candidates for the use in heterogeneous catalytic systems. Further, carbonaceous nanospheres, synthesized by the HTC process, have successfully demonstrated their surface reactivity by loading silver and palladium nanoparticles on their surfaces without using other reducing and stabilizing agents [72]. The surface redox reactions made the number of reducing functional groups decrease, as revealed by the FTIR spectra.

Another concept for the application of the reactivity of carbonaceous materials is the self-assembly process, based on the electrostatic attraction of surface charge. Nanoparticles with different surface charges can be easily self-assembled on the surface of carbonaceous materials. For example, citrate-stabilized gold nanoparticles have been assembled on the surface of carbonaceous spheres to fabricate a core–shell hybrid material [76]. The resulting nanocomposites have shown satisfactory chemical stability and good biocompatibility, and are expected to be promising templates for bimolecular immobilization and biosensor fabrication.

12.5.3. *In situ* HTC Nanocomposites

When noble-metal ions are firstly added into the HTC process, novel carbonaceous-encapsulated core–shell nanocomposites could be successfully

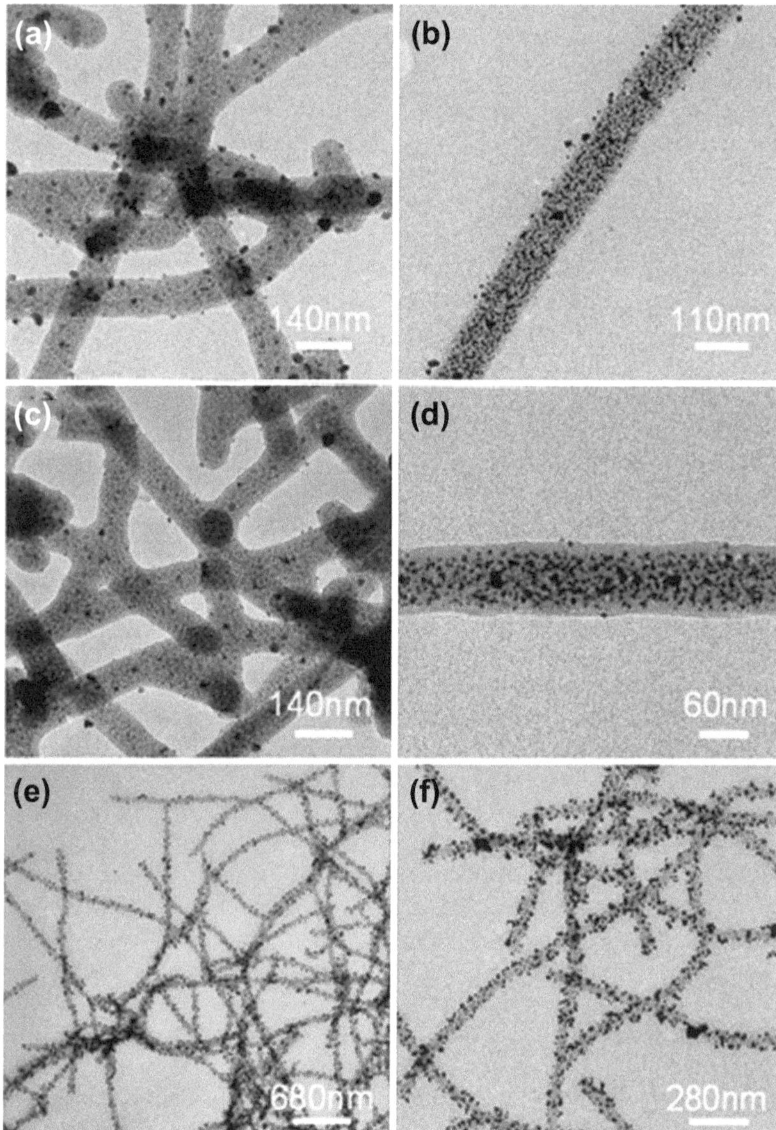

FIGURE 12.8 TEM images of noble metal nanoparticles loaded on the surfaces of carbonaceous nanofibers. (a-b) Pd nanoparticles with an average diameter of 6 nm; (c-d) Pt nanoparticles with an average diameter of 7 nm; and (e-f) Au nanoparticles ranging from 7 to 15 nm in diameter. *(Reproduced with permission from H. S. Qian, M. Antonietti, S. H. Yu, Adv. Funct. Mater. 2007, 17, 637.)*

produced in one-step HTC process [82–91]. The size of the resulting composites could be carefully controlled by changing the reagent ratios and reaction conditions. The advantages of this one-step HTC process include the following: (I) the carbonization process and the reduction and growth of noble-metal ions proceed simultaneously, (II) the noble-metal ions can catalyze the carbonization process, and (III) it is the one-step process which can potentially be upscaled.

One example is the synthesis of silver-carbonaceous nanocables [31]. These nanocables, with lengths up to 10 μm and diameters of ~200–500 nm, contain pentagonal-shaped silver nanowire cores and tend to branch or fuse with each other.

Poly(vinyl alcohol) (PVA)-assisted synthesis of coaxial nanocables has also been reported [92–96]. First, flexible silver@cross-linked PVA coaxial nanocables have been fabricated by the HTC process (Fig. 12.9a–c) [92]. Further, some glucose-based saccharide, including glucose, β-cyclodextrin, starch, and maltose, can take part in this PVA-assisted synthesis process, forming flexible silver@carbon-rich composite submicrocables (Fig. 12.9d–f) [93]. This PVA-assisted synthesis process can be extended to the synthesis of Cu@cross-linked PVA coaxial microcables [94,95]. Unique necklace-like Cu@cross-linked PVA coaxial microcables with strict wire-bead forms have been successfully synthesized (Fig. 12.9g–i) [94]. Further, Cu@carbonaceous submicrocables can be synthesized in the presence of some glucose-based saccharide, including glucose, β-cyclodextrin, starch, and maltose [96].

Pd@C core–shell nanoparticles have also been successfully synthesized by the one-step HTC process, [97] and shown selective catalytic capability for the batch partial hydrogenation of hydroxyl aromatic derivatives (Fig. 12.10a). In a similar fashion, using titanium isopropoxide as inorganic precursor and glucose as HTC precursor, mesoporous TiO_2/HTC composites were successfully synthesized (Fig. 12.10b). Such composites exhibited strong photocatalytic activity under visible light, which was not observed for the pure TiO_2 materials [98]. *In situ* coating of inorganic nanoparticles with a thin layer of conductive carbon was also successfully accomplished for $LiFePO_4$ (LFPs) mesocrystals (Fig. 12.10c). In this case, the formation of the inorganic mesocrystals and the carbonization process take place simultaneously. In order to achieve uniform coatings the concentration of carbon precursor needs to be maintained low in comparison to the one of the inorganic precursors. Such HTC-coated LFP mesocrystals showed improved performance as compared with their noncarbon-coated counterparts when used as positive electrodes in Li-ion batteries [99].

One-step HTC process can synthesize carbonaceous nanocomposites with more complex structures and specific properties, provided that the desired nanoparticles are already preformed [23,100,101]. For instance, core–shell Pt@C nanoparticles can embed in the porous channels of mesoporous silica (SBA-15) by the one-step HTC process [100]. After high-temperature

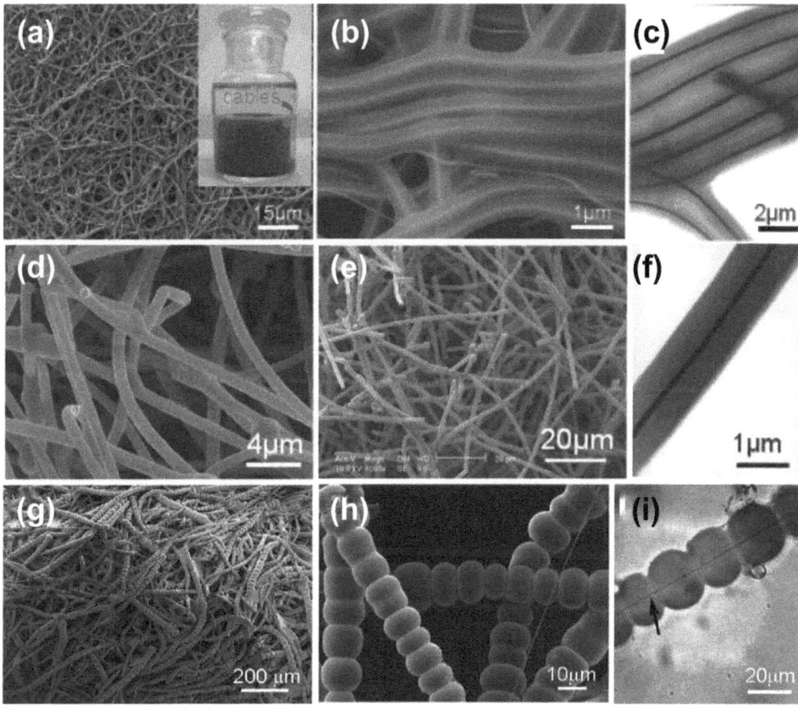

FIGURE 12.9 (a–c) SEM and TEM images of flexible silver@cross-linked PVA coaxial nanocables (reproduced with permission from L. B. Luo, S. H. Yu, H. S. Qian and T. Zhou, J. Am. Chem. Soc., 2005, 127, 2822). (d–f) SEM and TEM images of flexible silver@carbon-rich composite submicrocables (reproduced with permission from L. B. Luo, S. H. Yu, H. S. Qian and J. Y. Gong, Chem. Commun., 2006, 793). (g–i) SEM and TEM images of unique necklace-like Cu@cross-linked PVA coaxial microcables with strict wire-bead *(Reproduced with permission from Y. J. Zhan and S. H. Yu, J. Am. Chem. Soc., 2008, 130, 5650).*

carbonization and removal of SBA-15, the final complex nanocomposites have shown high catalytic capability for methanol-tolerant oxygen electroreduction in direct methanol fuel cell. Furthermore, one-step HTC process followed by removal of the carbon has successfully synthesized hollow core–shell mesospheres with aggregated crystalline SnO_2 nanoparticles [101] and mesoporous SnO_2 microspheres with nanoparticle assembly (Fig. 12.10d). Both these SnO_2 nanocomposites have shown a high capability for lithium-ion storage.

12.6. APPLICATIONS OF HTC

The most appealing feature of hydrothermal carbonization is the fact that it represents an easy, green, and scalable process allowing the production of various carbon and hybrid nanostructures with practical applications on a price

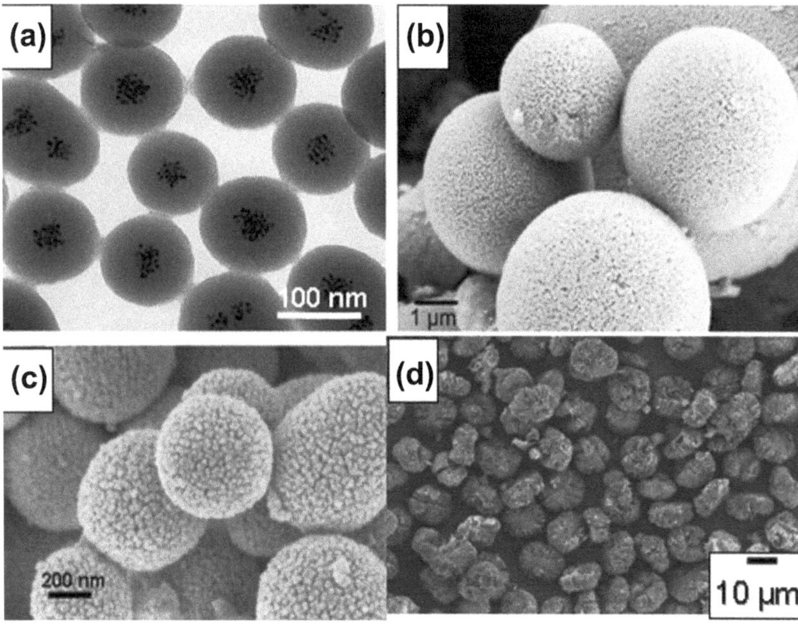

FIGURE 12.10 Figure 10 In situ HTC composites: (a) Pd/HTC core/shell particles *(Taken with permission from P. Makowski, R. D. Cakan, M. Antonietti, F. Goettmann and M.-M. Titirici, Chem. Commun., 2008, 999)*; (b) TiO2/ HTC composite *(Taken with permission from L. Zhao, X. Chen, X. Wang, Y. Zhang, W. Wei, Y. Sun, M. Antonietti and M.-M. Titirici, Adv. Mater., 2010, 22, 3317)*; and (c) SnO_2 mesoporous spheres obtained after removal of HTC from the resulting composite *(Taken with permission from R. Demir-Cakan, Y.-S. Hu, M. Antonietti, J. Maier and M.-M. Titirici, Chem. Mater., 2008, 20, 1227)*. (d) LFP/HTC composite.

base which is mostly well below any number of corresponding petrochemical processes. Even if only 10 years old, HTC materials have already found numerous applications in relevant fields of modern materials device manufacturing and the chemical industry such as soil enrichment, catalysis, water purification, energy storage, and CO_2 sequestration. In the following sections, we will summarize some of the most remarkable examples where HTC-based materials have proved not only to be sustainable, but also to possess extraordinary properties which, in some cases, surpass those of current "gold standards".

12.6.1. Agricultural Applications: Biochar

Distinguished from charcoal but similar in its chemical structure, biochar is produced with the intent to be applied to soil as a means to improve soil health, to filter and retain nutrients from percolating soil water, and to provide carbon storage. The advantage of converting biomass into biochar *via* the HTC process is that it can transform wet input material into carbonaceous solids at relatively high yields without the need for an energy-intensive drying before or during the process. This opens up the field of potential feedstock to a variety of nontraditional sources: wet animal manures, human waste, sewage sludge, municipal solid waste (MSW), as well as aquaculture and algal residues. These feedstocks represent large, continuously generated, renewable residual streams that require some degree of management, treatment, and/or processing to ensure protection to the environment.

In order to predict the behavior of different biochars in soil, it is very important to understand the final chemical structure of the employed biomass-derived carbon. Normally, pyrolysis is the classical way to produce biochar from biomass. Therefore, several groups have compared the chemical and physical properties of the materials obtained using the same precursor by pyrolysis or hydrothermal carbonization: swine manure [102], corn stover [103], or carbohydrates and cellulose in general [104].

Steinbeiss and Antonietti investigated the behavior of biochars in arable and forest soil in a greenhouse experiment in order to prove that these amendments can increase carbon storage in soils [105]. Two qualities of biochar were produced by hydrothermal pyrolysis from ^{13}C-labeled glucose (0% N) and yeast (5% N), respectively, and the respiratory losses of soil and biochar carbon were quantified. The mean residence times of the biochars were calculated using the isotopic label. Extraction of phospholipid fatty acids from soil at the beginning and after 4 months of incubation was used to quantify changes in microbial biomass and to identify microbial groups utilizing the biochars. Mean residence times varied between 4 and 29 years, depending on soil type and quality of biochar. Yeast-derived biochar promoted fungi in the soil, while glucose-derived biochar was utilized by Gram-negative bacteria. The results suggest that residence times of biochar in soils can be manipulated with the aim to "design" the best possible biochar for a given soil type.

Rillig and Titirici tested the effect of beetroot chip-derived HTC material on plant growth or on root-associated symbionts such as arbuscular mycorrhizal fungi [106]. Several studies were carried out and it was found that increasing concentrations of HTC material could be deleterious for plant growth of *Taraxacum*, starting at 10 vol% additions. Conversely, root colonization of the fungal symbiont was stimulated at an addition of 20 vol%. Soil pH changes occurring during the study could be traced by microbial reduction reactions, and these led to an increase in pH of the medium despite the quite acidic nature of the HTC material itself. The results suggest that HTC materials should be carefully tested and optimized to reduce negative effects on plant growth before applications in the field are undertaken, particularly at high addition rates.

12.6.2. Adsorption

12.6.2.1. Liquid Phase

Adsorption is by far the most encountered application of activated carbons. Contrary to activated carbons, hydrothermal carbons as produced at 180 °C do not have any microporosity unless further thermally treated or chemically activated. This is counterbalanced by the high number of oxygenated groups located at their surface, which can promote adsorption. Furthermore, these functionalities can be easily tuned by further surface modifications as previously described in Section 12.4. As mentioned before in Section 12.4.2, the addition of small amounts of vinyl organic monomers into the hydrothermal carbonization of glucose leads to the production of carbon materials rich in carboxyl functionalities [60]. The acrylic acid plays a double role in this process: on one hand, it brings in the functionality and, on the other hand it reacts with the compounds formed in the previous step of hydrothermal carbonization *via* cycloaddition reactions stabilizing the small droplets and leading to assembly into macroporous materials. The synthesized materials proved to be feasible for use in adsorption experiments for removal of heavy metals from aqueous solutions. The adsorption capacity was as high as 351.4 mg/g for Pb(II) and 88.8 mg/g for Cd(II), which is well beyond ordinary sorption capacities, proving the efficiency of the materials to bind and buffer ions, or more specifically to remove heavy metal pollutants. The possibility of water purification achieved with carbonaceous materials produced under mild conditions and with an overall low cost base is definitely an attractive alternative, especially for less developed areas of the world.

Another interesting example of adsorbents for water treatment is the production of carbons from lignocellulosic biomass using the HTC technology, especially having the advantage of using wet biomass as precursor [107]. Most precisely, materials were produced here from switchgrass at 300 °C in subcritical water and characterized using X-ray diffraction, Fourier transform infrared spectroscopy, scanning electron microscopy, and thermogravimetric analysis. The physiochemical properties indicated that they could serve as an excellent adsorbents to remove uranium from groundwater. A batch adsorption experiment at the natural pH indicated an H-type isotherm. The adsorption data were fitted using a Langmuir isotherm model and the sorption capacity was estimated to be ca. 2.12 mg of U g^{-1} of HTC. The adsorption process was highly dependent on the pH of the system. An increase towards neutral pH resulted in the maximum adsorption of ca. 4 mg U g^{-1} of HTC. The results of the column study indicate that such materials could be used as effective adsorbents for U(VI), as a reactive barrier medium.

Another study compared two types of materials, prepared from hydrothermal liquefaction of pinewood and rice husk for lead removal from aqueous solution. The results indicated that the resulting carbons contained a large amount of oxygen-containing groups on the surface, which were quite effective for lead removal with capacities of 4.25 and 2.40 mg/g for HTC pinewood and HTC rice husk, respectively. The adsorption equilibrium was achieved around 5 h. Higher temperature favored the removal capacity implying that the adsorption was an endothermic process [108].

Xing et al. also compared pyrolysis and hydrothermally produced materials for the adsorption of bisphenol A (BPA), 17a-ethinyl estradiol (EE2), and phenanthrene (Phen) from water [109]. The thermal chars were produced from feedstocks of poultry litter and wheat straw, respectively, through pyrolysis at 400 °C. The hydrothermal chars were made out of poultry litter and swine solids at 300 °C. X-ray diffraction and solid-state nuclear magnetic resonance results suggested that hydrothermal materials consisted of more amorphous aliphatic-C, possibly being responsible for their high sorption capacity of phenol. This study demonstrated that hydrothermal carbons could adsorb a wider spectrum of both polar and nonpolar organic contaminants than thermally produced biochars carbons, suggesting that HTC derived from poultry and animal waste is a potential sorbent for agricultural and environmental applications.

12.6.2.2. Gas Adsorption

12.6.2.2.1. CO_2 Capture on HTC Carbons

The mitigation of carbon dioxide emissions is a crucial issue as this gas is the main anthropogenic contributor to climate change. Among the possible strategies for CO_2 abatement, capture and storage have attracted keen interest.

The CO_2 capture performance of HTC-based activated carbons was recently investigated by Sevilla and Fuertes [110]. These authors analyzed the use as CO_2 adsorbents of porous carbons obtained by chemical activation of several HTC materials prepared from starch, cellulose, and eucalyptus sawdust. The hydrothermal carbonization of these materials was carried out at temperatures in the 230–250 °C range. The chemical activation was performed with KOH at temperatures in the 600–850 °C range, exhibiting BET surface areas between 1260 $m^2\,g^{-1}$ and 1940 $m^2\,g^{-1}$ characteristic of microporous materials.

The CO_2 adsorption uptake for several HTC activated carbons is listed in Table 12.2 for a pressure of 1 bar and three adsorption temperatures (0 °C, 25 °C, and 50 °C). It can be seen that the CO_2 capture capacities of the HTC-based porous carbons prepared from starch, cellulose, or eucalyptus sawdust are quite substantial and similar (about 5.5–5.8 mmol $CO_2\,g^{-1}$, 243–256 mg $CO_2\,g^{-1}$). This result is coherent with the fact that the pore characteristics of these materials are analogous.

For practical applications, in addition to a high CO_2 adsorption capacity, sorbents need to show fast adsorption kinetics, a high selectivity toward CO_2, and they must also be easy to regenerate. Sevilla and Fuertes showed that the CO_2 adsorption is very fast, around 95% of CO_2 uptake occurring in a span of

TABLE 12.2 CO_2 Capture Capacities of the Porous Carbons at Different Adsorption Temperatures and 1 atm

	Chemical Activation		CO_2 Uptake, mmol g^{-1} (mg g^{-1})		
HTC Precursor	T (°C)	KOH/HTC	0 °C	25 °C	50 °C
Starch	700	4	5.6 (247)	3.5 (152)	2.2 (196)
Cellulose	700	4	5.8 (256)	3.5 (155)	1.8 (79)
Eucalyptus sawdust	600	4	5.2 (230)	2.9 (128)	–
	700	4	5.5 (243)	2.9 (128)	1.8 (79)
	800	4	5.2 (227)	3.0 (130)	–
	600	2	6.1 (270)	4.8 (212)	3.6 (158)
	650	2	6.0 (262)	4.7 (206)	3.3 (145)
	700	2	6.6 (288)	4.3 (190)	2.6 (116)
	800	2	5.8 (255)	3.9 (170)	3.1 (136)

(Adapted from Ref. [110] with Permission)

2 min. By contrast, N_2 adsorption is slower, around 60 min being needed for the maximum adsorption uptake to take place. Then, the [CO_2/N_2] selectivity measured under equilibrium conditions was 5.4. Easy regeneration is another critical property that must be considered when designing CO_2 sorbents. In this respect, for the HTC-based carbons >95% of CO_2 was desorbed within 3 min under these conditions.

Another interesting approach in relation to the use of the HTC materials for the CO_2 capture is based on their functionalization with amine groups, which exhibit a high affinity to CO_2. In this respect, Titirici et al. [57] reported the CO_2 capture by means of an amine-rich HTC product. This material was prepared by a two-step process: (a) hydrothermal carbonization of glucose in the presence of small amounts of acrylic acid and (b) functionalization of carboxylic-rich HTC products with triethylamine. This aminated HTC material shows high CO_2 capture capacities (up to 4.3 mmol CO_2 g^{-1} at –20 °C). More importantly, these materials exhibit a very high [CO_2/N_2] selectivity at low (–20 °C) and high (70 °C) temperatures, up to 110 at 70 °C.

12.6.2.2.2. Methane Storage on HTC Carbons

Falco and Titirici in collaboration with Lozano-Castello at the University of Alicante examined the methane storage properties of two activated carbons derived from the hydrothermal carbonization of glucose (G) and rye straw (RS) at 240 °C, using KOH at 700° [45]. Herein very preliminary results regarding the comparison of their performance with the uptakes of other adsorbents are presented (Table 12.3). Both G-240 and RS-240 show CH_4 uptakes comparable with the top performing materials in methane storage applications (e.g. MOFs). These excellent preliminary results could be further improved by tailoring the PSD of the ACs to the optimum pore width value for methane adsorption (ca. 0.8–1.1 nm) [111]. Such specific material pore-size features could be achieved by tuning the KOH activation process parameters and the chemical structure of the parent HTC carbon as previously shown (i.e. changing HTC temperature).

12.6.2.2.3. Hydrogen Storage on HTC Carbons

Hydrogen storage is currently one of the main obstacles toward the commercial use of hydrogen in fuel cell systems. Among the large variety of materials investigated as carriers for hydrogen storage, porous carbons have generated great attention. Porous carbons with very large BET surface areas (>2000 $m^2\,g^{-1}$) and a porosity made up of micropores of around 1 nm have demonstrated to exhibit a large capacity to adsorb hydrogen [112].

Recently, Sevilla et al. reported the use as H_2 adsorbents of porous carbons produced by chemical activation (KOH) of HTC materials obtained by hydrothermal carbonization of glucose, starch, furfural, cellulose, and sawdust [113]. The main textural parameters for the HTC-based activated carbons, as well as their hydrogen storage capacity at 1 and 20 and at −196 °C, are given in Table 12.4. Independently on the type of precursor selected to produce the HTC material, all the activated samples exhibit similar textural properties: (i) large BET surface areas, around 2200 $m^2\,g^{-1}$, (ii) pore volume of ~1 $cm^3\,g^{-1}$, and (iii) a porosity made up mainly of micropores

TABLE 12.3 CH_4 **Uptake of Different Adsorbents at 298 K and 30 bar (if not differently stated in brackets)**

Adsorbent Type	CH_4 Uptake (mmol/g)
G-240 and RS-240	9.0 and 9.1
Commercially available ACs	3.0–10.6 (35 bar)
Coal-precursor-derived ACs	4.0–9.8
Petroleum pitch-based carbon-fiber-derived ACs	2.0–8.7
Microporous Organic Polymers (MOPs)	4.3–5.5 (36 bar)
Metal Organic Frameworks (MOFs)	6.5–10.3 (36 bar)

(<2 nm). Their hydrogen storage capacity is ~2.5 wt % at 1 bar and in the 5.3–5.6 wt % range at 20 bar. These hydrogen uptakes are in most cases superior to those obtained for other activated carbons with large surface area under similar conditions.

12.6.3. Heterogeneous Catalysis

It is well known that the performance of heterogeneous catalysts is highly affected by their supporting materials. It was a straightforward trial to explore biomass-derived HTC-carbon materials and colloids as supports for metal particles for different catalytic applications because of their high surface area and functionality. Subsequently a selection of catalytic applications from the literature is presented.

As already described in the nanocomposite section, Titirici et al. demonstrated for the first time the possibility of directly loading the HTC sphere with various nanoparticles in a one-step reaction [114]. The Pd-supported HTC spheres (Fig. 12.10a) proved to be successful catalysts for a broad range of reactions related to hydrogenation of hydroxy aromatic derivatives. The most important example was the selective hydrogenation of phenol to cyclohexanone, a reaction which is otherwise very difficult to achieve. The increased selectivity toward cyclohexanone as compared to cyclohexanol was attributed to the hydrophilic properties of HTC where once cyclohexanone is formed, being hydrophobic is repelled out of the hydrophilic reactive pocket and can no longer react [97].

TABLE 12.4 Hydrogen Uptake Capacity and Textural Properties of Chemically Activated Carbons Produced from HTC Materials

Precursor[a]	BET Surface Area ($m^2 g^{-1}$)	Pore Volume ($cm^3 g^{-1}$)	Micropore Volume ($cm^3 g^{-1}$)	H_2 Uptake (wt %)[b]
Furfural	2180	1.03	0.94	5.4 (2.5)
Glucose	2120	1.00	0.91	5.3 (2.4)
Starch	2190	1.01	0.92	5.4 (2.4)
Cellulose	2370	1.08	0.96	5.6 (2.5)
Wood	2250	1.03	0.91	5.6 (2.5)

[a]*The HTC products were obtained at a temperature of 230 °C (glucose and starch) and 250 °C (furfural, cellulose and eucalyptus sawdust - wood -). For the chemical activation of HTC materials a temperature of 700 °C and a KOH/HTC mass ratio of 4 were used.*
[b]*Hydrogen uptake capacity measured at −196 °C and 20 bar. In parenthesis, the hydrogen uptake capacity measured at −196 °C and 1 bar is given.*
(Adapted from Ref. [113] with permission)

The active nanoparticles can be postloaded onto the carbon support. Thus, using a nitrogen-doped carbon support obtained from the hydrothermal carbonization of seafood waste (lobster shells) loaded with Pt^{+2} nanoparticles were successfully tested as solid catalysts in the direct oxidation of methane in fuming sulfuric acid [77]. Remarkably, catalytic performance superior to the golden standard Periana system consisting of molecular dichlorobipyrimidyl platinum(II) complex [78] was achieved *via* the C–H activation, while the catalysts could be recycled over several runs without loss in activity. The remarkable catalytic performance together with further targeted material development will facilitate new strategies in process development to design viable methane utilization.

Pt nanoparticles supported on HTC spheres were used for methanol and ethanol electrooxidation in alkaline media. Experimental results demonstrate that, in comparison with the carbon black or hydrothermally synthesized carbon nanosphere support, carbon-nanosphere-supported Pt electrocatalyst showed an enhanced efficiency for both methanol and ethanol electrooxidation in terms of electrode conductivity, electrochemically active surface, oxidation peak current density, and onset potential [115].

The catalytic activity of a novel nanoparticulate Pt–Ru alloy supported on hydrothermal carbon for dry methane reforming and for the partial oxidation of methane was also studied. Upon removal of the carbon support by calcination under air, a mixture of Pt^0 and Ru oxide was obtained and their catalytic properties have been as well investigated for the same reactions. For dry methane reforming the initial activity in both samples was high but the catalysts were readily deactivated in time. Regarding the partial oxidation of methane, the Pt–Ru alloy supported on hydrothermal carbon showed a lower initial activity but a better catalytic stability than its oxidized analogs, suggesting that hydrothermal carbon-supported Pt–Ru alloys could be employed as a potential catalyst in hydrocarbon catalytic oxidations [116].

A facile and efficient one-pot route for the fabrication of Fe_xO_y@C spheres embedded with highly dispersed iron oxide nanoparticles by hydrothermal treatment of a glucose solution containing iron nitrate at a mild temperature has been also reported [117]. The catalytic study revealed the remarkable stability and selectivity of the reduced Fe_xO_y@C spheres in the Fischer–Tropsch synthesis reaction, which clearly exemplifies the confinement effect of the carbonaceous matter on the embedded nanoparticles imparted by the unique microstructure. The surrounding carbonaceous matter facilitated the formation of iron carbides during H_2 activation, which is beneficial for the formation of C_5+ hydrocarbons, and restricted the aggregation of the iron carbide nanoparticles during the activation and reaction processes.

Highly active carbonaceous nanofibers (CNFs) were prepared through a template-directed hydrothermal carbonization process and used as versatile nanoscale scaffolds for constructing macroscopic multifunctional membranes [118]. The broad applicability of the CNF scaffold was demonstrated by preparing a variety of CNF-based composite nanofibers including CNFs-Fe_3O_4, CNFs-TiO_2, CNFs-Ag, and CNFs-Au through various chemical routes (Fig. 12.11). Importantly, all of them inherit unique dimensionality (high aspect ratio) and mechanical properties (flexibility) of the original CNF scaffolds and thus can be assembled into macroscopic free-standing membranes through a simple casting process. The wide application potentials of these multifunctional composite membranes in magnetic actuation, antibiofouling filtration, and continuous-flow catalysis were also proven.

In all the examples presented thus far, HTC plays the role support for the nanosized catalytically active inorganic species. However, the carbon materials themselves can present intrinsic catalytic properties which will be provided by the various chemical functionalities.

Thus, imidazole-containing porous HTC materials were prepared by Titirici et al. combining the "hard templating" methodology with the "*in situ*" functionalization with small amounts of acrylic monomers as previously described. The resulting materials could be further functionalized to feature imidazolium

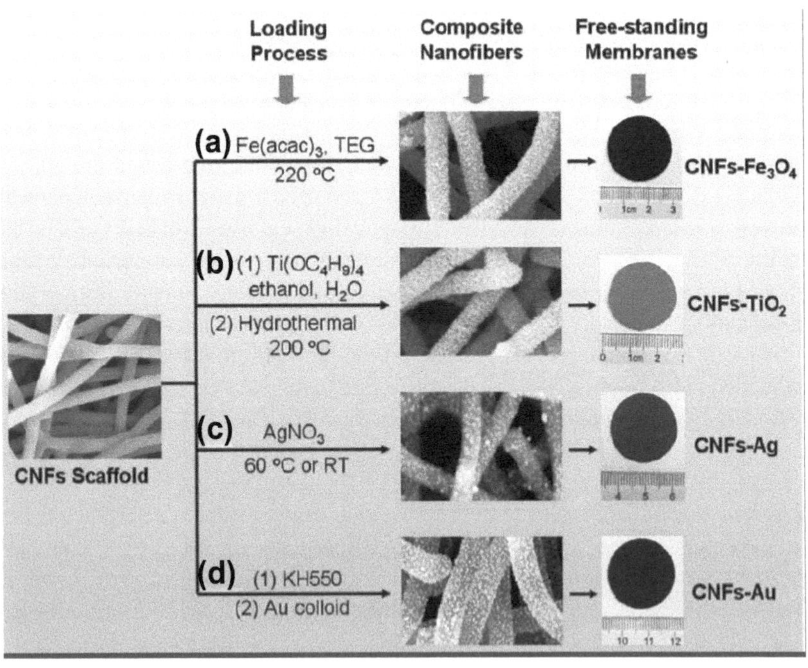

FIGURE 12.11 Schematic illustrations showing the broad versatility of the CNF scaffolds for constructing free-standing multifunctional membranes, including (a) CNFs-Fe_3O_4, (b) CNFs-TiO_2, (c) CNFs-Ag, and (d) CNFs-Au *(Taken with permission from W.-J. Zhang, H.-W. Liang, Y.-N. Ma, X. Cao, Q.-F. Guan, W.-P. Xu and S.-H. Yu, ACS Nano, 2011, 5, 8148).*

bromides on their surface and be used efficiently for various basic catalytic reactions such as transesterification and Knoevenagel and aldol reactions [61].

The main advantage of HTC is the easy tunability of the surface character. Thus, this can be turned from basic to highly acidic. In this respect, Sevilla et al. reported a route that involves the sulfonation of carbonaceous microspheres obtained by the hydrothermal carbonization of glucose. This synthetic strategy circumvents gas-phase pyrolysis, thereby avoiding the emission of harmful gases, and yields a solid acid comprising spherical particles of uniform, micrometer-regime size. These materials proved to be efficient catalysts toward the esterification of oleic acid with ethanol, a typical reaction in the synthesis of biodiesel. Moreover, the reaction rates and the intrinsic catalytic activity of the sulfonated microspheres showed better catalytic properties than other commercial solid acids, such as Amberlyst-15 [58].

Other authors had demonstrated that the functionalization of HTC with acidic groups for catalyzing various esterification reactions can also take place in one-step reaction *via* hydrothermal carbonization of furaldehyde and hydroxyethylsulfonic acid aqueous solution at 180 °C for 4 h [119]. Carbon exhibited high acidity and comparable activities to sulfuric acid for the traditional acid-catalyzed reactions.

12.6.4. Magnetic HTCs

A controlled aromatization and carbonization of glucose under hydrothermal reaction conditions enabled the preparation of carbonaceous surfactants that were used as shells for the coating of ferromagnetic Fe_3O_4 nanospheres with a uniform size distribution [120]. Under controlled experimental conditions, it was demonstrated that glucose could be employed as the carbon source in the preparation of continuous 1-D carbon nanoparticle chains with magnetic nanosphere inclusions (Fig. 12.12a,b). The functional groups on the carbon surface will facilitate the linkage of functional groups or catalytic species to the surface in future applications. The salient feature of the reported method was the assembly of magnetic nanospheres under hydrothermal reaction conditions in the absence of external fields. The magnetic behavior of Fe_3O_4/C nanochains, which is of importance for practical applications, was investigated as a comparison to that of pure Fe_3O_4 nanospheres. Figure 12.12c shows magnetic hysteresis curves, measured at room temperature, for the two samples. The hysteresis loop of the Fe_3O_4/C nanochains exhibited magnetization which was lower than that of the Fe_3O_4 nanospheres (92.8 emu/g), and this was believed to be due to the amorphous carbon shell in the chainlike structure at the interface providing a smaller magnetic torque per unit mass as opposed to that of the ferromagnetic cores. The magnetic separability of such magnetic chains was tested in water by placing a magnet near the glass bottle; the black chains were attracted toward the magnet within 20 s, clearly demonstrating the magnetic properties of the Fe_3O_4/C nanochains. Thus, an easy and efficient way to

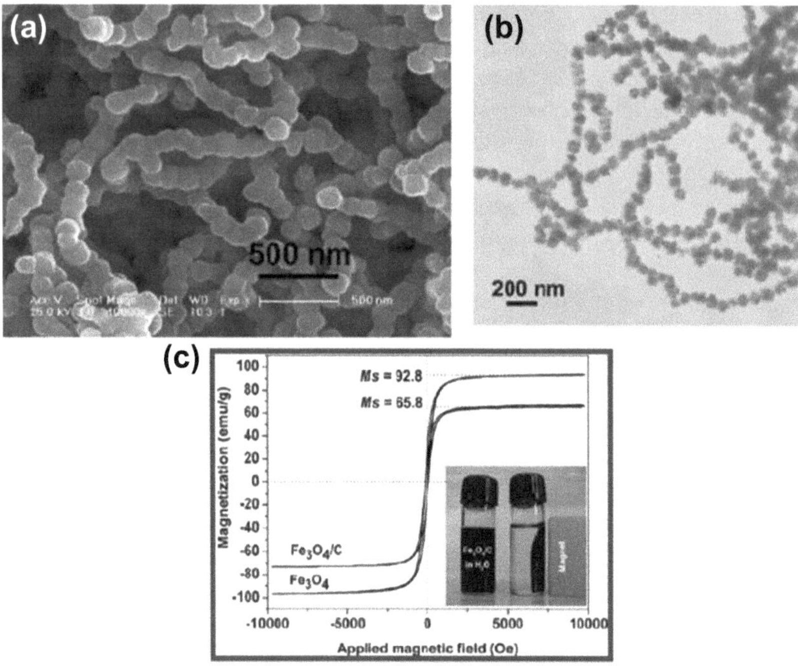

FIGURE 12.12 SEM image 11 (a) and TEM micrographs (b) of Fe3O4/C composites; (c) hysteresis curve and the magnetic separability of the Fe3O4/C nanochains at room temperature *(Taken with permission from H. Duan, Z. Zhang, S. Li and Y. Lin, Langmuir, 2010, 26(9), 6676).*

separate the Fe_3O_4/C nanochains from a suspension system is to use an external magnetic field.

Another example of self-assembly is the formation of porous Fe_3O_4/carbon core/shell nanorods fabricated *via* a three-step process. α-Fe_2O_3 nanorods were first obtained, and α-Fe_2O_3/carbon core/shell nanorods were subsequently fabricated using glucose as a carbon source by a hydrothermal method, in which the thickness of the carbon coating was about 3.5 nm. Fe_3O_4/carbon core/shell nanorods were synthesized after an annealing treatment of the product above under a mixture of Ar/H_2 flow. After the H_2 deoxidation process, the Fe_3O_4 core become porous while the thickness of the carbon shell was decreased to about 2.5 nm, and its degree of graphitization was enhanced. The interesting core/shell nanostructures are ferromagnetic at room temperature. The electromagnetic properties of the core/shell nanorod composite were investigated in detail. The maximum reflection loss was about −27.9 dB at 14.96 GHz for the composite with a thickness of 2.0 mm, and the absorption bandwidth with the reflection loss below −18 dB was up to 10.5 GHz for the absorber with the thickness of 2–5 mm. The excellent electromagnetic wave absorption properties of the porous Fe_3O_4/carbon core/shell nanorods were attributed to effective complementarities between the dielectric loss and the magnetic loss [121].

Rare-earth-doped gadolinium oxide (Gd_2O_3) hollow nanospheres have been successfully prepared on a large scale *via* a template-directed method using hydrothermal carbon spheres as sacrificed templates as previously reported by Titirici et al. [114]. By doping the rare-earth ions (Yb/Er) into the Gd_2O_3 host matrix, these NPs emitted bright multicolored upconversion emissions that can be fine-tuned from green to red by adjusting the co-doped Yb/Er ratio under 980 nm NIR laser excitation. The possibility of using these upconversion nanoparticles for optical imaging *in vivo* has been demonstrated. It was also shown that these Gd_2O_3 nanospheres brightened the T1-weighted images and enhanced the r1 relaxivity of water protons, which suggested that they could act as T1 contrast agents for magnetic resonance (MR) imaging. Moreover, these hollow spheres can be used as drug delivery host carriers, and drug storage/release properties were investigated using ibuprofen as the model drug. As a result, the so-prepared nanoscaled Gd_2O_3 hollow spheres bearing upconversion luminescence, MR imaging, and drug delivery capabilities could be potentially employed for simultaneous MR/fluorescent imaging and therapeutic applications.

12.6.5. Photocatalysis

TiO_2, as a cheap, nontoxic, and highly efficient photocatalyst, has been extensively studied and applied for degradation of organic pollutants, air purification, sterilization, and as a demister. Generally, TiO_2, with a bandgap of 3.2 eV, can only be excited by a small UV fraction of solar light, which accounts for only 3–5% of the solar energy. In the past decade, many works have been done to extend the photoresponse of the TiO_2 to the visible region. Among these attempts, TiO_2 doping, with either main group elements or transition metals, has been the most important approach for improving the photocatalytic performance of the catalyst.

The hydrothermal carbonization process has been used in this direction to produce carbon-doped TiO_2 with visible light activity (Fig. 12.10b) [98]. It was shown that the surface of nanometer-sized carbon materials can also show collective polarization modes and, therefore, these optical absorption transitions are feasible to sensitize TiO_2 which then acts as a novel "dyade"-type structure, with an improved TiO_2 hole reactivity, while the electron is taken up by the carbon component. This results in an improved photocatalytic activity over the complete spectral range. In order to avoid carbon from doping directly into bulk TiO_2 lattice our hybrid TiO_2/C is synthesized at low temperature under solvothermal conditions by a one-step carbonization of furfural in the presence of Ti-isopropoxide, allowing for the formation and co-assembly of carbon and TiO_2 into an interpenetrating C/TiO_2 nanoarchitecture containing 12.6% carbon.

UV–vis diffuse reflectance spectroscopy demonstrated that C@TiO_2 material can adsorb significantly more light in the 420–800 nm regions as compared with pure TiO_2. Therefore, the hydrothermally prepared C@TiO_2 exhibited a very high photocatalytic activity for degradation of methyl orange under the visible light. Furthermore, the C@TiO_2 material was indeed able to generate significant photocurrents under both UV and visible-light irradiation.

A typical n-type photocurrent was generated with UV ($\lambda > 320$ nm), while under visible-light irradiation the material exhibits both n-type and p-type currents. This was explained by the fact that UV-absorption also activates pure TiO_2 bands, while the ambipolar, biphasic nature of C@TiO_2 is activated under visible light. This allows both electrons and holes to contribute to the charge transport, which is crucial for photo-electrochemical applications.

Other authors also reported that the catalysts prepared under the same conditions also had high activity under the visible for the degradation of Acid Orange 7 and 2,4-dichlorophenol. It was also assumed that two kinds of sensitization processes, carbon sensitization and dye sensitization, are responsible for the visible-light-induced photocatalysis of TiO_2@C [122].

12.6.6. Energy Storage

Considering the energy shortage, environmental crisis, and the ever-increasing customer demands, the discovery of facile, low-cost, environmentally friendly, and nontoxic routes to produce novel valuable carbon materials is a key issue in today's society. In this respect, HTC offers viable solutions as it can convert biomass, a renewable and highly available resource, into interesting carbon materials, which can be used for energy storage obtained from nonconventional sources. Hydrothermal carbons and their nanocomposites have been so far intensively used as electrodes in Li^+ or Na^+ ion batteries, supercapacitors, or fuel cells.

12.6.6.1. Li/Na Ion Batteries

Lithium-ion batteries play a vital role as the prominent power sources for cell phones, laptop computers, digital cameras, power tools, electric-assist bicycles, and many consumer products, and are already well established and commercially available technologies.

Carbon materials are one of the mostly used negative electrodes in Li-ion batteries. Hydrothermal carbons have been also employed as electrodes in Li-ion batteries. The first study was performed by Huang et al. [123]. It was shown that the reversible lithium insertion/extraction capacity of this kind of material is much higher than the theoretical capacity of graphitized carbonaceous materials. However, such HTC materials had no significant porosity and only very small amounts of nanopores <1 nm. Titirici et al. investigated also the performance of HTC materials as anodes in Li^+ as well as Na^+ ion batteries. However, the morphology was different from the hard HTC spheres used by

Huang et al., namely hollow HTC nanospheres [55]. The reversible capacity at 1 C rate of hollow carbon nanospheres could reach up to 370 mAh g^{-1}, which is slightly higher than that of graphite as anodes in Li-ion batteries. Figure 12.13a shows the cycling performance of the hollow carbon nanospheres at 1 C rate. The capacity loss in the initial cycles stems from the SEI film stabilization and the irreversible lithium ion insertion. While previous reports suggested that nongraphic carbons show a continuous and progressive decay in capacity during cycling, here we found that the hollow carbon nanospheres showed excellent cycling performance at 1C rate. After 200 discharge/charge cycles a reversible capacity as high as 310 $mAhg^{-1}$ can still be retained, and the coulombic efficiency approaches almost 100%.

However, the real advantage of such hollow spheres, besides the high cyclability, is the outstanding rate capabilities. As shown in Fig. 12.13b, a specific charge capacity of 200 $mAhg^{-1}$ is reached at 5 C. When the current density is increased by from 1 C to 20 C, the capacity retention of hollow

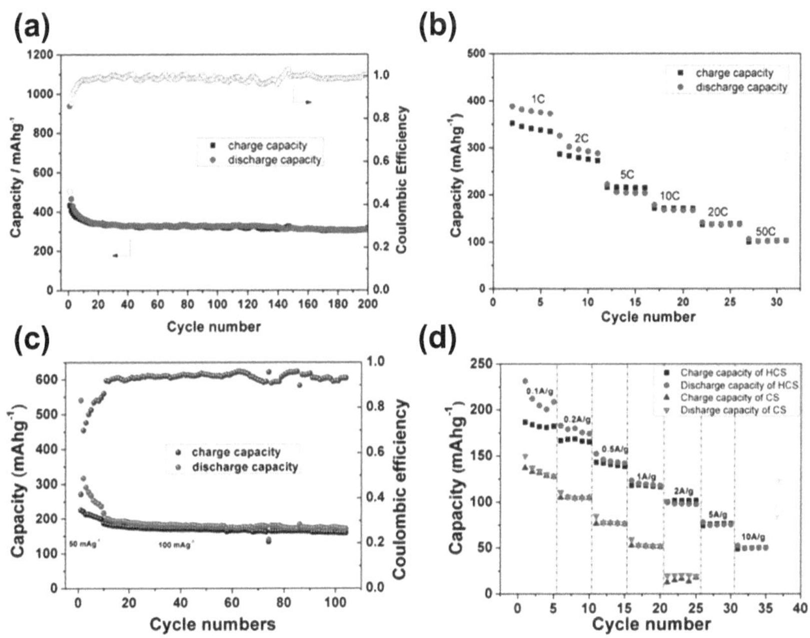

FIGURE 12.13 (a) Cycle performance of hollow carbon nanospheres cycled at a rate of 1C in Li Ion Batteries, (b) Rate performance of hollow carbon nanospheres at variant rates in Li ion batteries; *(Taken with permission from R. J. W. Kun Tang, Maria-Magdalena Titirici, Lijun Fu, Joachim Maier, ChemSus Chem **2012**, 5, 400)*, (c) Cycle performance of hollow carbon nanospheres as anodes in Na Ion Batteries, (d) Rate performance of hollow carbon nanospheres (HCS) and carbon spheres (CS) at variant rates as anodes in Na Ion Batteries. *(Taken with permission from L. F. Kun Tang, Robin J. White, Linghui Yu, Maria-Magdalena Titirici, Markus Antonietti, Joachim Maier Adv. Energ. Mater **2012**, 2, 873–877)*

carbon spheres can approach 50%. Even at the very high rate of 50 C (18.6 Ag^{-1}), a capacity of ~100 $mAhg^{-1}$ is still maintained. This value is much higher than traditional graphite electrode (almost neglectable at such high rate). To the best of our knowledge, it is almost the best rate performance ever measured on carbonaceous material. What should also be mentioned is that, the rate performance is much higher than those disordered carbon obtained from the similar precursor without template, indicating that hollow nanosphere morphology plays a vital role for the excellent electrochemical performance.

At the present time, lithium resources are confirmed to be unevenly distributed, located primarily in South America. The cost of lithium-based raw materials has roughly doubled from the first practical application in 1991 to now, and it may drastically increase when the demand for lithium increases as a result of commercialization of large-scale lithium-ion batteries in the future. However, sodium resources are nearly inexhaustible and unlimited around the world. Furthermore, the electrochemical equivalent and standard potential of sodium are the most advantageous for aprotic battery applications after lithium. Therefore, Titirici et al. have demonstrated for the first time that HTC materials, more precisely hollow spheres, can play a very important role as anode materials in Na ion batteries [124]. The advantage of the high rate capability of the hollow spheres over the standard hard HTC spheres obtained from the same precursor was also demonstrated. The first discharge/charge cycle delivered specific capacities of 537 and 223 mAh/g at a current rate of 50 mA/g. The reversible capacity of 200 mAh/g at a current density of 50 mA/g is significant, and only few reports achieved capacities more than 200 mAh/g for sodium anode materials obtained at much lower current densities. The cycling stability during sodium insertion/extraction in the hollow carbon nanospheres was also investigated at a current density of 50 mA/g for the first ten cycles and then 100 mAh/g for subsequent cycles (Fig. 12.13c). The capacity loss in the initial cycles stems from SEI film stabilization and irreversible sodium ion insertion. After 100 cycles, a reversible capacity of approximately 160 mAh/g is well maintained. The coulombic efficiency approaches around 94% after several cycles. Such a good cycle performance of more than 100 cycles was seldom achieved in previous reports for anode material of sodium ion battery. Remarkably, the reversible capacities are retained at 168, 142, 120, 100, and 75 mAh g^{-1} at current densities of 0.2 A/g, 0.5 A/g, 1 A/g, 2 A/g, and 5 A/g, respectively. Even at the very high current density of 10 A/g, a capacity of ~50 $mAhg^{-1}$ is still maintained, demonstrating the excellent rate performance of the hollow carbon nanosphere electrodes. To the best of our knowledge, the presented results indicate the best rate performance ever measured for anode materials of sodium ion battery so far (Fig. 12.13d).

Besides the use of pure HTC materials as electrodes in Li-ion batteries, great importance was given to the utilization of HTC/inorganic composites as electrode materials.

It is already known that certain elements such as Si and Sn have the great advantage of forming alloys with Li. Since about 4.4 Li atoms can be alloyed on one Si or Sn atoms, it is obvious that the capacity of such materials when used as negative electrodes in Li ion batteries will be greatly improved ($Li_{4.4}$ Si ~ 4200 $mAhg^{-1}$ as compared with 372 $mAhg^{-1}$ for LiC_6 in graphite). However, such alloying reactions are normally associated with severe volume changes during Li insertion–deinsertion, leading to very fast decay in capacity with cycling. Another associated problem is their low conductivity.

In order to overcome such problems, Titirici et al. have successfully coated commercially available silicon nanoparticles with a thin layer (25 wt%-10 nm) of HTC using the hydrothermal carbonization of glucose [69]. The resulting composite was further calcined to increase the conductivity of the carbon layer resulting in a $Si/SiO_X/C$ composite with an obvious improved cyclability compared with the pure Si. Especially when working in the presence of vinylene carbonate additive in the electrolyte, an excellent cycling performance was achieved. The reversible capacity is as high as 1100 $mAhg^{-1}$ at a current density of 150 mAg^{-1}, with no further decay of capacity even after 60 cycles. A large irreversible capacity was observed in the first discharge and charge process; however, after the initial cycles, the coulombic efficiency is above 99%. The irreversible capacity in the first initial cycles is an expected phenomenon in the Si-based electrodes in lithium batteries. The initial irreversible capacity loss could mainly originate from the reduction of the electrolyte, resulting in the formation of a solid electrolyte interphase (SEI) on the surface of the active particles and/or from irreversible lithium insertion into nanocomposites. This might be overcome by preforming an artificial SEI layer (through chemical modification) on the active particles and/or prelithiating the active particles.

Similarly, preformed SnO_2 nanoparticles were incorporated inside HTC spheres from glucose. The resulting composites can be calcined and used as electrodes in Li-ion batteries, or the carbon can be removed. Here, the hydrothermal carbonization acts as an assembly media of the initial nanoparticles into mesoporous SnO_2 microspheres with better packing density and superior Li insertion properties [23].

Cao et al. also prepared multilayered nanocrystalline SnO_2 hollow microspheres by chemically induced self-assembly in the hydrothermal environment [113]. First, a multilayered spherical SnO_2–C composite was produced through a condensation polymerization and carbonization of sucrose accompanied by hydrolysis of $SnCl_4$ in the hydrothermal reaction. Then, the HS–SnO_2 was obtained by removal of carbon. The as-prepared HS–SnO_2 exhibited good electrochemical performance as the anode material in lithium-ion batteries [125].

Besides such compounds forming alloys with Li, transition-metal oxides such as CoO, Co_3O_4, NiO, Cu_2O, CuO, FeO, Fe_2O_3, and MnO have attracted much attention. These oxides offer a promising carbon alternative to negative electrode materials in lithium-ion batteries, which is attributed to their high specific capacity and good cycling performance. A mechanism different from the classical Li insertion/deinsertion in carbonaceous compounds or Li-alloying processes in alloy electrode is proposed, which can be written as: $M_xO_y + 2y\ Li \leftrightarrow y\ Li_2O + xM$. Solid electrolyte interface (SEI) will also be formed during the discharge, but it will partially decompose during the subsequent charge process, which is attributed to the catalytic activity of metallic nanoparticles. The partially reversible formation/decomposition of SEI film will lead to an extra capacity. Thus, Tu et al. prepared spherical NiO–C composite by dispersing spherical NiO in glucose solution and subsequent carbonization under hydrothermal conditions at 180 °C [126]. The electrochemical properties of the NiO–C composite were investigated by galvanostatic discharge–charge test, cyclic voltammetric analysis (CV), and electrochemical impedance spectroscopy (EIS). The carbon coated on the surface, and also filled in the inner pores of the NiO sphere. The NiO–C composite exhibited higher initial coulombic efficiency (66.6%) than the NiO (56.4%) at a charging–discharging rate of 0.5 °C. Better cycling performance was also obtained for the NiO–C electrode. The specific capacity after 40 cycles for the NiO–C composite is 430 mAh g^{-1}, higher than that of NiO (200 mAh g^{-1}). These improvements can be attributed to the conductive carbon and its combination with the pores.

Besides negative electrodes in Li-ion batteries, the development of improved cathode materials is even more challenging. In this respect, olivine $LiFePO_4$ is considered one of the most promising cathode materials for Li-ion batteries. Titirici et al. developed a simple one-step, template-free, low-temperature solvothermal method for the synthesis of urchin-like hierarchical mesocrystals of pristine $LiFePO_4$ as well as carbon-coated $LiFePO_4$ composites. Each urchin-like mesocrystal consists of $LiFePO_4$ sheets self-assembled *via* a dipolar field in spheres during a solvothermal process under the influence of Cl^- anions (Fig. 12.9d). The obtained primary sheets of $LiFePO_4$ are single crystalline in nature and can be coated *in situ* with an amorphous nitrogen-doped carbonaceous layer several nanometers in thickness. To increase the conductivity of the carbon coating, the materials are subjected to further temperature treatment (700 °C) under an inert atmosphere. The lithium-storage performance of the pure $LiFePO_4$ is compared with that of its carbon-coated counterparts which proved superior [127].

12.6.6.2. Supercapacitors

Electrochemical capacitors, also usually called supercapacitors, are of great interest as high-power electrochemical energy storage devices due to the fact that they combine high-power density with long cycle life and wide operational temperature range, properties that are currently unattainable in Li-ion batteries. However, the energy density achieved by commercial products (about 5 Wh/kg) limits their use to few seconds of charge/discharge, hampering their utilization in energy-harvesting applications. Therefore, efforts are focused now on increasing the energy density of supercapacitors.

Based on the charge storage mechanism, electrochemical capacitors can be divided into: (i) electrochemical double-layer capacitors, where the energy storage is based on the electrostatic adsorption of electrolyte ions on the surface area of electrically conductive porous electrodes, and (ii) pseudocapacitors, where the energy is stored through redox reactions at the electrode/electrolyte interface [128].

HTC carbons are oxygen rich by nature, with the oxygen content being tuned through the operating conditions (temperature, time, precursor, and concentration). In addition, by selecting appropriately the initial carbon precursor, operating conditions, or additional additives, other functionalities can be introduced in the final carbonaceous materials. Unfortunately, as described in Section 12.3, HTC materials generally possess almost no open porosity and their heat treatment at elevated temperatures can only lead to moderate increase in surface area, which hampers their application in surface-area-sensitive applications, such as supercapacitors. However, this limitation has been circumvented recently by the chemical activation of HTC carbons, as thoroughly described in Section 12.3.1.

Wei et al. analyzed the supercapacitor performance in organic electrolyte (1 M $TEABF_4$ in acetonitrile) of HTC cellulose (C), starch (S), and sawdust (W) activated at 700 and 800 °C with a KOH/sample weight ratio = 4 [129]. The carbon materials possessed BET surface areas in the ~2100–3000 m^2/g range and pore volumes up to ~1.4 cm^3/g, with a relatively narrow pore-size distribution in the supermicropore range with virtually no pores >3 nm. The performance of those materials was spectacular, recording the highest capacitance ever reported for porous carbons in a symmetric two-electrode configuration using such electrolyte, i.e. 236 F/g (100 F/cm^3) at 1 mV/s (Fig. 12.14a). It exceeded the specific capacitance of commercial activated carbons optimized for EDLC applications. Among the tested carbons, the samples activated at the highest temperature (800 °C), which are the ones with the largest volume of small mesopores in the range of 2–3 nm, were the ones with better capacitance retention at high sweep rates in the CV measurements or higher current densities in the charging/discharging tests (Fig. 12.14b). This combination of very high specific and volumetric capacitance and good rate capability of the hydrothermally synthesized porous carbons is unmatched by state-of-the art activated carbons, carbon nanotubes, carbon onions, and graphene.

Wang et al. have also applied chemical activation to hydrothermal carbons, in this case using phosphoric acid as activating agent and rice husk as carbon precursor [130]. The authors varied the activation temperature between 300 and 700 °C and the weight ratio of phosphoric acid to hydrothermal carbons between 1 and 6. In this way, the obtained porous materials ranged from supermicroporous to mesoporous, with BET surface areas in the ~700–2700 m^2/g range and pore volumes up to ~2 cm^3/g. The specific capacitance of

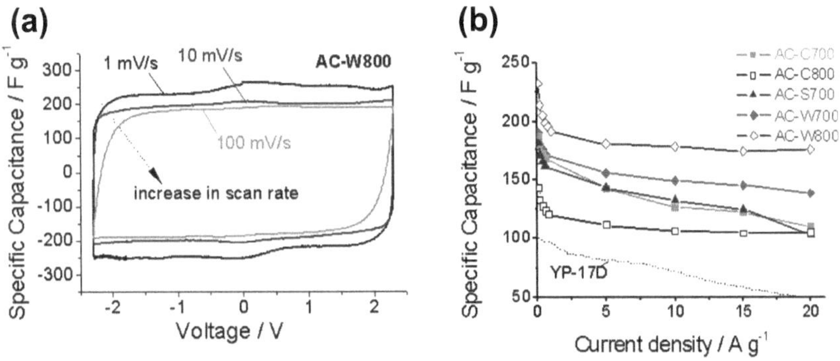

FIGURE 12.14 Electrochemical characterization of activated carbons derived from hydrothermally synthesized carbon materials in 1 M $TEABF_4$ solution in acetonitrile at room temperature: (a) cyclic voltammograms (CV) of the activated carbon obtained from sawdust at 800 °C with KOH/sample = 4 (AC-W800) and (b) capacitance retention with current density in comparison with that of commercially available YP-17D activated carbon *(Extracted from M. S. L. Wei, A. B. Fuertes, R. Mokaya, G. Yushin, Adv. Energ. Mater 2011, 1, 356. with permission).*

these activated carbons reached 130 F/g (measured at 2 mV/s) in an aqueous solution of KOH 6 M, a value quite below that measured for activated carbons with similar BET surface area but with narrower PSD in the micro–supermicropore range (235–286 F/g). This is due to the fact that, as shown by Chmiola et al., a good matching between the electrode pore size and the dimensions of the electrolyte ions is critical for an optimal performance of supercapacitors [131].

On the other hand, Zhao et al. have analyzed the capacitance behavior of nitrogen-containing hydrothermal carbons activated with KOH (weight ratio of KOH to hydrothermal carbons = 1–4, and $T = 600$ °C [22]). Although KOH was used as an activation agent, a high degree of activation was not achieved (surface area <600 m^2/g), which may be due to the low temperature used for the activation (600 °C). This was applied in order not to lose the structurally bound nitrogen at high temperatures of chemical activation. Despite the low surface area, the materials exhibited excellent electrochemical performance in KOH 6 M and H_2SO_4 1 M, achieving specific capacitances of up to 220 and 300 F/g at a current density of 0.1 A/g in the basic and acidic electrolyte, respectively. This superior capacitance is due to the combination of EDLC capacitance and pseudocapacitance arising from redox reactions of the nitrogen functionalities. It is worth noticing that the chemical activation process leads to a reduction of the nitrogen content of the materials, this reduction being larger with the hardening of the activation conditions (i.e. increase of the amount of KOH used). Nevertheless, the values of capacitance achieved with these materials are comparable to activated carbons reported in the literature with a much higher surface area (>3000 m^2/g) [132] and higher than that of a high surface area and high nitrogen-content carbon derived from polyacrylonitrile (160 F/g) [133]. Additionally, good capacitance retention at high current density (4 A/g) is observed which proves good conductivity and quick charge propagation in both acid and base electrolytes.

Recently Zhu et al. prepared electrodes for supercapacitors from hydrothermally carbonized fungi (HTC at 120 °C for 6 h) [134]. The product resulting from the HTC process was characterized by small particles (50–200 nm) fused together to give rise to sheets or tubular networks, the oxygen content being 13.4% (50% decrease from raw fungi). The value of the BET surface area was typical for hydrothermal carbons, ~14 m^2/g. With the aim of creating porosity, as well as improving the electronic conductivity, the authors pyrolyzed the hydrothermal carbon at 700 °C for 3 h. As a result, the oxygen content decreased to 5%, but the surface area increased to 80 m^2/g. Surprisingly, even though the oxygen content is not high and the surface area is really low, the material stored up to 196 F/g at a scan rate of 1 mV/s, a value comparable to high surface area carbons such as Maxsorb, or highly doped carbons like that obtained from seaweed. However, with the increase of the scan rate, the decrease of the capacitance was quite high (~40% decrease at 50 mV/s). The authors explain the large specific capacitance of that material on the basis of its small particle size, which reduces the ion and electron transport path and increases the material usage, as pseudocapacitors store charges in the first few nanometers below the surface.

12.6.6.3. Fuel Cells

The idea for the electrochemical conversion of the potential energy of coal into electrical energy originates from the late 19th century, when W. Jacques introduced the principle of the noncombustive, electrochemical oxidation of carbonaceous matter by air [135]. Carbon fuel cells are conceptually the most effective way of converting the chemical energy of carbon into electrical energy, especially compared to classical combustion processes. Paraknowitsch et al. used carbon colloids as a fuel, prepared by hydrothermal carbonization [136]. To evaluate the potential of HTC coal as a fuel, the oxidation properties of this carbon source compared to fossil coals are of main importance. Different samples of lignite and anthracite were applied as reference materials, using a simple Fe^{3+}-oxidation assay. The experiments revealed a significantly better capability of hydrothermal coal derived from glucose to get oxidized: it highly exceeds anthracite in its oxidation rate and it also shows better oxidation properties than lignite. Finally, using carbon colloids derived from hydrothermally treated biomass indeed yielded an indirect carbon fuel cell with an overall zero-emission balance concerning the green house gas CO_2.

The first to explore the use of an HTC-based carbon as electrocatalyst support in proton exchange membrane fuel cells was Yang et al. (R. Yang, X. Qiu, H. Zhang, J. Li, W. Zhu, Z. Wang, X. Huang, L. Chen, Carbon,

2005, 43, 11–16). They deposited 10 wt % Pt nanoparticles over carbon spherules obtained by HTC of sucrose at 190 °C and posttreatment at 1000 °C using two different methods, i.e. a polyol method and chemical reduction with $Na_2S_2O_4$. With regard to the electrooxidation of methanol, the catalyst prepared through the polyol method is the one exhibiting the highest current.

Sevilla et al. followed a different approach for the generation of graphitic structures from hydrothermal carbon [137] using a two-step process in which the synthesized hydrothermal carbon was impregnated with nickel nitrate and subjected to a heat treatment at 900 °C. As a result, highly crystalline, high surface area carbon nanocoils were formed. These nanocoils gather thus the key properties of electrocatalyst supports, i.e. relatively high and easily accessible surface area, and high crystallinity. This was confirmed by ther use as electrocatalyst supports toward the oxidation of methanol. Thus, Sevilla et al. deposited Pt/Ru nanoparticles [137] or Pt nanoparticles [138] over the graphitic carbon nanocoils and analyzed their behavior, comparing it with that of a carbon support widely used as electrocatalyst support, i.e. Vulcan. For both Pt/Ru and Pt nanoparticles, the electrocatalysts prepared with the carbon nanocoils exhibited a higher activity than that of Vulcan, which is a consequence of the fact that nanocoils combine good electrical conductivity and accessible surface that allows the diffusional resistances of reactants/products to be minimized.

12.6.7. Biological Applications

Recently, people began to explore the favorable bioapplications of HTC carbon materials such as bioimaging and drug delivery. Yu et al. synthesized green-luminescent monodisperse Ag@phenol PFR core–shell spheres in a facile one-step method [139]. By changing the concentration or the molar ratio of the starting materials, different architectures with tunable size can be obtained, such as monocore–shell, multicore–shell, and eccentric core–shell. These nanospheres give strong green emission under excitation at 340 nm. The fluorescent imaging showed that the particles could be internalized into living human lung cancer cells and exhibited almost no cytotoxic effects on the cells.

Kundu and co-workers [11] reported that the intrinsically fluorescent surface functional CSPs derived from the hydrothermal treatment of glucose can deliver a membrane-impermeable molecule into the mammalian cell nuclei and further modulate gene expression *in vivo*. The confocal laser scanning microscopy (CLSM) monitoring showed that the CSPs entered the nucleus within 3 h, with a further increasing amount over time. The fluorescent images and Raman spectrum also showed that the CSPs could cross the blood–brain barrier (a separation of circulating blood and cerebrospinal fluid) and penetrate into the brain. Thus, the CSPs are potential carriers to deliver drugs against brain tumors. The *in vivo* experiment further confirmed that the CSPs are nontoxic to animals and can be consumed over several days. CSPs can deliver a membrane-impermeable molecule, CTPB (N-(4-chloro-3-trifluoromethylphenyl)- 2-ethoxybenzamide), the only histone acetyltransferases (HATs) activator, into the cell nuclei.

12.7. CONCLUSIONS

In summary, carbons and carbonaceous materials synthesized by (HTC) from carbohydrates, organic molecules, and biomass have been discussed. The different experimental conditions and reaction mechanisms between the pure carbohydrates and lignocellulosic biomass as well as the dependence of the resulting chemical structure on the HTC temperature have been pointed out. Porosity can be easily introduced into such HTC materials using a variety of techniques from chemical activation to nanocasting and gelation agents which can interact with the HTC precursors. One important advantage of HTC materials is the retention of polar functionalities which makes the surface easily functionable either *via* post- or *in situ* modification. Depending on the desired application, the surface character can be easily tuned by either changing the HTC temperature or post-calcination under an inert atmosphere.

These carbonaceous materials can also be used to form nanocomposites with other materials and further serve as building blocks of hierarchical structures. Importantly, the as-synthesized carbon materials have already shown very broad practical applications in a variety of areas, such as environmental, catalytic, electronic, and biological applications. It is believed that the HTC process will play an increasingly important role in the fabrication of nanostructured carbon materials in the future. Heteroatoms (especially nitrogen) can be successfully incorporated within the framework of these materials and their chemical environment can also be controlled with the postannealing temperature. Such dopants confer very special electronic as well as catalytic properties to the HTC carbons.

Although the HTC process has achieved great success in a rather short time, there still remain some challenges and questions that should draw scientific attention: Can we find a proper catalyst to lower the reaction temperature even further, or shorten the reaction time? How to digest real waste biomass by the HTC to produce functional carbonaceous materials and nanocomposites with practical application? Solving these challenges and problems in the future will further facilitate and strengthen the capability for rational design of a variety of carbon materials and extended practical applications.

REFERENCES

[1] H.W. Kroto, J.R. Heath, S.C. Brien, R.F. Curl, S. R. E., Nature 318 (1985) 162.
[2] S. Iijima, Nature 354 (1991) 56.
[3] M.M. Titirici, A. Thomas, M. Antonietti, New J. Chem. 31 (2007) 787.

[4] S.H. Joo, S.J. Choi, I. Oh, J. Kwak, Z. Liu, O. Terasaki, R. Ryoo, Nature 412 (2001) 169.
[5] J.M. Planeix, N. Coustel, B. Coq, V. Brotons, P.S. Kumbhar, R. Dutartre, P. Geneste, P. Bernier, P.M. Ajayan, J. Am. Chem. Soc. 116 (1994) 7935.
[6] M.E. Davis, Nature 417 (2002) 813.
[7] S. Babel, T.A. Kurniawan, J. Hazard. Mater. 97 (2003) 219.
[8] A.C. Dillon, K.M. Jones, T.A. Bekkedahl, C.H. Kiang, D.S. Bethune, M.J. Heben, Nature 386 (1997) 377.
[9] E. Frackowiak, F. Beguin, Carbon 39 (2001) 937.
[10] K. Hemmes, M. Houwing, N. Woudstra, J. Fuel Cell Sci. Tech. 7 (2010) 051008.
[11] B.R. Selvi, D. Jagadeesan, B.S. Suma, G. Nagashankar, M. Arif, K. Balasubramanyam, M. Eswaramoorthy, T.K. Kundu, Nano Lett. 8 (2008) 3182.
[12] H. Dai, Acc. Chem. Res. 23 (2002) 1035.
[13] B.C. Satishkumar, A. Govindaraj, C.N.R. Rao, Chem. Phys. Lett. 307 (1999) 158.
[14] L.D. Schmidt, P.J. Dauenhauer, Nature 447 (2007) 914.
[15] M.M. Titirici, M. Antonietti, Chem. Soc. Rev. 39 (2010) 103.
[16] B. Hu, K. Wang, L. Wu, S.-H. Yu, M. Antonietti, M.M. Titirici, Adv. Mater. 22 (2010) 813.
[17] J.A. Libra, K.S. Ro, C. Kammann, A. Funke, N.D. Berge, Y. Neubauer, M-M. Titirici, C. Fühner, O. Bens, J. Kern, K.-H. Emmerich, Biofuels 2 (1) (2011) 89.
[18] M.M. Titirici, M. Antonietti, N. Baccile, Green Chem. 10 (2008) 1204.
[19] S. Kubo, R.J. White, Noriko Yoshizawa, M. Antonietti, M.M. Titirici, Chem. Mater. 23 (22) (2011) 4882.
[20] R.J. White, M. Antonietti, M.-M. Titirici, J. Mater. Chem. 19 (2009) 8645.
[21] R.J. White, N. Yoshizawa, M. Antonietti, M.-M. Titirici, Green Chem. 13 (2011) 2428.
[22] L. Zhao, L.-Z. Fan, M.-Q. Zhou, H. Guan, S. Qiao, M. Antonietti, M.-M. Titirici, Adv. Mater. 22 (2010) 5202.
[23] R. Demir-Cakan, Y.-S. Hu, M. Antonietti, J. Maier, M.-M. Titirici, Chem. Mater. 20 (2008) 1227.
[24] S. Kubo, I. Tan, R.J. White, M. Antonietti, M.-M. Titirici, Chem. Mater. 22 (2010) 6590.
[25] F.S. Asghari, H. Yoshida, Ind. Eng. Chem. Res. 46 (2007) 7703.
[26] Y. Yu, X. Lou, H. Wu, Energy Fuels 22 (2007) 46.
[27] J.N. Chheda, Y. Roman-Leshkov, J.A. Dumesic, Green Chem. 9 (2007) 342.
[28] X. Qi, M. Watanabe, T.M. Aida, R.L. Smith Jr., Catal. Commun. 9 (2008) 2244.
[29] Y. Roman-Leshkov, J.N. Chheda, J.A. Dumesic, Science 312 (2006) 1933.
[30] O. Sackur, Zeitschrift für Elektrochemie und angewandte physikalische Chemie 20 (1914) 260.
[31] S.H. Yu, X.J. Cui, L.L. Li, K. Li, B. Yu, M. Antonietti, H. Cölfen, Adv. Mater. 16 (2004) 1636.
[32] X. Cui, M. Antonietti, S.-H. Yu, Small 2 (2006) 756.
[33] M.M. Titirici, A. Thomas, S.-H. Yu, J.-O. Muller, M. Antonietti, Chem. Mater. 19 (2007) 4205.
[34] S.K.R. Patil, C.R.F. Lund, Energy Fuels 25 (2011) 4745.
[35] M. Sevilla, A.B. Fuertes, Carbon 47 (2009) 2281.
[36] M. Sevilla, A.B. Fuertes, Chem. Eur. J. 15 (2009) 4195.
[37] N. Baccile, G. Laurent, F. Babonneau, F. Fayon, M.M. Titirici, M. Antonietti, J. Phys. Chem. C 113 (2009) 9644.
[38] P. Kumar, D.M. Barrett, M.J. Delwiche, P. Stroeve, Ind. Eng. Chem. Res 48 (2009) 3713.
[39] D. Klemm, B. Heublein, H.-P. Fink, A. Bohn, Angew. Chem. 117 (2005) 3422.
[40] M.P. Pandey, C.S. Kim, Chem. Eng. Technol. 34 (1) (2011) 29.
[41] Y.S. Sun, X.B. Lu, S.T. Zhang, R. Zhang, X.Y. Wang, Bioresour. Technol. 102 (2011) 2936.

[42] H.D. Willauer, J.G. Huddleston, M. Li, R.D. Rogers, J. Chromatogr. B. 743 (1–2) (2000) 127.
[43] C. Falco, N. Baccile, M.M. Titirici, Green Chem. 13 (11) (2011) 3273.
[44] O. Ioannidou, A. Zabaniotou, Renew. Sust. Energy Rev. 11 (2007) 1966.
[45] C. Falco, D Lozano-Castello, M. M. Titirici, Carbon 2012-sumbitted.
[46] A. Linares-Solano, D. Lozano-Castello, M.A. Lillo-Rodenas, D. Cazorla-Amoros, Chemistry and Physics of Carbon, vol 30, Crc Press-Taylor & Francis Group, Boca Raton, 2008, pp. 1.
[47] C. Liang, Z. Li, S. Dai, Angew. Chem. Int. Ed. 47 (2008) 3696.
[48] D.Y. Zhao, J.L. Feng, Q.S. Huo, N. Melosh, G.H. Fredrickson, B.F. Chmelka, G.D. Stucky, Science 279 (1998) 548.
[49] S.H. Joo, R. Ryoo, M. Kruk, M. Jaroniec, J. Phys. Chem. B 106 (2002) 4640.
[50] A.H. Lu, W.C. Li, W. Schmidt, W. Kiefer, F. Schuth, Carbon 42 (2004) 2939.
[51] F. Zhang, Y. Meng, D. Gu, Y. Yan, C. Yu, B. Tu, D. Zhao, J. Am. Chem. Soc. 127 (2005) 13508.
[52] C. Liang, S. Dai, J. Am. Chem. Soc. 128 (2006) 5316.
[53] M.-M. Titirici, A. Thomas, M. Antonietti, Adv. Funct. Mater. 17 (2007) 1010.
[54] M.-M. Titirici, A. Thomas, M. Antonietti, J. Mater. Chem. 17 (2007) 3412.
[55] R.J. White, K. Tauer, M. Antonietti, M.-M. Titirici, J. Am. Chem. Soc. 132 (2011) 17360.
[56] N. Baccile, M. Antonietti, M.-M. Titirici, ChemSusChem 3 (2010) 246.
[57] L. Zhao, Z. Bacsik, N. Hedin, W. Wei, Y. Sun, M. Antonietti, M.-M. Titirici, ChemSusChem 3 (2010) 840.
[58] J.A. Macia-Agullo, M. Sevilla, M.A. Diez, A.B. Fuertes, ChemSusChem 3 (2010) 1352.
[59] Z. Chen, L. Ma, S. Li, J. Geng, Q. Song, J. Liu, C. Wang, H. Wang, J. Li, Z. Qin, S. Li, Appl. Surf Sci. 257 (2011) 8686.
[60] R. Demir-Cakan, N. Baccile, M. Antonietti, M.-M. Titirici, Chem. Mater. 21 (2009) 484.
[61] R. Demir-Cakan, P. , Makowski, M. Antonietti, F. Goettmann, M.-M. Titirici, Catal. Today 150 (2010) 115.
[62] Z. Li, N. Baccile, S. Gross, Z. Yuanjian, W. Wei, S. Yuhan, M. Antonietti, M.M. Titirici, Carbon 48 (2010) 378.
[63] N. Baccile, M. Antonietti, M.-M. Titirici, ChemSusChem 3 (2010) 246.
[64] N. Baccile, G. Laurent, C. Coelho, F. Babonneau, L. Zhao, M.-M. Titirici, J. Phys. Chem. C 115 (2011) 8976.
[65] S.H. Yu, K. Wang, L. Liu, X.W. Xu, Dalton Trans. (2008) 5414.
[66] H.S. Qian, S.H. Yu, L.B. Luo, J.Y. Gong, L.F. Fei, X.M. Liu, Chem. Mater. 18 (2006) 2102.
[67] Y. Wan, Y.L. Min, S.H. Yu, Langmuir 24 (2008) 5024.
[68] W.M. Zhang, X.L. Wu, J.S. Hu, Y.G. Guo, L.J. Wan, Adv. Funct. Mater. 18 (2008) 3941.
[69] Y.-S. Hu, R. Demir-Cakan, M.M. Titirici, J.O. Muller, R. Schlogl, M. Antonietti, J. Maier, Angew. Chem. Int. Ed. 47 (2008) 1645.
[70] X.W. Lou, C.M. Li, L.A. Archer, Adv. Mater. 21 (2009) 2536.
[71] W.M. Zhang, J.S. Hu, Y.G. Guo, S.F. Zheng, L.S. Zhong, W.G. Song, L.J. Wan, Adv. Mater. 20 (2008) 1160.
[72] X.M. Sun, Y.D. Li, Angew. Chem. Int. Ed. 43 (2004) 597.
[73] H.S. Qian, M. Antonietti, S.H. Yu, Adv. Funct. Mater. 17 (2007) 637.
[74] S.C. Tang, S. Vongehr, X.K. Meng, J. Phys. Chem. C 114 (2010) 977.
[75] B. Hu, Y. Zhao, H.Z. Zhu, S.H. Yu, ACS Nano 5 (2011) 3166.
[76] R.J. Cui, C. Liu, J.M. Shen, D. Gao, J.J. Zhu, H.Y. Chen, Adv. Funct. Mater. 18 (2008) 2197.
[77] M. Soorholtz, R.J. White, M.M. Titirici, M. Antonietti, R. Palkovits, F. Schüth, in preparation (2012).

[78] R.A. Periana, D.J. Taube, S. Gamble, H. Taube, T. Satoh, H. Fujii, Science 280 (1998) 560.
[79] D. Wolf, Angewandte Chemie-International Edition 37 (1998) 3351.
[80] B.L. Conley, W.J. Tenn, K.J.H. Young, S. Ganesh, S. Meier, V. Ziatdinov, O. Mironov, J. Oxgaard, J. Gonzales, W.A. Goddard, R.A. Periana, Methane Functionalization, Wiley-VCH Verlag GmbH & Co. KGaA, 2006.
[81] R.A. Periana, G. Bhalla, W.J. Tenn, K.J.H. Young, X.Y. Liu, O. Mironov, C.J. Jones, V.R. Ziatdinov, J. Mol. Catal. A Chem. 220 (2004) 7.
[82] X.M. Sun, Y.D. Li, Adv. Mater. 17 (2005) 2626.
[83] X.C. Song, Y. Zhao, Y.F. Zheng, E. Yang, J. Fu, Y. He, Cryst. Growth Des. 8 (2008) 1823.
[84] W.Z. Wang, S.L. Xiong, L.Y. Chen, B.J. Xi, H.Y. Zhou, Z.D. Zhang, Cryst. Growth Des. 6 (2006) 2422.
[85] Z. Fang, K.B. Tang, S.J. Lei, T.W. Li, Nanotechnology 17 (2006) 3008.
[86] J.C. Yu, X.L. Hu, Q. Li, L.Z. Zhang, Chem. Commun. (2005) 2704.
[87] W.Z. Wang, S. Qiu, B.J. Xi, L.Y. Chen, S.L. Xiong, Z.D. Zhang, Chem. Asian J. 3 (2008) 834.
[88] X.L. Hu, J.C. Yu, J.M. Gong, J. Phys. Chem. C 111 (2007) 5830.
[89] G.Y. Chen, B. Deng, G.B. Cai, W.F. Dong, W.X. Zhang, A.W. Xu, Cryst. Growth Des. 8 (2008) 2137.
[90] W.Z. Wang, L. Sun, Z. Fang, L.Y. Chen, Z.D. Zhang, Cryst. Growth Des. 9 (2009) 2117.
[91] B. Deng, A.W. Xu, G.Y. Chen, R.Q. Song, L.P. Chen, J. Phys. Chem. B 110 (2006) 11711.
[92] L.B. Luo, S.H. Yu, H.S. Qian, T. Zhou, J. Am. Chem. Soc. 127 (2005) 2822.
[93] L.B. Luo, S.H. Yu, H.S. Qian, J.Y. Gong, Chem. Commun. (2006) 793.
[94] Y.J. Zhan, S.H. Yu, J. Am. Chem. Soc. 130 (2008) 5650.
[95] J.Y. Gong, L.B. Luo, S.H. Yu, H.S. Qian, L.F. Fei, J. Mater. Chem. 16 (2006) 101.
[96] J.Y. Gong, S.H. Yu, H.S. Qian, L.B. Luo, T.W. Li, J. Phys. Chem. C 111 (2007) 2490.
[97] P. Makowski, R.D. Cakan, M. Antonietti, F. Goettmann, M.M. Titirici, Chem. Commun. (2008) 999.
[98] L. Zhao, X. Chen, X. Wang, Y. Zhang, W. Wei, Y. Sun, M. Antonietti, M.-M. Titirici, Adv. Mater. 22 (2010) 3317.
[99] J. Popovic, R. Demir-Cakan, J. Tornow, M. Morcrette, D.S. Su, R. Schloegl, M. Antonietti, M.-M. Titirici, Small 7 (2011) 1127.
[100] Z.H. Wen, J. Liu, J.H. Li, Adv. Mater. 20 (2008) 743.
[101] D. Deng, J.Y. Lee, Chem. Mater. 20 (2008) 1841.
[102] X. Cao, K.S. Ro, M. Chappell, Y. Li, J. Mao, Energy Fuels 25 (2010) 388.
[103] A.B. Fuertes, M.C. Arbestain, M. Sevilla, J.A. Macia-Agullo, S. Fiol, R. Lopez, R.J. Smernik, W.P. Aitkenhead, F. Arce, F. Macias, Aust. J. Soil Res. 48 (2010) 618.
[104] F. Perez Caballero, C. Falco, F. Babonneau, C. Gervais, M.M. Titirici, G. Laurent, N. Baccile, Langmuir 27 (23) (2011) 1446027 (23), (2011) 14460.
[105] S. Steinbeiss, G. Gleixner, M. Antonietti, Soil Biol. Biochem. 41 (2009) 1301.
[106] M.C. Rillig, M. Wagner, M. Salem, P.M. Antunes, C. George, H.-G. Ramke, M.-M. Titirici, Appl. Soil Ecol. 45 (2010) 238.
[107] V.A. Loganathan, Sandeep Kumar, R.B. Gupta, M.O. Barnett, J. Environ. Manage. 92 (2011) 2504.
[108] F.-S. Zhang, Z. Liu, J. Hazard. Mater. 167 (2009) 933.
[109] K. Ro, K. Sun, M. Guo, J. Novak, H.M. Mashayekhi, B. Xing, Bioresour. Technol. 102 (2011) 5757.
[110] M. Sevilla, A.B. Fuertes, Energy Environ. Sci. 4 (2011) 1765.
[111] J.W. Lee, M.S. Balathanigaimani, H.C. Kang, W.G. Shim, C. Kim, H. Moon, J. Chem. Eng. Data 52 (2007) 66.

[112] Y. Yürüm, A. Taralp, N. Veziroglu, Int. J. Hydrogen Energy 34 (2009) 3784.
[113] M. Sevilla, A.B. Fuertes, R. Mokaya, Energy Environ. Sci 4 (2011) 1400–1410.
[114] M.-M. Titirici, M. Antonietti, A. Thomas, Chem. Mater. 18 (2006) 3808.
[115] X. Wang, C. Hu, Y. Xiong, H. Liu, G. Du, X. He, J. Power Sources 196 (2011) 1904.
[116] J. Matos, M. Rosales, R. Demir-Cakan, M. M. Titirici, Appl. Catal. A. Gen. 386, 140.
[117] G. Yu, Y. Pei, B. Sun, S. Xie, S. Yan, M. Qiao, K. Fan, X. Zhang, B. Zong, J. Am. Chem. Soc. 132 (2009) 935.
[118] W.J. Zhang, H.W. Liang, Y.-N. Ma, X. Cao, Q. Guan, W. Xu, S.-H. Yu, ACS Nano 5 (2011) 8148.
[119] X. Liang, M. Zeng, C. Qi, Carbon 48 (1844).
[120] H. Duan Zhang, S. Li, Y. Lin, Langmuir 26 (9) (2010) 6676.
[121] G. Xiao Chen, T.-S. Wang, Q.-Y. Ouyang, Li-H. Qi, Y. Ma, P. Gao, C. Zhu, H.-B. Jin, J. Phys. Chem. C 115 (2011) 13603.
[122] J. Zhong, F. Chen, J. Zhang, J. Phys. Chem. C 114 (2010) 933–939.
[123] Q. Wang, L. Chen, H. Li, X. Huang, Carbon 39 (2001) 2211.
[124] K. Tang, J.R. White, L. Yu, M.-M. Titirici, M. Antonietti, J. Maier, Adv. Energ. Mater. 2 (2012) 873–877.
[125] M. Eswaramoorthy, K. Balasubramanyam, X. Yang, F. Jiang, Z.X. Chen, X.P. Ai, Y.L. Cao, J. Phys. Chem. C 111 (2007) 14067.
[126] X.H. Huang, J.P. Tu, C.Q. Zhang, X.T. Chen, Y.F. Yuan, H.M. Wu, Electrochim. Acta 52 (2007) 4177.
[127] J. Popovic, R.D. Cakan, J. Tornow, M. Morcrette, D. Su, R. Schlögl, M.-M. Titirici, Small 8 (2011) 1127.
[128] B.E. Conway, Electrochemical Supercapacitors: Scientific Fundamentals and Technological Applications, New York, 1999.
[129] L. Wei, M. Sevilla, A.B. Fuertes, R. Mokaya, G. Yushin, Adv. Energ. Mater. 1 (2011) 356.
[130] L. Wang, Y. Guo, B. Zou, C. Rong, X. Ma, Y. Qu, Y. Li, Z. Wang, Bioresour. Technol. 102 (2011) 1947.
[131] G.Y. Chmiola, Y. Gogotsi, C. Portet, P. Simon, P.L. Taberna, Science 313 (2006) 1760.
[132] E. Raymundo-Piñero, K. Kierzek, J. Machnikowski, F. Béguin, Carbon 44 (2006) 2498.
[133] E. Frackowiak, G. Lota, J. Machnikowski, C. Vix-Guterl, F. Béguin, Electrochim. Acta 51 (2006) 2209–2214.
[134] H. Zhu, X. Wang, F. Yang, X. Yang, Adv. Mater. 23 (2011) 2745.
[135] W. Jacques, Vol. 555511 (Ed.: U. patent), US, 1896.
[136] J.P. Paraknowitsch, A. Thomas, M. Antonietti, Chem. Mater. 21 (2009) 1170.
[137] M. Sevilla, G. Lota, A.B. Fuertes, J. Power Sources 171 (2007) 546.
[138] M. Sevilla, T. Valdés-Solís, E. Morallón, A.B. Fuertes, M. Sevilla, C. Sanchís, Electrochim. Acta 54 (2009) 1134.
[139] J.Y. Gong, S.R. Guo, P. Jiang, M. Wu, Y. Lu, S.H. Yu, Adv. Funct. Mater. 18 (2008) 872.

Chapter 13

Porosity and Adsorption Properties of Single-Wall Carbon Nanohorn

Shigenori Utsumi
Faculty of Systems Engineering, Tokyo University of Science, Suwa, Japan

Tomonori Ohba
Graduate School of Science, Chiba University, Chiba, Japan

Hideki Tanaka
Department of Chemical Engineering, Kyoto University, Kyoto, Japan

Koki Urita
Faculty of Engineering, Nagasaki University, Nagasaki, Japan

Katsumi Kaneko
Research Center for Exotic Nanocarbons, Shinshu University, Nagano, Japan

Chapter Outline

13.1. Single-Walled Carbon Nanohorn as One of Carbon Nanotubilites 402
13.2. Nanoporosity 407
13.2.1. Interaction Potential View of Single SWCNH Particle 407
13.2.2. Donation of Nanowindows to SWCNH 410
13.2.3. Accessible Sites of an SWCNH Assembly 411
13.2.4. Surface Chemistry of SWCNH 414
13.3. Adsorption of Molecules on SWCNH 415
13.3.1. Adsorption of Vapors 416
13.3.2. Adsorption of Supercritical Gases 419
13.4. Adsorption of Quantum Fluid 423
13.5. Nanoscale Materials Growth Fields 426
13.6. Conclusion 430
References 430

Novel Carbon Adsorbents. DOI: 10.1016/B978-0-08-097744-7.00013-2
Copyright © 2012 Elsevier Ltd. All rights reserved.

13.1. SINGLE-WALLED CARBON NANOHORN AS ONE OF CARBON NANOTUBILITES

The presence of single-walled carbon nanotube (SWCNT) was published by Iijima and Ichihashi [1], and Bethune et al. [2] with the high-resolution transmission electron microscopy (HR-TEM) in 1993. SWCNT has a rolled structure of graphene with caps at the both sides of the tube structure, as shown in Fig. 13.1. SWCNTs form the bundle structure of the hexagonal symmetry, being shown by X-ray diffraction except for so-called "super-growth" SWCNT produced by Hata et al. [3]. The cross-sectional TEM view of the SWCNT bundle is also shown in Fig. 13.1. The tube diameter of SWCNT produced by arc-discharge method and laser ablation method is in the range of 0.4–2 nm, while chemical vapor deposition (CVD) method often gives a bolder SWCNT whose diameter is in the range of 0.5–6 nm.

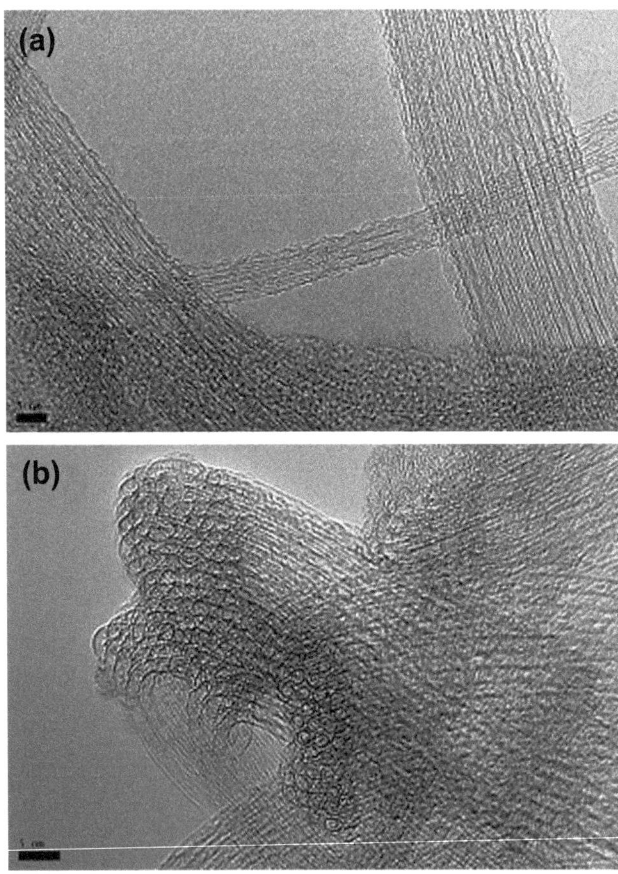

FIGURE 13.1 HR-TEM images of SWCNT bundles. (a) Side view of the well-aligned SWCNT bundle. (b) Cross section of the hexagonal bundle. Bar lengths are 5 nm.

The tube diameter d (nm) and the radial breathing mode (RBM) vibration frequency ω of Raman scattering spectrum observed below 800 cm^{-1} can be correlated with the electronic structure, which is given by Eqn (13.1) [4]:

$$d = \frac{248}{\omega} \tag{13.1}$$

Raman spectroscopy provides crystallinity of the graphene structure in addition to the tube diameter. Figure 13.2 shows Raman spectra of SWCNT prepared by different methods above 1000 cm^{-1}. SWCNTs prepared by the arc-discharge and the laser ablation methods give a sharp peak at 1600 cm^{-1}, which is called a G-band (graphite band). On the other hand, a weak peak is observed at 1350 cm^{-1}, which is named a D-band (defect band). The D-band can be observed, when the selection rule is broken due to an imperfection of the crystalline structure. Therefore, the intensity ratio of G and D, (I_G/I_D), is often used to show the crystallinity [5].

The I_G/I_D ratio is also associated with the crystallite size of the graphitic structure [6]. SWCNT prepared by CVD has strong D-band, being different from SWCNTs by the arc-discharge and the laser ablation methods. Although the CVD method can produce massively SWCNT, the crystallinity is not enough compared with that from the arc-discharge or laser ablation methods.

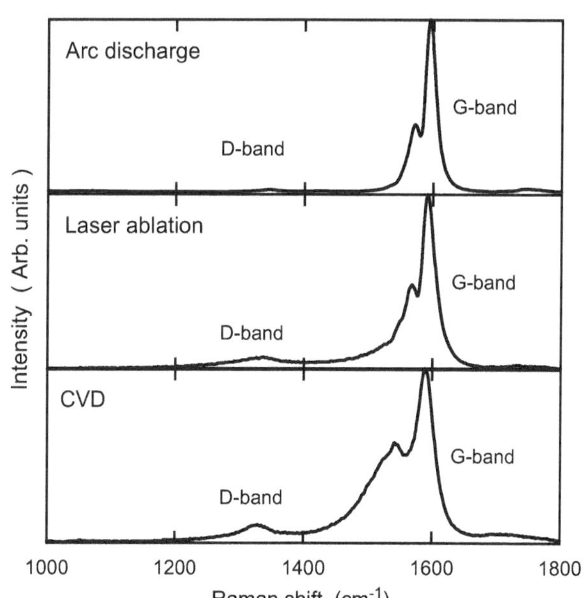

FIGURE 13.2 Raman spectra in the region of G- and D-bands of SWCNTs prepared by arc-discharge, laser ablation, and CVD methods.

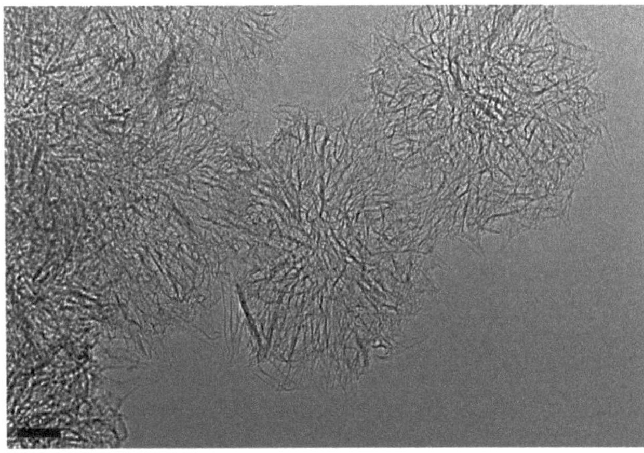

FIGURE 13.3 HR-TEM image of dahlia-flower-like SWCNH assembly. Bar length is 20 nm.

Harris et al. found a single-walled carbon nanohorn (SWCNH) structure in the electron-radiated activated carbon in 1994 [7]. Iijima et al. succeeded to produce pure SWCNH from graphite by CO_2-laser ablation [8]; the production rate was 10 g/h in 2000 and then SWCNH has a merit for chemical studies. Figure 13.3 shows an HR-TEM image of SWCNH. This SWCNH has a dahlia-flower-type assembly structure consisting of pencil-like particles. Although there is a bud-type SWCNH, the porosity is much less than the dahlia-flower-type SWCNH [9]. Here, only the dahlia-flower-type SWCNH is described. The schematic model of a primary SWCNH particle is shown in Fig. 13.4. The SWCNH is composed of tube and cone parts; the diameter and length of the tube are 2–6 nm and $\approx$40 nm, respectively. The SWCNH particle has only one edge structure, being called a cone structure. The cone structure is more defective than the cap of SWCNT, as mentioned later [10]. The Raman spectrum of SWCNH in the G–D band region is quite different from that of well-crystalline SWCNT [5]. Figure 13.5 shows the Raman spectra of SWCNH and activated carbon fiber (ACF) which are highly disordered carbon of huge porosity. The G-band of SWCNH is very broad and its D-band is strong. The feature of Raman spectrum of SWCNH is similar to that of ACF. Therefore, the

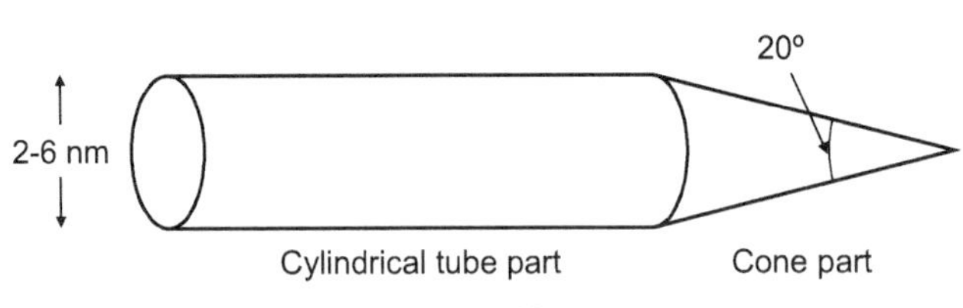

FIGURE 13.4 Schematic model of a primary SWCNH particle.

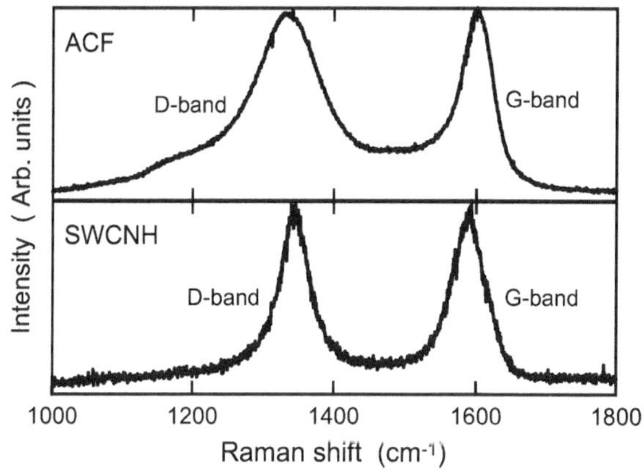

FIGURE 13.5 Raman spectra of SWCNH and ACF in the G- and D-band region.

fundamental carbon hexagon structure of SWCNH should be highly defective, being very different from SWCNT.

Although normal Raman scattering spectroscopy cannot provide more information except for the G- and D-band ratio, Fujimori et al. showed that

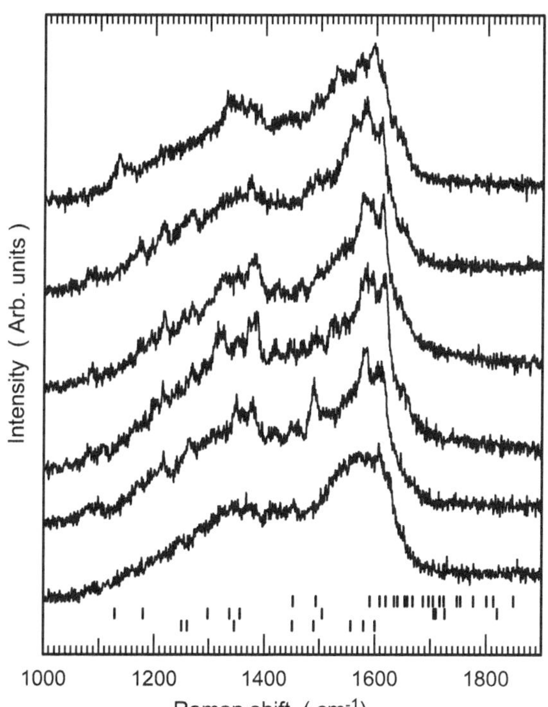

FIGURE 13.6 SERS spectra of SWCNH in the 1000–1900 cm^{-1} region obtained with 532-nm excitation. Each spectrum was obtained on the same focal point. Bars indicate the peak position of the calculated modes of SWCNT IMJs (top) [14], of the localized modes of graphene with a Stone–Wales defect (middle) [14], and of the calculated phonon modes of graphite (bottom) [15].

surface-enhanced Raman scattering (SERS) technique is efficient to detect pentagons and heptagons and the pentagon–heptagon pair (so-called Stone–Thrower–Wales (STW) defect) [11–13]. Figure 13.6 shows the SERS spectra of SWCNH obtained with 532-nm excitation. In addition to normal G- and D-bands, and RBM peaks, new peaks were observed in the wave number range of 200–1700 cm^{-1}, which are tentatively assigned to the vibrational mode due to topological defects such as pentagons and heptagons. Very recently, Fujimori et al. succeeded to show the presence of the hybrid structure of different sized caps in the cone site.

This highly defective nature of SWCNH leads to n-type semiconductivity in SWCNH. The cap structure of SWCNT needs pentagons in the cap. The cone site of SWCNH should contain many pentagons, as indicated by the SERS study. The pentagon has an electron donation nature according to the π-electron theory in organic chemistry. Therefore, SWCNH has highly concentrated quasi-free electrons than SWCNT and other carbon materials. Ordinary sp^2-based carbon materials are electron deficient, exhibiting p-type semiconductivity; the predominant carriers even in SWCNT are holes. The distinguishing of n- and p-types can be carried out by the response on

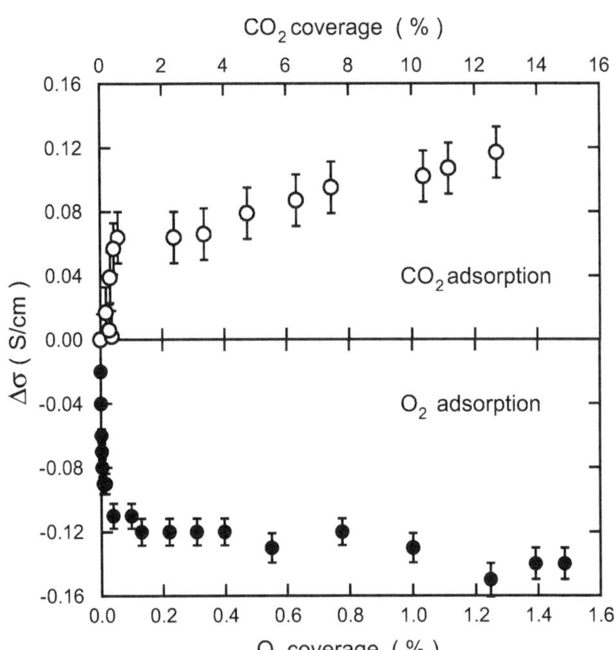

FIGURE 13.7 DC electrical conductivity changes of compressed SWCNH pellets on adsorption of O_2 and CO_2 at 303 K. Horizontal axes are expressed by O_2 or CO_2 coverage on SWCNHs, respectively.

exposing to O_2. As an O_2 molecule has a large electron affinity, O_2 molecules rob electrons from the solid on chemisorption. If the solid is p-type semiconductor, the solid becomes more electron deficient to increase positive hole carriers; the electrical conductivity increases on O_2 chemisorption. On the contrary, chemisorption of O_2 on the n-type solid consumes carrier electrons, decreasing the electrical conductivity. Figure 13.7 shows the DC electrical conductivity changes of compressed SWCNH pellets on adsorption of O_2 and CO_2. Adsorption of O_2 evidences a rapid reduction in the electrical conductivity, being a representative response of n-type semiconductor. On the other hand, adsorption of CO_2, being an electron donor, increases the electrical conductivity. Thus, SWCNH has a unique electronic property in carbon materials. On the other hand, oxidized SWCNH having many holes shows the p-type response on O_2 adsorption [16].

13.2. NANOPOROSITY

13.2.1. Interaction Potential View of Single SWCNH Particle

Single-walled carbons are unique solids. The frame structure consists of graphene which has the large surface area of 2630 m^2/g. The carbon atomic density in the single-walled carbons is high (38.2 $atom/nm^2$) due to the carbon hexagon network structure, being a great merit for interaction with atoms, molecules, and ions. The bent structure of graphene has two surfaces whose nanoscale curvature signs are opposite to each other, as shown in Fig. 13.8. The concave surface of the negative curvature has a stronger interaction potential than the convex surface of the positive curvature. All carbon atoms are exposed to both surfaces and there is no bulk atom in graphene. Each carbon atom can play a role in both the surface processes in a different way

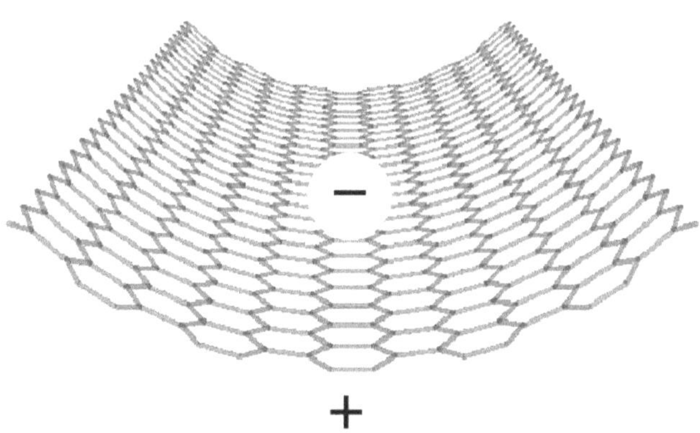

FIGURE 13.8 Schematic model of the bent graphene structure which has two surfaces with negative and positive curvatures.

depending on the curvature sign. Single-walled carbons are not bulk solids but "surface solids". Figure 13.9 shows the dispersion interaction potential profiles of N_2 and SWCNT surfaces of the negative (internal) and positive (external) curvatures [17]. Surprisingly, the interaction potential depends remarkably on the curvature sign. As the average molecule–carbon atom distance in the concave (negative) surface is shorter than that in the convex (positive) surface, the internal surface gives a much deeper interaction potential well than the external one. Here we use analytical potential functions for the interaction potential profile calculation. The interaction potential (Φ_{int}) of internal SWNCT is expressed as

$$\Phi_{\text{int}} = \pi^2 \rho_s \varepsilon_{\text{sf}} \sigma_{\text{sf}}^2 \left[\frac{63}{32} \frac{F(-4.5,\ -4.5,\ 1.0;\ \beta^2)}{\left[R^*(1-\beta^2)\right]^{10}} - 3\frac{F(-1.5,\ -1.5,\ 1.0;\ \beta^2)}{\left[R^*(1-\beta^2)\right]^{4}} \right] \tag{13.2}$$

Similarly, the interaction potential (Φ_{ext}) of external SWCNT is expressed as

$$\Phi_{\text{ext}} = \pi^2 \rho_s \varepsilon_{\text{sf}} \sigma_{\text{sf}}^2 \left[\frac{63}{32} \frac{F(-4.5,\ -4.5,\ 1.0;\ \delta^2)}{\left[R^*(1-\delta^2)\right]^{10}} \delta^{11} - 3\frac{F(-1.5,\ -1.5,\ 1.0;\ \delta^2)}{\left[R^*(1-\delta^2)\right]^{4}} \delta^5 \right] \tag{13.3}$$

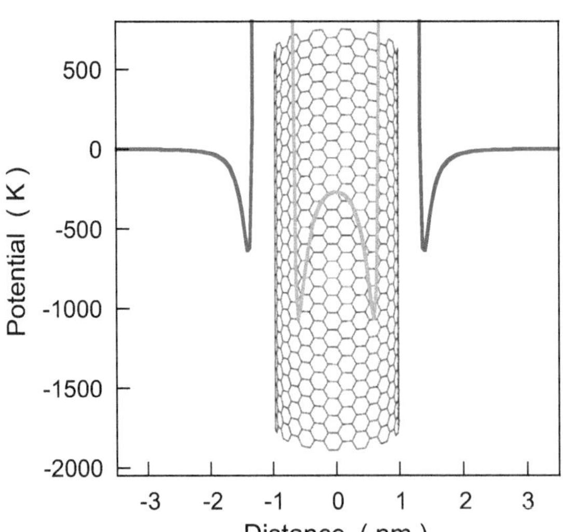

FIGURE 13.9 Interaction potential profile of an N_2 molecule with SWCNT surfaces of the negative (internal) and positive (external) curvatures.

Here, if r is the distance between an adsorbate and center of the tube, R the radius of SWCNT to the center of carbon atoms, we obtain $\beta = r/R$, $R^* = R/\sigma_{sf}$, $\delta = R/r$, and $\rho_s = 38.2\ \text{nm}^{-2}$. $F(a,b,c;d)$ is the hypergeometric function (a, b, c, and d are constants) [18]. The internal and external potential minima appear at the monolayer adsorption site. However, the potential energy at the internal monolayer positions is almost a double value of that at the external monolayer sites, and the potential energy difference becomes deeper with decreasing tube radius R. This is because the overlapping effect of the internal molecular potential received from a cylinder pore wall becomes remarkable in a thin tube. When the tube diameter equals an infinite value, the surface is completely flat and both side surfaces are mutually equivalent.

As mentioned above, SWCNH has a cone-shaped cap on one side. The cone structure can be approximated by the hybrid of caps of different curvatures. The interaction potential difference between the positive and negative curvature surfaces is more marked than that of the tube. Figure 13.10 shows the interaction between H_2 and the carbon cap [19]. As the interaction of H_2 with the internal cap is strong enough to adsorb supercritical H_2, the cap structure is promising even in adsorption of supercritical gases.

Also the cone part can be approximated by the regular cone structure. When we calculate the mutual interaction potentials between component carbon atoms and the target molecule (so-called atomistic calculation method), the reasonable interaction potential can be obtained to predict adsorption process, as shown in quantum molecular sieving of H_2 and D_2 on SWCNH. If we neglect the unknown structure of the opposite side of the SWCNH particle, there are four sites of internal cone space, internal tube space, external surface, and

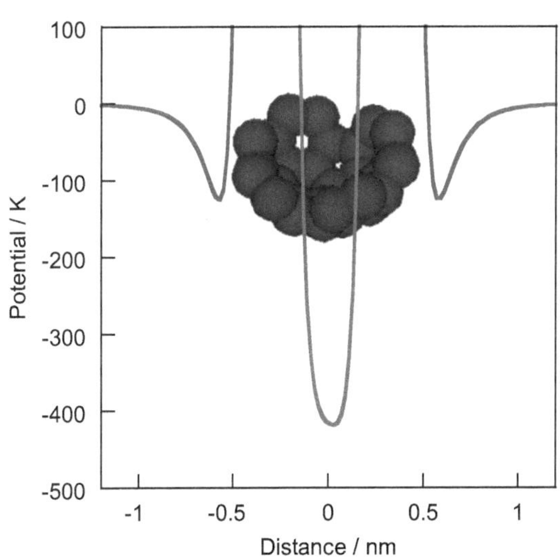

FIGURE 13.10 Interaction potential profile between H_2 and the carbon cap.

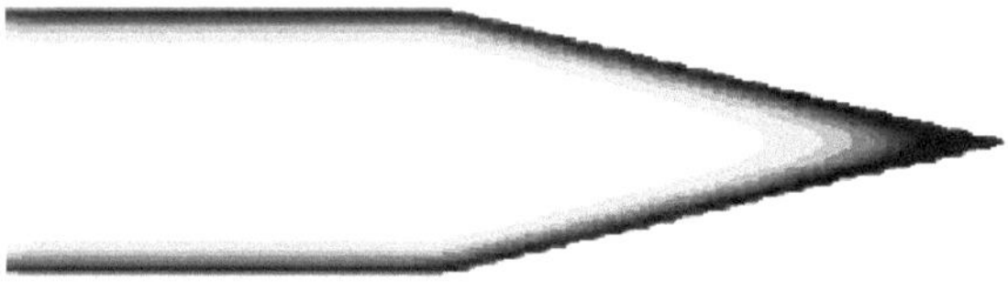

FIGURE 13.11 Schematic model of interaction potential distribution inside a primary SWCNH particle. The darker color indicates the deeper potential site.

external cone surface. The attractive interaction strength for a molecule is in the order of the internal cone surface, internal tube surface, external surface, and external cone surface. If the tube diameter is smaller than the trilayer size of the target molecule, the residual space after completion of monolayer adsorption on the internal wall surfaces of the cone and tube gives the stronger attractive interaction potential than the external surfaces. Figure 13.11 shows the distribution of interaction potential inside a primary SWCNH particle.

13.2.2. Donation of Nanowindows to SWCNH

Although we considered the internal space of an SWCNH particle, the graphene wall of SWCNH has no sufficient accessible windows (holes) for molecules. The almost close situation of SWCNH produced is evidenced by the particle density measurement using high-pressure buoyancy method with He gas up to 8 MPa [20]. Opening of the wall of SWCNH can be done by oxidation with O_2 or H_2O_2 for example. Utsumi et al. established a highly reproducible donation method of nanowindows in the SWCNH wall using oxidation with O_2 [21]. They controlled the oxidation temperature, the flow rate of O_2 gas, heating and cooling conditions, and so on. Here, they used the SWCNH assembly sample whose closing state was evidenced by the particle density method using buoyancy measurement. Figure 13.12 shows the BET specific surface area of oxidized SWCNH as a function of the oxidation temperature. The specific surface area increases with elevating the oxidation temperature up to 823 K, giving the maximum value of 1420 m^2/g which is four times larger than that of nonoxidized SWCNH. On the other hand, the surface area drops above 823 K, because the overoxidation loses pore walls, particularly the cone part having the greatest interaction potential. The increase in the surface area is brought by the addition of nanowindows on the SWCNH graphene wall. The pore-size distribution analyses and the comparison plot technique of the N_2 adsorption isotherms of SWCNH oxidized at different temperatures against that of nonoxidized SWCNH reveal that the minimum temperature of the nanowindow donation on the SWCNH wall is 626 K and the exact internal pore volume of SWCNH is 0.49 mg/L. Furthermore, TG-DTA experiments on SWCNHs oxidized at different temperatures clarified that SWCNH is composed of four carbon components: amorphous, defective, tubular, and graphitic carbons.

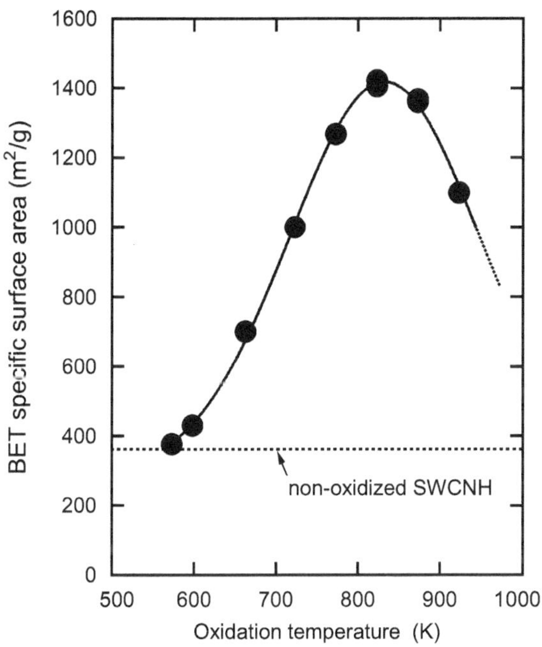

FIGURE 13.12 Relationship between the BET specific surface area of oxidized SWCNH and the oxidation temperature.

These carbon components were determined from the weight loss of TG profiles as 2.5 wt% of amorphous carbon, 15 wt% of defective carbon, 70 wt% of tubular carbon, and 12 wt% of graphitic carbon, respectively. Here, defective carbon indicates pentagons and heptagons situated at the cone cap and bending sites. It is clear that comparatively many defective carbons exist in an SWCNH assembly from the viewpoint of TGA, coinciding with the results of SERS Raman spectroscopy and TEM observations.

13.2.3. Accessible Sites of an SWCNH Assembly

SWCNH particles are associated with each other to form the dahlia-flower-type assembly. Hence, we must take into account interstitial porosity in the SWCNH assembly. The SWCNH assembly sample gives a diffuse X-ray diffraction peak corresponding to the mutually arrayed tube–tube structure. Ohba et al. proposed the following partially bundled SWCNH assembly structure, as shown in Fig. 13.13 [22]. They simulated the N_2 adsorption isotherm of the SWCNH assembly at 77 K with GCMC simulation, comparing it with the experimental adsorption isotherm. The fitting procedure indicates the presence of micropores as the interstitial pores. Then we can approximate the interstitial nanoporosity (0.11 mL/g) by the summation of microporosity of 0.096 mL/g and mesoporosity of 0.014 mL/g [20]. Consequently, there are two porosities of

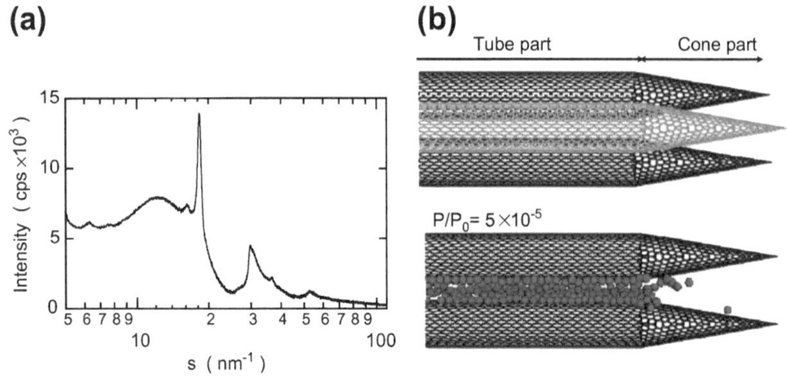

FIGURE 13.13 (a) XRD pattern of SWCNH. (b) Snapshots of N_2 molecules in interstitial pores of the nearest-neighbor tube wall-to-wall distance $D = 0.7$ nm. Interstitial pore is surrounded by three SWCNH particles, as shown in the top figure.

micropores and mesopores in addition to the external surface in the case of the nonoxidized SWCNH assembly. If we donate the nanowindows in the SWCNH wall, the internal nanoporosity (0.36 mL/g [23]) must be taken into account. Figure 13.14 shows the approximated model of interstitial nanoporosity of the nonoxidized SWCNH assembly.

The nanowindow-donated SWCNH by oxidation has four times larger nanopore volume than nonoxidized SWCNH. This is because the tube diameter is in the range of 2–6 nm. Table 13.1 summarizes the pore volume and the surface area of closed- and open-SWCNH samples [23]. In this article, SWCNH without opening treatment and SWCNH oxidized for opening are denoted SWCNH and open-SWCNH, respectively.

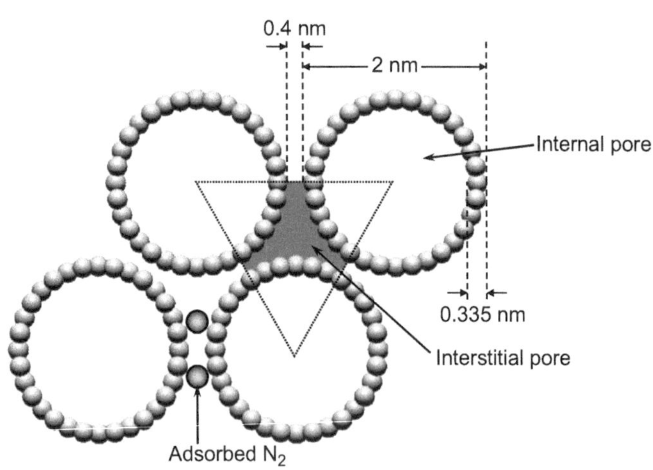

FIGURE 13.14 Approximated interstitial nanoporosity model of a closed SWCNH assembly.

TABLE 13.1 Nanopore Structural Parameters of SWCNH [23]

	Surface Area (m^2/g)	Nanopore Volume (mL/g)
SWCNH	200	0.11
Open-SWCNH	518	0.47

Another route for opening of SWCNH is possible. Mechanical grinding is a promising method for opening. Urita et al. [24] ground both closed- and open-SWCNH samples with an agate mortar in air for 3 min. The analysis of C1s XPS peak showed that sp^3 carbon atoms increased with the grinding procedures from 26% to 35% for closed SWCNH and from 32% to 60% for open-SWCNH on the nine-times grinding treatment. Therefore, grinding produces the dangling bonds, which react with O_2 to form surface oxygen functional groups. The N_2 adsorption and He buoyancy measurement of ground SWCNH provide changes of particle density and opening ratio of the SWCNH tubilites, as shown in Fig. 13.15. When we ground open-SWCNH, the particle density and opening ratio were 2.09 g/mL and 0.82, respectively. Thus, the mechanical treatment is efficient.

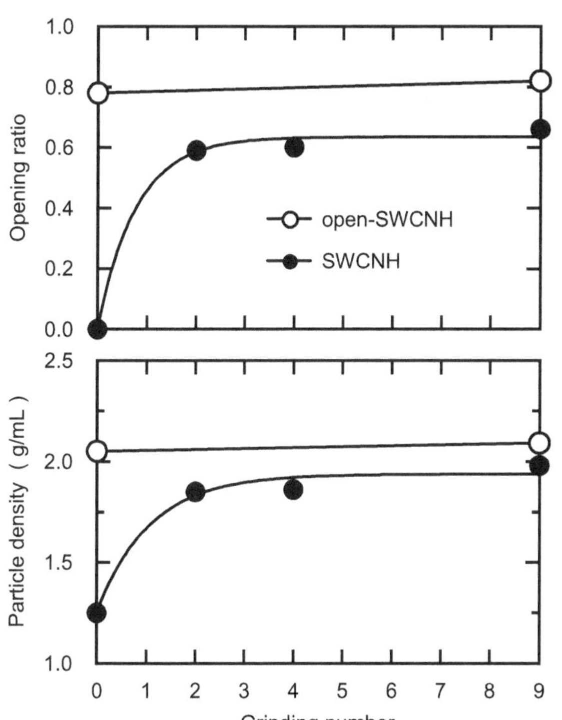

FIGURE 13.15 Changes in particle density and opening ratio of SWCNH and open-SWCNH depending on grinding number.

13.2.4. Surface Chemistry of SWCNH

SWCNH is produced from pure graphite without metallic catalysts, as mentioned before. Basically SWCNH samples have no metallic impurity. X-ray photoelectron spectroscopy (XPS) cannot detect any signals due to metallic impurities. We can detect O1s peak in the XPS of SWCNH even before oxidation treatment for donation of nanowindows, stemming from surface oxygen functional groups. The oxygen content increases after the oxidation treatment. Figure 13.16 shows the deconvolution results of C1s and O1s XPS spectra of SWCNH and open-SWCNH [5]. Some subpeaks assigned to oxygen-containing functional groups are observed in the relatively high binding energy region in C1s XPS spectrum of SWCNH, which are assigned to C–O at 286.1 eV and C = O at 287.0 eV. After oxidation treatment, a new peak assigned to the COO group at 289.0 eV is observed for open-SWCNH, indicating that the COO group is mainly produced by the oxidation treatment. Furthermore, O1s spectra of SWCNH and open-SWCNH consist of two peaks which are assigned to oxygen's single bonding (C–O; 533.8 eV) and double bonding (C = O; 532.4 eV) to carbon. The intensity ratio of C–O to C = O (C–O/C = O) increases by the oxidation treatment, which is consistent with the C1s results. Special attention should be paid to a subpeak at 285.5 eV, which is assigned to sp^3 single-bonding carbons (C–C). This peak intensity increases by the oxidation treatment, which is brought by the degradation of the C = C peak. Thus, the considerable amount of the C–C single bonding is present in SWNH to form a dahlia-flower-like assembly structure, accompanying with a strong D-band in Raman spectrum of SWNH.

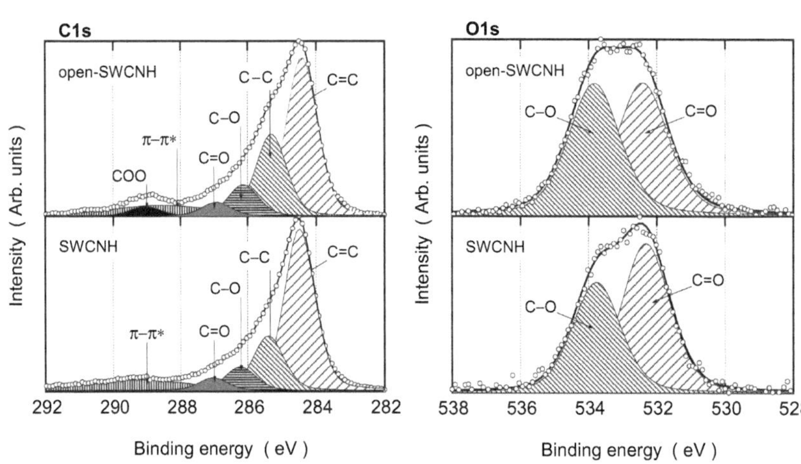

FIGURE 13.16 Deconvolved C1s (left) and O1s (right) XPS spectra of SWCNH and open-SWCNH. Open circles and solid lines indicate the experimental data and fitting curves, respectively.

Hattori et al. fluorinated the SWCNH to form semi-ionic and covalent C–F bonds to control the surface properties. Fluorination can introduce the perfect hydrophobicity to SWCNH samples and thereby molecular adsorptivities of the fluorinated SWCNH must be studied [25].

13.3. ADSORPTION OF MOLECULES ON SWCNH

Open-SWCNH has a large nanoporosity of different interaction potential for molecules. Molecules favor the pore which has the strongest attractive interaction potential; molecules are adsorbed at first in the pores of the strongest interaction potential and then adsorption occurs in other pores having weaker interaction potential. Here we describe physical adsorption which stems from the dispersion attractive interaction. In the case of physical adsorption, we need to distinguish vapors and supercritical gases, because the adsorption features of vapors and supercritical gases are very different from each other. The gas whose critical temperature T_C is lower than adsorption temperature is named a supercritical gas. Although there is a coexistent region between liquid and vapor below T_C, there is no coexistent region above T_C. Therefore, supercritical gas cannot equilibrate with liquid, and thereby the saturated vapor pressure P_0 is not defined. We measure adsorption isotherms of gases under the equilibrium conditions. We use the relative pressure P/P_0, where P is the equilibrium pressure, for description of the adsorption isotherm of vapors. The adsorption isotherms of vapors obey the IUPAC types. However, we cannot express the adsorption isotherms of supercritical gases with P/P_0. The attractive interaction between the supercritical gas and surface is not strong enough to induce a measurable adsorption below the atmospheric pressure. Hence, we apply the gas pressure higher than the atmospheric pressure to enhance surface adsorption. Application of high pressure induces both surface adsorption and the intermolecular interaction which varies the attractive and repulsive nature depending on the absolute pressure. Strictly speaking, we need to use the fugacity to take into account the intermolecular interaction factor instead of pressure. Consequently, the high-pressure adsorption isotherm of supercritical gases does not obey the IUPAC classification. Murata et al. [26] proposed three types of high-pressure adsorption isotherms, as shown in Fig. 13.17: Henry, Langmuir, and Maximum types; this order indicates the attractive interaction strength. For example, if we have the M-type isotherm, supercritical gas–solid surface interaction is strong enough to induce a considerable adsorption; the repulsive molecule–molecule interaction must be taken into account. Thus, we need to describe vapor adsorption and supercritical gas adsorption in a different way. Quantum nature must be considered in adsorption of light molecules at low temperature. An H_2 molecule is a light molecule, and the fluctuation extent at low temperature due to the quantum nature is a subnanometer order. In this case, we must treat the light H_2 molecule as quanta. The adsorption of quantum molecules must be described in terms of quantum interaction potential. In this

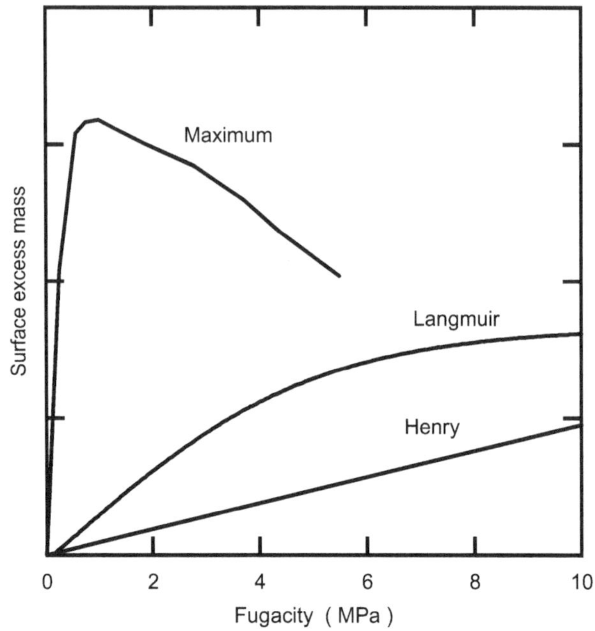

FIGURE 13.17 Three types of high-pressure adsorption isotherms of supercritical gases.

review, examples of adsorption of vapors, supercritical gases, and quantum molecules on SWCNH are described.

13.3.1. Adsorption of Vapors

Typical vapor adsorption isotherms of N_2 on SWCNH and open-SWCNH at 77 K are shown in Fig. 13.18 [21–23,27]. As the boiling temperature of N_2 is 77 K, the adsorption isotherm of N_2 at 77 K is a representative vapor adsorption isotherm. Both adsorption isotherms of SWCNH and open-SWCNH are close to IUPAC type II. However, the initial uptake is more predominant than ordinary type II isotherm. This is because both SWCNH samples have micropores and small mesopores. In particular, open-SWCNH has the large nanoporosity stemming from the internal tube spaces. Ohba et al. [22] simulated N_2 adsorption processes with GCMC method using the partially oriented model. N_2 molecules are adsorbed onto the cone sites and then onto the internal tube walls in the case of adsorption in the internal spaces, according to the snapshots, as shown in Fig. 13.19. The adsorption process predicted by GCMC simulation is quite reasonable. Liquid water and water vapor are coexistent under the ambient conditions. Then, the adsorption isotherm of water at the ambient temperature belongs to the vapor one. Figure 13.20 shows water vapor adsorption isotherms of SWCNH and open-SWCNH at 303 K [28]. Both

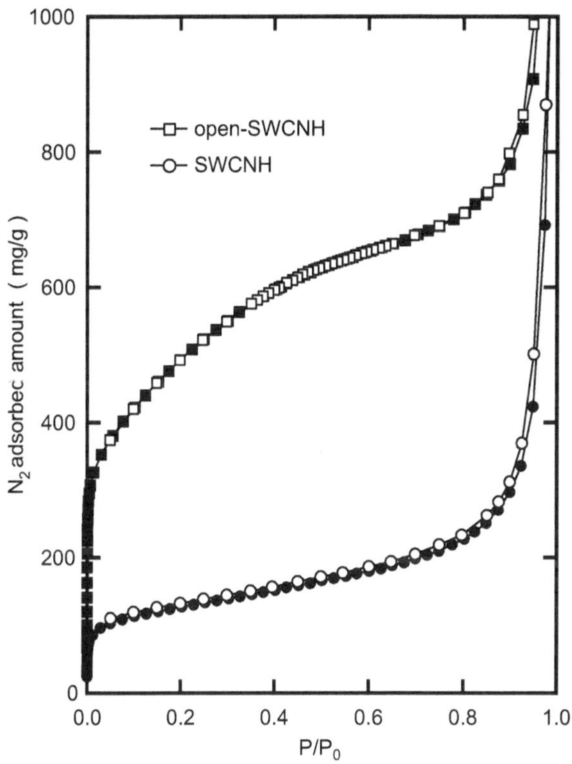

FIGURE 13.18 Typical vapor N_2 adsorption isotherms of SWCNH and open-SWCNH at 77 K.

isotherms have characteristics of the adsorption isotherm on the microporous carbons of hydrophobicity; adsorption below $P/P_0 = 0.3$ is negligible and predominant adsorption starts above $P/P_0 = 0.4$, accompanying with a clear adsorption hysteresis. If the micropore size is uniform, the adsorption rising and desorption dropping curves are almost vertical against the horizontal P/P_0

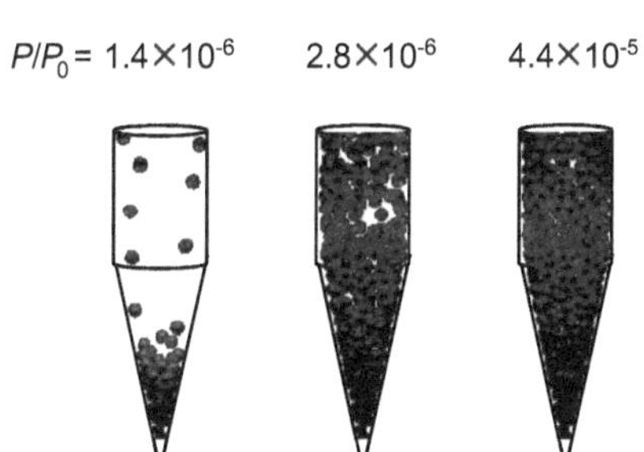

FIGURE 13.19 Snapshots of N_2 molecules adsorbed in an internal SWCNH space.

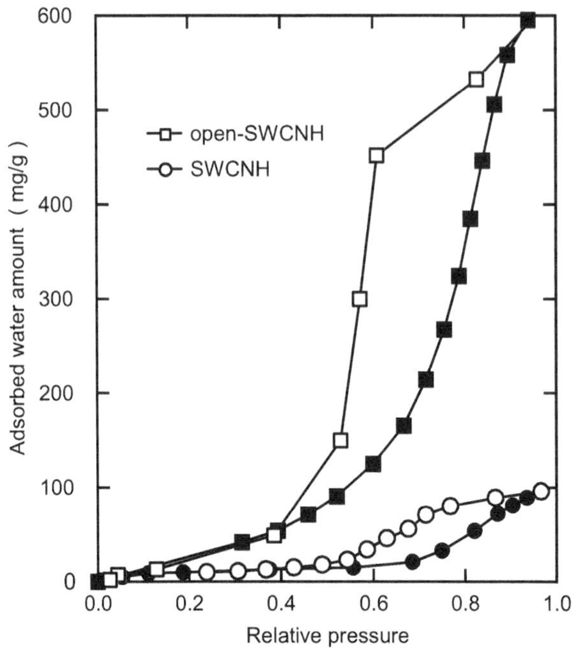

FIGURE 13.20 Vapor water adsorption isotherms of SWCNH and open-SWCNH at 303 K.

axis. The interstitial pores and internal pore spaces have considerable size distributions, inducing gradual adsorption and desorption above $P/P_0 = 0.5$. The pore volume of open-SWCNH is almost four times larger than that of SWCNH, and thereby water adsorption amount on open-SWCNH is much larger than that on SWCNH according to the porosity difference. The water vapor adsorption on SWCNH samples cannot be interpreted by classical capillary condensation theory originating from the Kelvin equation [29]. Ohba and Kaneko have elucidated the adsorption mechanism of water vapor on hydrophobic carbon micropores [30]. The water adsorption in hydrophobic carbon micropores proceeds through the molecular cluster-mediated micropore filling mechanism. Water molecules form hexamers or heptamers to be stabilized in the hydrophobic micropores, giving rise to the collective pore filling. The adsorption process is in the metastable state, whereas the desorption process is in the equilibrium. Nakamura et al. [31] examined the effect of equilibration on the adsorption hysteresis by changing the waiting time for measurement from 5 min to 16 h for single measurement point. The width of the adsorption hysteresis becomes narrower by waiting longer time for the adsorption measurement. The most important feature is that only adsorption branch shifts to a lower pressure side with increase of a longer equilibration time. This fact evidences that the adsorption branch stems from the metastable state. The metastable state consists of stable clusters and the stability is slightly different from the equilibrium adsorption state, which is observed in desorption process. There are several papers on this mechanism [29]. The water adsorption on SWCNH samples can be understood by the above molecular cluster-mediated micropore filling mechanism.

Very recently, unusual penetration process of water molecules through the nanowindow of open-SWCNH was found. Water molecules form the cluster even in the interstitial pores and thereby the nanowindow size must be larger than the stable cluster size of about 0.6 nm in order to induce smooth penetration. Ohba et al. controlled the window size to be 0.40 ± 0.05 nm. Astonishingly, water molecules can quickly penetrate into the internal tube spaces through the windows. Molecular dynamics predicts the transformation of the cluster to the one-dimensional chain. Thus, water molecules can adapt the different nanostructures of hydrophobicity by changing their molecular association structure [32].

13.3.2. Adsorption of Supercritical Gases

Adsorption behavior of supercritical CH_4 and H_2 on SWCNH samples is described here. The critical temperatures T_C of CH_4 and H_2 are 191 K and 33 K, respectively. The T_C of CH_4 is much higher than that of H_2. Adsorption of supercritical CH_4 is more predominant than that of supercritical H_2 near ambient temperature. Figure 13.21 shows high-pressure CH_4 adsorption isotherms of SWCNH samples up to 6 MPa at 303 K. The adsorption isotherms of SWCNH and open-SWCNH are of M-type, indicating the relatively strong

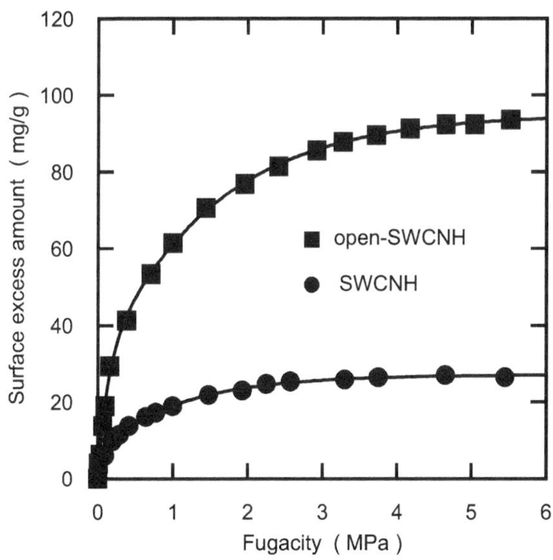

FIGURE 13.21 High-pressure CH_4 adsorption isotherms of SWCNH and open-SWCNH at 303 K.

interaction between CH_4 and pores of SWCNH. Of course, adsorption amount of CH_4 on open-SWCNH is larger than that on SWCNH owing to the larger micropore volume of open-SWCNH. We can evaluate the adsorbed density of CH_4 in micropores of SWCNH and open-SWCNH using the micropore volume from N_2 adsorption and the adsorption amount of CH_4. As SWCNH has the ultramicropores, the adsorbed density of CH_4 on SWCNH is higher than that on open-SWCNH. Interestingly, the adsorbed density of CH_4 on SWCNH (215 mg/cm^3) at 3.5 MPa and 303 K is about a half of to the bulk liquid density of 415 mg/cm^3 at boiling temperature of 112 K. This behavior is supported by the GCMC simulation [33].

Bekyarova et al. [33] produced the pellet samples of open-SWCNH by compression of SWCNH powders at 50 MPa. The compressed open-SWCNH has larger micropore volume than the powdered open-SWCNH due to mutual intricated assembly structure. This compressed SWCNH shows a very high adsorptivity for supercritical CH_4, as illustrated in Fig. 13.22. Here, the ordinate of the adsorption amount is expressed by the volume%. DOE gives the application guideline of 180 vol.% at 3 MPa and 298 K as the CH_4 storage. The compressed SWCNH has a 160 vol.% value, being close to the DOE target value. Figure 13.22 also shows the GCMC-simulated CH_4 adsorption isotherms for hexagonal and cubic arrays of SWCNTs [34]. The compressed SWCNH gives the CH_4 adsorptivity close to the SWCNT bundles. Murata et al. reported that dispersion of lanthanide nitrate enhanced the adsorbed CH_4 density by

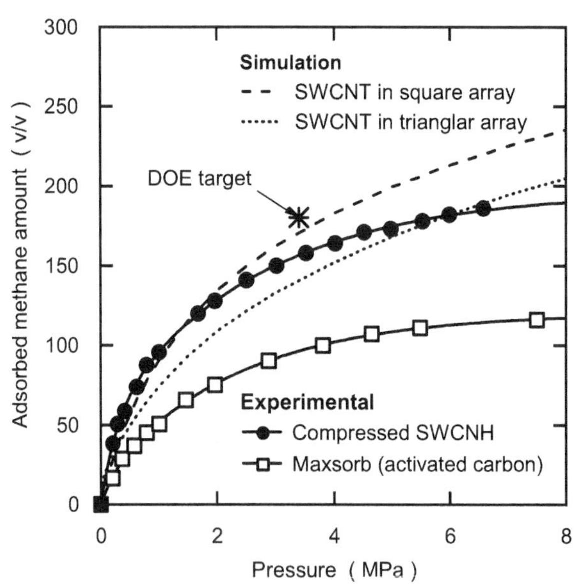

FIGURE 13.22 High-pressure CH_4 adsorption isotherms of compressed SWCNH and maxsorb (activated carbon), and the simulated isotherms of SWCNTs with square and triangle arrays at 303 K.

50% due to charge transfer effect [35]. Thus, SWCNH is a promising candidate for storage of supercritical CH_4.

Recently, CH_4 molecular state in the internal tube spaces of open-SWCNH was studied by rotational motion analysis of vibration spectra of CH_4 [36]. As IR beam can penetrate SWCNH, we can obtain the vibration–rotation band of CH_4 in the tube spaces of open-SWCNH. Although molecular motional states of CH_4 were determined under the vapor conditions due to enough adsorption amount near an atmospheric pressure, information on the adsorbed state can be referable to understand CH_4 molecules in micropores above the critical temperature. Figure 13.23 shows the difference spectra of anti-stretching mode between adsorbed CH_4 and gaseous CH_4 above 112 K. The gaseous CH_4 has fine rotational structures, so-called P, Q, and R branches at the boiling point (112 K). However, the rotational structure of adsorbed CH_4 almost disappears except for Q-branch at 112 K. With elevation of the measuring temperature, the shifted P and R branches appear slightly; the recovery of the P and R branches becomes clear at 140 K. Therefore, a severe suppression of rotational motions of CH_4 confines even above boiling temperature of the bulk CH_4, indicating the highly packed CH_4 molecular state. This molecular spectroscopic study explicitly shows the nanoconfinement effect on the molecular motion.

Supercritical H_2 adsorption on SWCNH leads to a new insight into the nanoconfinement. Figure 13.24 shows high-pressure adsorption isotherms of

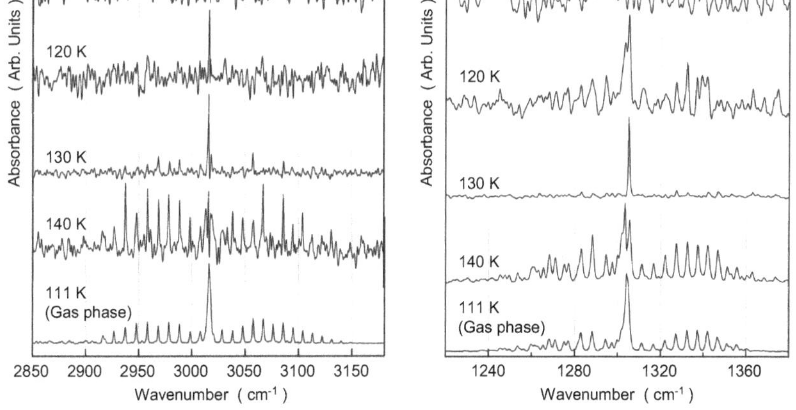

FIGURE 13.23 Difference spectra of anti-stretching mode of CH_4 between adsorbed CH_4 and gaseous CH_4 above 112 K.

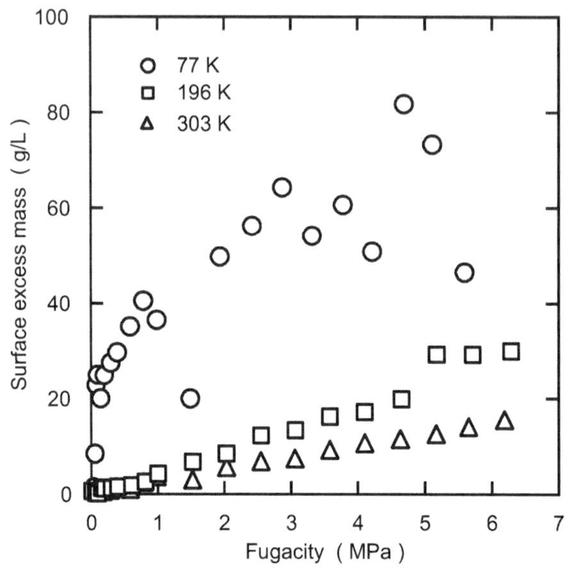

FIGURE 13.24 High-pressure adsorption isotherms of supercritical H_2 on SWCNH at 77 K, 196 K, and 303 K.

supercritical H_2 above 77 K [37]. The T_C of H_2 gas is 33 K and thereby H_2 gas at 77 K is still supercritical gas. These three adsorption isotherms have representative M- (at 77 K), L- (at 196 K), and H-types (at 303 K). Thus, the shape of the adsorption isotherm of H_2 sensitively varies with the measuring temperature. This is because the attractive interaction energy between H_2 and SWCNH is not remarkably large compared with the thermal energy at the adsorption temperature. The interaction energy at 77 K becomes large enough to induce the M-type isotherm. Murata and Kaneko analyzed these adsorption isotherms using DFT theory to obtain the H_2 intermolecular attractive and repulsive factors, and H_2 molecule-wall interaction energy [26,38]. The information on the H_2 adsorbed density at different temperature provides H_2 molecular states in the pores of SWCNH. Figure 13.25 shows the H_2 adsorbed density changes with the adsorption amount for SWCNH and open-SWCNH at 77 K [39]. The adsorbed density of SWCNH is slightly larger than that of open-SWCNH, which comes from higher ratio of the ultramicropores in SWCNH. The adsorbed density values of both SWCNH samples at 77 K are close to the bulk liquid density of H_2 (71 g/L) at 20 K. The above fundamental studies on supercritical CH_4 and H_2 adsorption on SWCNH are helpful to understand the nature of supercritical adsorption in nanoscale pores.

Open-SWCNH exhibits classical molecular sieving effect for gases whose molecular sizes are different. Murata et al. [40] examined molecular sieving effect using He, Ar, N_2, CH_4, SF_6, and C_{60} for open-SWCNH having different nanowindow sizes. Very recently, Ohba et al. [41] showed a splendid separation

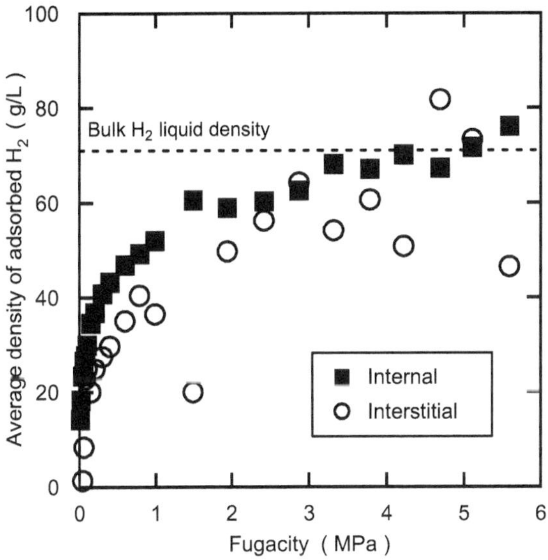

FIGURE 13.25 H_2 adsorbed density changes with the adsorption amount for interstitial and internal pores of SWCNH at 77 K. Broken line indicates the liquid density of bulk H_2.

ability of open-SWCNH having 0.4-nm windows for CH_4 and CO_2, being promising for biogas separation. These molecular sieving effects depend on the classical size of gas molecules. The classical molecular sieving is not efficient for isotopic separation, as shown below.

13.4. ADSORPTION OF QUANTUM FLUID

Steele indicated theoretically that He molecules in the second layer on the flat graphite surface interact strongly with the surface near 4 K, inducing an almost simultaneous monolayer and second-layer adsorption [42]. Kaneko et al. showed that He adsorbed in micropores had unusually large adsorbed density of 0.20–0.26 g/mL at 4.2 K, being an intermediated value between liquid density at 4.2 K (0.125 g/mL) and solid density at 19 K and 169 MPa (0.348 g/mL) [43–45]. Also Kaneko et al. suggested that He confined in the micropores has a larger effective size due to quantum nature in the restricted space [46]. Beenekker et al. proposed the concept of quantum molecular sieving for light molecules due to quantum nature using a simple model [47]. The de Broglie wavelengths of H_2 and D_2 at 77 K are 0.14 nm and 0.11 nm, respectively. The fluctuation of H_2 and D_2 molecules cannot be neglected in the nanoscale spaces. Johnson et al. studied the quantum molecular sieving effect for H_2 and D_2 on SWCNT with a path-integral grand canonical Monte Carlo (PI-GCMC) simulation [48,49]. Darkrim et al. showed the applicability of the Feynman–Hibbs (FH) effective potential to describe H_2 adsorption above 77 K [50]. Sesé gave the criterion for the applicability of the FH effective potential [51].

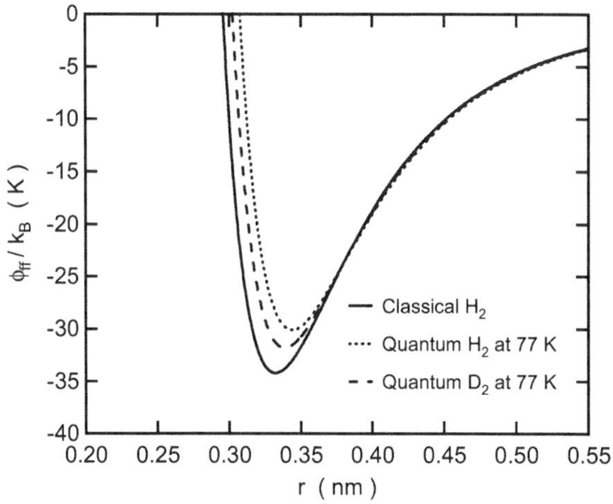

FIGURE 13.26 Interaction potential profiles for classical H_2 and quantum H_2 and D_2 at 77 K.

Although the quantum molecular sieving effect had gathered a great attention, there was no experimental study on adsorption of H_2 and D_2 onto SWCNT up to 2005 [52]. As we could not get pure SWCNT samples, SWCNH was used to obtain the evidence on the quantum molecular sieving effect.

Figure 13.26 shows intermolecular potential profiles of H_2 and D_2 quanta. The potential profile of classical H_2 is also shown for comparison. Here, the quantum interaction profiles are obtained using the FH effective potential function. Although quantum H_2 has a shallower potential well at a larger distance than quantum D_2, classical H_2 and D_2 give the same potential profiles,

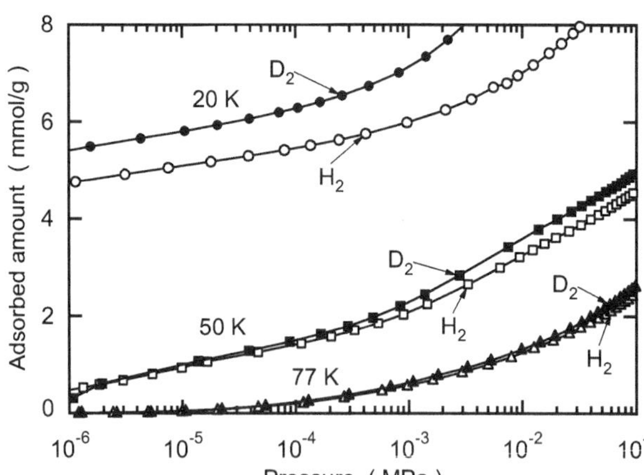

FIGURE 13.27 Adsorption isotherms of H_2 and D_2 on SWCNH at 20 K, 50 K, and 77 K.

leading to the deeper potential well at the smaller distance than quantum D_2. Therefore, classical theory cannot predict different adsorption behavior of H_2 and D_2; H_2 and D_2 quanta have larger molecular sizes than classical one from the FH effective potential; quantum H_2 and quantum D_2 can be distinguishable. Also, the intermolecular interaction of quantum H_2 is different from that of quantum D_2. Therefore, H_2 and D_2 quanta give different adsorption behaviors in the nanoscale pores. This is the reason for the quantum molecular sieving [52].

Figure 13.27 shows the adsorption isotherms of H_2 and D_2 on SWCNH having only interstitial porosity. As the de Broglie wavelength difference becomes more marked with lowering the adsorption temperature, adsorption difference between H_2 and D_2 is more evident at a lower temperature. Surely the adsorption difference at 20 K is about 20% at 1 kPa. The adsorption difference at 77 K is about 5%. The similar tendency was obtained for open-SWCNH. We need to describe the observed adsorption difference between H_2 and D_2 using quantum molecular simulation. Here, the FH effective potential was used for GCMC simulation of adsorption isotherms at 77 K. As the interparticle structure of SWCNH particles is not well defined, adsorption only in the internal tube space was simulated with the FH effective potential. The adsorption isotherms of H_2 and D_2 in the internal tube spaces are obtained by subtraction of adsorption on SWCNH from adsorption on open-SWCNH. Fig. 13.28 shows the experimental adsorption isotherms of H_2 and D_2 on open-SWCNH, SWCNH, and in-pore spaces at 77 K [52]. The explicit adsorption

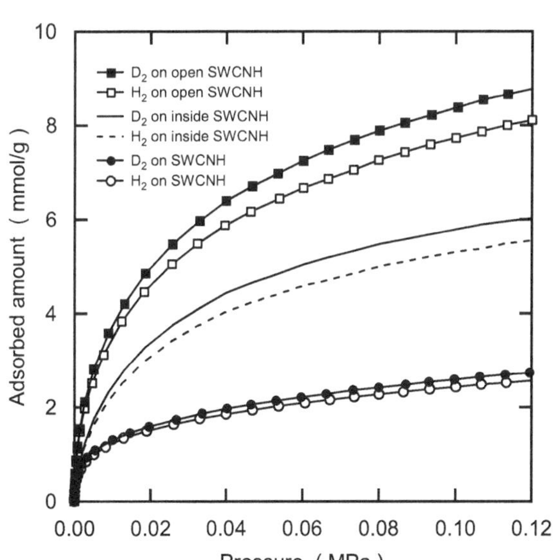

FIGURE 13.28 H_2 and D_2 adsorption isotherms for SWCNH, open-SWNH, and the internal tube space of SWCNH at 77 K.

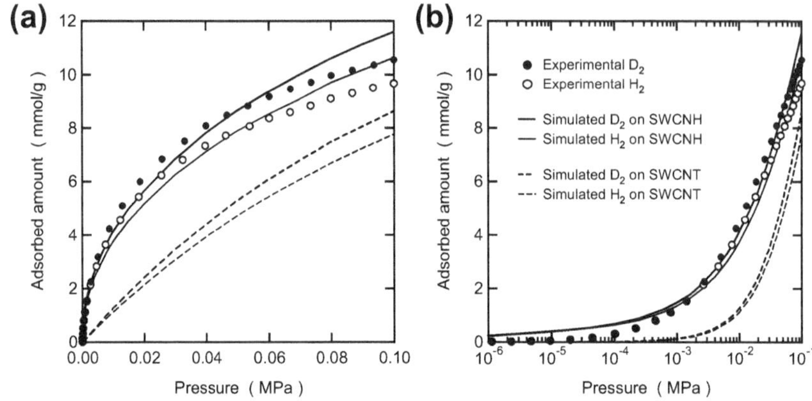

FIGURE 13.29 H_2 and D_2 adsorption isotherms at 77 K in a linear scale (a) and a logarithmic scale (b) of the pressure. The circles represent the experimental values for the internal space of SWNH. Solid and broken lines are prediction from the FH-GCMC simulations for SWCNH and SWCNT, respectively.

difference between H_2 and D_2 in the in-pore spaces is shown. The adsorption isotherms of H_2 and D_2 in the in-pore spaces are well described by the quantum simulation, as shown in Fig. 13.29. As the adsorption difference cannot be described by the classical simulation, the observed adsorption difference comes from the quantum nature of H_2 and D_2. Thus, the quantum molecular sieving effect predicted theoretically was evidenced by the above study using SWCNH samples.

13.5. NANOSCALE MATERIALS GROWTH FIELDS

SWCNH and open-SWCNH have an attractive nanoporosity, as mentioned above. Adsorption from solution on SWCNH samples is also important to control the porosity and electronic properties. Iodine adsorption on SWCNH gives rise to a marked increase in the electrical conductivity due to the charge transfer interaction [53]. Also, the nanowindow can work as an ion sieving. The capacitance measurement on SWCNH samples was done and the blocking effect of organic ions of different size by the nanowindows was evidenced [54]. Thus, there are many topics on interfacial phenomena on SWCNH. In the last section, embedding of magnetite and KI in the internal spaces and dispersion of metallic nanoparticles are briefly described.

Utsumi et al. [55] synthesized magnetite nanoparticles on open-SWCNH using Fe^{2+}–Fe^{3+} solution. Magnetite nanoparticles are doped in the internal tube spaces and dispersed on the assembly structure of open-SWCNH. The particle size is in the range of 2–9 nm, as shown in Fig. 13.30. These magnetite-doped SWCNH colloids show an intensive magnetic response. If we contact a magnet to the dispersion solution, all magnetite-doped SWCNH colloids are

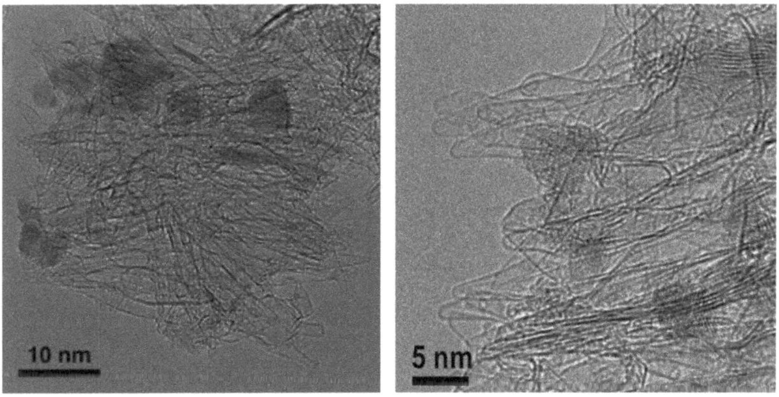

FIGURE 13.30 TEM images of magnetite nanoparticles-doped open-SWCNH.

attracted to the magnet. Figure 13.31 shows photographs of transformation of black magnetite-doped SWCNH dispersed solution into transparent solution by collection of magnetic SWCNH colloids by the magnet. The magnetization of the magnetite-doped SWCNH colloids is smaller than that of the bulk magnetite. Figure 13.32 shows the magnetization curves of magnetite-doped SWCNH colloids and the bulk magnetite at 1.9 K and 303 K. These magnetic behaviors indicate the hybrid property of ferrimagnetism of relatively large magnetite particles and superparamagnetism of small magnetite particles.

It is known that KI crystals transform the B1 phase of NaCl type into B2 phase of CsCl type at 1.9 GPa. Urita et al. [56] doped KI nanocrystals in open-SWCNH by heating the mixtures of SWCNH and KI crystals at 1073 K for 8 h below 0.1 MPa. The KI-doped SWCNH was examined by synchrotron X-ray

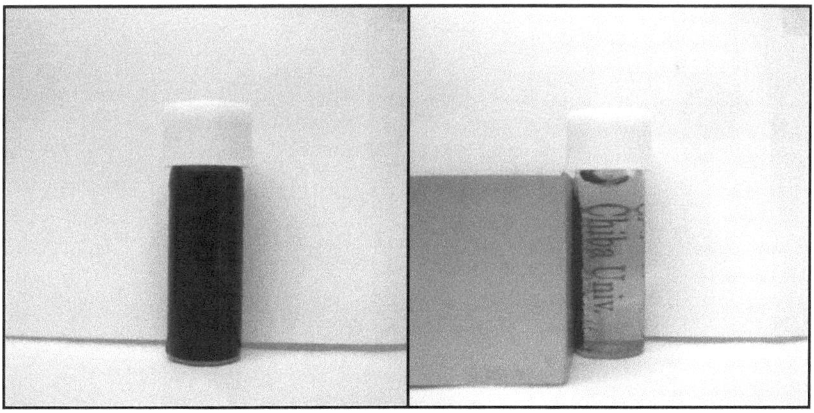

FIGURE 13.31 Photographs of transformation of black magnetite-doped SWCNH dispersed solution into transparent solution by collection of magnetic SWCNH colloids by the magnet.

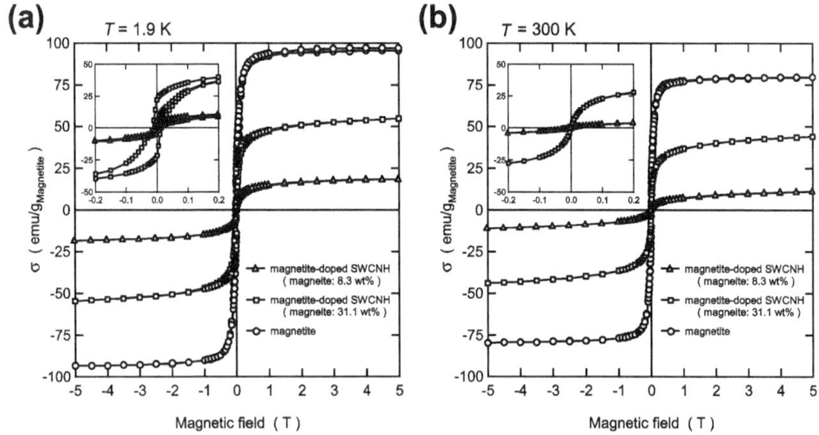

FIGURE 13.32 Magnetization curves of magnetite-doped SWCNH colloids and the bulk magnetite at 1.9 K (a) and 303 K (b), expressed as the magnetization per 1g of magnetite. Circles represent the magnetization of magnetite. Squares and triangles are the magnetization values of 8.3 wt% magnetite-doped SWCNH and 31.1 wt% magnetite-doped SWCNH, respectively. Insets are magnified figures in the magnetic field ranging from –0.2 to 0.2 T.

diffraction of 0.10029 nm wavelength and HR-TEM. Figure 13.33 shows TEM image of the B2-type KI structure in the internal tube space of open-SWCNH. Synchrotron X-ray diffraction gives more detailed structure information of the B2-type structure due to additional diffraction peaks which cannot be assigned to SWCNH and the B1 phase KI. Synchrotron XRD showed clearly the formation of the KI of the high-pressure B2 phase whose unit cell is slightly distorted from that of the bulk high-pressure B2 phase. This fact is really astonishing; confinement of KI nanocrystals in the internal tube spaces of open-SWCNH below 0.1 MPa provides the formation of the B2 phase which

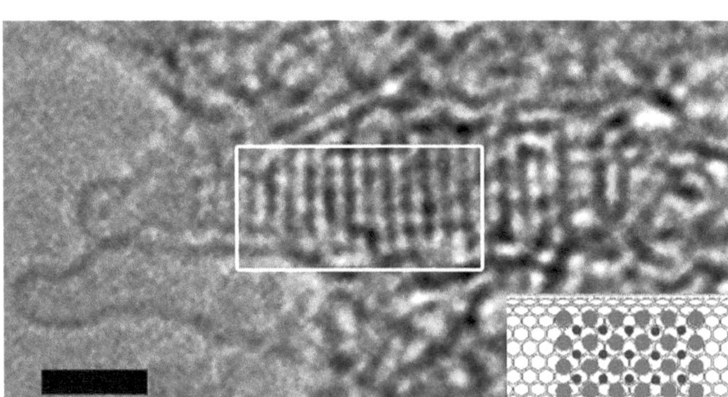

FIGURE 13.33 TEM image of the B2-type KI structure in the internal tube space of open-SWCNH. Bar scale is 2 nm. Insert is the model of the B2-type structure in SWCNH (Small black circle; potassium, large gray circle; iodine).

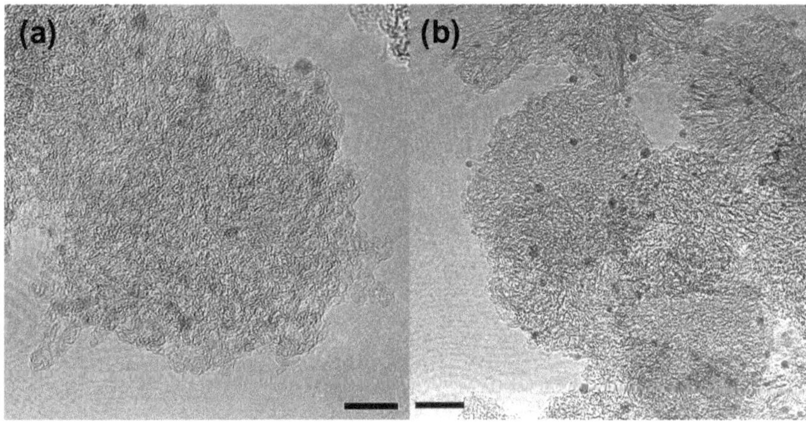

FIGURE 13.34 TEM images of (a) Pd-SWCNH (bar scale; 10 nm) and (b) Pd-open-SWCNH (bar scale; 20 nm).

needs the application of the pressure above 1.9 GPa. This high-pressure effect can be widely applicable to chemistry, physics, and materials science.

The interstitial pores and nanowindows can offer the stable sites for nanoparticles on the SWCNH samples. Considerably Pd nanoparticles of 2–3 nm can be highly dispersed on SWCNH samples, as shown in Fig. 13.34

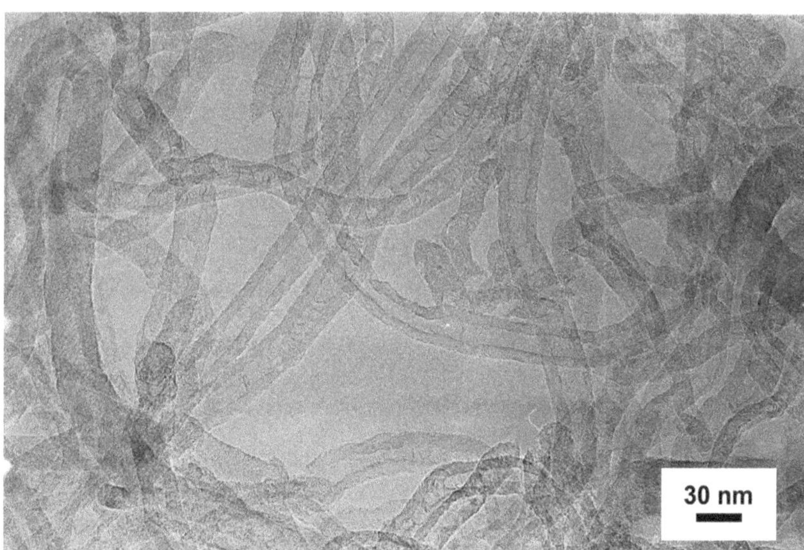

FIGURE 13.35 TEM image of multi-wall carbon nanotubes in Pd-open-SWCNH after CH_4 decomposition.

[57,58]. Pd nanoparticles are often observed in the interstitial sites and on the nanowindows. An XPS study indicated that there are two chemical species of metallic Pd as the majority and PdO_2. The nano-Pd-dispersed SWCNH exhibits high catalytic activity for CH_4 decomposition into carbon and H_2. Evolution of H_2 starts at 820 K and the produced carbon is a multi-walled carbon nanotube, as shown in Fig. 13.35 [58]. The nano-Pd-dispersed SWCNH also showed a high catalytic activity of water formation reaction from H_2 and O_2. The merit of SWCNH as the catalyst support is suggested.

13.6. CONCLUSION

SWCNH shows characteristic interfacial properties different from SWCNT, because SWCNH has unique dahlia-flower-type assembly structure and is highly defective, indicating the high surface activity. Although research on SWCNH has gradually extended, still only limited research groups can study SWCNH. A more active examination of SWCNH should be quite promising for new innovation in the fields of adsorbents, catalysts, and membranes.

ACKNOWLEDGMENT

K.K. has been supported by Exotic Nanocarbons, Japan Regional Innovation Strategy Program by the Excellence, Japan Science and Technology Agency.

REFERENCES

[1] S. Iijima, T. Ichihashi, Single-shell carbon nanotubes of 1-nm diameter, Nature 363 (1993) 603–605.

[2] D.S. Bethune, C.H. Klang, M.S. de Vries, G. Gorman, R. Savoy, J. Vazquez, R. Beyers, Cobalt-catalysed growth of carbon nanotubes with single-atomic-layer walls, Nature 363 (1993) 605–607.

[3] K. Hata, D.N. Futaba, K. Mizuno, T. Namai, M. Yumura, S. Iijima, Water-assisted highly efficient synthesis of impurity-free single-walled carbon nanotubes, Science 306 (2004) 1362–1364.

[4] H. Kataura, Y. Kumazawa, Y. Maniwa, I. Umezu, S. Suzuki, Y. Ohtsuka, Y. Achiba, Optical properties of single-wall carbon nanotube, Synth. Met. 103 (1999) 2555–2558.

[5] S. Utsumi, H. Honda, Y. Hattori, H. Kanoh, K. Takahashi, H. Sakai, M. Abe, M. Yudasaka, S. Iijima, K. Kaneko, Direct evidence on C–C single bonding in single-wall carbon nanohorn aggregates, J. Phys. Chem. C 111 (2007) 5572–5575.

[6] M.A. Pimenta, G. Dresselhaus, M.S. Dresselhaus, L.G. Cançado, A. Jorio, R. Saito, Studying disorder in graphite-based systems by Raman spectroscopy, Phys. Chem. Chem. Phys. 9 (2007) 1276–1290.

[7] P.J.F. Harris, S.C. Tsang, J.B. Claridge, M.L.H. Green, High-resolution electron microscopy studies of a microporous carbon produced by arc-evaporation, J. Chem. Soc. Faraday Trans. 90 (1994) 2799–2802.

[8] S. Iijima, M. Yudasaka, R. Yamada, S. Bandow, K. Suenaga, F. Kokai, K. Takahashi, Nano--aggregates of single-walled graphitic carbon nano-horns, Chem. Phys. Lett. 309 (1999) 165–170.

[9] E. Bekyarova, K. Kaneko, D. Kasuya, K. Murata, M. Yudasaka, S. Iijima, Oxidation and porosity evaluation of budlike single-wall carbon nanohorn aggregates, Langmuir 18 (2002) 4138–4141.

[10] M. Yudasaka, S. Iijima, V.H. Crespi, Single-wall carbon nanohorns and nanocones, in: Topics in Applied Physics 111, Carbon Nanotubes, Advanced Topics in the Synthesis, Structure, Properties and Application, Springer-Verlag, Berlin, 2008.

[11] T. Fujimori, K. Urita, Y. Aoki, H. Kanoh, T. Ohba, M. Yudasaka, S. Iijima, K. Kaneko, Fine nanostructure analysis of single-wall carbon nanohorns by surface-enhanced Raman scattering, J. Phys. Chem. C 112 (2008) 7552–7556.

[12] T. Fujimori, K. Urita, D. Tománek, T. Ohba, I. Moriguchi, M. Endo, K. Kaneko, Selective probe of the morphology and local vibrations at carbon nanoasperities, J. Chem. Phys. 136 (2012) 064505-1-064505-5.

[13] T. Fujimori, L.R. Radovic, A.B. Silva-Taoa, M. Endo, K. Kaneko, Structural importance of Stone-Thrower-Wales defects in rolled and flat graphenes from surface-enhanced Raman scattering, Carbon (in press), http://dx.doi.org/10.1016/j.carbon.2011.12.010.

[14] G. Wu, J. Dong, Raman characteristic peaks induced by the topological defects of carbon nanotube intramolecular junctions, Phys. Rev. B 73 (2006) 245414-1-245414-9.

[15] R. Saito, A. Jorio, A.G.S. Filho, A. Grueneis, M.A. Pimenta, D. Dresselhaus, M.S. Dresselhaus, Dispersive Raman spectra observed in graphite and single wall carbon nanotubes, Physica B 323 (2002) 100–106.

[16] K. Urita, S. Seki, S. Utsumi, D. Noguchi, H. Kanoh, H. Tanaka, Y. Hattori, Y. Ochiai, N. Aoki, M. Yudasaka, S. Iijima, K. Kaneko, Effects of gas adsorption on the electrical conductivity of single-wall carbon nanohorns, Nano Lett. 6 (2006) 1325–1328.

[17] T. Ohba, T. Matsumura, K. Hata, M. Yumura, S. Iijima, H. Kanoh, K. Kaneko, Nanoscale curvature effect on ordering of N_2 molecules adsorbed on single wall carbon nanotube, J. Phys. Chem. C 111 (2007) 15660–15663.

[18] H. Tanaka, D. Noguchi, A. Yuzawa, T. Kodaira, H. Kanoh, K. Kaneko, Quantum effects on hydrogen isotopes adsorption in nanopores, J. Low Temp. Phys. 157 (2009) 352–373.

[19] S. Lei, J. Miyamoto, T. Ohba, H. Kanoh, K. Kaneko, Novel nanostructures of porous carbon synthesized with zeolite LTA-template and methanol, J. Phys. Chem. C 111 (2007) 2459–2464.

[20] K. Murata, K. Kaneko, F. Kokai, K. Takahashi, M. Yudasaka, S. Iijima, Pore structure of single-wall carbon nanohorn aggregates, Chem. Phys. Lett. 331 (2000) 14–20.

[21] S. Utsumi, J. Miyawaki, H. Tanaka, Y. Hattori, T. Itoi, N. Ichikuni, H. Kanoh, M. Yudasaka, S. Iijima, K. Kaneko, Opening mechanism of internal nanoporosity of single-wall carbon nanohorn, J. Phys. Chem. B 109 (2005) 14319–14324.

[22] T. Ohba, K. Murata, K. Kaneko, W.A. Steele, F. Kokai, K. Takahashi, D. Kasuya, M. Yudasaka, S. Iijima, N_2 adsorption in an internal nanopore space of single-walled carbon nanohorn: GCMC simulation and experiment, Nano Lett. 1 (2001) 371–373.

[23] K. Murata, K. Kaneko, W.A. Steele, F. Kokai, K. Takahashi, D. Kasuya, M. Yudasaka, S. Iijima, Porosity evaluation of intrinsic intraparticle nanopores of single wall carbon nanohorn, Nano Lett. 1 (2001) 197–199.

[24] K. Urita, S. Seki, H. Tsuchiya, H. Honda, S. Utsumi, T. Ohba, H. Kanoh, M. Yudasaka, S. Iijima, K. Kaneko, Characterization of single wall carbon nanohorns restructured with mechanical treatment, in: Characterization of Porous Solid VIII, Royal Society of Chemistry Pub., London, 2009.

[25] Y. Hattori, H. Kanoh, F. Okino, H. Touhara, D. Kasuya, M. Yudasaka, S. Iijima, K. Kaneko, Direct thermal fluorination of single wall carbon nanohorns, J. Phys. Chem. B 108 (2004) 9614–9618.

[26] K. Murata, M. Yudasaka, S. Iijima, M. El-Merraoui, K. Kaneko, Classification of supercritical gas adsorption isotherms based on fluid–fluid interaction, J. Appl. Phys. 91 (2002) 10227–10229.

[27] K. Murata, K. Kaneko, W.A. Steele, F. Kokai, K. Takahashi, D. Kasuya, K. Hirahara, M. Yudasaka, S. Iijima, Molecular potential structures of heat-treated single-wall carbon nanohorn assemblies, J. Phys. Chem. B 105 (2001) 10210–10216.

[28] E. Bekyarova, Y. Hanzawa, K. Kaneko, J. Silvestre-Albero, A. Sepulveda-Escribano, F. Rodriguez-Reinoso, D. Kasuya, M. Yudasaka, S. Iijima, Cluster-mediated filling of water vapor in intratube and interstitial nanospaces of single-wall carbon nanohorns, Chem. Phys. Lett. 366 (2002) 463–468.

[29] T. Ohba, K. Kaneko, Cluster-induced adsorption of water in hydrophobic carbon nanopores by GCMC simulation-aided approach, in: Carbon Materials: Theory and Practice, Research Signpost, Kerala, 2008.

[30] T. Ohba, K. Kaneko, Cluster-associated filling of water molecules in slit-shaped graphitic nanopores, Mol. Phys. 105 (2007) 139–145.

[31] M. Nakamura, T. Ohba, P. Branton, H. Kanoh, K. Kaneko, Equilibration-time and pore-width dependent hysteresis of water adsorption isotherm on hydrophobic microporous carbons, Carbon 48 (2010) 305–308.

[32] T. Ohba, K. Kaneko, Kinetically forbidden transformations of water molecular assemblies in hydrophobic micropores, Langmuir 27 (2011) 7609–7613.

[33] E. Bekyarova, K. Kaneko, M. Yudasaka, K. Murata, D. Kasuya, S. Iijima, Micropore development and structure rearrangement of single-wall carbon nanohorn assemblies by compression, Adv. Mater. 14 (2002) 973–975.

[34] H. Tanaka, M. El-Merraoui, W.A. Steele, K. Kaneko, Methane adsorption on single-walled carbon nanotube: a density functional theory model, Chem. Phys. Lett. 352 (2002) 334–341.

[35] K. Murata, A. Hashimoto, M. Yudasaka, D. Kasuya, K. Kaneko, S. Iijima, The use of charge transfer to enhance the methane-storage capacity of single-walled, nanostructured carbon, Adv. Mater. 16 (2004) 1520–1522.

[36] S. Hashimoto, T. Fujimori, H. Tanaka, K. Urita, T. Ohba, H. Kanoh, T. Itoh, M. Asai, H. Sakamoto, S. Niimura, M. Endo, F. Rodriguez-Reinoso, K. Kaneko, Anomaly of CH_4 molecular assembly confined in single-wall carbon nanohorn spaces, J. Am. Chem. Soc. 133 (2011) 2022–2024.

[37] H. Tanaka, J. Miyawaki, K. Kaneko, K. Murata, D. Kasuya, M. Yudasaka, K. Kaneko, F. Kokai, K. Takahashi, D. Kasuya, M. Yudasaka, S. Iijima, Comparative study on physical adsorption of vapor and supercritical H_2 and CH_4 on SWNH and ACF, Mol. Cryst. Liq. Cryst. 388 (2002) 15–21.

[38] K. Murata, M. El-Merraoui, K. Kaneko, A new determination method of absolute adsorption isotherm of supercritical gases under high pressure with a special relevance to density-functional theory study, J. Chem. Phys. 114 (2001) 4196–4205.

[39] K. Murata, K. Kaneko, H. Kanoh, D. Kasuya, K. Takahashi, F. Kokai, M. Yudasaka, S. Iijima, Adsorption mechanism of supercritical hydrogen in internal and interstitial nanospaces of single-wall carbon nanohorn assembly, J. Phys. Chem. B 106 (2002) 11132–11138.

[40] K. Murata, K. Hirahara, M. Yudasaka, S. Iijima, D. Kasuya, K. Kaneko, Nanowindow-induced molecular sieving effect in a single-wall carbon nanohorn, J. Phys. Chem. B 106 (2002) 12668–12669.

[41] T. Ohba, H. Kanoh, K. Kaneko, Superuniform molecular nanogate fabrication on graphene sheets of single wall carbon nanohorns for selective molecular separation of CO_2 and CH_4, Chem. Lett. 40 (2011) 1089–1091.

[42] W.A. Steele, Concerning a theory of multilayer adsorption, with particular reference to adsorbed helium, J. Chem. Phys. 25 (1956) 819–823.

[43] H. Kuwabara, T. Suzuki, K. Kaneko, Ultramicropores in microporous carbon fibres evidenced by helium adsorption at 4.2 K, J. Chem. Soc. Faraday Trans. 87 (1991) 1915–1916.

[44] N. Setoyama, M. Ruike, T. Kasu, T. Suzuki, K. Kaneko, Surface characterization of microporous solids with helium adsorption and small angle x-ray scattering, Langmuir 9 (1993) 2612–2617.

[45] N. Setoyama, K. Kaneko, Density of He adsorbed in micropores at 4.2K, Adsorption 1 (1995) 165–173.

[46] K. Kaneko, N. Setoyama, T. Suzuki, Ultramicropore characterization by He adsorption. In: Characterization of Porous Solids III, Studies in Surf. Sci. Catal. Elsevier, 1994.

[47] J.J.M. Beenakker, V.D. Borman, S.Y. Krylov, Molecular transport in subnanometer pores: zero-point energy, reduced dimensionality and quantum sieving, Chem. Phys. Lett. 232 (1995) 379–382.

[48] Q.Y. Wang, J.K. Johnson, J.Q. Broughton, Path integral grand canonical Monte Carlo, J. Chem. Phys. 107 (1997) 5108–5117.

[49] Q.Y. Wang, J.K. Johnson, Molecular simulation of hydrogen adsorption in single-walled carbon nanotubes and idealized carbon slit pores, J. Chem. Phys. 110 (1999) 577–586.

[50] F. Darkrim, D. Levesque, High adsorptive property of opened carbon nanotubes at 77 K, J. Phys. Chem. B 104 (2000) 6773–6776.

[51] L.M. Sesé, Feynman-Hibbs potentials and path integrals for quantum Lennard-Jones systems: theory and Monte Carlo simulations, Mol. Phys. 85 (1995) 931–947.

[52] H. Tanaka, H. Kanoh, M. Yudasaka, S. Iijima, K. Kaneko, Quantum effects on hydrogen isotope adsorption on single-wall carbon nanohorns, J. Am. Chem. Soc. 127 (2005) 7511–7516.

[53] F. Khoerunnisa, T. Fujimori, T. Itoh, H. Kanoh, T. Ohba, M. Yudasaka, S. Iijima, K. Kaneko, Electronically modified single wall carbon nanohorns with iodine adsorption, Chem. Phys. Lett. 501 (2011) 485–490.

[54] C.M. Yang, Y.J. Kim, M. Endo, H. Kanoh, M. Yudasaka, S. Iijima, K. Kaneko, Nanowindow-regulated specific capacitance of supercapacitor electrodes of single-wall carbon nanohorns, J. Am. Chem. Soc. 129 (2007) 20–21.

[55] S. Utsumi, K. Urita, H. Kanoh, M. Yudasaka, K. Suenaga, S. Iijima, K. Kaneko, Preparing a magnetically responsive single-wall carbon nanohorn colloid by anchoring magnetite nanoparticles, J. Phys. Chem. B 110 (2006) 7165–7170.

[56] K. Urita, Y. Shiga, T. Fujimori, T. Iiyama, Y. Hattori, H. Kanoh, T. Ohba, H. Tanaka, M. Yudasaka, S. Iijima, I. Moriguchi, F. Okino, M. Endo, K. Kaneko, Confinement in carbon nanospace-induced production of KI nanocrystals of high-pressure phase, J. Am. Chem. Soc. 133 (2011) 10344–10347.

[57] T. Itoh, H. Danjo, W. Sasaki, K. Urita, E. Bekyarova, M. Arai, T. Imamoto, M. Yudasaka, S. Iijima, H. Kanoh, K. Kaneko, Catalytic activities of Pd-tailored single wall carbon nanohorns, Carbon 46 (2008) 172–175.

[58] Y. Aoki, K. Urita, D. Noguchi, T. Itoh, H. Kanoh, T. Ohba, M. Yudasaka, S. Iijima, K. Kaneko, Efficient production of H_2 and carbon nanotube from CH_4 over single wall carbon nanohorn, Chem. Phys. Lett. 482 (2009) 269–273.

Chapter 14

Adsorption Behaviors of Graphene and Graphene-related Materials

Soo-Jin Park and Seul-Yi Lee
Department of Chemistry, Inha University, Nam-gu, Incheon, South Korea

Chapter Outline

14.1. Introduction 435
14.2. Preparation Methods 437
14.2.1. Introduction 437
14.2.2. Epitaxial Growth and Chemical Vapor Deposition Technique 437
14.2.3. Micromechanical Exfoliation Technique 438
14.2.4. Chemical Reduction and Thermal Exfoliation 438
14.2.5. Others and so on 440
14.3. Adsorption Behaviors 445
14.3.1. Introduction 445
14.3.2. Gas Sensor 445
14.3.3. Hydrogen Storage 446
14.3.4. Carbon Dioxide Capture 456
14.3.5. Ammonia Adsorption 459
14.3.6. Surface Functionalization 460
14.4. Summary 461
References 461

14.1. INTRODUCTION

The discovery of graphene has been one of the most outstanding recent achievements. This material offers a unique combination of electrical, optical, thermal, and mechanical properties, and extensive research is being pursued on single- as well as bi- and few-layer graphene [1,2]. Graphene is one of the most promising materials in nanotechnology, as the mother of all graphitic materials, as presented Fig. 14.1 [3]. Graphene is the thinnest known material in the universe and the strongest ever measured.

Novel Carbon Adsorbents. DOI: 10.1016/B978-0-08-097744-7.00014-4
Copyright © 2012 Elsevier Ltd. All rights reserved.

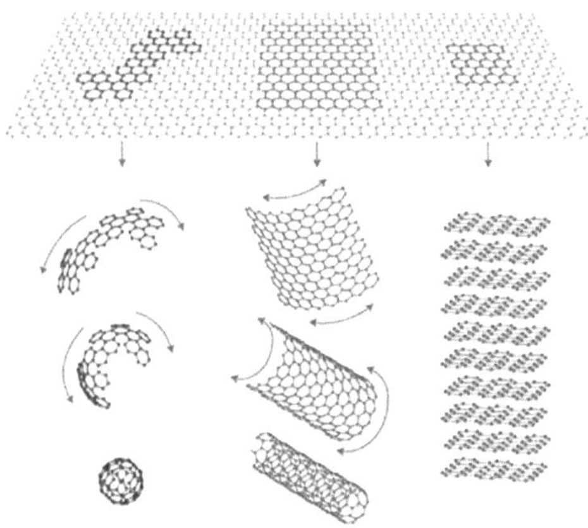

FIGURE 14.1 Schematic representation of graphene, which is the fundamental starting material for a variety of fullerene materials, buckyballs, carbon nanotubes, and graphite. *(With permission from Nature Materials (Ref. [3])).*

Graphene is the name given to a two-dimensional sheet of sp^2-hybridized carbon. Its extended hexagonal monolayer network is the basic building block of other important allotropes, which can be stacked to form 3D graphite, rolled to form 1D nanotubes, and wrapped to form 0D fullerenes. As shown in Table 14.1, it has superior thermal, mechanical, and electrical properties due to the long-range π-conjugation in graphene. More recently, the adsorption behavior of graphene and graphene-related materials has been of interest due to their high specific surface area and newly created textural properties.

It this chapter, it should be clarified that graphene is not confined to only the single-layer graphene nanosheet, which means it has two-, three-, or multi-layer nanosheets. Because adsorption behavior is mainly studied with graphene-derivative-related materials, the purpose of this chapter is to look at what is referred to as graphene in the literature and the adsorption behavior of graphene and graphene-related materials for gas sensors, hydrogen storage, carbon dioxide capture, and so on.

TABLE 14.1 Physical Properties of Single-layer Graphene at Room Temperature (With permission of Springer (Ref. [2]))

Content	Unit	Values	Ref.
C–C bond length in single-layer graphene	nm	0.142	[4]
Specific surface area	m^2/g	~2,630	[5]
Electron mobility	$cm^2/(V\ s)$	$\sim 1.5 \times 10^4$	[3]
Young's modulus	TPa	~1	[6]
Thermal conductivity	W/(m K)	$\sim 5.1 \times 10^3$	[7]

14.2. PREPARATION METHODS

14.2.1. Introduction

Graphene with varying number of layers can be synthesized by using different strategies, such as chemical vapor deposition, micromechanical exfoliation cleavage, chemical reduction of graphene oxide, and so on. In this section, we briefly overview the various synthetic methods and the properties of the graphene produced by each synthetic method.

14.2.2. Epitaxial Growth and Chemical Vapor Deposition Technique

For the development of large-scale synthesis of grown graphene, several methods have been studied by many researchers, including graphitization of SiC surfaces [8,9] and chemical vapor deposition (CVD) on transition metals [10–13].

At first, epitaxial graphene research evolved out of work on carbon nanotubes. It is well known that carbon nanotubes have a superior electronic property. However, the inability to manufacture well-controlled tubes and scale them up from single-tube transistors to large-scale integrated circuits has prevented serious technological applications [14].

In the 2000s, de Heer' et al. realized that two-dimensional graphene, essentially unrolled nanotubes, would have many of the same properties as carbon nanotubes, and this idea opened the way to a new approach to carbon electronics. With an epitaxial method, ultrathin epitaxial graphite was grown on single-crystal silicon carbide by vacuum graphitization at high temperature. The material can be patterned using standard nanolithography methods [13–15].

Epitaxially grown graphite has a number of differences related to physical properties in comparison with mechanically exfoliated graphene [16] due to the influence of interfacial effects in epitaxial graphene, which are strongly dependent on both the silicon carbide substrate and some growth parameters.

The second method is substrate-based growth of single layers by CVD technique. It has been reported that large and predominantly monolayer graphene of excellent quality can be synthesized by CVD on polycrystalline Ni, Cu, Co, Ru, and other transition metals [17,18], with the flow of hydrocarbon gases, such as methane, ethylene, acetylene, and benzene. In this procedure, the substrate is subjected to electron bombardment in ultrahigh vacuum at 1000 °C to remove oxide contaminants and then heated to temperatures ranging from 1250 °C to 1450 °C for a few minutes. Blakely et al. reported the formation of carbon films by the cooling of Ni foils saturated with carbon at high temperatures [19,20].

This is known to be relatively simple and economical, and has been used to produce graphene that can reach impressive sizes and can be easily transferred to other substrates [21,22] (You can be provided in detail and sufficient information from the ref. [13]).

Both epitaxial synthesis and CVD techniques take advantage of specially chosen platforms to encourage high-quality growth. They also have the prospect of producing a single sheet of graphene over an entire wafer.

Meanwhile, the following solution-based synthetic schemes by chemical reduction or exfoliation have no need for support substrates.

14.2.3. Micromechanical Exfoliation Technique

Geim et al. first observed a single-layer graphene in 2004 [1]. They isolated two-dimensional crystals from three-dimensional graphite by using the mechanical exfoliation technique. As a result, single- and few-layer flakes are locked up substanding by only van der Waals forces and can be made freestanding by etching the substrate [23–26]. This led to tremendous attention.

To exfoliate a single layer using mechanical exfoliation, the key is to overcome the van der Waals attraction [27,28] between exactly the first and second layers without disarranging any subsequent layers. However, this method is a typical top–down approach of graphene, which requires manual effort and produces unreliable results, while small areas hinder practical applications of graphene [29,30].

Meanwhile, it is interesting to note that mechanical exfoliation with cellophane tape from graphite is still considered best for producing the highest-performing samples [27], among the various methods mentioned in the following chapter.

14.2.4. Chemical Reduction and Thermal Exfoliation

In 1958, Hummers et al. [31] reported a method of preparing graphene oxide, which readily forms a stable colloidal suspension in water. This graphene oxide suspension is used to facilitate production of single-layer graphene oxide by ultrasonic treatment (under the hundreds of W and tens of kHz) [32–34].

By modifying the Hummers' methods, a large number of studies on the synthetic method of graphene production have been published, and the produced graphene is applied in a variety of applications. With improvement, gram quantities of graphene can also be obtained by a solvothermal procedure for preparing graphene oxide using sodium, potassium, ethanol, and so on [35,36]. Figure 14.2 is shown the chemical structure of graphene oxide.

To synthesize graphene, the reduction of graphene oxide by chemical methods has been carried out using reducing agents, such as hydrazine [38–41], dimethylhydrazine [42], hydroquinone [43], and $NaBH_4$ [44,45]. Figure 14.3 is presented a proposed reaction pathway for epoxide reduction with hydrazine in the process of synthesis of graphene.

Ruoff et al. reported a chemical reduction method to prepare graphene in which a graphene oxide dispersion in dimethlyformamide (DMF) is reduced with hydrazine hydrate [46], as presented in Fig. 14.4.

The chemical reduction method by graphene oxide in solution allows easy manipulation and transfer of graphene onto substrates.

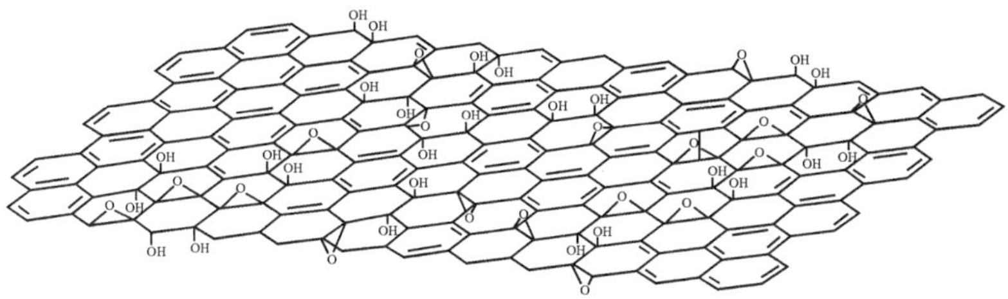

FIGURE 14.2 Chemical structure of graphene oxide. In this image, the carboxylic and carbonyl groups at the edges are omitted for clarity. *(With permission of (Ref. [37]))*.

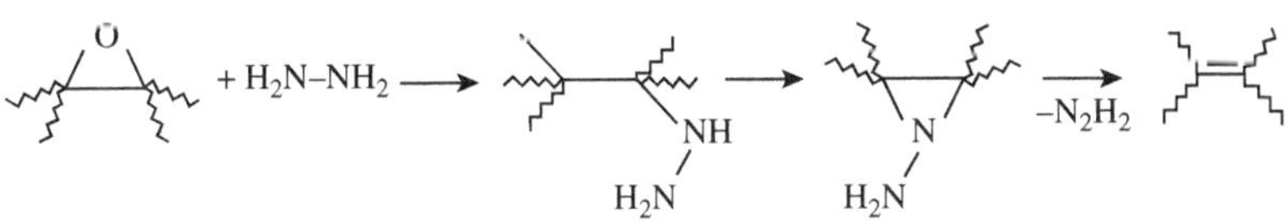

FIGURE 14.3 A proposed reaction pathway for epoxide reduction with hydrazine. *(With permission of (Ref. [38]))*.

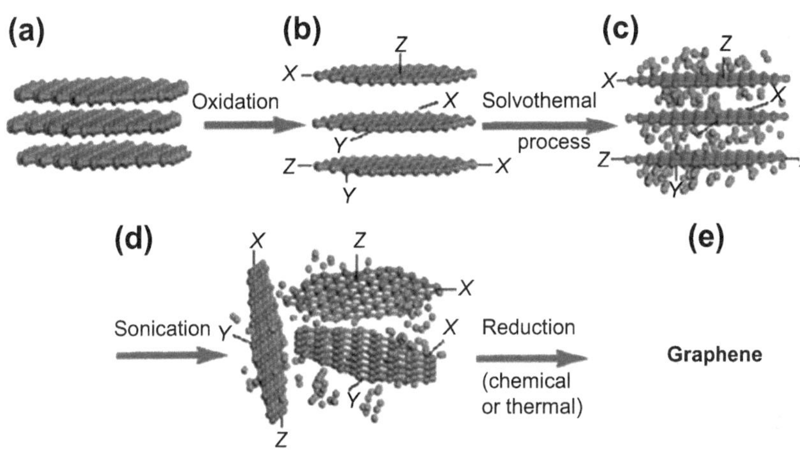

FIGURE 14.4 Schematic illustrating graphene through graphene oxide: (a) pristine graphite, (b) graphene oxide (X, Y, and Z are oxygen-containing groups), (c), graphene oxide containing solvent molecules between its sheets, (d) graphene oxide exfoliated by sonication and consisting of both single-layer graphene oxide and few-layers graphene oxide, and (e) reduced graphene. *(With permission of (Ref. [2]))*.

An important method to synthesize graphene is thermal exfoliation of graphene oxide at high temperatures [47–49]. In this method, chemically prepared graphene oxide is placed in a well-sealed long-type quartz tube and then purged with inert gas. Then, the tube containing graphene oxide is quickly inserted into a muffle or tubular furnace preheated to 1050 °C and maintained at that temperature for around 10 min. This is usually called exfoliated graphene.

FIGURE 14.5 SEM image of chemically modified exfoliated graphene (*Courtesy of Dr. Soo-Jin Park et al.*)

Recently, graphene and graphene related materials which are structurally similar to exfoliated graphene have been actively used in the field of adsorption in gas and liquid (Fig. 14.5).

14.2.5. Others and so on

Few-layer graphene is synthesized by microwave treatment, conversion of nanodiamonds, and arc discharge of graphite. The method of preparing graphene by microwave treatment is just a postprocess using chemically synthesized graphene oxide in large quantities in a short time [50,51]. Enoki et al. reported that graphene can be synthesized by heating nanodiamonds in an inert or reducing atmosphere [52,53]. Nanodiamond particles are treated by soaking them in concentrated hydrochloric acid to avoid contamination by magnetic impurities. Then, they are annealed in a graphite container, in a graphite furnace, in a helium atmosphere at temperatures ranging from 1600 to 2500 °C.

Rao et al. [54] reported that arc discharge of graphite in a hydrogen atmosphere yields graphenes exclusively containing 2–3 layers, although the flakes are relatively small (100–200 nm). It has been found that this process depends on the presence of hydrogen gas during the arc discharge which terminates the dangling carbon bonds with hydrogen, which prevents the formation of closed structures [55]. To prepare graphene by the arc discharge method, direct current arc discharge of graphite evaporation is carried out in a water-cooled stainless-steel chamber filled with a mixture of hydrogen and helium without any catalyst [56]. It has been reported that the best transition characteristics are exhibited by graphene synthesized by arc discharge of graphite in hydrogen.

The above-mentioned methods for preparing graphene influence the number of sheets in the graphene, which plays an important role in their characteristics. Table 14.2 lists the variety of chemically modified preparation from graphite, graphene oxide, and graphite compounds (Ref. [2]).

TABLE 14.2 Chemically Modified Preparation Methods Through GrO, Graphite, and Graphite Compound

Method or Substance	Solvent	GrO Concentration (mg/FL)	Plate Size (nm)	Particle or Plate Thickness (nm)	Procedure	Refs.
HM	Water, acetone, ethanol, 1-propanol, ethylene glycol, DMSO, DMF, MP, pyridine, THF	0.5	100–1000	1.0–1.4	In DMF, MP, THF, and ethylene glycol, GrO is stable in the form of suspension for a long time; GrO exists as SLGrO	Paredes et al., Langmuir., 24 (2008) 10560
HM	Water/methanol, acetone, acetonitrile, solvent mixtures	3–4	Several hundred	1.2	Reduction og GrO with $NaBH_4$ and hydrazine	Si et al., Nano Lett., 8 (2008) 1679
HM	Ethanol		Several hundred	2	Reduction of a TiO_2/GrO composite in ethanol	Williams et al., ACS Nano, 2 (2008) 1487
MHM	Water	1	–	–	Reduction of GrO in the presence of poly(sodium 4-styrene sulfonate)	Stankovich et al., J. Mater. Chem., 16 (2006) 155
MHM	Water(pH 10)	0.5	Several	1	Reduction of GrO with hydrazine	Li et al., Nat. Nanotechnol., 3 (2008) 101
MHM	Water	0.1	–	1.7	Reduction of GrO with hydrazine	Xu et al., J. Am. Chem. Soc., 130 (2008) 5856

(Continued)

TABLE 14.2 Chemically Modified Preparation Methods Through GrO, Graphite, and Graphite Compound—cont'd

Method or Substance	Solvent	GrO Concentration (mg/FL)	Plate Size (nm)	Particle or Plate Thickness (nm)	Procedure	Refs.
MHM	Water	7	Several hundred	1.7	Reduction of GrO with hydrazine in an alkaline medium	Park et al., Chem. Mater., 20 (2008) 6592
MHM	DMF, MP, DMSO, HMPA	1	560	1	–	Stankovich et al., Carbon, 44 (2006) 3342
MHM	DMF	1.5	200–400	0.6	Reduction of GrO with hydrazine	Tung et al., Nat. Nanotechnol., 4 (2008) 25
OR	DMF, THF, $CC1_4$, DCE	0.5	–	0.5–2.5	GrO modified with octadecylamine and reduced with $NaBH_4$	Niyogi et al., J. Am. Chem. Soc., 128 (2006) 7720
S	DMF, DMAC, MP	1	Several Hundred	1.8–2.2	Reduction with hydrazine	Lomeda et al., J. Am. Chem. Soc., 130 (2008) 16201
S	THF	<0.48	–	1–2	GrO modified with octadecylamine with $AuCl_4^-$ additions	Muszynski et al., J. Phys. Chem. C, 112 (2008) 5263

S	MP, DMF, DCB, THF, nitromethane	0.1	100–2500	Average 1.75	Thermal methods, reduction at ≈1050 °C	Schniepp et al., J. Phys. Chem., 110 (2007) 8535, McAllister et al., Chem. Mater., 19 (2007) 4396, Ramanathan et al., Nat. Nanotechnol., 3 (2008) 327.
Graphite powder	MP, DMAC, GBL, DMI	0.01	Several thousand	1–5	Suspension sonication (yield, 1–12%)	Hernandez et al., Nat. Nanotechnol., 3 (2008) 563
Graphite rods	DMF, MP, DMSO	1	500–700	11	Electrochemical processing	Liu et al., Adv. Funct. Mater., 18 (2008) 1518-25
GIC $K(THF)_xC_{24}$	MP	0.15	Several hundred (platelets or nanoribbons)	0.35–0.4	Suspension sensitive to air and stable after drying and deposition on a substrate	Valles et al., J. Am. Chem. Soc., 130 (2008) 15802
EGr prepared from a GIC at 1000 °C (1 min)	DCE	0.1	<10 (nanoribbon width)	1–1.8	Sonication in the presence of PmPv (SLG and FLG nanoribbons with adsorbed PmPv)	Li et al., Science, 319 (2008) 1229
EGr prepared from a GIC at 1000 °C (1 min)	DMF	–	250	1	90% yield of G coated with DSPE-mPEG	Li et al., Nat. Nanotechnol., 3 (2008) 538

(*Continued*)

TABLE 14.2 Chemically Modified Preparation Methods Through GrO, Graphite, and Graphite Compound—cont'd

Method or Substance	Solvent	GrO Concentration (mg/FL)	Plate Size (nm)	Particle or Plate Thickness (nm)	Procedure	Refs.
EGr prepared from a GIC at 1000 °C (1 min)	Water, DMF, DMSO	0.015–0.020	From several hundred nanometers to several microns	2–3 layered grapheme	TCNQ-coated G platelets	Hao et al., Chem. Commun., 48 (2008) 6576
Graphite fluoride $(CF_x)_n$	TMEDA, hexane, isopropanol, ethanol	0.002–0.54	1600	0.95	Reacted with R-Li (R = butyl, hexyl) to give$(R_y\text{-}CF_2)_n$; G obtained by calcination	Worsley et al., Chem. Phys. Lett., 445 (2007) 51
Natural graphite	DMF	–	–	Above 50% SLG	Sonication for more than 3 h, centrifugation at 13000 rpm for 10 min	Blake et al., Nano Lett., 8 (2008) 1704
Ethanol	Ethanol	–	–	–	Ethanol:Na = 1:1, autoclaving at 220 °C for 72 h; yield, 0.1 g/FL ethanol	Choucair et al., Nat. Nanotechnol., 4 (2009) 30

Note: OR, authors's original method; S, Staudenmaier method; DMF, dimethylformamide; DMAC, dimethylacetamide; DMSO, dimethylsulfoxide; MP, *N*-methyl-pyrrolidone; THF, tetrahydrofuran; MC, dichloromethane; DCE, 1,2-dichloroethane; DCB, 1,2-dichlorobenzene; HMPA, hexamethylphosphoramide; GIC, graphite intercalation compound; EGr, exfoliated graphite; MP, *N*-methylpyrrolidone; PmPv, poly(phenylenevinylene-co-2,5-dioctoxy-*p*-phenylenevinylene); DMF, dimethylformamide; DMAC, dimethylacetamide; DMSO, dimethyl sulfoxide; DCE, 1,2-dichloroetnane; HMPA, hexamethylphosphoramide; GBL, γ-butyrlactone; DMI, 1,3-dimethyl1-2-imidazolidinone; DSPE-mPEG, 1,2-distearoyl-sn-glycero-3-phosphoethanolamine-*N*-[methoxy(polyethylene glycol)-5000]; TCNQ, 7,7,8,8-teracyanoquinodimethane; TMEDA, tetramethyl-ethylenediamine.

14.3. ADSORPTION BEHAVIORS

14.3.1. Introduction

The unique chemical properties of graphene affecting its adsorption behavior have only recently attracted attention of many researchers; thus, it has not yet been studied in sufficient detail. Therefore, the adsorption behaviors of graphene and graphene-related materials will attempt to mention the following.

Adsorption behavior depends significantly on the unique structure and concentration of chemical species created by a process of preparing graphene. It is possible to possess a large high surface area, sufficient porosity, superior conductivity, broad potential window, and rich surface chemistry. In addition, the extended polyaromatic π-electron system and coordinatively unsaturated terminal carbon atoms have a decisive effect on the reactivity of graphene. The readily polarizable π-electron system of graphene is equally reactive with electro- and nucleophilic species, which makes the system sensitive even to radical species.

14.3.2. Gas Sensor

Theoretical studies of gas molecule adsorption on the graphene surface have been reported, and these studies showed that pristine graphene has the ability to physically adsorb gas molecules (NO_2, H_2O, NH_3, CO, NO, and so on) [57,58].

For the first time, the possibility of inducing charge carriers by the adsorption of various gases, including NH_3, H_2O, and NO_2, was experimentally found [8]. Novoselov et al. reported [59] that NH_3 and CO molecules can induce electrons (act as donors); on the other hand, H_2O and NO_2 molecules result in holes as charge carriers (act as acceptors), which has been demonstrated by Hall effect measurements. Thus, those gases have been detected at remarkably low concentrations, and particularly NO_2 has even been detected in the extreme limit of single molecules [59].

These observations give grounds to hope that graphene can be used as an important role for gas sensors.

Graphene nanoribbons, one of the materials derived from graphene, are quasi-one-dimensional materials, as though graphene can be patterned *via* standard lithographic techniques [60,61]. Graphene nanoribbons are advantageous in small volume and free reactive edges compared with graphene. Additionally, graphene nanoribbons have characteristics very similar to those of carbon nanotubes, such as energy gap dependence of widths and crystallographic orientation [62–64]. Also, the covalent attachment of chemical functional groups at graphene nanoribbons edges and molecules significantly influences their electronic properties due to dangling bond defects [65,66]. It is well known that dangling bond defects around vacancy sites or at the tips play a very important role in chemically inactive carbon materials such as carbon nanotubes [67–70]. However, unlike carbon nanotubes, graphene nanoribbons have long reactive edges; thus, they might be notably more accessible to doping and chemical modification as well as more susceptible to structural defects and impurities (Fig. 14.6).

Duan et al. [70] reported that the electronic and transport characteristics of graphene nanoribbons with armchair-shaped edges are sensitive to the adsorption of NH_3. The system of NH_3-adsorbed graphene nanoribbons reveals n-type semiconducting behavior because NH_3 molecule adsorption induces unoccupied local states in the conduction band, and the Fermi level is shifted into original conduction bands. In addition, partial charge density analysis demonstrated that the states near the Fermi level mainly come from the carbon atoms of the ribbons rather than the NH_3 molecules (Fig. 14.7).

Duan et al. have also studied the adsorption of other gas molecules, such as CO, NO, NO_2, O_2, N_2, and CO_2, on graphene nanoribbons. They demonstrated that other gas molecules have little effect on modifying the conductance properties of graphene nanoribbons, which are attributed to the adsorption geometries and electronic structure of graphene nanoribbons. Thus, they concluded that, on the basis of this characteristic, graphene nanoribbons could function as n-type transistors by NH_3 adsorption or as p-type transistors by CO_2 or O_2 adsorption on graphene nanoribbons.

Since Geim et al. [59] first reported the graphene Hall effect sensor device capable of detecting individual molecules of NO_2, the transport properties of graphene have been shown to be important to molecules adsorbed on the surface. However, the strength and character of the adsorption (chemical or physical) and the degree of charge transfer between the analyte and graphene are still under debate (Fig. 14.8).

14.3.3. Hydrogen Storage

The revised gravimetric capacity target for hydrogen storage systems of the US Department of Energy for 2015 is 5.5 wt.%. These targets have not yet been reproducibly met by any available storage technology. Solid carbon adsorbents, such activated carbons [75,76], fullerenes [77,78], carbon nanotubes [79–82], graphites, carbon nanofibers [83–85], ordered porous carbon [86,87], and carbon-based hybrid composites [88,89], had previously been considered to be promising candidates for hydrogen storage due to their large surface area, structural stability, and light mass. In particular, the adsorption mode of H_2 on carbon nanomaterials is reversible weak physisorption, which is a key factor in the operation of hydrogen/fuel-cell-powered systems.

To satisfy the increasing need for hydrogen storage, graphene has been attracting a lot of attention, as the newest carbon materials. It appears to be composed of a monolayer of carbon atoms packed into a dense honeycomb crystal structure. Single-layer graphene is theoretically predicted to have a large surface area of 2,600 m^2/g, while the specific surface area of few-layer graphenes prepared by various methods is in the range from 250 to 1,500 m^2/g.

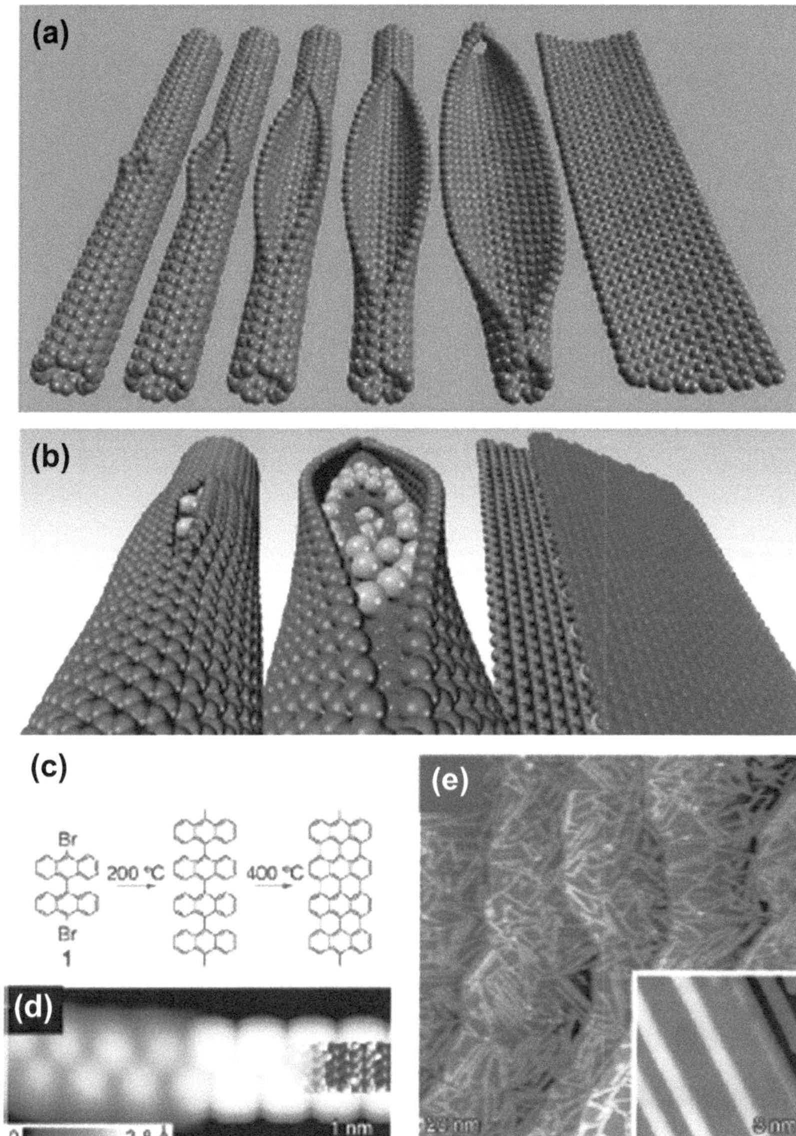

FIGURE 14.6 Graphene nanoribbons. (a) Oxidation and (b) reduction approaches to unzip CNTs into graphene nanoribbons. (c) Bottom-up synthesis of graphene nanoribbons from precursor 1 to straight graphene nanoribbons. STM images taken (d) after surface-assisted C–C coupling at 200 °C, showing a polyanthrylene chain, and (e) after the final cyclodehydrogenation step at 400 °C, showing a straight graphene nanoribbons structure. *(With permission from American Chemical Society © 2011 (Ref. [71–74])).*

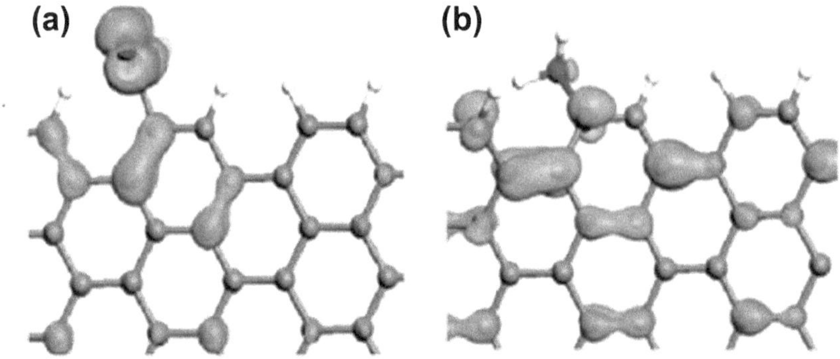

FIGURE 14.7 Isosurface plots of the partial charge density at the Γ point of the band crossing the Fermi level for the 10-armchair graphene nanoribbons with (a) CO_2 adsorption and (b) NH_3 adsorption. The isovalue is 0.01 e/Å^3. *(With permission from American Chemical Society © 2008 (Ref. [70])).*

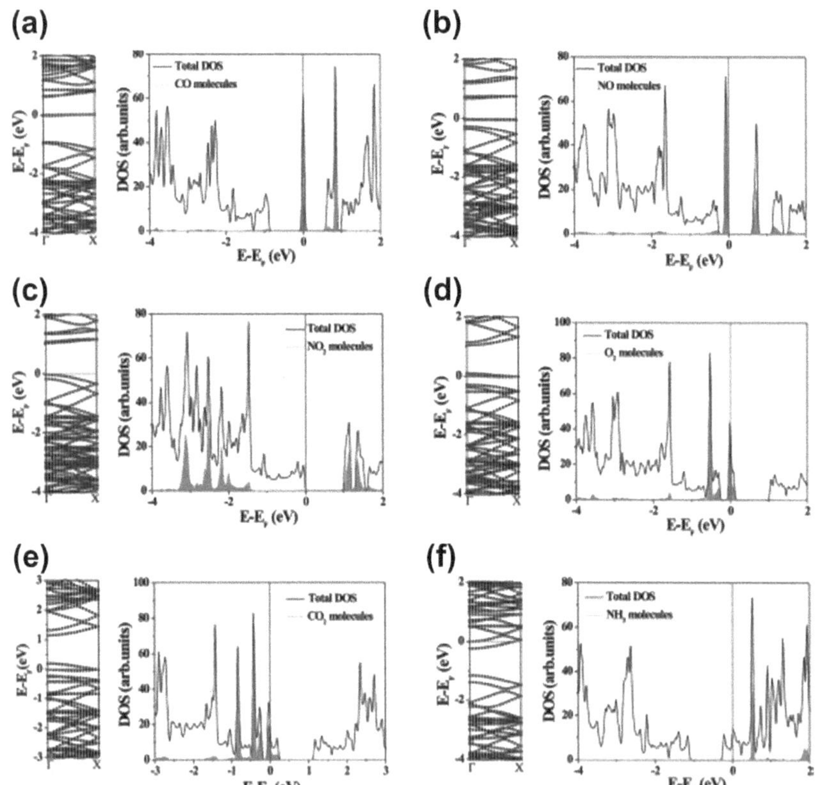

FIGURE 14.8 Band structure and density of states (DOS) of 10-armchair graphene nanoribbons with gas molecule adsorption: (a) CO, (b) NO, (c) NO_2, (d) O_2, (e) CO_2, and (f) NH_3. The local density of states (LDOS) of gas molecules is also plotted (red filled area under DOS curve). The Fermi level is set to zero. *(With permission from American Chemical Society © 2008 (Ref. [70])).*

Furthermore, recently, graphene-like nanosheets (in some cases, it has contained some oxygen functionalities; therefore, all of these graphene-like nanomaterials are called "graphene oxide") have been studied in relation to the adsorption of hydrogen. They are produced relatively easily *via* oxidative treatment of graphitic materials with strong oxidizing agents. Graphene oxide is a stacked structure similar to graphite but with a wider spacing between graphene oxide nanosheet layers with diverse range from 6 to 15 Å [90].

Psofogiannakis and Froudakis et al. [90] have studied new materials that are based on a previously simulated large surface area graphitic nanostructure, which is called pillared graphene oxide (PGO). It is composed of graphene-layered stacks that are joined by uniformly spaced carbon nanotubes. Their structures are shown in Fig. 14.9. It is suggested that their hydrogen storage capacity is influenced by the interplay between the graphene oxide and carbon nanotubes and the effect of extended doping on carbon surfaces with −OLi. As a simulated result, this new material allows the combination of (i) using carbon materials with a large and adjustable pore surface area and (ii) incorporating functionalities into the structure that increase the strength of interaction of hydrogen molecule with it.

Figure 14.10 shows the gravimetric hydrogen storage isotherms at 300 K of Li-doped PGO with a pore spacing of $d = 11$ Å and doping ratios of O/C = 0 (pristine PGO), 1:16, 1:8, and 1:4. The hydrogen storage capacity of Li-doped PGO, adding −OLi at an atomic ratio of O/C = 1:4, is significantly better than that of pristine PGO, for a final uptake of 3.0 wt.% at 100 bar. Li-doped PGO is

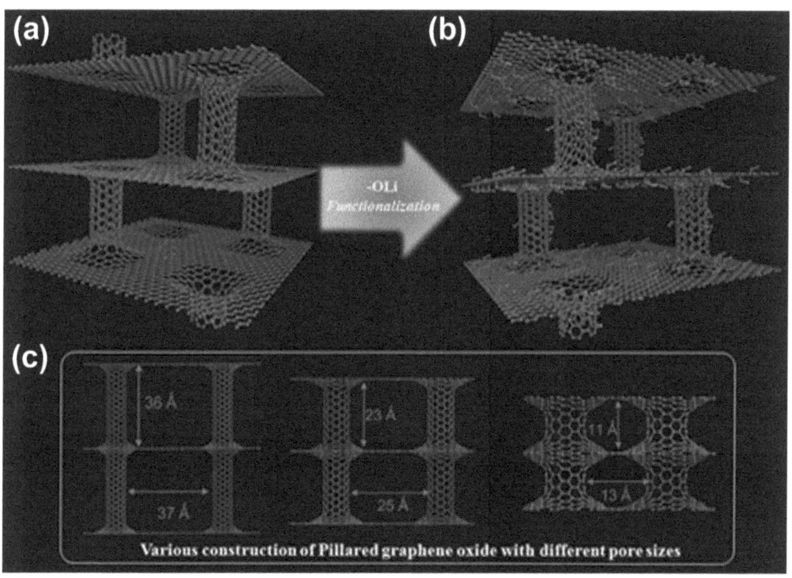

FIGURE 14.9 (a) Structure of pillared graphene composed of fused graphene layers and carbon nanotubes (PGO) and (b) Li-doped PGO; −OLi functionalities adsorbed on pristine PGO. *(With permission from American Chemical Society © 2010 (Ref. [90])).*

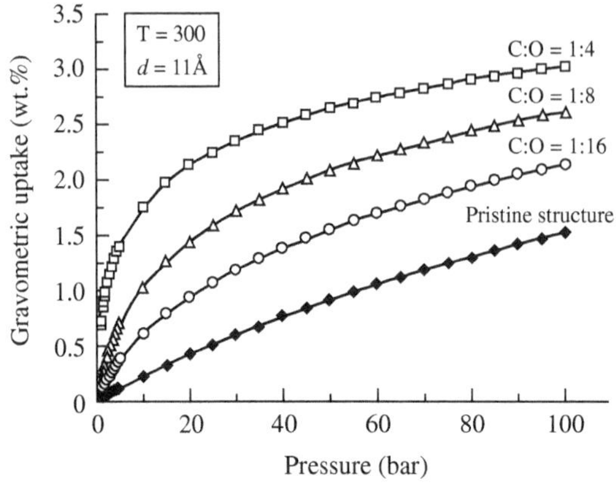

FIGURE 14.10 Gravimetric hydrogen storage capacity of Li-doped PGO. *(With permission from (Ref. [90]))*.

considered to be representative of a novel class of materials for hydrogen adsorption that can be formed from a variety of oxidized graphitic materials. From the Psofogiannakis and Froudakis' simulated study, the Li-doped PGO materials served as a model that helped to understand the mechanism of hydrogen physisorption on the existence of functionalities and effects of the pore volume and surface density of functionalities on hydrogen storage capacity.

G. Srinivas et al. [91] studied graphene oxide prepared by the reduction of a colloidal suspension of exfoliated graphite oxide for hydrogen storage capacity. As shown in Figs 14.11 and 14.12, the sample was found to be in

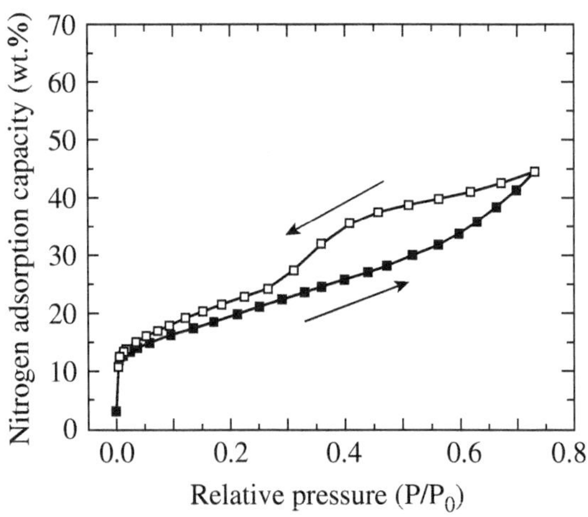

FIGURE 14.11 Nitrogen/77 K adsorption–desorption isotherms of graphene oxide prepared from Srinivas et al. *(With permission of (Ref. [91]))*.

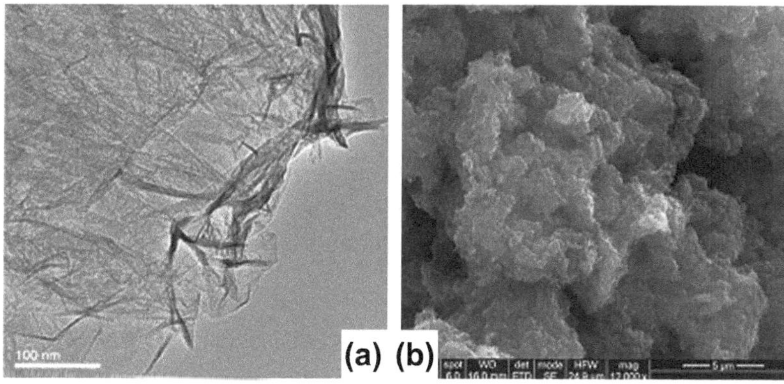

FIGURE 14.12 TEM (a) and SEM (b) images of graphene oxide prepared from G. Srinivas et al. *(With permission of (Ref. [91]))*.

a highly agglomerated state with many wrinkles, and it had a specific surface area of 640 m^2/g as measured by nitrogen adsorption at 77 K.

The isotherms exhibited a typical type-I curve at low relative pressure and a hysteresis loop at relative pressure from 0.4, indicating the presence of microporosity, mesoporosity, and some macroporosity [92]. They explained that the overlap and stacking of the exfoliated layers influenced the significant low textural characteristic.

In the Srinivas study, it was shown that the highest hydrogen storage capacities of a prepared sample were about 1.2 wt.% and 0.1 wt.% at pressures up to 10 bar at the temperatures of 77 K and 298 K, respectively. Furthermore, they predicted the adsorption capacities at 100 bar of ~3 wt.% and 0.72 wt.% at 77 K and 298 K, respectively, obtained by linear extrapolation using the model LangmuirEXT2 expression. The room temperature hydrogen adsorption capacity was higher than other recent experimental results on hydrogen storage in a pristine graphene sample [93], which might be attributable to the higher isosteric heat of adsorption values. Cheng et al. [93] reported that 0.4 wt.% and ~0.2 wt.% hydrogen uptake of graphene oxide were obtained at 77 K under 1 bar and at 298 K under 60 bar, respectively.

Ghosh et al. [94] found that high hydrogen adsorption capacity on graphene oxide obtained by thermal exfoliation or conversion of nanodiamond was 3.1 wt.% (in the case of single-layer graphene) at 298 K and 100 bar. As shown in Fig. 14.13, the hydrogen uptake varied linearly with the specific surface area, and the presence of more than one layer of graphene oxide in this study caused a decrease in the hydrogen uptake. In addition, it is noted that the specific surface area of graphene oxide depends on the number of layers, and not the pore effect [95,96].

It is interesting to note that the H_2–graphene interaction energies obtained in their study satisfy the optimum conditions for adsorptive storage of hydrogen

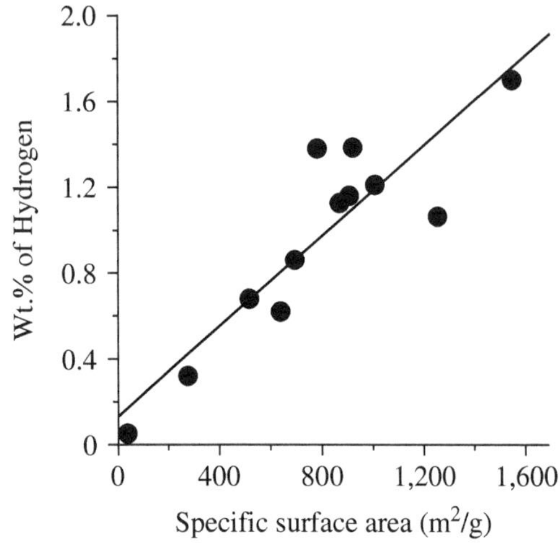

FIGURE 14.13 Linear relationship between the specific surface area and the hydrogen storage capacity at 77 K and 1 bar. *(With permission from American Chemical Society © 2008 (Ref. [94])).*

[97]. They also expected that higher values for hydrogen storage should be possible by preparing better samples and by reducing the average number of graphene layers.

The transition metal loaded/graphene oxide composites for hydrogen storage are being extensively researched. Zhang et al. [98] have investigated the structural properties of graphene oxide and the feasibility of using graphene oxide to anchor Ti for hydrogen storage with first-principles computations. They reported that on graphene oxide there exist stable structural motifs as the basic function units of oxygen functionalities. It is considered that graphene oxide could be an excellent substrate for Ti, but for hydrogen storage, it requires the hydrogenation of epoxy and open sp^2 sites. Upon hydrogenated graphene oxide, the Ti binding energy is increased to a level that is high enough to prevent Ti clustering. Dihydrogen binding energy is also favorable for room temperature storage. The theoretical gravimetric density was found to be 4.9 wt.%. Thus, Ti-anchored graphene oxide may offer a feasible solution for hydrogen storage (Fig. 14.14).

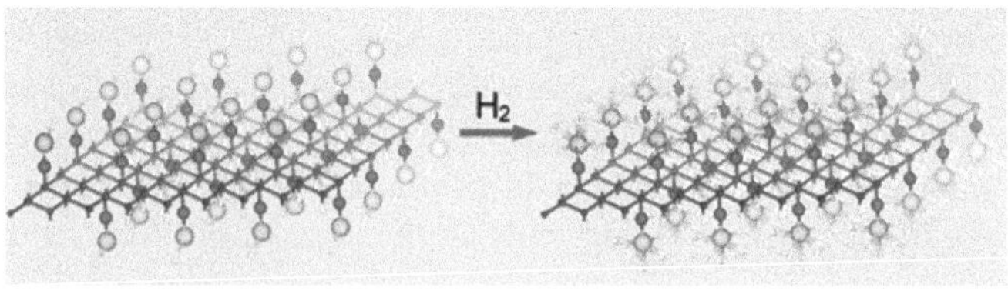

FIGURE 14.14 The multibinding hydrogen adsorption model of Ti-anchored graphene oxide. *(With permission from American Chemical Society © 2009 (Ref. [98])).*

Park et al. [99] have reported nickel/graphite hybrid materials which are produced by mixed acid treatment of graphite flakes for hydrogen storage. The graphene oxide samples were prepared by dipping graphite flake into a mixed acid solution (H_2SO_4 and H_2O_2) in a range of 10:1–5:5 at room temperature. The experimental conditions of mixed acid treatments and textural properties of produced graphene oxides are listed in Table 14.3.

Figure 14.15 is a schematic of the pore development of a graphite flake using the acidic method. The layer distance of each graphite sheet is nearly 0.354 nm, and it is impossible for hydrogen molecules to be adsorbed at this dimension. However, after the expansion of the layers, the layer distance widens and new adsorption sites become exposed for hydrogen molecules.

Furthermore, Ni nanoparticles were loaded onto graphene oxide surfaces as a function of Ni content in an effort to increase the final hydrogen uptake. In Fig. 14.16, the highest hydrogen storage result of the 8:2/Ni-0.05 sample was 4.48 wt.%, which was prepared by 0.05 wt.% of $NiSO_4$ per 1 g of graphene oxide.

However, it is important to note that the 10:1 sample showed lower hydrogen uptake than that of the 8:2 sample, although the former had a higher specific surface area. This result indicates that the 8:2 samples may have a more suitable pore dimension for the adsorption of hydrogen molecules. Although the 10:1 sample had a high specific surface area, it was checked by a nitrogen molecule, indicating that the area was not suitable for hydrogen molecules.

Park et al. suggested a new dipole-induced mechanism for hydrogen storage. As shown in Fig. 14.17, the first hydrogen molecules introduced can be reacted with metal particles due to the high oxidation power of metals on graphite surfaces due to the Kubas reaction [100,101].

TABLE 14.3 Specific Surface Area and Pore Volumes of Graphene Oxide as a Function of the Mixed Acid Ratio

Mixed Acid Ratio (H_2SO_4:H_2O_2)	Specific Surface Area (m^2/g)	Total Pore Volume (cm^3/g)
As-received	<5	–
10:1	145	0.056
8:2	110	0.042
6:4	80	0.031
5:5	65	0.025

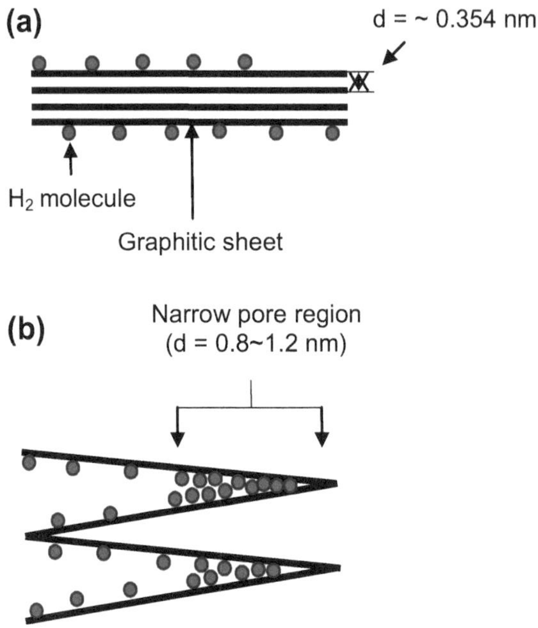

FIGURE 14.15 A schematic of hydrogen storage sites on graphite surfaces: (a) as-received graphite and (b) graphene oxide manufactured by mixed acid treatments in this study.

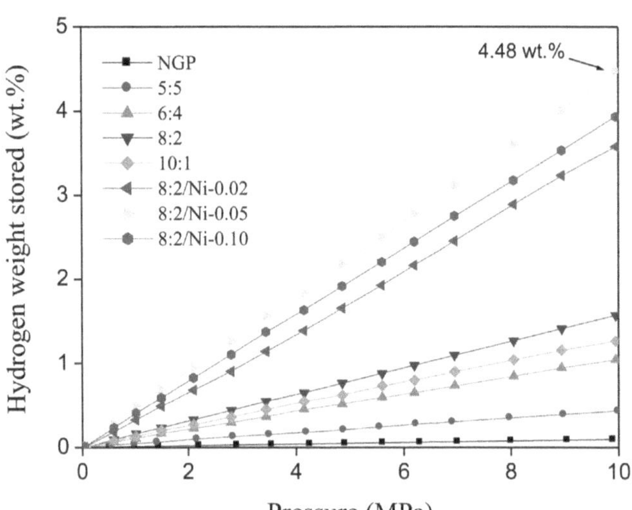

FIGURE 14.16 Hydrogen storage behavior of graphene oxide as a function of mixed acid ratio and nickel content.

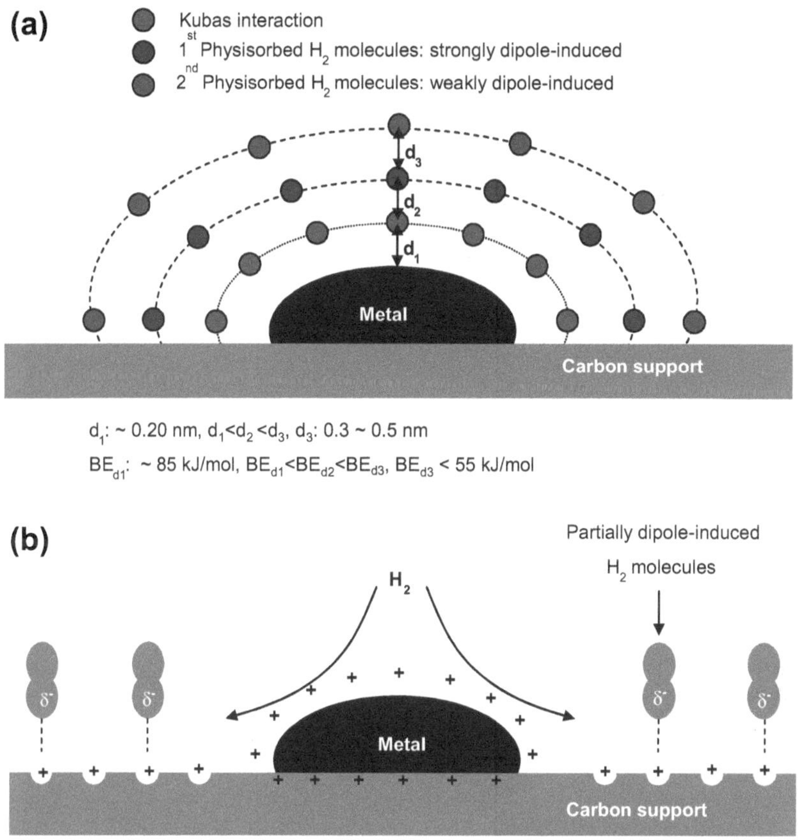

FIGURE 14.17 A schematic of the hydrogen storage mechanism on metal/carbon surfaces: (a) the accumulation of hydrogen molecules on metal/carbon surfaces, (b) the dispersion of hydrogen molecules from metals to carbon surfaces by dipole-induced effects. *(With permission from (Ref. [99]))*.

Here, unlike the chemical binding of hydrogen atoms to transition metal on carbon surfaces, lone transition metal atoms bind hydrogen *via* the so-called "Kubas interaction," in which the H–H bond undergoes some elongation rather than complete dissociation into atoms.

However, the amount of this initial reaction is very small. The binding energy and bond distance in this case are 85 kJ/mol and 0.2 nm, respectively, because the binding energy and bond distance of metal hydride are normally close to 85 kJ/mol and 0.2 nm, respectively. The second group of hydrogen molecules introduced cannot chemically react due to a site limitation caused by the initially introduced hydrogen molecules; however, they are physically adsorbed by dipole-induced effects. The hydrogen molecules are basically nonpolar, but the strong interaction of the metal particles leads to the dipole-inducing effects of the hydrogen molecules. The third group and any remaining hydrogen molecules are adsorbed by the same mechanism of high gas pressure, but this equilibrium can be broken when the interaction of oxidized graphite surfaces and dipole-induced hydrogen molecules is stronger than metal–hydrogen interactions [102–106]. This mechanism can be applied only under higher pressure, and graphite supports are charged strong electron acceptors.

These unique characteristics of graphene oxide can play a leading role in enhancing hydrogen storage capacity.

14.3.4. Carbon Dioxide Capture

Ramaprabhu et al. [107] suggested graphene oxide for CO_2 adsorption, which is prepared by hydrogen-induced thermal exfoliation of graphene oxide at 200 °C for possible large-scale production of graphene [108]. They confirmed the physical adsorption of CO_2 in graphene oxide using FT-IR spectroscopy.

As shown in Fig. 14.18, the band corresponding to –OH groups at 3435 cm^{-1} is quite prominent compared by the insignificant ratios of antisymmetric and symmetric $=CH_2$ vibrations for graphene oxide in this work. In addition, the intense peaks correspond to the carboxylic and carbonyl groups at 1726 (>C=C), 1563 (=CH–), and 1173 cm^{-1} (>C=O). An additional peak of 2324 cm^{-1} corresponds to the adsorbed molecular CO_2 in graphene oxide resulting from physical CO_2 adsorption. Ramaprabhu et al. reported that the maximum adsorption capacity was 21.6 mmol/g at 298 K and 11 bar. This value is higher than that of other adsorbents. In addition, they compared the CO_2 adsorption capacity of graphene oxide with those of other usual adsorbents, such as, carbon materials, MOF, and zeolite, as mentioned below.

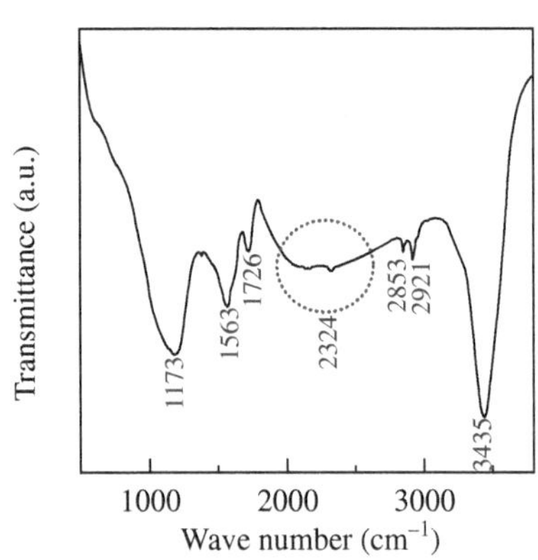

FIGURE 14.18 FT-IR spectrum of molecular adsorption of CO_2 in graphene oxide sample. Copyright © 2011, Author(s). This article is distributed under a Creative Commons Attribution 3.0 Unported license

Before everything else, Ramaprabhu's CO_2 adsorption capacity obtained by the study of multiwalled carbon nanotubes was observed to be 11.7, 8.3, and 7.0 mmol/g at 11 bar and 25 °C, 50 °C, and 100 °C, respectively [109]. Also, Siriwardane et al. [110] and Gensterblum et al. [111] reported 7 mmol/g and 6 mmol/g of CO_2 adsorption in activated carbon, respectively, at 45 °C and 11 bar. Zhang et al. [112] reported enhancement of around 20% in CO_2 uptake, which was achieved by modifying the AC with nitrogen at 298 K, and adsorption capacity was about 16 mmol/g adsorption capacity at 11 bar. Cavenati et al. reported 3.2 mmol/g of CO_2 adsorption in 13X zeolite at 298 K and 12 bar [113]. A high-pressure CO_2 adsorption study on various metal-organic frameworks by Millward and Yaghi [114] achieved CO_2 adsorption capacity ranging from 2 to 8 mmol/g under similar conditions.

In addition, Rao et al. reported CO_2 adsorption by graphene oxide [94]. They studied the CO_2 adsorption of graphene oxide at 195 K and 1 bar, which showed somewhat variable values of CO_2 uptake in the range from 10 to 38 wt.%. It is possible that low CO_2 uptake values can occur in graphene samples containing large particulates.

Next, we will discuss metal-loaded graphene composites for CO_2 adsorption. Guo et al. [115] investigated a first-principles study of CO_2 adsorbent materials consisting of calcium atoms and carbon-based nanostructures (Fig. 14.19).

They found that Ca-decorated graphene possesses unusually large CO_2 uptake capacities in the range from 0.1 to 0.4 g CO_2/g sorbent under a low-gas-pressure regime as a result of their unique topology and a strong interaction between the metal nanoparticles and CO_2 molecules. It is also reported that the strength of the Ca metal and CO_2 molecule interactions can be efficiently tuned as a function of the Ca loading content (Table 14.4).

Enhancement of the CO_2 reactivity properties of carbons in the presence of Ca atoms can be understood in terms of electronic structure, that is, overlapping of s,d-metallic and p-molecular states in the region near the Fermi level. Thus, by the use of first-principles simulation techniques, Guo et al. have confidently proposed that these Ca metal/carbon-based nanomaterial composites exhibit suitable adsorbents for CO_2 capture and sequestration application.

Magaña et al. [116] studied the adsorption of CO_2 on a Ti–graphene system with high metal coverage using density functional theory and molecular

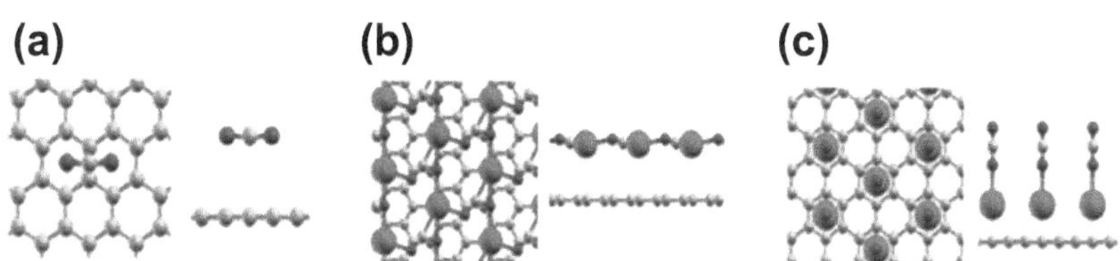

FIGURE 14.19 Carbon capture geometry optimized structure obtained; (a) pristine graphene, (b) Ca/graphene at 12.5%, and (c) Ca/graphene at 16.67% *(With permission of (Ref. [115]))*.

TABLE 14.4 Binding Energy (per CO_2 Molecule) and Corresponding Gas-adsorption Capacities of Ca/graphene as a Function of Ca Content (Ref. [115])

Ca Content (%)	Binding Energy (eV/molecule)	CO_2 Uptake (g CO_2/g sorbent)
0	−0.035	0.153
0–5	−1.022	0.134
5–15	−2.731	0.323
15–20	−0.101	0.392

dynamics simulation. Positively charged Ti atoms on graphene surfaces attract negatively charged oxygen atoms toward the graphene surfaces. This force is stronger than the initial repulsion on the C atom by the Ti atoms. When a CO_2 molecule approaches the graphene surfaces, the CO_2 molecule cannot be linear any longer. The O atoms in a bent CO_2 molecule are under different force fields. Thus, one O atom traps electronic charge from the Ti atoms of the upper plane and ends bonded to four Ti atoms, resulting in the CO_2 molecule interacting very strongly with the Ti atoms of the upper plane, as shown in Fig. 14.20.

To complete what Magaña et al. mentioned above [116], it is essential to place monodispersed Ti atoms on a graphene surface. Rosenberg et al. [117] reported that their experiments showed that adsorbed Ti atoms on a clean

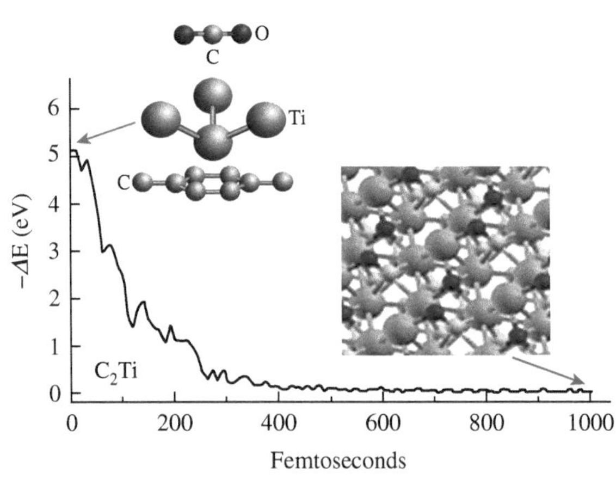

FIGURE 14.20 CO_2 adsorption onto Ti–graphene (C_2Ti). The final configuration shows the dissociation of the molecule in CO and O. The initial orientation of the molecule has no influence on the result.

graphite surface tend to form islands or some clusters. Meanwhile, Dai et al. [118] showed that not only the Ti coating atoms can be dispersed along single-walled carbon nanotubes, but also Al and Au nanoparticles are uniformly deposited on the pretreated nanotubes, which means that this procedure could facilitate the dispersion of Ti atoms on graphene surfaces.

14.3.5. Ammonia Adsorption

Eklund et al. [119] have reported studies of the interaction of NH_3 with a graphene field-effect transistor (FET) supported on Si/SiO_2 substrates to provide further insight into the nature of the molecule–graphene interaction mechanism. Here, the SiO_2 is used as a gate dielectric, and the heavily doped Si substrate is used as the bottom gate electrode. From the study, they elucidated that graphene FETs initially behave as a p-type system due to exposure to air, while these FETs can be rendered n-type by long-term vacuum degassing at 200 °C.

Leenaerts et al. [120] have investigated the adsorption process of NH_3 molecules on graphene through first-principle calculations and showed the presence of two main charge-transfer mechanisms. They considered two different orientations of NH_3 molecules with respect to the graphene surface: (i) the H atoms pointing away from the surface (*u*) and (ii) the H atoms pointing toward the surface (*d*). They showed the calculated adsorption energies and charge transfers (Table 14.5). It was found that the charge transfer is only determined by the orientation of the molecule and not by the adsorption sites. There is a small charge transfer from the molecule to the graphene surface of 0.03*e* in the *u* orientation, and there is almost no charge transfer in the *d* orientation.

TABLE 14.5 NH_3 on Graphene: the Adsorption Energy (E_a) and the Charge Transfer (ΔQ) from Graphene to the Molecule for Various Adsorption Sites and Orientations (Ref. [120])

Position	Orientation	E_a (meV)	ΔQ (e)
B	*u*	21	−0.026
T	*u*	20	−0.026
C	*u*	31	−0.027
B	*d*	15	−0.001
T	*d*	16	0.000
C	*d*	25	0.001

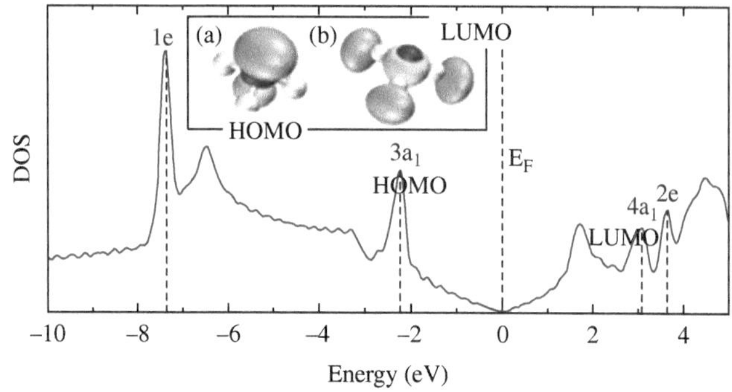

FIGURE 14.21 The density of states (DOS) of graphene with adsorbed NH_3 molecule calculated with a 60 × 60 × 1 MP grid and a Gaussian smearing of 0.14 eV. The dotted vertical lines show the molecular orbitals of NH_3.

As shown in Fig. 14.21, this dependence on the orientation can be understood in terms of the highest occupied and lowest unoccupied molecular orbital (HOMO and LUMO) of the molecule and the total DOS (density of states) of the system. The HOMO and LUMO are both close enough to the Dirac point to cause some charge transfer through orbital mixing (hybridization) with graphene orbital. In the *u* orientation, the NH_3 molecule acts as a donor, because the HOMO is the only orbital that can have a significant overlap with the graphene orbitals, and, as a consequence, acts as a donor in this orientation. Meanwhile, in the *d* orientation, both HOMO and LUMO of the NH_3 molecule are able to interact with graphene. This leads to competing charge transfers to and away from graphene, resulting in a total charge transfer close to zero. As observed experimentally in their study, the *u* orientation is energetically favored, which explains the donor character of NH_3 molecules.

14.3.6. Surface Functionalization

As it were the unique chemical properties of graphene, in addition to adsorption, a single-layer graphene is of interest as a giant flat molecule. Being a polyaromatic system, it is expected to attach reactive species to form covalent bonds.

In 2009, Novoselov et al. reported [121] that a new material, termed graphane, was obtained from graphene. Exposure of graphene to a hydrogen plasma leads to hydrogen attachment to graphene sheets (hydrogenation), and they remain flat. The process is reversible: heat treatment at 450 °C for 24 h leads to dehydrogenation and restores the π-electron system of graphene. Typical hydrogenated graphene (graphane) shows p-type behavior according to its electrical tests with an insulator transition. In contrast to graphene, a good conductor, graphane is shown to be dielectric [122–124] (You can be provided in detail and sufficient information from the ref. [123]).

It is well known that graphene is chemically inert and stable in air up to 200 °C [124]; thus, it requires relatively high energetic species to break its sp^2 bonds and initiate functionalization reactions. However, the above demonstrates that graphane derived from graphene is highly reactive. It is also expected that these characteristics from the defects or vacancies of graphane perform a role as adsorption sites to induce gas molecules to enhance donor–acceptor interactions in comparison with chemically inactive graphene.

14.4. SUMMARY

The field of graphene-related research has grown at a spectacular pace since a single-layer flake was first isolated in 2004. Graphene and graphene-related materials are promising candidates as components in various applications due to their superior properties.

Recently, the development of various methods for producing these materials has stimulated a vast amount of research, for example, epitaxial growth and chemical vapor deposition techniques, micromechanical exfoliation, chemical reduction, thermal exfoliation, microwave treatments, and conversion of nanodiamonds.

The main purpose of this chapter is to review the adsorption behaviors of graphene and graphene-related materials. We have also reported not only the nature of adsorption sites for various gases on their surfaces for high H_2 storage capacity and CO_2 capture, but also the sensor's ability for detecting individual gas molecules, such as CO, NO, NO_2, O_2, N_2, CO_2, and NH_3. The superior adsorption behaviors of graphene and graphene-related materials might be attributable to their large specific surface area and special interaction between electron-donor and -acceptor molecules.

REFERENCES

[1] K.S. Novoselov, A.K. Geim, S.V. Morozov, et al., Electric field effect in atomically thin carbon films, Science 306 (2004) 666–669.
[2] S.V. Tkachev, E.Y. Buslaeva, S.P. Gubin, Graphene: a novel carbon nanomaterial, Inorganic Mater. 47 (2011) 1–10.
[3] A.K. Geim, K.S. Novoselov, The rise of graphene, Nat. Mater. 6 (2007) 183–191.
[4] D.D.L. Chung, Review graphite, J. Mater. Sci. 37 (2002) 1475–1489.
[5] M.D. Stoller, S. Park, Y. Zhu, et al., Graphene-based ultracapacitors, Nano Lett. 8 (2008) 3498–3502.
[6] J.S. Bunch, V.D. Zande, A.M. Verbridge, et al., Electromechanical resonators from graphene sheets, Science 315 (2007) 490–493.
[7] A.A. Balandin, S. Ghosh, W. Bao, et al., Extremely high thermal conductivity of graphene: experimental study, Nano Lett. 8 (2008) 902–907.
[8] W.A. de Heer, C. Berger, X.S. Wu, et al., Epitaxial graphene, Solid State Commun. 143 (2007) 92–100.

[9] K.V. Emtsev, A. Bostwick, K. Horn, et al., Towards wafer-size graphene layers by atmospheric pressure graphitization of silicon carbide, Nat. Mater. 8 (2009) 203–207.
[10] K.S. Kim, Y. Zhao, H. Jang, et al., Large-scale pattern growth of graphene films for stretchable transparent electrodes, Nature 457 (2009) 706–710.
[11] A. Reina, X.T. Jia, J. Ho, et al., Large area, few-layer graphene films on arbitrary substrates by chemical vapor deposition, Nano Lett. 9 (2009) 30–35.
[12] Q.K. Yu, J. Lian, S. Siriponglert, et al., Graphene segregated on Ni surface and transferred to insulators, Appl. Phys. Lett. 93 (2008) 113103–113106.
[13] X.S. Li, W.W. Cai, J.H. An, et al., Large-area synthesis of high-quality and uniform graphene films on copper foils, Science 324 (2009) 1312–1314.
[14] C. Berger, Z. Song, X. Li, et al., Electronic confinement and coherence in patterned epitaxial graphene, Science 312 (2006) 1191–1196.
[15] J. Hass, W.A. de Heer, E.H. Conrad, The growth and morphology of epitaxial multilayer graphene, J. Phys. Condens. Matter. 20 (2008) 323202–323229.
[16] C. Berger, Z.M. Song, T.B. Li, et al., Ultrathin epitaxial graphite: 2D electron gas properties and a route toward graphene-based nanoelectronics, J. Phys. Chem. B 108 (2004) 19912–19916.
[17] L.A. Jauregui, H. Cao, W. Wu, et al., Electronic properties of grains and grain boundaries in graphene grown by chemical vapor deposition, Solid State Commun. 151 (2011) 1100–1104.
[18] A. Reina, X. Jia, J. Ho, et al., Large area, few-layer graphene films on arbitrary substrates by chemical vapor deposition, Nano Lett. 9 (2008) 30–35.
[19] M. Eizenberg, J.M. Blakely, Carbon monolayer phase condensation on Ni(111), Surf. Sci. 82 (1970) 228–236.
[20] L.C. Isett, J.M. Blakely, Segregation isosteres for carbon at the (100) surface of nickel, Surf. Sci. 58 (1976) 397–414.
[21] S. Bae, H. Kim, Y. Lee, et al., Roll-to-roll production of 30-inch graphene films for transparent electrodes, Nat. Nanotechnol. 5 (2010) 574–578.
[22] Y.P. Chen, Q.K. Yu, Nanomaterials graphene rolls off the press, Nat. Nanotechnol. 5 (2010) 559–560.
[23] J.C. Meyer, A.K. Geim, M.I. Katsnelson, et al., The structure of suspended graphene sheets, Nature 446 (2007) 60–63.
[24] K.I. Bolotin, K.J. Sikes, Z. Jiang, et al., Ultrahigh electron mobility in suspended graphene, Solid State Commun. 146 (2008) 351–355.
[25] K.I. Bolotin, K.J. Sikes, J. Hone, et al., Temperature-dependent transport in suspended graphene, Phys. Rev. Lett. 101 (2008) 096802–096805.
[26] K.S. Novoselov, D. Jiang, F. Schedin, et al., Two-dimensional atomic crystals, Proc. Natl. Acad. Sci. U.S.A 102 (2005) 10451–10453.
[27] S.J. Park, van der Waals interaction at surfaces, in: Encyclopedia of Surface and Colloid Science, Taylor & Francis, New York, 2006.
[28] S.J. Park, Long-range force contributions to surface dynamics, in: Interfacial Forces and Fields: Theory and Applications, Marcel Dekker, 1990. New York.
[29] X. Li, W. Cai, L. Colombo, et al., Evolution of graphene growth on Ni and cu by carbon isotope labeling, Nano Lett. 9 (2009) 4268–4272.
[30] S.J. Park, M.K. Seo, Interface Science and Composites, vol. 18, Elsevier, 2011.
[31] S. William, Hummers Jr., E.O. Richard, Preparation of graphitic oxide, J. Am. Chem. Soc. 80 (1958) 1339.

[32] M.J. Allen, V.C. Tung, R.B. Kaner, Honeycomb carbon: a review of graphene, Chem Rev. 110 (2009) 132–145.
[33] Y. Hernandez, V. Nicolosi, M. Lotya, et al., High-yield production of graphene by liquid-phase exfoliation of graphite, Nat. Nanotechnol. 3 (2008) 563–568.
[34] M. Lotya, Y. Hernandez, P.J. King, et al., Liquid phase production of graphene by exfoliation of graphite in surfactant/water solutions, J. Am. Chem. Soc. 131 (2009) 3611–3620.
[35] D.C. Marcano, D.V. Kosynkin, J.M. Berlin, et al., Improved synthesis of graphene oxide, ACS Nano 4 (2010) 4806–4814.
[36] K.S. Kim, S.J. Park, Influence of multi-walled carbon nanotubes on the electrochemical performance of graphene nanocomposites for supercapacitor electrodes, Electrochim. Acta 56 (2011) 1629–1635.
[37] H. He, J. Klinowski, M. Forster, et al., A new structural model for graphite oxide, Chem. Phys. Lett. 287 (1998) 53–56.
[38] S. Stankovich, D.A. Dikin, R.D. Piner, et al., Synthesis of graphene-based nanosheets via chemical reduction of exfoliated graphite oxide, Carbon 45 (2007) 1558–1565.
[39] J.R. Lomeda, C.D. Doyle, D.V. Kosynkin, et al., Diazonium functionalization of surfactant-wrapped chemically converted graphene sheets, J. Am. Chem. Soc. 130 (2008) 16201–16206.
[40] V.C. Tung, M.J. Allen, Y. Yang, et al., High-throughput solution processing of large-scale graphene, Nat Nanotechnol. 4 (2008) 25–29.
[41] P.G. Ren, D.X. Yan, X. Ji, et al., Temperature dependence of graphene oxide reduced by hydrazine hydrate, Nanotechnology 22 (2011) 055705–055713.
[42] S. Stankovich, D.A. Dikin, G.H.B. Dommett, et al., Graphene-based composite materials, Nature 442 (2006) 282–286.
[43] G. Wang, J. Yang, J. Park, et al., Facile synthesis and characterization of graphene nanosheets, J. Phys. Chem. C 112 (2008) 8192–8195.
[44] Y. Si, E.T. Samulski, Synthesis of water soluble graphene, Nano Lett. 8 (2008) 1679–1682.
[45] R. Muszynski, B. Seger, P.V. Kamat, Decorating graphene sheets with gold nanoparticles, J. Phys. Chem. C 112 (2008) 5263–5266.
[46] S. Park, J. An, I. Jung, et al., Colloidal suspensions of highly reduced graphene oxide in a wide variety of organic solvents, Nano Lett. 9 (2009) 1593–1597.
[47] M.J. McAllister, J.L. Li, D.H. Adamson, et al., Single sheet functionalized graphene by oxidation and thermal expansion of graphite, Chem. Mater. 19 (2007) 4396–4404.
[48] H.C. Schniepp, J.L. Li, M.J. McAllister, et al., Functionalized single graphene sheets derived from splitting graphite oxide, J. Phys. Chem. B 110 (2006) 8535–8539.
[49] K.S. Subrahmanyam, S.R.C. Vivekchand, A. Govindaraj, et al., A study of graphenes prepared by different methods: characterization, properties and solubilization, J. Mater. Chem. 18 (2008) 1517–1523.
[50] C.N.R. Rao, A.K. Sood, K.S. Subrahmanyam, et al., Graphene: the new two-dimensional nanomaterials, Angew. Chem. Int. Ed. 48 (2009) 7752–7777.
[51] G.M. Morales, P. Schifani, G. Ellis, et al., High-quality few layer graphene produced by electrochemical intercalation and microwave-assisted expansion of graphite, Carbon 49 (2011) 2809–2816.
[52] E.A. Odd, B.L.V. Prasad, H. Sato, et al., Structure and electronic properties of graphite nanoparticles, Phys. Rev. B 58 (1998) 16387–16395.

[53] B.L.V. Prasad, H. Sato, T. Enoki, et al., Heat-treatment effect on the nanosized graphite π-electron system during diamond to graphite conversion, Phys. Rev. B 62 (2000) 11209–11218.

[54] K.S. Subrahmanyam, L.S. Panchakarla, A. Govindaraj, et al., Simple method of preparing graphene flakes by an arc-discharge method, J. Phys. Chem. C 113 (2009) 4257–4259.

[55] C.N.R. Rao, K.S. Subrahmanyam, H.S.S.R. Matte, et al., A study of the synthetic methods and properties of graphenes, Sci. Technol. Adv. Mater. 11 (2010) 054502–054517.

[56] R. Seshadri, A. Govindaraj, H.N. Aiyer, et al., Investigations of carbon nanotubes, Curr. Sci 66 (1994) 839–847.

[57] T.O. Wehling, K.S. Novoselov, S.V. Morozov, et al., Molecular doping of graphene, Nano Lett. 8 (2008) 173–177.

[58] O. Leenaerts, B. Partoens, F.M. Peeters, Adsorption of H_2O, NH_3, CO, NO_2, and NO on graphene: A first-principles study, Phys. Rev. B 77 (2008) 125416–125422.

[59] F. Schedin, A.K. Geim, S.V. Morozov, et al., Detection of individual gas molecules adsorbed on graphene, Nat. Mater. 6 (2007) 652–655.

[60] M.Y. Han, B. Özyilmaz, Y. Zhang, et al., Energy band-gap engineering of graphene nanoribbons, Phys. Rev. Lett. 98 (2007) 206805–206809.

[61] Z.H. Chen, Y.M. Lin, M.J. Rooks, et al., Graphene nano-ribbon electronics, Physica E Low. Dimens. Syst. Nanostruct. 40 (2007) 228–232.

[62] I. Patla, S. Acharya, L. Zeiri, et al., Synthesis, two-dimensional assembly, and surface pressure-induced coalescence of ultranarrow PbS nanowires, Nano Lett. 7 (2007) 1459–1462.

[63] C. Li, S. Mitra, Processing of fullerene-single wall carbon nanotube complex for bulk heterojunction photovoltaic cells, Appl. Phys. Lett. 91 (2007) 253122–253125.

[64] B. Huang, F. Liu, J. Wu, et al., Suppression of spin polarization in graphene nanoribbons by edge defects and impurities, Phys. Rev. B 77 (2008) 153411–153415.

[65] Z.F. Wang, Q. Li, H. Zheng, et al., Tuning the electronic structure of graphene nanoribbons through chemical edge modification: A theoretical study, Phys. Rev. B 75 (2007) 113406–113410.

[66] F. Cervantes-Sodi, G. Csányi, S. Piscanec, et al., Edge-functionalized and substitutionally doped graphene nanoribbons: electronic and spin properties, Phys. Rev. B 77 (2008) 165427–165440.

[67] J.A. Robinson, E.S. Snow, S.C. Bădescu, et al., Role of defects in single-walled carbon nanotube chemical sensors, Nano Lett. 6 (2006) 1747–1751.

[68] J. Andzelma, N. Govinda, A. Maiti, Nanotube-based gas sensors-role of structural defects, J. Chem. Phys. 421 (2006) 58–62.

[69] V.A. Basiuk, Reactivity of carboxylic groups on armchair and zigzag carbon nanotube tips: a theoretical study of esterification with methanol, Nano Lett. 2 (2002) 835–839.

[70] B. Huang, Z. Li, Z. Liu, et al., Adsorption of gas molecules on graphene nanoribbons and its implication for nanoscale molecule sensor, J. Phys. Chem. C 112 (2008) 13442–13446.

[71] D.V. Kosynkin, A.L. Higginbotham, A. Sinitskii, et al., Longitudinal unzipping of carbon nanotubes to form graphene nanoribbons, Nature 458 (2009) 872–876.

[72] D.V. Kosynkin, W. Lu, A. Sinitskii, et al., Highly conductive graphene nanoribbons by longitudinal splitting of carbon nanotubes using potassium vapor, ACS Nano 5 (2011) 968–974.

[73] J. Cai, P. Ruffieux, R. Jaafar, et al., Atomically precise bottom–up fabrication of graphene nanoribbons, Nature 466 (2010) 470–473.

[74] Z.Z. Sun, D.K. James, J.M. Tour, Graphene chemistry: synthesis and manipulation, J. Phys. Chem. Lett. 2 (2011) 2425–2432.
[75] B.J. Kim, Y.S. Lee, S.J. Park, Novel porous carbons synthesized from polymeric precursors for hydrogen storage, Int. J. Hydrogen Energy 33 (2008) 2254–2259.
[76] B.J. Kim, S.J. Park, Influence of surface treatments on micropore structure and hydrogen adsorption behavior of nanoporous carbons, J. Colloid Interface Sci. 311 (2007) 619–621.
[77] D. Saha, S. Deng, Hydrogen adsorption on Pd- and Ru-doped C(60) fullerene at an ambient temperature, Langmuir 27 (2011) 6780–6786.
[78] J. Chen, F. Wu, Review of hydrogen storage in inorganic fullerene-like nanotubes, Appl. Phys. A 78 (2004) 989–994.
[79] S.Y. Lee, S.J. Park, Effect of temperature on activated carbon nanotubes for hydrogen storage behaviors, Int. J. Hydrogen Energy 35 (2010) 6757–6762.
[80] S.J. Park, S.Y. Lee, Hydrogen storage behaviors of platinum-supported multi-walled carbon nanotubes, Int. J. Hydrogen Energy 35 (2010) 13048–13054.
[81] S.Y. Lee, S.J. Park, Influence of CO_2 activation on hydrogen storage behaviors of platinum-loaded activated carbon nanotubes, J. Solid State Chem. 183 (2010) 2951–2956.
[82] S.Y. Lee, S.J. Park, Effect of chemical treatments on hydrogen storage behaviors of multi-walled carbon nanotubes, Mater. Chem. Phys. 124 (2010) 1011–1014.
[83] B.J. Kim, Y.S. Lee, S.J. Park, A study on the hydrogen storage capacity of Ni-plated porous carbon nanofibers, Int. J. Hydrogen Energy 33 (2008) 4112–4115.
[84] S.J. Park, B.J. Kim, Y.S. Lee, et al., Influence of copper electroplating on high pressure hydrogen-storage behaviors of activated carbon fibers, Int. J. Hydrogen Energy 33 (2008) 1706–1710.
[85] J.S. Im, O.S. Kwon, Y.H. Kim, et al., The effect of embedded vanadium catalyst on activated electrospun CFs for hydrogen storage, Microporous Mesoporous Mater. 115 (2008) 514–521.
[86] S.Y. Lee, S.J. Park, Preparation and characterization of ordered porous carbons for increasing hydrogen storage behaviors, J. Solid State Chem. 184 (2011) 2655–2660.
[87] J. Jiang, Q. Gao, Z. Zheng, et al., Enhanced room temperature hydrogen storage capacity of hollow nitrogen-containing carbon spheres, Int. J. Hydrogen Energy 35 (2010) 210–216.
[88] S.Y. Lee, S.J. Park, Effect of platinum doping of activated carbon on hydrogen storage behaviors of metal-organic frameworks-5, Int. J. Hydrogen Energy 36 (2011) 8381–8387.
[89] S.J. Park, S.Y. Lee, Hydrogen storage behaviors of carbon nanotubes/metal-organic frameworks-5, Carbon Lett 10 (2009) 19–22.
[90] G.M. Psofogiannakis, G.E. Froudakis, E. Tylianakis, Li-doped pillared graphene oxide: a graphene-based nanostructured material for hydrogen storage, J. Phys. Chem. Lett. 1 (2010) 2459–2464.
[91] G. Srinivas, Y.W. Zhu, R. Piner, et al., Synthesis of graphene-like nanosheets and their hydrogen adsorption capacity, Carbon 48 (2010) 630–635.
[92] A.B. Bourlinos, ThA. Steriotis, M. Karakassides, et al., Synthesis, characterization and gas sorption properties of a molecularly-derived graphite oxide like foam, Carbon 45 (2007) 852–857.
[93] L.P. Ma, Z.S. Wu, J. Li, et al., Hydrogen adsorption behavior of graphene above critical temperature, Int. J. Hydrogen Energy 34 (2009) 2329–2332.
[94] A. Ghosh, K.S. Subrahmanyam, K.S. Krishna, et al., Uptake of H_2 and CO_2 by graphene, J. Phys. Chem. C 112 (2008) 15704–15707.

[95] A. Peigney, Ch Laurent, E. Flahaut, et al., Specific surface area of carbon nanotubes and bundles of carbon nanotubes, Carbon 39 (2001) 507–514.
[96] K. Kaneko, C. Ishii, M. Ruike, et al., Origin of superhigh surface area and microcrystalline graphitic structures of activated carbons, Carbon 30 (1992) 1075–1088.
[97] S.K. Bhatia, A.L. Myers, Optimum conditions for adsorptive storage, Langmuir 22 (2006) 1688–1700.
[98] L. Wang, K. Lee, Y.Y. Sun, et al., Graphene oxide as an ideal substrate for hydrogen storage, ACS Nano 3 (2009) 2995–3000.
[99] B.J. Kim, S.J. Park, Optimization of the pore structure of nickel/graphite hybrid materials for hydrogen storage, Int. J. Hydrogen Energy 36 (2011) 648–653.
[100] G.J. Kubas, R.R. Ryan, B.I. Swanson, et al., Characterization of the first examples of isolable molecular hydrogen complexes, $M(CO)_3(PR_3)_2(H_2)$ ($M = Mo$, W; $R = C_y$, i-Pr). Evidence for a side-on bonded H_2 ligand, J. Am. Chem. Soc. 106 (1984) 451–452.
[101] G.J. Kubas, Molecular hydrogen complexes: coordination of a σ bond to transition metals, Acc. Chem. Res. 21 (1988) 120–128.
[102] T.K.A. Hoang, D.M. Antonelli, Exploiting the Kubas interaction in the design of hydrogen storage materials, Adv. Mater. 21 (2009) 1787–1800.
[103] A.K. Singh, A. Sadrzadeh, B.I. Yakobson, Metallacarboranes: toward promising hydrogen storage metal organic frameworks, J. Am. Chem. Soc. 132 (2010) 14126–14129.
[104] C.V.J. Skipper, A. Hamaed, D.M. Antonelli, Computational study of silica-supported transition metal fragments for Kubas-type hydrogen storage, J. Am. Chem. Soc. 132 (2010) 17296–17305.
[105] A. Hamaed, T.K.A. Hoang, G. Moula, et al., Hydride-induced amplification of performance and binding enthalpies in chromium hydrazide gels for Kubas-type hydrogen storage, J. Am. Chem. Soc. 133 (2011) 15434–15443.
[106] H. Zhu, Y. Chen, S. Li, et al., Novel sandwich-type dimetallocenes: toward promising candidate media for high-capacity hydrogen storage, Int. J. Hydrogen Energy 36 (2011) 11810–11814.
[107] A.K. Mishra, S. Ramaprabhu, Carbon dioxide adsorption in graphene sheets, AIP Adv 1 (2011) 032152–032155.
[108] A. Kaniyoor, T.T. Baby, S. Ramaprabhu, Graphene synthesis via hydrogen induced low temperature exfoliation of graphite oxide, J. Mater. Chem. 20 (2010) 8467–8469.
[109] A.K. Mishra, S. Ramaprabhu, Nano magnetite decorated multiwalled carbon nanotubes: a robust nanomaterial for enhanced carbon dioxide adsorption, Energy Environ. Sci. 4 (2011) 889–895.
[110] R.V. Siriwardane, M. Shen, E.P. Fisher, et al., Adsorption of CO_2 on molecular sieves and activated carbon, Energy Fuels 15 (2001) 279–284.
[111] Y. Gensterbluma, P. van Hemert, P. Billemont, et al., European inter-laboratory comparison of high pressure CO(2) sorption isotherms. I: activated carbon, Carbon 47 (2009) 2958–2969.
[112] Z. Zhang, M. Xu, H. Wang, et al., Enhancement of CO_2 adsorption on high surface area activated carbon modified by N_2, H_2 and ammonia, Chem. Eng. J. 160 (2010) 571–577.
[113] S. Cavenati, C.A. Grande, A.E. Rodrigues, Adsorption equilibrium of methane, carbon dioxide, and nitrogen on zeolite 13X at high pressures, J. Chem. Eng. Data 49 (2004) 1095–1101.
[114] A.R. Millward, O.M. Yaghi, Metal-organic frameworks with exceptionally high capacity for storage of carbon dioxide at room temperature, J. Am. Chem. Soc. 127 (2005) 17998–17999.
[115] C. Cazorla, S.A. Shevlin, Z.X. Guo, Calcium-based functionalization of carbon materials for CO_2 capture: A first-principles computational study, J. Phys. Chem. C 115 (2011) 10990–10995.

[116] C.E. Rangel, L.F. Magaña, Adsorption of carbon dioxide and methane on graphene with a high titanium coverage, Carbon 47 (2009) 2758–2760.
[117] M. Qing, R.A. Rosenberg, Interaction of Ti with the (0001) surface of highly oriented pyrolytic graphite, Phys. Rev. B 60 (1999) 2827–2832.
[118] Y. Zhang, N.W. Franklin, R.J. Chen, et al., Metal coating on suspended carbon nanotubes and its implication to metal-tube interaction, Chem. Phys. Lett. 331 (2000) 35–41.
[119] H.E. Romero, P. Joshi, A.K. Gupta, et al., Adsorption of ammonia on graphene, Nanotechnology 20 (2009) 245501–245509.
[120] O. Leenaerts, B. Partoens, F.M. Peeters, Adsorption of small molecules on graphene, Microelectronics J. 40 (2009) 860–862.
[121] D.C. Elias, R.R. Nair, T.M.G. Mohiuddin, et al., Control of graphene's properties by reversible hydrogenation: evidence for graphane, Science 323 (2009) 610–613.
[122] R. Yang, L. Zhang, Y. Wang, et al., Anisotropic etching effect in the graphene basal plane, Adv. Mater. 22 (2010) 4014–4019.
[123] R.E.F.A. Savchenko, Transforming graphene, Science 323 (2009) 589–590.
[124] S. Chen, L. Brown, M. Levendorf, et al., Oxidation resistance of graphene-coated Cu and Cu/Ni alloy, ACS Nano 5 (2011) 1321–1327.

Chapter 15

Porous Texture Versus Surface Chemistry in Applications of Adsorption by Carbons

José Luís Figueiredo and Manuel Fernando R. Pereira
Laboratório de Catálise e Materiais (LCM), Laboratório Associado LSRE/LCM, Departamento de Engenharia Química, Universidade do Porto, Porto, Portugal

Chapter Outline

15.1. Introduction 471
15.2. Carbon Materials 472
15.2.1. Structure and Properties 472
15.2.2. Strategies to Modify the Textural and Surface Properties of Carbon Materials 476
15.3. Adsorption of Organic Compounds 479
15.3.1. Effect of the Surface Chemistry 481
15.4. Adsorption of Inorganic Compounds 491
15.5. Conclusions 493
References 494

15.1. INTRODUCTION

The versatility of carbon materials in adsorption is well documented in the literature. Activated carbons, together with activated carbon fibers and cloths, are among the most important adsorbents for industrial applications. However, in recent years, there has been a steady growth in the number of reports on the use of nanostructured carbon materials (such as templated carbons, polymer-derived carbons, and carbon nanotubes).

Both the physical and the chemical surface properties of the carbon material have a significant impact on the adsorption of substances in the liquid phase, and structure–property relationships for carbon adsorbents in the liquid phase

Novel Carbon Adsorbents. DOI: 10.1016/B978-0-08-097744-7.00015-6
Copyright © 2012 Elsevier Ltd. All rights reserved.

have been discussed in previous reviews [1–3]. In particular, a carbon material immersed into an aqueous solution will develop a surface charge, which results from either the dissociation of surface groups or the adsorption of ions from the solution [1]. An electrical double layer is formed, which affects the adsorption of ions from the solution. The surface charge depends on both the solution pH and the surface chemistry of the carbon.

The pH of the solution controls the electrostatic interactions between the adsorbent and the adsorptive. Thus, the surface charge will be positive when $pH < pH_{PZC}$ (point of zero charge of the material), whereas it will be negative when $pH > pH_{PZC}$. On the other hand, the pH of the solution also controls the dissociation or ionization of the adsorptive; for instance, acidic organic solutes are dissociated at $pH > pKa$ [4,5]. Ions and polar molecules in solution will therefore be attracted by opposite charges on the surface [6].

Such electrostatic interactions do not occur in the gas phase; so the adsorption process is determined primarily by the textural properties of the carbon adsorbent, namely the surface area or micropore volume, with the exception of those applications where the carbon material acts as both an adsorbent and a catalyst, in which case its surface chemistry may play a decisive role in the oxidation–reduction of inorganic and organic pollutants [7,8].

In both cases (adsorption from the gas or liquid phase), the adsorption capacity will depend on the accessibility of the adsorptive to the inner surface of the adsorbent. Therefore, small molecules can access the micropores, whereas large molecules can adsorb only onto the larger pores.

Given the editorial limitations imposed on the size of this chapter, we will restrict this review to the adsorption of organic and inorganic compounds from dilute aqueous solutions. In order to clarify the roles of the porous texture and surface chemistry of the adsorbent, we will mainly discuss those results that were obtained on a series of carbon materials prepared from the same precursor with similar textural properties and different amounts of surface groups. In this way, it is possible to correlate adsorption properties with the surface chemistry of the carbons.

15.2. CARBON MATERIALS

15.2.1. Structure and Properties

The carbon materials used in adsorption present the graphitic structure. Active sites are expected to be found essentially at the edges of the graphene layers, where the unsaturated carbon atoms may chemisorb oxygen, water, or other compounds such as ammonia, originating surface groups such as those represented schematically in Fig. 15.1. Similar surface groups can be formed on basal plane defects. In addition, the π-electron system of the basal planes contributes to the carbon basicity, affecting its adsorption and catalytic properties.

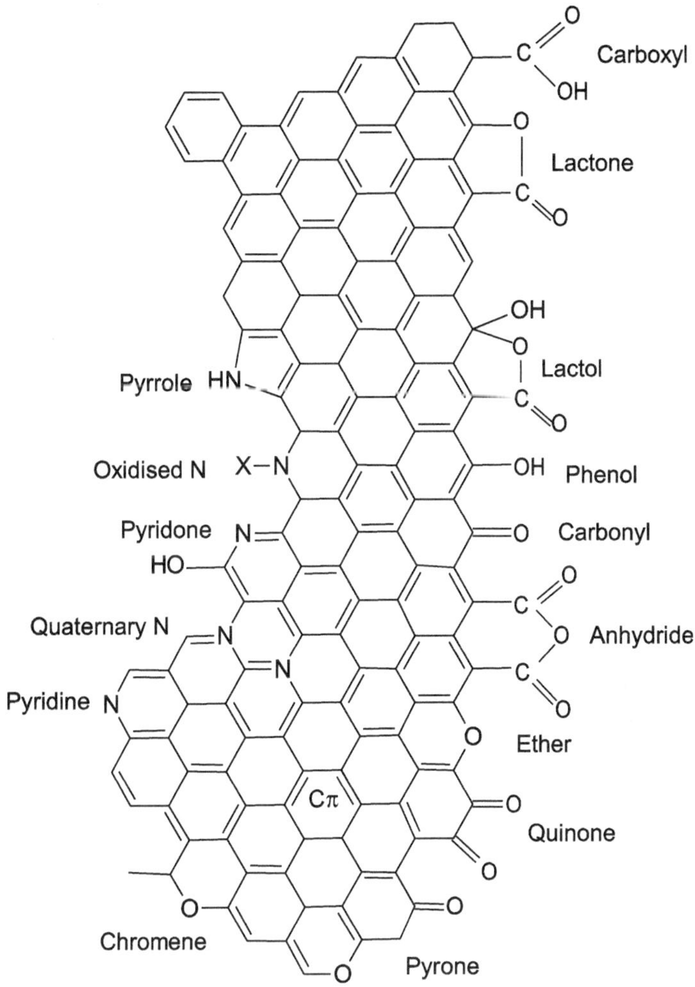

FIGURE 15.1 Nitrogen and oxygen surface groups on carbon. *(Reprinted with permission from Ref. [9])*

Activated carbons (ACs) and activated carbon fibers (ACFs) are the carbon materials which have traditionally been used in adsorption, but various types of nanostructured materials have been developed in recent years, which offer quite interesting properties for these applications. Thus, these new carbon materials will also be briefly described, highlighting their most relevant features.

Activated carbons are obtained from solid carbonaceous precursors (namely peat, coal, lignite, wood, and agricultural wastes such as fruit shells and stones, in particular coconut shells) by two main processes: physical and chemical activation. In physical (or steam) activation, the precursor is carbonized under inert gas at temperatures typically in the range of 600–800 °C, further development of the porous structure being achieved by partial gasification with steam (or other gasification agents, such as air or CO_2) at temperatures higher than 850 °C. In chemical activation, the precursor (generally wood or other lignocellulosic material) is impregnated with a chemical (ZnO, H_3PO_4, KOH, and NaOH) and carbonized at temperatures in the range of 400–600 °C. The chemical activation agent is then recovered and reused [10]. The activated carbons produced by both these methods contain micro- (<2 nm), meso- (2–50 nm), and macropores (>50 nm). The chemically activated carbons usually have a larger proportion of mesopores, while physically activated carbons are more microporous. The micropores can be considered slit-shaped [11], and can be further subdivided into narrow, or ultramicropores (up to 0.7 nm), and wide, or supermicropores (0.7–2.0 nm). The micropore volume of activated carbons may range from 0.15 to 0.50 cm^3/g, and accounts to at least 95% of the adsorption capacity; the mesopore volumes are generally lower than 0.10 cm^3/g, and account for a maximum of 5% of the adsorption capacity. The macropore volumes range from 0.2 to 0.5 cm^3/g, and the corresponding surface area (and adsorption capacity) is negligible (<2 m^2/g). Therefore, the meso- and macropores are mainly used as transport pores, providing for the access of the adsorptives to the micropores, where most of the adsorption occurs [12].

One disadvantage of these materials is that they are obtained from natural raw materials, and therefore they may contain relatively large amounts of mineral impurities (up to 20 wt%, depending on the precursor) which may catalyze undesirable reactions. Activated carbons are available in powder or in granular form (pellets, extrudates, and granules).

Activated carbon fibers are obtained by carbonization and activation of precursor fibers, including textile fibers (such as PAN, aramid, and novoloid), and pitch fibers obtained from coal liquids. They are available as strand, cloth, or felt, and can find similar applications as activated carbons. There are, however, a few differences between the textural and chemical properties of these materials: a) the micropores are open directly to the external surface of the fibers, whereas in activated carbons the micropores are found inside the particles, and can only be accessed through the larger pores; b) activated carbon fibers have generally lower ash contents, as they are produced from synthetic precursors; and c) the use of nitrogen-containing precursor fibers (such as polyacrylonitrile fibers, PAN) leads to carbon fibers containing nitrogen, which may be an asset for some applications, as will be discussed subsequently. As a result of these properties and their fiber shape, ACFs may achieve better performance than ACs in some applications, as recently reviewed [13].

Carbon nanofibers and carbon nanotubes fall under the classification of filamentous carbons [14], which includes materials produced from gaseous carbon compounds by the catalytic action of metal particles. Both materials have diameters lower than a few hundred nanometers, but they differ in the arrangement of the graphene layers: carbon nanotubes (CNTs) exhibit a tubular morphology with one or more concentric graphene layers, while carbon nanofibers (CNFs) are nontubular, and the graphene layers are usually at an angle with the fiber axis. Thus, CNTs can be functionalized only at defects or at the tube ends, while CNFs expose a large amount of edge sites, where functional groups can be attached easily. The individual filaments are nonporous, but they tend to agglomerate into bundles. Nitrogen adsorption isotherms therefore show some mesoporosity, which corresponds to the void space between filaments. A comprehensive review of the catalytic applications of CNTs and CNFs is available [15], and a short review on the use of CNTs for the adsorption of organic and inorganic pollutants was published more recently [16]. An overview of nanocarbons in the context of adsorption is included in the first chapter of this book.

Carbon gels are produced by carbonization of organic gels obtained by the sol–gel polycondensation of organic monomers [17], originating a network structure consisting of interconnected nanosized primary particles. In particular, resorcinol–formaldehyde aqueous gels lead to carbon materials whose texture can be adjusted within a very wide range of pore sizes (microporous, micro–mesoporous, or micro–macroporous) by adequate selection of the synthesis variables [18–20]. Depending on the method of drying the aqueous gels, the materials are designated as aerogels (supercritical drying), xerogels (conventional drying), and cryogels (freeze drying). As an example, Fig. 15.2 shows two carbon xerogels with quite different textural properties which were obtained under different synthesis conditions [21].

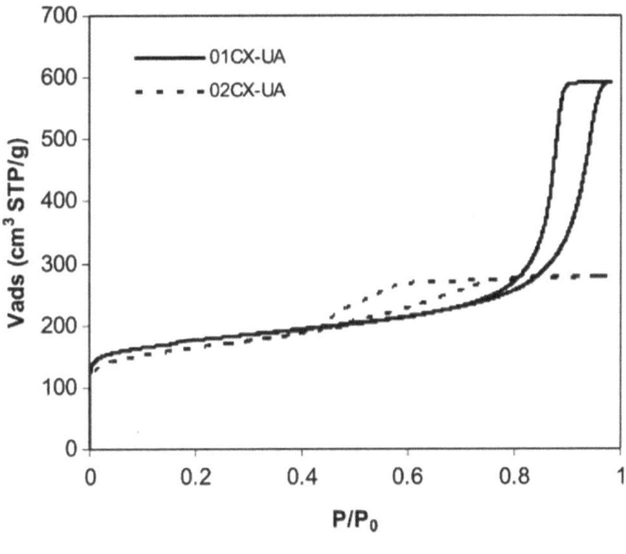

FIGURE 15.2 N_2 adsorption isotherms on carbon xerogels synthesized at different conditions, namely pH = 6.0, direct drying (01CX-UA), and pH = 6.85, drying after solvent exchange (02CX-UA). *(Reprinted with permission from Ref. [21])*

Nitrogen doping can be accomplished by the introduction of a nitrogen-containing precursor in the organic xerogel, such as urea or melamine [22]. A variety of surface nitrogen functionalities can be obtained after pyrolysis, including pyrrolic and pyridinic groups. In the case of pyridinic-N, the nitrogen atom contributes one p-electron to the aromatic π-system and has an isolated electron pair in the plane of the ring, while for pyrrolic-N the nitrogen atom contributes two p-electrons to the π-system, and a hydrogen atom is bound in the plane of the ring. Carbons with pyrrole nitrogen provide the highest charge mobility in a carbon matrix and the best donor–acceptor properties [23]. After carbonization, the carbon gels can be further activated to introduce oxygen functional groups [24]. Therefore, by proper tuning of the surface chemistry and texture, it is possible to design materials suitable for specific applications.

Ordered mesoporous carbons (or templated carbons) can be synthesized by a nanocasting technique, involving the impregnation of a suitable porous template (usually, an ordered mesoporous silica, which by itself is also obtained by a template approach) with a carbon precursor. After carbonization, the template is dissolved and a porous carbon material is obtained which is a negative replica of the original silica template. This method was first used by Ryoo et al. [25], who used an MCM-48 silica template impregnated with sucrose and sulfuric acid. The carbon materials obtained by this procedure (designated as CMK-1) have uniform mesopores of about 3 nm together with some micropores, and present high surface areas (up to 1800 m^2/g) and total pore volumes (up to 1.2 cm^3/g); however, the periodic structure of CMK-1 does not match exactly with that of the MCM-48 template. An exact replica of the template was synthesized later, using SBA-15 silica as a template. This material (CMK-3) has pores of 4.5 nm [26]. The SBA-15 silica templates can be synthesized with a range of wall thicknesses [27], which offers a means for tailoring the pore size of CMK-3 materials, while their surface chemistry can be modified by the procedures described in the following section.

15.2.2. Strategies to Modify the Textural and Surface Properties of Carbon Materials

The textural properties of the carbon materials described in the previous section are largely determined by their synthesis conditions; nevertheless, there are several postsynthesis treatments which can be used for finer tuning. Thermal treatments under inert atmosphere generally lead to shrinking of the porosity, especially when carried out at high temperatures. Oxidation treatments, especially in the gas phase, can be used to enlarge the pore sizes and to open up some of the eventual closed porosity. Catalytic carbon gasification is an alternative method that has been used to promote the development of mesoporosity in activated carbons, after impregnation with a suitable catalyst. On the other hand, controlled carbon deposition (for instance, by CVD from hydrocarbons) can be used to reduce the pore sizes [28]. The method can be controlled in such a way that carbon deposition occurs mainly at the pore entrances, thereby keeping most of the adsorption capacity of the original material unchanged [29]. As an example, Fig. 15.3 shows the nitrogen adsorption isotherms at 77 K obtained for a series of activated carbons derived from a common precursor (NORIT ROX 0.8) using these methodologies.

The nature and concentration of surface functional groups may be modified by thermal or chemical posttreatments. Oxidation with various agents (O_2, N_2O, H_2O_2, HNO_3, etc.) in the gas or liquid phase can be used to increase the concentration of surface oxygen groups, while heating under inert atmosphere may be used to selectively remove some of these functions. Carboxyl, carbonyl, phenol, and lactone groups have been identified on carbon surfaces by a variety of methods, including chemical and potentiometric titration, Fourier transform infrared spectroscopy (FTIR), X-ray photoelectron spectroscopy (XPS), and temperature-programmed desorption (TPD) [30]. Some of these methods allow to determine the amounts of each type of group, TPD being the most dependable one, especially when the carbon materials present small porosity [9,31,32]. Upon heating, carboxylic acids and lactones release CO_2, while phenol and carbonyl groups release CO; carboxylic anhydrides decompose by releasing both CO and CO_2.

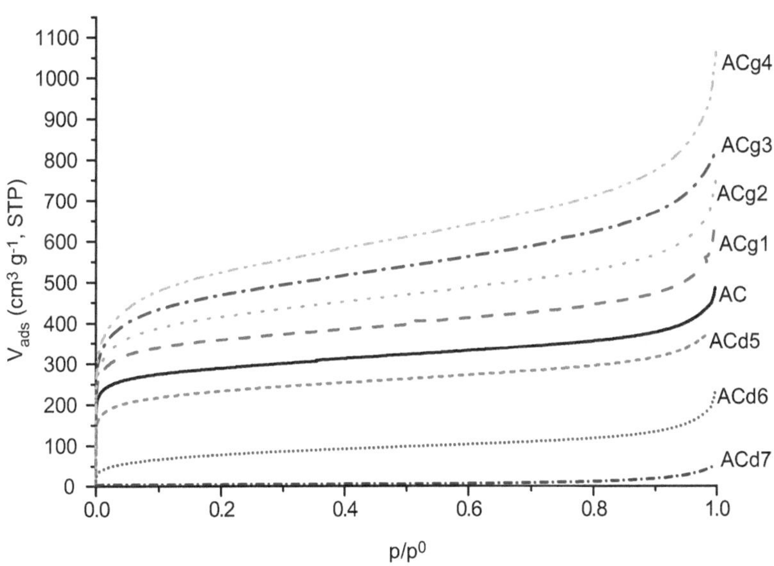

FIGURE 15.3 N_2 adsorption isotherms on modified activated carbons: AC: original sample; ACg1–ACg4, obtained by gasification of AC with CO_2 to different extents, with Co as catalyst; ACd5–ACd7 obtained by depositing different amounts of coke on AC, from pyrolysis of benzene at 825 °C. *(Adapted from Ref. [28])*

Figure 15.4 shows the TPD spectra obtained with carbon xerogels oxidized by different methods, namely oxygen plasma (70 W, for 10 min), treatment with HNO_3 7 M in a Soxhlet, under reflux, for 3 h, and oxidation in the gas phase with a mixture of 5% O_2 in N_2, at 415 °C for 30 h (17% burn-off) [33]. The total amounts of CO and CO_2 released, calculated from the corresponding TPD-MS spectra, are included in Table 15.1. It is clear that different activation processes introduce surface oxygen groups of different types and to different extents.

The TPD profiles of the oxygen plasma-treated sample are very similar in shape to those of the unactivated sample, with slight increases in the amounts of

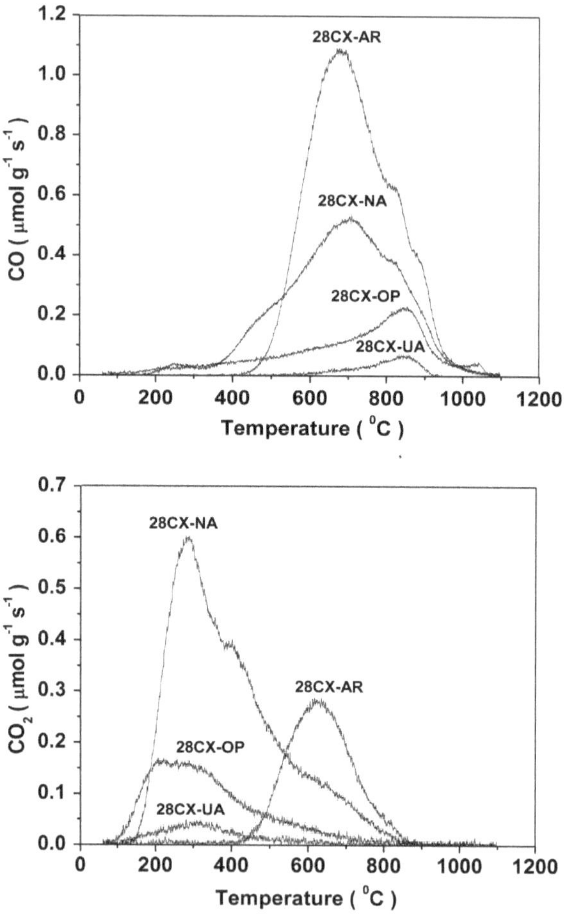

FIGURE 15.4 CO and CO_2 TPD spectra of activated and unactivated carbon xerogel samples: original sample (28CX-UA); sample treated by oxygen plasma (28CX-OP); sample treated in nitric acid (28CX-NA); and sample treated in oxygen (28CX-AR). *(Reprinted with permission from Ref. [33])*

TABLE 15.1 **Surface Oxygen Determined by TPD (as CO and CO_2) and by XPS on Different Carbon Xerogels: Original Sample (28CX-UA); Sample Treated By Oxygen Plasma (28CX-OP); Sample Treated in Nitric Acid (28CX-NA); and Sample Treated in Oxygen (28CX-AR)**

			Surface oxygen (wt.%)	
Sample	CO ($\mu mol\, g^{-1}$) (± 20)	CO_2 ($\mu mol\, g^{-1}$) (± 20)	TPD (± 0.1)	XPS (± 0.1)
28CX-UA	148	149	0.7	2.1
28CX-OP	877	644	3.5	34.1
28CX-NA	2255	1990	10.0	16.8
28CX-AR	3549	736	8.0	12.2

Adapted from Ref. [33]

evolved CO and CO_2. On the other hand, the profiles of nitric acid and air-activated samples are very different, showing large increases in the amounts of evolved CO and CO_2. XPS analysis shows that large amounts of oxygenated groups were introduced onto the external surface of the xerogel upon plasma treatment (Table 15.1). By comparing the increases in surface oxygen content determined by XPS and by TPD, it may be concluded that the effect of the plasma treatment is very selective to the external surface layers. On the other hand, nitric acid and air activation create significant amounts of oxygen groups on the outer surface as well as inside the pores. By deconvolution of the TPD spectra it is possible to estimate the amount of each individual contribution from specific oxygen surface groups [31,34]. Nitric acid activation is most suitable to create surface carboxylic acid groups. Activation by air creates mainly surface phenolic and carbonyl groups. Oxidation by plasma introduces mainly carboxylic acid groups onto the external surface of the particles.

The introduction of nitrogen groups onto the surface of porous carbons induces surface basicity. Moreover, the presence of nitrogen atoms in the carbon matrix has been shown to enhance the catalytic activity of carbon materials in oxidation reactions [35]. Nitrogen functional groups can be introduced onto carbon surfaces during their synthesis (as discussed in the previous section), or by posttreatments with HNO_3, diamine compounds, and ammonia, ammonia–air or ammonia–steam mixtures [36–38].

15.3. ADSORPTION OF ORGANIC COMPOUNDS

A large variety of aromatic compounds can be found in water for human consumption and in wastewater [39,40], including the emergent pollutants such as pharmaceuticals, personal care products, endocrine disruptors, and pesticides, which have recently attracted the attention of the public [41]. The noxious character of this type of compounds has been studied in detail [42], and different processes were developed for their elimination [43], with adsorption on carbon materials being one of the most important.

The adsorption capacity of carbon materials depends on quite different factors. Obviously, it depends on the activated carbon properties: texture (surface area and pore-size distributions), surface chemistry (surface functional groups), and ash content [44–46]. It also depends on the properties of the adsorptive: molecular weight, polarity, pK_a, molecular size, and functional groups. Finally, the solution condition is another factor to be taken into account: pH, adsorptive concentration, and the presence of other possible adsorptives [40,46–48].

The influence of the texture and surface chemistry of activated carbons in the adsorption of organic compounds has been studied for years, and many references on this topic can be found in the literature. The adsorption of organic molecules from aqueous solutions on activated carbons has been recently reviewed [1,49]. Since the pioneering works of Coughlin and Ezra [50] and Mattson et al. [51] to the others published more recently [30,52–56], it has been found that the adsorption capacity is significantly affected by the carbon surface chemistry. In the case of activated carbons, three components of the surface are usually considered: basal planes, heterogeneous surface groups (mainly oxygen-containing surface groups), and inorganic ash. For aromatic compounds, most of the adsorption sites are found on the basal planes, which correspond to about 90% of the carbon surface [54]. However, heterogeneous groups have a higher activity and define the chemical characteristics of the carbon surface.

Different studies have shown that acid oxygen-containing surface groups decrease the adsorption of organic compounds in aqueous solution, while their absence favors adsorption, independently of the polarity of the compounds [54,57,58]. This was justified by the fact that water is mainly adsorbed on the surface establishing hydrogen bonds with the oxygen surface groups, which can produce clusters that may block the passage of the adsorptive molecules into the micropores. Another explanation is based on the increase of the surface hydrophilicity with the increase of the oxygen-containing surface groups, decreasing the hydrophobic interactions. The attraction forces between the organic solutes and the carbon surface (both generally hydrophobic) are relevant.

There are two types of interactions between the adsorptive and the activated carbon: electrostatic and dispersive [1,46,49]. The former appears when the adsorptive is dissociated under the experimental conditions used; for the latter, three mechanisms have been proposed: π–π dispersion interaction mechanism, the hydrogen-bonding formation mechanism proposed by Coughlin and Ezra [50], and the electron donor–acceptor complex mechanism proposed by

Mattson et al. [51]. The π–π dispersion interaction mechanism has been the most widely accepted [1,49,56].

The surface chemistry and solution pH are the most important factors controlling the adsorption process [59]. Most of the aromatic pollutants are found in water solution in the molecular state for a broad range of pH values. In this case, dispersive interactions are predominant, mainly because of the attraction between the π-orbitals on the carbon basal planes and the electronic density in the adsorptive aromatic rings (π–π interactions). However, when the solution pH is very high or very low, ions may be present; so electrostatic interaction between them and charged functional groups on the carbon surface could be significant [46].

In order to highlight the effect of the surface chemistry on the adsorption of organic compounds, the best approach is to use a family of carbon materials, if possible starting from the same parent carbon, where only the surface chemistry is changed while maintaining the original textural properties as much as possible. This approach was followed for the adsorption of different classes of dyes on activated carbons [60–62] and carbon xerogels and templated carbons [63], of simple model aromatic compounds (phenol, aniline, nitrobenzene, sulfanilic acid, and benzenesulfonic acid) on activated carbon [64,65], and of the antibiotic ciprofloxacin on activated carbon, carbon xerogels, and carbon nanotubes [66,67]. Selected examples are discussed below.

15.3.1. Effect of the Surface Chemistry

A set of activated carbon samples were prepared using a NORIT GAC1240 PLUS (sample AC1). Acidic samples were prepared by oxidation with HNO_3 6 M at boiling temperature, for 3 h (sample AC2) and with H_2O_2 10 M at room temperature (sample AC3). Basic samples were prepared by heat treatment at 700 °C under a flow of H_2 (sample AC4) or N2 (sample AC5). Thermal treatments under a flow of H_2 generate higher amounts of basic sites on the carbon surface than thermal treatments under a flow of N_2 [68]. The textural characterization and the acidity/basicity of the AC samples are presented in Table 15.2, and the TPD results are given in Fig. 15.5. No drastic changes in the textural properties of the prepared samples were observed (cf. Table 15.2). Concerning the surface chemistry, it is clear that sample AC2 has a strong acid character while samples AC4 and AC5 have basic properties (cf. Table 15.2 and Fig. 15.5). Three dyes from different classes (C.I. basic red 14, C.I. reactive red 241, and C.I. acid blue 113) were selected to evaluate the adsorption capacity of the prepared materials. More details about the selected dyes and their characteristics can be found in [61].

Since no drastic changes were made in the textural properties of the adsorbents, any disparity in dye uptakes for the different activated carbons may be explained almost entirely by the differences in the chemical nature of

TABLE 15.2 Textural Characterization and Acidity/Basicity of the AC Samples

Sample	V_{micro} (cm^3/g)	S_{meso} (m^2/g)	W_{01} (cm^3/g)	W_{02} (cm^3/g)	L_1 (nm)	S_{BET} (m^2/g)	Acidity (μeq/g)	Basicity (μeq/g)	pH_{pzc}
AC1	0.367	125	0.312	0.058	1.1	972	149	302	9.7
AC2	0.353	91	0.303	0.048	1.1	909	716	34	2.7
AC3	0.365	110	0.311	0.054	1.2	949	358	153	5.4
AC4	0.375	103	0.318	0.058	1.2	972	89	437	10.8
AC5	0.356	114	0.301	0.050	1.2	946	179	365	9.9

V_{micro} – micropore volumes calculated by the t-method
S_{meso} – mesopore surface areas calculated by the t-method
W_{01} and W_{02} – volumes of the two micropore structures obtained by the Dubinin method
L_1 – average micropore width of the smaller pores obtained by the Stoeckli equation
S_{BET} - surface area
Adapted from Ref. [61]

the prepared samples. For basic dyes, no significant correlation could be established between the textural properties and color removal. For instance, with samples AC1, AC2, and AC3, the dye removal increased with the decrease in both V_{micro} and S_{meso}, as shown in Fig. 15.6. Similar behavior was observed for samples AC4 and AC5, in relation to S_{meso}. These observations demonstrate that the textural properties of the studied materials cannot explain the behavior of these adsorbents toward basic dyes. Nevertheless, the influence of the adsorbent texture in the dye uptake should not be completely discarded, as weak correlations between the adsorption capacity of the ACs and their textural properties may be found for acid and reactive dyes [61].

In contrast, strong correlations between acidity/basicity and maximum adsorption capacity for acid and reactive dyes are clearly shown in Fig. 15.7. Activated carbons are materials with amphoteric character; thus, depending on the pH of the solution, their surfaces might be positively or negatively charged. At $pH > pH_{pzc}$, the carbon surface becomes negatively charged favoring the adsorption of cationic species. On the other hand, adsorption of anionic species will be favored at $pH < pH_{pzc}$ [52]. Considering that the initial pH of the aqueous dyestuff solutions was set to 6–7, if the electrostatic interactions were the main adsorption mechanism, cationic (basic) dyes would be expected to show greater affinity for samples AC2 and AC3 ($pH_{pzc} < pH$) and anionic (reactive and acid) dyes would preferably adsorb on samples AC1, AC4, and AC5 ($pH_{pzc} > pH$). Nonetheless, two parallel adsorption mechanisms, one involving electrostatic interactions and a second one involving dispersive

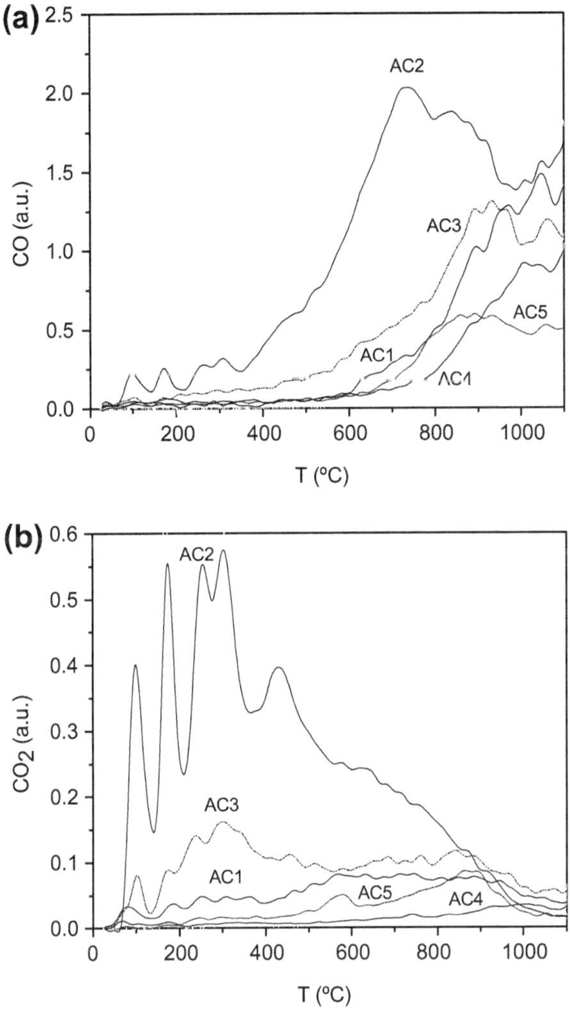

FIGURE 15.5 TPD spectra for AC samples: (a) CO evolution; (b) CO_2 evolution. *(Reprinted with permission from Ref. [61])*

interactions, were reported to explain the adsorption of organics in activated carbons [46]; thus, one cannot only account for the role of electrostatic interactions.

For the acid and reactive dyes, a strong correlation between the basicity of the AC samples and the dye uptakes is shown in Fig. 15.7, which is in agreement with the previous reasoning. Comparing both heat-treated samples, it is evident that sample AC4 has the best adsorption capacity for both acid and reactive dyes. According to Leon y Leon et al. [69], basic carbons are

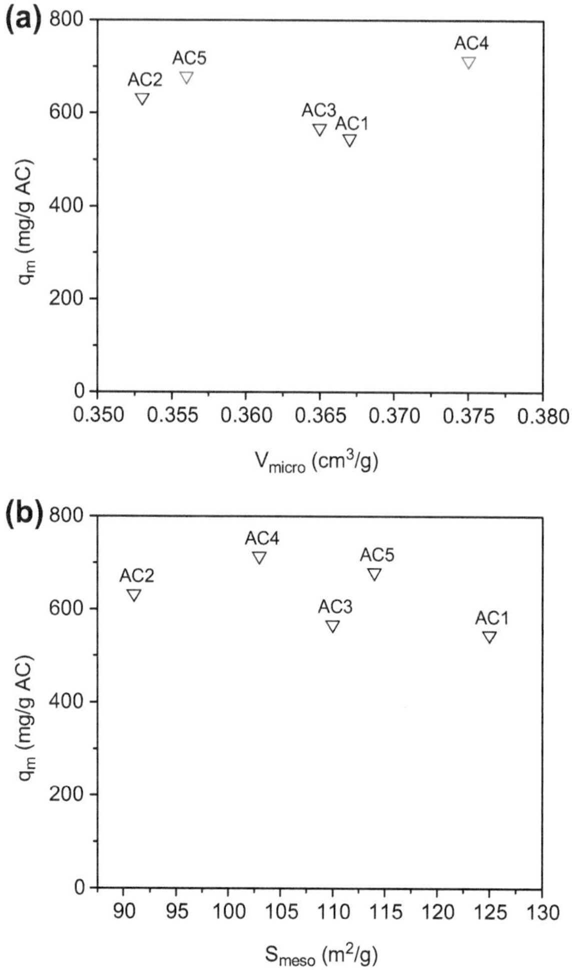

FIGURE 15.6 Maximum adsorption capacity for the basic dye *versus* V_{micro} (a) and S_{meso} (b) of the AC samples. *(Reprinted with permission from Ref. [61])*

characterized by a high content of electron-rich sites on their basal planes and a low concentration of electron-withdrawing groups. The interaction of the dye molecules and the activated carbon surface is expected to occur between the delocalized π-electrons of the oxygen-free Lewis basic sites and the free electrons of the dye molecule present in the aromatic rings and multiple bonds. The same authors have reported evidence for the protonation of basal plane sites on basic carbon. They state that oxygen-free carbon sites can adsorb protons from the solution, conferring a positively charged surface to the carbon. Thus, it is possible that negatively charged ions of the dyes also interact with these sites. The presence of oxygen-containing functional groups, which are

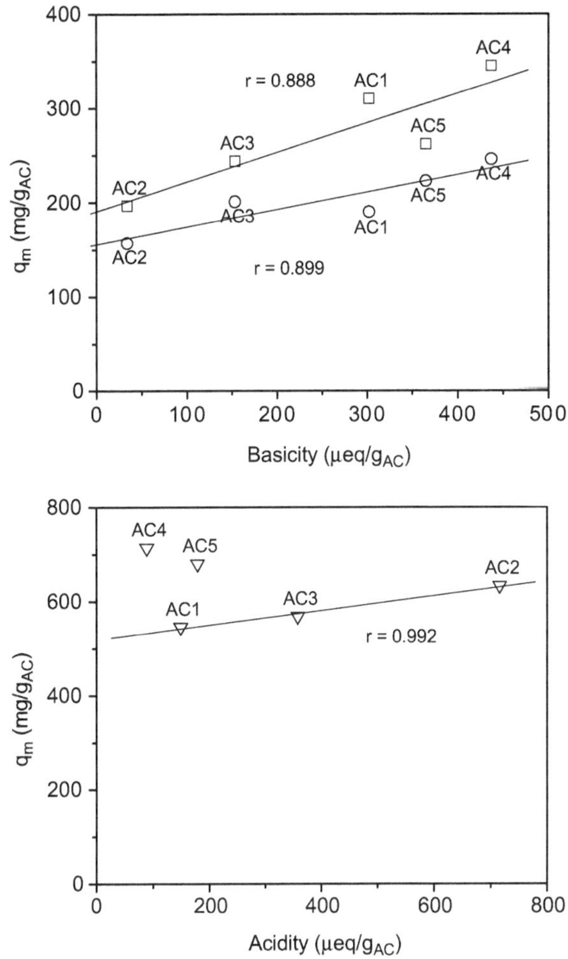

FIGURE 15.7 Correlation between samples acidity/basicity and maximum adsorption capacity for acid (□), reactive (○), and basic (▽) dyes. *(Reprinted with permission from Ref. [61])*

electron-withdrawing groups, has a negative effect on the adsorption of anionic species, as can be observed for sample AC2. These groups reduce the electron density on the surface of the carbon [46], thus decreasing the adsorption potential for the dye molecules. Simultaneously, there is an increase in repulsive electrostatic interactions between the anions of the dyes and the negatively charged surface of sample AC2.

When dissolved in water, basic dyes are positively charged; thus, it was expected that these dyes would preferably adsorb onto activated carbons with acid characteristics. In fact, when comparing the behavior of anionic and cationic dyes, we observe that sample AC2 shows better performance for the latter ones, relatively to AC1. This was expected as a result of the electrostatic interactions between the cations and the negatively charged surface of acid samples. For basic dyes, there is a strong correlation between the dye uptake and the increasing acidity of samples AC1, AC3, and AC2 (Fig. 15.7). However, samples AC4 and AC5 provide higher adsorption uptakes than the previous samples. This observation leads to the conclusion that, even though adsorption of cationic species on activated carbons is enhanced by the presence of surface acid groups, basic carbons still present better performances, indicating that dispersive interactions between the delocalized π-electrons on the surface of the basic activated carbons and the free electrons of the dye molecule present in the aromatic rings and multiple bonds play a dominant role in the adsorption mechanism.

In order to better assess the influence of the solution pH, samples equivalent to AC1, AC2, and AC4 were prepared and used in the adsorption of the dye C.I. reactive red 241 from aqueous solutions with different pH (2, 7, or 12) [62] and the most relevant results are presented in Fig. 15.8. It may be observed that the maximum adsorption capacity varies in the order AC2 < AC1 < AC4, independently of the pH value. This strongly correlates with the increasing basicity of the solid samples as discussed above. In addition, the pH of the solution has a remarkable effect on the performance of the adsorption process, as we can see in Fig. 15.8. At $pH > pH_{pzc}$, the carbon surface becomes negatively charged favoring the adsorption of cationic species. On the other hand, adsorption of anionic species, which is the case under discussion, will be favored at values of $pH < pH_{pzc}$ [1,49,52]. The surfaces of the activated carbon samples AC1 ($pH_{pzc} = 9.7$) and AC4 ($pH_{pzc} = 10.8$) are positively charged at pH values

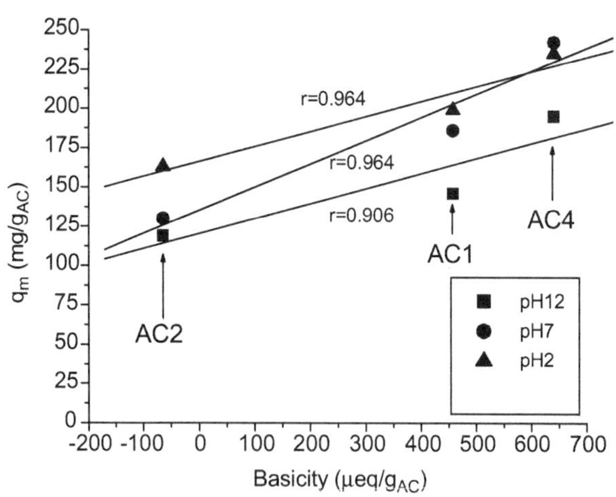

FIGURE 15.8 Correlations between samples basicity and maximum adsorption capacity. *(Reprinted with permission from Ref. [62])*

of 2 and 7 and negatively charged at pH = 12. The results presented in Fig. 15.8 are consistent with the previous discussion, as the maximum adsorption capacities at pH 2 and 7 are much higher than those corresponding to pH 12, since in the last case there are repulsive forces between the solute and the negatively charged carbon surface. On the contrary, the surface of sample AC2 ($pH_{pzc} = 2.7$) is negatively charged at pH 7 and 12. Accordingly (for example, comparing the adsorption equilibrium results at pH 7 with those obtained for sample AC1), a significant reduction of the uptake can be noticed, becoming very close to that obtained at pH 12. As expected, at pH 2 the activated carbon surface is electrically neutral, and the results are clearly better.

A similar approach was used with two types of high-surface-area carbon adsorbents, namely a carbon xerogel and a CMK-3 templated carbon, both subjected to appropriate treatments in order to modify their surface chemistries while keeping their textural properties relatively unchanged, which were evaluated in the adsorption of two anionic dyes (reactive red 241 and acid blue 113) at different pH values [63]. The effect of the surface chemistry was similar to that previously discussed. The only relevant point to discuss here is the comparison between these materials and the activated carbon. Although it has been reported that the use of mesoporous carbons may be advantageous for the adsorption of bulky molecules, such as textile dyes and some of the emergent pollutants [70–75], the results obtained showed that carbon xerogels do not seem to be a good option to remove color from the effluents, since their performances are worse or similar to activated carbons; on the other hand, templated carbons perform better than activated carbons in some cases, as a result of their higher mesopore surface areas.

Li et al. [76] presented a similar approach to study the adsorption of simple aromatics (aniline, pyridine, and phenol) from aqueous solutions on modified activated carbon fibers (ACFs). The conclusions reported were in line with the previous reasoning: the increase of the surface acidic groups of ACFs significantly reduced the adsorption capacity of ACF for aromatics because the π–π dispersive interaction between ACF and aromatic compounds decreased, competition of water adsorption increased because of increase of hydrophilic property of ACF, and more water clusters formed to block the access of the compounds to the micropores.

More recently, the adsorption of phenol was studied on carbon fibers activated by supercritical fluids (H_2O and CO_2) and conventionally activated carbon fibers (from the same precursor). Samples with different textural properties (S_{BET} ranging from 700 to 1900 m^2/g) and different concentrations of surface oxygen (from 0.44 to 2.1 $\mu mol/m^2$) were obtained [77].

A linear correlation between the amount of phenol adsorbed and the volume of small micropores (determined by adsorption of CO_2 at 0 °C) was found, but there was no apparent correlation between the phenol uptake (normalized by the micropore volume) and the surface oxygen content. However, further insights could be reached by subjecting the phenol-loaded ACFs to temperature-programmed desorption experiments with water at 250 bar. All TPD spectra showed two distinct peaks, corresponding to different types of interactions. According to Salvador et al. [78], the first (and larger) peak, in the zone of liquid water ($T < 374\,°C$), can be assigned to physisorbed phenol, while the second peak, in the zone of supercritical water ($T > 374\,°C$), corresponds to chemisorbed phenol. The corresponding amounts were determined from the areas of the two peaks, and correlated with the surface oxygen content as shown in Fig. 15.9. The results show that the physical adsorption of phenol over these ACFs is nearly independent of the surface oxygen concentration,

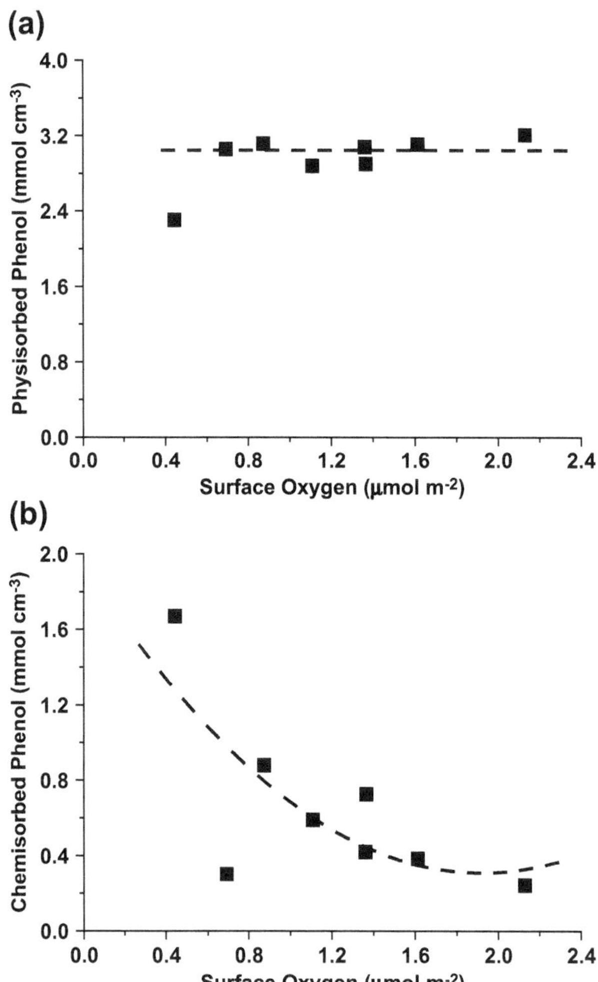

FIGURE 15.9 Dependence of physisorbed (a) and chemisorbed (b) phenol on surface oxygen concentration. *(Reprinted with permission from Ref. [77])*

while chemisorption is clearly inhibited by the presence of oxygen surface groups. These findings support the propositions of Radovic et al. [1], namely that the sites responsible for chemisorption are oxygen-free sites located at the edges of the graphene layers, whereas physisorption occurs all over the surface. Of course, one must recall that the ACFs used in this study were activated at temperatures higher than 700 °C; therefore, the amounts of surface oxygen, and in particular those corresponding to carboxylic groups, are relatively modest. When the fibers were oxidized in nitric acid, thereby introducing substantial amounts of surface oxygen (6.6 $\mu mol/m^2$) with a large proportion of carboxylic acid groups, the uptake of phenol was significantly reduced (by more than 60%) [77]. Acidic surface groups act as electron acceptors, decreasing the π-electron density in the carbon basal planes, which leads to a decrease in the amount adsorbed, as discussed before.

A comparison was made for the adsorption capacity of the antibiotic ciprofloxacin (CPX) on three types of carbon-based materials: activated carbons, carbon nanotubes, and carbon xerogels, used as received/prepared and functionalized through oxidation with nitric acid. Once again, the influence of the surface chemistry follows the same trend previously discussed for the adsorption on activated carbons [66]. The characterization of the parent carbon samples is presented in Table 15.3 and their performance on CPX adsorption is shown in Fig. 15.10. The commercial activated carbon used in these experiments (NORIT C-GRAN NC01-125) was prepared from wood, using a chemical activation process with phosphoric acid, and therefore presented a large amount of oxygen-containing surface groups (sample ACG). It can be seen that the maximum adsorption capacity (q_m), in absolute value, increases as CX < CNT < ACG (Fig. 15.10, left). The fact that ACG adsorbs much more CPX than the other materials is most likely related to its largest surface area (Table 15.3). Since CPX has an almost planar configuration, with approximate dimensions of $13.5 \times 3 \times 7.4$ Å, in principle it should be able to penetrate into the pores of ACG. As the micropores are normally slit shaped, the molecule should be able to enter "sideways." CPX can also enter into the pores of CX (which are cylindrical, but their dimensions are larger than those of the molecule) and into the CNT bundles. However, the surface area is not enough to explain all results, since CX, which has much larger surface area than CNT (Table 15.3), adsorbs less CPX (Fig. 15.9). This is most likely related to the fact that CNT has less surface groups (Table 15.3), and, as explained above, the presence of acidic surface groups is detrimental to adsorption. On the other hand, adsorption on basic (or less acidic) materials is favored, mainly due to the dispersive interactions between the delocalized π-electrons in the carbon basal planes and the free electrons in the CPX molecules (namely those of the aromatic rings). Figure 15.10 (right) shows the adsorption capacity normalized by the BET surface area of the materials. It can be seen then that CNT has a better performance, since it adsorbs more CPX per unit surface area. This can be related to the fact that this material has less surface groups, as explained above. Moreover, since CNT is not porous, unlike

TABLE 15.3 Description and Characterization of Carbon Samples

Sample	S_{BET} (m^2/g)	V_p (cm^3/g)	L (nm)	V_{micro} (cm^3/g)	$S_{external}$ (m^2/g)	pH_{PZC}	CO (μmol/g)	CO_2 (μmol/g)
ACG	1237	0.98	3.8	0.276	650	4.21	3347	1538
CX	617	1.30	31.3	~0	617	6.80	379	125
CNT	284	1.59	3.1	~0	284	7.10	189	65

S_{BET} – surface area; V_p – total pore volume; L – mode of the mesopore size distribution (BJH method); V_{micro} – micropore volume (t-method), $S_{external}$ – external area (t-method); pH_{PZC} – point of zero charge; CO and CO_2 desorbed (determined by TPD)
Adapted from Ref. [67]

ACG and CX (CNT pores result from the free space in the bundles), the access is easier and the π-electrons of CPX can easily interact with the surface by dispersive forces, the molecule being, most likely, lying down on the CNT walls. The performance of ACG is slightly better than CX (unlike what would be expected, since CX has less surface groups). A possible explanation for the better performance of ACG can lie in the influence of the electrostatic forces. In this case, it might be that the adsorption is not only π–π, but that there are also some interactions with the functional groups. In fact, at pH 5 (pH of water, at which the adsorption experiments were carried out) the cationic protonated form of CPX is more abundant. The surface of ACG contains carboxylic acid groups, and is

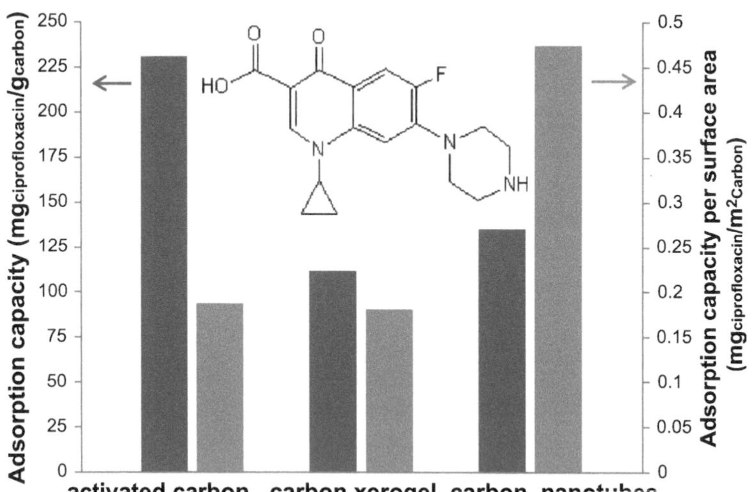

FIGURE 15.10 Adsorption capacity (q_m) of ciprofloxacin on carbon samples. *(Reprinted with permission from Ref. [67])*

negatively charged at the experimental pH around 5 ($pH > pH_{PZC}$), thus attracting CPX.

15.4. ADSORPTION OF INORGANIC COMPOUNDS

Adsorption on carbon materials is one of the most used technologies for the removal of inorganic compounds such as heavy metals in water and wastewater. Therefore, there is a large number of publications on this subject (cf. reviews [1,79,80]), with chromium, aluminum, magnesium, iron, cobalt, nickel, copper, zinc, cadmium, mercury, and lead being the most studied metals due to their toxicity. More recently, adsorption studies on pollutants such as sulfides, cyanates, nitrates, chlorides, and chlorates and bromates have been increasing.

The fundamental principles that govern the adsorption of inorganic compounds are the same as those mentioned above for the adsorption of organic compounds. However, the adsorption of inorganic compounds is mostly due to interactions of electrostatic nature. In addition, it is essential to take into account the different species under which the solute is present in solution to determine its electric charge, i.e. the solute speciation. Therefore, the adsorption results depend on the textural properties and surface chemistry of the adsorbents, the pH of the solution, and the characteristics of the inorganic pollutant. Here, the solution pH plays a key role because it determines directly the surface charge of the carbon materials and governs the chemistry of this type of pollutants in the solution, in particular their speciation. The presence of other electrolytes in solution that can compete with the inorganic compound for the same adsorption sites should also be considered. It is important to highlight here the role of the ionic strength of the solution. Electrostatic interactions (attractive or repulsive) can be reduced by increasing the ionic strength of the solution. Therefore, when the electrostatic interaction between the surface and the adsorptive is repulsive, or the surface concentration is sufficiently high, an increase in ionic strength will increase adsorption. Conversely, when the electrostatic interactions are attractive, or the surface concentration is sufficiently low, an increase of the ionic strength will diminish adsorption [49]. In a simplified manner, the behavior for the adsorption of inorganic compounds can be summarized by taking into account the concepts presented previously: for values of $pH < pH_{pzc}$, the carbon surface is positively charged, promoting the adsorption of anionic compounds; for values of $pH > pH_{pzc}$, the carbon surface is negatively charged, promoting the adsorption of cationic compounds.

The adsorption of chromium on carbon materials is selected to exemplify the relevance of the parameters discussed above. The first information we need to know is how the different chromium species appear in the solution. Chromium can be found as Cr (II) which is unstable and easily oxidized, as Cr (III), which is stable in a wide range of conditions or as Cr (VI) which is stable in strongly oxidizing conditions. In natural water streams, Cr (III) (as $Cr(OH)^{2+}$, $Cr(OH)_2^+$, and $Cr(OH)_3$) and Cr (VI) (as $HCrO_4^-$ and CrO_4^{2-}) are the most often encountered species [81,82]. Figure 15.11 shows the speciation diagram of Cr (III) [83]. The Cr (III) species depend on the pH value. At pH below 8, Cr (III) is present primarily as cationic species, and adsorption will be promoted for $pH > pH_{pzc}$. In this case, the use of carbon materials with acidic surface groups (carboxylic acids and anhydrides) enhances the electrostatic attraction forces between cations and the negatively charged surface. Adsorption of Cr (VI), normally present as anionic species, will be favored at $pH < pH_{pzc}$, therefore for carbon materials with basic properties. Nevertheless, it should be noted that Cr (VI) is reduced to Cr (III) under acidic conditions, and, in particular, in the presence of acidic surface groups, thus showing that the influence of the pH is more complex than it seems [1].

Rivera-Utrilla and Sánchez-Polo [84] studied the adsorption of Cr (III) at different pH values (see Fig. 15.12) on a series of activated carbons submitted to an oxidation treatment by ozonation to modify the contents of oxygenated surface groups (mainly acidic). The parent carbon (sample F) contains 1.82% of oxygen and has an S_{BET} of 1000 m^2/g. Samples F10 (obtained after 10 min of oxidation) and F120 (obtained after 120 min of oxidation) contain 4.3% and 24.9% of oxygen, respectively, and present S_{BET} of 1023 and 632 m^2/g, respectively. Therefore, in spite of decreasing the specific surface area, the adsorption of Cr (III) increases with the increase of the oxidation of the carbon materials (see Fig. 15.12), which demonstrates the predominant role played by the oxygen-containing surface groups in comparison to the textural properties. The maximum adsorption was reached, for all carbons, when the charge sign of the carbon surface was opposite to that of the Cr (III) species present at the pH of the experiment, which supports the predominance of electrostatic

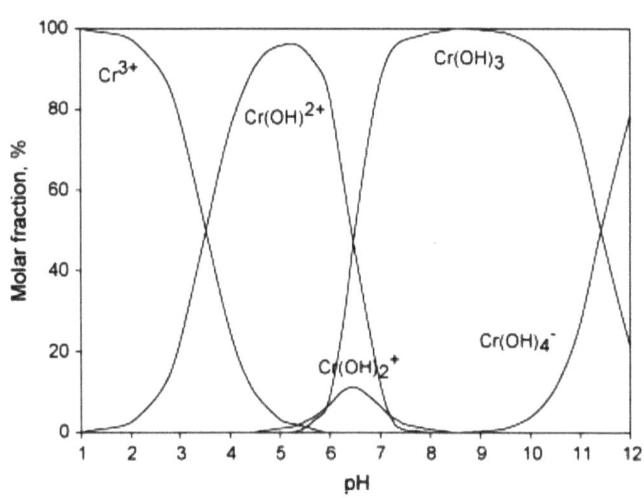

FIGURE 15.11 Speciation diagram for Cr (III) at 25 °C and 10 mg/L ion content. *(Adapted from Ref. [83])*

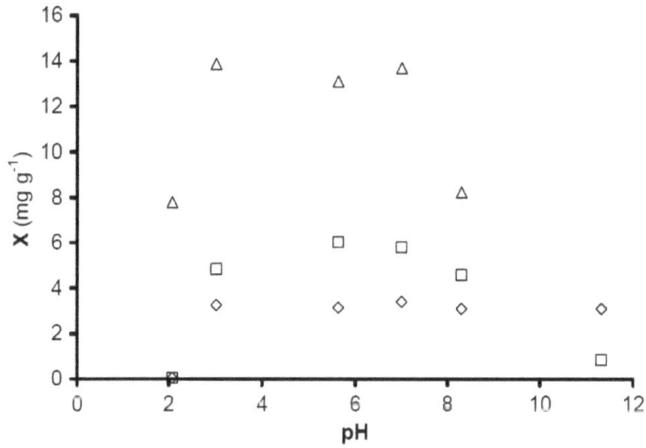

FIGURE 15.12 Influence of solution pH on Cr(III) adsorption on samples F ($pH_{pzc} = 8{,}82$) - ◇, F10 ($pH_{pzc} = 5{,}96$) - □, and F120 ($pH_{pzc} = 2{,}59$) - △. *(Reprinted with permission from Ref. [84])*

interactions in this adsorption process. With samples F10 and F120, the maximum values are obtained at pH ranges of 5.96–7, and 2.59–7, respectively (the surface charge is negative and the species present are Cr^{3+}, $Cr\,(OH)^{2+}$, and $Cr\,(OH)_2^+$. For values above 7, neutral ($Cr(OH)_3$) and anionic ($Cr\,(OH)_4^-$) species begin to appear, reducing the adsorption capacity of the samples due to the presence of repulsive adsorbent–adsorptive interactions. The amount adsorbed is virtually independent of the pH of the solution (except at pH 2) in the case of sample F. This is because the adsorbent and adsorptive have the same charge sign at all the values of pH studied. Thus, at pH values below 8.82, both carbon F and the Cr(III) species [$Cr(OH)^{2+}$ and $Cr(OH)^{2+}$] showed positive charge, and, at pH values above 8.82, both the carbon and the most common Cr(III) species, $Cr\,(OH)_4^-$, showed a negative charge. At pH 2, the amount of Cr(III) adsorbed on samples F and F10 was negligible, which is explained by the high concentration of protons in solution that prevents the ionic exchange between Cr^{3+} and protons on the carbon surface.

15.5. CONCLUSIONS

Relevant conclusions regarding the roles of texture and surface chemistry in carbon adsorption have been obtained by studying series of adsorbents prepared from the same carbon precursor (i) with similar textural properties and different amounts of surface groups and (ii) with similar surface chemistry and different textural properties.

Some very general guidelines can be inferred from these studies:

In the case of similar surface chemistries, the adsorption capacity can be generally correlated with the micropore volume, except for the case of bulky molecules (like those of some dyestuffs), in which case their access to the smaller pores may be difficult. For adsorption of small molecules, the best situation is to have a carbon adsorbent with small micropores (the smaller, the better).

In the case of identical textural properties, the surface chemistry, the adsorptive properties, and the solution properties (pH and ionic strength) become more relevant. For nonionic compounds, the dispersive interactions predominate, and an adsorbent with basic surface properties performs generally better. Electrostatic interactions may become more important in the case of ionic adsorptives.

A finer prediction of the adsorption behavior requires a more comprehensive analysis of the specific case under consideration.

ACKNOWLEDGMENTS

This work was partially supported by projects PEst-C/EQB/LA0020/2011 and NANO/NTec-CA/0122/2007, financed by FEDER through COMPETE - Programa Operacional Factores de Competitividade and by FCT - Fundação para a Ciência e a Tecnologia.

REFERENCES

[1] L.R. Radovic, C. Moreno-Castilla, J. Rivera-Utrilla, Carbon materials as adsorbents in aqueous solutions, in: L.R. Radovic (Ed.), Chem. Phys. Carbon, vol. 27, Marcel Dekker, New York, 2001, 227–405.

[2] A.P. Terzyk, Molecular properties and intermolecular forces—factors balancing the effect of carbon surface chemistry in adsorption of organics from dilute aqueous solutions, J. Colloid Interface Sci. 275 (2004) 9–29.

[3] C. Moreno-Castilla, Adsorption of organic solutes from dilute aqueous solutions, in: E.J. Bottani, J.M.D. Tascón (Eds.), Adsorption by Carbons, Elsevier, 2008, pp. 653–678.

[4] G. Müller, C.J. Radke, J.M. Prausnitz, Adsorption of weak electrolytes from dilute aqueous solution onto activated carbon. Part I. Single-solute systems, J. Colloid Interface Sci. 103 (1985) 466–483.

[5] G. Müller, C.J. Radke, J.M. Prausnitz, Adsorption of weak electrolytes from dilute aqueous solution onto activated carbon. Part II. Multisolute systems, J. Colloid Interface Sci. 103 (1985) 483–492.

[6] C.A. Leon y Leon, L.R. Radovic, Interfacial chemistry and electrochemistry of carbon surfaces, in: P.A. Thrower (Ed.), Chem. Phys. Carbon, vol. 24, Marcel Dekker, Inc., New York, 1994, pp. 213–310.

[7] T.J. Bandosz, Removal of inorganic gases and VOCs on activated carbons, in: E.J. Bottani, J.M.D. Tascón (Eds.), Adsorption by Carbons, Elsevier, 2008, pp. 533–564.

[8] J.L. Figueiredo, M.F.R. Pereira, Carbon as catalyst, in: P. Serp, J.L. Figueiredo (Eds.), Carbon Materials for Catalysis, John Wiley & Sons, Inc, 2009, pp. 177–217.

[9] J.L. Figueiredo, M.F.R. Pereira, The role of surface chemistry in catalysis with carbons, Catal. Today 150 (2010) 2–7.

[10] F. Rodríguez-Reinoso, Activated carbon. Structure, characterization, preparation and applications, in: H. Marsh, E.A. Heintz, F. Rodríguez-Reinoso (Eds.), Introduction to Carbon Technologies, Printer of the University of Alicante, 1997, pp. 35–102.

[11] H.F. Stoeckli, Microporous carbons and their characterization: The present state of the art, Carbon 28 (1990) 1–6.
[12] T. Wigmans, Fundamentals and practical implications of activated carbon production by partial gasification of carbonaceous materials, in: J.L. Figueiredo, J.A. Moulijn (Eds.), Carbon and Coal Gasification, Martinus Nijhoff Publishers, 1986, pp. 559–599.
[13] A. Linares-Solano, D. Cazorla-Amorós, in: E.J. Bottani, J.M.D. Tascón (Eds.), Adsorption by Carbons, Elsevier, 2008, pp. 431–454.
[14] E. Fitzer, K.-H. Kochling, H.P. Boehm, H. Marsh, Recomended terminology for the description of carbon as a solid, Pure Appl. Chem. 67 (1995) 473–506.
[15] P. Serp, Carbon nanotubes and nanofibers in catalysis, in: P. Serp, J.L. Figueiredo (Eds.), Carbon Materials for Catalysis, John Wiley & Sons, Inc, 2009, pp. 309–372.
[16] X. Ren, C. Chen, M. Nagatsu, X. Wang, Carbon nanotubes as adsorbents in environmental pollution management: a review, Chem. Eng. J. 170 (2011) 395–410.
[17] R.W. Pekala, Organic aerogels from the polycondensation of resorcinol with formaldehyde, J. Mater. Sci. 24 (1989) 3221–3227.
[18] T. Horikawa, J. Hayashi, M. Katsuhiko, Controllability of pore characteristics of resorcinol–formaldehyde carbon aerogel, Carbon 42 (2004) 1625–1633.
[19] N. Job, R. Pirard, J. Marien, J.-P. Pirard, Porous carbon xerogels with texture tailored by pH control during sol–gel process, Carbon 42 (2004) 619–628.
[20] N. Job, A. Théry, R. Pirard, J. Marien, L. Kocon, J.-N. Rouzaud, F. Béguin, J.-P. Pirard, Carbon aerogels, cryogels and xerogels: influence of the drying method on the textural properties of porous carbon materials, Carbon 43 (2005) 2481–2494.
[21] N. Mahata, A.R. Silva, M.F.R. Pereira, C. Freire, B.d. Castro, J.L. Figueiredo, Anchoring of a [Mn(salen)Cl] complex onto mesoporous carbon xerogels, J. Colloid Interface Sci. 311 (2007) 152–158.
[22] H.F. Gorgulho, F. Gonçalves, M.F.R. Pereira, J.L. Figueiredo, Synthesis and characterization of nitrogen-doped carbon xerogels, Carbon 47 (2009) 2032–2039.
[23] V.V. Strelko, V.S. Kuts, P.A. Thrower, On the mechanism of possible influence of heteroatoms of nitrogen, boron and phosphorus in a carbon matrix on the catalytic activity of carbons in electron transfer reactions, Carbon 38 (2000) 1499–1503.
[24] P.V. Samant, F. Gonçalves, M.M.A. Freitas, M.F.R. Pereira, J.L. Figueiredo, Surface activation of a polymer based carbon, Carbon 42 (2004) 1321–1325.
[25] R. Ryoo, S.H. Joo, S. Jun, Synthesis of highly ordered carbon molecular sieves via template-mediated structural transformation, J. Phys. Chem. B 103 (1999) 7743–7746.
[26] S. Jun, S.H. Joo, R. Ryoo, M. Kruk, M. Jaroniec, Z. Liu, T. Ohsuna, O. Terasaki, Synthesis of new, nanoporous carbon with hexagonally ordered mesostructure, J. Am. Chem. Soc. 122 (2000) 10712–10713.
[27] D. Zhao, Q. Huo, J. Feng, B.F. Chmelka, G.D. Stucky, Nonionic triblock and star diblock copolymer and oligomeric surfactant syntheses of highly ordered, hydrothermally stable, mesoporous silica structures, J. Am. Chem. Soc. (1998) 6024–6036.
[28] M.F.R. Pereira, J.J.M. Órfão, J.L. Figueiredo, Influence of the textural properties of an activated carbon catalyst on the oxidative dehydrogenation of ethylbenzene, Colloids Surf. A 241 (2004) 165–171.
[29] M.M.A. Freitas, J.L. Figueiredo, Preparation of carbon molecular sieves for gas separations by modification of the pore sizes of activated carbons, Fuel 80 (2001) 1–6.
[30] H.P. Boehm, Some aspects of the surface chemistry of carbon blacks and other carbons, Carbon 32 (1994) 759–769.

[31] J.L. Figueiredo, M.F.R. Pereira, M.M.A. Freitas, J.J.M. Órfão, Modification of the surface chemistry of activated carbons, Carbon 37 (1999) 1379–1389.
[32] H.F. Gorgulho, J.P. Mesquita, F. Gonçalves, M.F.R. Pereira, J.L. Figueiredo, Characterization of the surface chemistry of carbon materials by potentiometric titrations and temperature-programmed desorption, Carbon 46 (2008) 1544–1555.
[33] N. Mahata, M.F.R. Pereira, F. Suárez-García, A. Martínez-Alonso, J.M.D. Tascón, J.L. Figueiredo, Tuning of texture and surface chemistry of carbon xerogels, J. Colloid Interface Sci. 324 (2008) 150–155.
[34] J.L. Figueiredo, M.F.R. Pereira, M.M.A. Freitas, J.J.M. Órfão, Characterization of active sites on carbon catalysts, Ind. Eng. Chem. Res. 46 (2007) 4110–4115.
[35] B. Stöhr, H. Boehm, R. Schlögl, Enhancement of the catalytic activity of activated carbons in oxidation reactions by thermal treatment with ammonia or hydrogen cyanide and observation of a superoxide species as a possible intermediate, Carbon 29 (1991) 707–720.
[36] J. Lahaye, The chemistry of carbon surfaces, Fuel 77 (1998) 543–548.
[37] J. Przepiórski, M. Skrodzewicz, A. Morawski, High temperature ammonia treatment of activated carbon for enhancement of CO_2 adsorption, Appl. Surf. Sci. 225 (2004) 235–242.
[38] Y. El-Sayed, T.J. Bandosz, Role of surface oxygen groups in incorporation of nitrogen to activated carbons via ethylmethylamine adsorption, Langmuir 21 (2005) 1282–1289.
[39] V. Dubey, S. Madhusudhanan, R. Nath, N. Rao, B. Singh, Active carbon for removal of toxic chemicals from contaminated water, Carbon 34 (1996) 327–330.
[40] T. Karanfil, J.E. Kilduff, Role of granular activated carbon surface chemistry on the adsorption of organic compounds. 1. Priority pollutants, Environ. Sci. Technol. 33 (1999) 3217–3224.
[41] B. Kasprzyk-Hordern, R.M. Dinsdale, A.J. Guwy, The removal of pharmaceuticals, personal care products, endocrine disruptors and illicit drugs during wastewater treatment and its impact on the quality of receiving waters, Water Res. 43 (2009) 363.
[42] Agency for Toxic Substances and Disease Registry, Toxicological Profile for Aniline, Phenol, Nitrobenzene, U.S. Department of Health and Human Services, July, 1999.
[43] Metcalf, Eddy, Wastewater Engineering: Treatment, Disposal and Reuse, McGraw-Hill, 1991.
[44] R.C. Bansal, J.B. Donnet, F. Stoeckli, Active Carbon, Marcel Dekker, New York, 1988.
[45] S.J. Gregg, K.S.W. Sing, Adsorption, Surface Area and Porosity, Academic Press, London, 1982.
[46] L.R. Radovic, I.F. Silva, J.I. Ume, J.A. Menéndez, C.A. León y León, A.W. Scaroni, An experimental and theoretical study of the adsorption of aromatics possessing electron-withdrawing and electron-donating functional groups by chemically modified activated carbons, Carbon 35 (1997) 1339–1348.
[47] S. Nouri, F. Haghseresht, Adsorption of dissociating aromatic compounds by activated carbon: effects of ionization on the adsorption capacity, Adsorpt. Sci. Technol. 20 (2002) 417–432.
[48] F. Haghseresht, S. Nouri, J.J. Finnerty, G.Q. Lu, Effects of surface chemistry on aromatic compound adsorption from dilute aqueous solutions by activated carbon, J. Phys. Chem. B 106 (2002) 10935–10943.
[49] C. Moreno-Castilla, Adsorption of organic molecules from aqueous solutions on carbon materials, Carbon 42 (2004) 83–94.
[50] R. Coughlin, F.S. Ezra, Role of surface acidity in the adsorption of organic pollutants on the surface of carbon, Environ. Sci. Technol. 2 (1968) 291–297.
[51] J.S. Mattson, J.H.B. Mark, M.D. Malbin, J.W.J. Weber, J.C. Crittenden, Surface chemistry of activated carbon: specific adsorption of phenols, J. Colloid Interface Sci. 31 (1969) 116–130.

[52] F. Rodríguez-Reinoso, The role of carbon materials in heterogeneous catalysis, Carbon 36 (1998) 159–175.

[53] L. Li, P.A. Quinlivan, D.R.U. Knappe, Effects of activated carbon surface chemistry and pore structure on the adsorption of organic contaminants from aqueous solution, Carbon 40 (2002) 2085–2100.

[54] M. Franz, H.A. Arafat, N.G. Pinto, Effect of chemical surface heterogeneity on the adsorption mechanism of dissolved aromatics on activated carbon, Carbon 38 (2000) 1807–1819.

[55] C.C. Leng, N.G. Pinto, Effects of surface properties of activated carbons on adsorption behaviour of selected aromatics, Carbon 35 (1997) 1375–1385.

[56] S. Haydar, M.A. Ferro-Garcia, J. Rivera-Utrilla, J.P. Joly, Adsorption of p-nitrophenol on an activated carbon with different oxidations, Carbon 41 (2003) 387–395.

[57] P. Pendleton, S.H. Wong, R. Schumann, G. Levay, R. Denoyel, J. Rouquerol, Properties of activated carbon controlling 2-methylisoborneol adsorption, Carbon 35 (1997) 1141–1149.

[58] O.P. Mahajan, C. Moreno-Castilla, P.L. Walker Jr., Surface-treated activated carbon for removal of phenol from water, Sep. Sci. Technol. 15 (1980) 1733–1752.

[59] C. Moreno-Castilla, J. Rivera-Utrilla, M.V. López-Ramón, F. Carrasco-Marín, Adsorption of some substituted phenols on activated carbons from bituminous coal, Carbon 33 (1995) 845–851.

[60] M.F.R. Pereira, S.F. Soares, J.J.M. Órfão, J.L. Figueiredo, Adsorption of dyes on activated carbons: influence of surface chemical groups, Carbon 41 (2003) 811–821.

[61] P.C.C. Faria, J.J.M. Órfão, M.F.R. Pereira, Adsorption of anionic and cationic dyes on activated carbons with different surface chemistries, Water Res. 38 (2004) 2043–2052.

[62] J.J.M. Órfão, A.I.M. Silva, J.C.V. Pereira, S. Barata, I.M. Fonseca, M.F.R. Pereira, Adsorption of a reactive dye on chemically modified activated carbons - influence of pH, J. Colloid Interface Sci. 296 (2006) 480–489.

[63] J.L. Figueiredo, J.P.S. Sousa, C.A. Orge, M.F.R. Pereira, J.J.M. Órfão, Adsorption of dyes on carbon xerogels and templated carbons: influence of surface chemistry, Adsorption 17 (2011) 431–441.

[64] F. Villacañas, M.F.R. Pereira, J.J.M. Órfão, J.L. Figueiredo, Adsorption of simple aromatic compounds on activated carbons, J. Colloid Interface Sci. 293 (2006) 128–136.

[65] P.C.C. Faria, J.J.M. Órfão, J.L. Figueiredo, M.F.R. Pereira, Adsorption of aromatic compounds from the biodegradation of azo dyes on activated carbon, Appl. Surf. Sci. 254 (2008) 3497–3503.

[66] S.A.C. Carabineiro, T. Thavorn-Amornsri, M.F.R. Pereira, J.L. Figueiredo, Adsorption of ciprofloxacin on surface-modified carbon materials, Water Res. 45 (2011) 4583–4591.

[67] S.A.C. Carabineiro, T. Thavorn-Amornsri, M.F.R. Pereira, P. Serp, J.L. Figueiredo, Comparison between activated carbon, carbon xerogel and carbon nanotubes for the adsorption of the antibiotic ciprofloxacin. Catal. Today doi:10.1016/j.cattod.2011.08.020, in press.

[68] J.A. Menéndez, J. Phillips, B. Xia, L.R. Radovic, On the modification and characterization of chemical surface properties of activated carbon: in the search of carbons with stable basic properties, Langmuir 12 (1996) 4404–4410.

[69] C.A. Leon y Leon, J.M. Solar, V. Calemma, L.R. Radovic, Evidence for the protonation of basal plane sites on carbon, Carbon 30 (1992) 797–811.

[70] B.H. Han, D.H. Shin, S.Y. Cho, Graphite catalyzed reduction of aromatic and aliphatic nitro compounds with hydrazine hydrate, Tetrahedron Lett. 26 (1985) 6233–6234.

[71] Y. Xun, Z.S. Ping, X. Wei, C.H. You, D.X. Dong, L.X. Mei, Y.Z. Feng, Aqueous dye adsorption on ordered mesoporous carbons, J. Colloid Interface Sci. 310 (2007) 83–89.

[72] W. Yang, D. Wu, R. Fu, Effect of surface chemistry on the adsorption of basic dyes on carbon aerogels, Colloids Surf. A 312 (2008) 118–124.
[73] X. Zhuang, Y. Wan, C. Feng, Y. Shen, D. Zhao, Highly efficient adsorption of bulky dye molecules in wastewater on ordered mesoporous carbons, Chem. Mater. 21 (2009) 706–716.
[74] D.D. Asouhidou, K.S. Triantafyllidis, N.K. Lazaridis, K.A. Matis, S.S. Kim, T.J. Pinnavaia, Sorption of reactive dyes from aqueous solutions by ordered hexagonal and disordered mesoporous carbons, Micropor. Mesopor. Mat. 117 (2009) 257–267.
[75] A. Derylo-Marczewska, A.W. Marczewski, S. Winter, D. Sternik, Studies of adsorption equilibria and kinetics in the systems: Aqueous solution of dyes–mesoporous carbons, Appl. Surf. Sci. 256 (2010) 5164–5170.
[76] B. Li, Z. Lei, X. Zhang, Z. Huang, Adsorption of simple aromatics from aqueous solutions on modified activated carbon fibers, Catal. Today 158 (2010) 515–520.
[77] J.L. Figueiredo, N. Mahata, M.F.R. Pereira, M.J. Sánchez-Montero, J. Montero, F. Salvador, Adsorption of phenol on supercritically activated carbon fibres: Effect of texture and surface chemistry, J. Colloid Interface Sci. 357 (2011) 210–214.
[78] F. Salvador, M.J. Sánchez-Montero, A. Salvador, M.J. Martín, Study of the energetic heterogeneity of the adsorption of phenol onto activated carbons by TPD under supercritical conditions, Appl. Surf. Sci. 252 (2005) 641–646.
[79] S. Faust, O. Aly, Adsorption Processes for Water Treatment, Butterworths, Boston, 1987.
[80] C.P. Huang, Chemical interactions between inorganics and activated carbon, in: P. Cheremisinoff, F. Ellerbusch (Eds.), Carbon Adsorption Handbook, Ann Arbor Science, Ann Arbor, 1978, 282–329.
[81] R. Leyva-Ramos, L. Fuentes-Rubio, R. Guerrero-Coronado, J. Mendoza-Barron, Adsorption of trivalent chromium from aqueous solutions onto activated carbon, J. Chem. Technol. Biotechnol. 62 (1995) 64–67.
[82] R. Leyva-Ramos, A.J. Martinez, Adsorption of chromium (VI) from aqueous solutions on activated carbon, Water Sci. Technol. 30 (1994) 191–197.
[83] G. Blázquez, F. Hernáinz, M. Calero, M.A. Martín-Lara, G. Tenorio, The effect of pH on the biosorption of Cr (III) and Cr (VI) with olive stone, Chem. Eng. J. 148 (2009) 473–479.
[84] J. Rivera-Utrilla, M. Sánchez-Polo, Adsorption of Cr(III) on ozonised activated carbon. Importance of Cπ—cation interactions, Water Res. 37 (2003) 3335–3340.

Chapter 16

Catalytic Removal of Water-Solved Aromatic Compounds By Carbon-Based Materials

Eva Castillejos
Instituto de Catálisis y Petroleoquímica, CSIC, Campus Cantoblanco, Madrid, Spain

Inmaculada Rodríguez-Ramos
Instituto de Catálisis y Petroleoquímica, CSIC, Campus Cantoblanco, Madrid, Spain, and Unidad Asociada UNED-CSIC, Grupo de Diseño y Aplicación de Catalizadores Heterogeneos, Spain

Antonio Guerrero-Ruiz
Unidad Asociada UNED-CSIC, "Grupo de Diseño y Aplicación de Catalizadores Heterogeneos", Spain, and Dpto. Química Inorgánica y Química Técnica, Facultad de Ciencias, UNED, Spain

Chapter Outline

16.1. Introduction 500
16.2. Properties of Carbon Materials in Relation to their Applications in Heterogeneous Catalysis 501
16.2.1. Graphite Materials 502
16.2.2. Activated Carbon and Carbon Blacks 503
16.2.3. Carbon Nanotubes 505
16.2.4. Carbon Nanofibers 506
16.3. Applications of Carbon Materials for Removing Organic Aromatic Compounds Contained in Polluted Waters 507
16.3.1. The Most General Pollutants of Water 508
16.3.2. Adsorption of Aromatic Compounds on Novel Carbon Materials 509
16.3.3. Environmental Catalysis and Oxidation Reactions in Aqueous Media 510
16.3.4. Hydrogenation Reactions 512
16.3.5. Photocatalysts 513
16.4. Future Prospects and Outlook 514
References 515

Novel Carbon Adsorbents. DOI: 10.1016/B978-0-08-097744-7.00016-8
Copyright © 2012 Elsevier Ltd. All rights reserved.

16.1. INTRODUCTION

Wastewater treatments have become a major social and technological issue in the last decades. Depending on the pollutants' nature and their concentration, different strategies can be applied in order to remove the contamination (Scheme 16.1). Biological treatments are highly effective for the removal of most pollutants. Despite their success and low costs, these treatments are ineffective for highly concentrated organic pollutants and inappropriate for compounds that are toxic for the microorganisms. In addition, the sludge produced in the biological treatments, which are highly concentrated solutions with high chemical oxyg\en demand values, may be removed by burning (incineration) or landfilling.

Physical treatments, such as the adsorption, have been effectively applied in the case of high pollutant concentrations, and activated carbons are some of the most widespread adsorbents used for the removal of a wide variety of organic and inorganic pollutants from aqueous or gaseous media. These adsorbent materials are widely used due to their specific physical and chemical properties (specific surface functionalities, pore-size distribution, pore volume, confinement effect, and tuning surface) as well as their hydrophobicity that permit the retention of aromatic compounds and the water repellency. While the effectiveness of the adsorption by carbon materials is well documented [1–4], research in catalytic methods based on medium values of chemical oxygen demand and toxic compounds is still essential. The potentiality of using carbon materials as catalysts or as catalytic support has already been investigated and documented for a number of technological processes in the field of polluted water treatments [5–8]. The main objective of this review is to list and compare those catalytic methods where novel carbon materials are applied to remove organic pollutants from the aqueous

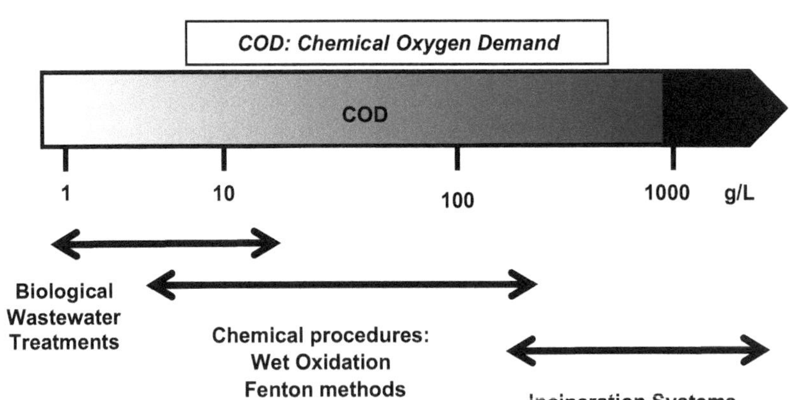

SCHEME 16.1 Applicability of different procedures as a function of the concentration of pollutants in wastewaters.

phase, particularly aromatic compounds. The methodology has been categorized into two broad groups: carbon materials and applications of carbon materials.

16.2. PROPERTIES OF CARBON MATERIALS IN RELATION TO THEIR APPLICATIONS IN HETEROGENEOUS CATALYSIS

Carbon solids such as activated carbons, carbon blacks, graphite, and graphitic derivatives are versatile and fascinating materials because of their multiple applications, many of them related with their surface properties. Nowadays, carbon materials can be classified in different families depending on graphene sheets' partial structural disposition or dimension: Carbon blacks (CBs), activated carbons (ACs), graphites including high surface area graphites (HSAGs), single-walled carbon nanotubes (SWCNTs), multi-walled carbon nanotubes (MWCNTs), and graphite nanofibers (GNFs). Many of these materials find a number of important applications [9] including separation science, heterogeneous catalysis, electronics, environmental pollution control (water purification), as well as in developing future areas, such as energy generation or nanoelectronics or nanomedicine. Novel carbon materials present important properties in relation to their porous structure (Table 16.1) and diversity in the adsorption properties into the different dimensions of their pores, which results in their potential applications. Each pore size range corresponds to a characteristic nitrogen adsorption mechanism, as evidenced by the isotherm shapes (five general types) [10]. In microporous materials, three-dimensional condensation of the adsorbates occurs inside a strong electromagnetic field induced by the narrow pore dimensions. In mesoporous materials, the formation of adsorbate layers completed with a capillary condensation phenomenon takes place. Finally, the macroporous materials cannot be filled by capillary condensation and present porous properties similar to conventional flat surfaces [11]. Importantly, small alterations to the specific surface area can significantly change the physical and chemical properties of these materials. Furthermore, the surface of carbon

TABLE 16.1 Specific Surface Areas

	A C	Graphite	SWCNTs	MWCNTs	GNFs
S_{BET} (m^2/g)	500–3000	1–300	400–900	150–450	10–250
Porosity (cm^3/g)	Microporous V_{micro}: 0.3–0.8	Mesoporous V_{meso}: 0.04–0.3	Microporous V_{micro}: 0.15–0.3	Mesoporous V_{meso}: 0.5–2	Mesoporous V_{meso}: 0.2–2

materials can be functionalized, for instance, by introduction of adatoms, which can also vary their surface properties [12]. In addition, the structural defects, for example, the graphene edges and corners, can contribute significantly to applications as adsorbents or to their catalytic properties. In this chapter, we will highlight the main features of the surface specific characteristics that make carbon attractive and will exemplify these with the most relevant studies of the literature.

16.2.1. Graphite Materials

Graphite (Fig. 16.1) is a semi-metal with a layered crystal structure where the adsorption proceeds mainly by dispersion forces. The graphite surface can be considered as a combination of basal planes and unsaturated sites at the edges of the graphene layers that constitute the carbon crystallites. The edge carbon atoms are frequently combined with other heteroatoms, such as oxygen, forming functional groups at the surface. In general, in crystalline graphite of low surface area, the edge proportion is small compared to the

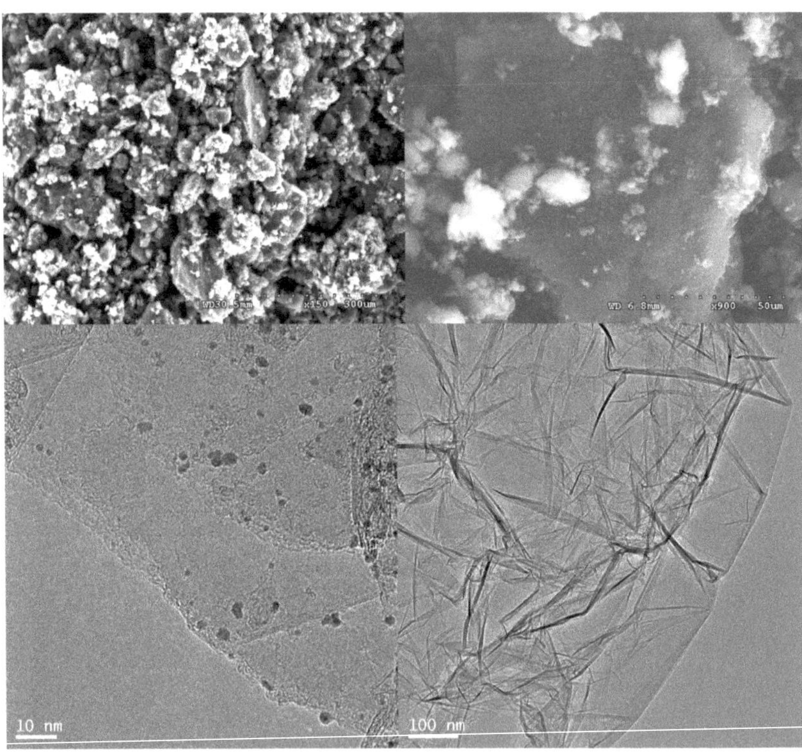

FIGURE 16.1 Scanning and transition electron microscopy of HSAGs and an expanded graphene oxide.

basal planes, and these materials do not exhibit significant oxygen chemisorption. However, HSAGs contain a high proportion of edges, then exhibit significant oxygen chemisorption and easy creation of surface functionalities [13]. Interestingly, HSAG can also suffer fine-tuning in order to introduce surface groups, which could be exploited in the design of catalytic systems. The specific interactions between the graphite surface and the adsorbate may be attributed in many cases to the functionalities exposed on carbon surfaces that mostly have acidic properties. In addition, these carbon surfaces can also contain basic sites that are thought to be oxygen-free sites of Lewis-type located at π electron-rich regions within the basal planes. In this case, nonionic organic compounds, particularly aromatic molecules, are able to originate specific interactions with the diverse types of sites exposed on the functionality-free surface of carbons. Moreover, electron-transfer bonds between the edges of the graphite layers and the aromatic molecules are possible [14]. Besides the chemical properties of graphite, these materials can have electrical, electronic, and electrochemical (combination of electrical and chemical activities) properties. The highly ordered pyrolytic graphite (HOPG) is known for its anisotropic feature in electrical, mechanical, and thermal properties associated with its atomic structure since electrical conductivity is much higher within the graphite layers [15]. Finally, in recent years the term 'graphene oxide' or 'graphene' has been introduced as an original and emerging issue; however, this new denomination refers to the classic oxidized graphite (graphitic oxide or graphitic acid) or exfoliated graphite widely used to describe properties of various carbon-based materials and studied for sixty years [16–18].

Graphene is the name given to a flat monolayer of carbon atoms packed into a two-dimensional honeycomb lattice and is the basic building block for graphitic materials. The significant attraction of this material can be attributed to its outstanding electrical, optical, electrochemical, and mechanical properties, especially in comparison to other carbon materials [19–23], and it should be a relevant material for adsorption processes or as a catalyst component [24]. On the other hand, expanded graphites have been used as adsorbents for treatment of industrial wastewater [25].

16.2.2. Activated Carbon and Carbon Blacks

Activated carbons (Fig. 16.2) are the most common and classical materials used as adsorbents due to their high surface area, high volume of pores, surface chemical properties, possibility to control the porosity and surface chemistry, resistance to acid/basic media [8] and, in the case of catalytic applications as support for noble metals, because of the easy recovery and high dispersion of the metal in the porous structure [26,27]. In addition, carbons materials offer unparalleled flexibility in tuning and tailoring properties to specific needs. The activated carbon structure consists of aromatic

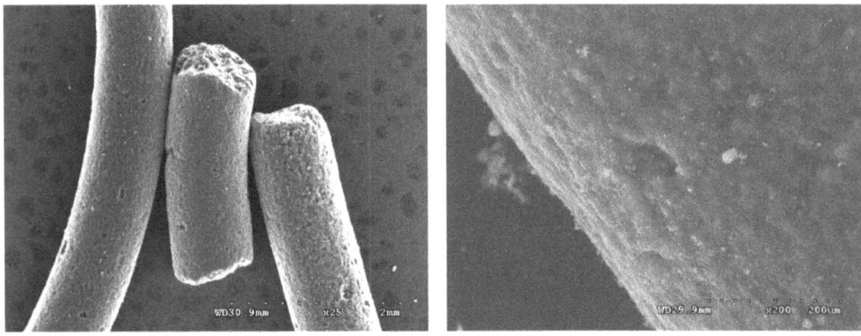

FIGURE 16.2 Scanning electron microscopy of commercial AC structures.

sheets and strips with variable gaps of molecular dimensions (turbostratic structures) between them, leading to micropores [28]. At the nanometer scale, sp^2 bonded carbons are intrinsically anisotropic and in the direction parallel to the basal plane, the strong aromatic C–C bond is dominant, but in the direction perpendicular to the basal plane, the weak van der Waals or π–π interaction are dominant. Bulk properties, such as electronic, electrical, or thermal ones, are a consequence of the carbon structure, while its anisotropic nanostructure is responsible for the behavior as a good conductor or an insulator [29]. Nitrogen doping of activated carbons has an effect on their electrical properties, and the electrical conductivity is higher than that of nitrogen-free carbons [30]. Further, surface properties (physical and chemical), which are responsible for catalytic and adsorbent applications, can be tailored by introduction of nitrogen ad-atoms [12]. In general, activated carbons have a porous structure, usually with a certain amount of surface chemically bonded heteroatoms, mainly oxygen. Although most of the adsorption takes place in the micropores of activated carbons, meso- and macropores play an important role because they serve as a passage for the adsorbate transport to the micropores. The adsorptive properties of activated carbons are determined not only by their porous structure but also by their chemical composition (mainly oxygenated surface groups). In microcrystalline carbons, such as activated carbon, the edge area is extremely extended due to a more disordered structure, which results in a larger propensity for oxygen chemisorption. Furthermore, carbon structures offer high flexibility to introduce oxygen surface groups not only in the edges but also within the basal planes, which contain defects where oxygen chemisorption is possible [31]. Among the different techniques (gas or solutions) that have been applied for surface oxidation, nitric acid solution treatments are the most common and most effective methods. These functionalities are not only responsible for the development of surface charge, but also for much of the chemical activity. In this sense, carbon is an amphoteric solid, which can be acidic or basic, and its surface charge is a function of the pH [32].

In the case of carbon blacks, their porous structure, surface area, and degree of structural order can be varied over a wide range by simple preparation tuning. Carbon blacks have small particle size and normally are presented in the form of robust pellets. Carbons blacks are conductors, which influence their performances as catalyst supports due to the shifting of electron density between them and the metal crystal. The level of conductivity depends on the structure of its particles. Carbon black with a larger structure shows excellent conductive property.

16.2.3. Carbon Nanotubes

CNTs are an ideal nanoscale material (Fig. 16.3) with enough chemical stability, excellent mechanical properties, high surface area, specific electronic properties, and adaptable thermal conductivity [7]. They may be viewed as a cylindrical structure formed from graphene sheets and closed by fullerenoid end-caps. There are single-wall carbon nanotubes (SWCNTs) and multiwall

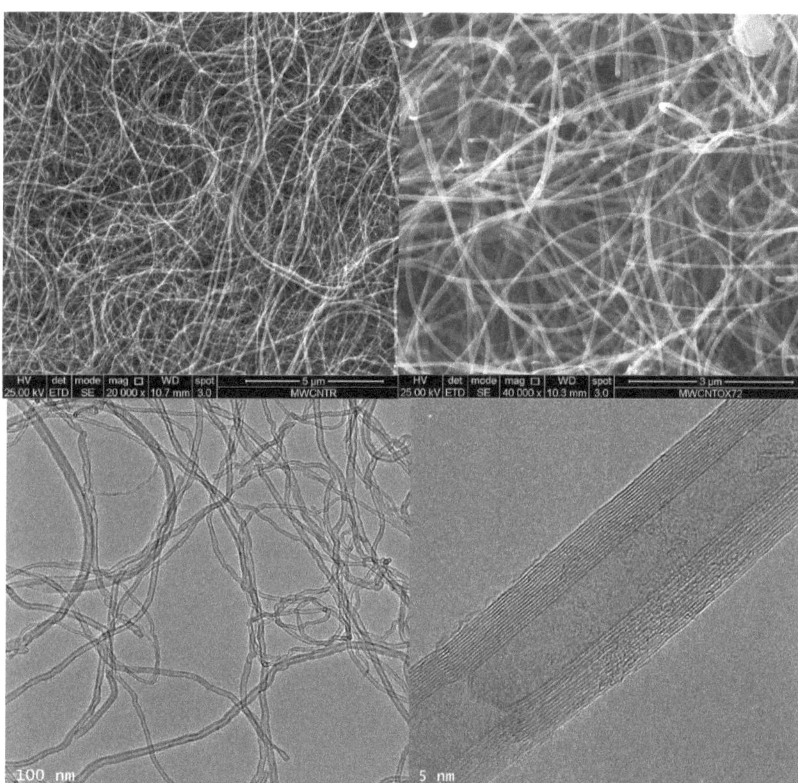

FIGURE 16.3 SEM and TEM micrographs of MWCNT structures.

carbon nanotubes (MWCNTs). These latter consist of several, nested co-axial single wall tubes that furthermore exhibit different surface properties. Due to the high energy of the carbon–carbon double bond, CNTs resist acid treatment quite well and are stable in all reductive atmospheres. In addition to the unique properties of CNTs, their nanometric dimensions and hollow structure result in increased surface/volume ratio, when suitably dispersed and with unimodal cylindrical pores. A detailed study of the CNTs' isotherms has shown that these materials have many potential adsorption sites namely, the outer surfaces of tubes, tube ends, intertube sites in an SWNT bundle, and the hollow cavity of open tubes. Nitrogen adsorption studies performed on CNTs have highlighted the porous nature of these materials. For SWCNTs, nitrogen isotherms clearly show the microporous nature of this type of CNTs. Most experiments show that the specific surface area of SWCNTs is often larger than that of MWCNTs (Table 16.1). For MWCNTs, pores can be mainly divided into inner hollow cavities of small diameter, external walls, and aggregated pores formed by interaction of isolated MWCNTs. Thus, MWCNTs are essentially a mesoporous material that also presents some macroporosity (pores >50 nm).

The adsorption sites can be decorated with a wide variety of atoms and molecules, where the nanostructure serves as a template to anchor or organize molecules and particles in a highly dispersed state. Due to their interaction with the surface of the nanotube substrate, the deposited metals, semiconductors, oxides, organic molecules, and biomolecules often present unique structure–property relationships. Thus, it has been observed that the curved surface of the SWNT modifies the interaction of the substrate with the coated material, thereby giving rise to new properties when compared with graphite [33].

The possibility of macroscopic shaping makes MWCNTs attractive as catalyst supports and as adsorbents in large-scale applications but only if their specificity justifies the extra cost of their production. Indeed, resistance to abrasion, thermal and dimensional stabilities, and specific adsorption properties are important factors that make the CNTs attractive systems. In addition, the MWCNTs' surface properties can be easily controlled. In particular, MWCNTs could replace activated carbon supports in liquid-phase reactions where the microporosity limits the mass transport [6]. In general, the electrical conductivity of CNTs as well as the presence of structural defects, the helicity, the curvature of the surface of these materials, and the presence of an inner cavity are expected to affect the adsorption, possibly in a differential manner if compared to activated carbon or graphite.

16.2.4. Carbon Nanofibers

GNFs have no inner cavity; thus, they are different from MWCNTs. However, they were largely studied in order to understand the behavior of the external carbon surface and because their easy conformation helps to form nets or clothes. The different types of GNFs are classified concerning the orientation of the graphite platelets with respect to the growth catalytic axes. They can be grouped in a ribbon-like structure (the graphite platelets in the carbon material are aligned parallel to the fiber axis), platelet-like structure (the graphite platelets are oriented perpendicular to the fiber axis), and herringbone-type structure or fishbone nanofibers (the graphitic layers are orientated at an angle to the fiber axis). The specific surface area of GNFs, which are nonporous materials, are generally lower than for MWCNTs. The diameters of GNFs are generally higher than those for MWCNTs and can easily reach 500 nm. GNFs show electronic and thermal conductivity properties, and these are of increasing interest for adsorbents and electrocatalyst support. GNFs present a surface area of easy access for adsorptive or reactive fluids and in general to their active surface sites. The peculiar 3-D mesoporous network formed by these materials provides an improved mass transport without mass transfer limitations. GNFs are often considered as conductive substrates that can exert electronic perturbations similar to those of graphite [34]. Furthermore, GNFs offer the possibility of cloth fabrication that permits to improve catalytic efficiency because of its lower mass-transfer resistance.

16.3. APPLICATIONS OF CARBON MATERIALS FOR REMOVING ORGANIC AROMATIC COMPOUNDS CONTAINED IN POLLUTED WATERS

Carbon materials attract a growing interest due to their specific characteristics in the field of wastewater treatments [35]. Especially, the novel carbon structures, such as graphite and carbon nanofibers, nanofilaments, and nanotubes, are very promising candidates for application in this area. The interest in carbon materials as support for catalysts has been gathered in several reviews [7,26,27].

Although carbon materials are much more frequently used as catalyst supports than as catalysts themselves, there are several reactions catalyzed by carbon materials [36–39]. Waters polluted by organic compounds, fertilizers, metals, or detergents are very frequently treated by means of adsorption processes [1] or by heterogeneous catalysis in which carbon materials (CNTs, GNFs, ACs, and graphite structures) act as both adsorbents or as a catalyst support for the active phase. In some cases, there is a cooperative action between the adsorbent sites, where the organic pollutants are concentrated, and the neighboring catalytic active sites, where the adsorbed pollutant molecules are decomposed. Richard et al. [40] have shown a three-step process to remove organic compounds that comprised: adsorption–concentration, catalytic hydrogenation, and regeneration of adsorbent.

16.3.1. The Most General Pollutants of Water

Many different organic chemicals are regarded as pollutants of water and can be classified depending on the activities where they are originated, considering their toxicities or taking into account their concentration. Some examples are given in Table 16.2. Pollutants can exist in water in different states: dissolved, in suspension, or absorbed by particles. All states of pollutants can travel great distances through water in many different ways.

Further, the so-called "emerging pollutants" or "micropollutants," which are contaminants presented at very low concentrations (μg/L), but with very high environmental impact should be considered. These compounds are a source of concern because they are used and released in large quantities, and their physical and chemical properties contribute to their widespread distribution into the environment. As representative examples are chlorobenzene derivatives, usually from industrial wastes, typical largely used pharmacological compounds, i.e. paracetamol (*N*-(4-hydroxyphenyl)ethanamide) or

TABLE 16.2 Organic Aromatic Pollutants in Wastewater: Activities where they are Produced and Typical Decontamination Treatments

Sources	Contaminants	Water treatment	Reference
Chemical	Phenolic compounds, surfactants	Catalytic wet oxidation, catalytic hydrogenation, adsorption	[41–44]
Pharmaceutical	Dyes, chlorophenol solvent, benzenes, plasticizers	Fenton process, catalytic wet oxidation, electrochemical oxidation processes, selective oxidation, oxidative dehydrogenation	[45–49]
Petrochemical	Phenolic compounds	Catalytic wet oxidation, adsorption	[50,51]
Textile industry	Dyes	Fenton process, persulfate oxidative degradation, catalytic ozonation, adsorption	[52–55]
Agricultural and agro-Industrial sector	Phenolic compounds, fertilizer, herbicides and pesticides	Catalytic wet oxidation, adsorption	[40,56–58]
Municipal waste	Surfactants and dyes	Adsorption and photocatalysis	[59,60]

ibuprofen ((*RS*)-2-(4-(2-methylpropyl)phenyl)propanoic acid) present in sewage from the cities, and agricultural commodities, pesticides, or herbicides, such as DDT (dichlorodiphenyltrichloroethane) or chloramben (3-amino-2,5-dichlorobenzoic acid).

16.3.2. Adsorption of Aromatic Compounds on Novel Carbon Materials

So many studies and reviews have been done in order to understand thoroughly the mechanisms involved in the adsorption processes [51,61–64]. The adsorption capacity of carbon materials depends on the nature of the adsorbent, the nature of the adsorbate, and the solution conditions (pH, temperature, and ionic strength). With respect to the graphitic surfaces, the main adsorption sites are located on: (i) the carbon basal planes, (ii) edges and crystal defects, (iii) oxygen surface groups, and (iv) ash impurities (i.e. metal oxides). The oxygen surface groups are mainly placed, in general, on the edges of the graphitic layers and also, in the case of graphene oxide, in the interlayer space on the graphitic basal planes. Despite being a small fraction of the overall carbon surface, the oxygen groups are however very active, exhibiting a significant influence on the adsorption capacity [65]. In the particular case of CNTs or GNFs, few studies distinguish between adsorption sites in the external or internal nanotube surfaces. However, the interaction of CNTs with adsorbed species on their either external or internal surfaces attracts an increasing attention, considering that the specific physico-chemical properties of the tubes can likely influence on the adsorption process and bearing in mind the opportunity of using these materials as selective adsorbents [66]. For MWCNTs, mesopores can be mainly divided into inner hollow cavities with cylindrical walls of small diameter and pores formed by interaction or aggregation of isolated MWCNTs. The adsorption of aromatic compounds by MWCNT might occur either on or inside the tubes, or in spaces between aggregated MWCNTs. Indeed, the rolling up of the graphene sheet to form the tube causes a rehybridization of the carbon orbital (nonplanar sp^2 configuration), thus leading to modifications of the π density of the graphene sheet.

Many attempts have been made to understand how the different ring substituents of aromatic compounds affect the adsorption [3,67]. Kaneko et al. [68] observed an adsorption trend with the compound's solubility, Deng et al. [69] related the uptake with the electron density of the aromatic ring, and Moustafa et al. [70] proposed a dependence on pK values. Although the nature of the adsorption driving forces for aromatic compounds is still under discussion, it seems that the hypothesis more accepted is a π–π interaction [3]. Then aromatic compounds (in their uncharged form) are adsorbed onto the carbon materials surface by dispersion forces between the π electrons of the aromatic ring and the π electrons of the graphene layers.

Notice also that the pH of the solution can affect the charge of phenolic compounds, and therefore electrostatic interactions between adsorbent and adsorbate can also occur.

16.3.3. Environmental Catalysis and Oxidation Reactions in Aqueous Media

Catalytic oxidation processes are currently applied for industrial wastewater treatments, and activated carbons have been widely used as the main component of catalysts due to their high adsorption capacity and stability in reaction media. Catalytic wet air oxidation (CWAO) of organic solutions and advanced oxidation processes (AOPs) such as catalytic ozonation, Fenton-derived methodologies, or catalytic photooxidation of pollutants, these latter being more appropriate for the removal of diluted compounds or with low adsorbability, are some of the most successful processes currently applied or under study. The sources of oxidants can be either air (dioxygen) and dissolved hydrogen peroxide or ozone.

CWAO is an emerging technology for contaminated waters with toxic and hazardous organic compounds, which aims at the total oxidation of the polluting species present in the effluent under relatively mild conditions (398–493 K, 2–50 bar), using air as oxidant. Depending on the reaction conditions and catalyst composition, either a total mineralization of pollutants yielding CO_2, N_2, and H_2O or an increase in the effluent biodegradability by forming easily biodegradable by-products, such as carboxylic acids, can be achieved. Adsorption of organic pollutants by activated carbons is a well-established technology [65], and its combination with catalytic oxidation represents an attractive alternative because of the synergetic action of this support, concentrating the reactants in close proximity of the catalytically active metal species. In this way, ruthenium supported on high surface area-activated carbon and promoted by ceria (Ru–CeO_2/AC) was demonstrated to be an active CWAO catalyst for phenol and acrylic acid in aqueous solutions. The enhanced activity was explained by the particular cooperative effect between the adsorption capacity of the support and selected ingredients of the catalyst [41]. In the same way, the CWAO of aniline solutions has been studied using Pt catalysts supported on carbon xerogel, MWCNT, and AC [71]. More recently, it has been demonstrated that certain carbon nanostructured materials directly exhibit catalytic activities for phenol oxidation in aqueous phase [72], in agreement with previous studies over MWCNTs chemically modified using HCl or HNO_3–H_2SO_4 [73]. Thus, novel carbon materials can exhibit high activity and good stability for the phenol CWAO. On the other hand, Garcia et al. [74] investigated the effect of ruthenium precursor and the surface chemistry of MWCNT on the performance of Ru/MWCNT for the CWAO of aniline. In addition, the same authors [75] found that Pt/oxidized MWCNT catalysts lead to a simpler oxidation pathway in the aniline CWAO compared with the bare support. The same behavior was achieved with Pt/MWCNTs in the treatment of textile dyes (monoazo dye Chromotrope 2R, diazo dye Erionyl Red B and triazo dye Solophenyl Green BLE) [75]. In short, as reviewed in [76], the use of carbon materials is opening up new perspectives in CWAO due to the potential specific properties of some novel solid carbon structures.

The AOPs are emerging technologies with high efficiency in the degradation of persistent organic micropollutants in the aqueous phase by oxidation with hydroxyl radicals [77]. AOPs are characterized by the production of $^{\cdot}OH$ radicals, which are powerful and unselective oxidants that can oxidize and mineralize resistant organic pollutants in water. In the specific case of Fenton processes, hydroxyl and perhydroxyl radicals are produced by the reaction between Fe ions, which act as homogeneous catalysts, and the reagent hydrogen peroxide. The Fenton process can be also carried out under heterogeneous conditions, by immobilizing the iron catalysts on a support, such as activated carbons [78,79]. On the other hand, ozonation, which is a typical treatment of drinking waters, is limited by chemical kinetics and by the generation of degradation products. In order to overcome these limitations, some modifications by adding supported metal catalysts have been postulated. Under the conditions of catalytic ozonation, an enhancement in the removal of highly refractory compounds is obtained by the transformation of ozone into the more reactive $^{\cdot}OH$ radicals [54,80]. Several reported studies have shown that composites of metal oxides and activated carbon have better performance for mineralization of organic pollutants than the individual components, increasing the catalytic activity of the composites with the amount of activated carbon introduced.

A critical point in the use of carbon materials as catalysts for oxidation processes in water media is the possibility of active phase lixiviation under the reaction conditions. To overcome this difficulty, an adequate anchoring of active catalytic species on the carbon surface is required. Among the different strategies to conduce this anchoring, the carbon surface functionalization of GNFs by acetylacetonate groups, before reaction with Fe ions, has been demonstrated to produce chelate bidentate complexes, which are very active and stable (several reusing cycles) in the phenol CWAO [72]. On the other hand, for economic and environmental reasons, spent adsorbents/catalysts are not disposed of but undergo several cycles of regeneration. The regeneration of carbon materials is at present conducted by three procedures: thermal regeneration at near 800 °C, desorption by hot water at temperatures up to 180 °C, and catalytic regeneration by oxidation at 200–300 °C or by reduction at 30–50 °C [81]. Thus, for the latter catalytic procedure it is important to incorporate active and very stable catalytic ingredients in the adsorbents/catalysts with the final purpose of their regeneration. The molecular anchoring of metal oxides can be carried out over previously functionalized carbon surfaces *via* covalent bonds that are required for stable oxidation catalysts. In the case of reductive conditions, the metallic particles are stabilized on the carbon surface by interaction with graphitic edges of the carbon structures, giving a stable composite during the adsorption–reduction process [82].

One of the more interesting technologies for wastewater treatments is the electrochemical advanced oxidation processes (EAOP) [83]. The most popular EAOP is the electro-Fenton (EF) process, in which H_2O_2 is continuously electrogenerated in the anode with the acid media where oxygen and two electrons are injected ($O_{2\,(g)} + 2H^+ + 2e^- \rightarrow H_2O_2$), whereas Fe^{2+} ions are added as a catalyst to react with this generated H_2O_2 originating homogeneous hydroxyl radicals ($^{\cdot}OH$) and Fe^{3+} ions from Fenton's reaction ($Fe^{2+} + H_2O_2 \rightarrow Fe^{3+} + {}^{\cdot}OH + OH^-$). Cathodes usually consist of carbon felts, which favor the regeneration of Fe^{2+} from the cathodic reduction of Fe^{3+}. Formed $^{\cdot}OH$s in the water are strong oxidants with high standard reduction potential ($E^0(^{\cdot}OH/H_2O) = 2.80$ V), which permit a nonselective reaction with organics *via* dehydrogenation and/or hydroxylation until total mineralization (conversion into CO_2, water, and inorganic ions). Some studies of EAOP have been reported for decontamination of wastewaters containing aromatic compounds, such as chlorophenoxy acids, typical herbicides [84,85], and dyes [86]. From the novel carbon materials point of view, a key point is the modification of the cathodes constituted by carbon felts based on GNF through the incorporation of MWCNTs. These latter materials bring: (i) a high electronic conductivity, (ii) a high mesoporosity to attain high metal dispersion at reasonable loading, and (iii) a good hydrophobicity for water management. The particular structure and properties of the different carbon materials have a direct impact on electrocatalyst performance. Therefore, the unique properties of MWCNTs make them promising building blocks for nanoscale electronic devices with enhanced performance.

16.3.4. Hydrogenation Reactions

Reductive processes for wastewater treatment are more expensive and less popular than oxidation methods, but, in special cases, they are advantageous. Although convectional and advance oxidative degradation processes are efficiently used for the disposal of many pollutants, they were found to be of limited effectiveness for certain groups of pollutants, such as halogenated organics. The oxidative processes are potentially prone to produce hazardous by-products, such as phosgene, dioxins, and chlorine. In addition, as indicated above, certain catalytic reduction treatments are applied for the regeneration of spent adsorbents. The majority of these use Pd as a catalytic active component [81]. Both these reductive regeneration processes as well as the reduction treatments of industrial wastewaters are applied for effluents contaminated by halogenated organic compounds (such as trichlorobenzene or chlorophenol). Furthermore, these hydrodehalogenation (or dechlorination) reactions are usually performed with Pd catalysts, and in some cases carbon materials have been tested as supports for these procedures. In contrast with oxidative processes, catalytic hydrodechlorination cannot produce ultimate harmless products such as water and carbon dioxide, but leads to break carbon–chlorine bonds giving place to innocuous organic compounds without chlorine and without aromatic functions [87].

Another advantage of the hydrogenation–hydrodechlorination methods in comparison with the oxidation ones in aqueous media is the higher solubility of hydrogen, under hydrodehalogenation operation conditions, in comparison with the oxygen solubility in water during CWAO. It should also be noticed that the hydrogenation reactions take place at temperatures and pressures much lower than that required for CWAO processes. In the hydrochlorination reactions, sometimes Pd has been supported on carbon supports. Early studies demonstrated that the reaction rates are higher on Pd/AC than on Pd/alumina, and this effect was interpreted as due to an enhanced adsorption of the reactant molecules on the carbon support, facilitating their presence in contact with the Pd catalytic sites, and yielding significantly faster reaction rates [88].

On the other hand, catalytic hydrogenation treatments have been proposed for purification of wastewater from the olive oil industry [40] and even for phenol elimination from industrial polluted water [43]. It should be indicated that the wastewater of the olive oil extraction industry (so-called alpechin) is composed of numerous organic pollutants, which can be assumed to be mainly phenolic compounds, such as tyrosol (4-(2-hydroxyethyl)phenol). Catalytic hydrogenation of phenolic compounds leads to dearomatized organic compound formation, which can be subsequently degraded by standard biological processes or recovered as valuable organic compounds.

16.3.5. Photocatalysts

Photocatalysis is considered an effective method for the mineralization of many organic compounds through the photogeneration of oxidizing radicals, such as $^{\cdot}OH$ and $^{\cdot}O_2^-$. These processes are usually oxidation reactions but are considered independently in this section taking into account their potential use in a variety of conditions and specific geographical locations. Photocatalysis has important applications but one of the more relevant is the degradation of environmental pollutants in aqueous wastewater, these treatments being more adequate because of their on-place use and because they do not consume extra energy. TiO_2 is the reference for oxide semiconductors because of its specific photocatalytic properties, but it should be noted that it suffers from low energetic efficiency and a narrow light response range. Hence, in recent years many efforts have been focused on developing new composite photocatalysts. Actually, it has been showed that the presence of carbon materials (i.e. AC) improves the efficiency and the activity of TiO_2 heterogeneous photocatalysts in processes where considerable attention has been paid to the unique and controllable structural and electrical properties of these composites. Thus, Matos et al. [89–94] examined the synergy effects between TiO_2 and AC for the photocatalytic degradation of model aromatic pollutants. The combination of adsorption and photocatalytic treatments greatly improves the depuration of residual waters, by using AC associated with TiO_2 in order to activate the metal oxide. Moreover, the presence of AC in different percentages acts as an efficient template for the TiO_2 nanoparticles, which can be prepared by a novel alkoxide sol–gel precipitation [95]. As an example, the photodegradation of different azo dyes has been also studied by using TiO_2–AC composite catalysts [96]. More examples of the use of TiO_2 and AC can be found in a general overview in a brief book chapter published by Faria et al. [97]. Finally, it must be remarked that the carbon–TiO_2 composite, when optimized in composition, can exhibit photocatalytic activities under visible light range irradiation [98].

On the other hand, CNTs show remarkable potential to contribute increasing photocatalytic activity of TiO_2; therefore, the number of studies has been increasing over the past decade [60]. It was shown that the presence of MWNTs in the composite permits: (i) to increase the effective surface area of the photocatalyst, (ii) to stabilize the nanoparticles of TiO_2 in the active state, and (iii) a strong interface interaction that favors adsorption [99] and electron transfer. In these reactions, MWNTs promote the transport of electron/hole pairs along the tubes, thus decreasing their recombination rate and improving the photocatalytic activity of TiO_2. It has also been shown that an optimum amount of CNTs is needed to obtain a significant improvement in the catalyst efficiency. Above this critical amount, the photocatalytic activity decreases due to the increased absorbing and scattering of photons by the carbon support. Nanotube–titania composites have been studied for the photodegradation of phenol [100,101] and azo dyes [102].

Further, the newly developed graphene–TiO_2 nanocomposites have received significant attention, with the intention to harness the redox properties of graphene for creating new-generation photocatalytic systems with enhanced activity. Thus, graphene–semiconductor composites act as photocatalysts, with a mechanism scheme where the TiO_2 nanoparticles adsorb the light-producing electrons, and the reduced graphene oxide captures the electrons and shuttles them to the adsorbed species, inducing the oxidation reaction [103]. Recently, examples, synthesis routes, and future applications about these composites have been summarized and discussed by Leary et al. [60].

16.4. FUTURE PROSPECTS AND OUTLOOK

Emerging pollutants or micropollutants are contaminants present at very low concentrations (μg/L) but with very high environmental impact. Chemicals, such as flame retardants, surfactants, chloroalkanes, pesticides, or drugs (and their metabolic derivatives), are included in this family of pollutants. Most of these compounds are demonstrated to be very dangerous for living organisms (because of their toxic character, endocrine disruptor properties, etc.). The treatment of these pollutants is extremely difficult because of their low concentration. Therefore, although these pollutants are organic molecules, their concentration is too low for the application of well-established technologies, such as wet air oxidation, and other advanced processes are required (electrochemical, catalytic, photocatalytic, etc.). Because of this fact, those processes leading to a concentration of the pollutant are needed as a preliminary strategy for the treatment of these streams. Therefore, the much diluted initial streams are ideally transformed into streams with a pollutant concentration several orders of magnitude higher. Adsorption is the most promising approach for doing this, since it can be accomplished at ambient temperatures and pressures. In addition, adsorption operations can be considered as a continuous operation when working at this low concentration. As alternative adsorbents to classical activated carbons, several research studies have showed the ability of carbon nanotubes to adsorb different pollutants from various aqueous samples [104]. This outstanding ability of carbon nanotubes compared with other adsorbents is due to strong interaction between the carbon surface and the pollutants due to the unique structure of nanocarbon as a result of the delocalized π-electrons on the hexagonal arrays of carbon atoms in graphene sheets. Despite the fact that the cost of carbon nanotubes is rather high compared with other conventional treatment methods, the production cost is getting lower and the unique properties of these tubes have made them a promising material for environmental remediation. In addition, carbon nanotubes have shown exceptional adsorption capability and high adsorption efficiency for various organic contaminants, such as dioxin, benzene, 1,2-dichlorobenzene, trihalomethanes, uremic toxins, polycyclic aromatic hydrocarbons (PAHs), etc. [105].

REFERENCES

[1] J. Rivera-Utrilla, M. Sanchez-Polo, V. Gomez-Serrano, P.M. Alvarez, M.C.M. Alvim-Ferraz, J.M.J. Dias, Activated carbon modifications to enhance its water treatment applications. An overview, Hazar. Mater. 187 (2011) 1–23.

[2] S. Furmaniak, P.A. Gauden, A.P. Terzyk, G. Rychficki, Water adsorption on carbons – critical review of the most popular analytical approaches, Adv. Colloid Interface Sci. 137 (2008) 82–143.

[3] L.R. Radovic, C. Moreno-Castilla, Rivera-Utrilla, Carbon materials as adsorbents in aqueous solutions, Chem. Phys. Carbon (2001).

[4] V.K.K. Upadhyayula, S.G. Deng, M.C. Mitchell, G.B. Smith, Application of carbon nanotube technology for removal of contaminants in drinking water: a review, Sci. Total Environ. 408 (2009) 1–13.

[5] K.P. Jong de, J.W. Geus, Carbon nanofibers: catalytic synthesis and applications, Catal. Rev. Sci. Eng 42 (2000) 481–510.

[6] P. Serp, M. Corrias, P. Kalck, Carbon nanotubes and nanofibers in catalysis, Review. Appl. Catal. A Gen. 253 (2003) 337–358.

[7] P. Serp, J.L. Figueiredo, Carbon Materials for Catalysis, J. Wiley & Sons, Hoboken (NJ), 2009.

[8] F. Rodriguez-Reinoso, The role of carbon materials in heterogeneous catalysis, Carbon 36 (1998) 159–175.

[9] R.J. White, V. Budarin, R. Luque, J.H. Clark, D.J. Macquarrie, Tuneable porous carbonaceous materials from renewable resources, Chem. Soc. Rev. 28 (2009) 3401–3418.

[10] R.H. Perry, D.W. Green, J.O. Maloney, Chemical Engineer's Handbook, sixth ed. McGraw-Hill, New York, 1984.
[11] S.J. Greg, K.S. Sing, Adsorption, Surface Area and Porosity, Academic Press, London, New York, Paris, Tokyo, 1982.
[12] A. Guerrero-Ruiz, I. Rodriguez-Ramos, F. Rodriguez-Reinoso, C. Moreno-Castilla, J.D. López-González, The role of nitrogen and oxygen surface groups in the behavior of carbon-supported iron and ruthenium catalysts, Carbon 26 (1988) 417–423.
[13] A.B. Dongil, B. Bachiller-Baeza, A. Guerrero-Ruiz, I. Rodríguez-Ramos, A. Martínez-Alonso, J.M.D. Tascón, Surface chemical modifications induced on high surface area graphite and carbon nanofibers using different oxidation and functionalization treatments, J. Collo. Interface Sci. 355 (2011) 179–189.
[14] E. Castillejos-López, D.M. Nevskaia, V. Muñoz, I. Rodríguez-Ramos, A. Guerrero-Ruiz, Specific interactions between aromatic electrons of organic compounds and graphite surfaces as detected by immersion calorimetry, Langmuir 20 (2004) 1013–1015.
[15] R.L. McCreery, Electroanalytical Chemistry. A.J. Bard, Marcel Dekker, New York, 1991.
[16] P.R. Wallace, The band theory of graphite, Phys. Rev. 71 (1947) 622–634.
[17] J.W. McClure, Diamagnetism of graphite, Phys. Rev. 104 (1956) 666–671.
[18] E.L. Evans, J.D.D. Lopez Gonzalez, A. Martin Rodriguez, F. Rodriguez Reinoso, Kinetic of formation of graphite oxide, Carbon 13 (1975) 461–465.
[19] A.H. Castro Neto, F. Guinea, N.M.R. Peres, K.S. Novoselov, A.K. Geim, The electronic properties of graphene, Rev. Mod. Phys. 81 (2009) 109–162.
[20] S. Stankovich, D.A. Dikin, G.H.B. Dommett, K.M. Kohlhaas, E.J. Zimney, E.A. Stach, R.D. Piner, S.T. Nguyen, R.S. Ruoff, Graphene-based composite materials, Nature 442 (2006) 282–286.
[21] R. Arsat, M. Breedon, M. Shafiei, P.G. Spizziri, S. Gilje, R.B. Kaner, K. Kalantar-zadeh, W. Wlodarski, Graphene-like nano-sheets for surface acoustic wave gas sensor applications, Chem. Phys. Lett. 467 (2009) 344–347.
[22] F. Scarpa, S. Adhikari, A.S. Phani, Effective elastic mechanical properties of single layer graphene sheets, Nanotechnology 20 (2009) 065709.
[23] T.G. Pedersen, C. Flindt, J. Pedersen, A.P. Jauho, N.A. Mortensen, K. Pedersen, Optical properties of graphene antidot lattices, Phys. Rev. B 77 (2008) 245431.
[24] W. Gao, M. Majumder, L.B. Alemany, T.N. Narayanan, M.A. Ibarra, B.K. Pradhan, P.M. Ajayan, Engineered graphite oxide materials for application in water purification, Appl. Mater. Interface 3 (2011) 1821–1826.
[25] M. Toyoda, M. Inagaki, Heavy oil sorption using exfoliated graphite: new application of exfoliated graphite to protect heavy oil pollution, Carbon 38 (2000) 199–210.
[26] E. Auer, A. Freund, J. Pietsch, T. Takke, Carbons as supports for industrial precious metal catalysts, Appl. Catal. A Gen 173 (1998) 259–271.
[27] R. Schlögl, in: G. Ertl, H. Knözinger, J. Weitkamp (Eds.), Preparation of Solid Catalysts, Wiley-VCH, Weinheim, 1999.
[28] R.C. Bansal, J.B. Donnet, H.F. Stoeckli, Active Carbon, Marcel Dekker, New York, 1988.
[29] S. Mrozowski, Semiconductivity and diamagnetism of polycrystalline graphite and condensed ring systems, Phys. Rev. 85 (1952) 609–620.
[30] D.P. Kim, C.L. Lin, T. Mihalisin, P. Heynei, M.M. Labes, Electronic properties of nitrogen-doped graphite flakes, Chem. Mater. 3 (1991) 686.
[31] L.R. Radovic, F. Rodriguez-Reinoso, in: P.A. Thrower (Ed.), Chemistry and Physics of Carbon, Marcel Dekker, New York, 1997.
[32] L.R. Radovic, in: J.A. Schwarz, C.I. Contescu (Eds.), Surface of Nanoparticles and Porous Materials, Marcel Dekker, New York, 1999.

[33] M. Menon, A.N. Andriotis, G.E. Froudakis, Curvature dependence of the metal catalyst atom interaction with carbon nanotubes walls, Chem. Phys. Lett. 320 (2000) 425–434.
[34] C.A. Bessel, K. Laubernds, N.M. Rodriguez, R.T.K. Baker, Graphite nanofibers as an electrode for fuel cell applications, J. Phys. Chem. B 105 (2001) 1115–1118.
[35] O. Hernadez- Ramirez, S.M. Holmes, Novel and modified materials for wastewater treatment applications, J. Mater. Chem. 18 (2008) 2751–2761.
[36] D.L. Trimm, Catalysis, The Royal Society of Chemistry, London, U. K, 1981.
[37] J. Zhang, X. Liu, R. Blume, A. Zhang, R. Schlögl, D.S. Su, Surface-modified carbon nanotubes catalyze oxidative dehydrogenation of n-butane, Science 322 (2008) 73–77.
[38] G.C. Grunewald, R.S. Drago, Carbon molecular sieves as catalysts and catalyst supports, J. Am. Chem. Soc. 113 (1991) 1636–1639.
[39] F. Löcking, H. Köser, M. Jank, A. Ritter, Iron powder, graphite and activated carbon as catalysts for the oxidation of 4-chlorophenol with hydrogen peroxide in aqueous solution, Wat. Res. 32 (1998) 2607–2614.
[40] D. Richard, M.L. Delgado-Nuñez, Kinetics of the degradation by catalytic hydrogenation of tyrosol, a model molecule present in olive oil waste waters, J. Chem. Technol. Biotechnol. 78 (2003) 927–934.
[41] L. Oliviero, J. Barbier, D. Duprez, A. Guerrero-Ruiz, B. Bachiller-Baeza, I. Rodriguez-Ramos, Catalytic wet air oxidation of phenol and acrylic acid over Ru/C and Ru–CeO_2/C catalysts, Appl. Catal. B Environ. 25 (2000) 267–275.
[42] A. Rey, M. Faraldos, J.A. Casas, J.A. Zazo, A. Bahamonde, J.J. Rodríguez, Catalytic wet peroxide oxidation of phenol over Fe/AC catalysts: influence of iron precursor and activated carbon surface, Appl. Catal. B Environ. 86 (2009) 69–77.
[43] D. Richard, L. Delgado-Nuñez, C. de Bellefon, D. Schweich, Depullation of waters contaminated by phenol and chlorophenols, in: Environmental Chemistry-Green Chemistry and Pollutants in Ecosystems, Springer, 2004.
[44] B. Petrova, T. Budinova, B. Tsyntsarski, V. Kochkodan, Z. Shkavro, N. Petrov, Removal of aromatic hydrocarbons from water by activated carbon from apricot stones, Chem. Eng. J. 165 (2010) 258–264.
[45] F. Duarte, F.J. Maldonado-hódar, M.L. Madeira, Appl. Catal. B Env. 103 (2011) 109–115.
[46] H.T. Gomes, S.M. Miranda, M.J. Sampaio, A.M.T. Silva, J.L. Faria, Activated carbons treated with sulphuric acid: catalysts for catalytic wet peroxide oxidation, Catal. Today 151 (2010) 153–158.
[47] H. Wang, Z.Y. Bian, D.Z. Sun, Degradation mechanism of 4-chlorophenol with electro-generated hydrogen peroxide on a Pd/C gas-diffusion electrode, Water Sci. Technol. 63 (2011) 484–490.
[48] L. Kesavan, R. Tiruvalam, M.H.A. Rahim, M.I.B. Saiman, D.I. Enache, R.L. Jenkins, N. Dimitratos, J.A. Lopez-Sanchez, S.H. Taylor, D.W. Knight, C.J. Kiely, G.J. Hutchings, Solvent-free oxidation of primary carbon–hydrogen bonds in toluene suing Au–Pd alloy nanoparticles, Science 133 (2011) 195–199.
[49] G. Mestl, N.I. Maksimova, N. Keller, V.V. Roddatis, R. Schlögl, Carbon nanofilaments in heterogeneous catalysis: an industrial application for new carbon materials? Angew. Chem. Int. Ed. 40 (2001) 2066–2068.
[50] E. Castillejos-Lopez, A. Maroto-Valiente, D.M. Nevskaia, V. Munoz, I. Rodriguez-Ramos, A. Guerrero-Ruiz, Comparative study of support effects in ruthenium catalysts applied for wet air oxidation of aromatic compounds, Catal.Today 143 (2009) 355–363.

[51] D.M. Nevskaia, E. Castillejos-Lopez, V. Munoz, A. Guerrero-Ruiz, Adsorption of aromatic compounds from water by treated carbon materials, Env. Sci. Technol. 38 (2004) 5786–5796.
[52] A. Rodriguez, G. Ovejero, J.L. Sotelo, M. Mestanza-Garcia, Heterogeneous Fenton catalyst supports screening for mono azo dye degradation in contaminated wastewaters, J. Ind. Eng. Chem. Res. 49 (2010) 498–505.
[53] S. Yang, X. Yang, X. Shao, R. Niu, L. Wang, Activated carbon catalyzed persulfate oxidation of azo dye acid orange 7 at ambient temperature, J. Hazard. Mater. 186 (2011) 659–666.
[54] C.A. Orge, J.J.M. Orfao, M.F.R. Pereira, Catalytic ozonation of organic pollutans in the presence of cerium oxide- carbon composites, Appli. Catal. B Environ. 102 (2011) 539–546.
[55] G. Cimino, R.M. Cappello, C. Caristi, G. Toscano, Characterization of carbons from olive cake by sorption of wastewater pollutants, Chemosphere 61 (2005) 947–955.
[56] S. Morales-Torres, A.M.T. Silva, A.F. Pérez-Cadenas, J.L. Faria, F.J. Maldonado-Hódar, J.L. Figueiredo, F. Carrasco-Marín, Wet air oxidation of trinitrophenol with activated carbon catalysts: effect of textural properties on the mechanism of degradation, Appl. Catal. B Environ. 100 (2010) 310–317.
[57] A. Bhatnagar, Removal of bromophenols from water using industrial wastes as low cost adsorbents, J. Hazard. Mater. 139 (2007) 93–102.
[58] M. Anbia, A. Ghaffari, Adsorption of phenolic compounds from aqueous solutions using carbon nanoporous adsorbent coated with polymer, Appl. Surf. Sci. 255 (2009) 9487–9492.
[59] K. Verschueren, Handbook of Environmental Data on Organic Chemicals, fourth ed., John Wiley & Sons, Inc., 2001.
[60] R. Leary, A. Westwood, Carbonaceous nanomaterials for the enhancement of TiO_2 photocatalysis, Review. Carbon 49 (2011) 741–772.
[61] F.L. Slejko, Adsorption Technology: a Step-By-Step Approach to Process Evaluation and Application, Marcel Dekker, New York, 1985.
[62] S.D. Faust, O.M. Aly, Adsorption Processes for Water Treatment, Butterworth Publishers, London, 1987.
[63] N.P. Cheremishinoff, P.N. Cheremishinoff, Carbon Adsorption for Pollution Control, Prentice Hall, Upper Saddle River, NJ, 1993.
[64] D.M. Nevskaia, E. Castillejos-Lopez, A. Guerrero-Ruiz, V. Munoz, Effects of the surface chemistry of carbon materials on the adsorption of phenol–aniline mixtures from water, Carbon 42 (2004) 653–665.
[65] C. Leon y Leon, L. Radovic, Chemistry and Physics of Carbon, Marcel Dekker, New York, 1994.
[66] R.Q. Long, R.T. Yang, Carbon nanotubes as superior sorbent for dioxin removal, J. Am. Chem. Soc. 123 (2001) 2058–2059.
[67] R.W. Coughlin, F.S. Ezra, R.N. Tan, Influence of chemisorbed oxygen in adsorption onto carbon from aqueous solution, J. Coll. Inter. Sci. 28 (1968) 386–396.
[68] Y. Kaneko, M. Abe, K. Ogino, Adsorption characteristics of organic compounds dissolved in water on surface-improved activated carbon fibers, Colloids Surf. 37 (1989) 211–222.
[69] X. Deng, Y. Yue, Z. Gao, Preparation and characterization of active carbon adsorbents for wastewater treatment from elutrilithe, J. Colloid Interf. Sci. 192 (1997) 475–480.
[70] M.R. Moustafa, S.E. Samra, A.M. Youssef, Removal of organic pollutants from aqueous solution. Part 1. Adsorption of phenols by activated carbons, Ind. J. Chem. 28 (1989) 946–948.
[71] H.T. Gomes, P.V. Samant, P. Serp, P. Kalck, J.L. Figueiredo, J.L. Faria, Carbon nanotubes and xerogels as supports of well-dispersed Pt catalysts for environmental applications, Appl. Catal. B Environ. 54 (2004) 175–182.

[72] M. Soria-Sanchez, A. Maroto-Valiente, J. Alvarez-Rodriguez, V. Muñoz-Andrés, I. Rodríguez-Ramos, A. Guerrero-Ruíz, Carbon nanostrutured materials as direct catalysts for phenol oxidation in aqueous phase, Appl. Catal. B Environ. 104 (2011) 101–109.
[73] S.X. Yang, X. Li, W.P. Zhu, J.B. Wang, C. Descorme, Catalytic activity, stability and structure of multi-walled carbon nanotubes in the wet air oxidation of phenol, Carbon 46 (2008) 445–452.
[74] J. Garcia, H.T. Gomes, P. Serp, P. Kalck, J.L. Figueiredo, J.L. Faria, Carbon nanotube supported ruthenium catalysts for the treatment of high strength wastewater with aniline using wet air oxidation, Carbon 44 (2006) 2384–2391.
[75] J. Garcia, H.T. Gomes, P. Serp, P. Kalck, J.L. Figueiredo, J.L. Faria, Platinum catalysts supported on MWNT for catalytic wet air oxidation of nitrogen containing compounds, Catal. Today 102 (2005) 101–109.
[76] F. Stuber, J. Font, A. Fortuny, C. Bengoa, A. Eftaxias, A. Fabregat, Carbon materials and catalytic wet air oxidation of organic pollutants in wastewater, Top. Catal. 33 (2005) 3–50.
[77] S. Esplugas, J. Jiménez, S. Contreras, E. Pascual, M. Rodríguez, Comparison of different advanced oxidation processes for phenol degradation, Water Res. 36 (2002) 1034–1042.
[78] F. Duarte, F.J. Maldonado-Hódar, A.F. Pérez-Cadenas, L.M. Madeira, Fenton-like degradation of azo-dye orange II catalyzed by transition metals on carbon aerogels, Appl. Catal. B Environ. 85 (2009) 139–147.
[79] M.A. Fontecha-Cámara, M.A. Álvarez-Merino, F. Carrasco-Marín, M.V. López-Ramón, C. Moreno-Castilla, Heterogeneous and homogeneous Fenton processes using activated carbon for the removal of the herbicide amitrole from water, Appl. Catal. B Environ. 101 (2011) 425–430.
[80] M. Sánchez-Polo, J. Rivera-Utrilla, Ozonation of 1,3,6-naphthalenetrisulfonic acid in presence of heavy metals, J. Chem. Technol. Biotechnol. 79 (2004) 902–909.
[81] Yurii I. Matatov-Meytal, S. Moshe, Catalytic abatement of water pollutants, Ind. Eng. Chem. Res. 37 (1998) 309–326.
[82] A. Guerrero-Ruiz, P. Badenes, I. Rodríguez-Ramos, Study of some factors affecting the Ru and Pt dispersions over high surface area graphite-supported catalysts, Appl. Catal. A Gen. 173 (1998) 179–187.
[83] E. Brillas, I. Sires, M.A. Oturan, Electro-Fenton process and related electrochemical technologies based on Fenton's reaction chemistry, Chem. Rev. 109 (2009) 6570–6631.
[84] E. Brillas, B. Boye, I. Sirés, J.A. Garrido, R.M. Rodríguez, C. Arias, P.L. Cabot, C. Comninellis, Electrochemical destruction of chlorophenoxy herbicides by anodic oxidation and electro-Fenton using a boron-doped diamond electrode, Electrochim. Acta 49 (2004) 4487–4496.
[85] E. Brillas, M.A. Baños, J.A. Garrido, Mineralization of herbicide 3,6-dichloro-2-methoxybenzoic acid in aqueous medium by anodic oxidation, electro-Fenton and photoelectro-Fenton, Electrochim. Acta 48 (2003) 1697–1705.
[86] C. Flox, S. Ammar, C. Arias, E. Brillas, A.V. Vargas-Zavala, R. Abdelhedi, Electro-Fenton and photoelectro-Fenton degradation of indigo carmine in acidic aqueous medium, Appl. Catal. B Environ. 67 (2006) 93–104.
[87] G. Yuan, M.A. Keane, Liquid phase catalytic hydrodechlorination of chlorophenols at 273 K, Catal. Commun. 4 (2003) 195–201.
[88] S. Kovenklioglu, Z. Cao, D. Shah, R.J. Farrauto, E.N. Balko, Hydrodechlorination of toxic organics in wastewater, AIChE. J. 38 (1992) 1003–1012.

[89] T. Cordero, J.M. Chovelon, C. Duchamp, C. Ferronato, J. Matos, Surface nano-aggregation and photocatalytic activity of TiO_2 on H-type activated carbons, Appl. Catal. B 73 (2007) 227–235.

[90] J. Matos, J. Laine, J.M. Herrmann, D. Uzcategui, J.L. Brito, Influence of activated carbon upon titania on aqueous photocatalytic consecutive runs of phenol photodegradation, Appl. Catal. B Environ. 70 (2007) 461–469.

[91] J. Matos, J. Laine, J.M. Herrmann, Effect of the type of activated carbons on the photocatalytic degradation of aqueous organic pollutants by UV-irradiated titania, J. Catal. 200 (2001) 10–20.

[92] J.M. Herrmann, J. Matos, J. Disdier, C. Guillard, J. Laine, S. Malato, J. Blanco, Solar photocatalytic degradation of 4-chlorophenol using the synergistic effect between titania and activated carbon in aqueous suspension, Catal. Today 54 (1999) 255–265.

[93] J. Matos, J. Laine, J.M. Herrmann, Association of AC of different origins with titania in the photocatalytic purification of water, Carbon 37 (1999) 1870–1877.

[94] T. Cordero, C. Duchamp, J.M. Chovelon, C.J. Ferronato, Matos, Influence of L-type activated carbons on photocatalytic activity of TiO_2 in 4-chlorophenol photodegradation, J. Photochem. Photobiol. A 191 (2007) 122–131.

[95] G. Colón, M.C. Hidalgo, M. Macías, J.A. Navío, J.M. Doña, Influence of residual carbon on the photocatalytic activity of TiO_2/C samples for phenol oxidation, Appl. Catal. B Environ. 43 (2003) 163–173.

[96] W. Wang, Gomes C. Silva, J.L. Faria, Photocatalytic degradation of chromotrope 2R using nanocrystalline TiO_2/activated-carbon composite catalysts, Appl. Catal. B Environ. 70 (2007) 470–478.

[97] J.L. Faria, W. Wang, Carbon materials in photocatalysis, in: P. Serp, J.L. Figueiredo (Eds.), Carbon Materials for Catalysis, vol. 13, J. Wiley & Sons, Hoboken (NJ), 2009, pp. 481–499.

[98] J. Matos, A. García, L. Zhao, M.M. Titirici, Solvothermal carbon-doped TiO_2 photocatalyst for the enhanced methylene blue degradation under visible light, Appl. Catal. A Gen. 390 (2010) 175–182.

[99] W. Wang, P. Serp, P. Kalck, J.L. Faria, Photocatalytic degradation of phenol on MWNT and titania composite catalysts prepared by a modified sol–gel method, Appl. Catal. B 56 (2005) 305–312.

[100] W. Wang, P. Serp, P. Kalck, J.L. Faria, Visible light photodegradation of phenol on MWNT–TiO_2 composite catalysts prepared by a modified sol–gel method, J. Mol. Catal. 235 (2005) 194–199.

[101] Z. Kulesius, E. Valatka, Synthesis, characterization and photocatalytic activity of some TiO_2/carbon composites, Polish. J. Chem. 80 (2006) 335–344.

[102] Y. Yu, J.C. Yu, C.Y. Chan, Y.K. Che, J.C. Zhao, L. Ding, W.K. Ge, P.K. Wong, Enhancement of adsorption and photocatalytic activity of TiO_2 by using carbon nanotubes for the treatment of azo dye, Appl. Catal. B 61 (2005) 1–11.

[103] H. Zhang, X. Lv, Y. Li, Y. Wang, J. Li, P25-graphene composite as a high performance photocatalyst, ACS Nano. 4 (2009) 380–386.

[104] Q.Wang, J. Li, C. Chen, X. Ren, J. Hu, X. Wang, Removal of cobalt from aqueous solution by magnetic multiwalled carbon nanotube/iron oxide composites. Chem. Eng. J., 174 (2011) 126–133. doi:10.1016/j.cej.2011.08.059.

[105] C. Ye, Q.M. Gong, F.P. Lu, J. Liang, Adsorption of uraemic toxins on carbon nanotubes, Sep. Purif. Technol. 58 (2007) 2–6.

Chapter 17

Photochemical Behavior of Carbon Adsorbents

Conchi O. Ania, Leticia F. Velasco and Teresa Valdés-Solís
Instituto Nacional del Carbón, INCAR-CSIC, Oviedo, Spain

Chapter Outline

17.1. Introduction 521
17.2. Fundamentals of Semiconductor Photocatalysis 523
17.3. Role of Carbon on Photocatalysis 525
17.3.1. Carbon/TiO_2 Photocatalysts 526
17.3.2. Carbon as Photocatalyst 532
17.3.3. Carbon/ZnO Photocatalysts 535
17.4. Conclusions 539
References 539

17.1. INTRODUCTION

According to IUPAC, photochemical reactions are those caused by absorption of UV, visible, or infrared radiation [1]. Most of them proceed through the formation of thermodynamically instable compounds (e.g., radicals) that overcome large activation barriers during short periods, thus allowing reactivity otherwise inaccessible by conventional methods. Many molecules and atoms are capable of absorbing light and as a result, undergo chemical changes. A light-initiated reaction takes place by the conversion of the light energy into chemical energy, generating excited state species usually more reactive than the ground sate. When a catalyst is involved in the photochemical reaction, the process is known as photocatalysis. Many photochemical reactions also take place in nature, such as photosynthesis (conversion of CO_2 into polysaccharides and O_2 in plants), human skin production of vitamin D, ozone formation and dissociation in the atmosphere, or photochemical smog (air pollution due to the reaction of nitrogen oxides and some volatile organic compounds with sunlight).

Novel Carbon Adsorbents. DOI: 10.1016/B978-0-08-097744-7.00017-X
Copyright © 2012 Elsevier Ltd. All rights reserved.

The use of light energy (and particularly trapping of solar light) in non-biological photochemical processes has been long researched by scientists for a number of years, and many ground-state reactions, such as photo-cycloadditions, photodecarbonylations, photo-enolizations, photo-isomerizations, photo-oxidations, and photosubstitutions, have been studied. In environmental chemistry, photochemical reactions are useful to the trace chemistry of atmospheric pollution, the degradation of different pollutants (since the excitation of electronic molecular states at energies corresponding to the ultraviolet may induce chemical bond breaking), and in the field of solar energy storage and conversion [2–6]. More recently, emerging applications include water splitting for hydrogen generation, carbon dioxide remediation, self-cleaning surfaces, light harvesting for solar energy production, and air purification and wastewater remediation among the most representative [3–6].

Photocatalytic reactions are based on the use of semiconductors as catalysts; among them, titanium oxide has been extensively studied in different applications after the first demonstration of the photo-electrochemical water splitting on a TiO_2 electrode [7]. Compared to other semiconductors, TiO_2 has quite a few advantages including biological and chemical inertness, non-toxicity, low cost and availability, large stability against photo- and chemical corrosion, and high efficiency under UV light [5,7–9]. However, major drawbacks of titanium oxide are related to its low efficiency under visible light, the high recombination rate of photogenerated electron−hole pairs, and technological limitations associated to the separation (filtration), recovery, and reutilization of the fine titania powders – an important challenge that prevents the large-scale implementation of photocatalytic processes [4–6]. To overcome these cornerstones different approaches are being considered, including doping with transition metal and nonmetal ions, surface sensitization of dyes, design of advanced hybrid nanostructured photocatalysts, and synthesis of porous semiconductors or immobilization on porous supports [10–16].

Despite carbons being strong light absorbing materials, extensive work has been carried out on the improvement of the catalytic activity of photo-active materials upon immobilization on different carbon supports [14,15]. Indeed, carbon–titania composites have shown quite high efficiencies for the photodegradation of a variety of pollutants [14–28]. Moreover, although it is hard to understand how light penetrates the carbon particles to reach the immobilized photocatalyst, recent studies have demonstrated that certain carbon materials show an anomalous photochemical response when exposed to UV light, resulting in an improved photo-oxidation of phenol in aqueous solution [26–29] compared to the photolytic breakdown (i.e., non-catalyzed reaction).

Summarizing, the purpose of this chapter is to provide a global view of the role of carbon in photocatalysis, covering the use as both photocatalyst itself and support in semiconductor/carbon composites (mainly based on TiO_2 and ZnO). An overview of the results available in the literature concerning the photochemical response of carbon-containing catalysts and their efficiency for the degradation of different pollutants will be presented and discussed. Particular emphasis will also be drawn on the recently reported photochemical behavior of carbons under UV irradiation in the absence of semiconductor additives.

17.2. FUNDAMENTALS OF SEMICONDUCTOR PHOTOCATALYSIS

After the pioneering studies in 1977 reporting the performance of titania powders to decompose cyanides in solutions [30,31], the interest of semiconductor photocatalysis for environmental applications has largely increased and nowadays it has become one of the most popular and promising advanced oxidation processes (AOPs) for the degradation of recalcitrant pollutants in air, water, and wastewater [4,5,15,32].

The success of heterogeneous photocatalysis as an efficient environmental remediation tool is mainly related to the choice of titanium oxide as the photoactive material, although other transition metal oxides and sulfides such as ZnO, CdS, ZnS, and WO_3 have been also investigated as semiconductors. The mechanism of light absorption in a wide bandgap n-type semiconductor – such as titanium oxide – is based on direct interband electron transitions. A series of chain oxidative-reduction reactions have been postulated and are summarized in Fig. 17.1.

When titania particles are irradiated with photons with energy equal to or higher than its bandgap (i.e., 3.2 eV in the case of anatase powders), a charge separation takes place: electrons are excited from the valence band (VB) into the unoccupied conduction band (CB), whereas positive holes are created in the valence band. The photogenerated charge carriers (electron/hole pairs, e^-/h^+) can recombine radiatively (photoluminescence) or non-radiatively

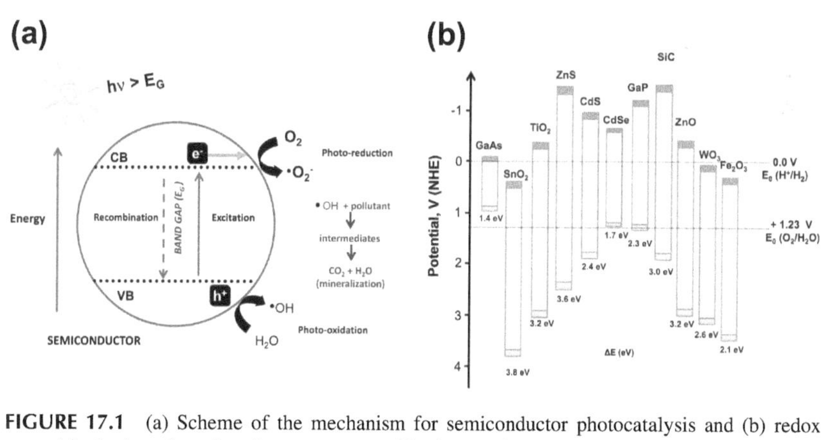

FIGURE 17.1 (a) Scheme of the mechanism for semiconductor photocatalysis and (b) redox potential of selected semiconductors vs normal hydrogen electrode (NHE).

(dissipating the energy as heat), or migrate to the surface reaction sites and get trapped (scavenging) to react with electron donors or acceptors adsorbed onto the photocatalyst surface. The competition between these processes determines the overall efficiency of the photocatalytic process, since it has been demonstrated that the photo-redox chemistry occurring at the particle surface emanates from trapped electrons/holes rather than from free valences [33]. Spectroscopic studies have also demonstrated that the electron–hole trapping or recombination rates are extremely fast (of the order of 10^{-6}–10^{-15} s) [34] and therefore can largely affect the photocatalytic efficiency of the catalyst.

Subsequent chemical events at the gas or liquid–solid interface between the e^-/h^+ pairs and the species in solution also determine the efficiency of the photocatalytic reaction. The redox potential of the photogenerated holes is +2.53 V versus standard hydrogen electrode (SHE) [7,35], positive enough to directly oxidize the pollutant and to react with water molecules or hydroxyl groups to produce hydroxyl radicals (Eo = +2.81 V vs SHE), which are also powerful oxidizing agents (Fig. 17.1). These reactive species may undergo further decomposition reactions, increasing the photodegradation yield, and leading to complete mineralization eventually (see Eqn (17.1)):

$$\text{Pollutant} + h\nu\ (TiO_2) \rightarrow \text{intermediates} \rightarrow CO_2 + H_2O \qquad (17.1)$$

Similarly, the redox potential of the conduction band electrons is −0.52 V vs SHE, negative enough to form superoxide O_2^- (Eo = +0.89 V vs SHE) or hydrogen peroxide H_2O_2 (Eo = + 1.78V vs SHE) when oxygen is present in the reaction medium as an electron acceptor. All these species are also powerful oxidizing agents that might enhance the yield of the photocatalytic reaction. Thus, photo-oxidation of a compound can occur by either direct electron transfer to an active hole or indirect reaction with the intermediate oxidizing species also formed.

Another aspect to be considered in heterogeneous photocatalysis is the degradation rate. In terms of reaction rate, several steps can be distinguished in the photocatalytic process [36], accounting for i) the mass transfer of the pollutant from the bulk solution to the catalyst surface, ii) followed by adsorption on the catalyst, iii) photocatalytic reaction, iv) desorption of the degradation intermediate, and v) mass transfer of the intermediate from the catalyst surface to the solution. Based on general knowledge on reaction kinetics, the overall rate of a given reaction is controlled by the slowest step. The photocatalytic reaction itself (i.e., photon absorption by the photocatalysts) is typically very fast; in addition, in a stirred reaction vessel the mass transfer (external diffusion) is also very fast, the internal diffusion and the adsorption/desorption of the compounds in solution being the most common rate-limiting steps. Thus, surface contact of the pollutant with the active sites of the photocatalyst is a critical parameter defining the overall rate in heterogeneous photocatalytic reactions.

Summarizing, heterogeneous photocatalytic reactions depend on several factors governing the activity of the photocatalyst and the rate of the degradation process. As a general rule, the activity of a photocatalyst mainly depends on the bandgap of the material, the extent of surface recombination of the photogenerated e^-/h^+ pairs, and the nature of the chemical surface reactions occurring. In this regard, several approaches are currently being investigated to enhance the activity of photocatalysts; these include mainly photocatalyst bandgap tuning via chemical modification, doping of the TiO_2 lattice for reducing surface recombination, applying nanosized catalysts for an efficient charge separation, increasing the crystallinity, and so forth [37–39]. On the other hand, the degradation rate is connected to the surface contact between the active sites in the photocatalyst and the compounds in solution, for which most research efforts on this axis focus on incorporating porous catalysts, either via synthesis of porous semiconductors or immobilization on porous supports [5,14,15].

Nevertheless, some other operating conditions, such as catalysts surface area and composition, solution pH, external diffusion control, catalyst concentration, and dispersion of titania particles, must also be carefully considered as they can control the success or failure of the photocatalytic process.

17.3. ROLE OF CARBON ON PHOTOCATALYSIS

Triggered by the rising interest in heterogeneous photocatalysis for environmental remediation, numerous efforts have been made in the last decades to improve the photocatalytic activity of semiconductors. Aside from tuning their nanostructure and chemical composition, novel hybrid materials prepared by immobilization of the photoactive semiconductor on appropriate substrates have been explored due to the significantly improved performance observed on such composites [40–43]. Among them, the potential role of carbon adsorbents as supports has been investigated in a number of photocatalytic degradation reactions both in liquid and gas phase [14,15,44, and references therein].

Recent advances in the design of carbon materials have contributed to an unforeseen control of their electronic, physicochemical, and structural properties (particle size and shapes, uniformity in porous void, and flexibility). The versatility of such highly featured carbons offers unexpected opportunities in different fields beyond traditional uses, including photocatalysis. Indeed, different types and forms of carbons ranging from conventional (i.e., activated carbon, carbon black, and graphites) to novel carbons (i.e., nanotubes, graphene, fullerenes, etc.) have been used in photocatalytic applications, mainly as additives and supports in carbon/semiconductor composite photocatalysts [14,15,45,46].

The enhancement on the photocatalytic response of carbon/TiO_2 composites has been attributed to either single or collective factors traditionally associated to visible light absorption, textural features (i.e. surface area), and strong interfacial electronic effects [14,15,44,47]. Most recently, the photochemical activity of certain carbon materials under UV has been demonstrated in the absence of conventional semiconductors [26–28]. An overview of the photochemical response of carbon-containing photocatalysts and their efficiency for the degradation of different pollutants will be presented in the next sections of this chapter. Special attention will be paid to photocatalytic degradation of C/TiO_2 and C/ZnO systems and those based on carbons themselves as photocatalysts.

17.3.1. Carbon/TiO_2 Photocatalysts

The majority of the studies in the literature deal with the use of TiO_2 as a photoactive material, mainly due to its low cost and high photo-activity and stability [5,9,47]. The photocatalytic performance of carbon-doped TiO_2 or hybrid C/TiO_2 composites has been reported over a variety of carbon sources, forms, and morphologies (powders, fibers, foams, and cloths), using different synthetic routes for both components of the hybrid catalyst material (varying the carbon and titania precursors) and for the catalyst preparation (mechanical mixture, coating by liquid impregnation, hydrothermal process, chemical vapor deposition, and so forth) [14,15,18]. Moreover, photocatalytic techniques based on carbon-containing catalysts have been employed for water splitting into H_2 and O_2, and the gas-phase and liquid-phase degradation of a vast number of pollutants; phenol and related aromatic compounds, dyes, pesticides, and herbicides are among most representative. Owing to space limitations, only the basis of the role of carbon on the photocatalytic performance of C/TiO_2 composites is presented in this section. More detailed information on the experimental procedures for the preparation of the carbon-containing photocatalysts and their photocatalytic performance can be found in the literature [14,15,18, and references therein presented].

The enhanced photo-activity of carbon/titania composites has been discussed by different mechanisms, depending on the nature of the carbon matrix itself. It seems generally accepted that the role of carbon on the overall photocatalytic enhancement differs greatly for porous carbons (including activated carbons, carbon fibers and nanofibers, carbon foams, and so forth) compared to other forms of nanostructured carbons, such as carbon nanotubes (CNT), fullerenes, and graphene. All these aspects are addressed and discussed in the following sections.

17.3.1.1. Carbon-Doped TiO_2

TiO_2 doping is nowadays considered a promising route to extend the optical absorption of this material toward the visible spectral region, allowing the use of sunlight in photochemistry and photocatalysis. Indeed, many attempts to improve the photo-activity of titanium oxide have focused on using metallic (Fe, Cr, Ag, and Ce) and nonmetallic (N, C, S, B, and F) doping [48–60]. Comparatively, non-metallic TiO_2 doping is considered to be more effective than transition metal doping due to their high photostability and lack of photocorrosion of the resulting catalyst.

Nitrogen and most recently carbon have been widely studied and considered to be more effective than other non-metal heteroatoms for the degradation of organic pollutants [54–59]. The effect of carbon doping seems to be more beneficial, compared to nitrogen doping, although the synthesis and structural and electronic control of carbon-doped titania are usually more complex.

The origin of the improved photocatalytic performance of carbon-doped titanium oxide is still uncertain, although most authors agree to explain it in terms of the shift in the absorption properties of carbon-doped TiO_2 toward the visible region and the modifications in the electronic band structure of titanium oxide associated to bandgap narrowing and localized midgap levels [55–58]. However, the exact role of the carbon atoms on the bandgap narrowing remains rather unclear and controversial [53,54]. For instance, a decreased direct oxidation ability (involving e^-/h^+ pairs) for carbon-doped TiO_2 has been reported, along with an improvement in the indirect oxidation properties via reactions with surface intermediates (due to reduced recombination fate of the holes trapped at the carbon doping sites) [59].

17.3.1.2. Porous Carbon/TiO_2 Composites

Besides semiconductor doping, the incorporation of porous supports in hybrid carbon/TiO_2 composites is another promising alternative to prepare more efficient photocatalysts, overcoming the operational drawbacks associated to the use of bare (naked) nanosized titania powders that hinder the application in continuous flow systems due to limited recovery and reuse of the catalyst [13,60,61]. In this sense, the deposition of titanium oxide on a granular carbon support has been reported to improve the practical efficiency of the composite – defined in terms of the decantability of the catalyst and its recovery from the reaction media – through a faster sedimentation and therefore a lower turbidity of the final solution [22,62].

Hence, the immobilization of titanium oxide on porous supports, such as alumina, zeolite, silica, or glass beads, has attracted much attention. In particular, porous carbon materials (mainly activated carbons, fibers and nanofibers, and foams) have been widely investigated due to their large specific surface areas, high pore volume, and well-developed micro-mesoporosity [14–29,44,47,62]. Most of the studies report an enhanced photoconversion of pollutants when a porous carbon is incorporated to the catalyst, which has been explained in terms of a synergistic effect in the composite. The idea, firstly introduced by Matos et al. [21], relies on the combination of the confinement of the target species in the porosity of the catalyst via adsorption, and the photo-activity of the immobilized semiconductor. This has been well documented in the literature [14,15,22–29,44,62–69].

The increased photoconversion yield on porous carbon/titania composites has been ascribed to the enhanced mass transfer of the adsorbed pollutant from the bulk solution to the photoactive TiO_2 particles through the interface between the two catalyst components and to the interactions between the substrate and the immobilized catalyst [62,72]. Preferential adsorption and surface concentration of the pollutant onto the carbon porosity are followed by a spontaneous transfer from the support to the titania surface, where it is more rapidly decomposed due to the large concentration gradient between the two solid phases. This is important because the photogenerated oxidizing species (•OH) do not migrate far from the active centers on TiO_2, with catalytic degradation taking place essentially at the catalyst/water interface [72]. Furthermore, simultaneously to the adsorption/desorption of the studied pollutant, the degradation intermediates formed during the photochemical reaction are also expected to be adsorbed/desorbed from the porous matrix of the catalysts.

A synergistic factor has been defined as the ratio between the [66] apparent rate constants of the photoconversion reactions using supported and bare titania as catalysts. Different synergy factors have been reported for porous carbon/titania composites, which suggest that the carbon material itself plays a significant role in the photo-activity of titania powders. However, it should be pointed out that the conversion observed for a given pollutant does not only depend on the catalyst and the pollutant as there are many other operating parameters that control the reaction rate and yield (photoreactor geometry, illumination source, mass-transfer limitations, diffusion or convection control, chemical structure of the pollutant, temperature, etc.). For this reason, comparison of the synergy factors obtained from different studies is not straightforward, although it seems reliable to compare a set of experiments performed under similar conditions.

On the other hand, the photo-activity of carbon/titania composites strongly depends on many other parameters such as the nature of the carbon, the titania/carbon loading, the synthetic route chosen for infiltrating TiO_2 nanoparticles inside the porous carbon, or the porosity of the carbon adsorbent. For instance, as a general rule large surface areas and pore volumes are usually required in the carbon material for achieving a high dispersion of the semiconductor. However, the porosity of the carbon is generally largely underexploited [26,70], mostly due to the difficulty in obtaining an efficient infiltration of the small pores (i.e., narrow micropores) and the lack of precise control on the size of TiO_2 nanoparticles that potentially remain on the outer surface of the carbon [71,72]. Besides microporous carbons, mesoporous supports have been employed to immobilize TiO_2 nanoparticles [63,64], obtaining interesting materials when large molecules (i.e., dyes) are to be degraded.

Carbon-coating TiO_2 particles have also been extensively explored. Along with the increased adsorption rate of the pollutant in the porosity created by the carbon layer, increased photocatalytic activity is attributed to the effect of carbon on stabilizing the anatase structure (suppressing rutile transformation at high temperature) and inhibiting the sintering of titania (anatase) crystals [71]. More detailed information on the experimental procedures for the preparation of the carbon–titania composites and the effects of the synthesis conditions on their photocatalytic activity can be found in the literature [14,15,18,71,73].

Contrary to porosity, the effect of the carbon surface chemistry has been scarcely studied. Ocampo-Perez et al. [74] have investigated the photocatalytic activity of a series of carbon/titania composites, where the carbon phase was chemically modified by wet oxidation treatments. An improved photocatalytic yield of activated carbons/titania composites was observed after oxidation of the adsorbents. The authors speculated that this behavior is attributed to interactions between the surface functionalities of the carbon and the radical species generated from the TiO_2 particles. More specifically, the catalytic activity showed a dependence with the concentration of carboxylic acid groups, which would act as catalytic centers capturing the electrons from the photo-activation of TiO_2. This reaction would produce the reduction of carboxylic groups to ketones generating H_2O_2 which would be decomposed into HO• radicals and, subsequently, generated ketones would be transformed into an alcoholic superficial group generating additional HO• radicals. The authors support this theory based on the slight increase in the phenolic and lactone groups detected in the carbons after the photocatalytic experiments. This contrasts with recent investigations carried out in our group [27], where no modifications of the surface chemistry of acidic carbons were observed after UV irradiation. Thus, so far the role of the surface functional groups still remains unclear and needs to be further investigated.

Recent works on the photocatalytic degradation of several organic compounds have also demonstrated that the presence of a carbon matrix in the catalyst composition actually changes the catalytic behavior of TiO_2 beyond the above-mentioned synergistic effect on the degradation kinetics [22,26,27,62]. Compared to bare titania, different degradation intermediates have been detected using activated carbon/titania composites as photocatalysts, confirming the role of the carbon matrix on the photodegradation mechanism of the target pollutant [26,27,62,64,71]. FTIR studies have revealed some changes in the interactions occurring between the pollutant and the photocatalyst surface in the presence of the carbon layer, which would explain the modification of the photo-oxidation pathway observed for the organic pollutants [62]. Moreover, our recent studies suggest that the activated carbon support itself is capable of a significant level of self-photo-activity under UV irradiation [26–29]. This issue will be further discussed in section 3.2.

17.3.1.3. Composites Based on Other Carbon Nanostructures

For carbon/titania composites with none or small porosity, the above-mentioned synergistic effect associated to the increased degradation rate via adsorption does not count since the contribution of adsorption is negligible; that is the case for instance of carbon nanotubes, fullerenes, graphenes, and other low porous sophisticated carbon nanostructures. Two other mechanisms have been proposed to explain the photocatalytic performance obtained for such C/TiO_2 composites (Fig. 17.2), based on 1) the separation of the photogenerated carriers in the presence of the carbon network and 2) the role of carbon as a photosensitizer [8,75,76].

The concept of the separation of the photogenerated species is based on the model proposed by Woan et al. [75] for semiconductor–metal or semiconductor–semiconductor junctions, where electrons flow from the higher to lower Fermi level material (until Fermi levels are aligned), forming a space charge separation layer at the interface of both materials. The barrier formed (so-called Schottky barrier) serves as an efficient electron trap, enlarging the time for the electron–hole recombination and thus enhancing the photocatalytic efficiency of a given catalyst.

This idea has been extended to explain the photochemical behavior of certain carbon/TiO_2 composites, where the carbon material exhibits metallic conductivity (similar to the metal–semiconductor junctions). After the generation of the e^-/h^+ pair, the quantum efficiency of the photochemical reaction is controlled either by the competition between recombination and trapping of the

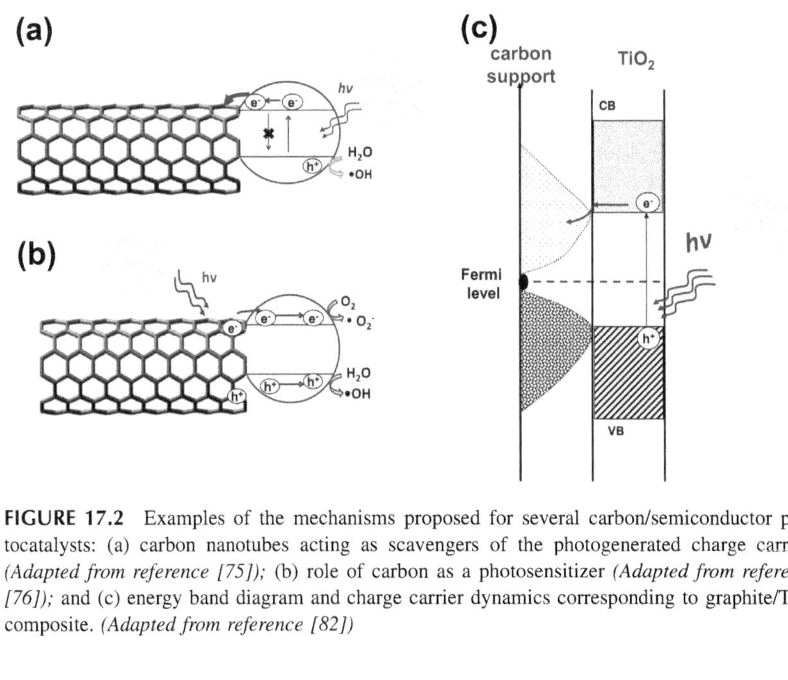

FIGURE 17.2 Examples of the mechanisms proposed for several carbon/semiconductor photocatalysts: (a) carbon nanotubes acting as scavengers of the photogenerated charge carriers *(Adapted from reference [75])*; (b) role of carbon as a photosensitizer *(Adapted from reference [76])*; and (c) energy band diagram and charge carrier dynamics corresponding to graphite/TiO_2 composite. *(Adapted from reference [82])*

charge carrier or by the competition between the trapped carrier recombination and interfacial charge transfer [75–80]. Due to their structural organization and electronic properties, the graphene sheets of carbon materials can be regarded as electron-storage reservoirs (i.e., one π-electron per aromatic ring). Thus, in the presence of carbon/titania composites, photogenerated electrons are transferred to the carbon structure where they are free to flow within the graphene layers of the carbon/Ti substrate to the external circuit. This freedom hinders the recombination of the e^-/h^+ pairs, leaving the holes also free to move to the outer surface of the catalyst and therefore ready to participate in oxidative reactions [75] or become trapped in other processes. A similar behavior based on the increase in the carriers income coupled to a drop in the carriers expenditure has been reported for semiconductor heterostructures that incorporate photonic effects [81].

Studies on the charge carrier dynamics on carbon/TiO_2 nanofilms have demonstrated that an optimal interfacial interaction between the carbon film and the semiconductor promotes an efficient charge carrier leakage from the TiO_2 nanoparticles to the carbon, reducing the radiative recombination. In contrast, the incorporation of an insulator film (ca. SiO_2) between the carbon and titania layers acts as a barrier, suppressing the charge transfer across the interface [82].

The second mechanism considers that the carbon material acts as a photosensitizer [76] by injecting electrons into the conduction band of the TiO_2, and allowing the formation of reactive species (i.e., superoxide and hydroxyl radicals). The increase in the formation of highly reactive radicals (hydroxyl superoxide or hydroperoxy radicals) in carbon/titania composites has been confirmed by electron paramagnetic resonance (EPR) [83,84]. At the same time, the positively charged carbon would remove an electron from the titania valence band leaving a hole, which is also capable of reacting with adsorbed water to form hydroxyl radicals or directly react with the pollutant. Albeit considered rather unlikely, some experimental evidence on direct oxidation via the valence-band hole has been reported in the literature for tightly bound electron-donor compounds [85].

The photoluminescence response of different catalysts has revealed that some CNTs act as dispersing agents for titania nanoparticles, favoring the nucleation of small-sized TiO_2 crystallites [86]. However, the low photocatalytic efficiencies reported for small crystallites indicate that the nanosizing effect on the titania particles is not the most important factor accounting for a high photodegradation performance [47,87–89]. More recently, carbon quantum dots immobilized over titanium oxide have also been reported to increase the photocatalytic activity of the resulting composite; again the charge separation and increased recombination lifetime of the photogenerated species [90] seem to be more critical for the photochemical response than the nanosize effect of the carbon quantum dots.

The absorption properties of carbon/TiO_2 composites show a gradual shift in the spectrum toward higher wavelengths compared to that of TiO_2, pointing out the role of the surface electric charge of the multiwall carbon nanotubes (MWCNT) in the photochemical response of the catalysts [88,89]. A correlation between the UV–Vis spectrum and the carbon content has been reported for MWCNT which contrasts with activated carbon/TiO_2 systems [14,15,22,62] where no correlation has been found.

Graphene-based composite materials have also been postulated as candidates to anchor semiconductor particles, with improved performance in optoelectronic and energy conversion devices [91–99]. As in the case of fullerenes and CNT, electrons stored in TiO_2 nanoparticles are readily scavenged by graphene nanostructures, as demonstrated by flash photolysis [93–98]. However, despite the unique structural and electronic properties of graphene itself, graphene/titania composites are considered in essence the same as other carbon/titania materials [97].

The synthetic route used for the composite preparation also controls its photocatalytic activity by defining the recombination lifetime and interfacial interactions between the two phases (i.e., due to crystal defects and surface morphologies) [88,89]. An elegant way to prepare carbon/titania catalysts has been reported using hydrothermal carbonization methods [100]; core–shell carbon/titania composites with controlled morphology and topology of the active phase can be prepared, showing enhanced photocatalytic activity for degradation of pollutants [101,102]. However, the electronic structure (i.e., electron density) and structural defects of the carbon material seem to be much more important than its chemical composition or any chemical bond between carbon and TiO_2 [75,103].

17.3.2. Carbon as Photocatalyst

Heretofore, knowledge regarding carbon/semiconductor photocatalysis is based mainly on the role of carbons as non-photo-active additives and supports [21,99]. The enhancement in the photocatalytic performance of carbon/TiO_2 composites has been attributed to either single or collective factors associated to visible light absorption, textural features of the carbon support, and strong interfacial electronic effects [14,15,47] (see earlier sections).

Recently, the self-photochemical activity of activated carbons under UV irradiation has been demonstrated in the absence of semiconductor additives [26–29]. An activated carbon showed an improved photo-oxidation of phenol in aqueous solution compared to both the photolytic reaction and the photodegradation using bare or carbon-immobilized titania. The results pointed out that beyond the synergistic effect of carbon/TiO_2 composites, the activated carbon alone was capable of a significant level of self-photo-activity under UV irradiation [26–29]. Similar to carbon/TiO_2, when the activated carbon alone was used as photocatalyst the photodegradation pathway of phenol was modified, through the preferential photo-oxidation of phenol toward the catechol route [26,71]. For carbon/TiO_2 composites, this observation had been linked to weak interactions occurring between the carbon material and the semiconductor particles [22,26,62]. The fact that the same is observed when carbon alone is used as a photocatalyst confirmed the role of carbon in the photo-oxidation mechanism of the pollutant.

To discriminate whether this behavior was a singular characteristic of that particular activated carbon or rather an intrinsic property of carbon materials, an in-depth investigation into the photochemical behavior of several porous activated carbons under UV irradiation was undertaken [27,29,104,105]. A selection of various carbonaceous adsorbents obtained from different precursors and showing varied chemical and textural features was considered. The yield of the photocatalytic degradation of phenol was evaluated from a twofold viewpoint; due to their textural features, the contribution of the amount removed via adsorption (dark conditions) was also taken into account. Under dark conditions, the rate of phenol adsorption followed the expected trend considering the physicochemical characteristics of the investigated carbons, with fast adsorption rates and a sharp decrease in the amount adsorbed in carbons of acidic nature [27,104,105]. Upon UV irradiation, the overall rate of phenol disappearance increased remarkably for all the carbons. This trend was particularly more evident at the initial stage of the reaction and also particularly more evident for the activated carbon of an acidic nature (where phenol removal via adsorption is hindered).

The different photochemical behavior (rate and intermediates speciation) obtained for the studied activated carbons under dark and UV irradiation cannot be explained if the carbons were considered nonphotoactive and photolytic reaction of the pollutant is the only process occurring upon UV irradiation. Moreover, the differences observed in the photocatalytic performance of carbons with similar porous features demonstrated that this behavior was not attributed to the confinement of the pollutant in the porosity of the activated carbons (i.e., concentration effect on photolysis). In fact, the relative abundance of the degradation intermediates detected in solution depended on the characteristics of the activated carbons. However, this singular photocatalytic behavior does not apply for all types of carbon adsorbents, and although the effect has not yet been fully understood, several hypotheses and scenarios have been postulated based on the carbons' physicochemical and structural features [27,28].

Elucidating the mechanism of the photo-activity of porous carbons is a complex task, since various processes (photo-oxidation and adsorption) coexist simultaneously. To investigate the catalytic activity inside carbon materials disregarding the effects of adsorption kinetics and solution photolysis, irradiation was performed on carbon samples preloaded with the target pollutant [27,28]. The extent of photodegradation (if any) inside the carbons was evaluated by extracting the irradiated carbons in an adequate solvent. For certain activated carbons, varied amounts of the pollutant and its photo-oxidation intermediates were detected in the extracts. Since no leaching out of any compound was detected upon irradiation, the results demonstrated the photodegradation of the pollutant adsorbed inside the porosity of the carbon, when exposed to UV irradiation.

Additionally, photodegradation yields for certain activated carbons were larger than or similar to those of the photolytic reaction. This is most remarkable since the incident photo flux arriving at the molecules adsorbed inside the carbon is expected to be small compared to the photolytic reaction, due to the strong absorption of the carbon matrix and light scattering by the catalyst particles suspended in solution.

In an attempt to quantify the photochemical response of the activated carbons for the photodegradation of a given pollutant, and to compare it with the photochemical quantum yield of the photolytic reaction (Φ [1]), an estimation of an apparent or pseudo quantum yield (Φ_{ps}) was proposed, considering that all photons are absorbed by the solution [27,28,104]. Due to the difficulties in evaluating the fraction of light absorbed by the carbon, the estimation of the actual Φ for carbons is not accurate; the approximation of Φ_{ps} – although cannot stricto sensu be considered a quantum yield – would account for the minimum limit of the actual quantum yield, allowing the comparison of the different carbons. The retrieved Φ_{ps} values for different carbon materials were higher than or similar to that of photolysis (Fig. 17.3). The results do not demonstrate whether the UV light penetrates inside the carbon porosity or carbon–photons interactions occur at the external carbon surface and are subsequently propagated

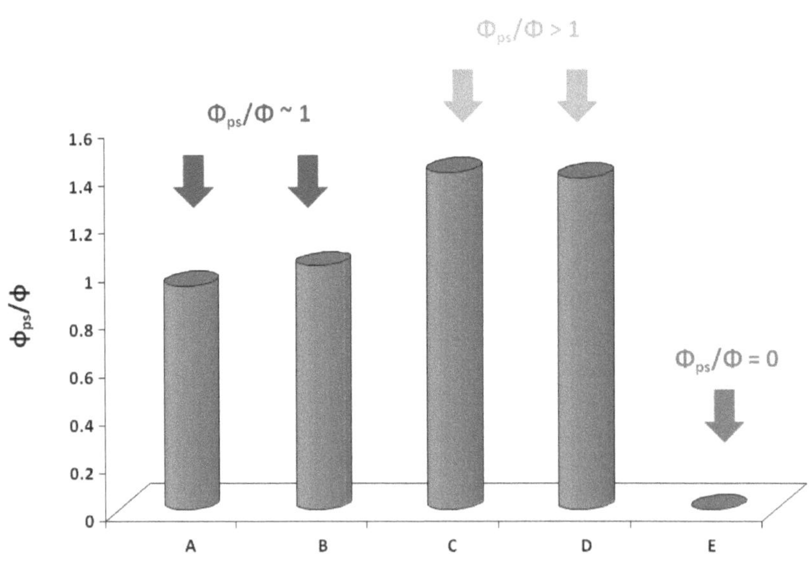

FIGURE 17.3 Correlation between phenol photochemical quantum yields evaluated for the preloaded activated carbons (Φ_{ps}) and the photolytic reaction (Φ).

through the graphene sheets, but the self-activity of activated carbons and its application to the degradation of pollutants was confirmed.

This effect has not yet been fully understood, and further research is currently ongoing to comprehend the nature of the photochemical behavior of carbon adsorbents under UV light, analyzing different aspects of carbon materials [27,28]. Based on the mechanisms already proposed for carbon nanotubes and other carbon nanostructures, several hypotheses and most likely mechanisms have been postulated to explain this behavior [27,28]. When the carbon is illuminated under UV light, a fraction of the incident photon flux is absorbed by the carbon matrix while another fraction generates charge carriers, which are expected to diffuse randomly through the graphene sheets. The role of the graphitic sheets of carbons is important for the migration of the photogenerated electrons, thus minimizing recombination and favoring the electron transfer to the adsorbed species (i.e., pollutant and water).

As for the carbon surface chemistry, the presence of oxygen-containing functionalities on the edges of the graphene sheets does not seem to have a negative effect on the photochemical behavior of the carbon [27–29,104,105]. Surface functionalities could either act as charge injectors upon UV excitation or stabilize the photogenerated charge carriers, as already reported for other carbon materials [106]. These photogenerated carriers seem to have enough redox potential to generate more reactive species (•OH and O_2^{2-}) and/or directly oxidize phenol. The formation of hydroxyl radicals (•OH) can be expected as water is simultaneously co-entrapped in the porosity of the carbon during the adsorption [107], whereas the occurrence of superoxide anions (O_2^-) is also suspected as the photo-oxidation yield is reduced under oxygen-free conditions [29]. In this regard, preliminary data on EPR of activated carbons have confirmed their ability to generate a similar or even higher quantity of hydroxyl radicals than commercial P25 [105], when exposed to UV irradiation in an aqueous system. Additionally, although the contribution of the mineral matter of the carbons to catalytic reactions cannot be disregarded [108], studies on de-ashed adsorbents have demonstrated that photo-oxidation of phenol still occurs for demineralized carbons [29,104].

The self-photo-activity of MWCNT has also been suggested in the visible range [14,109] and has been attributed to the presence of structural defects and vacancies, accompanied by local lattice reordering and inter-tube reorientation during the synthesis of the nanotubes. A general interest in the modification of the electronic properties of carbon nanostructures and further research on the photochemical response of activated carbons may contribute to progress regarding self-photocatalytic properties of carbon materials.

17.3.3. Carbon/ZnO Photocatalysts

Although TiO_2 was the first effective photocatalyst to be applied, many other semiconductors have been also been investigated in environmental applications. Among them, zinc dioxide appears to be a suitable alternative since its photodegradation mechanism is similar to that of TiO_2, and it has been reported to be more active than titanium oxide in some processes, such as bleaching wastewater, phenol treatment, and the degradation of some dyes [110–113]. Furthermore, the reported optimal pH value for ZnO process is close to the neutral one, whereas the optimum pH for TiO_2 mostly lies in the acidic region [112].

Zinc oxide is an n-type semiconductor with a wide bandgap of 3.17 eV [112]. As in the case of titanium oxide, the potential of water and oxygen exists within the bandgap of ZnO; thus, the photogenerated holes in the valence band can oxidize water to produce hydroxyl radicals and the photogenerated electrons in the conduction band can reduce oxygen to form superoxide anions (Fig. 17.1). All these species assist the oxidation of the pollutant, resulting in an increase in the photocatalytic activity of ZnO.

As already pointed out for titanium oxide, at the level of the material's surface an important limitation of photocatalysts from a practical point of view is the decrease in the efficiency due to recombination reactions of the photogenerated e^-/h^+ pairs. Based on the photochemical response observed in carbon/titania composites, immobilization of ZnO on different supports has also emerged as an interesting approach to improve the photocatalytic yield of ZnO systems. Scarcely studied when compared to titania/carbon systems, several carbon materials (activated carbons, carbon nanotubes, fullerenes, etc.) have been investigated and the photocatalytic efficiency of the resulting carbon/ZnO composites evaluated for environmental photocatalytic applications.

Activated carbon is by far the most widely used carbon material for the immobilization of ZnO [110–118], due to the versatility of morphologies and structures. Several synthetic routes have been described in the literature for the preparation of carbon/ZnO composites, obtaining materials with very different structures and composition, whose efficiency has been investigated for a variety of pollutants (both from the liquid and gas phase). Some contradictory results can be found in the literature concerning the role of carbon in these carbon/ZnO composites, reporting either that the photocatalytic performance of the composite is not related to the carbon matrix (which would act as an inert support) [110], or an enhanced stability of the catalyst due to the presence of a carbon phase [115]. In some other cases, a synergistic effect similar to that described for carbon/titania materials has been reported when ZnO is immobilized on (porous) activated carbons. The shift of the absorption properties of carbon/ZnO toward the visible range of the spectrum, thus improving the quantum efficiency under solar irradiation, has also been reported [115–117]. Furthermore, the role of carbon has also been associated to the extended lifetime of the catalysts and their reutilization over several cycles [114].

Novel strategies to enhance the photocatalytic performance of ZnO photocatalysts have been proposed based on coupling ZnO with other inorganic semiconductors (i.e., ZnS), to facilitate the separation of photogenerated hole–electron pairs and reduce recombination. The immobilization of these binary photoactive systems on carbon materials shows high photodegradation efficiencies of dyes from the aqueous phase [117]. For instance, carbon/ZnO and carbon/ZnO–ZnS show active under visible irradiation, the latter being considerably more efficient (about 5.3 times) than the former. This has been attributed to the formation of ZnO–ZnS heterojunctions, leading to a higher hole–electron pair generation and photostability (suppressing photocorrosion).

A different approach to Zn–C materials is the use of carbon as an additive, either coating the ZnO surface or in close contact with it; catalysts with complex morphologies (nanoflowers and nanorods) have been synthesized using various carbon precursors [119–123], and used for the degradation of various pollutants (dyes and formaldehyde) under UV and visible light irradiation. The photocatalytic activity of these materials increased after carbon doping, outperforming nitrogen-doped titania systems under visible light irradiation [123]. The enhanced effect of carbon-doped ZnO materials compared to ZnO or TiO_2 has been attributed to the synergy effect of carbon via nanoconfinement of the target pollutant in the porosity of the carbon phase (adsorption), the shift toward the visible range of the spectrum, and reduced recombination [120,123]. The presence of a carbon layer coating the ZnO particles also minimizes the sintering of ZnO and avoids the nanoparticle growth due to agglomeration and coalescence processes, while preserving the surface area, shape, and crystallinity of the catalysts [119,122]. Additional benefits of C-doped ZnO materials are associated with the increased stability of the catalyst at extreme pH values, inhibiting photocorrosion.

Some other interesting innovative morphologies for carbon/ZnO composites can be found in the literature. For instance, Wang et al. [124] synthesized a number of ZnO-C composites (carbon-core/ZnO shell) using colloidal carbon spheres (CCSs) as template. Varying the synthesis conditions (basically with the selection of additives), the authors prepared ZnO dandelions over carbon spheres or CCS@ZnO nanospheres. Both materials showed different photocatalytic activity toward methylene blue degradation; whereas the activity of CCS@ZnO nanospheres was similar to that of commercial titania, ZnO dandelions showed an enhanced activity and degradation rate. These differences were mainly associated to the morphology of the samples, whereas the role of the carbon layer was not addressed.

Carbon nanotubes (single walled or multiwalled), nanofibers, fullerenes, and graphenes have been extensively investigated as support for ZnO-based photocatalysts. The use of these carbon nanostructures has become rather popular in the last decade based on their electronic structure and the ability of semiconductor–metal junctions to enhance the photocatalytic efficiency by facilitating interfacial electron transfer processes [125–138]. Due to their performance and characteristics, CNT, fullerenes, and graphene-based semiconductor nanocomposites appear to be promising materials with potential applications in photocatalysis; however, none of them seems to be suitable for a large-scale application, due to their current high price [125].

First reports about the photocatalytic behavior of ZnO-CNT account from 2005 [126], and since then quite a large number of publications have focused on describing and understanding the photochemical response of CNT/ZnO catalysts [127–135]. Carbon nanotubes have excellent electron acceptor properties [136], while ZnO can be considered as good electron donor under UV illumination [137]. Thus, MWCNTs would act as scavengers of the photogenerated electrons arisen from the immobilized ZnO nanoparticles, improving the fast electron transfer at the interface between the carbon and the semiconductor layers. This behavior has been observed even when the catalysts are obtained by physical mixture of the two components [126,128].

Photoluminescence studies have also revealed a blue shift change in the absorption features of the CNT/ZnO composites, [139–141], pointing out a stronger oxidation/reduction potential of the CNT/ZnO nanocomposites compared to bare ZnO. The immobilization on CNT also enhanced the crystallinity of ZnO nanoparticles while reducing their size [140,141]. Quantum size effects can increase the energy levels of the conduction and valence band edges, which would improve the redox potentials to utilize UV light in photocatalysis [142,143]. Similar behaviors have already been reported in ZnO nanocrystallites.

Hybrid C60/ZnO composites have shown higher photocatalytic activity than pure ZnO for the degradation of methylene blue, even for low amounts of fullerene (i.e., up to 2.5 wt.%, the optimum being at 1.5 wt.%) [144]. An additional benefit of the incorporation of C60 as additive (besides the efficient charge transfer and decreased recombination) seems to be the inhibition of photocorrosion in the C60/ZnO composite, compared to naked ZnO. In the case of graphene, synthesized graphene/ZnO nanocomposites showed layered structures with ZnO nanoparticles inserted between the layers of the reduced graphite oxide used as a graphene precursor. The higher photocatalytic performance of the graphene/ZnO composite was once more attributed to the electrical conductivity of the carbon matrix, acting as an acceptor of the photo-induced electrons and efficiently reducing the recombination rate of the charge carrier [125,145].

The development of visible light-active photocatalysts is also attracting considerable attention in the quest to utilize solar energy effectively [146]. A large number of undoped, single-phase semiconductor photocatalysts with absorption properties under visible light have been developed based on binary oxides of transition metals, such as $BiVO_4$, Bi_2WO_6, and $CaBi_2O_4$ [147–150]. Among them, $BiVO_4$ has been reported to be an efficient photocatalyst under visible light irradiation, although it suffers from poor adsorption ability and fast recombination of photo-induced charge carriers [150]. To enhance the photocatalytic activity of $BiVO_4$, immobilization on porous carbon supports has also been investigated. The carbon/$BiVO_4$ composite showed higher photodegradation efficiency of an organic dye [151]; this was explained on the basis of the improved optical absorption properties of the composite due to the formation on carbon nano-islands on the surface of the semiconductor and the enhanced charge carrier separation. In this case, the thickness of the carbon layer and the carbon loading are critical parameters, since a fall in the photocatalytic activity can be observed if the optimal carbon loading is exceeded due to the ability of the carbon layer to absorb visible light.

17.4. CONCLUSIONS

The application of carbon/semiconductor composites and carbons themselves as photocatalysts has become an important area of research for the last decades. Substantial advances have been reported in the synthesis of more efficient photocatalysts, as well as in the comprehension of the mechanisms governing the photochemical response of these hybrid catalysts, particularly in the understanding of the role of carbon.

The enhanced photoconversion yield of carbon/semiconductor composites has been attributed to either single or collective factors based on the combination of the photo-activity of the immobilized semiconductor and the structural, textural, and electronic features of the carbon support. Several mechanisms on the role of carbon have been postulated, mainly associated to i) modification of the absorption properties of the composite (typically band-gap narrowing and visible light shift), ii) increased reaction rate due to the confinement of the pollutant in the porosity of the carbon support via adsorption, iii) strong interfacial electronic effects between the carbon film and the semiconductor leading to efficient charge carrier transfer and separation that hinder recombination of the e^-/h^+ pairs, iv) carbon acting as a photosensitizer by injecting electrons into the conduction band of the semiconductor, and v) increased stability of the catalyst inhibiting photocorrosion.

Most recently, the self-photochemical activity of certain carbon materials under UV irradiation in the absence of conventional semiconductors has been demonstrated. The results marked a starting point for further research in this field, offering the possibility of coupling photocatalytic degradation of refractory pollutants with classic and highly skilled adsorption technologies based on carbon materials.

ACKNOWLEDGMENTS

The authors thank the financial support of the Spanish MINECO and FICYT (grants CTM2008/01956, CTM2011/02338 and PC10-002).

REFERENCES

[1] S.E. Braslavsky, Glossary of terms used in photochemistry, Pure Appl. Chem. 79 (2007) 293–465.
[2] H. Okabe, Photochemistry of Small Molecules, John Wiley & Sons, 1978.
[3] R.P. Wayne, Principles and Applications of Photochemistry, Oxford Science Publications, 1988.

[4] D.F. Ollis, H. Al-Ekabi, Photocatalytic Purification and Treatment of Water and Air, Elsevier, 1993.
[5] E. Pelizzetti, N. Serpone, Photocatalysis: Fundamental and Applications, Wiley-Interscience, 1989.
[6] M. Kaneko, I. Okura, Photocatalysis: Science and Technology, Springer, 2002.
[7] A. Fujishima, K. Honda, Electrochemical photolysis of water at a semiconductor electrode, Nature 238 (1972) 37–38.
[8] M.R. Hoffmann, S.T. Martin, W. Choi, D.W. Bahnemann, Environmental applications of semiconductor photocatalysis, Chem. Rev. 95 (1995) 69–96.
[9] A.L. Linsebigler, L. Guangquan, J.T. Yate, Photocatalysis on TiO_2 surfaces: principles, mechanisms and selected results, Chem. Rev. 95 (1995) 735–758.
[10] H. Choi, E. Stathatos, D. Dionysiou, Sol–gel preparation of mesoporous photocatalytic TiO_2 films and TiO_2/Al_2O_3 composite membranes for environmental applications, Appl. Catal. B 63 (2006) 60–67.
[11] L. Erdeia, N. Arecrachakula, S. Vigneswaran, A combined photocatalytic slurry reactor-immersed membrane module system for advanced wastewater treatment, Sep. Purif. Technol. 62 (2008) 382–388.
[12] F. Sunada, A. Heller, Effects of water, salt water, and silicone overcoating of the TiO_2 photocatalyst on the rates and products of photocatalytic oxidation of liquid 3-octanol and 3-octanone, Environ. Sci. Technol. 32 (1998) 282–286.
[13] A. Fernández, G. Lassaletta, V.M. Jiménez, A. Justo, A.R. González-Elipe, J.M. Herrmann, H. Tahiri, Y. Ait-Ichou, Preparation and characterization of TiO_2 photocatalysts supported on various rigid supports (glass, quartz and stainless steel). Comparative studies of photocatalytic activity in water purification, Appl. Catal. B Environ. 7 (1995) 49–63.
[14] R. Leary, A. Westwood, Carbonaceous nanomaterials for the enhancement of TiO_2 photocatalysis, Carbon 49 (2011) 741–772.
[15] J.L. Faria, W. Wang, Carbon Materials in Photocatalysis (chapter 13), in: Ph. Serp, J.L. Figueiredo (Eds.), Carbon Materials for Catalysis, Wiley & Sons, 2009.
[16] B. Tryba, A.W. Morawski, M. Inagaki, Application of TiO_2-mounted activated carbon to the removal of phenol from water, Appl. Catal. B 41 (2003) 427–433.
[17] M. Toyoda, Y. Nanbu, T. Kito, M. Hirano, M. Inagaki, Preparation and performance of anatase-loaded porous carbons for water purification, Desalination 159 (2003) 273–282.
[18] G.L. Puma, A. Bono, D. Krishnaiah, J.G. Collin, Preparation of titanium dioxide photocatalyst loaded onto activated carbon support using chemical vapor deposition: a review paper, J. Hazard. Mater. 157 (2008) 209–219.
[19] N. Keller, G. Rebmann, E. Barraud, O. Zahraa, V. Keller, Macroscopic carbon nanofibers for use as photocatalyst support, Catal. Today 101 (2005) 323–329.
[20] T. Torimoto, S. Ito, S. Kuwabata, H. Yoneyama, Effects of adsorbents used as supports for titanium dioxide loading on photocatalytic degradation of propyzamide, Environ. Sci. Technol. 30 (1996) 1275–1281.
[21] J. Matos, J. Laine, J.M. Herrmann, Synergy effect in the photocatalytic degradation of phenol on a suspended mixture of titania and activated carbon, Appl. Catal. B Environ. 18 (1998) 281–291.
[22] J. Araña, J.M. Doña-Rodríguez, E. Tello–Rendón, C. Garriga i Cabo, O. González-Díaz, J.A. Herrera-Melián, J. Pérez-Peña, G. Colón, J.A. Navío, TiO_2 activation by using activated carbon as a support. Part I. Surface characterisation and decantability study, Appl. Catal. B Environ. 44 (2003) 161–172.

[23] X. Zhang, M. Zhou, L. Lei, TiO_2 photocatalyst deposition by MOCVD on activated carbon, Carbon 44 (2006) 325–333.
[24] C.G. Silva, W. Wang, J.L. Faria, Photocatalytic and photochemical degradation of mono-, di- and tri-azo dyes in aqueous solution under UV irradiation, J. Photochem. Photobiol. A Chem. 181 (2006) 314–324.
[25] N. Takeda, N. Iwata, T. Torimoto, H. Yoneyama, Influence of carbon blacks as an adsorbent used in photocatalyst films on photodegradation behaviors of propyzamide, J. Catal 177 (1998) 240–246.
[26] L.F. Velasco, J.B. Parra, C.O. Ania, Role of activated carbon features on the photocatalytic degradation of phenol, Appl. Surf. Sci. 256 (2010) 5254–5258.
[27] L.F. Velasco, I.M. Fonseca, J.B. Parra, J.C. Lima, C.O. Ania, Photochemical behavior of activated carbons under UV irradiation, Carbon 50 (2012) 249–258.
[28] L.F. Velasco, C.O. Ania, Materiales de carbono en fotocatálisis, Bol. Grupo Español del Carbón 21 (2011) 2–5.
[29] L.F. Velasco, Dissertation for MSc degree. Estudio de la fotodegradación de fenol sobre materiales carbonosos. University of Oviedo, Spain, 2010.
[30] S.N. Frank, A.J. Bard, Heterogeneous photocatalytic oxidation of cyanide ion in aqueous solution at TiO_2 powder, J. Am. Chem. Soc. 99 (1977) 303–304.
[31] S.N. Frank, A.J. Bard, Heterogeneous photocatalytic oxidation of cyanide and sulfite in aqueous solutions at semiconductor powders, J. Phys. Chem. 81 (1977) 1484–1488.
[32] S. Parsons, Advanced Oxidation Processes for Water and Wastewater Treatment, IWA Publishing, 2004.
[33] N. Serpone, D. Lawless, R. Khairutdinov, E.J. Pelizzetti, Subnanosecond relaxation dynamics in TiO_2 colloidal sols (Particle Sizes Rp = 1.0–13.4 nm). Relevance to heterogeneous photocatalysis, J. Phys.Chem. 99 (1995) 16655–16661.
[34] D.P. Colombo Jr., R.M. Bowman, Does interfacial charge transfer compete with charge carrier recombination? A femtosecond diffuse reflectance investigation of TiO_2 nanoparticles, J. Phys. Chem. 100 (1996) 18445–18449.
[35] A. Fujishima, T.N. Rao, D.A. Tryk, Titanium dioxide photocatalysis, J. Photochem. Photobiol. C Photochem. Rev. 1 (2000) 1–21.
[36] J.M. Herrmann, Heterogeneous photocatalysis: fundamentals and applications to the removal of various types of aqueous pollutants, Catal. Today 53 (1999) 115–129.
[37] A. Kudo, Y. Miseki, Heterogeneous photocatalyst materials for water splitting, Chem. Soc. Rev. 38 (2009) 253–278.
[38] X. Hu, G. Li, J.C. Yu, Design, fabrication, and modification of nanostructured semiconductor materials for environmental and energy applications, Langmuir 26 (2009) 3031–3039.
[39] J.C. Yu, J.J. Yu, W. Ho, L. Zhang, Preparation of highly photocatalytic active nano-sized TiO_2 particles via ultrasonic irradiation, Chem. Commun. (2001) 1942–1943.
[40] W. Ho, J.C. Yu, S. Lee, Low-temperature hydrothermal synthesis of S-doped TiO_2 with visible light photocatalytic activity, J. Solid State Chem. 179 (2006) 1171–1176.
[41] J.G. Yu, J.C. Yu, B. Cheng, S.K. Hark, K. Iu, The effect of F^--doping and temperature on the structural and textural evolution of mesoporous TiO_2 powders, J. Solid State Chem. 174 (2003) 372–380.
[42] S. Yin, Y. Aita, M. Komatsu, J. Wang, Q. Tang, T. Sato, Synthesis of excellent visible-light responsive $TiO_{2-x}N_y$ photocatalyst by a homogeneous precipitation-solvothermal process, J. Mater. Chem. 15 (2005) 674–682.

[43] H. Luo, T. Takata, Y. Lee, J. Zhao, K. Domen, Y. Yan, Photocatalytic Activity Enhancing for Titanium Dioxide by Co-doping with Bromine and Chlorine, Chem. Mater. 16 (2004) 846–849.
[44] J. Matos, Environmental green chemistry applications of nanoporous carbons, in: T. Rufford, D. Hulicova-Jurkahova (Eds.), Green Carbons: Advances and Applications, PAN Stanford Pulbishing, 2012 (Chapter 13).
[45] B. Hu, K. Wang, L.H. Wu, S.H. Yu, M. Antonietti, M.M. Titirici, Engineering carbon materials from the hydrothermal carbonization process of biomass, Adv. Mater. 2 (2002) 813–828.
[46] M. Inagaki, K.C. Park, M. Endo, Carbonization under pressure, New Carbon Mater. 25 (2010) 409–420.
[47] M.A. Henderson, A surface science perspective on TiO_2 photocatalysis, A surface science perspective on TiO_2 photocatalysis, Surf. Sci. Rep. 66 (2011) 185–297.
[48] B. Llano, G. Restrepo, J.M. Marin, J.A. Navio, M.C. Hidalgo, Characterisation and photocatalytic properties of titania-silica mixed oxides doped with Ag and Pt, Appl. Catal. A Gen. 387 (2010) 135–140.
[49] M.I. Litter, J.A. Navio, Photocatalytic properties of iron-doped titania semiconductors, J. Photochem. Photobiol. A, Chem. 98 (1996) 171–181.
[50] C. Minero, G. Mariella, V. Maurino, E. Pelizzetti, Photocatalytic transformation of organic compounds in the presence of inorganic anions. 1. Hydroxyl-mediated and direct electron-transfer reactions of phenol on a titanium dioxide-fluoride system, Langmuir 16 (2000) 2632–2641.
[51] K. Lv, X. Li, K. Deng, J. Sun, X. Li, M. Li, Effect of phase structures on the photocatalytic activity of surface fluorinated TiO_2, Appl. Catal. B Environ 95 (2010) 383–392.
[52] E.M. Rockafellow, L.K. Stewart, W.S. Jenks, Is sulfur-doped TiO_2 an effective visible light photocatalyst for remediation? Appl. Catal. B Environ 91 (2009) 554–562.
[53] S.U.M. Khan, M. Al-Shahry, WB. Ingler Jr., Efficient photochemical water splitting by a chemically modified n-TiO_2, Science 297 (2002) 2243–2245.
[54] R. Asahi, T. Morikawa, T. Ohwaki, K. Aoki, Y. Taga, Visible-light photocatalysis in nitrogen-doped titanium oxides, Science 293 (2001) 269–271.
[55] S. Sakthivel, H. Kisch, Daylight photocatalysis by carbon-modified titanium dioxide, Angew. Chem. Int. Ed. 42 (2003) 4908–4911.
[56] S. Sakthivel, M. Janczarek, H.J. Kisch, Visible light activity and photoelectrochemical properties of nitrogen-doped TiO_2, Phys. Chem. B 108 (2004) 19384–19387.
[57] Y. Choi, T. Umebayashi, M.J. Yoshikawa, Fabrication and characterization of C-doped anatase TiO_2 photocatalysts, Mater. Sci 39 (2004) 1837–1839.
[58] X. Nie, S. Zhuo, G. Maeng, K. Sohlberg, Doping of TiO_2 polymorphs for altered optical and photocatalytic properties, Int. J. Photoenergy (2009) 294042.
[59] C. Di Valentin, G. Pacchioni, A. Selloni, Theory of carbon doping of titanium dioxide, Chem. Mater. 17 (2005) 6656–6665.
[60] M. Mohseni, Gas phase trichloroethylene (TCE) photooxidation and byproduct formation: photolysis vs. titania/silica based photocatalysis, Chemosphere 59 (2005) 335–342.
[61] M. Bideau, B. Claudel, C. Dubien, L. Faure, H. Kazouan, On the "immobilization" of titanium dioxide in the photocatalytic oxidation of spent waters, J. Photochem. Photobiol. A 91 (1995) 137–144.
[62] J. Araña, J.M. Doña-Rodríguez, E. Tello Rendón, C. Garriga i Cabo, O. González-Díaz, J.A. Herrera-Melián, J. Pérez-Peña, G. Colón, J.A. Navio, TiO_2 activation by using activated carbon as a support: Part II. Photoreactivity and FTIR study, Appl. Catal. B 44 (2003b) 153–160.

[63] X. Qian, Y. Wan, Y. Wena, N. Jia, H. Li, D. Zhao, Synthesis of ordered mesoporous crystalline carbon–anatase composites with high titania contents, J. Colloid Interf. Sci. 328 (2008) 367–373.
[64] S. Schlienger, J. Parmentier, R. Gadiou, C. Vix-Guterl, L.F. Velasco, J.B. Parra, C.O. Ania, Phenol Photodegradation in Aqueous Solution by Ordered Mesoporous TiO_2/C Nanocomposites Prepared by Evaporation Induced Self-assembly Process, Presented at the 3rd Int. conference on Carbon for Energy Storage and Environmental Protection (CESEP'11), Vichy, France, 2011.
[65] C.-H. Huang, D. Gu, D. Zhao, R.-A. Doong, Direct synthesis of controllable microstructures of thermally stable and ordered mesoporous crystalline titanium oxides and carbide/carbon composites, Chem. Mater. 22 (2010) 1760–1767.
[66] J. Matos, J. Laine, J.M. Herrmann, Association of activated carbons of different origins with titania in the photocatalytic purification of water, Carbon 37 (1999) 1870–1872.
[67] J. Matos, J. Laine, J.M. Herrmann, Effect of the type of activated carbons on the photocatalytic degradation of aqueous organic pollutants by UV-irradiated titania, J. Catal. 200 (2001) 10–20.
[68] J. Matos, J. Laine, J.M. Herrmann, D. Uzcategui, J.L. Brito, Influence of activated carbon upon titania on aqueous photocatalytic consecutive runs of phenol photodegradation, Appl. Catal. B 70 (2007) 461–469.
[69] B. Tryba, A.W. Morawski, T. Tsumura, M. Toyoda, M. Inagaki, Hybridization of adsorptivity with photocatalytic activity–carbon-coated anatase, J. Photochem. Photobiol. A: Chem. 167 (2004) 27–135.
[70] L.F. Velasco, B. Tsyntsarski, B. Petrova, T. Budinova, N. Petrov, J.B. Parra, C.O. Ania, Carbon foams as catalyst supports for phenol photodegradation, J. Hazard. Mater. 184 (2010) 843–848.
[71] B. Tryba, Increase of the Photocatalytic Activity of TiO_2 by carbon and iron modifications, J. Photoenergy, ID 721824 (2008) 1–15.
[72] C. Minero, F. Catozzo, E. Pelizzetti, Role of adsorption in photocatalyzed reactions of organic molecules in aqueous TiO_2 suspensions, Langmuir 8 (1992) 481–486.
[73] D.K. Lee, S.C. Kim, S.J. Kim, I.S. Chung, S.W. Kim, Photocatalytic oxidation of microcystin-LR with TiO_2-coated activated carbon, Chem. Eng. J. 102 (2004) 93–98.
[74] R. Ocampo-Pérez, M. Sánchez-Polo, J. Rivera-Utrilla, R. Leyva-Ramos, Enhancement of the catalytic activity of TiO_2 by using activated carbon in the photocatalytic degradation of cytarabine, Appl. Catal. B 104 (2011) 177–184.
[75] K. Woan, G. Pyrgiotakis, W. Sigmund, Photocatalytic carbon-nanotube–TiO_2 composites, Adv. Mater. 21 (2009) 2233–2239.
[76] W.D. Wang, Ph. Serp, P. Kalck, J.L. Faria, Visible light photodegradation of phenol on MWNT-TiO_2 composite catalysts prepared by a modified sol–gel method, J. Mol. Catal. A Chem. 235 (2005) 194–199.
[77] H.T. Yu, X. Quan, S. Chen, H. Zhao, TiO_2-multiwalled carbon nanotube heterojunction arrays and their charge separation capability, J. Phys. Chem. C 35 (2007) 12987–12991.
[78] X.H. Xia, Z.H. Jia, Y. Yu, Y. Liang, Z. Wang, L.L. Ma, Preparation of multi-walled carbon nanotube supported TiO_2 and its photocatalytic activity in the reduction of CO_2 with H_2O, Carbon 45 (2007) 717–721.
[79] Y. Yao, G. Li, S. Ciston, R.M. Lueptow, K.A. Gray, Photoreactive TiO_2/carbon nanotube composites: synthesis and reactivity, Environ. Sci. Technol 42 (2008) 4952–4957.
[80] H.T. Yu, X. Quan, S. Chen, H.M. Zhao, Y.B. Zhang, TiO_2-carbon nanotube heterojunction arrays with a controllable thickness of TiO_2 layer and their first application in photocatalysis, J. Photochem. Photobiol. A 200 (2008) 301–306.

[81] H. Chen, S. Chen, X. Quan, Y. Zhang, Structuring a TiO_2-based photonic crystal photocatalyst with schottky junction for efficient photocatalysis, Environ. Sci. Technol. 44 (2010) 451–455.
[82] R. Sellappan, A. Galeckas, V. Venkatachalapathy, A.Y. Kuznetsov, D. Chakarov, On the mechanism of enhanced photocatalytic activity of composite TiO_2/carbon Nanofilms, Appl. Catal. B Environ. 106 (2011) 337–342.
[83] R. Morelli, I.R. Bellobono, C.M. Chiodarolib, S. Alborghetti, EPR spin-trapping of hydroxyl radicals onto photocatalytic membranes immobilizing titanium dioxide, and spin adduct competition, as a probe of reactivity with aqueous organic micropollutants, J. Photochem. Photobiol. A Chem. 112 (1998) 271–276.
[84] D.C. Hurum, A.G. Agrios, S.E. Crist, K.A. Gray, T. Rajh, M.C. Thurnauer, Probing reaction mechanisms in mixed phase TiO_2 by EPR, J. Electron Spectros. Relat. Phenomena 150 (2006) 155–163.
[85] E.R. Carraway, A.J. Hoffman, M.R. Hoffmann, Photocatalytic production of H_2O_2 and organic acids on quantum-sized semiconductor colloids, Environ. Sci. Technol. 28 (1994) 786–793.
[86] R. Sellappan, J. Zhu, H. Fredriksson, R.S. Martins, M. Zäch, D. Chakarov, Preparation and characterization of TiO_2/carbon composite thin films with enhanced photocatalytic activity, J. Mol. Catal. A Chem. 335 (2011) 136–144.
[87] B. Gao, G.Z. Chen, G.L. Puma, Carbon nanotubes/titanium dioxide (CNTs/TiO_2) nanocomposites prepared by conventional and novel surfactant wrapping sol-gel methods exhibiting enhanced photocatalytic activity, Appl. Catal. B 89 (2009) 503–509.
[88] C.G. Silva, J.L. Faria, Photocatalytic oxidation of phenolic compounds using carbon nanotube-TiO_2 composite catalyst, ChemSusChem 5 (2010) 609–618.
[89] W. Wang, Ph. Serp, P. Kalck, C.G. Silva, J.L. Faria, Preparation and characterization of nanostrustured MWCNT-TiO_2 composite materials for photocatalytic water treatment application, Mater. Res. Bull. 43 (2008) 958–967.
[90] H. Li, X. He, Z. Kang, H. Huang, Y. Liu, J. Liu, S. Lian, C.H.A. Tsang, X. Yang, S.T. Lee, Water-soluble fluorescent carbon quantum dots and photocatalyst design, Angew. Chem. Int. Ed. 49 (2010) 4430–4434.
[91] S. Stankovich, D.A. Dikin, G.H.B. Dommett, K.M. Kohlhaas, E.J. Zimney, E.A. Stach, R.D. Piner, S.T. Nguyen, R.S. Ruoff, Graphene-based composite materials, Nature 442 (2006) 282–286.
[92] A.P. Yu, P. Ramesh, M.E. Itkis, E. Bekyarova, R.C. Haddon, Graphite nanoplatelet-epoxy composite thermal interface materials, J. Phys. Chem. C 111 (2007) 7565–7569.
[93] G. Williams, B. Seger, P.V. Kamat, TiO_2-graphene nanocomposites. UV-assisted photocatalytic reduction of graphene oxide, ACS Nano 2 (2008) 1487–1491.
[94] P.V. Kamat, I. Bedja, S. Hotchandani, Photoinduced charge transfer between carbon and semiconductor clusters. One-electron reduction of C60 in Colloidal TiO_2 semiconductor suspensions, J. Phys. Chem. 98 (1994) 9137–9142.
[95] A. Kongkanand, P.V. Kamat, Electron Storage in single wall carbon nanotubes. fermi level equilibration in semiconductor-SWCNT suspensions, ACS Nano 1 (2010) 13–21.
[96] Y. Zhang, Z.R. Tang, X. Fu, Y.-J. Xu, TiO_2−graphene nanocomposites for gas-phase photocatalytic degradation of volatile aromatic pollutant: Is TiO_2−graphene truly different from other TiO_2−carbon composite materials? ACS Nano 4 (2010) 7303–7314.
[97] H. Zhang, X. Lv, Y. Li, Y. Wang, J. Li, P25-Graphene composite as a high performance photocatalyst, ACS Nano 4 (2010) 380–386.

[98] S. Wang, D. Choi, J. Li, Z. Yang, Z. Nie, R. Kou, D. Hu, C. Wang, L.F. Saraf, J. Zhang, I.A. Aksay, J. Liu, Self-assembled TiO_2–graphene hybrid nanostructures for enhanced Li-ion insertion, ACS Nano 3 (2009) 907–991.

[99] J. Matos, J.M. Chovelon, T. Cordero, C. Ferronato, Influence of surface properties of activated carbon on photocatalytic activity of TiO_2 in 4-chlorophenol degradation, Open Environ. Eng. J. 2 (2009) 21–29.

[100] X.M. Sun, Y.D. Li, Colloidal carbon spheres and their core/shell structures with noble-metal nanoparticles, Angew. Chem., Int. Ed. 43 (2004) 597–601.

[101] L.F. Velasco, J. Dentzer, G. Gadiou, J. Parmentier, C. Vix-Guterl, C.O. Ania, Synthesis of TiO_2/core shell composites for the photodegradation of organic compounds (2011). Proc. Carbon International Conference 2011, n. 143, Shanghai.

[102] J. Matos, A. Garcia, M.M. Titirici, Solvothermal carbon-doped TiO_2 photocatalyst for the enhanced methylene blue degradation under visible light, Appl. Catal. A: Gen. 390 (2010) 175–182.

[103] G. Pyrgiotakis, S.H. Lee, W.R. Sigmund, Advanced Photocatalysis with Anatase Nano-Coated Multi-Walled Carbon Nanotubes (2005). Presented at MRS Spring Meeting, San Francisco, CA 2005.

[104] L.F. Velasco, I.M. Fonseca, J.B. Parra, J.C. Lima, C.O. Ania, Evidences of Photocatalytic Activity of Activated Carbons under UV Irradiation (2011). Proc. Carbon International Conference 2011, n. 162, Shanghai.

[105] L.F. Velasco, V. Maurino, E. Laurenti, C. Minero, C.O. Ania, Unraveling the mechanisms of carbon photochemical behaviour under UV Light by EPR, Chem. Sus. Chem. (2012) (submitted for publication).

[106] S. Lu, B. Panchapakesan, Photoconductivity in single wall carbon nanotube sheets, Nanotechnology 17 (2006) 1843–1850.

[107] L.F. Velasco, C.O. Ania, Understanding phenol adsorption mechanisms on activated carbons, Adsorption 17 (2011) 247–254.

[108] F. Rodríguez-Reinoso, C. Moreno-Castilla, A. Guerrero-Ruiz, I. Rodríguez-Ramos, J.D. López-González, The effect of inorganic constituents of the support on the characteristics of carbon-supported platinum catalysts, Appl. Catal. 15 (1985) 293–300.

[109] Y. Luo, Y. Heng, X. Dai, W. Chen, J. Li, Preparation and photocatalytic ability of highly defective carbon nanotubes, J. Solid State Chem. 182 (2009) 2521–2525.

[110] E. Pulido-Melián, O. González-Díaz, J.M. Doña-Rodriguez, G. Colon, J. Araña, J. Herrera-Melián, J.A. Navío, J. Pérez-Peña, ZnO activation by using activated carbon as a support: Characterisation and photoreactivity, Appl. Catal. A: Gen 364 (2009) 174–181.

[111] N. Gokon, N. Hasegawa, H. Kaneko, H. Aoki, Y. Tamaura, M. Kitamura, Photocatalytic effect of ZnO on carbon gasification with CO_2 for high temperature solar thermochemistry, Sol. Energ. Mat. Sol. C. 80 (2003) 335–341.

[112] J. Matos, E. García-López, L. Palmisano, A. García, G. Marcì, Influence of activated carbon in TiO_2 and ZnO mediated photo-assisted degradation of 2-propanol in gas-solid regime, Appl. Catal. B Environ. 99 (2010) 170–180.

[113] K.M. Joshi', V.S. Shrivastava, Removal of hazardous textile dyes from aqueous solution by using commercial activated carbon with TiO_2 and ZnO as photocatalyst, Int. J. Chem. Tech. Res. 2 (2010) 427–435.

[114] N. Sobana, M. Swaminathan, Combination effect of ZnO and activated carbon for solar assisted photocatalytic degradation of Direct Blue 53, Sol. Energ. Mat. Sol. C 91 (2007) 727–734.

[115] N. Sobana, M. Muruganandam, M. Swaminathan, Characterization of AC-ZnO catalyst and its photocatalytic activity on 4-acetylphenol degradation, Catal. Commun. 9 (2008) 262–268.
[116] K. Byrappa, A.K. Subramani, S. Ananda, K.M.L. Rai, M.H. Sunitha, B. Basavalingu, K. Soga, Impregnation of ZnO onto activated carbon under hydrothermal conditions and its photocatalytic properties, J. Mater. Sci. 41 (2006) 1355–1362.
[117] H. Ma, J. Han, Y. Fu, Y. Song, C. Yu, X. Dong, Synthesis of visible light responsive ZnO–ZnS/C photocatalyst by simple carbothermal reduction, Appl. Catal. B Environ. 102 (2011) 417–423.
[118] F.F. Brites, V.S. Santana, N.R.C. Fernandes-Machado, Effect of support on the photocatalytic degradation of textile effluents using Nb_2O_5 and ZnO: Photocatalytic degradation of textile dye, Top. Catal. 54 (2011) 264–269.
[119] S. Liu, C. Li, J. Yu, Q. Xiang, Improved visible-light photocatalytic activity of porous carbon self-doped ZnO nanosheet-assembled flowers, CrystEngComm 13 (2011) 2533–2541.
[120] Y. Guo, H. Wang, C. He, L. Qiu, X. Cao, Uniform carbon-coated ZnO nanorods: Microwave-assisted preparation, cytotoxicity, and photocatalytic activity, Langmuir 25 (2009) 4678–4684.
[121] S. Cho, J.W. Jang, J.S. Lee, K.H. Lee, Carbon-doped ZnO nanostructures synthesized using vitamin C for visible light photocatalysis, CrystEngComm 12 (2010) 3929–3935.
[122] L. Zhang, H. Cheng, R. Zong, Y. Zhu, Photocorrosion suppression of ZnO nanoparticles via hybridization with graphite-like carbon and enhanced photocatalytic activity, J. Phys. Chem. C. 113 (2009) 2368–2374.
[123] X. Zhou, Y. Li, T. Peng, W. Xie, X. Zhao, Synthesis, characterization and its visible-light-induced photocatalytic property of carbon doped ZnO, Mater. Lett. 63 (2009) 1747–1749.
[124] X. Wang, P. Hu, Y. Fangli, L. Yu, Preparation and Characterization of ZnO hollow spheres and ZnO-carbon composite materials using colloidal carbon spheres as templates, J. Phys. Chem. C 111 (2007) 6706–6712.
[125] X. Chen, Y. He, Q. Zhang, L. Li, D. Hu, T. Yin, Fabrication of sandwich-structured ZnO/reduced graphite oxide composite and its photocatalytic properties, J. Mater. Sci. 45 (2010) 953–960.
[126] L. Jiang, L. Gao, Fabrication and characterization of ZnO-coated multi-walled carbon nanotubes with enhanced photocatalytic activity, Mater. Chem. Phys. 91 (2005) 313–316.
[127] X. Wang, S. Yao, X. Li, Sol-gel preparation of CNT/ZnO nanocomposite and its photocatalytic property, Chin. J. Chem. 27 (2009) 1317–1320.
[128] J. Xu, X.J. Song, X.W. Wei, Study on photocatalytic decomposition of azo-dyes by ZnO/carbon nanotubes composites by UV–vis spectroscopy, Spectrosc. Spect. Anal. 27 (2007) 2510–2513.
[129] L.P. Zhu, G.H. Liao, W.Y. Huang, L.L. Ma, Y. Yang, Y. Yu, S.Y. Fu, Preparation, characterization and photocatalytic properties of ZnO-coated multi-walled carbon nanotubes, Mat. Sci. Eng. B Adv. 163 (2009) 194–198.
[130] J. Chang, J.H. Lee, C.K. Najeeb, G.H. Nam, M. Lee, J.H. Kim, Area-selective growth of ZnO nanorod arrays on single-walled carbon nanotube patterns, Scripta Mater. 63 (2010) 520–523.
[131] Q. Wang, X. Li, W. Nie, Y. Xia, J. Dai, The preparation and photocatalytic activity of SWNTs doped ZnO/TiO_2 compound films, Adv. Mater. Res. 97–101 (2010) 1611–1615.
[132] F. Vietmeyer, B. Seger, P.V. Kamat, Anchoring ZnO particles on functionalized single wall carbon nanotubes. Excited state interactions and charge collection, Adv. Mater. 19 (2007) 2935–2940.
[133] J.G. Ok, S.H. Tawfick, K.A. Juggernauth, K. Sun, Y. Zhang, A.J. Hart, Electrically addressable hybrid architectures of zinc oxide nanowires grown on aligned carbon nanotubes, Adv. Funct. Mater. 20 (2010) 2470–2480.

[134] C.C. Lin, B.T.T. Chu, G. Tobias, S. Sahakalkan, S. Roth, M.L.H. Green, S.Y. Chen, Electron transport behavior of individual zinc oxide coated single-walled carbon nanotubes, Nanotechnology 20 (2009) 105703.

[135] K. Byrappa, A.S. Dayananda, C.P. Sajan, B. Basavalingu, M.B. Shayan, K. Soga, M. Yoshimura, Hydrothermal preparation of ZnO: CNT and TiO_2:CNT composites and their photocatalytic applications, J. Mater. Sci. 43 (2008) 2348–2355.

[136] S. Kment, Z. Hubicka, H. Kmentova, P. Kluson, J. Krysa, I. Gregora, M. Morozova, M. Cada, D. Petras, P. Dytrych, M. Slater, L. Jastrabik, Photoelectrochemical properties of hierarchical nanocomposite structure: carbon nanofibers/ TiO_2/ZnO thin films, Catal. Today 161 (2011) 8–14.

[137] K. Zhang, W.C. Oh, Photocatalytic degradation of methylene blue by ACF/ TiO_2 and ACF/ ZnO composites under UV light, Kor. J. Mater. Res. 20 (2010) 31–36.

[138] T.A. Saleh, M.A. Gondal, Q.A. Drmosh, Z.H. Yamani, A. AL-yamani, Enhancement in photocatalytic activity for acetaldehyde removal by embedding ZnO nano particles on multiwall carbon nanotubes, Chem. Eng. J. 166 (2011) 407–412.

[139] W.D. Zhang, L.C. Jiang, J.S. Ye, Photoelectrochemical study on charge transfer properties of ZnO nanowires promoted by carbon nanotubes, J. Phys. Chem. C 113 (2009) 16247–16253.

[140] Y. Yan, T. Chang, P. Wei, S.Z. Kang, J. Mu, Photocatalytic activity of nanocomposites of ZnO and multi-walled carbon nanotubes for dye degradation, J. Dispers. Sci. Technol. 30 (2009) 197–202.

[141] T.A. Saleh, M.A. Gondal, Q.A. Drmosh, Preparation of a MWCNT/ZnO nanocomposite and its photocatalytic activity for the removal of cyanide from water using a laser, Nanotechnology 21 (2010) 495705.

[142] L.C. Jiang, W.D. Zhang, Y.X. Yu, J. Wang, Preparation and charge transfer properties of carbon nanotubes supported CdS/ZnO-NWs shell/core heterojunction, Electrochem. Commun. 13 (2011) 627–630.

[143] N. Hobeika, L.F. Velasco, R. Schneider, L. Balan, C.O. Ania, Photocatalytic activity of ZnO quantum dots immobilized on a carbon support (2011). Presented at the 3rd Int. conference on Carbon for Energy Storage and Environmental Protection (CESEP'11), Vichy, France.

[144] H. Fu, T. Xu, S. Zhu, Y. Zhu, Photocorrosion inhibition and enhancement of photocatalytic activity for ZnO via hybridization with C60, Environ. Sci. Technol. 42 (2008) 8064–8069.

[145] T. Lv, L. Pan, X. Liu, T. Lu, G. Zhu, Z. Sun, Enhanced photocatalytic degradation of methylene blue by ZnO-reduced graphene oxide composite synthesized via microwave-assisted reaction, J. Alloy. Compds 509 (2011) 10086–10091.

[146] H.G. Kim, P.H. Borse, W. Choi, J.S. Lee, Photocatalytic nanodiodes for visible-light photocatalysis, Angew. Chem. Int. Ed. 44 (2005) 4585–4589.

[147] A. Kudo, K. Ueda, H. Kato, I. Mikami, Photocatalytic O_2 evolution under visible light irradiation on BiVO4 in aqueous $AgNO_3$ solution, Catal. Lett. 53 (1998) 229–230.

[148] A. Kudo, S. Hijii, H_2 or O_2 evolution from aqueous solutions on layered oxide photocatalysts consisting of Bi^{3+} with $6s^2$ configuration and d0 transition metal ions, Chem. Lett. 28 (1999) 1103–1104.

[149] J.W. Tang, Z.G. Zou, J.H. Ye, Efficient photocatalytic decomposition of organic contaminants over $CaBi_2O_4$ under visible-light irradiation, Angew. Chem. Int. Ed. 43 (2004) 4463–4466.

[150] M. Long, W. Cai, J. Cai, B. Zhou, X. Chai, Y. Wu, Efficient photocatalytic degradation of phenol over $Co_3O_4/BiVO_4$ composite under visible light irradiation, J. Phys. Chem. B 110 (2006) 20211–20216.

[151] D.K. Lee, I.S. Cho, S. Lee, S.T. Bae, J.H. Noh, D.W. Kim, K.S. Hong, Effects of carbon content on the photocatalytic activity of $C/BiVO_4$ composites under visible light irradiation, Mater. Chem. Phys. 119 (2010) 106–111.

Chapter 18

Carbon-based Catalyst Support in Fuel Cell Applications

Ji Liang, Shi Zhang Qiao, Gao Qing (Max) Lu and Denisa Hulicova-Jurcakova

ARC Centre of Excellence for Functional Nanomaterials, Australian Institute for Bioengineering and Nanotechnology, University of Queensland, St Lucia, QLD, Australia

Chapter Outline

18.1. Introduction: Fuel Cells and Carbons: Why Carbon is Indispensable in Fuel Cells? 549

18.2. Conventional Carbons in Fuel Cells 551

18.2.1. Carbon Black and Electrochemical Corrosion of Carbon in Fuel Cells 551

18.2.2. Graphitized Carbons and Graphite 554

18.2.3. Surface-modified Carbons 556

18.2.4. Catalyst-Loading Techniques 558

18.3. Novel Carbon Materials: Mesoporous Carbon and Carbon Nanomaterials 559

18.3.1. Mesoporous Carbons 559

18.3.2. Carbon Nanomaterials 561

18.4. Heteroatom-doped Carbons and Carbon-based Materials 568

18.4.1. Nitrogen-Doped Carbons 569

18.4.2. Boron-Doped Carbons (BDCs) 573

18.5. Summary, Perspectives, and Further Directions 573

References 574

18.1. INTRODUCTION: FUEL CELLS AND CARBONS: WHY CARBON IS INDISPENSABLE IN FUEL CELLS?

Sustainable energy is a key area of human challenges and opportunities for research and development in the next century. Among various energy

Novel Carbon Adsorbents. DOI: 10.1016/B978-0-08-097744-7.00018-1
Copyright © 2012 Elsevier Ltd. All rights reserved.

conversion methods, fuel cell is a very efficient process and does not generate pollution [1,2].

Fuels of different types can be converted into the electrical energy through redox reactions in fuel cells (Fig. 18.1). The general concept of a fuel cell involves oxidation of fuel on the anode and the reduction of oxygen on the cathode. The hydrogen fuel is commonly used in proton exchange membrane fuel cells (PEMFCs) and phosphoric acid fuel cells (PAFCs), while methanol and ethanol are the fuels for direct methanol/ethanol fuel cells (DMFCs/DEFCs). These fuel cells are also known as low-temperature fuel cells since their operation temperature is below 200 °C. The high-temperature fuel cells represented by solid oxide fuel cells (SOFCs) and molten carbonate fuel cells (MOFCs) operate at temperatures higher than 600 °C and hydrogen, natural gas, hydrocarbon, or coals can be utilized as fuels. The low-temperature fuel cells are particularly attractive to the consumer market due to their mild working condition, high safety factor, and moderate mobility [2]. However, due to the low operation temperatures, fuels cannot be oxidized without the presence of catalyst. Catalyst is also required at the cathode side for the reduction of oxygen. Noble metals including Pt and Pd or their alloys Pt/Pd and Pt/Ru have been traditionally employed as catalysts in fuel cells [4–6]. However, their high cost and low reserve are hindering the wide application and commercialization of fuel cells and driving researchers to make the utmost of the catalyst. In this regard, the major effort has been paid toward nanoscaling of the catalyst particles to form more active sites per mass unit. In this case, the nanosized metallic catalyst particles must be loaded on certain supports preventing them from aggregation and/or sintering during the synthesis and fuel cell operation. The morphology, structure, and activity of the catalyst, and correspondingly the whole lifetime of a cell, thus strongly depend on the catalyst support [7].

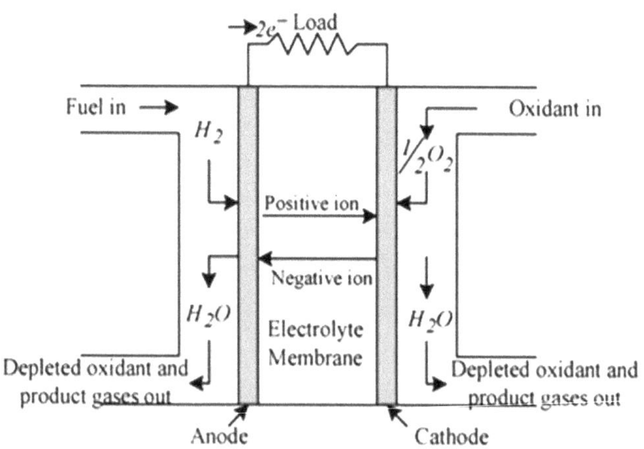

FIGURE 18.1 Typical scheme of a fuel cell consuming hydrogen [3].

Catalyst support not only functions as a matrix for even dispersion of metal particles but also influences its electronic structure, which has a great impact on the catalytic activity [8]. The shape and/or the size of catalyst particles can also be influenced by the surface geometry of the support, especially by porous support materials. Hence, finding a proper support for the catalyst is, to some extent, as important as finding a new and highly active catalyst.

The critical characteristics for catalyst support include high surface area for high catalyst loadings and good dispersion, suitable pore size for smooth mass transfer of the fuel and its oxidized form, high electrical conductivity for good electron transport through the electrode, as well as electrochemical corrosion resistance in harsh fuel cell environments. Among different schools of materials (polymers, metals, and inorganic–nonmetallic materials/ceramics), carbon or carbon-based materials are the most widely applied as they meet nearly all of above-mentioned criteria [9]. The research on using carbon black-loaded Pt particles for PAFCs dates back to 40 years ago and, until the 1990s, the nearly exclusively available catalyst support was carbon black. As a result of the progresses in carbon material research, more types of carbon or carbon-based materials have been developed, such as mesoporous carbon, carbon nanomaterials (graphitized carbon, carbon nanotube, graphene, and carbon nanocoils), and heteroatom-doped carbons (nitrogen- and boron-doped carbons). In this chapter, an overview of different carbon-based materials applied as catalyst supports in fuel cells is presented. The latest developments in designing novel carbon materials including heteroatom-doped carbons are also summarized.

18.2. CONVENTIONAL CARBONS IN FUEL CELLS

Using carbon as a Pt catalyst support in fuel cells appeared for the first time in the 1970s by United Technologies Corporation [10]. Even after nearly 40 years of development and new carbon materials emerging, carbon black and graphite remain as the most widely used and successfully commercialized fuel cell catalyst supports today.

18.2.1. Carbon Black and Electrochemical Corrosion of Carbon in Fuel Cells

Standard procedure for synthesis of carbon black is a thermal cracking or partial combustion of carbon-containing precursors, mostly hydrocarbons such as natural gas at elevated temperatures and in an inert atmosphere or inadequate air supply. To date, three main synthesis routes have been developed and these include furnace process, thermal process, and channel process, with most of the carbon black being produced through the first route [11].

TABLE 18.1 Physical Properties of Several Commercially Available Carbon Blacks for Fuel Cells

Carbon Blacks	Supplier	Synthesis Route	Surface Area (m^2/g)	Particle Size (nm)
Vulcan XC-72	Cabot	Furnace Black	250–260	20–50 [12]
Black Pearl 2000	Cabot	Furnace Black	1475–1500	15 [8,13]
Denka Black	Denka	Acetylene Black	58–65	40 [8,13]
Shawinigan Black	Chevron	Acetylene Black	70–90	40–50 [8,14]
Conductex 975 Ultra	Columbian	Furnace Black	250	24 [13]
3250/3750/3950	Mitsubishi	–	240/800/1500	28/28/16 [15]
Ketjen EC-300 J	Akzo Nobel	Furnace Black	800 [8]	30 [16]
Ketjen EC-600 JD	Akzo Nobel	Furnace Black	1270	30 [17]*

* *Ketjen EC-600 JD produced by Cabot Corp has an average particle size of 34 nm [18]*

Table 18.1 summarizes different types of commercially available carbon blacks with basic physical properties. The most widely used carbon black product in research and commercial fuel cells (up to 80% in total) is the Vulcan XC-72 carbon black produced by a furnace method using oil fracture as a raw material [8]. Besides Vulcan XC-72, other carbon blacks for fuel cells have been commercially available, such as Black Pearl 2000, Ketjen EC series, and Denka Black.

In spite of their high popularity and moderate performance in fuel cells, carbon blacks suffer from a poorly defined porous structure. In other words, their particle sizes and pore widths are distributed in a wide range. Taking Vulcan XC-72 as an example, only half of the total pore volume belongs to macropores or mesopores. The rest are micropores of mostly less than 1 nm [20] as shown in Fig. 18.2, which have been mutually believed to contribute little to the catalytic performance because of their inaccessibility to the fuels [21]. Apart from these drawbacks, the amorphous nature of carbon blacks makes them vulnerable to the electrochemical corrosion under the fuel cell operation conditions, especially in the presence of highly active metallic catalyst particles [22,23]. The working conditions for carbon supports in fuel cells, even in the low-temperature types, are extremely harsh, especially for the

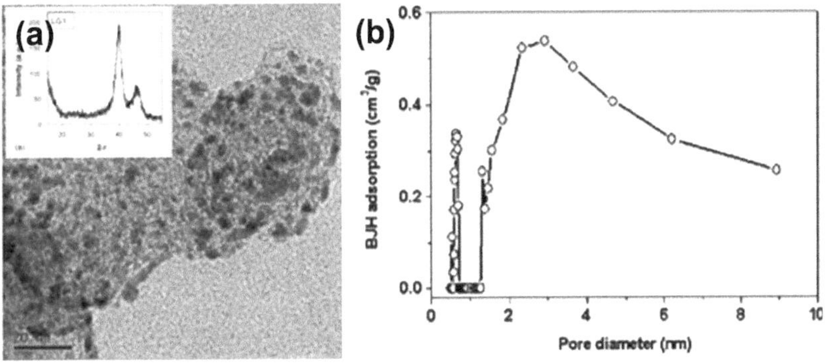

FIGURE 18.2 Transmission electron microscopy (TEM) image of metallic Pt particles loaded on Vulcan XC-72 (a) and corresponding X-ray diffraction spectra (inset of (a)) [19]. (b) Pore-size distribution of Vulcan XC-72 determined by the Barrett–Joyner–Halenda (BJH) method [20].

cathode where oxygen is reduced. In most cases, the electrochemical corrosion of carbon catalyst support can be referred to as carbon oxidation or combustion. When this process begins, the support loses surface area, leading to a catalyst detachment or agglomeration and the performance of a fuel cell is thus significantly compromised. Such electrode (carbon support) corrosion effect has been attributed to several factors as described below.

The presence of high Pt loading as well as small Pt particle size (3–8 nm) on carbon support not only enhance the efficiency of favorable redox electrode reactions, but also facilitate the carbon electrooxidation [24,25]. Pt-catalyzed carbon combustion has been observed at a temperature as low as 125 °C in dry air, and the oxidation was further enhanced with increasing temperature. On the other hand, pristine carbon black with no Pt loading is stable up to 400 °C in air [26]. Generally, the higher the activity of catalyst, the more severe the carbon combustion comes along, as shown in Fig. 18.3.

Another factor causing corrosion of the carbon black support is the low potential for carbon oxidation. It has been confirmed that carbon black can be converted to carbon monoxide at a potential higher than 0.2 V versus reversible hydrogen electrode (RHE) from a thermodynamic point of view [27]; that is to say, whenever a potential higher than that is applied, carbon begins to "burn off". Unfortunately, increasing of an output voltage of fuel cell is the major means to improve its power performance [28], which inevitably exposes electrodes (carbon supports) to a much higher potential than 0.2 V versus RHE [8].

For most fuel cells, a certain amount of water in both electrodes, in either liquid or vapor phase, is necessary for electron and proton transfer, especially in low-temperature PEMFCs. However, the presence of water has serious detrimental effect on carbon black. In PEMFCs where water is present as liquid due to the low operation temperature, carbon is likely to be

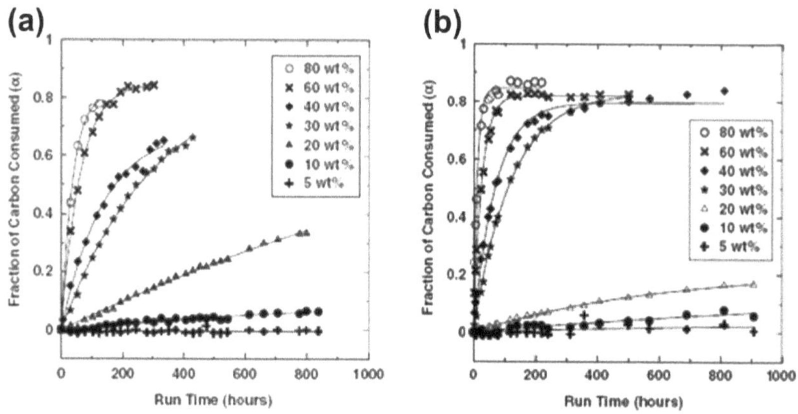

FIGURE 18.3 Isothermal weight loss of carbon blacks with different Pt loading at 173 °C (a) and 195 °C (b), the symbols are indentified in the legend in both figures [26].

directly oxidized into carbon dioxide [29], while carbon monoxide is formed by the oxidation of carbon black with water vapor at higher temperature in PAFCs [30].

In summary, the electrochemical corrosion of carbon blacks accelerates with decreasing Pt particles, increasing Pt loadings, operation temperature, cathode potential, and amorphous nature of carbon blacks. Hence, the search for a more durable and stable substitute for carbon blacks has been of current research interest.

18.2.2. Graphitized Carbons and Graphite

The issues of corrosion propensities of carbon blacks can be overcome by selection of carbon with high graphitic degree. It has been well known that highly crystallized carbon or, in other words, carbon with high graphitic degree has higher resistance against chemical and electrochemical oxidation [24]. A representative carbon with a high degree of crystallinity is graphite. Graphite can be obtained through a high-temperature treatment of carbon, the so-called graphitization of amorphous carbon at temperatures greater than 2000 °C in an inert atmosphere or a high vacuum. Several commercially available carbon blacks have been graphitized, including Vulcan XC-72 [31] and other carbon blacks [32]; the increment in crystallinity was clearly demonstrated by Raman spectroscopy by an order of magnitude decrease of I_D/I_G ratio as shown in Fig. 18.4a.

During the graphitization, carbon atoms rearrange in order to reduce the system energy, which results in higher crystallinity and order. However, these changes cause some significant structural changes such as shrinkage of pores, micropores closing, and decrease of surface areas. Commonly, the loss of

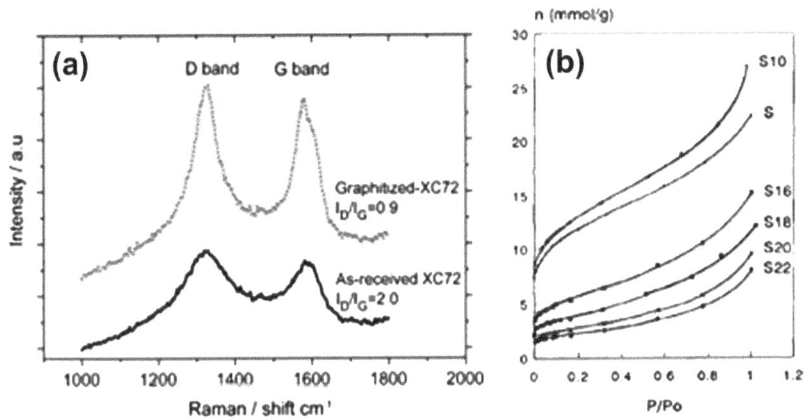

FIGURE 18.4 Feature changes of carbon black after graphitization. (a) Decrease of I_D/I_G of graphitized Vulcan XC-72 treated at 2200 °C for 1 h in Ar [31]. (b) Changes of N_2 adsorption properties of carbon blacks treated at different temperatures (S10, 16, 18, 20, and 22 refer to 1000, 1600, 1800, 2000, 2200 °C, respectively in He for 1 h) compared with the original sample (S) [32].

surface area can reach up to 60% for commercial carbon blacks. As mentioned previously, the porosity of carbon support is one of the crucial characteristics for high and uniform catalyst loading and for achieving sufficient flow of fuel through the electrode.

The issue of maintaining high surface area while improving the graphitic degree has been addressed by two ways. One is the direct ball milling of graphite into fine particles to gain high surface area, thus avoiding the high-temperature treatment and shrinkage of the pores. Figure 18.5a shows SEM image of natural flake graphite ball-milled for 50 h. The BET surface area increased from 7 m^2/g to 580 m^2/g upon treatment [33].

The increase of surface area with the ball-milling time is accompanied with ineligible disturbance to the crystallinity of graphite as depicted in Fig. 18.5b.

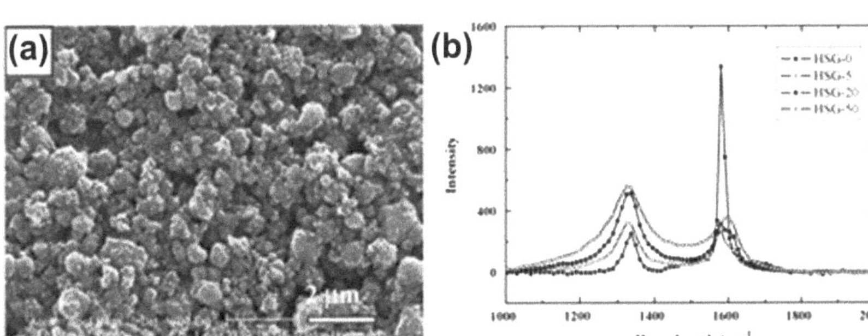

FIGURE 18.5 Morphology of 50-h ball-milled natural graphite (a), and (b) Raman spectra of graphite ball-milled for different periods of time (HSG-0–HSG-50 represents samples grinded for 0–50 h) [33].

The 50-h milled graphite, although having a surface area only twice than that of Vulcan XC-75 carbon black XC-72 of ~250 m^2/g, possesses nearly 200 times higher I_D/I_G than original carbon black sample [33]. In addition, the particle size and pore-size distribution were difficult to control during the ball milling. The process is also time consuming and requires considerable amount of energy for the long-term milling.

Another solution for increasing the surface area loss during the high-temperature treatment is to enlarge the pore size. By this, the pores, although shrunk during the heat treatment, do not disappear, and the surface area thus remains the same [34,35]. This will be discussed later in the following section of this chapter.

18.2.3. Surface-modified Carbons

The durability of a fuel cell largely depends on (i) the dispersion and fineness of metal particles and (ii) the stability of metal catalyst loaded on the support against detachment, agglomeration, and growth. The former determines the total amount of initial active sites on metal catalyst surface, while the latter greatly affects the speed of active site loss during the fuel cell operation. Although the decrease of particle size may seem to be the effective way for increasing the rate of catalytic process, it also has some disadvantages according to the thermodynamic principles in this case, which is to say, the finer the catalyst particles are, the easier they agglomerate to lower the system entropy. Therefore, good anchoring of metal particles on catalyst support is critical for both better dispersion and longer durability. For this purpose, carbon supports are usually functionalized to increase their affinity with the catalyst metal particles.

Traditionally, carbon can be functionalized through oxidation [36] and thermal treatment [37,38]. For the oxidation treatment, carbon is treated by different oxidants such as O_2, O_3, H_2O_2, HNO_3, H_2SO_4, and $KMnO_4$. As a result, different oxygen functional groups can be grafted on the carbon surface, including carbonyl (C=O), phenol (C–OH), carboxyl (C–COOH), etc. (Fig. 18.6).

Oxidation enhances the dispersion of metal particles and overall performance of the catalyst in two aspects [39]: (1) Oxygen groups, especially the strong ones such as carboxyl groups, can change the carbon surface from hydrophobic into hydrophilic, thus facilitating better impregnation of metal precursor solution, and (2) the relatively weak functional groups, e.g. C–O group, can increase the interaction between the metal precursor and carbon surface, leading to higher sintering resistance as shown in Fig. 18.7.

However, there are still some reports showing different or even contradictory results. Some research shows that during the reduction of metal precursors at high temperatures following the impregnation process, oxygen-containing

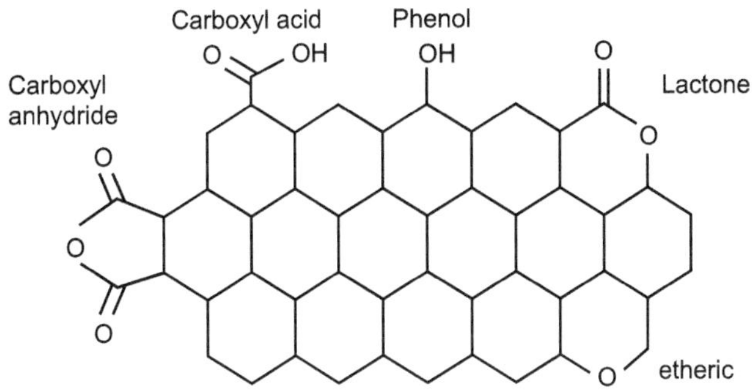

FIGURE 18.6 Different functional groups containing oxygen grafted on carbon surface by oxidative treatment [37].

functional groups tend to decompose, resulting in the decrease of Pt anchoring sites [40]. Others attribute the worsened Pt dispersion to the diminishment of basic sites, which are believed favorable for adsorption of $[PtCL_6]^{2-}$ on carbon during the acid treatment [41,42].

Heat treatment at high temperature is another method that can alter the surface properties of carbon [32], besides increasing the crystallinity as aforementioned. The enhanced properties are attributed to the formation of basic carbon surface during the treatment. It should be noted that such basicity is not the result of incorporation of basic groups, but the interaction between π-bonds on carbon and water. Such basicity is favorable for metal precursor anchoring.

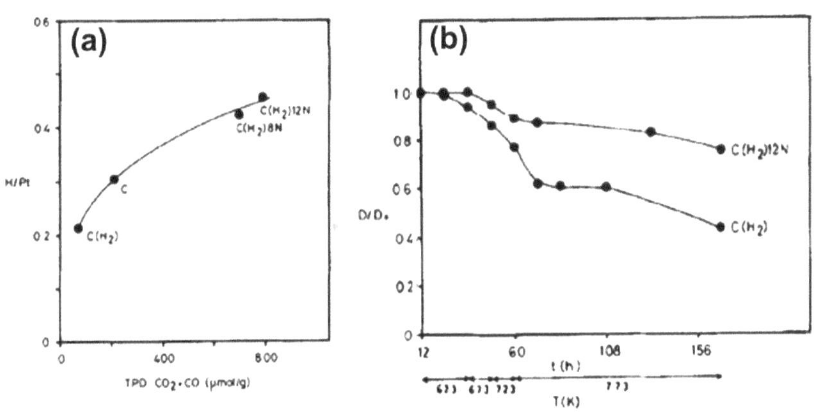

FIGURE 18.7 Relation between metal particle properties and amount of functional groups. (a) The Pt dispersion and total amount of oxygen-containing groups, denoted as CO for weak ones and CO_2 for strong ones from TPD analysis; (b) antisintering properties in accordance with the amount of functional groups [39].

TABLE 18.2 Novel Carbon Surface Treatment Methods for Applications in Fuel Cells [37]

Surface Treatment	Effect on Catalyst Performance
Chemical treatment with Sn–Pd chloride solution [43]	Pt dispersion and electrocatalyst activity is increased
C_2F_6 radio frequency plasma treatment [44]	Enhanced catalytic activity of Pt/C for ORR by CF_3 groups
Chemical treatment in basic (NaOH) and neutral (C_6H_6) media [45]	Enhanced activity for oxygen reduction reaction
Intercalation of $CuCl_2$ and $PdCl_2$ [46]	Enhanced activity for oxygen reduction reaction

Recently, a series of novel surface treatments of carbons for fuel cells have been reported, which are summarized by Yu and Ye [37] in Table 18.2.

18.2.4. Catalyst-Loading Techniques

Two methods are widely used to load Pt on carbon support: an impregnation method and a colloidal method.

For the impregnation method, the metal precursor is first loaded on carbon by simple impregnation of the carbon with the metal precursor solution followed by solvent removal. Then the metal precursor is usually reduced or decomposed at elevated temperatures or by reductants such as hydrazine and $NaBH_4$ to form the metal particles on carbon support. This method is particularly suitable to load the catalyst particles within the pores. A modified impregnation method, called incipient-wetness impregnation, has been developed to specifically target synthesis of catalyst particles inside the carbon support pores. The method utilizes the same amount of precursor solution as is the carbon pore volume, thus avoiding the catalyst loading outside the pores.

In the case of the colloidal method, metal particles are first prepared to form a colloid and then the carbon support is mixed with this colloid to adsorb the catalyst on its surface. This method can provide catalyst particles with uniform and small size, especially when the particles in colloid are stabilized by proper surfactants.

18.3. NOVEL CARBON MATERIALS: MESOPOROUS CARBON AND CARBON NANOMATERIALS

As described in the previous section, conventional carbons, although cheap and widely available, suffer from several drawbacks such as wide pore-size distribution and poor graphitic degree. Hence, alternative carbon substitutes have been developed and tested as a catalyst support for fuel cells. In this section, novel carbon materials, including mesoporous carbons and carbon nanomaterials, will be discussed.

18.3.1. Mesoporous Carbons

It has been generally agreed that ultramicropores (<1.0 nm) of carbon support are too narrow for sufficient fuel transport during the fuel cell operation. In the case of commercial carbon black, micropores contribute to more than half of the total BET surface area and such a surface of the catalyst support is thus difficult to be utilized. Hence, it is necessary to enlarge the pore size and narrow down the pore-size distribution of the carbon support for a more facile fuel supply. For this purpose, mesoporous carbons have been developed as the catalyst support for fuel cells.

Mesoporous carbons are commonly divided into two groups according to their structures: ordered mesoporous carbons (OMCs), with highly ordered pore structure and uniform pore size, nonordered mesoporous carbons with irregular pores.

The success in synthesis of mesoporous silica greatly facilitated the development of OMCs. The first successful synthesis of OMC is the carbon replica of a mesoporous silica molecular sieve MCM-48, denoted as CMK-1 (Fig. 18.8a) [47]. However, the most popular OMC for fuel cells is the replication of SBA-15, denoted as CMK-3 (Fig. 18.8d), because of its more favorable large pores. The synthesis involves sucrose infiltration using a mesoporous silica as a hard template, carbonization in inert atmosphere, followed by silica removal [48,49]. When SBA-15 is used as a template, synthesized OMC is built up with hexagonally ordered arrayed carbon rods (Fig. 18.8c) which are cross-linked by carbon nanorods (the replica of the micropores in the walls of SBA-15, Fig. 18.8e). Other forms of OMCs have also been successfully fabricated using mesoporous silica as a hard template with both 2-dimensional [49,50] and 3-dimensional ordered pore structures [51,52]. Organic–organic nanocomposite was also used to obtain OMC with similar structure as CMK-3, which avoided the use of SBA-15 [53]. Ia3d structured OMC (denoted as C-FDU-14) was also obtained from this polymerization–carbonization route [54].

Mesoporous carbon with nonordered pores could also be easily prepared. The wide availability of commercial silica colloids with different particle diameters and different types of carbon precursors makes it facile to obtain

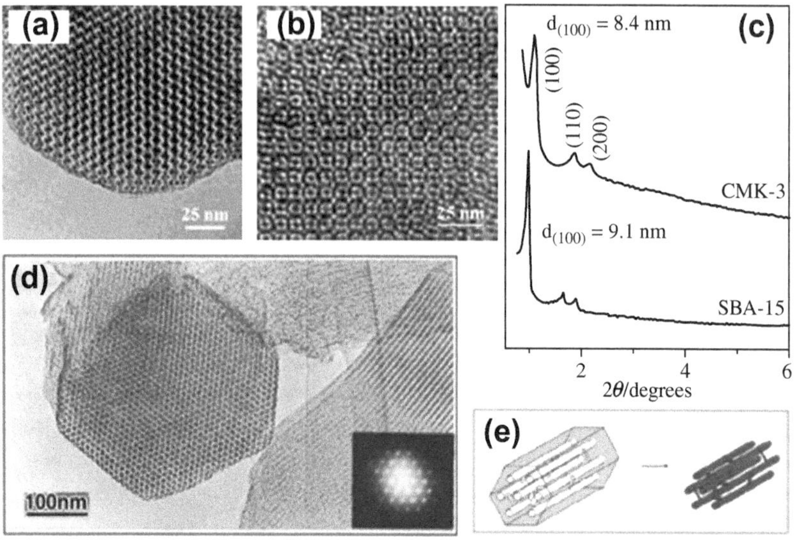

FIGURE 18.8 Ordered mesoporous carbons. (a and b) TEM images of CMK-1 and CMK-2 from 111 direction. [49] (c, d and e) Small-angle X-ray diffraction patterns of CMK-3 and parent template SBA-15, TEM image of CMK-3 [48] and schematic formation of CMK-3 from SBA-15 [49].

mesoporous carbon with a wide range of pore structures [55–57]. Another type of mesoporous carbon is carbon aerogels from carbonization of organic aerogels prepared from sol–gel polycondensation of organic monomers [58]. Such carbon aerogels vary from both the microscopic and the macroscopic level. From the macroscopic point of view, such materials can be prepared in forms of monoliths, beads, powders, and thin films.

Pt was first loaded on hexagonal CMK-5 mesoporous carbon, the tubular-like carbon arrays [49]. Narror Pt particles (size <10 nm) were obtained on this mesoporous carbon compared with much larger Pt particles up to 30 nm on carbon blacks. The electrocatalytic mass activities of Pt/OMC with different Pt loading from 20 to 50 wt % were all higher than the Pt/carbon black with similar Pt loading. CMK-3 is another widely used mesoporous carbon support for catalyst loading [59]. Pt could be loaded through a simple impregnation method, or a two-step process. In the later route, Pt precursor was firstly coated on the inner walls of SBA-15 mesoporous silica and then carbon was formed inside the pores of silica template. Nanosized Pt particles were formed through both routes, but smaller particles were observed on the Pt/OMC prepared by the later way (1–5 nm compared with 8–10 nm). The finer Pt probably resulted from the hindered aggregation of metal precursors during the polymerization/carbonization of carbon precursors. In addition, this strategy produced more exposed facets (100), which not only facilitated the oxygen reduction, but also enhanced the methanol tolerance. Following that, various metal particles have been deposited on OMCs as well as other types of disordered mesoporous carbons, including Pt or Pt/X (X stands for other metal), etc. [60,61]. For these metal particles, the size was well controlled between 3 and 5 nm and they were also highly dispersed.

Generally, catalysts supported on mesoporous carbons show higher catalytic activities toward both fuel oxidation and oxygen reduction, compared with the catalyst loaded on conventional carbon blacks. The enhancement in activity is attributed to two aspects: (1) the high surface area of mesoporous carbons and the small metal particle sizes and (2) the uniform pores and carbon networks. The former can provide enough active sites for the electrode reaction, while the latter makes the mass transfer easier inside the channels. Such features are favored for good metal dispersion and fuel/oxidant transfer.

However, some studies also indicated a declined activity on the catalysts loaded on OMCs [60]. For instance, Pt–Ru/CMK-3 prepared by colloidal method with the metal particles small (0.8–2.6 nm) and well dispersed on the carbon support showed inferior performance when compared with commercially available 20 wt. % Pt–Ru/XC-72. The authors ascribed this to the less metal loading. Interestingly, in the same study, Pt with larger particle size (~15 nm) showed even better performance compared with commercial E-TEK Pt/C catalyst. This clearly indicates the excellent properties of mesoporous carbons, but the intrinsic kinetics and mechanism of the activity enhancement are still not clear and further studies are necessary.

18.3.2. Carbon Nanomaterials

Carbon nanomaterials are a new class of carbons, nanoscaled at least in one dimension, including one-dimensional carbon nanotubes (CNTs) and carbon nanofibers (CNFs), two-dimensional carbon atom layers (graphene), and three-dimensional carbon nanohorns (CNHs), and carbon nanocoils (CNCs) as schematically depicted in Fig. 18.9.

The unique nanodimensions give these carbons superior properties to the conventional "macroscopic" carbons, and such materials have been used as novel catalyst supports for fuel cells as described in the next sections.

18.3.2.1. Carbon Nanotubes

Since their discovery in 1991, carbon nanotubes (CNTs) have drawn much attention for numerous applications including the fuel cell. CNTs can be divided into two categories: single-walled carbon nanotubes (SWNTs), composed of a single rolled graphene sheet, and multiwalled carbon nanotubes (MWNTs), with several graphene sheets rolled around the central axis. The diameters of CNTs range from several nanometers for single- and double-walled

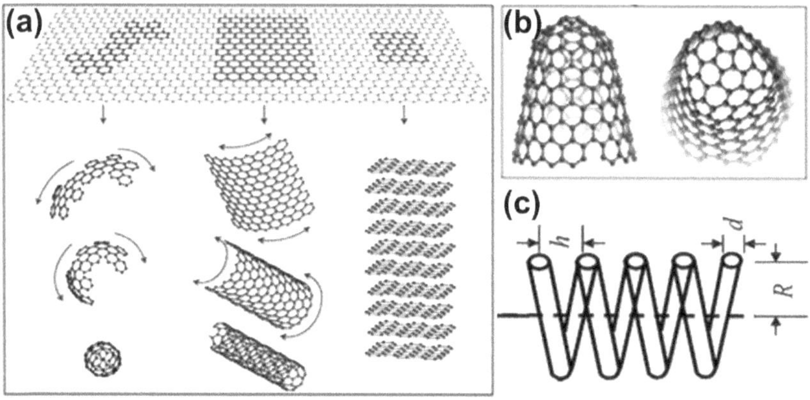

FIGURE 18.9 Schematic formation of different dimensional carbon materials: (a) graphene (2D), carbon nanotubes (1D) and bulkyball (0D) [62], (b) carbon nanohorns [63], and (c) carbon nanocoil [64].

CNTs to several hundred nanometers for MWNTs, while their length can be up to several microns [63].

The excellent properties of CNTs come from their unique structure. High crystallinity of CNTs is the feature important for applications in fuel cells since it provides: (1) high conductivity (up to 10^4 S/m [65]) compared with carbon blacks (4 S/m [66]) and (2) fewer surface defects and higher resistance against electrochemical corrosion [67]. At the same time, however, the defect-free surface of CNTs makes it difficult to deposit catalyst particles, and finding a way how to disperse and attach metal particles on CNT surface is more important and difficult than on other carbon materials. Surface modification is one of the possible ways. The most frequently adopted method for CNT surface modification is oxidization leading to introduction of oxygen surface functional groups to the CNTs external walls. The same approach is used in modification of other carbon materials as described previously in section 2.3. The difference between CNTs and other carbons is that extreme oxidation conditions are required due to high chemical resistance of CNT walls. Normally, CNTs are refluxed at a high temperature between 100 and 140 °C with concentrated sulfuric acid, nitric acid, or their mixture [11]. This treatment results in the introduction of several different functional groups with a ratio of 4(–OH):2(–COOH):1(–CO) [68]. More gentle oxidization conditions, such as diluted HNO_3, lead to the presence of carboxylic groups at the CNT surface [69]. The catalyst loaded on functionalized CNTs generally shows higher dispersion even with Pt loading up to 30 wt. %, lower particle size (3–5 nm), and higher electrode reaction catalytic activity than metal loaded on pristine CNTs [70,71].

To obtain even higher Pt loadings, smaller particle size, and better dispersion on CNTs, another method has been developed. This involves oxidization of CNTs followed by sensitization–activation process by $SnCl_2$ and $PdCl_2$ and

Pt loading [72]. CNTs were first oxidized by conventional reflux in HNO_3 at 140 °C, mixture of HNO_3/H_2SO_4 at 140 °C, or $K_2Cr_2O_7/H_2SO_4$ at 60 °C. Then the oxidized CNTs were sensitized in 0.1 M $SnCl_2$/0.1 M HCl solution for 1 h. The sensitized CNTs were activated in a $PdCl_2$/HCl solution, forming Pd–Sn nuclei. Pt was then deposited on these nuclei from $PtCl_2$ precursor. The catalyst on thus-prepared CNTs shows better activity for oxygen reduction than its counterparts loaded on pristine CNTs or oxidized CNTs.

CNTs can be also functionalized simultaneously with Pt deposition. Glacial acetic acid was used for decorating CNTs with monodispersed Pt nanoparticles in high loadings. The particle size varied between 2 and 4 nm depending on different Pt precursor concentrations [73]. Well-dispersed Pt particles were observed with catalyst loading up to 0.42 mg/cm^2. Glacial acetic acid played dual role in the process: (1) acted as a reducing agent for the Pt precursor and (2) created additional surface functional groups on the CNTs in combination with the acid pretreatment. Thus, prepared Pt–CNTs showed significant improvement when compared with commercial E-Tek 30 wt% carbon black-supported Pt, both in activity per specific surface area and per Pt mass.

Pt can also be electrochemically deposited on CNTs in the electrolyte-containing Pt precursor through a static potential (−0.25 V vs. SCE) or a pulsed potentiostatic method (potential stepped from 700 to 300 mV vs. SCE) in a traditional three-electrode cell [74,75]. Large Pt particles or their aggregations (60–80 nm) were obtained using nonfunctionalized CNTs as support. The particle size was significantly reduced after decorating the CNT surface with 4-aminobenzene monolayer. The thus-produced material possessed higher performance than Pt loaded on graphite. Functionalization of CNTs with 4-aminobenzene groups before electrical deposition provided better particle dispersion as well as stronger Pt–CNT binding.

Ultrasonic technique has also been employed to replace conventional mechanical stirring, and this process is named as sonochemical process. In this synthesis, CNTs were ultrasonically treated in a mixture of HNO_3 and H_2SO_4 at 60 °C, instead of conventional mechanical stirring, followed by Pt reduction in ethylene glycol–water solution [76]. The catalyst prepared in this way showed excellent dispersion on CNT surface (Fig. 18.10), and this is attributed to the more surface functional groups. The electrochemical adsorption and desorption of hydrogen on such Pt/CNT were two times as high as those on commercial Pt/carbon black as a result of unique structure of CNTs and the interactions between the metal particles and the CNT surface. The enhanced electrochemical activity on the Pt/CNT prepared in this way implied its potential application for PEMFCs.

Pt could also be prepared in vapor phase via sputtering at 10 mA for 30 s on CNTs, which were grown on carbon cloths by a CVD method [77]. Conventional liquid-phase Pt deposition was also carried out as comparison. The sputtering fabricated Pt showed smaller particle size (2 nm) compared with the

FIGURE 18.10 Catalyst with different Pt loadings (10, 20, 30 wt % from left to right) supported on sonochemically treated CNTs [76].

ones obtained in liquid phase (2–5 nm). This method also provides a simplified route for preparing the membrane electrode assembly for fuel cells.

18.3.2.2. Carbon Nanofibers and Activated Carbon Fibers

Carbon nanofibers (CNFs) are fibrous carbon nanomaterials somewhat similar to CNTs. They are composed of stacked carbon atom layers, which are normally in different direction with the fiber axis, while CNTs have coaxially distributed tubular graphite layers. CNFs can be fabricated through decomposition of carbon-containing precursors in the presence of transition metal catalysts [78,79]. CNFs can be divided into three types according to the graphite stack orientation [80]: platelet, ribbon, and herringbone structure as depicted in Fig. 18.11.

As shown in Fig. 18.11, the edges of graphite planes are exposed on the CNF surface, which potentially provides active sites for catalyst particle anchoring.

All three types of nanofibers have been studied as catalyst supports for fuel cells. Pt particles with the loading of 5 wt. % have been deposited on "platelet" and "ribbon" type of CNFs and they exhibited similar activities as the 25 wt. % Pt/Vulcan XC-72 [80]. However higher CO poisoning resistance was observed on these Pt/CNFs catalysts. Such performance enhancement was attributed to the special crystallographic orientation of Pt particles dispersed on the CNF surface. Another Pt–Ru bimetallic catalyst has been prepared on the surface of herringbone-structured CNFs [81,82]. By comparing with the same catalyst loaded on SWNTs and MWNTs, Pt–Ru/CNFs showed the highest performance in a DMFC of 0.23 A/cm^2 at 0.4 V. This power output value is 64% higher than that of unsupported metal particles. Again, the surface geometry played an important role by possibly modifying the crystal line structure of metal particles.

Another fibrous carbon material is porous carbon fiber, normally termed as activated carbon fibers (ACFs). ACFs can be prepared by activation of carbon fibers obtained from raw fibers (cellulose or viscose based raw fiber) at the

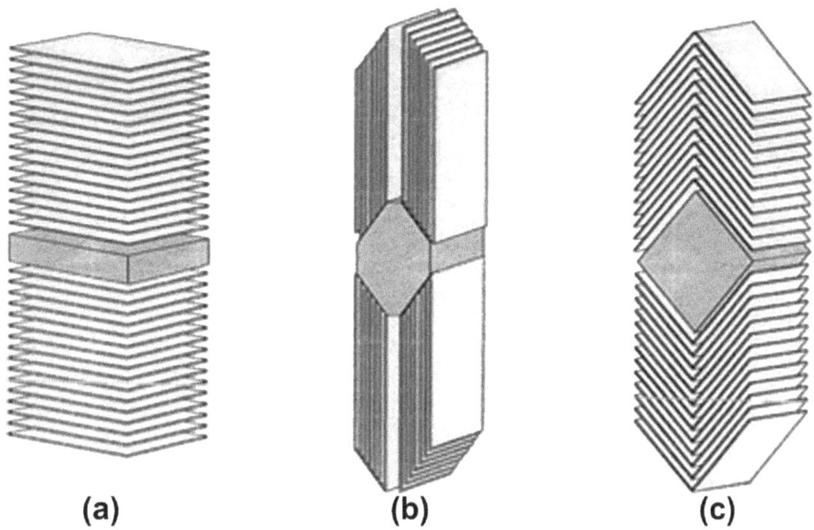

FIGURE 18.11 Different types of carbon nanofibers sorted by the orientation of graphite stacks. (a) Platelet; (b) ribbon; (c) herringbone [80].

carbonization process [83,84]. This material has the potential for catalyst support due to its unique structural features: (1) high specific surface area (>1000 m^2/g), (2) rich surface in oxygen functional groups, (3) relatively more mesopores compared with carbon blacks [84], and (4) reduction propensities for reduction of Pt^{4+} into Pt^{2+} [85] or Pd^{2+} into Pd^0 [86].

Pt has been loaded on ACFs by refluxing with chloroplatinic acid in ethylene glycol in the presence of NaOH [84]. Due to the abundance of surface functional groups, excellent Pt dispersion on ACFs was obtained. Electrochemical oxidation activity of methanol on Pt/AFC showed 2.4 times improvement compared with Pt/Vulcan XC-72. Remarkably, the Pt particles on AFCs maintained their structures after long-term test (1800 cycles by cyclic voltammetry), while those on Vulcan XC-72 suffered both particle growth and aggregation.

18.3.2.3. Graphene, Carbon Nanohorn, and Carbon Nanocoils

Graphene is a novel material with a single layer of hexagonally close-packed carbon atoms. This unique two-dimensional nanostructure gives graphene excellent properties such as high electrical and thermal conductivity, mechanical strength, and flexibility. These characteristics make it interesting candidate for fuel cell applications.

Pt has been deposited on graphenes via various methods. Using a colloidal Pt loading method described previously in this chapter, Pt^{4+} was reduced to metallic nanoparticles and at the same time oxidized graphene was reduced to graphene [87]. This route produced Pt/graphene electrode with the catalyst loading of 45 wt %. The material was tested for methanol oxidation and showed twice the peak current and better durability than commercial Pt/carbon black. Similarly prepared Pt/graphene showed 80% electrochemically active surface area enhancement after further reduction using hydrazine. Single hydrogen fuel cell test showed that such electrode had higher power output than unsupported Pt (161 mW/cm^2 vs. 95 mW/cm^2).

Similar to CNTs, graphene can also be functionalized prior to Pt deposition for better metal dispersion. Functionalized single but wrinkled graphene sheet was prepared by a rapid thermal expansion treatment of oxidized graphite/graphene [88]. The purpose of this thermal treatment was to achieve higher conductivity as well as exfoliation of graphene nanosheets [88,89]. Pt was loaded on as-prepared graphene via impregnation with a H_2PtCl_6 solution followed by reduction in hydrogen at 300 °C. Such electrode material showed better oxygen reduction catalytic performance than Pt/Vulcan XC-72 and excellent long-term durability. The enhanced performance was attributed to the small and stable Pt particles anchored on functionalized graphene. Poly(diallyldimethylammonium chloride) (PDDA) has also been employed to modify the surface of graphene [90]. PDDA is long-chain positively charged polyelectrolyte and can irreversibly adhere on graphene surface through π–π interactions. This modification can significantly improve the metal particle dispersion and decrease the particle size by a stronger interaction between the graphene surface and $PtCl_6^{2-}$, which has been previously confirmed for CNTs [91]. The oxygen reduction property of Pt on PDDA functionalized graphene was similar to Pt on CNTs or commercial Pt/carbon black, but the Pt/graphene exhibited 2–3 times higher durability compared with those two electrodes.

Pt clusters with different thicknesses were prepared on graphene through magnetron sputtering [92]. Thus synthesized Pt/graphene electrode showed the highest electrocatalytic activity toward methanol oxidation when the metal loading was the lowest. The improved property compared with large Pt clusters was attributed to a network of highly intercoupled Pt nanocrystals of a monolayer thickness (2–4 nm) combined with the high conductivity of graphene nanoflakes. Remarkably, the graphene in this catalyst was produced through microwave plasma chemical vapor deposition on silicon wafers, which has the potential for fabrication of microdirect methanol fuel cell by using micro-electromechanical systems technology, as proposed by the authors.

Apart from Pt, other metals or alloys have also been loaded on graphene. Pd nanoparticles were uniformly dispersed on graphene even at the loadings as high as 64 wt %. [93]. Figure 18.12 shows TEM images of these catalysts. The performance test analysis revealed that Pd/graphene showed higher oxygen reduction activity per mass or per surface area, compared with Pt/graphene, especially at high metal loadings.

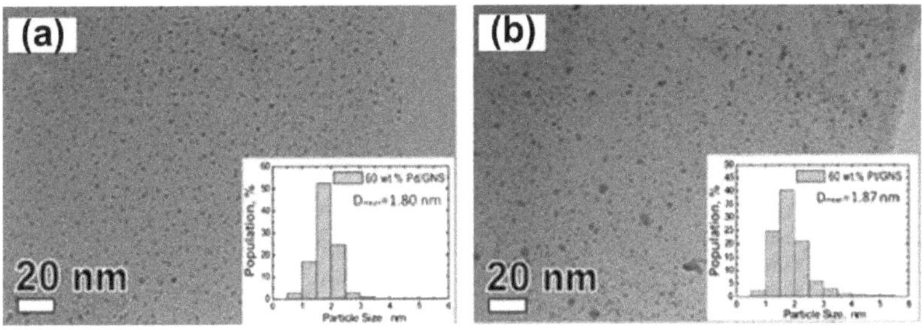

FIGURE 18.12 TEM images of Pd (a) and Pt (b) nanoparticles on graphene [93].

Pt–Ru was deposited on graphene through a simple colloidal route and 80 wt. % of metal loading was obtained [94]. Pt–Ru was also prepared on Vulcan XC-72 via a similar way as comparative sample. Tests on methanol oxidation revealed that Pt–Ru on graphene showed better methanol oxidation behavior than its counterpart on Vulcan XC-72. It was also concluded that Ru significantly reduces the CO poisoning effect on Pt, making it a suitable candidate for methanol and ethanol electrooxidations [95].

Pt–Pd core/shell structured metal particles were also deposited on graphene by a one-step microwave heating method [96]. Graphene-supported Pt–Pd showed 80% activity enhancement compared with unsupported Pt–Pd, indicating the better utilization of metal particles by the introduction of graphene.

Carbon nanohorns (CNHs) and carbon nanocoils (CNCs) are three-dimensional carbon nanomaterials with the structure depicted in Fig. 18.9b and c. These materials have also been studied as catalyst support for fuel cells.

CNHs are composed of horn-shaped sheaths of single-wall graphene sheets. A single CNH is 2–3 nm thick, and they usually form an aggregation with a diameter of about 80 nm [97]. According to the structure of their aggregations, CNHs can be divided into two groups: “dahlia-like” or “bud-like” form as shown in Fig. 18.13. These two kinds of CNH aggregations can be prepared by CO_2 laser vaporization of carbon in different atmospheres. “Dahlia-like” CNHs aggregates were produced using Ar as the buffer gas at 760 Torr, while “bud-like” SWNH aggregates were formed in either He or N_2 at similar condition.

Only a few works have been reported on the preparation of metal catalysts on CNHs for the fuel cell applications. One example is Pt/CNHs prepared by one-step synthesis employing the so-called ‘arc in liquid nitrogen’ method using Pt-contained graphite anode [98]. Pt nanoparticles with the diameters less than 5 nm were evenly dispersed on CNHs. The particle size was controlled by changing the concentration of Pt in the graphite anode. A simple colloidal method has also been adopted to load Pt particles on CNHs [99]. Smaller Pt particles were obtained via this method (~2 nm), which is less than half

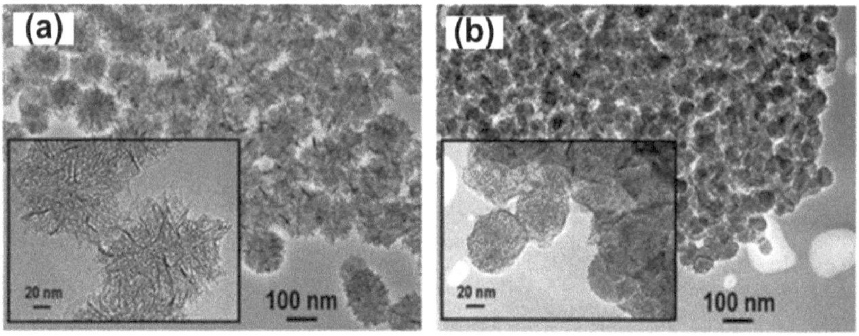

FIGURE 18.13 TEM images of different types of CNH aggregation. (a) "dahlia-like" CNHs; (b) "bud-like" CNHs [97].

compared with Pt on carbon blacks. The current density of the fuel cell using Pt/CNHs was larger than that using Pt/carbon black, which demonstrates the potential application of the CNHs as a catalyst support for fuel cells.

CNCs can be simply considered as helically shaped MWNTs. Pt–Ru nanoparticles have been deposited on this material through the colloidal method and particle size remained as small as 2.3 nm even at 60 wt. % metal loading [100]. Pt–Ru/CNC, Pt–Ru/VulcanXC-72, and commercial E-TEK catalyst were tested in a direct methanol fuel cell and the maximum power density of Pt–Ru/CNC was 146% and 180% higher than that of Pt–Ru/Vulcan XC-72 and E-TEK catalyst, respectively. The power density of Pt–Ru/CNC maintained stable up to 100 h in the test. The superior performance of Pt–Ru/CNC compared with standard catalysts was also obtained in methanol oxidation tests and it was attributed to the combination of good electrical conductivity of CNC that originated in their graphitic structure and a suitable porosity allowing a smooth diffusion of reactants [101].

18.4. HETEROATOM-DOPED CARBONS AND CARBON-BASED MATERIALS

As described in previous sections, noble metals are used as catalyst in fuel cells and in most cases carbons are treated under extreme conditions to create surface functional groups for easier metal particle deposition.

The main drawbacks of such technology are as follows: 1) high price of the noble metal often required in large quantities and 2) carbon surface functionalization always introduces defects to carbon lattice, which on a positive side facilitates better metal deposition, but on a negative side significantly reduce the mechanical and electrical properties of carbons, as well as their electrochemical stability in fuel cell environments. Doping carbons with other atoms has been recently developed as an effective way for producing electrochemically active carbon surface without deterioration of the mechanical and electrical properties of carbon supports [102–105]. The heteroatom active sites not only serve as metal anchoring points, but also show metal-free catalytic activity for electrochemical reduction of oxygen [106,107]. In this section, the preparation method of nitrogen-doped carbons and boron-doped carbons as well as their application in fuel cells will be introduced.

18.4.1. Nitrogen-Doped Carbons

It has been widely accepted that nitrogen-doped carbon shows an *n*-type semiconductive behavior and is expected to have greater electron mobility compared with pristine nitrogen-free carbon [108]. Moreover, nitrogen doping efficiently introduces chemically active sites, promoting metal particle anchoring. In this section, nitrogen-doped carbon nanotubes (NCNTs), carbon nanofibers (NCNFs), graphene, carbon nanofibers (NCNFs), carbon nanocoils (NCNCs), graphite, and mesoporous carbons (NMCs) will be presented and discussed. Nitrogen can be doped into carbon structure during the carbon synthesis or at the post-treatments in the presence of nitrogen sources.

18.4.1.1. NCTNs

The simplest and most economical way to prepare NCNT is an *in situ* introduction of nitrogen into carbon lattice during the chemical vapor deposition (CVD) synthesis of CNTs, which is also the most popular way to fabricate CNTs. Conventional CVD has been employed to prepare NCNTs using transition metal powder (Fe, Co on Al_2O_3) catalyst and pyridine as a nitrogen-containing carbon precursor in nitrogen atmosphere at 550–950 °C [109]. NCNTs in this process grew through the vapor–liquid–solid mechanism and had the highest and lowest nitrogen contents of 8.8 wt. % and 2.7 wt. % when prepared at 550 °C and 850 °C, respectively. Pt particles sized 3–9 nm were supported on such NCNTs (3–5 wt. % nitrogen content) without any further treatment using colloidal method [104]. The electrocatalytic activity of this Pt/NCNT was tested for methanol oxidation, and a 30 mV lower overpotential compared with Pt on pristine CNTs was observed, confirming the enhanced performance of Pt on NCNTs.

Pt–Ru particles (2.5–3.5 nm) have also been decorated on NCNTs using microwave-assisted colloidal method [110]. This bimetallic catalyst showed higher CO tolerance and long-term durability towards methanol oxidation when compared with the above-mentioned Pt/NCNTs and significantly higher electrochemical activity when compared with commercial Pt–Ru/C catalyst. Easy fabrication and excellent performance make these electrodes promising substitutes for conventional Pt/C or Pt–Ru/C catalysts.

Microwave plasma-enhanced CVD has been used to grow vertically aligned NCNTs (VANCNTs, also called NCNT arrays) on Si substrates using CH_4 and

N_2 as precursors and H_2 as an atmosphere [111]. Pt was loaded on such VANCNTs by DC sputtering with an average size of 2 nm, and this integrated composite catalyst comprised Pt, NCNTs and Si showed promising potential in micro-DMFCs. Similarly structured VANCNTs were also prepared via nano-casting technique using porous anodic alumina membrane as a template and polyvinylpyrrolidone as a carbon precursor [102]. Pt was well dispersed onto the inner walls of such NCNTs through impregnation followed by reduction in H_2 at 550 °C. Such catalyst achieved both the higher catalytic activity for methanol oxidation, and better durability compared with the commercial E-TEK catalyst. A large quantity of multilayered N-graphene could also be prepared on a large scale through arc discharge of graphite rods in ammonia [112] or by reducing graphite oxide in ammonia [113].

Nitrogen can also be incorporated in the carbon structure through post-treatments. Nitrogen plasma treatment of conventional CVD prepared CNTs introduced up to almost 4 wt. % of nitrogen into the CNTs [114]. Pt–Ru bimetallic nanoparticles were dispersed on such NCNTs as well as acid-treated CNTs through the impregnation method. The Pt–Ru/NCNT showed significantly higher activity towards methanol oxidation and CO durability, compared with Pt–Ru/CNT of E-TEK Pt–Ru/C. Pyridinic and pyrrolic nitrogen species located at the periphery of the graphene sheets have been confirmed as acting anchoring sites for catalyst. This work also suggests the critical importance of the type of nitrogen functionalities introduced to the carbon surface rather than high concentrations of nitrogen.

Similar treatments have also been applied to carbon nanocoils in order to stabilize colloid-prepared Pt nanoparticles on their surfaces [115]. Single hydrogen PEM fuel cell test showed a maximum power density of 550 mV/cm^2, compared with 490 mW/cm^2 obtained on Pt / NCNCs.

18.4.1.2. NMCs

Ordered mesoporous carbons (OMCs) have also been doped with nitrogen and afterward used as a support for methanol oxidation [116]. In this one-step synthesis, nitrogen-containing species were first introduced into the silica template, and then transferred into carbon structure during the carbonization. Hexachloroplatinic acid was dispersed into the carbon precursor prior to carbonization and thus the Pt was formed simultaneously with carbon. This Pt/N-OMCs with 1.8 wt% of nitrogen in the structure showed superior electro-catalytic activity and the tolerance to methanol crossover compared to Pt/OMC, which was attributed to the dispersion and unique nanostructure of Pt on N-OMCs and the mesoporous structure of N-OMCs.

Apart from Pt-based metallic catalysts, non-noble transition metals loaded on N-doped carbons have also been studied as potential catalysts for fuel cell applications. The development in this field has been review by Shao as summarized in Table 18.3 [117].

TABLE 18.3 Non-noble Metals Loaded on NCs [117]

Methods	Catalysts	Precursors	N Contents
HT	NA	NH_3	NA [118]
Pyrolysis	Fe	Melamine ($C_3H_9N_6$)	<7 [119]
Pyrolysis	Ferrocene ($Fe(C_5H_5)_2$)	Melamine ($C_3H_9N_6$)	10 [120]
AA-CVD	Ferrocene ($Fe(C_5H_5)_2$)	Benzylamine ($C_6H_5CH_2NH_2$)	<5 [121]
AA-CVD	Fe^0 from Fe(III) (acetylacetonate)$_3$	Acetonitrile (CH_3CN), tetrahydrofuran (C_4H_8O)	20 [122]
FC-CVD	Ferrocene ($Fe(C_5H_5)_2$)	Pyridine/xylene/NH_3	10 [123]
MPE-CVD	Fe	CH_4/N_2	15–17 [124]
Pyrolysis	Fe	Dimethylformamide ($HOCN(CH_3)_2$)	2–16 [125]
Pyrolysis	Co	2-Amino-4,6-dichloro-*s*-triazine ($C_3N_3Cl_2NH_2$)	1–2 [126]
HT	No	NH_3	0.7 [127]
HT	Fe	NH_3	2–4 [128]
FC-CVD	Ferrocene ($Fe(C_5H_5)_2$)	C_5H_5N	2.6 [129]
Pyrolysis	Fe	Melamine	4–5 [130]
Pyrolysis	Ni/Co	Acetonitrile	0.5–1.2 [131]
Pyrolysis	Co	Pyridine	2 [132]
Pyrolysis	Ferrocene ($Fe(C_5H_5)_2$)	NH_3	4.8–8.8 [133]
CVD	Ferrocene ($Fe(C_5H_5)_2$)	Pyridine (C_5H_5N)/ pyrimidine ($C_4H_4N_2$)	1–3 [134]
Pyrolysis	NA	FePc/CoPc/NiPc	0.7–7.8 [135]
CVD	NA	Acetonitrile	8 [136]
Pyrolysis	NA	C_3H_6/CH_3CN	3.2–3.5 [137]

HT, heat treatment; AA-CVD, aerosol-assisted CVD; FC-CVD, floating catalyst CVD; MPE-CVD, microwave plasma-enhanced CVD

18.4.1.3. Metal-free Nitrogen-Doped Carbons

In order to explain the enhanced performance of N-doped carbon, quantum mechanics calculations have been employed. The results showed that nitrogen heteroatoms in carbon structures produce a substantially higher positive charge density on the carbon atoms around them to counterbalance the strong electronic affinity of the nitrogen atom [106]. This charge delocalization changes the O_2 adsorption on carbon surface from end-on type to side-on type, which effectively weakens the O–O bonding and oxygen reduction is thus facilitated. This work also revealed that nitrogen-doped carbons can also act as a metal-free catalyst for oxygen reduction.

VANCNTs were grown on silica substrate using nitrogen-containing precursors. These NCNTs show excellent catalytic activity toward oxygen reduction, including much higher electrocatalytic activity, lower overpotential, smaller fuel crossover poisoning effect, and better long-term operation stability than commercially available E-TEK catalysts. Remarkably, these NCNTs had a high onset potential very close to that of Pt/C, showing the easy oxygen reduction on them [106]. NCNTs with the nitrogen content of 8.4 wt. % were also prepared by a floating catalyst CVD method [138]. Thus obtained nitrogen-doped MWNTs had large diameters of 100 nm as shown in Fig. 18.14a and showed higher activity toward ORR than commercial Ag/C catalyst, but still inferior to the Pt/C catalyst.

Metal-free nitrogen-doped graphene (N-graphene) was synthesized by in-situ nitrogen doping via the CVD of methane in ammonia (Fig. 18.14b) [139]. Such material showed a high efficiency for electron ORR pathway, indicating direct reduction of oxygen into water without the hydrogen peroxide intermediate. The N-graphene also showed high tolerance against CO poisoning.

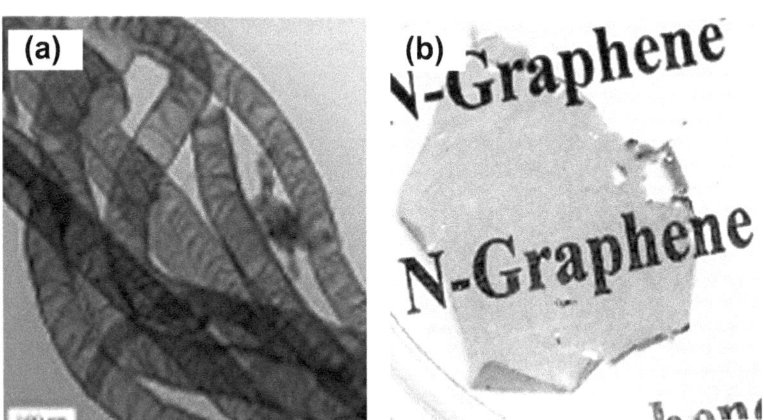

FIGURE 18.14 Nitrogen-doped nanostructures. (a) NCNTs produced by floating catalyst CVD method [138]; (b) optical image of N-graphene [139].

18.4.2. Boron-Doped Carbons (BDCs)

18.4.2.1. Boron-Doped Diamonds (BDDs)

The innovative BDDs have been proposed as promising carbon-based catalyst support for metal nanoparticles. BDDs possess much higher structural stability and electrochemical corrosion resistance in both acidic and alkaline medium compared with its sp^2 carbon counterparts due to their diamond allotropic nature [105,140]. Boron can be doped into diamond lattice either through post-treatment of diamond powder [141] or simultaneously during the deposition of diamond films, thus producing various structures of BDDs [142–145]. The initial research in this area showed the possibility of metal deposition (Pt, Pd, and Hg) on BDDs and after that a series of metals on BDDs have been prepared and tested as fuel cell catalysts.

Pt was deposited using the impregnation method on honeycomb BDDs prepared by high-pressure plasma-assisted CVD on n-Si(111) substrates [145]. As-prepared Pt/BDDs showed high catalytic activity for hydrogen adsorption and oxidation of several alcohols. BDDs were also prepared in a hot filament CVD process, in which a gaseous mixture of carbon precursor (usually methane) and boron source (usually trimethylboron) was thermally decomposed at 2440–2560 °C in H_2 atmosphere. This process produced BDD films [144] and several metals were loaded on such films including Pt, Pt–RuO_2, and Pt–RuO_2–RhO_2 [144,146]. Among all, Pt–RuO_2–RhO_2/BDD showed the lowest overpotential for methanol oxidation as well as higher poisoning resistance during the methanol and the ethanol oxidation reactions.

18.4.2.2. Boron-Doped Carbons (BDCs)

Boron-doped sp^2 carbon for fuel cell application was prepared via a conventional low-temperature CVD method using boron trichloride as a boron source and benzene as a carbon source at 1000 °C, respectively [147]. The Pt nanoparticles showed a slightly stronger interaction with boron-doped carbon than with the boron-free carbon. Nitrogen and boron could be co-doped into carbon by the pyrolysis of polymers containing melamine and BF_3, and these polymers also acted as nitrogen and boron precursors [148]. Such carbon showed an ORR catalytic activity, but inferior to the Pt/C catalyst.

The comprehensive comparison of boron-doped carbons with metal/carbon catalyst cannot be done at this stage since the number of research reports is still rather limited.

18.5. SUMMARY, PERSPECTIVES, AND FURTHER DIRECTIONS

Carbon materials have been extensively studied as catalyst supports or novel metal-free catalysts for fuel cells. However, these materials are still far from ideal as they do not meet all the requirements for fuel cell applications: high surface area for more catalyst loading and better dispersion, suitable pore size for smooth mass transfer, high conductivity for electron pathway, and electrochemical corrosion resistance in fuel cell environments. Although carbon blacks have been widely used in both research and industry as a support for catalyst nanoparticles, their drawbacks are obvious: irregular pore structures, low crystallinity, insufficient anchoring sites, and vulnerability to electrochemical corrosion.

New carbon-based materials have been developed including high-surface-area graphite, mesoporous carbon, carbon nanomaterials, and heteroatom-doped carbons. These novel supports showed significantly better performance in fuel cells in most case studies than the commercial carbon blacks, thanks to their unique structures. Carbon nanomaterials are promising future catalyst supports for fuel cells as they are prepared in a "bottom–up" way; thus, they can theoretically have both high surface area and crystallinity. The disadvantage of these materials is their "perfect" defect-free surface which makes the catalyst loading challenging. One possible way is to chemically introduce surface defects critical for stable attachments of catalyst metal particles. At the same time, however, such defects have a negative effect on their corrosion resistance and electric conductivity.

The higher price of these carbon nanomaterials compared with carbon blacks also hinders their wide application in fuel cells. Therefore, the development of new and economical ways for their synthesis is of high priority.

Metal-free heteroatom-doped carbons for ORR have paved the way for further development in the area of low cost and highly efficient catalyst for fuel cells. Although their performances in fuel cells are still to be evaluated more comprehensively, they are promising for a much better long-term stability and fuel poisoning resistance when compared with Pt-based catalysts.

The latest research has shown similar higher catalytic activity on nitrogen-doped carbon nanotubes than on Pt-based catalyst, indicating a great potential for the next-generation fuel cell catalyst. However, similar results have not been reported for other types of nitrogen-doped carbon nanomaterials. It is therefore essential to make further studies and to enhance the activity of these materials.

REFERENCES

[1] R. Dillon, et al., International activities in DMFC R&D: status of technologies and potential applications, J. Power Sources 127 (1–2) (2004) 112–126.

[2] P. Costamagna, S. Srinivasan, Quantum jumps in the PEMFC science and technology from the 1960s to the year 2000: part I. Fundamental scientific aspects, J. Power Sources 102 (1–2) (2001) 242–252.

[3] A. Kirubakaran, S. Jain, R.K. Nema, A review on fuel cell technologies and power electronic interface, Renew. Sustain. Energy Rev. 13 (9) (2009) 2430–2440.

[4] M. Neergat, D. Leveratto, U. Stimming, Catalysts for direct methanol fuel cells, Fuel Cells 2 (1) (2002) 25–30.

[5] X. Ren, M.S. Wilson, S. Gottesfeld, High performance direct methanol polymer electrolyte fuel cells, J. Electrochem. Soc. 143 (1) (1996) L12–L15.

[6] N.V. Long, M. Ohtaki, M. Nogami, Synthesis and characterization of Pt–Pd alloy and core-shell bimetallic nanoparticles for direct methanol fuel cells (DMFCs): enhanced electro-catalytic properties of well-shaped core-shell morphologies and nanostructures. Int. J. Hydrogen Energy, in press, Corrected Proof.

[7] D. Mirabile Gattia, et al., Study of different nanostructured carbon supports for fuel cell catalysts, J. Power Sources 194 (1) (2009) 243–251.

[8] S. Tang, et al., Review of new carbon materials as catalyst supports in direct alcohol fuel cells, Chin. J. Catal. 31 (1) (2010) 12–17.

[9] D.A. Landsman, F.J. Luczak, Catalyst studies and coating technologies. Handbook of Fuel Cells: John Wiley & Sons, Ltd.

[10] D.A. Landsman, F.J. Luczak, Catalyst Studies and Coating Technologies. Handbook of Fuel Cells, John Wiley & Sons, Ltd, 2010.

[11] E. Antolini, Carbon supports for low-temperature fuel cell catalysts, Appl. Catal. B Environ. 88 (1–2) (2009) 1–24.

[12] D. Santiago, et al., Platinum electrodeposition at high surface area carbon Vulcan-XC-72R material using a rotating disk-slurry electrode technique, J. Electrochem. Soc. 157 (12) (2010) F189–F195.

[13] M. Uchida, et al., Influences of both carbon supports and heat-treatment of supported catalyst on electrochemical oxidation of methanol, J. Electrochem. Soc. 142 (8) (1995) 2572–2576.

[14] E. Antolini, R.R. Passos, E.A. Ticianelli, Effects of the carbon powder characteristics in the cathode gas diffusion layer on the performance of polymer electrolyte fuel cells, J. Power Sources 109 (2) (2002) 477–482.

[15] K. Kanno, et al., Effects of carbon black addition on the carbonization of mesophase pitch, Carbon 32 (5) (1994) 801–807.

[16] G. Wang, et al., Improving the DMFC performance with Ketjen Black EC 300J as the additive in the cathode catalyst layer, J. Power Sources 180 (1) (2008) 176–180.

[17] K. Sakanishi, et al., Preparation of highly dispersed NiMo catalysts supported on hollow spherical carbon black particles, Energy Fuels 9 (6) (1995) 995–998.

[18] J. Yu, Y. Yoshikawa, T. Matsuura, M.N. Islam, M. Hori, Preparing gas-diffusion layers of PEMFCs with a dry deposition technique, Electrochem. Solid State Lett. 8 (3) (2005) A152–A155.

[19] M. Kim, et al., The preparation of Pt/C catalysts using various carbon materials for the cathode of PEMFC, J. Power Sources 163 (1) (2006) 93–97.

[20] V. Raghuveer, A. Manthiram, Mesoporous carbon with larger pore diameter as an electrocatalyst support for methanol oxidation, Electrochem. Solid State Lett. 7 (10) (2004) A336–A339.

[21] D.R. Rolison, Catalytic nanoarchitectures – the importance of nothing and the unimportance of periodicity, Science 299 (5613) (2003) 1698–1701.

[22] J. Willsau, J. Heitbaum, The influence of pt-activation on the corrosion of carbon in gas diffusion electrodes – a DEMS study, J. Electroanal. Chem. 161 (1) (1984) 93–101.

[23] L.M. Roen, C.H. Paik, T.D. Jarvi, Electrocatalytic Corrosion of carbon support in PEMFC cathodes, Electrochem. Solid State Lett. 7 (1) (2004) A19–A22.

[24] E. Passalacqua, P.L. Antonucci, M. Vivaldi, A. Patti, V. Antonucci, N. Giordano, K. Kinoshita, The influence of Pt on the electrooxidation behaviour of carbon in phosphoric acid, Electrochim. Acta 37 (15) (1992) 2725–2730.
[25] D.A. Stevens, et al., Ex situ and in situ stability studies of PEMFC catalysts, J. Electrochem. Soc. 152 (12) (2005) A2309–A2315.
[26] D.A. Stevens, J.R. Dahn, Thermal degradation of the support in carbon-supported platinum electrocatalysts for PEM fuel cells, Carbon 43 (1) (2005) 179–188.
[27] K. Kinoshita, Carbon: electrochemical and physicochemical properties (1988).
[28] K.I. Ota, et al., Improvement of cathode materials for polymer electrolyte fuel cell, J. New Mater. Electrochem. Syst. 8 (1) (2005) 25–35.
[29] J.M. Thomas, Reactivity of carbon: some current problems and trends, Carbon 8 (4) (1970) 413–421.
[30] K. Kinoshita, J. Bett, Electrochemical oxidation of carbon black in concentrated phosphoric acid at 135 °C, Carbon 11 (3) (1973) 237–247.
[31] C.-C. Hung, et al., Corrosion of carbon support for PEM fuel cells by electrochemical quartz crystal microbalance, J. Power Sources 196 (1) (2011) 140–146.
[32] F. Coloma, A. Sepulveda-Escribano, F. Rodriguez-Reinoso, Heat-treated carbon-blacks as supports for platinum catalysts, J. Catal. 154 (2) (1995) 299–305.
[33] H.-Q. Li, et al., A competitive candidate material for aqueous supercapacitors: high surface-area graphite, J. Power Sources 185 (2) (2008) 1557–1562.
[34] S.B. Yoon, et al., Graphitized pitch-based carbons with ordered nanopores synthesized by using colloidal crystals as templates, J. Am. Chem. Soc. 127 (12) (2005) 4188–4189.
[35] Z. Li, et al., High surface area graphitized carbon with uniform mesopores synthesised by a colloidal imprinting method, Chem. Commun. 13 (2002) 1346–1347.
[36] A. Guha, et al., Surface-modified carbons as platinum catalyst support for PEM fuel cells, Carbon 45 (7) (2007) 1506–1517.
[37] X. Yu, S. Ye, Recent advances in activity and durability enhancement of Pt/C catalytic cathode in PEMFC: part I. Physico-chemical and electronic interaction between Pt and carbon support, and activity enhancement of Pt/C catalyst, J. Power Sources 172 (1) (2007) 133–144.
[38] X. Yu, S. Ye, Recent advances in activity and durability enhancement of Pt/C catalytic cathode in PEMFC: Part II: degradation mechanism and durability enhancement of carbon supported platinum catalyst, J. Power Sources 172 (1) (2007) 145–154.
[39] C. Prado-Burguete, A. Linares-Solano, F. Rodríguez-Reinoso, C. Salinas-Martínez de Lecea, The effect of oxygen surface groups of the support on platinum dispersion in Pt/carbon catalysts, J. Catal. 115 (1) (1989) 98–106.
[40] F. Coloma, et al., Preparation of platinum supported on pregraphitized carbon blacks, Langmuir 10 (3) (1994) 750–755.
[41] M.A. Fraga, et al., Properties of carbon-supported platinum catalysts: role of carbon surface sites, J. Catal. 209 (2) (2002) 355–364.
[42] A. Guerrero-Ruiz, P. Badenes, I. Rodriguez-Ramos, Study of some factors affecting the Ru and Pt dispersions over high surface area graphite-supported catalysts, Appl. Catal. A Gen 173 (2) (1998) 313–321.
[43] T.D. Tran, S.H. Langer, Graphite pre-treatment for deposition of platinum catalysts, Electrochim. Acta 38 (11) (1993) 1551–1554.
[44] H. Shioyama, et al., C2F6 plasma treatment of a carbon support for a PEM fuel cell electrocatalyst, J. Power Sources 161 (2) (2006) 836–838.
[45] S. Kim, S.-J. Park, Effects of chemical treatment of carbon supports on electrochemical behaviors for platinum catalysts of fuel cells, J. Power Sources 159 (1) (2006) 42–45.

[46] H. Shioyama, et al., Graphite intercalation compounds as PEMFC electrocatalyst supports, Carbon 43 (11) (2005) 2374–2378.
[47] R. Ryoo, S.H. Joo, S. Jun, Synthesis of highly ordered carbon molecular sieves via template-mediated structural transformation, J. Phys. Chem. B 103 (37) (1999) 7743–7746.
[48] S. Jun, et al., Synthesis of new, nanoporous carbon with hexagonally ordered mesostructure, J. Am. Chem. Soc. 122 (43) (2000) 10712–10713.
[49] R. Ryoo, et al., Ordered mesoporous carbons, Adv. Mater. 13 (9) (2001) 677–681.
[50] S.H. Joo, et al., Ordered nanoporous arrays of carbon supporting high dispersions of platinum nanoparticles, Nature 412 (6843) (2001) 169–172.
[51] J. Lee, et al., Development of a new mesoporous carbon using an HMS aluminosilicate template, Adv. Mater. 12 (5) (2000) 359–362.
[52] F. Kleitz, S. Hei Choi, R. Ryoo, Cubic Ia3d large mesoporous silica: synthesis and replication to platinum nanowires, carbon nanorods and carbon nanotubes, Chem. Commun. (17) (2003) 2136–2137.
[53] S. Tanaka, et al., Synthesis of ordered mesoporous carbons with channel structure from an organic-organic nanocomposite, Chem. Commun.(16) (2005) 2125–2127.
[54] F. Zhang, et al., A facile aqueous route to synthesize highly ordered mesoporous polymers and carbon frameworks with Ia3d bicontinuous cubic structure, J. Am. Chem. Soc. 127 (39) (2005) 13508–13509.
[55] H.I. Lee, et al., Spontaneous phase separation mediated synthesis of 3D mesoporous carbon with controllable cage and window size, Adv. Mater. 23 (20) (2011) 2357–2361.
[56] Z. Li, M. Jaroniec, Synthesis and adsorption properties of colloid-imprinted carbons with surface and volume mesoporosity, Chem. Mater. 15 (6) (2003) 1327–1333.
[57] Z. Li, M. Jaroniec, Colloidal imprinting: a novel approach to the synthesis of mesoporous carbons, J. Am. Chem. Soc. 123 (37) (2001) 9208–9209.
[58] C. Moreno-Castilla, F.J. Maldonado-Hódar, Carbon aerogels for catalysis applications: an overview, Carbon 43 (3) (2005) 455–465.
[59] W.C. Choi, et al., Platinum nanoclusters studded in the microporous nanowalls of ordered mesoporous carbon, Adv. Mater. 17 (4) (2005) 446–451.
[60] J. Ding, et al., Platinum and platinum-ruthenium nanoparticles supported on ordered mesoporous carbon and their electrocatalytic performance for fuel cell reactions, Electrochim. Acta 50 (15) (2005) 3131–3141.
[61] H. Yamada, et al., A highly active Pt catalyst fabricated on 3D porous carbon, J. Power Sources 164 (2) (2007) 538–543.
[62] A.K. Geim, K.S. Novoselov, The rise of graphene, Nat Mater 6 (3) (2007) 183–191.
[63] S. Iijima, Carbon nanotubes: past, present, and future, Physica B Condensed Mat. 323 (1–4) (2002) 1–5.
[64] X. Chen, et al., Mechanics of a carbon nanocoil, Nano Lett. 3 (9) (2003) 1299–1304.
[65] A. Thess, et al., Crystalline ropes of metallic carbon nanotubes, Science 273 (5274) (1996) 483–487.
[66] W. Li, et al., Preparation and characterization of multiwalled carbon nanotube-supported platinum for cathode catalysts of direct methanol fuel cells, J. Phys. Chem. B 107 (26) (2003) 6292–6299.
[67] X. Wang, et al., Durability investigation of carbon nanotube as catalyst support for proton exchange membrane fuel cell, J. Power Sources 158 (1) (2006) 154–159.
[68] H. Hiura, T.W. Ebbesen, K. Tanigaki, Opening and purification of carbon nanotubes in high yields, Adv. Mat. 7 (3) (1995) 275–276.

[69] V. Lordi, N. Yao, J. Wei, Method for supporting platinum on single-walled carbon nanotubes for a selective hydrogenation catalyst, Chem. Mater. 13 (3) (2001) 733–737.
[70] J. Prabhuram, et al., Multiwalled carbon nanotube supported PtRu for the anode of direct methanol fuel cells, J. Phys. Chem. B 110 (11) (2006) 5245–5252.
[71] Z.D. Wei, et al., Spontaneous reduction of Pt(IV) onto the sidewalls of functionalized multiwalled carbon nanotubes as catalysts for oxygen reduction reaction in PEMFCs, J. Phys. Chem. C 112 (7) (2008) 2671–2677.
[72] Z. Liu, et al., Preparation and characterization of platinum-based electrocatalysts on multiwalled carbon nanotubes for proton exchange membrane fuel cells, Langmuir 18 (10) (2002) 4054–4060.
[73] M.S. Saha, R. Li, X. Sun, High loading and monodispersed Pt nanoparticles on multiwalled carbon nanotubes for high performance proton exchange membrane fuel cells, J. Power Sources 177 (2) (2008) 314–322.
[74] Z. He, et al., Deposition and electrocatalytic properties of platinum nanoparticals on carbon nanotubes for methanol electrooxidation, Mater. Chem. Phys. 85 (2–3) (2004) 396–401.
[75] D.-J. Guo, H.-L. Li, High Dispersion and Electrocatalytic properties of platinum on functional multi-walled carbon nanotubes, Electroanalysis 17 (10) (2005) 869–872.
[76] Y. Xing, Synthesis and electrochemical characterization of uniformly-dispersed high loading Pt nanoparticles on sonochemically-treated carbon nanotubes, J. Phys. Chem. B 108 (50) (2004) 19255–19259.
[77] C.-C. Chen, et al., Growth and characteristics of carbon nanotubes on carbon cloth as electrodes, Diam. Relat. Mater. 14 (3–7) (2005) 770–773.
[78] N.M. Rodriguez, M.S. Kim, R.T.K. Baker, Promotional effect of carbon monoxide on the decomposition of ethylene over an iron catalyst, J. Catal. 144 (1) (1993) 93–108.
[79] N. Krishnankutty, et al., The effect of copper on the structural characteristics of carbon filaments produced from iron catalyzed decomposition of ethylene, Catal. Today 37 (3) (1997) 295–307.
[80] C.A. Bessel, et al., Graphite nanofibers as an electrode for fuel cell applications, J. Phys. Chem. B 105 (6) (2001) 1115–1118.
[81] E.S. Steigerwalt, et al., A Pt–Ru/graphitic carbon nanofiber nanocomposite exhibiting high relative performance as a direct-methanol fuel cell anode catalyst, J. Phys. Chem. B 105 (34) (2001) 8097–8101.
[82] E.S. Steigerwalt, G.A. Deluga, C.M. Lukehart, Pt-Ru/carbon fiber nanocomposites: synthesis, characterization, and performance as anode catalysts of direct methanol fuel cells. A search for exceptional performance, J. Phys. Chem. B 106 (4) (2002) 760–766.
[83] L.L. Vasiliev, L.E. Kanonchik, Activated carbon fibres and composites on its base for high performance hydrogen storage system. Chem. Eng. Sci. 65(8) 2586–2595.
[84] H.X. Huang, S.X. Chen, C.E. Yuan, Platinum nanoparticles supported on activated carbon fiber as catalyst for methanol oxidation, J. Power Sources 175 (1) (2008) 166–174.
[85] D.-J. Guo, et al., Novel hollow PtRu nanospheres supported on multi-walled carbon nanotube for methanol electrooxidation, J. Power Sources 177 (2) (2008) 334–338.
[86] H.T. Zheng, et al., Effect of support on the activity of Pd electrocatalyst for ethanol oxidation, J. Power Sources 163 (1) (2006) 371–375.
[87] Y. Li, L. Tang, J. Li, Preparation and electrochemical performance for methanol oxidation of Pt/graphene nanocomposites, Electrochem. Commun. 11 (4) (2009) 846–849.
[88] R. Kou, et al., Enhanced activity and stability of Pt catalysts on functionalized graphene sheets for electrocatalytic oxygen reduction, Electrochem. Commun. 11 (5) (2009) 954–957.

[89] M.J. McAllister, et al., Single sheet functionalized graphene by oxidation and thermal expansion of graphite, Chem. Mater. 19 (18) (2007) 4396–4404.
[90] Y. Shao, et al., Highly durable graphene nanoplatelets supported Pt nanocatalysts for oxygen reduction, J. Power Sources 195 (15) (2010) 4600–4605.
[91] D.-Q. Yang, J.-F. Rochette, E. Sacher, Spectroscopic evidence for π–π interaction between poly(diallyl dimethylammonium) chloride and multiwalled carbon nanotubes, J. Phys. Chem. B 109 (10) (2005) 4481–4484.
[92] N. Shang, et al., Platinum integrated graphene for methanol fuel cells, J. Phys. Chem. C 114 (37) (2010) 15837–15841.
[93] M.H. Seo, et al., The graphene-supported Pd and Pt catalysts for highly active oxygen reduction reaction in an alkaline condition, Electrochem. Commun. 13 (2) (2011) 182–185.
[94] S. Bong, et al., Graphene supported electrocatalysts for methanol oxidation, Electrochem. Commun. 12 (1) (2010) 129–131.
[95] L. Dong, et al., Graphene-supported platinum and platinum–ruthenium nanoparticles with high electrocatalytic activity for methanol and ethanol oxidation, Carbon 48 (3) (2010) 781–787.
[96] H. Zhang, et al., Microwave-assisted synthesis of graphene-supported Pd1Pt3 nanostructures and their electrocatalytic activity for methanol oxidation, Electrochim. Acta 56 (20) (2011) 7064–7070.
[97] D. Kasuya, et al., Selective production of single-wall carbon nanohorn aggregates and their formation mechanism, J. Phys. Chem. B 106 (19) (2002) 4947–4951.
[98] N. Sano, S.-i. Ukita, One-step synthesis of Pt-supported carbon nanohorns for fuel cell electrode by arc plasma in liquid nitrogen, Mater. Chem. Phys. 99 (2–3) (2006) 447–450.
[99] T. Yoshitake, et al., Preparation of fine platinum catalyst supported on single-wall carbon nanohorns for fuel cell application, Physica B Condensed Mat. 323 (1–4) (2002) 124–126.
[100] T. Hyeon, et al., High-performance direct methanol fuel cell electrodes using solid-phase-synthesized carbon nanocoils, Angew. Chem. Int. Ed. 42 (36) (2003) 4352–4356.
[101] M. Sevilla, G. Lota, A.B. Fuertes, Saccharide-based graphitic carbon nanocoils as supports for PtRu nanoparticles for methanol electrooxidation, J. Power Sources 171 (2) (2007) 546–551.
[102] T. Maiyalagan, B. Viswanathan, U.V. Varadaraju, Nitrogen containing carbon nanotubes as supports for Pt – alternate anodes for fuel cell applications, Electrochem. Commun. 7 (9) (2005) 905–912.
[103] S. Jiang, et al., Facile construction of Pt–Co/CNx nanotube electrocatalysts and their application to the oxygen reduction reaction, Adv. Mater. 21 (48) (2009) 4953–4956.
[104] B. Yue, et al., CNx nanotubes as catalyst support to immobilize platinum nanoparticles for methanol oxidation, J. Mater. Chem. 18 (15) (2008) 1747–1750.
[105] Q. Chen, et al., Morphological and microstructural stability of boron-doped diamond thin film electrodes in an acidic chloride medium at high anodic current densities, J. Electrochem. Soc. 144 (11) (1997) 3806–3812.
[106] K. Gong, et al., Nitrogen-doped carbon nanotube arrays with high electrocatalytic activity for oxygen reduction, Science 323 (5915) (2009) 760–764.
[107] Qu, L., et al., Nitrogen-doped graphene as efficient metal-free electrocatalyst for oxygen reduction in fuel cells. ACS Nano. 4(3) 1321–1326.
[108] K. Xiao, et al., n-Type field-effect transistors made of an individual nitrogen-doped multiwalled carbon nanotube, J. Am. Chem. Soc. 127 (24) (2005) 8614–8617.
[109] H. Chen, et al., Synergism of C5N Six-Membered Ring and vapor–liquid–solid growth of CNx nanotubes with pyridine precursor, J. Phys. Chem. B 110 (33) (2006) 16422–16427.

[110] S. Jiang, et al., Direct immobilization of Pt–Ru alloy nanoparticles on nitrogen-doped carbon nanotubes with superior electrocatalytic performance. J. Power Sources. 195(22) 7578–7582.

[111] C.-L. Sun, et al., Ultrafine platinum nanoparticles uniformly dispersed on arrayed CNx nanotubes with high electrochemical activity, Chem. Mater. 17 (14) (2005) 3749–3753.

[112] N. Li, et al., Large scale synthesis of N-doped multi-layered graphene sheets by simple arc-discharge method, Carbon 48 (1) (2010) 255–259.

[113] X. Li, et al., Simultaneous nitrogen doping and reduction of graphene oxide, J. Am. Chem. Soc. 131 (43) (2009) 15939–15944.

[114] R. Chetty, et al., PtRu nanoparticles supported on nitrogen-doped multiwalled carbon nanotubes as catalyst for methanol electrooxidation, Electrochim. Acta 54 (17) (2009) 4208–4215.

[115] R.I. Jafri, N. Rajalakshmi, S. Ramaprabhu, Nitrogen-doped multi-walled carbon nanocoils as catalyst support for oxygen reduction reaction in proton exchange membrane fuel cell. J. Power Sources. 195(24): 8080–8083.

[116] S.-H. Liu, J.-R. Wu, Nitrogen-doped ordered mesoporous carbons as electrocatalysts for methanol-tolerant oxygen reduction in acid solution, Int. J. Hydrogen Energy 36 (1) (2010) 87–93.

[117] Y. Shao, et al., Nitrogen-doped carbon nanostructures and their composites as catalytic materials for proton exchange membrane fuel cell, Appl. Catal. B Environ. 79 (1) (2008) 89–99.

[118] R.A. Sidik, et al., O_2 reduction on graphite and nitrogen-doped graphite: experiment and theory, J. Phys. Chem. B 110 (4) (2006) 1787–1793.

[119] M. Terrones, et al., Carbon nitride nanocomposites: formation of aligned CxNy nanofibers, Adv. Mater. 11 (8) (1999) 655–658.

[120] M. Terrones, et al., Efficient route to large arrays of CN(x) nanofibers by pyrolysis of ferrocene/melamine mixtures, Appl. Phys. Lett. 75 (25) (1999) 3932–3934.

[121] M. Terrones, et al., Novel nanoscale gas containers: encapsulation of N in CN nanotubes, Chem. Commun. 23 (2000) 2335–2336.

[122] M. Glerup, et al., Synthesis of highly nitrogen-doped multi-walled carbon nanotubes, Chem. Commun. 20 (2003) 2542–2543.

[123] S. Maldonado, S. Morin, K.J. Stevenson, Structure, composition, and chemical reactivity of carbon nanotubes by selective nitrogen doping, Carbon 44 (8) (2006) 1429–1437.

[124] E.-G. Wang, A new development in covalently bonded carbon nitride and related materials, Adv. Mater. 11 (13) (1999) 1129–1133.

[125] C. Tang, et al., Structure and nitrogen incorporation of carbon nanotubes synthesized by catalytic pyrolysis of dimethylformamide, Carbon 42 (12–13) (2004) 2625–2633.

[126] M. Terrones, et al., Controlled production of aligned-nanotube bundles, Nature 388 (6637) (1997) 52–55.

[127] S.C. Roy, et al., Spectroelectrochemical study of the role played by carbon functionality in fuel cell electrodes, J. Electrochem. Soc. 144 (7) (1997) 2323–2328.

[128] F. Jaouen, et al., Heat-treated Fe/N/C catalysts for O_2 electroreduction: Are active sites hosted in micropores? J. Phys. Chem. B 110 (11) (2006) 5553–5558.

[129] D. Qian, et al., Low-temperature synthesis of large-area CNx nanotube arrays, J. Nanosci. Nanotechno. 3 (1–2) (2003) 93–97.

[130] S. Trasobares, et al., Compartmentalized CNx nanotubes: chemistry, morphology, and growth, J. Chem. Phys. 116 (20) (2002) 8966–8972.

[131] A.G. Kudashov, et al., Influence of Ni–Co catalyst composition on nitrogen content in carbon nanotubes, J. Phys. Chem. B 108 (26) (2004) 9048–9053.

[132] R. Sen, et al., B–C–N, C–N and B–N nanotubes produced by the pyrolysis of precursor molecules over Co catalysts, Chem. Phys. Lett. 287 (5–6) (1998) 671–676.
[133] J. Liu, S. Webster, D.L. Carroll, Temperature and flow rate of NH_3 effects on nitrogen content and doping environments of carbon nanotubes grown by injection CVD method, J. Phys. Chem. B 109 (33) (2005) 15769–15774.
[134] J. Liu, R. Czerw, D.L. Carroll, Large-scale synthesis of highly aligned nitrogen doped carbon nanotubes by injection chemical vapor deposition methods, J. Mater. Res. 20 (2) (2005) 538–543.
[135] H.C. Choi, J. Park, B. Kim, Distribution and structure of N atoms in multiwalled carbon nanotubes using variable-energy X-ray photoelectron spectroscopy, J. Phys. Chem. B 109 (10) (2005) 4333–4340.
[136] Y. Xia, R. Mokaya, Generalized and facile synthesis approach to N-doped highly graphitic mesoporous carbon materials, Chem. Mater. 17 (6) (2005) 1553–1560.
[137] W. Xu, et al., Synthesis of aligned carbon nanotubes with double coaxial structure of nitrogen-doped and undoped multiwalls, Adv. Mater. 15 (13) (2003) 1087–1090.
[138] H. Li, et al., Nitrogen-doped carbon nanotubes with high activity for oxygen reduction in alkaline media, Int. J. Hydrogen Energy 36 (3) (2010) 2258–2265.
[139] L. Qu, et al., Nitrogen-doped graphene as efficient metal-free electrocatalyst for oxygen reduction in fuel cells, ACS Nano 4 (3) (2010) 1321–1326.
[140] G.M. Swain, The susceptibility to surface corrosion in acidic fluoride media: a comparison of diamond, HOPG, and glassy carbon electrodes, J. Electrochem. Soc. 141 (12) (1994) 3382–3393.
[141] A.E. Fischer, G.M. Swain, Preparation and characterization of boron-doped diamond powder, J. Electrochem. Soc. 152 (9) (2005) B369–B375.
[142] G.R. Salazar-Banda, K.I.B. Eguiluz, L.A. Avaca, Boron-doped diamond powder as catalyst support for fuel cell applications, Electrochem. Commun. 9 (1) (2007) 59–64.
[143] G.R. Salazar-Banda, et al., Methanol and ethanol electro-oxidation on Pt-SnO_2 and Pt-Ta_2O_5 sol–gel-modified boron-doped diamond surfaces, Mater. Chem. Phys. 117 (2–3) (2009) 434–442.
[144] G.R. Salazar-Banda, et al., Sol–gel-modified boron-doped diamond surfaces for methanol and ethanol electro-oxidation in acid medium, J. Power Sources 162 (1) (2006) 9–20.
[145] K. Honda, et al., Electrochemical properties of Pt-modified nano-honeycomb diamond electrodes, J. Electroanal. Chem. 514 (1–2) (2001) 35–50.
[146] H.B. Suffredini, et al., Electro-oxidation of methanol and ethanol using a Pt–RuO_2/C composite prepared by the sol–gel technique and supported on boron-doped diamond, J. Power Sources 158 (1) (2006) 124–128.
[147] C.K. Acharya, et al., Effect of boron doping in the carbon support on platinum nanoparticles and carbon corrosion, J. Power Sources 192 (2) (2009) 324–329.
[148] J.-i. Ozaki, et al., Simultaneous doping of boron and nitrogen into a carbon to enhance its oxygen reduction activity in proton exchange membrane fuel cells, Carbon 44 (15) (2006) 3358–3361.

Chapter 19

Novel Carbon Materials for CO_2 Adsorption

Joaquín Silvestre-Albero and Francisco Rodríguez-Reinoso
Instituto Universitario de Materiales, Departamento de Química Inorgánica, Universidad de Alicante, Alicante, Spain

Chapter Outline

19.1. Introduction 583
19.2. Porous Carbon Materials as CO_2 Adsorbents 586
19.2.1. Effect of the Surface Chemistry 586
19.2.2. Effect of the Porous Structure 589
19.2.3. Selective CO_2 Adsorbents 593
19.2.4. High-Pressure Performance 595
19.2.5. Regeneration 597
19.3. Final Remarks 598
References 599

19.1. INTRODUCTION

Large-scale greenhouse gas emissions to the atmosphere are forcing the different environmental agencies (e.g. EPA) to encourage the development of new cost-effective technologies to separate and capture the gases that are mainly responsible for the climate change [1]. Among them, CO_2 is the most problematic gas not because of its power, i.e. CH_4 is 21 times a more powerful greenhouse gas than CO_2, but because of its large-scale emissions worldwide. CO_2 emissions mainly come from fossil fuel combustion which accounts, as primary energy source, for 90% of the global energy consumption. The actual concentration of CO_2 in the atmosphere is as high as 392 ppm (compared to its preindustrial level of *ca.* 280 ppm) and it is expected to continue growing if anthropogenic sources are used at the actual rate [2]. The main three sources where CO_2 capture technologies are required include flue-gas streams (separation of CO_2 from N_2), natural gas (separation of CO_2 from CH_4), and precombustion gas mixtures (separation of CO_2 from H_2) [3,4]. Apart from

Novel Carbon Adsorbents. DOI: 10.1016/B978-0-08-097744-7.00019-3
Copyright © 2012 Elsevier Ltd. All rights reserved.

these gases, other components (e.g. moisture) can be present in these gas mixtures and must be taken into account in the development of these technologies. Considering that around 75–80% of the cost of capturing and sequestering 90% of the CO_2 generated in a power plant are due to the capture and compression stages, the development of low-cost technologies is mandatory [5,6].

Actual technologies for CO_2 abatement are mainly based on chemical reaction processes in basic liquid media [7]. The acid gas (CO_2) is bubbled through the liquid solution containing the basic media, e.g. aqueous amine solutions or other fluids with basic character such as ammonia; these compounds react with CO_2 to form water-soluble compounds (carbamates and bicarbonates). This technology is widely applied for streams with low CO_2 partial pressure. Primary (monoethanolamine (MEA)), secondary (diethanolamine (DEA)), or tertiary (methyldiethanolamine (MDEA)) amines have been widely applied (e.g. Econamine FGSM, Kerr-McGee/ABB Lumus Crest MEA) [7,8]. Although this technology is the most energy efficient, it has important capital and running costs associated with the subsequent regeneration of the solvent (heat treatment of the solvent well above 100 °C), together with some associated corrosion and environmental problems (a large amount of water is needed to minimize corrosion and air flow problems). Thus, the typical energy penalty incurred by the operation of the MEA capture process is an estimated 15–37% of the net power output of the plant [9]. Furthermore, the CO_2 absorption capacity of these systems is rather limited.

Other technologies available for CO_2 separation and capture from other light gases include cryogenic distillation, membrane purification, absorption using ionic liquids, and adsorption using porous solids [3,4,10]. Although cryogenic distillation is a well-known technology for the separation of gas mixtures, it is economically unfeasible for the separation of CO_2 from other light gases. Concerning the use of membranes, they are highly efficient for the separation of CO_2 from large concentration streams but they fail for diluted systems [11]. Ionic liquids are a broad category of salts which are stable at high temperature, thus offering the possibility of recovering CO_2 by dissolution even at the high temperature of the flue gas [12,13]. However, ionic liquids are still at laboratory scale and the large-scale production is still limited by the current unit cost and the high viscosity of these liquids. Finally, adsorption by means of pressure (PSA), temperature (VSA), or vacuum (VSA) swing systems and using inorganic and/or hybrid solids as adsorbents is a growing technology which is based on the design of nanoporous solids with the appropriate textural and chemical properties to selectively adsorb CO_2, the other gases being excluded [10,14]. Zeolites [15,16], carbon materials (mainly activated carbons and carbon molecular sieves) [17–20], alkali metal carbonates [21], covalent organic frameworks (COFs) [22], and metal-organic frameworks (MOFs) [23–25] are the solids with the best performance for this application. Successful materials for CO_2 separation and capture must combine not only a large adsorption capacity but also the appropriate porous structure (target pore-size entrance) and/or surface chemistry to facilitate the selective adsorption of CO_2 from the mixture of gases. Taking into account these requirements, carbon materials exhibit certain advantages with respect to other inorganic and/or hybrid materials described above, such as the possibility to design their textural and chemical properties by selecting simple pre- and postsynthesis treatments, higher hydrophobicity, higher resistance to both alkaline and basic media, relative ease of regeneration, and a low cost [26].

As mentioned above, CO_2 capture using amine-based aqueous solutions suffers from important limitations. In order to overcome these drawbacks associated with handling basic liquid solvents, research has been devoted in the last few years to the development of economical amine-enriched, imine-enriched, and/or nitrogen-enriched sorbents by immobilizing these organic compounds on a high-surface-area material. These modified sorbents exhibit an improved adsorbate–adsorbent interaction, thus improving their selective behavior compared to the unmodified porous sorbents, even at the high temperatures of the power plant flue gases (80–150 °C) [27–32]. Among them, carbon has attracted much attention due to the easy incorporation of nitrogen-based functionalities either in the raw material and/or by postsynthesis treatments.

From these results it is clear that CO_2 capture on porous solids (e.g. carbon materials) can be performed using two main strategies: i) through a pure physical adsorption process using a porous carbon with a target porous structure able to adsorb CO_2 and exclude other molecules or ii) through a chemical process (acid–base reaction or chemical adsorption) working with modified carbon materials, i.e. carbon materials with a basic surface chemistry, able to interact preferentially with CO_2. Concerning the modified carbon materials, surface basicity is mainly introduced through nitrogen functionalities. These nitrogen groups can be incorporated by selecting nitrogen-containing carbon precursors, by reaction of the synthesized carbons with a nitrogen reagent at high temperature (e.g. ammonia), or by impregnation of the parent carbon with basic chemicals (e.g. amines, imines, and amides). Modification of the carbon material to exhibit improved adsorbent–adsorbate interactions may be somewhat inconvenient from an economical point of view, even more when these materials are postimpregnated with corrosive chemicals. Consequently, the challenge in the design of an optimum sorbent for CO_2 capture is to have a perfect control of the porous structure in order to tune the pore dimensions and the specific pore volume for an optimum selective adsorption behavior, avoiding undesired posttreatments. This chapter will be devoted to the analysis of the state of the art in the different approaches in which porous carbon materials can be used for a selective adsorption of CO_2 from the mixture of gases in which it is present.

19.2. POROUS CARBON MATERIALS AS CO_2 ADSORBENTS

19.2.1. Effect of the Surface Chemistry

The adsorption properties of porous carbons materials are defined not only by the porous structure but also by the presence of surface chemical functionalities, i.e. the presence of heteroatoms on the carbon surface such as oxygen, nitrogen, sulfur, and phosphorous. The amount and nature of these functionalities are defined either by the nature of the starting material and/or the preparation conditions applied, i.e. activation process performed, and/or by the application of postsynthesis treatments. In any case, the type of surface heteroatoms and the amount of delocalized π-electrons on the basal planes of the graphene layers of the carbon material will define the acidic or basic character of the synthesized activated carbons and, consequently, the adsorption behavior. Among these groups, oxygen surface functionalities are mainly responsible for the acidic properties of the carbon surface, whereas nitrogen functionalities, i.e. nitrogen-containing groups, are mainly associated with the basicity of the carbon [26,33].

It has been shown that the introduction of basic functionalities on the carbon surface (nitrogen-containing groups) produces a beneficial effect on the adsorption properties of these materials for CO_2 [27,29,34–41]. Nitrogen functionalities can be introduced by selecting nitrogen-containing carbon precursors, by reaction of the synthesized material with a nitrogen-containing reagent at high temperature (e.g. ammonia), or by impregnation of the parent carbon with basic chemicals (e.g. amines, imines, and amides).

Among the different alternatives to obtain carbons with basic surface functionalities, the most trivial one consists of the preparation of carbon materials starting from nitrogen-rich precursors. The use of either natural resources (e.g. soybean) or high-nitrogen-content polymers (e.g. urea–formaldehyde and melamine–formaldehyde resins) has been proposed in the literature as an excellent approach to develop CO_2 adsorbents [35,39]. Thote et al. showed that the nitrogen contained in the soybean (40 g/100 g, in the form of protein) could be retained in the carbon matrix after a carbonization in the form of amines, amides, imines, and pyridinic groups [39]. These carbons with significant nitrogen content (up to 0.64%) exhibited an improved adsorption behavior toward an acidic molecule such as CO_2 (up to 4.1 wt.% CO_2 at 30 °C). Recent studies described in the literature have shown that the appropriate selection of the nitrogen-rich precursor (e.g. polypyrrole) [31,41] and the appropriate selection of the carbonization procedure (e.g., pyrolysis of the copolymer of resorcinol, formaldehyde, and a basic amino acid such as lysine used as a nitrogen source and polymerization catalyst) [42] can produce carbon materials with a very large adsorption capacity for CO_2 (16.9 wt.%, 18.9 wt.%, and 13.7 wt.%, respectively, at 25 °C and 1 atm). Polycondensation of deep eutectic solvents (DESs), a new class of ionic liquids, constitutes another approach allowing the preparation of hierarchical carbon structures with a relatively high surface area (up to 700 m^2/g), which contain nitrogen functionalities in the microporous carbon framework [36]. The use of DESs (resorcinol, 3-hydroxypyridine, and choline chloride) is particularly attractive because DESs constitute i) the liquid medium ensuring reagent homogenization, ii) the structure-directing agents responsible of the hierarchical structure, and iii) the source of carbon and nitrogen of the solid sorbent. Carbon materials prepared using this approach exhibit high nitrogen content (5–12 at.%) and high CO_2 adsorption capacity (14 wt.% at 25 °C). The same CO_2 adsorption capacity was reported for porous carbon monoliths prepared by polycondensation of resorcinol, but in this case using a block copolymer (Pluronic F127) as structure-directing agent [43].

Since sometimes the original surface chemistry of the activated carbon is not sufficient to enhance the specific adsorbate–adsorbent interactions, basic functionalities have to be incorporated by postsynthesis treatments. Among them, reaction of carbon materials with gaseous nitrogen-containing compounds at high temperature, i.e. incorporation of nitrogen functionalities in the carbon structure, has been widely described in the literature. Arenilla et al. reported an improved adsorption behavior of a series of polymer-based activated carbons after incorporation of nitrogen functionalities in the carbon structure by reaction of the carbon precursor (polyethylene terephthalate) with nitrogen-containing reactants (acridine, carbazole, and urea) at high temperature [34]. The total CO_2 adsorption capacity of these modified samples was highly sensible to the nature of the surface groups and not to the total amount of nitrogen functionalities incorporated. Reaction with gaseous ammonia at high temperature has also been used to incorporate nitrogen groups into the carbon structure [44]. Ammonia decomposition takes place at high temperature on carboxylic groups, thus producing NH_2, NH, and H radicals which, upon reaction with the carbon surface, lead to the formation of $-NH_2$, –CN, pyridinic, pyrrolic, and quaternary nitrogen functional groups [45,46]. The amount of nitrogen incorporated highly depends on the presence of surface oxygen groups [44], whereas their nature highly depends on the temperature of the ammonia treatment used [44,47]. Amides, imides, imines, aromatic amines, nitriles, and protonated amides dominated at low temperature (<600 °C), whereas thermally stable aromatic rings such as pyrrole and pyridinic-like functionalities dominated at high temperatures (600–800 °C). Adsorption analyses show that these groups increase the total adsorption capacity for CO_2, the improvement being higher for samples heat treated with ammonia at high temperatures (up to 9.6 wt.% at 25 °C), i.e. the adsorption capacity is highly sensible to the nature of the nitrogen groups present. Interestingly, the adsorption capacity of these nitrogen-doped carbon materials decreases with an increase in the adsorption temperature, this being the behavior one would expect in a typical physical adsorption process [18,34,44].

Another approach to improve the adsorption behavior consists of the anchoring of basic compounds on the surface of carbons materials. Although impregnation produces *a priori* a blocking of the microporous structure, this "apparent" detrimental effect is usually offset by an improved CO_2 adsorption behavior, mainly at high temperatures (80–100 °C) [27,29,37]. Only carbon materials with narrow porosity exhibit a negative effect on the capture efficiency after impregnation with basic chemicals [18,48]. This is due to the presence of an adequate porosity (pores smaller than 1 nm) already in the unmodified carbon where physical adsorption of CO_2 is highly promoted, this porosity being somewhat inaccessible after incorporation of the chemicals. Besides these disadvantages, it is noteworthy to emphasize the great advantage of handling impregnated solids in relation to the unsupported basic solvent, in addition to the lower capital cost, lower pressure for gas recovery, and lower energy consumption for regeneration [49].

Gray et al. modified fly-ash carbon by impregnation with a basic compound such as 3-chloropropylamine–hydrochloride and reported a large improvement in the CO_2 adsorption capacity [29]. The promoting effect of the chemically attached amino compound in the CO_2 adsorption capacity was attributed to the formation of carbamate–amine species as observed using DRIFTS. A similar improvement in the adsorption capacity of fly-ash carbons (up to 6.8 wt.% at 30 °C) was reported after impregnation with amines [37] and imines [27]. In general, physical adsorption of CO_2 on carbon materials is an exothermic process, that is, the adsorption capacity decreases with an increase in the adsorption temperature (see, for instance, sample CC in Fig. 19.1). However, the situation is reversed on impregnated samples [27,37]. As shown in

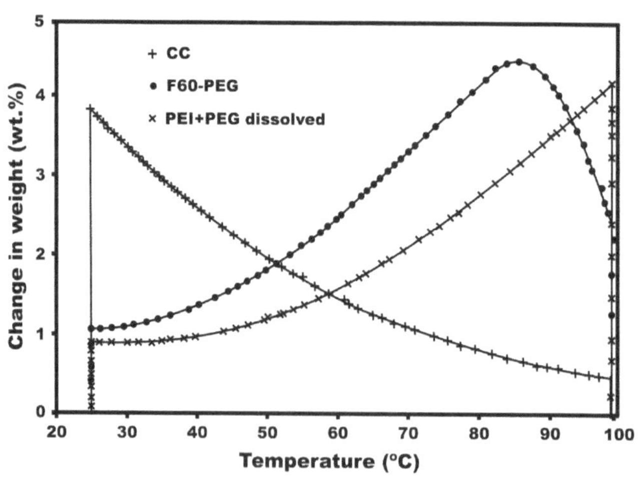

FIGURE 19.1 Temperature-programmed CO_2 adsorption tests for a fly-ash carbon F60-PEG, a commercial activated carbon (CC), and a mixture of PEI and PEG dissolved in methanol. *(Reprinted with permission from Ref. [27])*

Fig. 19.1, the adsorption capacity of a fly-ash carbon impregnated with polyethylenimine (PEI) and modified with polyethylene glycol (PEG), i.e. sample F60-PEG, is low at room temperature and increases with the adsorption temperature up to a maximum at 85 °C. Interestingly, the behavior of the impregnated sample resembles that of the unsupported liquid solvent, although with a slightly lower stability (see PEI + PEG dissolved in Fig. 19.1). Although the explanation for the improvement with temperature is still unclear, it must be attributed to a different adsorption mechanism on amine/imine-impregnated carbon materials, i.e. chemical interactions which are promoted at higher temperatures. However, Xu et al. proposed that at low temperatures the imines exist as nanosized particles in such a way that only the external sites are able to interact with CO_2 [32]. Consequently, higher temperatures are required to facilitate the diffusion of CO_2 into the inner sites of the particle, that is, it is a kinetically diffusion-controlled process. Consequently, while conventional carbon materials are good sorbents for CO_2 adsorption at room temperature, carbon materials impregnated with amines and/or imines constitute a better choice at the typical temperatures of the flue gases (up to 120 °C). In any case, temperatures above 80–120 °C become detrimental for the CO_2 adsorption on these impregnated systems due to the decomposition and/or volatilization of the nitrogen-based compounds used [27,29,37].

19.2.2. Effect of the Porous Structure

One of the main characteristics of adsorbents based on carbon materials is the possibility to tailor their porous structure for a given application. The type of precursor, the synthesis conditions (carbonization and activation temperature, activating agent, and so on), and the application of postsynthesis treatments are crucial parameters controlling the final properties of the synthesized carbons. These possibilities are extremely important for physical adsorption processes where the adsorbate–adsorbent interactions have to be improved in order to achieve an excellent adsorption behavior, in terms of total capacity and selectivity. This is the case when dealing with light molecules such as H_2, O_2, N_2, CH_4, and CO_2. Tuning the pore structure of the carbon adsorbent, i.e. tailoring the size and volume of micro-, meso-, and macropores, has been described in the literature as an efficient approach to improve the adsorption behavior of carbon materials for H_2 and CH_4 storage [50–53]. The improved behavior was attributed to the presence of narrow micropores (<1 nm) which provide an optimized storage capacity; the overlapping adsorption potential at the walls of narrow pores results in the enhancement of adsorption forces.

CO_2 is an acidic molecule with a kinetic diameter of 0.33 nm that has been widely used as a probe molecule for the characterization of porous solids. Whereas N_2 adsorption at cryogenic temperatures (–196 °C) and atmospheric pressure ($p/p_0 \sim 1.0$) measures the whole range of microporosity (pores below 2 nm), CO_2 adsorption at 0 °C and atmospheric pressure ($p/p_0 \sim 0.03$) was proposed some years ago as a complement to nitrogen for the characterization of the narrow microporosity (pores below 0.7 nm) in carbon materials [54,55]. However, recent studies described in the literature have shown that CO_2 adsorption may be kinetically restricted in some narrow pores (<0.3–0.4 nm), in spite of the relatively high temperature of the adsorption measurement (0 °C) [56]. The restricted accessibility of CO_2 at 0 °C on samples with narrow constrictions is clearly illustrated in Fig. 19.2 for a conventional activated carbon prepared by physical activation of olive stones with steam, but with a low degree of activation [57]. High equilibration times are required for CO_2 to access the narrow microporosity (above 300 s). Furthermore, the isotherm carried out at 25 °C exhibits higher uptake at low relative pressure than at 0 °C, thus showing that the adsorption at the later temperature is kinetically restricted. Consequently, these results anticipate the crucial role of the porosity, i.e. pore size, in the design of an adsorbent for CO_2 abatement under dynamic conditions. Micropores are required in order to enhance the adsorption potential for CO_2 adsorption [58], although care must be taken when narrow constrictions are expected.

Mathematical simulation studies using grand canonical Monte Carlo (GCMC) and nonlocal density functional theory (NLDFT) have shown that CO_2 forms a dense monolayer in pores wider than 0.3 nm, these pores being filled at $p \sim 5 \times 10^{-3}$ atm [59]. Additionally, these models predict a transition from a one-layer to a two-layer adsorbed structure for pores between 0.65 nm and 0.8 nm, whereas wider pores exhibit only two adsorbed monolayers on the opposite pore walls, with the central part of the pores unfilled at

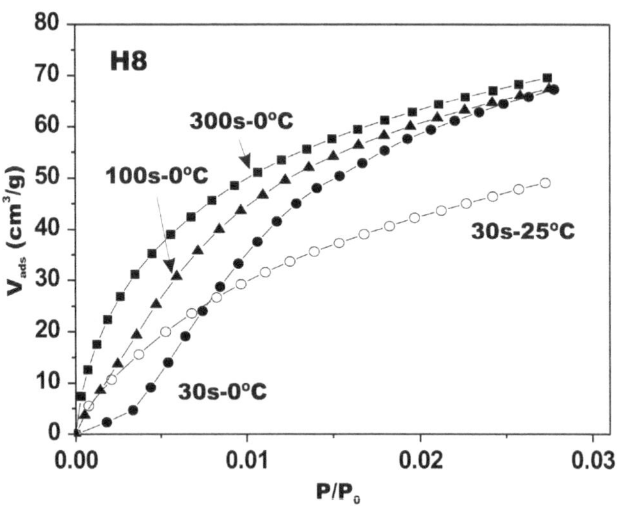

FIGURE 19.2 CO_2 adsorption isotherms at different equilibrium times (30, 100, and 300 s) for activated carbon H8 at 0 °C and 25 °C. *(Adapted from Ref. [56])*

subatmospheric pressures. Consequently, theoretical calculations predict the need of a pore size less than ~1 nm in order to design a carbon sorbent with optimized physical adsorption properties (selectivity and capacity) for CO_2. This observation confirms our assumption postulated in 1987 about the use of CO_2 adsorption at 0 °C as a complement to N_2 adsorption at −195 °C for the correct characterization of the narrow porosity (pores below 0.7–0.8 nm) in carbon materials [54].

Experimental studies using activated anthracites have shown that the total CO_2 adsorption capacity at atmospheric pressure and 30 °C does not relate to the BET surface area nor to the total micropore volume, thus suggesting that not all the surface area or the pore volume of these samples contributes to the adsorption of CO_2 [18]. Even the raw anthracites, without any activation, exhibit a large adsorption capacity (up to 4 wt.% at 30 °C) in spite of the low BET surface area of these samples, measured using N_2 adsorption at cryogenic temperature. Once again these results suggest the critical role of the narrow microporosity (pores not accessible to N_2 at −195 °C due to kinetic restrictions at so low a temperature, but accessible to a slightly smaller molecule such as CO_2 at a much higher adsorption temperature, 0 °C) in the adsorption of CO_2 at room temperature. A similar observation was described by Sevilla et al. on activated carbons prepared by hydrothermal treatment of polysaccharides and biomass [60]. The total CO_2 adsorption capacity was mainly independent of the carbon precursor used, but it was highly sensitive to the preparation conditions used, i.e. it was higher (up to 21 wt.% at 25 °C) on samples prepared under mild activation conditions (a lower amount of activating agent – KOH – and lower activation temperature). The correlation of these data with the textural parameters, determined by N_2 adsorption at −196 °C, suggested that narrow micropores (<1 nm) have the largest contribution to the total adsorption capacity for CO_2 at normal conditions, as confirmed for other precursors [61]. A larger activation degree led to carbon materials containing wider micropores and some mesopores, with a worse CO_2 adsorption behavior. The isosteric heat of adsorption was calculated to be ~20 ± 2 kJ/mol.

A more detailed analysis of the effect of the textural properties on the CO_2 adsorption capacity was described by Presser et al. using TiC–CDC (carbide-derived carbon) samples prepared using different TiC sources (different grain size precursors) and different annealing/activation conditions [58]. The total amount adsorbed on these samples does not relate to the BET surface area, the pore size, the total pore volume, or the volume of pores >1 nm. However, as shown in Fig. 19.3, the authors observed a correlation with the total volume of specific pores depending on the adsorption conditions (atmospheric or subatmospheric pressure) while keeping constant the adsorption temperature (0 °C). As it can be observed, the amount adsorbed at atmospheric pressure (1 bar) exhibits a better correlation when plotted as a function of the total volume of pores <0.8 nm (Fig. 19.3c), whereas the amount adsorbed at lower pressures (0.1 atm) fits better with the volume of pores <0.5 nm (Fig. 19.3h).

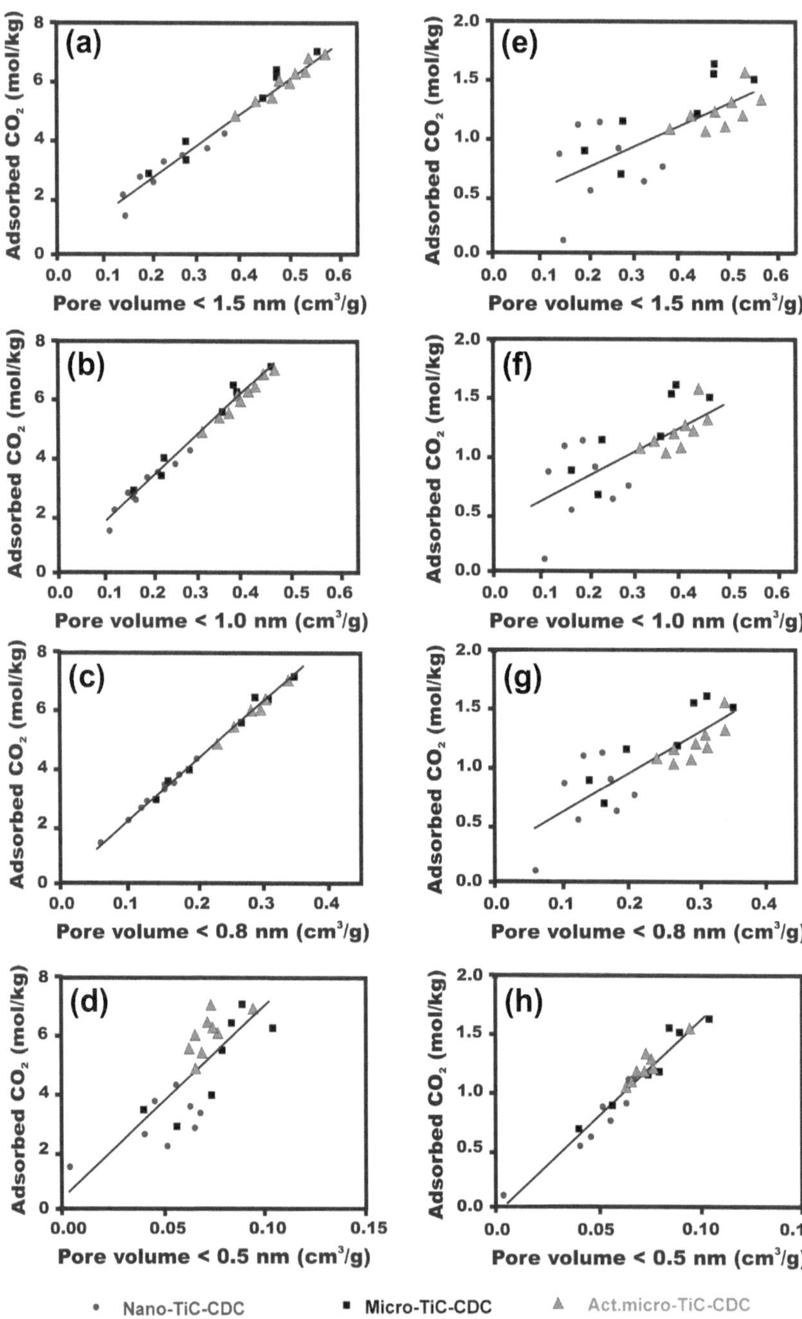

FIGURE 19.3 Correlation between the CO_2 uptake at 0 °C and 1.0 bar (a–d) and 0.1 bar (e–h) and the volume of pores smaller than 1.5 nm (a,e), 1.0 nm (b,f), 0.8 nm (c,g), and 0.5 nm (d,h) for a series of TiC–CDC samples prepared using different grain sizes (nano- and microsize) and different preparation conditions (chlorination temperature, annealing temperature, and so on). *(Reprinted with permission from Ref. [58])*

Consequently, angstrom-sized pore volume (volume of pores smaller than 0.5 nm or less) may play a major role for the adsorption of CO_2 at low pressures, whereas the volume of "larger" micropores (volume of pores smaller than 0.8 nm) may play a major role at atmospheric pressure, in close agreement with previous observations [59,60]. In any case, the activation or graphitization degree of the synthesized TiC–CDC had a minor effect on the total amount of CO_2 adsorbed compared to the total volume of specific pores. A similar linear correlation between the amount of CO_2 adsorbed and the volume of narrow micropores (those below 0.7–0.8 nm) was described by Vargas et al. for activated carbon monoliths at atmospheric pressure and 0 °C [62].

19.2.3. Selective CO_2 Adsorbents

For practical application, an appropriate carbon adsorbent must exhibit fast adsorption kinetics and a high selectivity toward CO_2 adsorption, i.e. molecular sieving effect, in addition to a high CO_2 adsorption capacity. The two main approaches to promote the preferential adsorption of CO_2 are i) to tune the porous structure, i.e. to design carbon molecular sieves, or ii) to modify the surface chemistry in order to promote specific CO_2–adsorbent interactions.

Carbon molecular sieves (CMSs) are carbonaceous materials with a narrow pore-size distribution, i.e. effective pore diameter in the nanoscale. The discrimination among different molecules on CMSs can be performed on the basis of a difference in adsorption equilibrium or on a difference in adsorption rate (size selectivity and/or kinetic selectivity). However, the achievement of the target pore size for a given application constitutes a difficult task because it requires a careful control of the preparation conditions used. This drawback can be overcome by performing pre- and postsynthesis treatments able to adjust the pore size to a given target, e.g. (i) control activation of the carbonized material, (ii) chemical vapor deposition (CVD) of the vapor of a organic molecule, (iii) controlled modification of the surface chemistry, and (iv) thermal contraction [63–67].

A carbon molecular sieve for CO_2 capture must fulfill several requirements: (i) a large adsorption capacity, (ii) a high selectivity to discriminate CO_2 from other gases, (iii) fast adsorption kinetics, and (iv) an easy regeneration. Previous studies described in the literature have shown that CO_2 adsorption at atmospheric pressure takes place mainly in micropores up to 0.8 nm, i.e. the larger the volume of pores <0.8 nm, the higher the adsorption capacity [58]. Concerning the second requirement, CO_2, with a kinetic diameter of 0.33 nm, can be found in combination with other gases of similar molecular dimension such as CH_4 (0.39 nm) and N_2 (0.36 nm). Consequently, a proper adsorbent for CO_2 separation must exhibit a well-defined pore structure (pore size around 0.4–0.5 nm), together with a narrow pore-size distribution. This statement requires a compromise between an optimum adsorption capacity, which requires pores ~0.65–0.8 nm [59], where an optimum CO_2 packing density will take place, and a large selectivity toward CO_2 that requires pores around ~0.4–0.5 nm, able to refuse other gases such as CH_4 and N_2. Pores below this limit (narrow constrictions) must be minimized to avoid kinetic restrictions [56].

Wahby et al. recently reported high selectivity for CO_2/CH_4 and CO_2/N_2 separation on activated carbons prepared from petroleum pitch, in either powder or monolith configurations, and using KOH as activating agent [20]. The nature of the petroleum residue, the mesophase content, and the final conformation (either powder or monolith) were reported as critical parameters defining the final porous structure and, consequently, the adsorption capacity and selectivity. In this sense, the appropriate combination of these parameters can provide samples (e.g. VR-93-M) with an extremely large adsorption capacity (~20 wt.% at 25 °C), fast adsorption kinetics, and a large CO_2/CH_4 (∞) and CO_2/N_2 (~14) selectivity (see Fig. 19.4). In accordance with previous statements described above, gas adsorption and immersion calorimetry measurements confirm that this sample exhibits an extremely large surface area (~2900 m^2/g) and, more importantly, a very large volume of narrow micropores (pore dimension below 0.55 nm). Interestingly, an activated carbon monolith prepared using the same experimental conditions but using a different petroleum residue (sample DO-88-M, Fig. 19.4) exhibited slightly wider micropores (~0.7–0.8 nm, determined by immersion calorimetry), and the consequence is that the separation ability drastically drops due to the considerable adsorption of CH_4 and N_2. An improvement in selectivity was observed when comparing powder and monolith forms of common samples due to the slight narrowing of the microporosity upon pressure conforming. Consequently, these results constitute a clear demonstration of the narrow window of textural properties required for an optimum sorbent for CO_2 capture. A similar behavior in terms of adsorption capacity and selectivity (S (CO_2/N_2) ~ 5.4) was also recently described by Sevilla et al. for activated carbons prepared from sawdust and using KOH as activating agent [60]. Interestingly, all these carbon materials also fulfill the last requirement: an easy regeneration. As described later on in

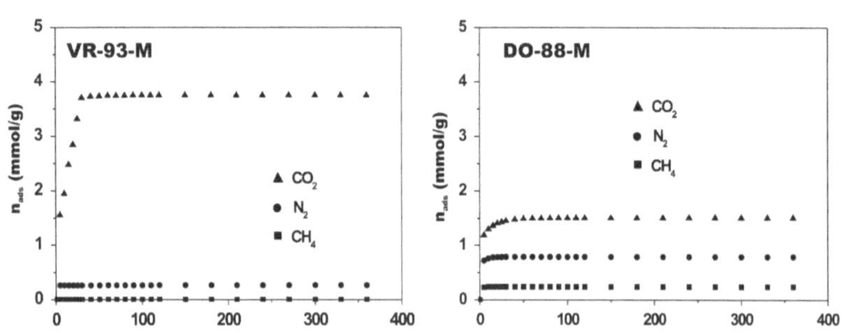

FIGURE 19.4 Adsorption kinetics for CO_2, N_2, and CH_4 at 25 °C for two activated carbon monoliths (VR-93-M and DO-88-M) prepared from different mesophase pitches. *(Adapted from Ref. [20])*

this chapter, CO_2 can be completely regenerated due to the presence of weak interactions between the adsorbate and the carbon surface (heat of adsorption ~20 kJ/mol).

The incorporation of specific groups on the carbon surface (e.g. nitrogen functionalities) can be considered as another approach to improve the preferential interaction of adsorbate (e.g. CO_2) with the carbon surface and, consequently, to improve the adsorption selectivity toward the desired molecule. Following this approach, large selectivity values for CO_2/CH_4 separation were reported for acrylic monolithic carbon molecular sieves, the surface of which was functionalized with nitrogen. However, these excellent values were associated to a low total adsorption capacity (4.6 wt.% at 25 °C) [68]. A high selectivity for CO_2/N_2 separation (around 5.3) together with a large adsorption capacity for CO_2 (up to 16.9 wt.% at 25 °C), fast adsorption kinetics, and an easy regeneration were reported in the literature for activated carbons prepared from a nitrogen-rich polymer (polypyrrole) and using KOH as activating agent [31]. Even larger selectivity values were reported for porous carbons, either powder or monoliths, prepared by polycondensation of resorcinol using as a structure-directing agent, a block copolymer (e.g. Pluronic F127) or deep eutectic solvents (CO_2/N_2 selectivity values ranging from 6 to 28) [36,43]. The excellent selective properties of these materials for CO_2 capture were attributed to a combination of a target porous structure (narrow micropores with enhanced adsorption potential) together with the presence of a large number of surface nitrogen functional groups, which lead to an increase in the isosteric heat of adsorption of an acidic gas molecule such as CO_2, i.e. an improved interaction between the large quadrupole moment of the CO_2 molecule (13.4×10^{-40} C/m^2) [3] and the polar sites associated with N groups. The isosteric heat of adsorption increased from ~20 kJ/mol on pure carbon [60] to values around 31.5–32.0 kJ/mol for N-doped samples [31,36]. At this point, it is noteworthy to mention that, in general, the selectivity values reported above for CO_2/CH_4 using novel activated carbon adsorbents are comparable or even larger, in terms of separation ability, kinetics, and capacity, compared to other porous solids such as zeolites, periodic mesoporous silicas, and conventional carbon materials [69,70].

In summary, the tailoring of the porous structure by pre- and postsynthesis treatments and the modification of the surface chemistry to improve the adsorbent–adsorbate interaction can be postulated as important factors to take into account when designing an adsorbent able to discriminate CO_2 from other gases and to adsorb the CO_2 in a very large amount. The accomplishment of any of these criteria will provide an adsorbent with an excellent performance for CO_2 capture in flue gases by means of swing adsorption systems operating at short time cycles (down to a few seconds).

19.2.4. High-Pressure Performance

Although most studies reported in the literature concern CO_2 adsorption at atmospheric pressure and room temperature, high-pressure studies are also very useful for application of porous sorbents in swing adsorption systems and also for storage and transportation. Unfortunately, high-pressure CO_2 analyses on carbon materials are very scarce in the literature. In the last few years, there has been a growing number of publications dealing with the apparently exceptional behavior of a relatively new class of porous materials, MOFs (metal-organic framework materials), in the adsorption of CO_2 at high pressure (up to 50 bar) [25]. Values up to 250 wt.% at 55 bar have been reported for MOF materials with an estimated surface area above 4500 m^2/g [71]. In contrast to the excellent adsorption behavior at high pressure, these hybrid systems fail at low pressures, i.e. they exhibit a low adsorption capacity at atmospheric pressure, at least those that outperform at high pressure (e.g. MOF-177 and MOF-200). This behavior is due to the appearance of an S-shape characteristic adsorption isotherm. Another important drawback when using MOF materials compared to other conventional sorbents (carbon materials or zeolites) may be the lack of high thermal and mechanical stability under certain environments.

The adsorption behavior of a carbon material, in terms of CO_2 adsorption capacity, highly depends on both the porous structure and the surface chemistry. Concerning the porous structure, the total adsorption capacity depends on the total volume of a specific pore, the target porosity being dependent on the pressure applied, i.e. at atmospheric pressure, total volume of pores below 0.8 nm contributes to the total CO_2 uptake, whereas at sub-atmospheric pressure, only pores below 0.5 nm must be considered [58]. Consequently, a carbon sorbent able to outperform MOF materials at high pressure must combine narrow micropores, wide micropores, and, possibly, some small mesopores.

Although the number of studies dealing with CO_2 adsorption at high pressure is scarce, very recently, Silvestre-Albero et al. have shown that the appropriate design of the porous structure can lead to adsorption values very close to the best MOFs reported in the literature [19]. The appropriate selection of the carbon precursor (e.g. vacuum residue (VR) pitch), the activating agent (KOH), and the activating ratio (3:1, KOH:pitch) provides materials (e.g. VR-93) with a total adsorption capacity at 50 bar of 207 wt.%, a value close to the best MOF material (MOF-200) with a total amount adsorbed of 239 wt.% (all these measurements at 25 °C) (see Fig. 19.5). Consequently, at high pressures, MOF materials exhibit a superior adsorption capacity compared to carbon materials, at least on a gravimetric basis (mg/g). However, for practical applications a porous solid for CO_2 storage should have not only a high adsorption capacity on a gravimetric basis but also a high adsorption capacity per unit volume of adsorbent. The volumetric adsorption capacity highly depends on the final bulk density of the synthesized material. A comparison of the excess adsorption isotherms for CO_2 on a volumetric basis (cm^3 adsorbed CO_2/cm^3 of sorbent) shows that now carbon molecular sieves exhibit a higher adsorption capacity compared to the best MOFs reported in the literature [19]. Sample VR-93 exhibits an adsorption capacity as high as 500 cm^3 STP/cm^3. Compared to these MOF materials, carbon molecular sieves can be envisaged

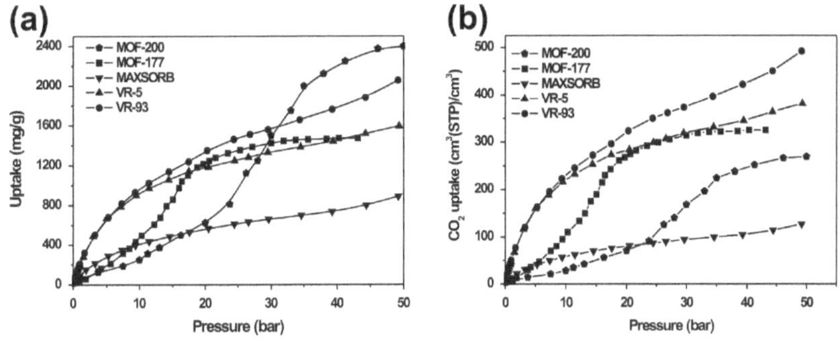

FIGURE 19.5 CO_2 adsorption isotherms in a (a) gravimetric (mg/g) and (b) volumetric (cm^3/cm^3) basis on VR carbon molecular sieves samples at 25 °C and up to 50 bar. MOF materials (MOF-177 and MOF-200) with exceptional adsorption properties and commercial activated carbon MAXSORB are included for the sake of comparison. *(Adapted from Ref. [19])*

as excellent alternatives for high-pressure applications due to their higher packing density and, consequently, higher volumetric adsorption capacity.

19.2.5. Regeneration

Besides achieving an excellent adsorption behavior at both low (atmospheric) and high pressures, another requirement for a final practical application of carbon materials is the ease of the regeneration step. Considering that the isosteric heat of adsorption of CO_2 on carbon materials is quite low (around 20 kJ/mol), the weak adsorbate–adsorbent interaction could anticipate an easy regeneration compared to other porous solids (e.g. zeolites). In this sense, Silvestre-Albero et al. studied the behavior of carbon molecular sieve monoliths under repeated cycles from vacuum to high pressure (50 bar) at room temperature (25 °C) [19]. As shown in Fig. 19.6, CO_2 adsorption on carbon

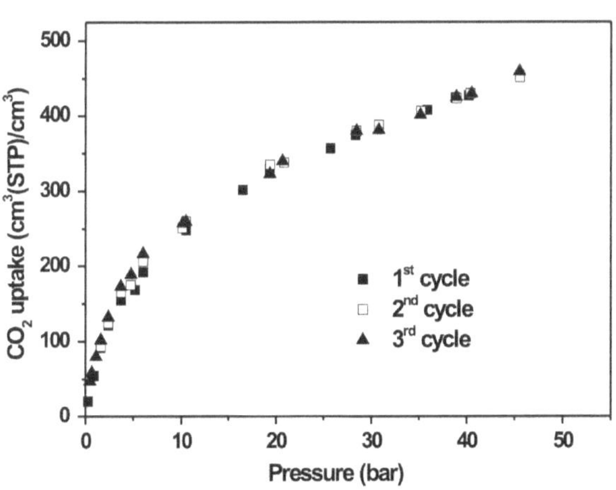

FIGURE 19.6 CO_2 isotherms for activated carbon monolith VR-93 at 25 °C and up to 50 bar after different adsorption–desorption cycles. *(Adapted from Ref. [19])*

materials is completely reversible, with no loss of adsorption capacity over the whole pressure range even after 3 cycles. A similar behavior in terms of regenerability was described in the literature for sawdust-based and N-doped porous carbons with no noticeable change in the kinetics of desorption and CO_2 uptake at atmospheric pressure and room temperature, even after six cycles [31,60].

19.3. FINAL REMARKS

The results described in this chapter show that carbon materials are excellent candidates for the adsorption of CO_2 at both low and high pressures, albeit a perfectly designed porous structure and surface chemistry is mandatory. Table 19.1 summarizes the best results described in the literature up to now following any of these strategies.

Adsorption of CO_2 can be achieved either chemically or physically. When physical adsorption is the predominant mechanism, i.e. when using carbon materials without nitrogen functionalities, the adsorbent–adsorbate interaction

TABLE 19.1 Comparison of CO_2 Adsorption Uptakes at Two Temperatures (0 °C And 25 °C) and 1 atm for the Best Carbon Materials Reported in the Literature

Material	CO_2 Uptake (mg/g) $^{0\,°C}$	CO_2 Uptake (mg/g) $^{25\,°C}$	Reference
N-doped hierarchical carbons	–	145	[36]
Zeolite-templated N-doped carbons	303	194	[72]
N-doped polypyrrole–graphene composites	–	189	[41]
N-doped polypyrrole-based carbons	271	169	[31]
Purified SWNTs	224	–	[17]
Micro-TiC–CDC	312	–	[58]
Petroleum pitch-based activated carbon monoliths	380	207	[20]
Sawdust-based porous carbon	–	212	[60]

is weak (van der Waals forces) and the amount adsorbed drastically decreases at high temperatures. For the physical adsorption process, the optimum pore size is always in the microporous range (below ~1 nm), although the target value highly depends on the final pressure applied.

Concerning the surface chemistry, the incorporation of nitrogen functionalities in the carbon surface and/or structure, i.e. an increase in the basic character of the carbon, gives rise to an improvement in the adsorbate–adsorbent interactions and, consequently, an increase in the adsorption capacity for CO_2. Nitrogen functionalities can be incorporated either using a nitrogen-rich carbon precursor or by postsynthesis treatments. Another approach consists of the impregnation of the carbon material with basic chemicals (amines, imines, and so on). In all these cases, chemisorption is the predominant mechanism and results in the formation of strong bonds between the adsorbate and the adsorbent. Concerning the effect of the adsorption temperature, for samples impregnated with chemicals (amines, imines, and so on) the adsorption capacity increases with the adsorption temperature up to a maximum around 80–100 °C, above which the adsorption capacity drops suddenly, as corresponds to a pure chemical process. For samples with nitrogen functionalities incorporated in the carbon matrix by reaction with a nitrogen-based chemical at high temperature, the adsorption capacity decreases with the adsorption temperature, although to a lower extent compared to the unmodified samples, i.e. hybrid behavior, suggesting a combination of a physical and chemical adsorption processes.

In general, CO_2 adsorption is highly reversible in carbon materials, exhibits fast adsorption kinetics, and, for certain samples described here, even outperforms, at both low and high pressures, other promising porous materials such as MOFs. These unique characteristics make carbon materials excellent candidates for industrial applications based on CO_2 capture.

REFERENCES

[1] DOE/NETL, Carbon Dioxide Capture and Storage RD&D Roadmap, U.S. Department of Energy. www.netl.doe.gov/technologies/carbon_seq, 2010.

[2] Trends in Atmospheric Carbon Dioxide, U.S. Department of Commerce. National Oceanic & Atmospheric Administration. www.esrl.noaa.gov/gmd/ccgg/trends, 2011.

[3] D.M. D'Alessandro, B. Smit, J.R. Long, Carbon dioxide capture: prospects for new materials, Angew. Chem. Int. Ed. 49 (2010) 6058–6082.

[4] J.D. Figueroa, T. Fout, S. Plasynski, H. McIlvried, R.D. Srivastava, Advances in CO_2 capture technology–the U.S. Department of Energy's carbon sequestration program, Int. J. Greenhouse Gas Control 2 (2008) 9–20.

[5] J. David, Economic evaluation of leading technology options for sequestration of carbon dioxide, in: Master of Science in Technology and Policy, Massachusetts Institute of Technology, 2000.

[6] C. Song, Global challenges and strategies for control, conversion and utilization of CO_2 for sustainable development involving energy, catalysis, adsorption and chemical processing, Catal. Today 115 (2006) 2–32.

[7] C.L. Leci, Financial implications on power generation costs resulting from the parasitic effect of CO_2 capture using liquid scrubbing technology from power station flue gases, Energy Convers. Manage 37 (1996) 915–921.

[8] D.G. Chapel, C.L. Mariz, J. Ernest, Recovery of CO_2 from flue gases: commercial trends, in: First National Conference on Carbon Sequestration, NETL Publications, 2001.

[9] H.J. Herzog, E.M. Drake, Greenhouse gas R&D program, IEA/93/OE6 (1993).

[10] N. Hedin, L. Chen, A. Laaksonen, Sorbents for CO_2 capture from flue gas-aspects from materials and theoretical chemistry, Nanoscale 2 (2010) 1819–1841.

[11] J. Zou, W.S.W. Ho, CO_2-selective polymeric membranes containing amines in crosslinked poly(vinyl alcohol), J. Membr. Sci. 286 (2006) 310–321.

[12] L.A. Blanchard, D. Hancu, E.J. Beckman, J.F. Brennecke, Green processing using ionic liquids and CO_2, Nature 399 (1999) 28–29.

[13] C. Wang, H. Luo, D.-e. Jiang, H. Li, S. Dai, Carbon dioxide capture by superbase-derived protic ionic liquids, Angew. Chem. Int. Ed. 49 (2010) 5978–5981.

[14] Q. Wang, J. Luo, Z. Zhong, A. Borgna, CO_2 capture by solid adsorbents and their applications: current status and new trends, Energy Environ. Sci. 4 (2011) 42–55.

[15] P.J.E. Harlick, F.H. Tezel, An experimental adsorbent screening study for CO_2 removal from N_2, Microp. Mesop. Mater. 76 (2004) 71–79.

[16] R.V. Siriwardane, M.-S. Shen, E.P. Fisher, J.A. Poston, Adsorption of CO_2 on molecular sieves and activated carbon, Energy Fuels 15 (2001) 279–284.

[17] M. Cinke, J. Li, C.W. Bauschlicher, A. Ricca, M. Meyyappan, CO_2 adsorption in single-walled carbon nanotubes, Chem. Phys. Lett. 376 (2003) 761–766.

[18] M.M. Maroto-Valer, Z. Tang, Y. Zhang, CO_2 capture by activated and impregnated anthracites, Fuel Process. Technol. 86 (2005) 1487–1502.

[19] J. Silvestre-Albero, A. Wahby, A. Sepúlveda-Escribano, M. Martínez-Escandell, K. Kaneko, F. Rodríguez-Reinoso, Ultrahigh CO_2 adsorption capacity on carbon molecular sieves at room temperature, Chem. Commun. 47 (2011) 6840–6842.

[20] A. Wahby, J.M. Ramos-Fernández, M. Martínez-Escandell, A. Sepúlveda-Escribano, J. Silvestre-Albero, F. Rodríguez-Reinoso, High-surface-area carbon molecular sieves for selective CO_2 adsorption, ChemSusChem 3 (2010) 974–981.

[21] Y. Liang, D.P. Harrison, R.P. Gupta, D.A. Green, W.J. McMichael, Carbon dioxide capture using dry sodium-based sorbents, Energy Fuels 18 (2004) 569–575.

[22] H. Furukawa, O.M. Yaghi, Storage of hydrogen, methane and carbon dioxide in highly porous covalent organic frameworks for clean energy applications, J. Am. Chem. Soc. 131 (2009) 8875–8883.

[23] P.L. Llewellyn, et al., High uptake of CO_2 and CH_4 in mesoporous metal-organic frameworks MIL-100 and MIL-101, Langmuir 24 (2008) 7245–7250.

[24] A.R. Millward, O.M. Yaghi, Metal-organic frameworks with exceptionally high capacity for storage of carbon dioxide at room temperature, J. Am. Chem. Soc. 127 (2005) 17998–17999.

[25] A.Ö Yazaydin, et al., Screening of metal-organic frameworks for carbon dioxide capture from flue gas using a combined experimental and modeling approach, J. Am. Chem. Soc. 131 (2009) 18198–18199.

[26] H. Marsh, F. Rodríguez-Reinoso, Activated Carbon, Elsevier, 2006.

[27] A. Arenillas, K.M. Smith, T.C. Drage, C.E. Snape, CO_2 capture using some fly ash-derived carbon materials, Fuel 84 (2005) 2204–2210.

[28] A. Demessence, D.M. D'Alessandro, M.L. Foo, J.R. Long, Strong CO_2 binding in a water-stable, triazolate-bridged metal-organic framework functionalized with ethylenediamine, J. Am. Chem. Soc. 131 (2009) 8784–8786.

[29] M.L. Gray, Y. Soong, K.J. Champagne, J. Baltrus, R.W. Stevens, P. Toochinda, S.S.C. Chuang, CO_2 capture by amine-enriched fly ash carbon sorbents, Separ. Purif. Technol. 35 (2004) 31–36.

[30] C. Lu, H. Bai, B. Wu, F. Su, J.F. Hwang, Comparative study of CO_2 capture by carbon nanotubes, activated carbons, and zeolites, Energy Fuels 22 (2008) 3050–3056.

[31] M. Sevilla, P. Valle-Vigón, A.B. Fuertes, N-doped polypyrrole-based porous carbons for CO_2 capture, Adv. Funct. Mater. 21 (2011) 2781–2787.

[32] X. Xu, C. Song, J.M. Andresen, B.G. Miller, A.W. Scaroni, Preparation and characterization of novel CO_2 "molecular basket" adsorbents based on polymer-modified mesoporous molecular sieve MCM-41, Microp. Mesop. Mater. 62 (2003) 29–45.

[33] J.L. Figueiredo, M.F.R. Pereira, M.M.A. Freitas, J.J.M. Órfao, Modification of the surface chemistry of activated carbons, Carbon 37 (1999) 1379–1389.

[34] A. Arenillas, F. Rubiera, J.B. Parra, C.O. Ania, J.J. Pis, Surface modification of low cost carbons for their application in the environmental protection, Appl. Surf. Sci. 252 (2005) 619–624.

[35] T.C. Drage, A. Arenillas, K.M. Smith, C. Pevida, S. Piippo, C.E. Snape, Preparation of carbon dioxide adsorbents from the chemical activation of urea-formaldehyde and melamine-formaldehyde resins, Fuel 86 (2007) 22–31.

[36] M.C. Gutiérrez, D. Carriazo, C.O. Ania, J.B. Parra, M.L. Ferrer, F. del Monte, Deep eutectic solvents as both precursors and structure directing agents in the synthesis of nitrogen doped hierarchical carbons highly suitable for CO_2 capture, Energy Environ. Sci. 4 (2011) 3535–3544.

[37] M.M. Maroto-Valer, Z. Lu, Y. Zhang, Z. Tang, Sorbents for CO_2 capture from high carbon fly ashes, Waste Manage. 28 (2008) 2320–2328.

[38] M.S. Shafeeyan, W.M.A.W. Daud, A. Houshmand, A. Shamiri, A review on surface modification of activated carbon for carbon dioxide adsorption, J. Anal. Appl. Pyrol. 89 (2010) 143–151.

[39] J.A. Thote, K.S. Iyer, R. Chatti, N.K. Labhsetwar, R.B. Biniwale, S.S. Rayalu, In situ nitrogen enriched carbon for carbon dioxide capture, Carbon 48 (2010) 396–402.

[40] H.A. Zinnen, T.A. Nalette, W. Papale, T. Filburn, Carbon dioxide removal using aminated carbon molecular sieves, US patent No. 4,810,266 (1989).

[41] V. Chandra, S.U. Yu, S.H. Kim, Y.S. Yoon, D.Y. Kim, A.H. Kwon, M. Meyyappan, K.S. Kim, Highly selective CO_2 capture on N-doped carbon produced by chemical activation of polypyrrole functionalized graphene sheets, Chem. Commun. 48 (2012) 735–737.

[42] G.-P. Hao, W.-C. Li, D. Qian, A.-H. Lu, Rapid synthesis of nitrogen-doped porous carbons monoliths for CO_2 capture, Adv. Mater. 22 (2010) 853–857.

[43] G.-P. Hao, et al., Structurally designed synthesis of mechanically stable poly(benzoxazine-co-resol)-based porous carbon monoliths and their application as high-performance CO_2 capture sorbents, J. Am. Chem. Soc. 133 (2011) 11378–11388.

[44] C. Pevida, M.G. Plaza, B. Arias, J. Fermoso, F. Rubiera, J.J. Pis, Surface modification of activated carbons for CO_2 capture, Appl. Surf. Sci. 254 (2008) 7165–7172.

[45] H.P. Boehm, G. Mair, T. Stoehr, A.R. De Rincón, B. Tereczki, Carbon as a catalyst in oxidation reactions and hydrogen halide elimination reactions, Fuel 63 (1984) 1061–1063.

[46] B. Stöhr, H.P. Boehm, R. Schlögl, Enhancement of the catalytic activity of activated carbons in oxidation reactions by thermal treatment with ammonia or hydrogen cyanide and observation of a superoxide species as a possible intermediate, Carbon 29 (1991) 707–720.

[47] C.L. Mangun, K.R. Benak, J. Economy, K.L. Foster, Surface chemistry, pore sizes and adsorption properties of activated carbon fibers and precursors treated with ammonia, Carbon 39 (2001) 1809–1820.

[48] M.G. Plaza, C. Pevida, A. Arenillas, F. Rubiera, J.J. Pis, CO_2 capture by adsorption with nitrogen enriched carbons, Fuel 86 (2007) 2204–2212.
[49] R.A. Khatri, S.S.C. Chuang, Y. Soong, M. Gray, Thermal and chemical stability of regenerable solid amine sorbent for CO_2 capture, Energy Fuels 20 (2006) 1514–1520.
[50] C. Almansa, M. Molina-Sabio, F. Rodríguez-Reinoso, Adsorption of methane into $ZnCl_2$-activated carbon derived discs, Microp. Mesop. Mater. 76 (2004) 185–191.
[51] Y. Gogotsi, R.K. Dash, G. Yushin, T. Yildirim, G. Laudisio, J.E. Fischer, Tailoring of nanoscale porosity in carbide-derived carbons for hydrogen storage, J. Am. Chem. Soc. 127 (2005) 16006–16007.
[52] M. Molina-Sabio, C. Almansa, F. Rodríguez-Reinoso, Phosphoric acid activated carbon discs for methane adsorption, Carbon 41 (2003) 2113–2119.
[53] G. Yushin, R. Dash, J. Jagiello, J.E. Fischer, Y. Gogotsi, Carbide-derived carbons: effect of pore size on hydrogen uptake and heat of adsorption, Adv. Funct. Mater. 16 (2006) 2288–2293.
[54] J. Garrido, A. Linares-Solano, J.M. Martín-Martínez, M. Molina-Sabio, F. Rodríguez-Reinoso, R. Torregrosa, Use of nitrogen vs. carbon dioxide in the characterization of activated carbons, Langmuir 3 (1987) 76–81.
[55] F. Rodríguez-Reinoso, J. Garrido, J.M. Martín-Martínez, M. Molina-Sabio, R. Torregrosa, The combined use of different approaches in the characterization of microporous carbons, Carbon 27 (1989) 23–32.
[56] R.V.R.A. Rios, J. Silvestre-Albero, A. Sepúlveda-Escribano, M. Molina-Sabio, F. Rodríguez-Reinoso, Kinetic restrictions in the characterization of narrow microporosity in carbon materials, J. Phys. Chem. C 111 (2007) 3803–3805.
[57] M. Molina-Sabio, M.T. Gonzalez, F. Rodríguez-Reinoso, A. Sepúlveda-Escribano, Effect of steam and carbon dioxide activation in the micropore size distribution of activated carbon, Carbon 34 (1996) 505–509.
[58] V. Presser, J. McDonough, S.-H. Yeon, Y. Gogotsi, Effect of pore size on carbon dioxide sorption by carbide derived carbon, Energy Environ. Sci. 4 (2011) 3059–3066.
[59] A. Vishnyakov, P.I. Ravikovitch, A.V. Neimark, Molecular level models for CO_2 sorption in nanopores, Langmuir 15 (1999) 8736–8742.
[60] M. Sevilla, A.B. Fuertes, Sustainable porous carbons with a superior performance for CO_2 capture, Energy Environ. Sci. 4 (2011) 1765–1771.
[61] J.M. Ramos-Fernández, M. Martínez-Escandell, F. Rodríguez-Reinoso, Production of binderless activated carbon monoliths by KOH activation of carbon mesophase materials, Carbon 46 (2008) 384–386.
[62] D.P. Vargas, L. Giraldo, J. Silvestre-Albero, J.C. Moreno-Piraján, CO_2 adsorption on binderless activated carbon monoliths, Adsorption 17 (2011) 497–504.
[63] G. Bello, R. García, R. Arriagada, A. Sepúlveda-Escribano, F. Rodríguez-Reinoso, Carbon molecular sieves from Eucalyptus globulus charcoal, Microp. Mesop. Mater. 56 (2002) 139–145.
[64] C. Gómez-de-Salazar, A. Sepúlveda-Escribano, F. Rodríguez-Reinoso, Preparation of carbon molecular sieves by controlled oxidation treatments, Carbon 38 (2000) 1889–1892.
[65] T. Horikawa, J. Hayashi, K. Muroyama, Preparation of molecular sieving carbon from waste resin by chemical vapor deposition, Carbon 40 (2002) 709–714.
[66] K. Miura, J. Hayashi, K. Hashimoto, Production of molecular sieving carbon through carbonization of coal modified by organic additives, Carbon 29 (1991) 653–660.
[67] A. Silvestre-Albero, A. Wahby, J. Silvestre-Albero, F. Rodríguez-Reinoso, W. Betz, Carbon molecular sieves prepared from polymeric precursors: porous structure and hydrogen adsorption properties, Ind. Eng. Chem. Res. 48 (2009) 7125–7131.

[68] J.M. Valente Nabais, P.J.M. Carrott, M.M.L. Ribeiro Carrott, A.M. Padre-Eterno, J.A. Menéndez, A. Dominguez, A.L. Ortiz, New acrylic monolithic carbon molecular sieves for O_2/N_2 and CO_2/CH_4 separations, Carbon 44 (2006) 1158–1165.
[69] Y. Belmabkhout, G. De Weireld, A. Sayari, Amine-bearing mesoporous silica for CO_2 and H_2S removal from natural gas and biogas, Langmuir 25 (2009) 13275–13278.
[70] Y. Belmabkhout, A. Sayari, Adsorption of CO_2 from dry gases on MCM-41 silica at ambient temperature and high pressure. 2: adsorption of CO_2/N_2, CO_2/CH_4 and CO_2/H_2 binary mixtures, Chem. Eng. Sci. 64 (2009) 3729–3735.
[71] H. Furukawa, et al., Ultrahigh porosity in metal-organic frameworks, Science 329 (2010) 424–428.
[72] Y. Xia, R. Mokaya, G.S. Walker, Y. Zhu, Superior CO_2 adsorption capacity on N-doped, high-surface-area, microporous carbons templated from zeolite, Adv. Energ. Mater. 1 (2011) 678–683.

Chapter 20

Nonenvironmental Industrial Applications of Activated Carbon Adsorption

Meenakshi Goyal
University Institute of Chemical Engineering and Technology, Punjab University, Chandigarh, India

Chapter Outline

20.1. Activated Carbon as a Decolorizer and a Purifier 607
- 20.1.1. Sugar and Sweetener Industries 607
- 20.1.2. Oils and Fats Industry 609
- 20.1.3. Alcoholic Beverage Industries 610
- 20.1.4. Chemical Industry 611
- 20.1.5. Activated Carbon Applications in Pharmaceutical Industry 612
- 20.1.6. Electroplating Industries for Purifying Electrolytic Baths 613

20.2. Activated Carbon as a Deodorizer 614
20.3. Activated Carbon Adsorption in Nuclear Technology 614
20.4. Activated Carbon in Vacuum Technology 617
20.5. Activated Carbon in Gas Chromatography 617
20.6. Application of Activated Carbon as Ion Exchanger 618
20.7. Activated Carbons as Porous Electrodes 619
20.8. Activated Carbon Application in the Refining of Mineral Oil and Gasoline 620
20.9. Activated Carbon Applications for Human Body Protection 621
20.10. Activated Carbon for Recovery 624
- 20.10.1. Recovery of Organic Solvents 624

Novel Carbon Adsorbents. DOI: 10.1016/B978-0-08-097744-7.00020-X
Copyright © 2012 Elsevier Ltd. All rights reserved.

20.10.2. Recovery of Gold 625
20.10.3. Recovery of Sulfur 626
20.11. Activated Carbon Applications in Wastewater Treatment 628
20.12. Activated Carbon Applications in Catalysis 629
20.12.1. Oxidation Reactions 630
20.12.2. Catalytic Hydrogenations 631
20.12.3. Activated Carbon Catalyzed C–C Coupling Reactions 632
20.12.4. Basic Activated Carbon Catalysts 632
20.12.5. Acid-doped Activated Carbon Catalysts 633
20.12.6. Combination Reactions 633
References 635

Activated carbons are unique and versatile adsorbents because they have a large surface area, highly microporous character, and a high degree of surface reactivity. They are universal adsorbents and can adsorb varying amounts of organic and inorganic compounds. The most widely used carbon adsorbents have surface areas between 800 and 1500 m^2/g. This surface area is contained predominantly within micropores with effective pore width smaller than 2 nm. In fact, activated carbons consist of a network of pores which have been classified into micropores with diameters less than 2 nm, mesopores with diameters between 2 and 50 nm, and macropores with diameters greater than 50 nm. Most of the adsorption (between 90% and 95%) generally takes place in the micropores while mesopores contribute to rest of the adsorption. Macropores do not contribute significantly toward surface area and act as conduits for the adsorbate into the micro- and macropores. Attempts, however, are underway to prepare active carbons with increased mesoporosity for the adsorption of larger molecules.

The random ordering of the aromatic sheets in activated carbons causes a variation in the arrangement of electron clouds in the carbon skeleton and results in the creation of unpaired electrons and incompletely saturated valencies, which would undoubtedly influence the adsorption properties of these materials. Activated carbons are invariably associated with certain amounts of oxygen and hydrogen bonded at the edges and corners of the aromatic sheets in the form of carbon–oxygen and carbon–hydrogen surface functional groups. Since these edges constitute the main adsorbing surface, these surface groups profoundly influence the adsorption properties of activated carbons. Besides, the active carbon surface has active sites in the form of edges, dislocations, and discontinuities, which determine the surface reactions and catalytic properties of active carbons.

The important applications of active carbons include decolorization, deodorization, dechlorination, separation, concentration in order to permit recovery, air purification in inhabited places such as restaurants, food processing, chemical and pharmaceutical industries, to filter, modify and separate harmful constituents from gases and liquid solutions, precious metal recovery, solvent recovery and in respirators and gas masks for work under hostile environment conditions. They are also used in medicine and for health applications to cure certain types of bacterial diseases and as catalysts and catalyst supports. They also find wide applications in the treatment of flue gases and hazardous waste remediation.

A comprehensive discussion of all the active carbon applications is beyond the scope of this chapter. Only some of the more important applications shall be briefly discussed.

20.1. ACTIVATED CARBON AS A DECOLORIZER AND A PURIFIER

Adsorption by activated carbon finds numerous applications in several food processing industries to remove unwanted color and to improve the quality and consumability of the food material. The more important industries where activated carbons have applications are the sugar industry for decolorization of sugar syrups, in the preparation of alcoholic beverages for removing unwanted compounds to improve the taste, color, and other properties, in fats and oils industry for decolorization, and in the chemical and pharmaceutical industries for decolorization and purification. The demand of activated carbons in this sector is ever on the increase because of the continuous expansion of these industries.

20.1.1. Sugar and Sweetener Industries

Activated carbons are used worldwide for the decolorization and purification of sugar syrups. Bone charcoal which was used in the nineteenth century was replaced by activated carbons at the turn of the twentieth century when its commercial production was started in Great Britain and Germany in 1911. At about the same time Norit also built up an activated carbon plant. The activated carbons with better decolorizing properties were prepared by carbonization of wood and other high-carbon-content carbonaceous raw materials in the presence of zinc chloride as the activating agent. Soon after, there was a tremendous upsurge in the production of active carbons using different source raw materials, preparation procedures, and different production conditions. Several adsorbents with ion-exchange properties and several synthetic ion exchangers were developed for decolorizing sugars but many of these increased the pH of the sugar syrup, which interfered with the crystallization process.

The more important coloring materials which need to be removed are (i) the melanoidines formed by reactions of reducing sugars with amines, (ii) the caramels which are nitrogen-free dyes produced by the partial thermal decomposition of sugars containing phenols and quinoid groups, and (iii) iron-containing phenolic complexes. These coloring substances can be present as both dissociated as well as associated molecules. The dissociated ionic forms are more common. It may, however, be mentioned that substances with molecular masses between 8000 and 15000 occurring in the colloidal form produce the most intense color. Technologically activated carbon improves both the color of the sugar syrup as well as the properties of the product. The removal of colloidal substances enhances the surface tension of the sugar solution and decreases its viscosity. Both these factors increase the rate of crystallization of the sugar solution and improve separation of crystals in the centrifuge. The metal salt impurities present in sugar solutions are also adsorbed by the activated carbon, thereby decreasing the amount of molasses formed as well.

Purification of sugar involves several steps where decolorization by active carbons is the last step and is carried out before the sugar solution is boiled to produce white mother liquor from which white sugar can be crystallized out.

Active carbon for decolorization of sugar solutions can be used in several forms such as powder, granule, and sometimes as decolorizing ion exchanger. However, better results can be obtained by using a combination of adsorbents at different stages of the production process.

Decolorization with powdered activated carbon can be carried out using a batch process or a fixed-bed process. In the case of the batch process, a given amount of the active carbon is added to a known volume of the sugar syrup in a container. The suspension is kept at a temperature between 80 and 90 °C for about half an hour to attain the adsorption equilibrium and then filtered and the syrup is separated from the active carbon. In the fixed-bed process, the sugar solution is passed through a fixed bed of active carbon. As the decolorizing capacity of the carbon bed continues to decrease with time, a system consisting of multiple filters is generally used. This ensures syrup decolorization to a constant degree.

When granulated active carbons are used, the decolorization can be carried out by fixed-bed method, the fluidized bed method, or the continuous countercurrent method. In the stationary-bed method, the sugar solution is heated to about 80 °C and then passed through a fixed carbon bed. The system consists of three sections where adsorptive decolorization of the sugar solution, washing of the active carbon, and its regeneration are carried out. In the fluidized-bed method, a certain amount of the spent-up carbon is removed from the bottom of the filter and a fresh amount of the carbon is added at the top of the bed. In the countercurrent method, the sugar solution flows upward through the activated carbon layer which moves down the adsorption column. This results in the exposure of the entire layer of the carbon surface to the sugar solution. The actual plant consists of several adsorption columns, which also include a column for the separation of sugar from the spent-up carbon, a system for removing water from the carbon, and a rotating carbon regeneration furnace.

The regenerated carbon is continuously returned to the decolorizing process. The continuous countercurrent method is the most economical because the sorption capacity of the carbon is fully utilized and the requirements of space, buildings, and the amount of sweet water are reduced to a minimum. It also reduces sugar losses and the consumption of the fuel.

The decolorizing capacity of an active carbon depends on both the porous structure which involves pore size and surface area of the carbon and the nature of the carbon surface which is due to the type and the concentration of the carbon–oxygen surface groups associated with the carbon surface, as well as on the nature of the coloring matter present and pH of the sugar solution. The carbon–oxygen surface groups make the carbon surface acidic or basic in character [1–3]. Sulfitation of the sugar solution before treatment with active carbon improves decolorization because sulfitation reduces the alkalinity of the sugar solution. Active carbons are also effective for the decolorization of several other natural and synthetic sweeteners such as glucose and fructose, corn syrups, dextrine, mannitol, and red and white vinegar. The decolorization of vinegar is considerably enhanced by increasing the mesoporosity of the activated carbon [4].

20.1.2. Oils and Fats Industry

Active carbons are very effective materials for the decolorization of oils and fats. They are generally used in combination with certain bleaching clays because clays are less expensive and the combination can produce good results. The composition of the mixture of the active carbon and the clay depends upon the nature of the oil or the fat and the nature of the coloring matter. A small amount of the activated carbon can reduce the amount of the clay considerably for decolorization of a given amount of the oil or the fat. The correct dose of the mixed adsorbent and the processing conditions depend upon the intensity of the color of the oil. The decolorization is usually carried out in a batch process. The oil or the fat is likely to be oxidized during the decolorization process due to the presence of acids in the fat or the oil. The oil or the fat is therefore first deacidified. A suitable dose of the mixed adsorbent is then added. The decolorization is then carried out in a vacuum vessel in the temperature range 100–120 °C. The charge in the vacuum vessel is vigorously stirred. The decolorized oil is separated from the carbon in a filter press. The small amount of oil remaining in the filter cake can be separated by washing the filter with petrol or some other solvent. In case the fat or the oil contains vitamins which need to be protected, the decolorization is carried out at room temperature. In this case, the oil is first thinned by the addition of a solvent before the decolorization.

Refined corn oil is obtained from crude oil through several steps which include incubating with alkali to remove fat, followed by washing, dewatering, decolorization, and deodorization. Activated carbon helps to improve the odor and the color of the corn oil [5].

20.1.3. Alcoholic Beverage Industries

Activated carbons have several functions in the manufacture of alcoholic beverages. The treatment with activated carbon promotes taste, color, and flavor, eliminates off-odors, delays oxidation, purifies and detoxifies water used, and supplies minerals. The minerals present in active carbon promote quality of the beverage and enhance its nutritional value. Several types of carbons such as charcoals, activated carbons impregnated with silver, and nonwoven fabrics impregnated with activated carbon are used.

Activated carbons are used at several stages of the brewing of beer. When used directly in the brewing process, activated carbons modify the color and improve the taste and odors produced by phenols, and by the autolysis of yeast and infections [6,7]. They are also used for the purification of water, air, and carbon dioxide used for brewing beer. The purification of beer is generally carried out before bottling. But if the quality of the beer is not good, about 2–2.5 g of activated carbon is added to the beer in the cask several days before bottling. Being universal adsorbents, activated carbons can also remove some useful components from the beer. Therefore, only minimum amounts of activated carbons should be used. These amounts can be determined through laboratory tests.

Vodka is prepared from a mixture of alcohol, water, and alcoholic extracts of certain seaweeds. According to Russian patent [8] 10,000 liters of vodka formulation contains 100–200 L of aqueous-alcoholic extract of *Unidaria pinnatifida*, 100–200 L of alcoholic extract of *Costaria costata* seaweeds, and rest of aqueous-alcoholic mixture. The seaweed extracts are obtained from a fresh raw material with 40% aqueous alcohol (ethanol). The resulting mixture is then cleared by passing through activated carbon filters and bottled. Before mixing alcohol with water in vodka manufacture, the water is purified from free organic acids [9] by the addition of suspension of alkali or alkaline earth metal, oxides, or hydroxides, and then cooled and filtered through active carbon filters which have already been treated with potassium permanganate. Vodka is finally treated with activated carbon impregnated with silver. Activated carbons can also remove fulvic acid present in water [10] and can reduce the amount of oxides and peroxides [11].

Vodka with a high nutrient value has been prepared [12] by mixing soya bean extract, vanillin, sucrose, fructose, and citric acid in suitable proportions. The contents are stirred thoroughly and then purified by passing through activated carbon filters and bottled. The amounts of organic impurities [13,14] and undesirable substances in water–alcohol solutions used in the preparation of vodka can be measured in terms of its optical density. The treatment with activated carbon has been found to reduce the optical density. The best-quality vodka prepared using distilled water and treated with activated carbon has an optical density at 260 nm and can be used as an index of its properties.

Rectified spirit is prepared by the rectification of crude spirit when three different fractions are collected: the light fraction, the medium fraction, and the heavy fraction. The light fraction contains low-boiling aldehydes and many other oxygen-containing organic compounds, while the heavy fraction is mainly fusal oil. The middle fraction is the rectified spirit which always contains small amounts of the fusal oil which gives it an unpleasant taste and odor. This fusal oil can be removed by filtering the rectified spirit through a bed of activated carbon. After the adsorptive removal of fusal oil, the rectified spirit is of good quality and can be used for preparing alcoholic beverages.

Brandies pick up fusal oil, furfural, and tannins during their manufacture, which give them undesirable taste and flavor. Activated carbons can help to remove these flavors by reducing the amounts of these impurities and certain other components in the raw distillate and accelerate maturing of the brandies. Addition of small amounts of activated carbons to the extent of 5 g/L produces good-quality brandies; however, to produce a distinctive taste and flavor, about 30 g/L of the activated carbon has to be added.

In the production of wines, the active carbons must conform to specific requirements. The activated carbon should be able to adjust the color and remove from it undesirable components originating from molds, cork, yeast, and the container. It should also not remove the components of the wine which give wines desirable characteristics. Several activated carbon manufacturers have developed and marketed specialty activated carbons that are more suitable for applications in the manufacture of specific wines. Detailed studies into the removal of particular substances from wines using activated carbons have been carried out [15] using a large number of different varieties of activated carbons. It is, therefore, essential to carry out laboratory analysis of wine to select a suitable active carbon and its amount which can give the best results. Usually, the amount of carbon varies between 0.5 and 1.0 g of carbon for each liter of wine or the must, but to avoid loss of quality the smallest effective dose should be determined in the laboratory. The activated carbon is added to the batch of wine in a vat in the form of its suspension in a small volume of wine and stirring the contents of the vat or by bubbling air fed at the bottom of the vat. This process takes several hours after which the carbon is separated by sedimentation followed by filtration on a filter press or in a centrifuge. Although activated carbons are insufficient in removing the wine components, which may sometimes reduce the quality of wine, the application of activated carbon in improving the taste and color of wines constitutes an important step in the preparation of wine. It is expected that the preparation of new grades of carbons with more selectivity will bring further progress in the production of better-quality wines.

20.1.4. Chemical Industry

Activated carbons have been and are widely used for the decolorization and clarification of solutions in many chemical industries. Active carbons can help in the removal of polymeric impurities and colloidal substances. The former impurities tend to make the product darker in color and cause several technological problems, especially during crystallization. The latter substances render the solution translucent or opaque by flocculation. The decolorization and clarification of the chemical solutions are generally carried out by adding a certain amount of a suitable grade of an active carbon and allowing the system to equilibrate. The active carbon clarifies the solution by flocculating the colloidal substances and decolorizes it by adsorption of the coloring matter. The carbon is then separated from the clarified and decolorized solution using filtration or centrifugation. The operation is usually carried out at room temperature. In the case of viscous solutions, either higher temperatures or dilution with suitable solvents is carried out to decrease the viscosity of the solution before decolorization or clarification of the solution.

The more important factors that determine the effectiveness of an active carbon for decolorization and clarification are the nature of the solvent, the nature of the carbon surface, the amount of ash content in the carbon, and the pH of the solution. Usually, the polarity of the solvent determines its effectiveness and is in the order water > methanol > ethanol > ethyl acetate > acetone > chloroform. The pH of the solution determines the amounts of different species present in the solution and results in the ionization of carbon–oxygen surface functional groups present on the carbon surface. The ionization of the carbon–oxygen surface groups gives the surface a charge. This carbon surface charge results in electrostatic interactions between the carbon surface and the species present in the solution. The ash present in the active carbon contains minerals which can contaminate the product being purified. Active carbons can also promote catalytic oxidations of certain organic compounds and may reduce yields. These oxidation reactions can be reduced by adding certain reducing agents such as sulfites or SO_2.

20.1.5. Activated Carbon Applications in Pharmaceutical Industry

Activated carbons have countless and varied applications in pharmaceutical industry. They are used for the adsorptive removal of coloring matter and bacterial toxins [16,17] which cannot be removed by filtration or destroyed by sterilization. They have applications in the production of vitamins, intravenous solutions, antibiotics, antivirals, analgesics, enzymes, and many pharmaceutical intermediates. Purification of glycerol, lactic acid, glutomic acid, tartaric acid and their salts, and the isolation of vitamins and penicillin from their fermentation broths are also carried out using activated carbons. Pharmaceutical industry uses only high-purity activated carbons because the concentration of the soluble impurities has to be very small. The activated carbon has to be used in small amounts relative to the product being treated. The usual activated carbon dose is from 0.1 to 0.5 percent which after shaking with the solution for a little time (~10 min) is separated by filtration on a glass filter paper. Filtration in this case is a very critical procedure because it is difficult to remove the last traces of carbon. In some cases, the impurities present in the carbon influence the pH of the product being treated. In such cases, corrected pH active carbons are used. Thus, the more important factors in the selection of the active carbon for pharmaceutical industry are the types of the impurities present in the solution, the basic treatment technology, the type of the activated carbon (powdered or granulated), and the purity level of the carbon. The varied applications of active carbons in pharmaceutical industry suggest that the global demand of these specialty carbons is likely to increase by about 5% every year.

An important application of activated carbon in the pharmaceutical industry is the depyrogenation [16–18] of solutions for hypodermic infections. These solutions are contaminated by bacteria which are not eliminated by filtration or destroyed by sterilization. The active carbon when used has a depyrogenating effect and is therefore directly used for this purpose. Activated carbons can also be used for the decolorization and deodorization of proteins and peptide solutions as well as for the removal of lipid denatured proteins [19].

20.1.6. Electroplating Industries for Purifying Electrolytic Baths

Electroplating is a useful surface coating process which protects metal surfaces from corrosion and decorates them. Recently, plastic surfaces are also given electroplating treatment to produce lightweight materials with shining metal-like surfaces. The electroplating process is carried out in an electrolytic bath which generally contains emulsified oils and fats, wetting agents, lacquers, and brighteners. After electroplating for several times, impurities such as residues of lacquers, decomposition products of wetting agents and brighteners, and impurities obtained from the surface accumulate in the bath which needs to be replaced. The insoluble impurities can be removed by filtration, while the soluble and emulsified impurities can be removed by activated carbon treatment.

The cleansing of the electrolytic baths can be carried out using batch process or a continuous process. In the batch process, a certain amount of the activated carbon is added into the bath, stirred for 20–25 min and then filtered. The process can also be carried out by filtering the bath solution through a fabric coated with an activated carbon. In the continuous process, the bath solution is purified by circulating it through a bed of activated carbon. The frequency of filtration and the amount of carbon used depends on the number of electroplating operations carried out and the extent of the contamination.

The first bath purification was carried out in the case of nickel-plating baths which produced bright nickel plating. Later on good results were also obtained after purification of acid copper plating, alkaline tin plating, and cyanide baths. Activated carbons can remove certain useful and expensive components of the bath such as wetting agents and brighteners which have to be added to the bath for further electroplating. It is many times preferable to purify a part of the bath solution and mix this with the other part of the solution. The activated carbons can also catalyze some decomposition or oxidation processes. Thus, only a required amount of the activated carbon should be used. Usually, this amount varies between 0.6 and 1.2 kg per 100 L of the bath solution [1].

20.2. ACTIVATED CARBON AS A DEODORIZER

Activated carbon filters can be effectively used for deodorization of inhabited places such as the social and occupational quarters, the kitchens of restaurants where the offensive gases and vapors can leak into the dining hall or other parts of the main building, and the laboratories attached to different chemical and pharmaceutical production units. Such places usually have activated carbon filters placed in ventilating system which can effectively remove the stable offensive and toxic vapors and supply fresh air to maintain a pleasant atmosphere to breathe. Ozone from ultraviolet lamps placed in ducts or in the rooms has been used for the oxidative decomposition of the offending vapors. However, ozone even in small amounts affects the mucous membranes and, therefore, the process is useful only for places of food storage. The activated carbon filters are, therefore, used widely. These filters contain activated carbon pellets or coarse crushed carbon fractions. The adsorption capacity of the carbon can be enhanced by impregnation with certain metal salts or by mixing them with other kinds of activated carbons and sometimes by mixing them with ion-exchange resins [1]. The effective life of an activated carbon filter is determined by the temperature and the humidity of the air as well as by the nature and the amount of the offensive contaminants to be removed. The average life of the filters is usually several months and they can be regenerated by activation with steam.

Deodorization activated carbon filters can be very conveniently used on ventilation of cold storages in the food and fruit storerooms where they can prevent the contamination of one product from the odors of other products. Activated carbon filters reduce the concentration of ethylene in the atmosphere which retards the ripening of apples and pears and enhances their shelf life. In addition, these filters can also be used in the ventilation system of public places such as schools, cinemas, and public buildings where large public gatherings are present. In big cities where air pollution is ever on the increase, deodorization of the outside air for ventilation is also becoming important. In Los Angeles, for instance, activated carbon deodorization filters have been successfully used for antismog protection. The offensive and toxic gases and vapors are removed almost completely by the activated carbon filters, and the filtered air provides an acceptable atmosphere even when the outside pollution level is very high.

20.3. ACTIVATED CARBON ADSORPTION IN NUCLEAR TECHNOLOGY

Helium is used as a protective gas in a nuclear reactor cooled and moderated by heavy water [20]. Helium forms a protective layer above the surface of heavy water and carries away deuterium and oxygen which are formed by radiolysis. These gases can form an explosive mixture. In addition, argon gets mixed up with the helium from the air which enters the system during exchange of fuel elements. Consequently, it is necessary to decontaminate helium and the exhaust gases.

The mixture of helium and exhaust gases is collected by a blower and passed over a reactivator in which deuterium and oxygen are converted into heavy water over a catalyst. The gases are then cooled and passed through an adsorber containing activated carbon cooled by liquid nitrogen. Activated carbon adsorbs argon and nitrogen and possibly also radioactive krypton and xenon which leak into the protective gas from faulty fuel elements [21]. The loaded activated carbon can be regenerated by raising its temperature and evacuation. The temperature is raised slowly so that different desorbed gases can be collected separately for disposal. The protective gas helium is collected first and recirculated into the reactor while the products of the radioactive decay are compressed into pressure vessels for disposal.

Activated carbon is also used in nuclear technology as a filter in the emergency ventilation system of the reactor building which is switched on automatically during breakdown of the reactor.

Among the various fission products released during nuclear plant operations, radioiodine is considered as the most important because it produces radioactive iodine isotopes I^{131} and I^{133}. I^{131} is the most abundant isotope and combines with hydrocarbons to produce CH_3I^{131} which is highly poisonous. Activated carbons can be used for the removal of elemental radioactive iodine and its components. However, when activated carbon is used as such its efficiency is very small because of the adsorption of water. Therefore, activated carbons are impregnated with different chemical reagents such as KI or triethylene diamine (TEDA) [22–24]. The impregnation with TEDA enhances CH_3I^{131} adsorption efficiency under high humidity conditions through specific dispersive interactions between methyl iodide and TEDA to form stable ammonium salts which remain fixed within the adsorbent porous structure.

Gonzalea-Garcia et al. [25] studied the adsorption of radioactive methyl iodine on a series of activated carbons impregnated with TEDA at 5 and 10 wt % and observed that the adsorption efficiency was markedly enhanced by the addition of the impregnant. The adsorption involved both the physical adsorption through nonspecific interactions and chemisorption through specific interactions between the methyl iodide and TEDA. The trapping efficiencies up to 98.1% were obtained. The amount of chemisorption was stoichiometrically proportional to the amount of the impregnant [26]. Billinge et al. [22], while comparing the adsorption efficiencies of KI impregnated coconut-based and coal-based activated carbons for radioactive methyl iodide, observed that the coal-based activated carbon was more efficient than the coconut-based charcoal. The adsorption of methyl iodide was related to the amount of carbon–oxygen surface groups associated with the carbon surface. Impregnated KI reacted with the acidic carbon–oxygen surface groups and modified their desorption behavior which enhanced the efficiency of the activated carbon for the removal of CH_3I.

The adsorption efficiency of charcoals impregnated with different pyridines has also been studied [23]. Different pyridines showed varying efficiencies for the removal of CH_3I, with pyridine being the most effective. The impregnation of the same charcoal with different amounts of 4-vinyl and 4-aminopyridine showed that while the retention of methyl iodide increases from 53 to 75% with increase in the concentration of 4-aminopyridine on the carbon surface, the retention remains more or less unchanged (~75%) with increase in the concentration of 4-vinylpyridine. Hoffmann [27] has prepared filters made from KI impregnated active carbons for the removal of iodine from off-gas from nuclear technology units. The off-gas is first passed over a noble-metal catalyst which decomposes the radioactive iodine organic compounds. The elemental iodine so produced is adsorbed in activated carbon filters.

Toshiaki and Takashi [28] studied the cleaning of laundry wastewater which contains organic compounds such as surfactants from washing clothes of nuclear plant workers using activated carbon filters. The wastewater is circulated through activated carbon filters. This treatment is capable of reducing the total organic carbon concentration from 100 ppm to 10 ppm. The spent-up carbon is then regenerated by heating in vacuum.

The preservation of atmospheric basin in the vicinity of a nuclear facility is of considerable importance. Aleshin et al. [29] has developed a special technology measure and a gas purification system. This system contains a radiochromatograph system and an activated carbon which adsorbs and purifies the gaseous wastes from short-lived radionuclides through dynamic adsorption and decay of short-lived radioactive gases.

During its operation, a nuclear reactor produces a number of highly radioactive gases which cannot be vented directly into the atmosphere. Usually, these gases are collected and stocked over the decay period. However, these gases can be decontaminated through active carbon filters and then vented to the atmosphere. As the carrier gas containing these radioactive gases is passed through the bed of activated carbon the adsorption on the surface of the carbon reduces the concentration of these gases in the carrier gas so that, in a sufficiently long active carbon bed, the activity of the gases is substantially reduced to be vented to atmosphere. In this system, oxygen gas is used as the carrier gas.

Conditions of adsorption on activated carbon used in the nuclear technology are very different and strict. The process is very expensive because only special-quality active carbons can be used and the adsorption is carried out at low temperature. As the gases coming out of the nuclear reactor have a high temperature, they are required to be passed through heat exchangers for cooling before passing over the carbon.

20.4. ACTIVATED CARBON IN VACUUM TECHNOLOGY

Clean vacuum systems are becoming more and more important for carrying out several processes. When mercury or oil diffusion pumps are used to produce vacuum in the laboratory, some oil or mercury vapors enter the evacuated space and interfere with the processes to be carried out in the evacuated system. It is quite common to see that the tubes connecting the mercury diffusion pump to the evacuated vessel appear black after some time due to the condensation of mercury vapors on the cooler portions of the tubes. Adsorption of gases and vapors from reactor vessels can be used as a method to produce clean evacuated system. The production of clean vacuum system using activated carbon does not need an elaborate equipment. The simple system involves a glass vessel filled with granulated activated carbon cooled with liquid air or nitrogen in a Dewar flask. The activated carbon is first evacuated with a rotary oil pump preferably at 350 °C to remove physisorbed gases and vapors, cooled and then connected to the system to be evacuated. The vacuum so obtained depends upon the size of the apparatus and the width of the inlet and lead tubes. It has been claimed [30] that pressures as low as 10^{-7} mm of mercury can be obtained with 30 L/sec as the rate of evacuation.

Because of the reliability and large adsorption capacity of active carbons, they can be used as complements to diffusion pumps or alone. They can also be used for degassing systems overnight without any supervision unlike diffusion pumps. The adsorption capacity of the activated carbon can be restored merely by raising the temperature of the saturated carbon to room temperature or slightly higher temperatures where the gases and vapors are desorbed.

Activated carbons can also be used in the laboratory for the separation or purification of gaseous mixtures when the gases have different adsorbabilities at a given temperature. Liquid air-cooled activated carbon can remove impurities from helium caused by air leak into the system. The method can also be used for the purification of helium enriched with He^3 isotope used in cyclotron. A mixture of neon and helium can be successfully separated by activated carbon cooled at −228 °C. Jech [31] used activated carbon as a reservoir of radon in a closed system. Radon was desorbed from the carbon on heating and entered the system. After completing the experiment, the vessel containing the active carbon was cooled to normal temperature when radon was readsorbed by the carbon.

20.5. ACTIVATED CARBON IN GAS CHROMATOGRAPHY

Active carbons are now increasingly being used in adsorption chromatography for the analysis of rare and permanent gases and light hydrocarbons. The analysis is generally carried out using activated carbons with a uniformly sized fraction preferably between 0.25 and 0.5 mm granular size. The coarse or the finer size of the activated carbon granules creates problems with the measurements. The more coarse particle size increases turbulent diffusion while the finer particles enhance the resistance of the chromatographic column. More important separations which can be carried out using activated carbon columns are the components of natural gas, water gas, generator gas, carbonization and coke oven gas [32], ventilation air from mines, mixtures of rare gases, and many more. A mixture of oxygen and nitrogen can be separated using activated carbon columns several meters long.

Several mixtures which cannot be separated using active carbon alone can be separated by using its combination with other adsorbents. Krejci [33] separated a mixture of oxygen, nitrogen, methane, and krypton using a column containing activated carbon and a molecular sieve 5A. Complex mixtures of hydrocarbons and permanent gases have been separated using two columns filled with another adsorbent. The adsorption chromatographic analysis or separation is carried out using elution technique which is better than the frontal or displacement technique for analytical purposes.

20.6. APPLICATION OF ACTIVATED CARBON AS ION EXCHANGER

Ion-exchange properties of activated carbons have been well recognized for the last several decades. They are in many ways better than many of the synthetic and mineral ion exchangers, and their ion-exchange properties can be suitably modified and can be tailor made for a particular application. Activated carbons owe their ion-exchange properties to the presence of different types of functional groups such as carbon–oxygen, carbon–nitrogen, and carbon–sulfur on their surface.

The carbon–oxygen surface groups are by far the most important in determining the ion-exchange properties of carbons. They are of two types: acidic and nonacidic surface groups. The ion-exchange properties of activated carbons are due to the presence of acidic surface groups which have been identified as carboxyls, lactones, and phenols. These groups in small amounts are always present on the carbon surface but their amounts can be considerably enhanced on oxidation of the carbons with HNO_3, H_2O_2, and ammonium persulfate in the solution phase and with oxygen gas at 350–400 °C. These modified activated carbons find widespread ion-exchange application in several technological processing units. The more important applications are in the removal of trace amounts of toxic heavy metals from drinking water, purification of inorganic salt solutions, recovery of small amounts of valuable metals, and concentration of micro-admixtures before their analytical analysis.

It may be worthwhile mentioning here that the oxidation of carbons with nitric acid is a very drastic treatment and may result in the loss of some carbon and may also produce large amounts of humins which may create problems with the ion-exchange properties of carbons. These, however, can be removed by heat treatment of the oxidized carbon in vacuum at 300–350 °C. Similarly in oxidation with oxygen gas if the treatment is carried out above 400 °C, then it results in a significant burning of the carbon. These treatments should, therefore, be carried out under milder conditions.

Electrolytic baths used for the preparation of caustic soda and chlorine gas by the electrolytes of sodium chloride using mercury contain several micro-impurities such as Mo, V, and Cr. These metallic impurities can poison the electrolytic process even when present in very small amounts. These impurities can be removed by oxidized activated carbon ion exchangers very effectively. The spent-up carbon can be regenerated by washing with hydrochloric acid (HCl) and can be reused. These carbon ion exchangers can also be used for the purification of the caustic soda solution. In this respect, they are better than the synthetic or mineral ion exchangers which are not stable to high concentrations of sodium hydroxide.

Activated carbon ion exchangers (oxidized carbons) are very useful when chemical compounds of high purity (specialty materials) are to be prepared. The more important of these specialty materials are the alkali metal hydrogen phosphates which are raw materials for the production of specialty glass, the rare earth metals which are raw materials for the production of cathode luminiphores such as KCl, $ZnSO_4$, and $(NH_4)_2CO_3$.

Ion-exchange activated carbons are also useful in several analytic chemistry procedures for enhancing their sensitivity by preconcentration of the component to be analyzed and when small amounts of additives are to be analyzed against a strong background of the main component. The preparation and quality control of high-purity materials for the microelectronic and nuclear industries, the manufacture of optical and luminescent materials, and many inorganic technology branches also utilize activated carbon ion exchangers. The activated carbons used in such sensitive technologies should be free of ash and other mineral impurities because they can contaminate the solution of the substance to be analyzed.

The treatment of oxidized carbons with ammonia results in the formation of amide, imide, and ammonium salt groups on the carbon surface. Such carbons are good anion exchangers. The formation of sulfur groups on the carbon surface also gives them ion-exchange properties, while phosphorizing the carbon surface with phosphoric acid or phosphorous trichloride at high temperatures results in the formation of cation exchangers. Although not fully explored, activated carbon ion exchangers have a great potential and are likely to be valuable supplement to synthetic and mineral ion exchangers.

20.7. ACTIVATED CARBONS AS POROUS ELECTRODES

Activated carbons are widely used as electron transfer materials in chemical devices such as fuel cells in the form of porous gas electrodes because they have a well-developed surface area, high resistance to acids and bases, and relatively large electrical conductivity. The electrode process on these electrodes involves a solid phase which is the carbon material itself, a liquid phase which is the electrolyte solution, and a gaseous phase which can be oxygen or air. The preparation of a successful carbon diffusion oxygen electrode with reproducible parameters requires a good understanding of the kinetics and mechanisms of the electrode processes, the transport micro-kinetics, and the design and the manufacturing technology of the electrode [34,35]. As activated carbon is inert to acids and bases, the carbon electrodes are very stable and durable and can be kept in contact with the electrolyte solution without any harm for a long time. The important parameter for the carbon electrode material for fuel cells is the porosity of the carbon which ensures large effective electrode surface and high rates of transport processes.

The development of activated carbon electrodes has generated a considerable interest in electrochemical reduction of oxygen on the surface of carbon materials. The rate of the reaction and the sorption properties of the electrode material depend on the chemical nature of its surface. The mechanism of the electrochemical reduction of oxygen on the carbon electrode involves diffusion of oxygen and its adsorption on the carbon surface and the transfer of an electron from the electrode to the adsorbed oxygen molecule which is converted into oxygen ion and reacts with water molecules [36]. These reactions produce peroxide ions. The decomposition of hydrogen peroxide results in the oxidation of the carbon surface which significantly influences the current–voltage characteristics of the carbon electrode. In order to remove H_2O_2 from the reaction medium and to enhance the electro-catalytical activity of the carbon–oxygen electrode in fuel cells, several catalysts such as silver and its salts, platinum, and phthalocynines and polyphthalocynines of transition metals have been suggested [35]. The diffusion oxygen electrodes can also be used as indicator electrodes in the electrochemical sensors of oxygen analyzers.

It may be worthwhile to mention here that although the mechanism and several other problems associated with electrochemical reduction of oxygen on porous carbon electrodes have been successfully solved, the mass production of carbon diffusion electrodes has not been carried out. This is probably due to the fact that the surface properties of activated carbons cannot be completely reproduced. The future, however, is bright for the carbon diffusion electrodes because specialty carbon materials with well-defined surface characteristics are now being prepared.

20.8. ACTIVATED CARBON APPLICATION IN THE REFINING OF MINERAL OIL AND GASOLINE

Activated carbons are effective adsorbents for the refining of mineral oils and gasoline. In the case of mineral oils, activated carbon is not used alone because of its cost. It is used in combination with certain bleaching clays, silica gel, and several other polar adsorbents. However, it is often used alone for the production of specialty products such as vaseline, paraffin, and several oils used in medicine.

Gasoline obtained from primary distillation of crude oil contains several types of sulfur compounds including mercaptans. These sulfur compounds can be removed from gasoline by passing through an adsorber containing a bed of granulated activated carbon impregnated with sodium hydroxide. The mercaptans are removed partly by adsorption on the carbon surface and partly by a chemical reaction with sodium hydroxide:

$$RSH + NaOH \rightarrow NaRS + H_2O$$

The bed can be regenerated by steam treatment until there is no more mercaptan odor. The regenerated bed can be reused after recharging with sodium hydroxide. Between 400 and 500 kg of gasoline can be treated with 1 kg of activated carbon depending upon the activated carbon grade.

20.9. ACTIVATED CARBON APPLICATIONS FOR HUMAN BODY PROTECTION

Protection against toxic and hazardous gases has been one of the important applications of activated carbons since World War I when in 1915 the German army released chlorine gas on allied forces. Protective clothing at that time consisted of impermeable rubber. However, wearing of such clothes for longer periods of time was very uncomfortable, especially in summer months. An immediate research was initiated to develop new lighter and more efficient and effective materials for protection against chemical warfare gases. The effort for more sophisticated design of adsorbents and respirators was further intensified with the development of new chemical and biological weapons which release more toxic and hazardous gases. Activated carbon, because of its large surface area and high chemical reactivity of its surface, has been considered as a potential adsorbent for preparing respirators and human body clothing because most of these gases can be adsorbed by the activated carbon and prevented from reaching the human body.

Activated carbon itself can provide sufficient protection against gases which can be easily adsorbed or for war gases with low chemical reactivity. The chemical reactivity of the activated carbon surface is enhanced by impregnation with suitable chemical reagents. The activated carbons act as both adsorbents and a porous support for the impregnated material. Although the physical adsorption of toxic gases by the carbons decreases on impregnation, it is compensated by their removal due to chemical reactions between the gas and the impregnated reagent. The protection of human body by the impregnated carbons involves adsorption as in the case of chloropicrin, chemical reactions with the impregnant such as neutralization, complex formation, hydrolysis as in the case of phosgene and hydrogen cyanide, and catalytic reactions such as oxidation in the case of arsine.

The effectiveness and the efficiency of a respirator or a protective clothing depend upon the porous and the chemical structure of the carbon surface, the nature of the impregnant, and the method of impregnation. Usually, carbons with sufficient mesoporosity and with low water adsorption capacity are used because the mesoporosity provides enough place for the deposition of the impregnant and moisture is detrimental and reduces the protection. The chemical substances used as impregnants are generally compounds of copper, chromium, zinc, cobalt, silver, manganese, vanadium, and molybdenum, as well as several organic compounds such as pyridine or some aromatic amines. The method of impregnation has a considerable influence on the protection of the adsorbent. Impregnation in several steps using dilute solutions is better than in a single step. This has been attributed to the deposition of the impregnant in the form of fine particles in a multistep impregnation. The composition of the impregnants and the procedure for impregnation are closely guarded military and industrial secrets.

Activated carbons impregnated with a combination of metals such as Cu^{2+}, Ag^{+}, Cr^{6+}, NH_4^+, and CO_3^{2-} have been found to be very effective adsorbates for Cyanogen chloride, hydrogen cyanide, phosgene, arsene, and chloroform [37–40]. In most cases, the adsorption on impregnated carbons involves chemisorption and decomposition. For example, in the case of an activated carbon impregnated with Cu^{2+}, Cr^{6+}, Ag^{+}, and NH_4^+ in a given ratio, the Cyanogen chloride was chemisorbed irreversibly and was given out as CO_2 on heating the loaded carbon sample. The surface of the impregnate was regenerated and more Cyanogen chloride could be chemisorbed. Comparison of the adsorption behavior of BPL activated carbons with ASC whetlerites prepared by impregnation with Cu^{2+}, Cr^{4+}, and Ag^{+}, and of ASB carbons prepared by impregnation with Cu^{2+}, and BO_3^{3-}, for the adsorption of phosgene, Cyanogen chloride, hydrogen cyanide, and chloroform [37,38] showed that the adsorption of phosgene and chloroform increased by about 17% on the ASB impregnated carbons while the adsorption of Cyanogen chloride increased by about 20% on ASC whetlerite. Both physisorption and chemisorption were observed on the impregnated carbons, although chemisorptions were more pronounced in the case of phosgene, Cyanogen chloride, and hydrogen cyanide. The breakthrough curves of phosgene mixed with dry and wet air through a NBC canister made by impregnating a coconut shell activated carbon with Cu, Cr, Ag, sodium hydroxide, and pyridine [39] showed that, at 4.0 mg/L phosgene concentration, the canister was able to provide protection for 84 min under wet and only 37 min under dry conditions at 30 °C and 30 L/min flow rate. The protection mainly involved physisorption, hydrolysis, and a chemical reaction between the impregnant and phosgene. These processes can be represented as

$)C + COCl_2$	$\longrightarrow$	$)C\text{-}COCl_2$	Adsorption
$)C + COCl_2 + H_2O$	$\longrightarrow$	$)C + CO_2 + 2\,HCl$	Hydrolysis
$)C\text{-}CuO + COCl_2$	$\longrightarrow$	$)C + CuCl_2 + H_2O$	Chemical reaction
Impregnant		Carbon surface	

An activated carbon cloth retained about five times more phosphine when impregnated with $AgNO_3$ and three times larger when it was impregnated with $Cu(NO_3)_2 \cdot 3H_2O$ [41] as compared to the nonimpregnated carbon. Active carbons impregnated with silver and copper compounds are particularly useful for the removal of chlorine, phosgene, arsine, and hydrogen cyanide from air [39]. The addition of zinc oxide into the impregnant enhanced the activity of the carbon for phosgene and diphosgene. The chromium copper adsorbent prepared by impregnating activated carbon in a solution of tetramine Cu (II) carbonate, ammonium chromate, and silver salts exhibited good protection against HCN, CNCl, and AsH_3. The protection was better against CNCl when the proportion of chromate was more on the adsorbent and better against arsine when it contained more of copper. Protection against hazardous gases such as sulfur mustard can also be carried out using impregnated activated carbons [42].

Activated carbons impregnated with certain organic compounds such as pyridine, 4-vinylpyridine, 4-aminopyridine, 4-cynopyridine, and 4-n-propylpyridine have also been used for the adsorptive removal of Cyanogen chloride [23]. The activated carbon impregnated with 4-vinylpyridine showed higher adsorption capacity compared to the pyridine-impregnated carbon. The reactivity of the impregnated carbon increases with increase in the amount of the impregnant, although it varies from one pyridine to another.

The respirator filters for the protection of upper respiratory tract against toxic gases generally contain an antismoke screen which traps the biological aerosols, radioactive dust and irritating smokes, and a layer of activated carbon which is held between two sheets of wire gauge for the removal of toxic gases. In complete body suits for the protection of whole body from nerve gases which are generally physically adsorbed, the activated carbons used should have a high surface area and a high pore volume. Thus, spherical carbons which provide a higher surface area and permit rapid and effective adsorption are used. These carbon spheres are fixed on a fabric with the help of suitable adhesives.

Such modified carbons in the form of filters are also suitable for cleaning air in underground shelters and army transport vehicles. They can also be used in several chemical and process industries which produce hazardous and toxic gases and vapors and can endanger the life of workers especially during emergencies. In industrial protective respirators, the activated carbons are generally impregnated with potassium salts for acidic pollutants, with zinc salts for protection against ammonia and H_2S, with copper salts for HCN, with potassium bicarbonate against SO_2, with potassium iodide for catalytic oxidation of PH_3, and with iodine or sulfur compounds for protection against mercury vapors.

20.10. ACTIVATED CARBON FOR RECOVERY

Activated carbons are versatile adsorbents for the recovery of several organic and inorganic compounds which are of considerable commercial importance. The more important compounds which are recovered using activated carbons are the organic solvents, precious metals, and sulfur.

20.10.1. Recovery of Organic Solvents

Recovery of organic solvents from industrial waste gases is an important application of activated carbons. This has significance from the commercial point of view because of the recovery of valuable materials as well as protects the natural environment from air pollution. Some of the components of the waste gas are inflammable and can result in fires. The emissions from power plants can combine with NO in the presence of sunlight to produce ozone. Ozone, although useful in the upper atmosphere and prevents UV radiations reaching the Earth's surface, can result in acute respiratory problem, decrease lung capacity, and harm the immune defense system.

Recovery of solvents using activated carbons is very cost effective. In well-designed units the cost of recovery is less than 20% of the price of the solvent because the recovery efficiency of the adsorption process is very high (92–98%). The recovery costs, however, are determined by the process conditions, the size and design of the unit, and the composition of the waste gas. The first solvent recovery using activated carbon prepared by activation with $ZnCl_2$ was carried out in 1916 [1].

The main solvents recovered by adsorption on activated carbons are benzene, toluene, xylene, petrol, lower alcohols, carbon disulfide, and halocarbons such as chloroform, methylene chloride, carbon tetrachloride, and dichloroethane. These solvents are recovered from waste gases of printing and paint manufacturing units and from dry cleaning, polymers, adhesives, rayon, rubber, and several other industries. The concentration of these solvent vapors in the waste gases is usually of the order of 1–20 g/m^3. The recovery of the solvent is carried out by passing the waste gas through a bed of activated carbon. The organic solvent vapors are adsorbed by the activated carbon and almost pure air passes out as effluent. The solvent from the saturated carbon bed is usually recovered by desorption with steam and separating the condensed vapors. The active carbon in turn is regenerated for reuse. The high efficiency of the activated carbon in the recovery of solvents present in small concentrations is very important from the point of view of safety because the ventilation system can be so arranged to maintain a constant composition of the air–vapor mixture well below the explosive limit. The application of activated carbon in solvent recovery plants is going to increase considerably in view of the fact that ever-increasing amounts of various solvents are being used in several industrial processes and because of the ever-increasing prices of the solvents.

20.10.2. Recovery of Gold

Activated carbons can selectively adsorb gold and silver from solutions containing base metals such as copper, iron, cobalt, nickel, and antimony. Consequently, these are abundantly used for the extraction of gold from chlorine and cyanide leach solutions. The first attempt to recover gold from chlorine leach solutions was carried out as early as 1880 [43] using wood charcoal but later on extended to cyanide leach solutions. However, the carbons available at that time had smaller pore volume and there was no process available to obtain gold from the carbon surface except to burn it and to melt the gold. But this produced a low-quality gold. Zadra [44], working with US Bureau of Mines, in 1950 developed a direct carbon-in-pulp (CIP) process which formed the basis for the first CIP commercial plant in USA by air Homestake Mining Company [45] for the large-scale production of gold using activated carbon adsorption. By that time good-quality activated carbons with a large surface area and pore volume which were more suitable for the recovery of gold were available. Consequently, several gold recovery plants based on activated carbon adsorption were set up in South Africa where zinc concentration process was being used. In the CIP process, the adsorption of gold by the carbon is not influenced by the presence of other baser metals in the leach solution; the activated carbon can therefore be directly added to the leach solution avoiding any filtration and clarification steps and making the CIP process more economical. Furthermore, the gold could be recovered from the carbon surface by elution which also regenerated the activated carbon. With the availability of activated carbons with surface characteristics more suitable for the adsorption of gold, it will be possible to recover gold from the gold plant effluents.

The mechanisms of gold adsorption by activated carbons have been differently explained by different groups of workers, the more important being ion–ion pair adsorption theory, the aurocyamnide anion adsorption theory, the cluster compound adsorption theory, and the reduction theory. Similarly, several methods have been developed to strip gold from the carbon surface. These methods involve high-temperature desorption of gold using dilute solutions of certain electrolytes or the selective desorption of gold from the carbon surface using certain organic solvents. The detailed discussion of the mechanisms of adsorption of gold by activated carbons and the procedures for elution of gold from the activated carbon surface is beyond the scope of this chapter and have been discussed elsewhere [2,3].

20.10.3. Recovery of Sulfur

Industrial waste exhaust gases contain large amounts of toxic sulfurized gases such as SO_2, H_2S, CS_2, and sulfur-containing organic compounds. Removal of these gases from exhaust gases is one of the major applications of activated carbons. The removal helps to protect the environment from pollution and provides commercial benefits in terms of the recovery of a valuable component sulfur.

Sulfur dioxide mainly enters the atmosphere through waste gases from coal-fired power plants, from metallurgical industries using ferrous and nonferrous sulfide ores, many chemical industries, and from petrochemical-based units. Sulfur contained in fuels and sulfide ores is usually given out as SO_2 during their combustion. Although attempts are made to remove sulfur from the fuels, these have proved to be more or less ineffective. The emission of SO_2 to atmosphere in highly industrialized countries is of the order of 20–30 million tons per year. These large amounts of SO_2 are recovered by conversion into sulfuric acid.

There are several methods by which this recovery can be carried out, but the method used depends upon the composition of the flue gas. When the flue gas does not contain any oxygen or water vapor, the removal of SO_2 involves physical adsorption on the carbon surface. The adsorbed SO_2 can be removed from the carbon surface by evacuation or by passing another gas over the carbon surface. However, most often, the flue gases are associated with appreciable amounts of oxygen and water vapors. In such a case, both physical adsorption into the micropores and oxidation of SO_2 to SO_3 take place resulting in the formation of sulfuric acid. Some of the sulfur is also chemisorbed on the carbon surface as carbon–sulfur surface groups. The amount of sulfur removed by each of these processes depends upon the temperature and the composition of the flue gas as well as on the nature of the carbon surface. At higher temperatures, the oxidation of SO_2 to SO_3 and chemisorption are predominant while at lower temperatures physisorption of SO_2 is larger. The chemisorbed sulfur is difficult to be removed because carbon–sulfur compounds are highly stable and can be removed only as H_2S on treatment with hydrogen gas. Therefore, activated carbons associated with larger micropores or mesopores and having smaller amounts of carbon–oxygen surface groups are preferable. In such cases, the SO_2 will preferably involve physisorption.

The purification of flue gases can be carried out using several methods which include stationary or fluidized-bed adsorbers. The most commonly used recovery processes are the Sumitomo process developed in Japan and the Beagban–Forschung (BF) process developed in Germany and then developed in USA. The details of these recovery plants together with the conditions of operation are available in the literature [46–48]. Briefly speaking, the flue gases are first passed through a dust catcher and then through a battery of adsorbers containing activated carbon. The purification of the flue gas and the recovery of SO_2 are carried out between 100 and 150 °C. The adsorbers are then washed first with concentrated and then with dilute sulfuric acid followed by washing with water. The liquid washings are collected which contain about 50% of sulfuric acid which can be further concentrated by heating. In fact, several sets of adsorbers are used. While adsorption is carried out in one set, the washings simultaneously take place in the second set. The hot flue gases after purification are passed through a third set of adsorbers which contain washed but still wet activated carbon. This helps to dry the carbon and removes last traces of SO_2 from the flue gas. The process is therefore continuous.

Activated carbon fibers prepared from polyacrylonitrile, coal tar, and petroleum pitches have been found to be very effective for the conversion of SO_2 into H_2SO_4 [49]. The catalytic activity of these carbon fibers has been found to increase with increase in the heat treatment temperature, the extent of increase depending on the type of the carbon fiber, the heat treatment temperature, and the experimental conditions of its preparation [50].

Carbon disulfide and hydrogen sulfide are two other obnoxious sulfur compounds which are present in waste gases emitted from the rayon industry. The amount of H_2S and CS_2 can be as large as 2000 mg/m^3 and 8000 mg/m^3, respectively. H_2S has been removed by absorption in some solutions but the removal has not been complete. CS_2, on the other hand, can be effectively removed by adsorption using activated carbons. During the last three decades, attempts have been made to develop technologies where both CS_2 and H_2S can be removed simultaneously by using activated carbons. In one such technology, the flue gases containing CS_2 and H_2S are passed through a single static bed of a mesoporous activated carbon impregnated with iron. H_2S present in the flue gases is oxidized in the front layer of the carbon bed while the adsorption of CS_2 takes place afterward. The elemental sulfur produced by the oxidation of H_2S and the adsorbed CS_2 are recovered from the carbon surface in a multistep regeneration cycle [46–48].

In another method, the flue gases are passed at 20–50 °C through two stationary beds containing two different types of activated carbons. One of the beds contains a mesoporous carbon with pores in the 40–80-nm range usually impregnated with iodine. H_2S is oxidized on the surface of the impregnated carbon. The elemental sulfur so produced is deposited in the mesopores. In the second stationary bed, which contains a microporous carbon, adsorption of CS_2 takes place. The adsorbed CS_2 is removed by passing super heated steam through the bed and the elemental sulfur is extracted with liquid CS_2. Small amounts of H_2SO_4 formed as a result of partial oxidation of H_2S can be removed as ammonium sulfate by passing gaseous ammonia through the carbon bed and can be easily washed from the carbon surface with water in the regeneration cycle. The catalytic oxidation of H_2S on the activated carbon surface also finds applications in the desulfurization of several industrial gases such as water gas or coke oven gas.

20.11. ACTIVATED CARBON APPLICATIONS IN WASTEWATER TREATMENT

Drinking water contains small amounts of a large number of organic and inorganic compounds such as metals, phenols, pesticides, aliphatic and aromatic hydrocarbons, dyes, surfactants, halomethanes, organic sulfur compounds, amines, nitro compounds, and many more. These compounds enter surface water and groundwater through industrial and municipal discharge, urban and rural runoff, natural decomposition of animal and vegetable matter, and from water and wastewater chlorination practices. Many of these chemical compounds are carcinogenic and many others cause several other ailments of varying intensity and character. Several physical and chemical methods such as coagulation, oxidation, aeration, and ion exchange have been used for the removal of these chemical pollutants from water. These methods have achieved only a limited success because the amount and the variety of these chemicals in wastewater are always increasing. Many studies which include laboratory tests and field operations have shown that the adsorption by activated carbons is perhaps the best broad-spectrum control technology at the present moment.

The adsorption capacity of an activated carbon for organic and inorganic compounds from aqueous solutions generally depends upon the porous characteristics of the carbon surface which includes surface area and pore-size distribution, the chemistry of the carbon surface, the electrokinetic properties, and the nature and the dimension of the pollutants. In the case of metal ions, the more important factors that influence and determine the adsorption are the carbon–oxygen groups present on the carbon surface and the pH of the solution. These parameters determine the nature and concentration of the ionic and molecular species in the solution as well as the nature and concentration of the carbon surface charge. The carbon surface can have a positive or a negative charge depending upon the pH of the carbon-solution system. The positive charge on the carbon surface is due to the presence of basic surface groups, the excessive protonation of the carbon surface at low pH values, and to graphene layers that act as Lewis bases. These Lewis bases result in the formation of acceptor–donor complexes important for the adsorption of many organic compounds. At higher pH values, the carbon surface has a negative charge due to the ionization of acidic carbon–oxygen surface groups.

The adsorption of metals mainly involves electrostatic interactions between the positively charged metal ionic species present in the solution and the negative sites created by the ionization of the acidic carbon–oxygen surface groups involving ion-exchange mechanism [51–62]. The dispersive interactions between the ionic species in the solution and the graphene layers and the surface area of the carbon play only a smaller role in the adsorption of inorganic compounds.

In the case of the organic compounds which can be polar or nonpolar, the adsorption will involve both the electrostatic and dispersive interactions. In addition, hydrogen bonding can also be an important consideration in the adsorption of certain polar organic molecules. The porous structure of the activated carbon shall also have an important consideration for the adsorption of essentially nonpolar organic molecules because a certain proportion of the microporosity may not be accessible to very large organic molecules. The discussion of the adsorption processes involved in the adsorption of individual organic and inorganic compounds and the mechanisms involved shall be the subject matter of a separate chapter in the book.

20.12. ACTIVATED CARBON APPLICATIONS IN CATALYSIS

Besides being good adsorbents, activated carbons can promote a variety of surface reactions. They can be used directly as catalysts as well as suitable catalyst supports. The role of activated carbon as a catalyst is, however, different in different reactions. While the surface area and porous structure of activated carbons play a dominant role in some reactions, the concentration of the active sites and the carbon surface chemistry become extremely important in some other reactions. The carbon surface is invariably associated with certain heteroatoms such as oxygen, hydrogen, nitrogen, and sulfur which are present in the form of functional groups. The carbon–oxygen surface functional groups are by far the most important surface groups which considerably influence the catalytic properties of these materials. These carbon–oxygen surface groups have been identified as acidic functional groups which are carboxyls, lactones and phenols, and nonacidic functional groups which have been postulated as quinones. The acidic functional groups render the carbon surface polar and hydrophobic in character which enhances the adsorption of metal cations and polar organic compounds. The nonacidic surface groups make the carbon surface hydrophobic and enhance the adsorption of nonpolar organic molecules. Thus, both the porous and the chemical structure of the carbon surface are important parameters in catalytic applications of activated carbons. Furthermore, these parameters can be modified by suitable treatments to satisfy any specific requirements during carbon catalyst preparation so that the catalyst surface is accessible to precursor materials. In this section, some catalytic reaction where activated carbon itself acts as a catalyst or as a catalyst support or both as a catalyst and as a catalyst support shall be discussed. These catalytic reactions have been classified as oxidation reactions, reduction reactions, combination reactions, and reactions catalyzed by acid or base supported on activated carbons. A brief discussion of such a type of reaction will be given because a detailed discussion is beyond the scope of this chapter.

A comprehensive review on activated carbons as catalyst support has already been published [63].

20.12.1. Oxidation Reactions

Activated carbon-supported metals can catalyze a large number of oxidation reactions used in the preparation of industrial products and intermediates. In the synthesis of many fine chemicals and pharmaceuticals, they act as selective catalysts to prepare individual products with a high degree of selectivity. Several workers have carried out selective oxidations using gold and platinum supported on activated carbons as catalysts. Similar catalysts have been used for the preparation of glyceric acid [64,65] with 100% selectivity by the oxidation of glycerol and selective liquid-phase oxidation of ethylene glycol to glycolate [66]. Several other selective conversions, such as carbohydrates and aldehydes to corresponding carboxylic acids [67], glyoxal into glyoxalic acid have been carried out using gold- and palladium-supported carbon catalysts. Diethyl carbonate, which is an additive to gasoline and diesel fuels to reduce pollutant emissions, has been prepared by air oxidative carbonylation of ethanol in gas phase with high selectivity using $CuCl_2$ or $PdCl_2$ supported on activated carbons [68–70]. Comparatively pure hydrogen gas free from oxides of carbon has been prepared by oxidative decomposition of methane over activated carbon-supported iron catalysts [71]. Gold-supported activated carbon catalysts [72] have also been used for selective oxidation of CO in hydrogen-rich gas to eliminate CO impurities. This hydrogen can be used as a source of energy for stationary power plants, moving vehicles, and fuel cell applications [73,74]. Several metal and metal compounds supported on activated carbons [75–77] have also been used as catalysts for the oxidation of H_2S form flue gases.

Activated carbons as such also catalyze a number of oxidation reactions because they are capable of accepting electrons due to the existence of unpaired electrons on the edges of the crystallites. They are being increasingly used for the removal of toxic and hazardous materials [78–81] under mild conditions using wet oxidation method. Hydrogen peroxide is the most common oxidizing agent because it is environment friendly. The hydroxyl radicals generated by the decomposition of H_2O_2 can be resonance stabilized on the carbon surface and can cause the oxidation of heterocyclic organo-sulfur compounds [82,83]. Such an oxidative desulfurization of diesel fuels with H_2O_2 using activated carbons as catalysts has been studied [84,85]. The desulfurization was found to depend on the adsorption capacity of the carbon and was larger at pH values less than 2. The desulfurization also depended upon the amount of the acidic carbon–oxygen surface groups. It increased with increase in the concentration of these groups on the carbon surface. The sulfur content of a commercial diesel oil was reduced from 800 ppm by weight to almost 142 ppm. The addition of formic acid to the system further decreased the sulfur content of the diesel. The recovery of the oil was about 97% with sulfur removal to the extent of 98%. The resulting oil contained about 16 ppm by weight of sulfur without any negative effect in the fuel quality. The activated carbon could be reused several times. These gases could be used for upgrading biomethane into true biobased synthetic natural gas without adding fossil components.

Several other oxidation reactions such as the oxidation of oxalic acid [86], mercaptans [87,88], SO_2 and H_2S [89,90], creatinine, and ferrous ions have been carried out using activated carbons as a catalyst.

20.12.2. Catalytic Hydrogenations

Metal catalysts supported on oxide surfaces have been used to carry out several reduction processes such as hydrogenation of aliphatic and aromatic hydrocarbons and the synthesis and manufacture of industrial chemicals, hydrocraking of higher hydrocarbons into lighter ones, and for the desulfurization, dechlorination, and denitrogenation of impurities from hydrocarbon streams. These metals activate hydrogen under milder experimental conditions which carries out the reduction process. During the last several decades, the oxide supports are increasingly being replaced by activated carbons.

The hydrogenation of aliphatic and aromatic hydrocarbons, aldehydes, esters, nitro compounds, nitriles, and imines can be carried out by using metal-supported activated carbons. However, the hydrogenation of aromatic ring requires more severe conditions as compared to the hydrogenation of other functional groups. The hydrogenation of phenols to cyclohexanols [91] in aqueous solutions on palladium supported on activated carbon can be used for the treatment of wastewater and for the preparation of intermediates for the fragrance and pharmaceutical industry. Precious metal (Pt, Pd, Ru, or Rh) supported activated carbons [92–94] have been used to reduce esters, carboxylic acids, and aldehydes to their respective alcohols. Nickel-supported activated carbons are useful for the preparation of compounds used in the synthesis of biologically active compounds and natural products. Amines which are basic raw materials for dyes, rubber chemicals, urethane, and several pharmaceutical products can be prepared by the catalytic hydrogenation of corresponding nitro compounds over palladium-supported activated carbons.

Dehydrochlorination of chlorofluorocarbons [95,96], which are major source of chlorine in the upper atmosphere and which can cause ozone depletion and of chlorinated hydrocarbons which are pollutants in drinking water [97,98], has been carried out by activated carbon-supported palladium.

A few studies [99] have also used nickel supported on activated carbon catalysts for hydrodechlorination of chlorofluorohydrocarbons. The selective catalytic reduction of nitrogen oxides in the industrial waste gases can be carried out using an activated carbon-supported vanadium catalyst. The reductant in this case usually is ammonia or urea.

Drinking water contains nitrates, nitrites, and ammonia which are undesirable. The biological denitrification of this water is carried out by heterotrophic denitrifying bacteria. The culture, however, always requires an activated carbon catalyst which acts as an electron donor for the reduction of nitrate and nitrite. The activated carbon has a large adsorption capacity and an irregular shape which acts as a shelter for bacteria.

20.12.3. Activated Carbon Catalyzed C–C Coupling Reactions

Several C–C coupling reactions have been carried out on palladium supported on activated carbon catalysts. These coupling reactions lead to the production of several intermediates used in the pharmaceutical industry. The coupling reactions between aryl halides and alkenes (Heck reaction) or between aryl halides are catalyzed by palladium supported on activated carbons [100,101]. Mukhopadhyay et al. [102] carried out coupling of substituted halo-benzenes to respective biphenyls in the presence of a reducing agent and a weak base Na_2CO_3 using tetrabutylammonium bromide and palladium supported on the surface of an activated carbon. Palladium supported on activated carbon has also been used as a catalyst for the reaction between aryl or vinyl boronic acid and aryl or vinyl halide. This reaction is widely used in the synthesis of polyolefines, styrenes, and substituted biophenyls [103–105]. Several fine and bulk chemicals can be prepared by carbonylation reactions involving the insertion of C=O group into alkenes, alkynes, and alcohols using Ni or Rh supported on activated carbons [106,107]. The oxidative carbonylation of methane [108] has been carried out using rhodium chalcogen halide-supported activated carbon catalysts, while the vapor-phase carbonylation of methane can be carried out over $NiCl_2$–$CuCl_2$ catalysts supported on activated carbon [109].

20.12.4. Basic Activated Carbon Catalysts

More recently, basic solid catalysts are being increasingly used for several preparation and synthetic reactions. These catalysts can be basic activated carbons or activated carbon impregnated or doped with alkalis. The major advantage of such catalytic reactions is the easy separation and recovery of the catalyst and the products. Several investigators [110–112] used alkali-doped activated carbons for the catalysis of several organic compounds using sonography. The activated carbons used were doped with sodium and cesium compounds. The use of sonography increased the rate of the reaction, reduced the induction period, and enhanced the catalytic efficiency of the catalyst. They synthesized α,β–unsaturated nitriles by sonochemical activation of carbonylic compounds with malononitrile using two activated carbons doped with sodium and cesium salts. These doped activated carbons activated by ultrasound waves have also been used for the selective synthesis of n-alkylated imidazolic rings with medium-chain and long-chain alkyl halides which are important intermediates in the synthesis of several pharmaceutical drugs. The ultrasound waves are adsorbed by the activated carbon which converts these waves into heat energy, thereby reducing the reaction temperature and modifying the selectivity.

20.12.5. Acid-doped Activated Carbon Catalysts

The development of acid solid catalysts for the preparation of pharmaceutical and agricultural chemicals has been the subject matter of numerous researches. As a result, several activated carbon-supported catalysts have been prepared. Cu (II) acetylacetonate complex loaded on activated carbon has been used for the aziridination of styrene [113]. These aziridines are important intermediates for the synthesis of several pharmaceutical and agricultural chemical compounds. Acetals and ketals have been synthesized using tungstosilicic acid supported on activated carbon [114]. Activated carbons can entrap heteropolyacids [114] which are very good catalysts for pharmaceutical as well as food and flavor industries. Tungstophosphoric acid supported on activated carbon has been used [115] for the preparation of biodiesel oil from low-quality canola oil containing up to 20% free fatty acids.

20.12.6. Combination Reactions

Activated carbons have also been used as catalysts for several combination, decomposition, elimination, and chlorination reactions. The combination of hydrogen and bromine has been carried out using several activated carbons having different surface areas and associated with varying amounts of the carbon–oxygen surface groups. The reaction efficiency was found to be between 80 and 100% at a temperature between 80 and 150 °C in the presence of the activated carbon [116,117]. The reaction did not take place in the absence of the carbon. The catalytic activity was related to the surface area of the carbon while it was retarded by the presence of carbon–oxygen surface groups on the carbon surface. It has been suggested [118,119] that both hydrogen and bromine are adsorbed on the carbon surface at the neighboring sites. The activated carbon surface probably activates the bromine molecule in a manner similar to its activation by light. The activated bromine then reacts with the hydrogen to give HBr which is desorbed, making the active sites available and starting a chain reaction.

The combination of phosgene and formaldehyde for the formation of dichloromethane in the temperature range 180–200 °C is also catalyzed by activated carbons [120]. There is almost complete conversion of the reactants below 200 °C in the presence of activated carbons. However, when the reaction is carried out above 200 °C, the percentage conversion into dichloromethane is considerably decreased. Temperature-controlled desorption studies indicated that both phosgene and formaldehyde were adsorbed on the carbon surface and then reacted to produce an intermediate compound which then decomposed into dichloromethane and CO_2.

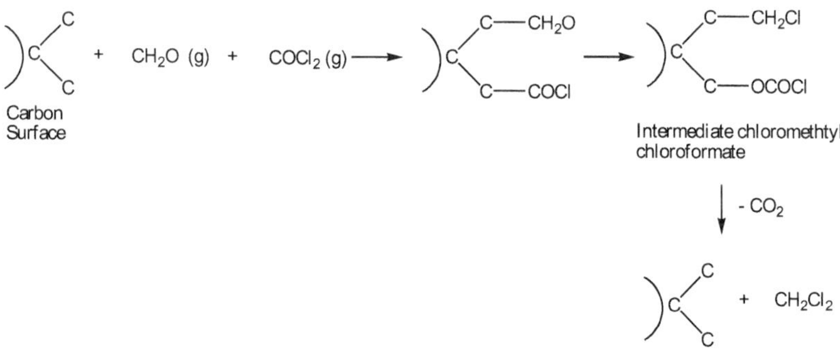

Several decomposition reactions are catalyzed by activated carbons. The decomposition of H_2O_2 has been studied as a function of the concentration of the solution, the temperature, as well as the nature of the carbon surface. The decomposition increases with concentration of the solution and the temperature. The catalytic activity of a carbon is considerably enhanced in the presence of degassed carbon samples. In fact, the carbons which are associated with carbon–oxygen surface groups inhibit the decomposition of H_2O_2 [121–123]. The results show that although the acidic medium is conducive to the stability of H_2O_2, an alkaline medium is conducive to its decomposition. The function of the carbon is simply to provide an environment of a particular degree of acidity or alkalinity. Thus, while acidic activated carbons inhibit the decomposition, the basic carbons enhance the decomposition. The decomposition has been attributed to the existence of chromene groups [124] which initiate a chain reaction for the decomposition. At low pH values, these chromene groups are oxidized to carbonium ions and thus the carbon loses its ability to decompose H_2O_2.

The catalytic decomposition of benzoyl peroxide in organic solutions has been investigated by several workers. The decomposition in benzene solutions increases considerably with concentration of the solution, the temperature, and the amount of the carbon used. The decomposition produces benzoic acid and CO_2. The amount of benzoyl peroxide decomposed is related linearly to the surface area of the activated carbon. The decomposition activity of the carbon catalyst is considerably reduced when the same carbon was reused in a second cycle. This is attributed to the fact that some of the carbon surface has been covered by the chemisorbed free radical benzoyl species $C_6H_5COO^-$.

REFERENCES

[1] M. Smisek, S. Cerny, Active Carbon: Manufacture, Properties and Applications, Elsevier Publishing Company, New York, 1970.

[2] R.C. Bansal, J.B. Donnet, F. Stoeckli, Active Carbon, Marcel Dekker, New York, 1988.

[3] R.C. Bansal, M. Goyal, Activated Carbon Adsorption, Taylor & Francis, CRC group, New York, 2005.

[4] L. Francisco, M. Francisco, P. Marin, G. Carme, J. Colloid Interface Sci. 257 (2003) 173.

[5] Q.Y. He, Y.P. Zhang, W.F. Yu, Hebei Keji Daxue Xuebao 24 (2003) 66.

[6] G. Nawak, Brauwelt 95 (1955) 760.

[7] K. Kaiser, Brauwelt 97 (1957) 1203.

[8] L.N. Potishuk,V.G. Belkin, T.K. Kalenik, N.A. Poroshin, Russian Patent RUXXE7 RU 2383612 C2 20100310, 2010.

[9] Z.M. Maiboroda, Russian Patent RUXXE7 RU 2275421 C2 20060427, 2008.

[10] N.I. Glyantsev, V.V. Kotov, N.M. Stekolnikova, Ekologiya Promyshlennogo Proizvodstva 4 (2005) 55.

[11] S.M. Zenovich, B.Kh. Strelets, V.A. Polyakov, I.I. Burachevskii, Russian Patent RUXXE7 RU 2243996 C2 20050110, 2005.

[12] V.A. Frenkel, N.P. Fedulaev, E.A. Erantseva, V.G. Lopyreva, Russian Patent RUXXE7 RU 2181763 C1 20020427, 2002.

[13] E.T. Podlubnaya, V.E. Stepnaya, T.V. Sushinskaya, N.I. Slavutskaya, Fermentnaya i Spirtovaya Promyshlennost 1 (1985) 9.

[14] E.T. Podlubnaya, V.E. Stepnaya, N.I. Slavutskaya, G.I. Zaikanova, Fermentnaya i Spirtovaya Promyshlennost 7 (1984) 9.

[15] V.L. Singleton, D.E. Draper, Am. J. Enol. Viticult. 13 (1962) 114.

[16] F. Howard, E.C.R. Spooner, Chem. Ind. (London) (1964) 186.

[17] S. Rajkowski, K. Nowak, T. Piechocpki, Pharma Zentralhalle 97 (1958) 1.

[18] R. Charonnat, P. Lechat, Pharma France 8 (1950) 171.

[19] P.T. Nedkov, S.B. Tzokov, P.G. Dalev, Bulg. Chem. Commun. 28 (1) (1995) 133.

[20] O. Lehmer, Atomwirtsch 5 (1960) 356.

[21] U. Kohrt, Chem. Eng. Technol. 33 (1961) 135.

[22] B.H.M. Billinge, J.B. Docherty, M.J. Bevan, Carbon 22 (1984) 83.

[23] J.A. Baker, E.J. Poziomek, Carbon 12 (1974) 45.

[24] D.A. Collins, L.R. Taylor, R. Taylor, in: , Proc. Ninth AEC Air Cleaning Conf., CONF. 660904, Vol. I, 1957, p. 59.

[25] C.M. González-García, J.F. González, S. Román, Fuel Process Technol. 92 (2011) 247.

[26] S.W. Park, W.K. Lee, Sep. Technol. 3 (1993) 133.

[27] W. Hoffmann, German Patent GWXXBX DE 3347118 A1 19850711, 1985.

[28] M. Toshiaki, N. Takashi, Carbon 38 (5) (2000) 709.

[29] A.A. Aleshin, I.V. Golikov, V.A. Kazakov, D.V. Ochkin, Radiatsion naya Bezopasnosti Zashchita AES 11 (1986) 95.

[30] B.G. Lazarev, M.F. Fedorova, Zh. Techn. Phys. 29 (1950) 862.

[31] C. Jech, in: R.C. Extermann (Ed.), Radioisotopes in Scientific Research (Proc. Inter Conf. UNESCO in Paris 1957), Vol. II, Pergamon Press, New York, 1958, p. 491.

[32] J. Janak, Chem. Listy 47 (1953) 837.

[33] M. Krejci, Sbornik Praci O aktivnium Uhli (Collection of papers on active carbon), Cs Vedtechn. Spol., Prague (1959) 199.

[34] W. Tomassi, S. Neffe, Biul. WAT 22 (1973) 256.

[35] H. Jankowska, H.A. Swiatkowski, J. Choma, in: Active Carbon, Ellis Horwood, New York, 1991.
[36] N. Jankowska, S. Neffe, Prezem. Chem. 59 (1980) 591.
[37] Z. Barnir, C. Aharoni, Carbon 13 (1975) 363.
[38] C.T. Chiou, P.J. Reucroft, Carbon 15 (1977) 49.
[39] P.J. Reucroft, C.T. Chiou, Carbon 15 (1977) 285.
[40] B. Singh, A. Saxena, D. Sharda, S.S. Yadav, D. Chintamani, D. Pande, K. Shekh, in: G.N. Mathur, V.S. Tripathi, T.L. Dhami, O.P. Bahl (Eds.), Proc. Indo Carbon 2003 Conf, DMSRDE, Kanpur, Indian Carbon Soc. Publ., 2003, p. 33.
[41] P.G. Hall, P.M. Gittins, J.M. Winn, J. Robertson, Carbon 23 (1985) 353.
[42] G.K. Prasad, B. Singh, J. Hazard. Mater. 116 (2004) 213.
[43] W.N. Davis, U. S. Patent 227963, 1880.
[44] J.B. Zadra, U. S. Bureau of Mines Washington DC RI No. 11672 (1950) and RI No. 4843, 1952.
[45] K.B. Hall, World Min. 27 (1974) 44.
[46] N.V. Keltsev, in Podstawi techniki adsorpcyjnej (Principles of Adsorption Technology) WNT, Warsaw, 1980.
[47] H. Kienle, E. Bader, Aktivkohle und ihre industrielle Anwendung, Ferdinand Enke Verlag Stuttgart, 1980.
[48] H. Juentgen, Carbon 15 (1977) 273.
[49] Y. Fei, Y.N. Sun, E. Givens, F. Derbyshire, ACS FCD 40 (1995) 1051.
[50] R. Andrews, E. Raymundo-Pincro, F. Derbyshire, in Proc. Carbon'98 Strasbourg, France, 1988, p. 275.
[51] D. Aggarwal, M. Goyal, R.C. Bansal, Carbon 37 (1999) 1989.
[52] M. Goyal, V.K. Rattan, D. Aggarwal, R.C. Bansal, Colloids Surf. A 190 (2001) 229.
[53] M. Goyal, V.K. Rattan, R.C. Bansal, Indian J. Chem. Technol 6 (1999) 305.
[54] L.R. Radovic, in: L.R. Radovic (Ed.), Chemistry and Physics of Carbon, Vol. 27, Marcel Dekker, New York, 2001.
[55] M. Goyal, M. Bhagat, R. Dhawan, J. Hazard. Mater. 171 (2009) 1009.
[56] M. Goyal, R. Amutha, Carbon Science 7 (2006) 97.
[57] D. Aggarwal, M. Goyal, J. Ind. Pollut. Control 22 (2006) 329.
[58] R.C. Bansal, D. Aggarwal, M. Goyal, B.C. Kaistha, Indian J. Chem. Technol. 9 (2002) 290.
[59] M. Goyal, M. Bhagat, R. Dhawan, Sep. Sci. Technol. 45 (2010) 1265.
[60] M. Goyal, Carbon Science 5 (2004) 55–61.
[61] D. Aggarwal, M. Goyal, Res. J. Chem. Environ. 10 (2006) 28–36.
[62] P. Aggarwal, S.K. Kapoor, J.C. Kapoor, A.K. Bhalla, R.C. Bansal, Indian J. Chem. Technol. 3 (1996) 187.
[63] V. Calvino-Casilda, A.J. Lopez-Peinodo, C.J. Duran-Valle, R.M. Martin-Aranda, Catal. Rev. Sci. Eng. 52 (2010) 325.
[64] S. Carrettin, P. Mc Morn, P. Johnston, K. Griffin, G.J. Hutchings, Chem. Commun. 7 (2002) 696.
[65] L. Chen, M. Ding, X. Bau, J. Phys. Chem. 111 (2007) 2229.
[66] L. Prati, F. Porta, Appl. Catal. A 291 (2005) 199.
[67] A. Deffernez, S. Hermans, M. Devillers, J. Phys. Chem. 111 (2007) 9448.
[68] N. Roh, B.C. Dunn, E.M. Eyring, R.J. Pugmire, H.L.C. Meuzelaar, Fuel Process Technol. 83 (2003) 27.
[69] Z. Zhang, X. Ma, P. Zhang, Y. Li, S. Wang, J. Mol. Catal. A Chem. 266 (2007) 202.
[70] J. Song, T. Zhao, Y. Du, Chin. J. Catal. 27 (2006) 386.

[71] N. Muradov, F. Smith, C. Huang, A.T. Raissi, Catal. Today 116 (2006) 281.
[72] F. Porta, M. Rossi, J. Mol. Catal A Chem. 204 (2003) 553.
[73] M. Okumura, S. Tsubota, M. Haruta, J. Mol. Catal. A Chem. 199 (2003) 73.
[74] F. Wang, G. Lu, Catal. Lett. 115 (2007) 46.
[75] V. Meeyoo, J.H. Lee, D.L. Trimm, N.W. Cant, Catal. Today 44 (1998) 67.
[76] A.K. Dalai, M.T. Cundall, M. De, Can. J. Chem. Eng. 86 (2008) 768.
[77] L. Ma, P. Ning, Y. Zhang, X. Wang, Chem. Eng. J. 137 (2008) 471.
[78] A. Fortuny, J. Font, A. Fabregat, Appl. Catal. B Environ. 19 (1998) 165.
[79] A. Fortuny, C. Miro, J. Font, A. Fabregat, Catal. Today 48 (1999) 323.
[80] S. Imamura, Ind. Eng. Chem. Res. 38 (1999) 1743.
[81] F. Luck, Catal. Today 27 (1996) 195.
[82] H.P. Boehm, Carbon 32 (1994) 759.
[83] T. Hirai, Y. Shiraishi, K. Ogawa, I. Komasawa, Ind. Eng. Chem. Res. 36 (1997) 530.
[84] G. Yu, S. Lu, H. Chen, Z. Zhu, Carbon 43 (2005) 2285.
[85] D. Joachim, A. Fastabend, A. Greve, V. Heil, M. Juricev-Spiric, A. Kraft, M. Krzanowski, K. Meller, A. Menne, C. Unger, W. Urban, J. Danzig, A. Fastabend, Mineraloeltechnik 54 (2009) 1.
[86] R. Kurth, B. Tereczki, H.P. Boehm, in 15th Bien. Conf. Carbon, Philadelphia, 1981, pp. 244.
[87] C. Ishizaki, J.T. Cookson, in: A.J. Rubin (Ed.), Chemistry of Water Supply, Treatment and Distribution, Ann Arbor Sci. Publ., Ann Arbor, Mich., 1980, p. 201.
[88] J.T. Cookson Jr., in: I.H. Suffet, M.J. Mc Guire (Eds.), Activated Carbon Adsorption, Ann Arbor Sci. Publ., Ann Arbor, Mich, 1981, p. 379.
[89] B.R. Puri, B. Kumar, K.C. Kalra, Ind. J. Chem. 9 (1971) 970.
[90] O.C. Cariaso, P.L. Waker Jr., Carbon 13 (1975) 233.
[91] C. Park, M.A. Keane, J. Colloid Interface Sci. 266 (2003) 183.
[92] M.A. Fraga, M.J. Mendes, E. Jordao, J. Mol. Catal A Chem. 179 (2002) 243.
[93] Z. Zhang, J.E. Jackson, D.J. Miller, Appl. Catal A Gen 219 (2001) 89.
[94] J.P. Breen, R. Burch, K. Griffin, C. Hardacre, M. Hayes, X. Huang, S.D. O'Brien, J. Catal. 236 (2005) 270.
[95] D.J. Moon, M.J. Chung, K.Y. Park, S.I. Hong, Carbon 37 (1999) 123.
[96] M. Makkee, A. Wiersma, E.J.A.X. van de Sandt, H. van Bekkum, J.A. Moulijn, Catal. Today 55 (2000) 125.
[97] S. Ordonez, H. Sastre, F.V. Diez, Appl. Catal. B Environ 25 (2000) 49.
[98] N.C. Concibido, T. Okuda, W. Nishijima, M. Okada, Appl. Catal. B 71 (2007) 64.
[99] A. Morato, C. Alonso, F. Medina, P. Salagre, J.E. Sueiras, R. Terrado, A. Giralt, Appl. Catal. B 23 (1999) 175.
[100] G. Dyker, A. Kellner, J. Organomet. Chem. 551 (1998) 141.
[101] M. Gruber, S. Chouzier, K. Koehler, L. Djakovitch, Appl. Catal. A 265 (2004) 161.
[102] S. Mukhopadhyay, G. Rothenberg, N. Qafisheh, Y. Sasson, Tetrahedron Lett. (2001) 6117.
[103] A. Corma, H. Garcia, A. Leyva, J. Mol. Catal. A 230 (2005) 97.
[104] R. Narayanan, M.A. El-Sayed, J. Catal. 234 (2005) 348.
[105] M. Seki, J. Synth. Org. Chem. 64 (2006) 853.
[106] M.E. Halttunen, M.K. Niemela, A.O.I. Krause, T. Vaara, A.I. Vuori, Appl. Catal. A. 205 (2001) 37.
[107] J. Liu, Q. Zhang, H. Yao, Chem. React. Eng. Technol. 23 (2007) 244.
[108] G.R. Kosmambetova, P.E. Strizhak, V.I. Gritsenko, S.V. Volkov, L.B. Kharkova, O.G. Yanko, O.M. Korduban, J. Nat. Gas. Chem. 17 (2008) 1.
[109] F. Peng, X.B. Fu, Catal. Today 93 (2004) 451.

[110] C.J. Duran-Valle, I.M. Fonseca, V. Calvino-Casilda, M. Picallo, A.J. Lopez-Peinado, R.M. Martin-Aranda, Catal. Today 107 (2005) 500.
[111] C.J. Duran-Valle, S. Ferrera-Escudero, V. Calvino-Casilda, J. Diaz-Teran, R.M. Martin-Aranda, Appl. Surf. Sci. 238 (2004) 97.
[112] L. Costarrosa, V. Calvino-Casilda, S. Ferrera-Escudero, C.J. Duran-Valle, R.M. Martin-Aranda, Appl. Surf. Sci. 252 (2006) 6089.
[113] A.R. Silva, J.L. Figueiredo, C. Freire, B. de Castro, Catal. Today 102 (2005) 154.
[114] Y. Shui-jin, D. Xin-Xian, H. Lan, S. Ju-Tang, J. Zhejiang. Univ. Sci. 6B (5) (2005) 373.
[115] M.G. Kulkarni, R. Gopinath, L.C. Meher, A.K. Dalai, Green Chem. 8 (2006) 1056.
[116] B. Bruns, O. Zarubina, Zh. Fiz. Khim 11 (1938) 300.
[117] H.P. Boehm, U. Hoffman, A. Clauss, in: Proc. 3rd Carbon Conf, Pergamon Press, Oxford, 1959, p. 241.
[118] J. Tijssen, J.F.M. Kaptein, J. Feijen, A. Bantjes, A.W.J. VanDoorn, in: R.M. Kenedi, et al. (Eds.), Artificial Organs, Macmillan Press, London, 1977, p. 158.
[119] B.R. Puri, K.C. Kalra, Indian J. Chem. 5 (1967) 638.
[120] T.A. Ryan, M.H. Stacey, Fuel 63 (1984) 1101.
[121] E.C. Larson, J.H. Walton, J. Phys. Chem. 44 (1940) 70.
[122] R.C. Bansal, Ph. D. thesis, Panjab University, Chandigarh, India, 1965.
[123] V.N. Latyarva, O.N. Nuretdinove, Tr. Khim Technol. 3 (1960) 110.
[124] V.A. Garten, D.E. Weiss, Rev. Pure Appl. Chem. 7 (1957) 69.

Chapter 21

Biomedical Applications of Carbon Adsorbents

Sergey V. Mikhalovsky
University of Brighton, Brighton, UK
Nazarbayev University, Astana, Kazakhstan

Susan R. Sandeman, Carol A. Howell and Gary J. Phillips
University of Brighton, Brighton, UK

Vladimir G. Nikolaev
Kavetsky Institute of Experimental Pathology, Oncology and Radiobiology (IEPOR), Kiev, Ukraine

Chapter Outline

21.1. Introduction 639
21.2. Oral Activated Carbon (Enterosorbent) in Poisoning and Beyond 640
21.3. Activated Carbon in Hemoperfusion 645
21.3.1. The History of Activated Carbon in Hemoperfusion 645
21.3.2. The Principles of Hemoperfusion 647
21.3.3. Complications of Hemoperfusion 650
21.3.4. AC Containing Hemoperfusion Devices 651
21.3.5. AC Containing Combined Extracorporeal Devices 652
21.4. Activated Carbon Adsorbents for Use in CBRN Incidents 655
21.5. Activated Carbon Materials for Wound Dressings 657
21.6. Conclusions 662
References 663

21.1. INTRODUCTION

In modern times, the pharmaceutical industry has become one of the largest, most important, and financially profitable of human activities. It is among the top

Novel Carbon Adsorbents. DOI: 10.1016/B978-0-08-097744-7.00021-1
Copyright © 2012 Elsevier Ltd. All rights reserved.

performing high-tech sectors with significant R&D investment, since only one or two substances out of every 10,000 tested will reach the market [1]. Yet, there is one substance, or rather a group of substances with the same name of activated carbon (AC) or charcoal, which occupies a uniquely universal place among the multitude of medicines, as it can be used as an oral antidote against a large number of toxic substances and poisons, for extracorporeal purification of bodily fluids, most importantly, blood, and in wounds and burns treatment.

In this chapter, medical applications of activated carbon are discussed in the light of recent advances in carbon materials science.

21.2. ORAL ACTIVATED CARBON (ENTEROSORBENT) IN POISONING AND BEYOND

Activated carbon, in a range of forms and made from diverse starting materials, has a long history of oral use for medical applications [2,3]. Its use as a treatment for food poisoning was first recorded on Egyptian papyrus in 1550 BC, and anecdotal evidence suggests carbon preparations were periodically recommended for the treatment of bilious complaints and malodor throughout the Middle Ages. Modern activated carbons for medical use as well as in other applications came into production in the 1900s when activation processes were developed to greatly increase the adsorptive capacity of the derived carbons, and useful resources for pyrolysis were better defined for a more superior product, sourcing from peat, wood char, coconut shells, and petroleum products among others. An in-depth description of the historical evidence, on which the efficacy of oral carbon use for the treatment of poisoning is based, can be found in Cooney [2]. However, it is worth noting that no systematic clinical trials were ever carried out prior to the widespread clinical acceptance of oral carbon use for poisoning and drug overdose. The precedent for human use was initially based on a number of striking, single-subject demonstrations in the 1800s, one of which used a 15 g dose of carbon to negate the effect of 10 times the lethal dose of strychnine. Since that time, the high adsorption capacity of oral carbons has been examined and refined for a range of applications, particularly in former Soviet countries, but remains limited to clinical treatment of poisoning in the West [4]. In the former Soviet countries, the word "enterosorbent" is commonly used to indicate oral administration of a sorbent. Originally, the term "enterosorption" had been proposed to define a method of sorption therapy based on oral administration of "enterosorbents", i.e. activated synthetic carbon [5].

In any case, activated carbon, or charcoal, occupies a rather unique position among other medicines, as was underlined in the court case: Forsyth vs Eli Lilly, by a medical expert, David Healy: "In the case of strychnine and activated charcoal, the end point is so clear-cut that there can be no doubt about the answer… Its occurrence … is so well-established that no investigator has ever felt the need to conduct a randomized placebo-controlled trial to establish this fact" [6]. This argument was accepted by the court. It is therefore highly unlikely that anyone will be prepared to spend large sums of money on a clinical trial to confirm the therapeutic action of activated carbon.

Activated carbon alone is the preferred treatment for drug overdose and acute poisoning, although some reticence exists over the clinical impact of its use in some cases [3,7,8]. Its use in combination with forced emesis and gastric lavage is no longer recommended since these treatments were found to offer little benefit over the use of carbon alone and may have reduced carbon efficacy, increased the risk of aspiration, and promoted more rapid duodenal drug absorption. The latest position statement published by the American Academy of Clinical Toxicology and the European Association of Poison Centres and Clinical Toxicologists is cautious and does not recommend the continuous routine use of single-dose activated carbon because of the absence of clear evidence from multicenter, randomized controlled clinical trials in favor of its use [9–13]. However, a recent review suggests that the variables involved in such studies mean that there may never be clear evidence available to completely define the parameters that should govern the clinical use of activated carbon in overdose [14]. A large 2008 controlled clinical trial comparing multiple-dose, single-dose activated carbon treatment, and conservative treatment of self-poisoning highlights the point. No mortality difference was shown between the three groups indicating no benefit of AC treatment in these cases [15]. However, the trial was based in rural Sri Lanka where supportive care may vary and time to administration of treatment was often delayed, potentially influencing the outcome [14]. The use of activated carbon is currently recommended in a subset of patients where treatment can be administered in less than 1 h after ingestion of toxin and at a dose of 1 g/kg for children under 12 and 25–100 g for adults. Application can be single dose or repeated dose over time and preparations may be made more palatable with sweeteners [3,16]. Later use as a decontamination agent may benefit cases involving slow-release drug overdose and to reduce vascular drug concentration through impact on enterohepatic and enterovascular drug recirculation. Already absorbed drug, present in the enteric vasculature or processed by the liver and present in bile secretions, may be adsorbed by carbon present in the gut following drug desorption back into the intestine. A thorough description of current, oral-activated carbon formulations and listed classes of toxins and drugs adsorbed can be found in [4]. The superior adsorptive capacity of activated carbon is predominantly for organic molecules, the size of which depends on the pore size distribution of the carbon itself. Most low molecular-weight organic molecules are adsorbed within micropores of less than 2 nm, formed and expanded to increase surface area by a physical or chemical activation step following pyrolysis. The removal of high molecular weight molecules for applications other than poisoning requires the presence of larger pores above 2 nm and can be achieved using high-purity synthetically sourced materials, regulation of the pore former and by control of the activation process. The mechanisms by which the therapeutic action of activated carbon occurs are poorly understood and can be rather complex. It has been suggested that the main role of activated carbon as an oral adsorbent is to remove from the gastrointestinal tract: (i) exogenous toxins, i.e. substances consumed with food; (ii) substances diffused into intestines from the blood; and (iii) secondary toxic metabolites produced as a result of digestion [17]. The removal of toxic substances occurs through both reversible physical and irreversible chemical binding, depending on the presence of surface functional groups. Other possible mechanisms include modification of the diet, transfer of bioactive substances such as bile acids adsorbed onto the AC surface, and change in the volume of the gut contents similar to the effect of fibrous food. Inorganic molecules and ions are poorly removed by activated carbon although its ion exchange capacity can be improved by adding ionic surface functional groups. Microporous oral carbon slurry preparations are currently used to treat drug and poison overdose in the West and are also marketed in tablet form for the treatment of diarrhea, indigestion, and flatulence. While treatments are marketed under different brand names internationally from different suppliers, the majority are Norit-based formulations from peat or coconut-shell precursors [4].

In addition to naturally sourced carbons such as KAU made from fruit stones, oral carbons from porous synthetic polymer resins such as polystyrene (SUGS and SCS) or polyvinylpyridine (SCN, or SKN) resin have an established history of use for a much broader spectrum of applications in former Soviet countries [18–34]. The benefit of using polymer-pyrolyzed AC includes a consistent spherical bead product that is mechanically strong, does not release microparticulates, has controlled and very low levels of impurities, and can be more palatable for oral use. A high surface area and a pore size extending beyond 50 nm are the possible parameters which allow the removal of high-molecular-weight biological toxins. SCN or SKN, synthesized from porous poly(4-vinylpyridine) cross-linked with styrene–divinylbenzene, was reported as the most commonly used oral activated bead carbon in the former USSR. Prior to the development of SCN in the 1970s, AC use had diminished since the high doses of powdered carbon required for efficacy were poorly tolerated due to nausea, vomiting, and constipation. High dose treatment of 20–80 g per day of SCN given in 2–4 doses in between meals was found to reduce these side-effects to negligible levels [31]. SCN oral therapy has been in clinical use for over 30 years with evidence indicating benefit for the treatment of liver cirrhosis, intestinal infection, chronic kidney disease, autoimmune disease, hypercholesterolemia, urticaria, and in the reduction of side effects associated with chemotherapy [18–34]. The mechanism of benefit is based on the porosity of the source carbons and is broadly described by the removal of toxins and metabolites synthesized within or introduced into the GI tract, the removal of toxins from the blood, modification of lipid and amino acid metabolism through selective removal of aromatic amino acids and long chain fatty acids, and the transfer of agents from the upper intestinal tract which can play a role in neutralizing toxins lower in the intestinal tract [30,31]. Oral adsorbent impact on blood toxin levels may occur indirectly. In the case of chronic kidney disease associated with high serum creatinine levels, gastrointestinal adsorption by low-dose activated carbon AST-120 did not affect serum creatinine levels directly but only as a result of improved glomerular filtration suggesting no transluminal adsorption from blood occurs [35]. Activated carbon carriage of inflammatory molecules through the intestinal tract may simply alleviate pressure on the mucosal lining and reduce toxin overloading of the normal processing strategies that become dysfunctional in disease states where treatment with oral SCN has been shown to alleviate symptoms.

One further oral carbon in clinical use is the microporous spherical carbon microbeads called AST-120 or Kremazin (Kureha Corporation), made from petroleum-based pitch, which has been marketed in Japan since 1991 as a drug to slow down the progression of CKD [36]. Its mechanism of action is similar to that described in the Russian literature reporting the clinical use of the carbon enterosorbent SCN for the treatment of CKD in the 1980s [18]. The drug is postulated to work by binding to dietary metabolites, such as advanced glycation end products, phenols, and indoles present in the gut, limiting their absorption, systemic build up, and negative impact on endothelial and renal physiology. Impact is most clearly described in studies investigating the protein-bound uremic toxin indoxyl sulfate (IS). IS has been studied using cell, animal, and human models with respect to removal of the dietary indole precursor by AST-120 in the gut, subsequent reduction in plasma indoxyl sulfate levels and in markers of oxidative stress, vascular calcification, renal interstitial fibrosis, nephrovascular senescence, and glomerular sclerosis [35–43]. A number of clinical studies have reported some efficacy in the treatment [43]. However, no definitive results from large, multicenter clinical trials have been obtained to date. The latest reported results from a one-year study using a 6 g per day dose of AST-120 showed no significant impact on primary end-point events indicating event-free survival in a multicenter, randomized, controlled trial involving predialysis CKD patients [35]. Two phase III randomized, placebo-controlled clinical studies investigating the prevention of CKD progression in patients with moderate to severe CKD are in process. These studies will hopefully provide more definite answers on the clinical efficacy of AST-120 in the treatment of CKD. It is worth noting that the dosage used in these trials was much lower than that reported in Eastern European clinical practice with a dose of 6–9 g per day [35,36,43]. If such a treatment were proven successful, it would lead to a delay, by several months to years, in putting CKD patients on life-long dialysis, the latter being more costly and limiting quality of life.

Recently, a new technology for manufacturing microbeaded AC from porous phenol–formaldehyde resin has been developed by MAST Carbon International Ltd (Basingstoke, Hants, UK) [44] (Fig. 21.1). In preliminary *in vitro* and *ex vivo* evaluation of its adsorption properties toward clinically relevant substances, this

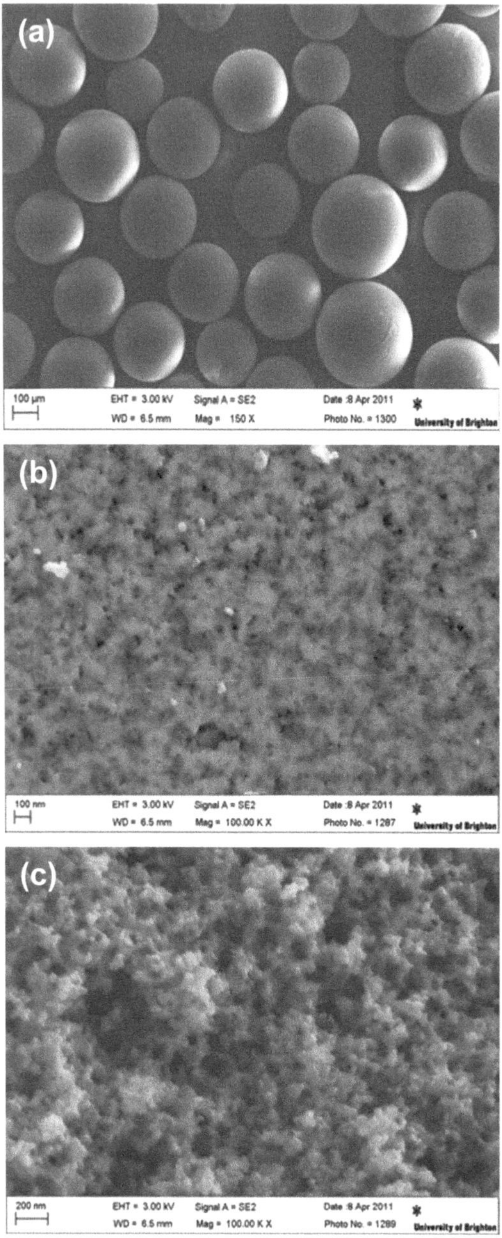

FIGURE 21.1 Scanning electron micrographs of phenol–formaldehyde-derived activated carbon (MAST Carbon International Ltd). a) AC microbeads at 150x magnification. b) External surface at 100,000x magnification. c) Internal surface at 100,000x magnification.

AC has shown results comparable to the best SCN series. Its main advantage is bimodal pore size distribution with peaks in the range of micropores and mesopores [45]. The mesopore radius can be varied in the range from 10 to 50 nm, which allows adsorption of large biomolecular toxins.

The dissolution of the Soviet Union has brought about improved collaborative links with Eastern European clinical research communities and insight into previously undisclosed technology and clinical practice. In addition, there is a greater understanding of the role that changes in intestinal microbial colonization and integrity play in infection related to chronic disease progression and a recognized need to find alternative, broadly acting, cost-effective therapies that reduce toxin load and address the growing incidence of virulent strain resistance to antibiotics. Such developments open up new possibilities for the use of next-generation activated carbon adsorbents to treat pathologies influenced by altered gastrointestinal physiology beyond poisoning.

21.3. ACTIVATED CARBON IN HEMOPERFUSION

21.3.1. The History of Activated Carbon in Hemoperfusion

There are three main types of extracorporeal treatment, hemoperfusion, hemodialysis, and hemofiltration, which differ by their physicochemical mechanism of action (Table 21.1).

Hemoperfusion (HP), or hemoadsorption, is a medical treatment used to remove toxic substances from the patient's blood. The process requires large volumes of the patient's anticoagulated blood to be passed over an adsorbent material in order to remove toxic substances. HP is an extracorporeal form of treatment because the blood is taken out of an artery/vein, pumped through the adsorbent material contained within a medical device situated outside the patient's body, and the cleansed blood is returned to the patient's body (Fig. 21.2).

TABLE 21.1 Mechanism of Extracorporeal Blood Cleansing Using Different Techniques

Substance Removed	Hemodialysis	Hemofiltration	Hemoperfusion Over AC
Water	Ultrafiltration/high	Ultrafiltration/high	None
Small-molecular-weight solutes	Diffusion	Convection	Molecular adsorption
Ionic solutes	Diffusion	Convection	Ion exchange
Large and medium molecular weight biomolecules	Diffusion/molecular adsorption	Convection/ molecular adsorption	Molecular adsorption

Adapted from [46].

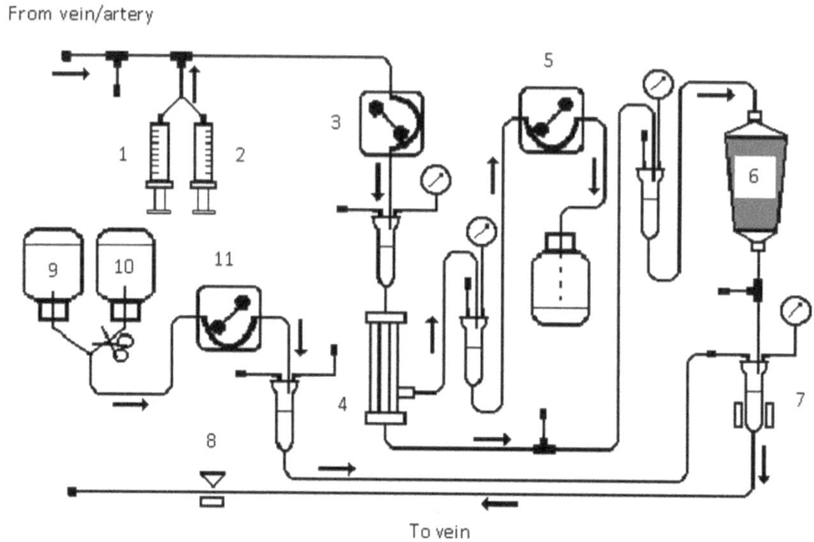

FIGURE 21.2 Schematic of the extracorporeal hemoperfusion circuit. Anticoagulation solutions: 1 – heparin and 2 – citrate; 3 – blood pump; 4 – ultrafilter; 5 – ultrafiltrate pump; 6 – HP cartridge; 7 – air trap; 8 – alarm clamp; 9 – replacement fluid; 10 – purified albumin solution; and 11 – replacement fluid pump.

The use of HP for the treatment of uremia was first examined as early as 1948, by Muirhead and Reid who conducted animal experiments using cation and anion exchange resin filters to show successful removal of urea, although the animals died [47]. Later in 1958, Schreiner treated a pentobarbital-overdose patient with 2 × 15 min sessions of HP on a lactated anion resin column, and showed some removal, although at the time they encountered massive hemolysis, electrolyte derangement, and death, which were cited as the main obstacles to the use of this therapy [48]. In 1964, Yatzidis *et al.* were the first investigators to use activated carbon produced from coconut shells for hemoperfusion treatment in 5 × 1 h sessions, to treat 2 patients with barbiturate overdoses. Both patients regained consciousness, although they suffered anaphylactoid side-effects and a 50% drop in their platelet counts following the procedures [49,50]. Chang *et al.* demonstrated in initial investigations comparing hemoperfusion and hemodialysis that although overall uremic toxin removal was greater with HP using coated AC microcapsules, the technique could not remove urea in clinically relevant quantities [51]. Hemoperfusion was introduced clinically in the 1970–80s in the West for the treatment of poisoning, yet despite its availability and its long-term use, it is only used infrequently (Fig. 21.3).

Hemoperfusion has been considered over the years for a number of medical uses including: a method to increase the efficiency of dialysis in patients with

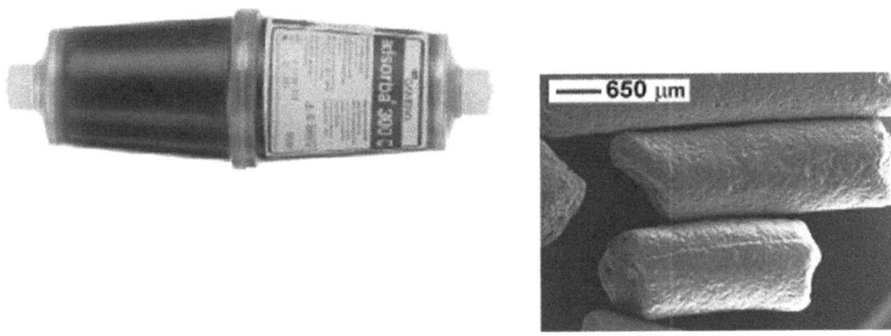

FIGURE 21.3 Left: HP cartridge Adsorba® 300C (Gambro, Sweden), one of the most common HP cartridges in West. Right: Electron micrograph of coated AC used in Adsorba® cartridges.

kidney failure, to treat acute drug and chemical intoxication, and as a treatment for fulminant hepatic encephalopathy. More recently, the use of AC as a sorbent system has found favor as a means of regeneration of dialysate, in wearable kidney systems [52].

21.3.2. The Principles of Hemoperfusion

The HP device usually consists of a polymer housing containing the sorbent particulate material through which the blood flows directly in a laminar flow [53]. The type of sorbent used within the hemoperfusion device can vary, although the most commonly employed to date have been AC and either ion-exchange or nonionic macroporous resins. The particle size is larger than that in adsorbents used for oral administration. Microbeaded ACs for direct HP have particle diameters in the range from 0.5 to 1.0 mm, whereas microspheres for oral administration are in the range from 0.1 to 0.25 mm.

Clinical hemoperfusion requires: a device with both inlet and outlet lines, vascular access to the patient, a blood pump that can maintain a blood flow of between 100 and 500 ml/min, pressure gauges to detect clotting within the device, the use of a suitable anticoagulant (most commonly heparin), and monitoring of safety and efficacy of the device (Fig. 21.2) [54]. There are different geometric configurations and sizes of HP cartridge available in the market, although it is doubtful that these parameters have ever been optimized. The cartridge volume varies from 50 to 500 ml mostly packed with granular adsorbent leaving ca. 20% of the volume as free space usually filled with sterile saline. Hemoperfusion over coated AC and porous polymer resins is known to be effective at clearing small, lipid-soluble drug molecules, but it demonstrates variable effectiveness at removing small water-soluble molecules. The adsorption and thus removal from blood of any given molecule depends not only on its size but also on the affinity of the particular sorbent for that molecule [55,56].

The key requirement for adsorbents used in HP is their sufficient hemocompatibility, which is the most important consideration when designing a medical device that is in direct contact with blood. There must be a direct flow of blood through the sorbent material contained within the device without any notable destruction of blood components or excessive stimulation of the inflammatory response. Hemocompatibility of sorbent materials is often tested *in vitro* in terms of protein adsorption, platelet retention/activation, and hemolysis. Initial AC materials showed poor compatibility, which prompted the addition of semipermeable biocompatible polymeric coatings to AC to try to overcome this failing and related clinical complications. However, such a coating hampered the adsorption properties of the AC and restricted the use of HP in comparison to other extracorporeal blood purification techniques [57]. During hemoperfusion with a device containing uncoated AC, the diffusion of molecules from blood to the inner surface of pores (up to 2000 m^2/g) is the rate-limiting step, whereas, in coated AC devices, the rate-limiting step is diffusion of the molecule through the biopolymer coating. Thus, diffusion of substances with low molecular mass, such as creatinine, is only slightly affected by the coating, but diffusion of molecules with molecular mass of 500 Da or larger is significantly delayed or restricted, and the higher the size of the solute the lesser is the accessibility of the adsorbent surface underneath the coating [58]. Not surprisingly, the use of coated AC for hemoperfusion proved to be efficient mainly in the treatment of acute poisoning with low-molecular-mass substances. Coated ACs used in hemoperfusion are predominantly microporous, as the permeability of membrane coatings to larger solutes is very limited.

The effect of a coating on adsorption of even a small molecular solute, such as methylene blue (MW 320 Da), is shown in Table 21.2.

TABLE 21.2 Adsorption of Methylene Blue on Coated and Uncoated AC in Cyclic Experiments[a]

AC	Cartridge/ Company	Mass of AC/ Column Volume	$\tau_{50\%}$, min	$\tau_{95\%}$, min
Cellophane-coated, 3–5 μm, Norit RBX, peat	Adsorba® 300C/ Gambro	300 g/600 mL	39	154
SCN	Ukrainian Academy of Sciences	130 g/450 mL	5	14

[a]*Experimental conditions: initial concentration C_0 of methylene blue – 500 mg/l, solution volume 6 l, flow rate 125 ml/min, 25 °C. $\tau_{50\%}$ and $\tau_{95\%}$ is the time by which C_0 reduced by 50% and 95%, respectively [59]*

A major advancement in hemoperfusion was made when uncoated ACs for direct contact with blood were developed. It means that a new generation of uncoated AC were created that showed hemocompatibility equal to or better than coated AC used in HP.

This was achieved by both introducing microbeaded polymer-pyrolyzed AC, which retains high mechanical strength even at high degree of activation (burn-off up to 85–90%), and surface pretreatment of AC prior to use in direct HP. The latter is necessary to ensure that AC in contact with blood does not shift the pH balance or violate the mineral composition of blood and to remove microparticulates adhering to the carbon granule (Figs 21.4 and 21.5).

The effect that ion balancing of the AC surface has on pH and other ions, changes in contact with activated carbon. Unbalanced ACs have a significant effect on these parameters, whereas ion-balanced AC does not induce any changes in Tyrode's solution.

The most efficient removal of microparticulates from the AC surface was achieved using a surfactant-containing saline, such as heparin [60].

As AC is an electrochemically active substance, its biocompatibility has been related to the electrochemical potential at the surface–blood interface. It was shown that the magnitude of pH shift of saline in contact with AC

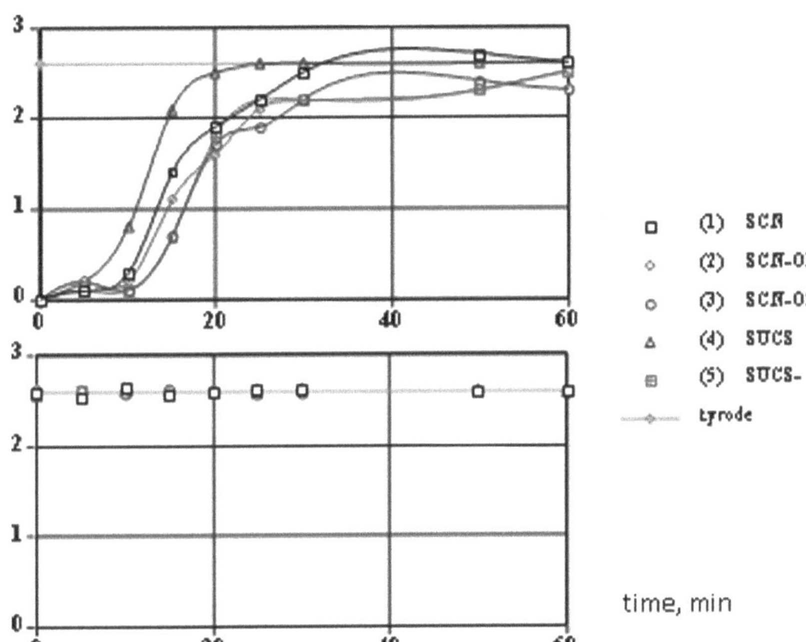

FIGURE 21.4 Breakthrough curves of K^+ in Tyrode's solution perfused over polymer-pyrolysed AC SCN, oxidized SCN (SCN-O1 and SCN-O2), SUCS and oxidized SUCS (SUCS-O2). Upper graph – ion-unbalanced AC, lower graph – ion-balanced AC. Tyrode's solution is a solution mimicking the mineral composition and pH of blood.

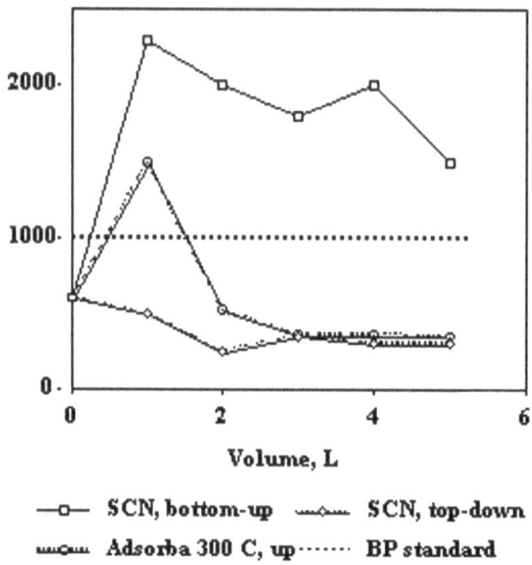

FIGURE 21.5 Number of microparticles, per cm^3, released from AC after washing with saline. Bottom-up and top-down indicate the direction of the saline flow [59].

correlated with the shift of the electrochemical potential of the latter vs its zero charge potential [60].

The use of uncoated hemocompatible ACs has broadened the possible medical applications of HP. In addition manipulating the porous structure of AC, particularly its mesopore size distribution, improves the size specificity for larger biological molecules, such as inflammatory cytokines and lipopolysaccharide (endotoxin) [61,62]. The molecules are separated by their ability to penetrate the porous structure. Thus, to target middle-molecular-weight toxins the pores must be tailored to a size to enable their selection while restricting larger proteins, such as albumin (66 kDa), from orientating within the porous structure [63].

21.3.3. Complications of Hemoperfusion

Activated carbon use in hemoperfusion encountered a number of serious problems in its early use. Microparticle release from carbons with low mechanical strength led to particle embolization, a serious clinical side-effect [64]. Blood proteins act as surfactants and are capable of releasing the small microparticles that are loosely attached to the carbon surface. Even initial experiments using very extensive washing of the charcoal bed still resulted in continued fine particle release [65]. Later work has shown that this side effect can be reduced by introducing a washing stage on the commercial scale, the use of polymer coatings, and the introduction of AC with greater mechanical strength, all of which have reduced microparticulate release to acceptable infusion fluid limits according to the British Pharmacopoeia [4]. In addition, considerable blood cell losses were experienced after passage through the early uncoated AC containing hemoperfusion devices. Excessive platelet deposition in particular was seen [66–68]. Transient leukopenia (loss of leukocytes) was also measured, caused as a direct result of blood complement activation by the foreign surface of the sorbent [48]. Current devices have overcome blood cell loss by microencapsulation techniques and experience platelet losses ranging from 0 to 30% [69,70]. During hemoperfusion, adsorption or activation of coagulation factors has been detected, including fibrinogen and fibronectin depletion and iC3b adsorption [71–73]. Changes in the mineral composition of the blood after contact with the AC sorbent were also noted [50], although these varied greatly between materials and were not considered to be significant if the device was first equilibrated in an isotonic buffer solution. In order to reduce the tendency for the HP device to generate blood clots, an anticoagulant such as sodium heparin was recommended to increase the intra-circuit whole blood clotting time [56]. Interestingly, there were initial concerns over whether heparin adsorption by the AC within the device would be a significant problem and reduce the effectiveness of the anticoagulant. However, studies using four different HP columns determined that heparin adsorption never reached higher than 7% of the inlet amount and thus would not be an issue [74]. Even with the introduction of more biocompatible AC and anticoagulation regimes, there are still issues associated with the use of AC in hemoperfusion, such as hypotension likely due to platelet activation and release of vasoactive amines and the unselective removal of factors, such as calcium, glucose, hormones, coagulation factors, and trace elements.

21.3.4. AC Containing Hemoperfusion Devices

During the 1970s various companies developed and commercially sold a number of HP devices containing AC as the adsorbent. The columns contained AC produced from peat or pitch and were coated with biocompatible cellulose, polyHEMA, acrylic hydrogel, or heparin hydrogel membranes. The columns were produced in Europe, Japan, and the United States by companies, such as Gambro Lundia AB (Adsorba cartridge, Sweden), Asahi Medical (Hemosorba, Japan), Clark R&D (Clark Biocompatible HP system, USA), Becton Dickinson (Hemodetoxifier, USA), and Smith & Nephew (Haemocol, UK) [2,46]. The columns were marketed mainly for acute poisoning, although some (Asahi & Gambro) gained US FDA approval for treatment of liver failure; but later clinical trials indicated no benefit over standard therapy, attributed mainly to complications in the treated group [75]. Antibody- or antigen-coated AC was also developed as a treatment for specific autoimmune states; although these demonstrated some clinical success [76,77], their use has been limited. Since this time, hemoperfusion use in both Europe and the USA has been on the decline. At present, only the Adsorba® C cartridges (Adsorba 150 & 300) associated with the Prismaflex® HP system produced by

Gambro Lundia AB are still commercially available in Europe and the US. The system is marketed for removing drugs and toxins including both lipophilic and hydrophilic drugs, by AC hemoperfusion using cellulose-coated Norit RBX charcoal, which reduces the undesired deposition of blood components and protects against fine particle release. This overall decline in the use of HP in the West can be explained by a number of factors, not least by the limited usefulness of coated AC for blood cleansing.

Hemoperfusion using uncoated AC appears to be much more popular in the former USSR. For example, a spherical carbon called FAS manufactured by Neorganika Ltd. from a furfurol precursor (Electrostal, Moscow region, Russia) is used for HP applications in the treatment of a range of conditions ranging from peritonitis, nephritic and hepatic insufficiency, autoimmune and skin diseases, and sepsis. In recent years, a new type of AC produced by pyrolysis of nitrogen-containing synthetic (4-vinylpyridine-styrene-divinylbenzene copolymer) resins, called HSGD (HaemoSorbent Granulated Deliganding), has been developed for hemoperfusion applications at IEPOR, Kiev, Ukraine. Uncoated HSGD in an extracorporeal system is used clinically in Ukraine, Russia, and Uzbekistan for the treatment of hepatic, liver, and multiorgan insufficiency. In particular, it is used to treat multiorgan failure in children with death-cup fungi mycotoxin poisoning and has been clinically shown to decrease mortality from 58 to 10%. HSGD has also demonstrated a greater capacity (10–20-fold higher) for unconjugated bilirubin than commercial hemoadsorbents from Gambro and Asahi and conventional granulated carbons [78,79].

Switching from HP over coated AC to uncoated AC therefore has a high potential for development and applications, particularly for treatment of patients with septic shock and multiorgan failure, conditions in which mortality remains high with little reduction in the last 40–50 years.

In the UK, a novel, 3-dimensional, AC monolith containing large transport channels allowing blood flow through the device has been developed for hemoperfusion applications by MAST Carbon International Ltd using a Novolac phenol–formaldehyde resin (Fig. 21.6) [80]. The monolith structure is uncoated and has demonstrated its ability to remove inflammatory cytokines and uremic toxins and is undergoing further development for a range of possible hemoperfusion applications to treat conditions such as systemic inflammatory response syndrome (SIRS) related to sepsis, and to augment current hemodialysis by adsorbing protein-bound uremic toxins not currently removed [81,82].

21.3.5. AC Containing Combined Extracorporeal Devices

When the biocompatibility of the AC is not as important an issue as with direct contact with blood, uncoated AC has been used for regeneration and recycling of ultrafiltrate or dialysate to reduce the volume of replacement fluid required and

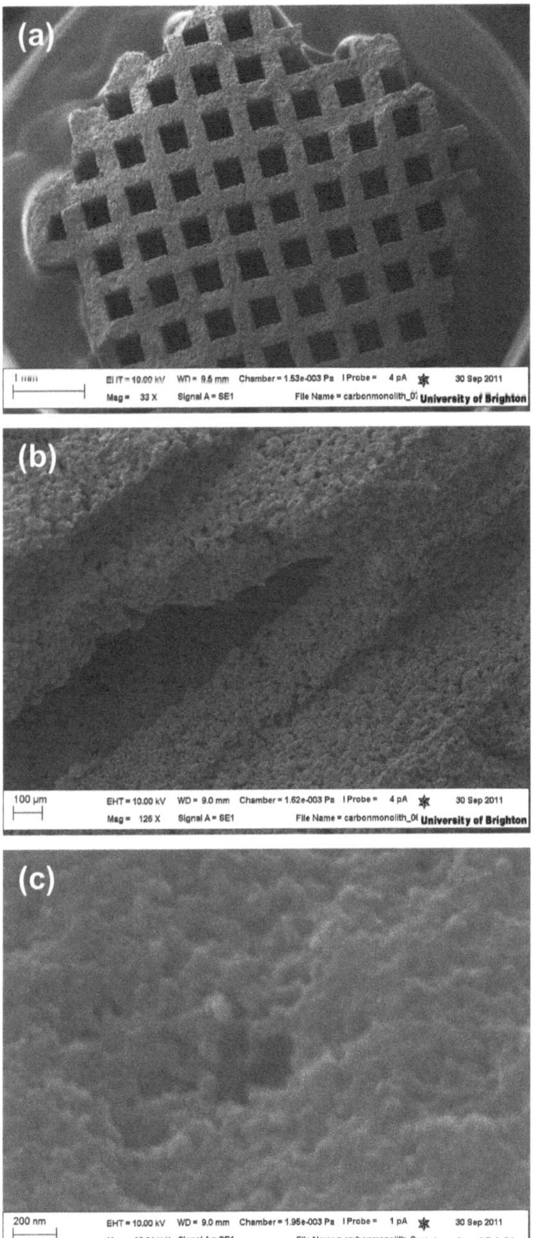

FIGURE 21.6 Scanning electron micrographs of a carbon monolith (MAST Carbon International Ltd). a) external view of the channel structure; b) internal view of the channels; c) porous structure of the channel walls.

thus limit costs of the extracorporeal treatment. As early as 1977, the Liver Dialysis Unit (originally called the BioLogic-DT®, HemoCleanse Inc, USA) was the first artificial liver device with FDA approval for the treatment of encephalopathy due to liver failure [83]. The device used a sorbent suspension of powdered charcoal (Norit A Supra, American Norit Company, USA) and cation exchangers and was the prototype for the Molecular Adsorbent Recirculating System (MARS®, Teraklin, Germany) and Prometheus® (Fresenius Medical Care) and BioLogic-DTPF (HemoCleanse Inc) devices [84]. The Liver Dialysis Unit is not currently marketed in the USA and is being redesigned using an immobilized powdered charcoal in a block system, which will enable regeneration of dialysate, albumin dialysate, or plasma. Studies have shown improved toxin binding by the carbon block compared with granular carbon, for bilirubin, cytokines, and other small toxins, although the system will be limited in its capacity for strongly protein-bound toxins [85].

The MARS system removes protein-bound and water-soluble toxins using albumin dialysis. The system runs blood through the dialyzer, where water-soluble and protein-bound toxins in the blood can pass through the membrane due to the albumin dialysate on the other side. This toxin-enriched albumin solution is then passed firstly through a dialyzer to remove water-soluble toxins, then through two adsorber cartridges, one filled with AC, the other with an anion exchanger, to remove the protein-bound toxins. The diaMARS® AC250 cartridge is filled with vapor-activated carbon that is uncoated and especially suited for elimination of low-molecular-weight, nonpolar compounds, such as aromatic polycyclic hydrocarbons or fatty acids. Although the system has been developed for the treatment of chronic liver conditions [86] and has a CE mark to allow sale in Europe, at the present time in the US, *the MARS® system is cleared for use in the treatment of drug overdose and poisonings only.*

The sorbent column-dependent dialysate regeneration and recirculation system (REDY system, Renal Solutions Inc, USA) was developed in the 1970s to regenerate dialysis fluid for the renal home dialysis market. The system used AC in combination with zirconium phosphate, zirconia, and alumina to regenerate dialysate, but by the 1990s, the REDY had disappeared from clinical use [87]. Following on from this technology, the Allient Sorbent Hemodialysis System (Renal Solutions Inc) was developed for acute and chronic renal failure including an improved sorbent column. The Sorbent Management for Advanced Renal Replacement Therapy (SMARRT™) technology enables the use of just 6 l of dialysate for one treatment with a sorbent cartridge composed of an AC and several inorganic layers necessary to convert urea into ammonia using immobilized enzyme urease followed by binding of the ammonia.

There has been a recent resurgence of interest in home hemodialysis in which sorbent systems play an integral part, with the ultimate goal being a wearable system and which has been made possible by the advent of micro-technologies [88]. The development of a wearable artificial kidney (WAK) has recently undergone two successful human trials in dialysis patients [52].

The system regenerates dialysate by continuous recirculation through a series of cartridges containing AC, immobilized urease, zirconium hydroxide, and zirconium phosphate and holds the potential to change the lives of patients with end-stage renal failure.

The future for AC use as uncoated sorbents in hemoperfusion or other extracorporeal devices looks very promising. As continuous improvements and developments are made to the physicochemical and structural properties of AC, this enables the AC to be tailored for the particular application, whether that be in powdered, spherical bead, carbon block, or monolith form. Increasing AC hardness or 3-dimensional structure to prevent fine generation, introduction of pores of selected dimensions, and modifying surface chemistry are just some of the ways by which ACs have been adapted in recent years to improve their suitability for medical applications.

21.4. ACTIVATED CARBON ADSORBENTS FOR USE IN CBRN INCIDENTS

The XXI century brought a serious threat of terrorism using weapons of mass destruction, such as chemical, biological, radiological, and nuclear (CBRN) warfare. Although so far most attackers used conventional weapons, well-known cases of anthrax crisis in the USA in autumn 2001 and the use of sarin by extremist groups in Matsumoto and Tokyo, Japan, in 1994 and 1995, suggest that the use of CBRN agents in terrorist events is likely and, according to some forecasts, even imminent [89]. The most recent terrorist attack in Mumbai, India, in November 2008, killed almost 200 people. These terrorists did not use any CBRN agents, but the US Congress hearings in the immediate aftermath have concluded that 'Terrorists are likely to use a weapon of mass destruction somewhere in the world in the next five years'. Even if the crisis management procedures are well established and first-aid responders are properly trained, it is likely that terrorists will use CBRN agents in a covert rather than overt way, and the first signs of exposure will be revealed with a delay of hours if not days, with the exact nature of the agent used yet to be determined [90]. This is particularly relevant to the use of biological agents, but even results of exposure to chemical and radiological agents may be delayed if, for example, the chemical poisoning occurred *via* contaminated food, or water, or a low-level-radioactivity radiological dispersal device (RDD), or a 'dirty bomb', as it is commonly known, was used.

With the exact nature of the agent used still unknown, the use of AC as an oral adsorbent or for HP treatment becomes well justified due to its ability to adsorb a large number of substances with different molecular mass and of different chemical origin. This statement is based on the experience of using AC for the treatment of so-called 'liquidators', or military staff who participated in the cleanup of the Chernobyl zone following an accident at the nuclear power plant in 1986. The nature of this event allows a comparison with a powerful 'dirty bomb' on a large scale, and most of about 800,000 liquidators were continuously exposed to low-level doses of radioactivity except for firemen, who received high doses. Incorporation of radionuclides in the human body may have long-lasting health effects caused by ionizing radiation. The main medical concept that was exploited to prevent negative health effects of low-level radioactivity was to accelerate removal of radionuclides from the body or prevent their accumulation in human body by the prophylactic use of AC.

In earlier experiments, HP over activated carbon SUGS was successfully used for the treatment of acute radiation syndrome induced in dogs, by passing the blood of irradiated animals through a hemoperfusion column (Table 21.3) [91].

TABLE 21.3 Effect of Hemoperfusion (HP) Over Activated Carbon SUGS on Acute Radiation Syndrome (ARS) in Dogs[a]

		Number of Animals That Died		
Group	Survival Rate, %	5–9 days	10–14 days	15–30 days
Control ($n=31$)	3.2	9	20	1
HP 2 hours after irradiation ($n=19$)	68.4	–	1	5
HP 24 hours after irradiation ($n=19$)	62.4	–	2	5

[a]*Experimental conditions: Mongrel dogs, 8–12 kg weight, exposed to 5.25 Gy external radiation dose with X-ray; this dose causes acute radiation sickness. All animals received antibiotics and human serum albumin transfusion as a basic conventional therapy. The 3rd group additionally had one session of massive saline transfusion, following diuretic furosemide prescription. Those animals that survived after 30 days, lived for another 1–2 years*

In these experiments, there was no incorporated radioactivity; therefore, the therapeutic action of hemoperfusion cannot be assigned to direct adsorption/removal of radionuclides. Instead, activated carbon adsorbed products of radioactive destruction of natural proteins, DNA, and other molecules. These products known as radiotoxins are toxic to the organism and cause systemic endotoxicoses, severe inflammation, and multiple organ failure [92]. This mechanism was further confirmed by measuring viability of marrow cells, blood cell count, etc. Radiotoxins can be large and middle molecular mass species, which are adsorbed into the mesopores.

Hemoperfusion over activated carbon was successfully used in conjunction with external radiation treatment of cancer patients. Such a treatment is always a balance between a positive effect of radiation killing cancer cells and a damaging effect on the rest of the body. By purifying the blood of irradiated patients with activated carbon, negative effects of radiation were significantly reduced and higher doses of radiation could be applied increasing efficiency of such a treatment [93]. It remains practically unknown in the West that a large group of patients ($n = 51$) diagnosed with acute radiation syndrome (ARS) was treated at the Institute of Oncology and Radiology, in Kiev, Ukraine, and all patients survived [94]. The Kiev group treated acute radiation syndrome as a type of acute poisoning and used decorporation techniques, such as lavage, in combination with extracorporeal techniques for blood purification, dialysis and hemoperfusion over activated carbon.

In the case of liquidators exposed to low-level radioactivity, oral administration of activated carbon SCN for two weeks while they stayed in the Chernobyl zone, led to a significant reduction in the incorporated radionuclides in their bodies, compared to the control group [3,95]. In total, the level of radionuclides in the personnel who used oral activated carbon was by almost an order of magnitude lower than in the control group. In comparison with the control group, radioactivity of feces was significantly higher and radioactivity of blood and urine significantly lower in the treated group. These results confirmed that activated carbon bound radionuclides. Importantly, removal of radioactivity through the gastrointestinal tract also led to reduction in radioactivity in the blood and other body compartments, which suggests that oral adsorption is an efficient way of reducing radioactive load in the whole body. In the group of personnel treated with oral adsorbents, the biochemical parameters of blood normalized faster than in the control group. Lipid peroxidation, as an indicator of the presence of reactive oxygen species, is one of the processes enhanced by radiation. Parameters of lipid peroxidation system in the blood of personnel given activated carbon orally were significantly different from those in the control group and closer to normal.

21.5. ACTIVATED CARBON MATERIALS FOR WOUND DRESSINGS

Knowledge surrounding the wound-healing process and the factors that impact the non-healing wound has expanded significantly in the last 50 years [96,97]. In addition, the number and types of dressings developed to treat the different stages of wound healing and combat the non-healing wound have increased. Depending on the physiology of the wound bed and stage of healing, a wound dressing should maintain a moist barrier to infection, allow gas exchange, absorb exudate and microorganisms, promote debridement, be non-allergenic, and have a low cost/benefit ratio [98]. A large number of advanced moist wound-healing dressings have been designed to promote rapid wound healing, including foams, alginates, hydrocolloids, and hydrogels [99]. However, it has been estimated that half of all wound-care dressings remain traditional gauze-based varieties [86]. Recent advances in wound-care management have sought to consider the physiology of the wound bed, particularly with regard to the non-healing wound where biochemical imbalance leads to heightened proteolysis, increased microbiological load, increased cell senescence cues, and the release of inflammatory species [100]. Antimicrobial dressings that control wound bioburden, address biochemical imbalance, and absorb exudates have been developed to address some of the issues associated with the non-healing wound. However, the treatment of non-healing wounds remains a serious clinical issue with little clear evidence from large, controlled clinical trials of improved healing rates using more expensive advanced dressings [101].

The treatment of infected wounds centers on the control of infecting pathogens, the removal of exudate, and the reduction of odor. Wound odor is particularly associated with venous ulcers, fungating carcinomas, and surgical wounds where bacterial metabolism of amino acids derived from proteins in wound exudate or necrotic tissue results in the release of ammonia, amines, and sulfurous compounds [102]. The resulting smell emitted is foul and can provide a serious negative impact on the quality of life of patients and their carers. Topical or systemic antibiotics are used to treat infection. However, efficacy may be limited in some wounds by a lack of therapeutic dose at the wound site and a detrimental effect on wound healing [103,104]. Additionally, there is an increase in pathogens resistant to antibiotic therapy requiring the development of alternative treatment options. Honey-based dressings are also used to treat infected and malodorous wounds. These control infection and malodor *via* the slow-release of low concentrations of hydrogen peroxide, a hyperosmotic effect in addition to the presence of plant-derived antimicrobial properties. Bacterial metabolism of glucose in preference to amino acids reduces the release of malodorous compounds in favor of lactic acid [102]. While some clinical efficacy has been observed in surgical and burn wounds studies, more evidence from randomized controlled studies is required, and these dressings are not routinely used by the wound-care community.

Activated carbon dressings are most commonly incorporated into dressings and marketed for the treatment of malodorous wounds [97,105]. These are effective because they offer a highly adsorbent surface area for small gas molecules. The use of an activated carbon cloth dressing derived from cellulose was first reported in the 1970s for the treatment of wound malodor using an activated cloth developed by the Ministry of Defence (UK). Similarly, the use of activated carbon cloth in combined dressings was proposed in the former USSR to reduce infection, remove toxins, and increase re-epithelialization in burns treatment [106,107]. Subsequently, a range of products have been produced that include activated carbon cloth produced by pyrolysis of cellulosic material and activation at high temperatures [108–113]. While activated carbon is applied to wounds for malodor adsorption, activation can produce a large surface area and pore volume to allow adsorption of bacteria, bacterial

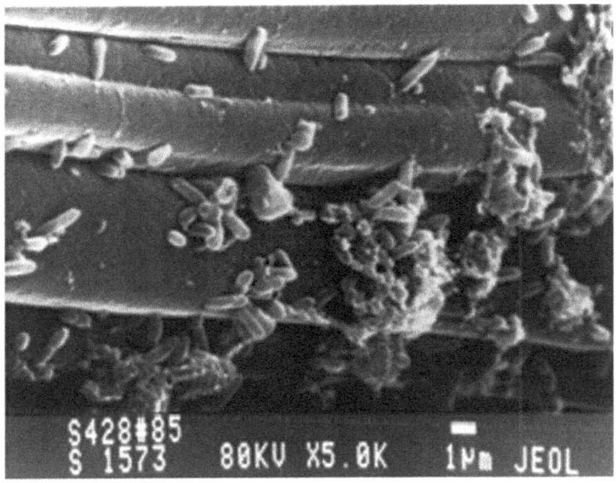

FIGURE 21.7 Scanning electron micrograph showing Bacteria (*Bacillus, Pseudomonas* and *Streptococcus* sp.) fixed to Zorflex® fibers.

toxins, and larger wound-degradation products in addition to the small volatile chemicals that produce malodor (Fig. 21.7). Numerous dressings incorporate activated carbon cloth and are clinically recommended predominantly as an odor and exudate-adsorbing component [105]. However, evidence indicates that the activated carbon itself contributes to the reduction in wound bioburden and healing of chronic, ulcerated wounds [114].

A number of multicomponent dressings containing activated carbon cloth are marketed for use under the trademark names Actisorb (Systagenix Wound Management), Carboflex (ConvaTec UK), Carbonet (Smith & Nephew), and Clinisorb (Clinimed) [98]. The majority of these dressings incorporate Zorflex® Medical cloth produced by Chemviron Carbon, a subsidiary of Calgon Carbon Corporation and the world's leading manufacturer of activated carbon cloth for wound dressings and other applications (Personal comm., J Taylor, Chemviron). Zorflex® is a microporous activated carbon cloth synthesized from viscose rayon and used in medical devices certified to ISO 13485 standard. The cloth is carbonized and activated by controlled heating to 800 °C to produce a highly microporous material with a superior surface area (1,000–2,000 m^2/g) and rapid adsorption kinetics through electrostatic bonds formed within the micropores. The company has also developed Zorflex® VB and Zorflex® VB Plus, which when tested by the Health Protection Agency using the MS2 Coliphage surrogate test virus showed high virus kill rates after immobilization. While Zorflex has inherent antimicrobial properties, a unique pre-impregnation step with silver salts prior to carbonization produces a stable metallic silver coating that is non-cytotoxic and does not leach ionic silver into the wound bed. Zorflex® medical activated carbon cloth finds application in pressure sores, fungating lesions, venous and arterial ulcers, minor burns, donor sites, and surgical wounds in addition to other applications, such as in ostomy filters, medical and personal care products, and biological and chemical warfare equipment. Actisorb® products, incorporating the Zorflex® Medical cloth, initially manufactured by Johnson & Johnson Medical Limited, now known as Systagenix Wound Management, were the first to be commercially available for wound-dressing applications. The range includes activated carbon cloth with silver (Actisorb® 220), an antimicrobial dressing containing silver bound to carbon fibers within a spun bonded nylon envelope. The Actisorb® range is marketed for the management of a wide range of chronic wounds including malodorous, infected wounds, fungating lesions, fecal fistulae, infected pressure sores, and heavily exuding leg ulcers.

One of the primary targets for chronic wound care is a reduction in wound bioburden since high levels of bacteria in nonhealing wounds have been directly linked to poor wound-closure rates. High microbial levels in the wound bed can disrupt the normal wound-healing process, form antimicrobial resistant biofilms, and enhance tissue necrosis through toxin secretion and the heightened activation of pro-inflammatory stimulatory pathways. Silver-incorporating dressings have been developed to combat wound infection. Silver is a well-established, broadly acting antimicrobial that damages cell membrane and membrane receptor function, interferes with electron transport mechanisms to disrupt bacterial metabolic pathways, and binds to and denatures bacterial DNA inhibiting cell respiration [115,116]. Dressings coated with nanocrystalline silver or silver salts release silver ions into the wound bed to reduce bioburden and speed up the wound-healing rate. While *in vitro* studies show biocidal activity in the presence of silver-containing dressings, evidence for their use in the clinical setting remains unclear [117–122]. The Actisorb® 220 dressings, as for the Zorflex® silver-activated carbon cloths, incorporate silver in a processing step prior to carbonization. The silver is not leached into the wound but has a bactericidal effect on microbial contaminants drawn out of the wound bed by the powerful adsorptive capacity of the activated carbon matrix [99]. Concerns over silver cytotoxicity from silver release dressings, particularly in immunocompromized patients, suggest that a carbon fiber matrix with stably incorporated silver, which adsorbs microbial species and toxins from the wound bed but does not release silver ions may be a safer approach to antimicrobial wound management. Conclusive evidence from large, randomized, multicenter clinical studies investigating the impact of silver (all species)-containing dressings on wound closure rates is lacking [123]. However, evidence to date indicates that silver dressings do induce a more rapid reduction in wound area [114,124]. In addition, recent data indicate that highly activated carbon cloth dressings alone may be appropriate for antimicrobial wound management to improve healing of chronic and ulcerated wounds. *In vitro* studies comparing the bactericidal effect of Zorflex® compared to silver carbon cloth indicated no significant difference in bacterial colony formation between the dressings. A 2011 Medica-Compamed press release presented evidence from individual clinical case studies carried out

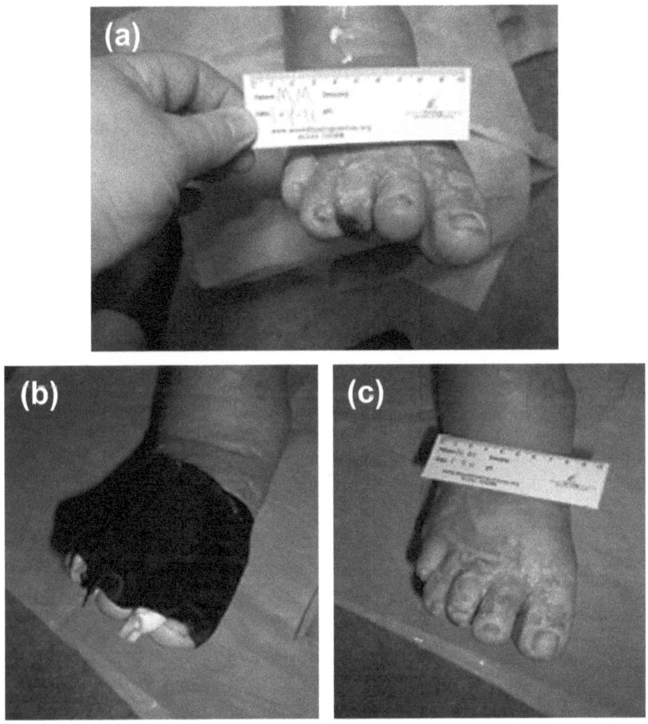

FIGURE 21.8 Images showing treatment of a diabetic foot ulcer with Zorflex® a. Preapplication of Zorflex® b. Dressing *in-situ* c. After 4 days' use of Zorflex® showing less inflamed tissue and softening of the necrotic tissue on the third toe. *(Courtesy of Chemviron Carbon Cloth in association with The Wound Healing Centres, UK)*

by Chemviron Carbon Cloth Division in association with Wound Healing Centres Ltd. (UK). The data showed increased granulation tissue and increased wound margin epithelialization in the wound bed of a mixed etiology leg ulcer and venous leg ulcer respectively, after 3 days' application of Zorflex® and reduced inflammation and necrotic tissue softening in a diabetic foot ulcer after 4 days' application of Zorflex® dressing (Fig. 21.8). Larger controlled clinical studies are certainly necessary but the early evidence suggests the cost-effective prospect that highly adsorbent microporous carbon dressings alone without more costly silver incorporation adsorb bacteria, bacterial products, inflammatory molecules, and exudate in sufficient quantity to alter the wound-bed environment in favor of wound resolution.

The incorporation of a mesoporous component into carbon wound-dressing technology may also be used to enhance dressing efficacy. Highly microporous dressings using Zorflex® maintain a strong electrostatic charge for binding of small molecules and immobilizing microbial species (Fig. 21.7). The introduction of mesopores may increase the adsorptive capacity of carbon dressings for high-molecular-weight proteins and bacterial toxins, as is observed in carbon adsorbents for extracorporeal applications, but may also reduce molecular binding strength in wound-dressing applications. Microporous activated carbon cloth is made by pretreating a cellulose-based starting material with a Lewis acid prior to pyrolysis and activation of the char to produce activated carbon cloth with micropores of up to 2 nm in width [98]. Methods to incorporate mesopores with a width of 2 to 50 nm were developed by impregnation with boric acid and at least one alkali metal prior to activation [111,125]. A method to increase mesopore volume and fiber physical strength was additionally developed by cellulose impregnation with a solution of a phosphate and a Lewis acid containing a metal halide, such as aluminum or zinc chloride, prior to carbonization and activation. *In vitro* experiments comparing these highly mesoporous dressings with Actisorb dressings indicated significantly increased adsorption of representative high-molecular-mass proteins cytochrome C (12.5 KDa) and bovine serum albumin (66 KDa) [112].

While a range of dressings exist containing activated carbon in both microporous and mesoporous forms, the data confirming clinical efficacy remain unclear with conflicting results found in reports of *in vitro*, *in vivo*, and clinical studies. It appears that specifically engineered 'next generation' activated carbon materials have yet to be fully exploited for wound-dressing applications. However, there is great potential for the production of further optimized dressings drawing from evidence of carbon antibacterial properties and biotoxin adsorption in other applications. These would combine biotoxin adsorption with antimicrobial, anti-odor and exudate absorbing properties to augment reduction in wound-bed bioburden and a subsequent reduction in mediators of inflammatory perturbation commonly found in the nonhealing wound.

21.6. CONCLUSIONS

Throughout its long history of use in medical applications, activated carbon has been in and out of fashion. Recent advances in designing novel carbon materials with controlled porosity and in better understanding their performance as medicinal products and in medical devices have opened up new horizons. New opportunities have been created for wider applications in the treatment of acute and chronic poisoning of an exogenous and endogenous nature, in the treatment of organ failure, wound care, and in providing protection against deliberate or accidental use of chemical, biological, and radiological warfare.

ACKNOWLEDGMENTS

This work has been supported by 7th Framework Programme (EU), grants FP7-PEOPLE-IRSES MEAD-ET and FP7-PEOPLE-IAPP ACROBAT.

The authors are grateful to Prof. SR Tennison, Dr. OP Kozynchenko and Mr AP Rawlinson (MAST Carbon International Ltd), and Dr. J Taylor (Chemviron Carbon Cloth Division) for providing samples of activated carbon materials and useful discussions of experimental data and properties of activated carbon.

REFERENCES

[1] The Pharmaceutical Industry in Figures, European Federation of Pharmaceutical Industries and Associations, Brussels, 2009.
[2] D.O. Cooney, Activated Charcoal in Medical Applications, Marcel Dekker Inc., New York, 1995.
[3] B.G. Katona, E.G. Siegel, R.J. Cluxton Jr., The new black magic: activated charcoal and new therapeutic uses, J. Emerg. Med. 5 (1) (1987) 9–18.
[4] S.V. Mikhalovsky, V.G. Nikolaev, Activated carbons as medical adsorbents, in: T.J. Bandosz (Ed.), Activated Carbon Surfaces in Environmental Remediation, Elsevier Press, 2006.
[5] V. Nikolaev, V. Strelko, Yu. Korovin, Theoretical basis and practical use of enterosorption method, in: Sorption Methods of Detoxication and Immunocorrection in Medicine, 1982, pp. 112–114. Kharkiv, Ukraine (In Russian).
[6] D. Healy, Let Them Eat Prozac, Canadian Association of University Teachers, J Lorimer & Co Ltd. 2003, http://www.healyprozac.com/Trials/Forsyth/Transcripts/default.htm
[7] T.E. Albertson, R.W. Derlet, G.E. Foulke, M.C. Minguillon, S.R. Tharratt, Superiority of activated charcoal alone compared with ipecac and activated charcoal in the treatment of acute toxic ingestions, Ann. Emerg. Med. 18 (1) (1989) 56–59.
[8] G.R. Bond, The role of activated charcoal and gastric emptying in gastrointestinal decontamination: a state-of-the-art review, Ann. Emerg. Med. 39 (3) (2002) 273–286.
[9] P.A. Chyka, D. Seger, E.P. Krenzelok, J.A. Vale, Position paper: single-dose activated charcoal, Clin. Toxicol. (Phila) 43 (2) (2005) 61–87.
[10] American Academy of Clinical Toxicology, European Association of Poisons Centres and Clinical Toxicologists, Position statement and practice guidelines on the use of multi-dose activated charcoal in the treatment of acute poisoning, J. Toxicol. Clin. Toxicol. 37 (6) (1999) 731–751.
[11] P.A. Chyka, D. Seger, Position statement: single-dose activated charcoal. American Academy of Clinical Toxicology; European Association of Poisons Centres and Clinical Toxicologists, J. Toxicol. Clin. Toxicol. 35 (7) (1997) 721–741.
[12] G.K. Isbister, V.V. Kumar, Indications for single-dose activated charcoal administration in acute overdose, Curr. Opin. Crit. Care. 17 (4) (2011) 351–357.
[13] T.E. Albertson, K.P. Owen, M.E. Sutter, A.L. Chan, Gastrointestinal decontamination in the acutely poisoned patient, Int. J. Emerg. Med. 4 (2011) 65.
[14] K.R. Olson, Activated charcoal for acute poisoning: one toxicologist's journey, J. Med. Toxicol. 6 (2) (2010) 190–198.
[15] M. Eddleston , E. Juszczak, N.A. Buckley, L. Senarathna, F. Mohamed, W. Dissanayake, A. Hittarage, S. Azher, K. Jeganathan, S. Jayamanne, M.H.R. Sheriff, D.A. Warrell, Ox-Col Poisoning Study Collaborators, Multiple-dose activated charcoal in acute self-poisoning: a randomised controlled trial, Lancet 371 (9612) (2008) 579–587.
[16] P. Gaudreault, Activated charcoal revisited, Clin. Ped. Emerg. Med. 6 (2005) 76–80.
[17] V. Nikolaev, Peroral application of synthetic activated charcoal in USSR, Biomat. Art. Cells Art. Org. 4 (1990) 555–568.

[18] B.G. Lukichev, G.D. Shostka, V.V. Strelko, T.S. Azizova, Iu.R. Kavrăskĭ, I.Iu Panina, 10-years' experience in using enterosorption for treating chronic kidney failure (in Russian), Ter. Arkh. 64 (8) (1992) 52–56.
[19] E.S. Ryss, S.I. Riabov, M.B. Lutoshkin, A comparative evaluation of the efficacy of the clinical use of SKN-4M-, SKT-6A and polifepan-type sorbents in treating patients with chronic kidney failure (clinical and experimental studies) (in Russian), Ter. Arkh. 68 (8) (1996) 39–43.
[20] E.S. Ryss, M.B. Lutoshkin, The effect of enterosorbents on stomach functional activity and the parameters of nitrogen metabolism in patients with nephrogenous gastropathies (in Russian), Ter. Arkh. 63 (7) (1991) 112–116.
[21] I.K. Sledzevskaia, V.I. Bul'da, K.D. Babov, Enterosorbent SKN in the combined rehabilitative treatment of patients who have had a myocardial infarct. (in Russian), Vrach Delo 9 (1991) 87–89.
[22] S.I. Riabov, V.I. Tsiura, B.G. Lukichev, G.D. Shostka, Iu.V. Natochin, Reduced experimental blood creatinine as affected by the SKN enterosorbent (in Russian), Urol. Nefrol. (Moscow) 2 (1987) 49–52.
[23] V.I. Davydov, S.S. Stavitskaia, V.I. Galinskaia, N.V. Gerasimenko, V.V. Strelko, The possibility of using carbon enterosorbents for normalizing cholesterol metabolism (in Russian), Biokhimiia 59 (2) (1994) 304–312.
[24] E.A. Friedman, E.I. Feinstein, M.M. Beyer, R.S. Galonsky, S.R. Hirsch, Charcoal-induced lipid reduction in uremia, Kidney Int. Suppl 8 (1978) S170–S176.
[25] T.D. Nykula, V.V. Bondur, K.H. Pluzhnyk, The effect of the SKN charcoal enterosorbent on the clinico-laboratory indices of patients with chronic circulatory failure (in Ukrainian), Lik Sprava 7–9 (1996) 106–109.
[26] A.I. Neimark, Ia.V. Iakovets, Changes in the hemostatic system indices of patients with chronic kidney failure under the influence of enterosorption and plasmapheresis (in Russian), Ter Arkh 66 (8) (1994) 57–60.
[27] M.A. Andreichyn, O.L. Ivakhiv, Enterosorption in the combined treatment of patients with salmonellosis (in Ukrainian), Lik Sprava 8 (1992) 81–84.
[28] M.L. Tarakhovskii, A.G. Tsypkun, V.P. Sergeev, T.D. Zadorozhnaia, V.K. Tishchenko, V.F. Litvinov, Efficiency of enterosorbents and detoxication mechanisms in immature rats with simulated hepatitis (in Russian), Fiziol Zh 37 (6) (1991) 48–55.
[29] I.V. Grin'ko, A.A. Krivchik, O.V. Ivankova, Functional state of the liver during development of extrahepatic cholestasis and during use of enterosorption (in Russian), Vopr Med Khim 37 (3) (1991) 26–28.
[30] V.G. Nikolaev, Peroral application of synthetic activated charcoal in USSR, Biomater. Artif. Cells Artif. Organs 18 (4) (1990) 555–568.
[31] V. Nikolaev, Enterosorption, in: T.M.S. Chang, B.-L. He (Eds.), Proceedings of the Fifth International Symposium on Hemoperfusion and Artificial Organs, China Academic Publishers, 1983, pp. 87–99.
[32] L.P. Bobkova, o V.A. Beĭk, P.V. Grishilo, A.V. Shevchuk, V.V. Glinskĭ, S.V. Mikhalovsky, V.V. Strelko, N.M. Berezhnaia, Enterosorption in the treatment of patients with chronic urticaria (in Russian), Vrach Delo 7 (1990) 87–90.
[33] G.V. Muravskaya, V.G. Nikolaev, V.P. Sergeev, N.I. Krutilina, L.V. Bonatskaya, V.N. Klevtsov, V.V. Surovikina, V.V. Sinajko, Enterosorption in oncotherapy, Biomater. Artif. Cells Immob. Biotechnol. 19 (1) (1991) 167–174.
[34] V.G. Nikolaev, V.V. Sarnatskaya, V.L. Sigal, V.N. Klevtsov, K.E. Makhorin, L.A. Yushko, High-porosity activated carbons for bilirubin removal, Int. J. Artif. Organs 14 (3) (1991) 179–185.

[35] G. Schulman, A nexus of progression of chronic kidney disease: tryptophan, profibrotic cytokines, and charcoal, J. Ren. Nutr. 22 (1) (2012) 107–113.
[36] T. Niwa, Role of indoxyl sulfate in the progression of chronic kidney disease and cardiovascular disease: experimental and clinical effects of oral sorbent AST-120, Ther. Apher. Dial. 15 (2) (2011) 120–124.
[37] H. Fujii, K. Nakai, M. Fukagawa, Role of oxidative stress and indoxyl sulfate in progression of cardiovascular disease in chronic kidney disease, Ther. Apher. Dial. 15 (2) (2011) 125–128.
[38] H. Fujii, F. Nishijima, S. Goto, M. Sugano, H. Yamato, R. Kitazawa, S. Kitazawa, M. Fukagawa, Oral charcoal adsorbent (AST-120) prevents progression of cardiac damage in chronic kidney disease through suppression of oxidative stress, Nephrol. Dial. Transplant 24 (7) (2009) 2089–2095.
[39] T. Niwa, Y. Emoto, K. Maeda, Y. Uehara, N. Yamada, M. Shibata, Oral sorbent suppresses accumulation of albumin-bound indoxyl sulphate in serum of haemodialysis patients, Nephrol. Dial. Transplant 6 (2) (1991) 105–109.
[40] Y.K. Lee, S.J. Moon, H.R. An, J.K. Kim, S.C. Bae, B.S. Kim, H.C. Park, S.K. Ha, Effects of oral adsorbent AST-120 (Kremazin) on the progression of chronic kidney disease, Korean J. Nephrol. 29 (2010) 450–457.
[41] S. Owada, T. Maeba, Y. Sugano, A. Hirayama, A. Ueda, S. Nagase, S. Goto, F. Nishijima, K. Bannai, H. Yamato, Spherical carbon adsorbent (AST-120) protects deterioration of renal function in chronic kidney disease rats through inhibition of reactive oxygen species production from mitochondria and reduction of serum lipid peroxidation, Nephron. Exp. Nephrol. 115 (4) (2010) e101–111.
[42] T. Niwa, H. Shimizu, Indoxyl sulfate induces nephrovascular senescence, J. Ren. Nutr. 22 (1) (2012) 102–106.
[43] T. Akizawa, Y. Asano, S. Morita, T. Wakita, Y. Onishi, S. Fukuhara, F. Gejyo, S. Matsuo, N. Yorioka, K. Kurokawa, CAP-KD Study Group, Effect of a carbonaceous oral adsorbent on the progression of CKD: a multicenter, randomized, controlled trial, Am. J. Kidney Dis. 54 (3) (2009) 459–467.
[44] S.R. Tennison, O.P. Kozynchenko, V.V. Strelko, A.J. Blackburn, Porous carbons. (2011). US PA 20110097583.
[45] M. Melillo V.M. Gun'ko, L.I. Mikhalovska, G.J. Phillips, J.G. Davies, A.W. Lloyd, S.R. Tennison, O.P. Kozynchenko, D.J. Malik, M. Streat, S.V. Mikhalovsky, Influence of structural characteristics of activated carbons on ibuprofen adsorption affected by bovine serum albumin, Langmuir 20 (7) (2004) 2837–2851.
[46] S.V. Mikhalovsky, Emerging technologies in extracorporeal treatment: focus on adsorption, Perfusion 18 (Suppl. 1) (2003) 47–54.
[47] E.E. Muirhead, A.F. Reid, A resin artificial kidney, J. Lab. Clin. Med. 33 (7) (1948) 841–844.
[48] G.E. Schreiner, The role of hemodialysis (artificial kidney) in acute poisoning, AMA Arch. Intern. Med. 102 (6) (1958) 896–913.
[49] H. Yatzidis, S. Voudiclari, D. Oreopoulos, D. Triantaphyllidis, N. Tsaparas, C. C. Gavras, A. Stavroulaki, Treatment of severe barbiturate poisoning, Lancet 286 (7405) (1965) 216–217.
[50] H. Yatzidis, D. Oreopoulos, Early clinical trials with sorbents, Kidney Int. (Suppl. 7) (1976) S215–S217.
[51] T.M. Chang, M. Migchelsen, J.F. Coffey, A. Stark, Serum middle molecule levels in uremia during long term intermittent hemoperfusions with the ACAC (coated charcoal) microcapsule artificial kidney, Trans. Am. Soc. Artif. Intern. Organs 20A (1974) 364–371.

[52] A. Davenport, V. Gura, C. Ronco, M. Beizai, C. Ezon, E. Rambod, A wearable haemodialysis device for patients with end-stage renal failure: a pilot study, Lancet 370 (9604) (2007) 2005–2010.

[53] H.D. Polaschegg, C. Ronco, M. Soli, Characterization of flow-dynamic pattern in a new sorbent cartridge for combined hemoperfusion-hemodialysis, Contrib. Nephrol. 133 (2001) 154–165.

[54] J.F. Winchester, Dialysis techniques: hemoperfusion, in: W.H. Horl, R.M. Lindsay, C. Ronco, J.F. Winchester (Eds.), Replacement of Renal Function by Dialysis, fifth ed., Kluwer Academic Publ., 2004.

[55] J. Botella, P.M. Ghezzi, C. Sanz-Moreno, Adsorption in hemodialysis, Kidney Int. 58 (2000) S60–S65.

[56] J.F. Winchester, A. Boldur, C. Oleru, C. Kitiyakara, Use of dialysis and hemoperfusion in treatment of poisoning, in: J.T. Daugirdas, P.G. Blake, T.S. Ing (Eds.), Handbook of Dialysis, fourth ed. Lippincott Willliams & Wilkins, Philadelphia, 2007, p. 300.

[57] E. Denti, J.M. Walker, Activated carbon: properties, selection and evaluation, in: C. Giordano (Ed.), Sorbents and Their Clinical Applications, Academic Press, New York, 1980.

[58] J.F. Winchester, C. Ronco, J.A. Brady, J. Clemmer, T.E. Muller, V. Davankov, M. Tsyurupa, L. Pavlova, M. Pavlov, J. Clemmer, H.D. Polaschegg, T.E. Muller, G. La Greca, N.W. Levin, Rationale for combined hemoperfusion/hemodialysis in uremia, Contrib. Nephrol. 133 (2001) 174–179.

[59] S.V. Mikhalovsky, T.A. Alexeeva, E.A. Fesenko, N.T. Kartel, V.V. Strelko, Biocompatibility of activated carbons, in: Extended Abstracts, Carbon 2001, 2001. Lexington, Ky.

[60] N.T. Kartel, Biocompatibility of carbonaceous haemosorbents, Efferent Ther. 4 (4) (1998) 3–9.

[61] C.A. Howell, S.R. Sandeman, G.J. Phillips, J.G. Davies, A.W. Lloyd, S.V. Mikhalovsky, S.R. Tennison, A.P. Rawlinson, O.P. Kozynchenko, H.L.H. Owen, J.D.S. Gaylor, J.J. Rouse, J.M. Courtney, The *in vitro* adsorption of cytokines by polymer-pyrolyzed carbon, Biomaterials 27 (30) (2006) 5286–5291.

[62] S. Yachamaneni, G. Yushin, S.H. Yeon, Y. Gogotsi, C. Howell, S. Sandeman, G. Phillips, S. Mikhalovsky, Mesoporous carbide-derived carbon for cytokine removal from blood plasma, Biomaterials 31 (18) (2010) 4789–4794.

[63] V. Davankov, L. Pavlova, M. Tsyurupa, J. Brady, M. Balasamo, E. Yousha, Polymeric adsorbent for removing toxic proteins from blood of patients with kidney failure, J. Chromatogr. B Biomed. Sci. Appl. 739 (1) (2000) 73–80.

[64] K.E. Hagstam, L.E. Larsson, H. Thysell, Experimental studies on charcoal haemoperfusion in phenobarbital intoxication and uraemia, including histopathologic findings, Acta. Med. Scand. 180 (5) (1966) 593–603.

[65] J.D. Andrade, K. Kunitomo, R. van Wagenen, B. Kastiger, D. Gough, W.J. Kolff, Coated adsorbents for direct blood perfusion II, Trans. Am. Soc. Artif. Intern. Organs 18 (1972) 473–483. 485.

[66] G. Dunea, W.J. Kolff, Clinical experience with the Yatzidis charcoal artificial kidney, Trans. Am. Soc. Artif. Intern. Organs 11 (1965) 178–182.

[67] T.M.S. Chang, Semipermeable aqueous microcapsules ("artificial cells"): with emphasis on experiments in an extracorporeal shunt system, Trans. Am. Soc. Artif. Intern. Organs 12 (1966) 13–19.

[68] J.D. Andrade, K. Kumitomo, R. van Wagenen, B. Kastigir, D. Gough, W.J. Kolff, Coated adsorbents for direct blood perfusion: HEMA-activated carbon, Trans. Am. Soc. Artif. Intern. Organs 17 (1971) 222–228.

[69] C. Ronco, A. Brendolan, J.F. Winchester, E. Golds, J. Clemmer, H.D. Polaschegg, T.E. Muller, G. La Greca, N.W. Levin, First clinical experience with an adjunctive hemoperfusion device designed specifically to remove beta(2)-microglobulin in hemodialysis, Blood Purif. 19 (2) (2001) 260–263.

[70] T. Chang, Microcapsule artificial kidney in replacement of renal function. With emphasis on adsorbent hemoperfusion, in: W. Drukker, F.M. Parsons, J.F. Maher (Eds.), Replacement of Renal Function by Dialysis, first ed. Martinus Nijhoff, 1978.

[71] J.F. Winchester, Hemoperfusion, in: J.F. Maher (Ed.), Replacement of Renal Function by Dialysis, third ed. Kluwer, Dordrecht, 1988.

[72] G. Pott, B. Voss, J. Lohmann, P. Kundorf, Loss of fibronectin plasma of patients with shock and septicaemia and after haemoperfusion in patients with severe poisoning. Fibronectin concentrations in plasma in human diseases. II, J. Clin. Chem. Clin. Biochem. 20 (6) (1982) 333–335.

[73] S.R. Sandeman, C.A. Howell, G.J. Phillips, J.G. Davies, A.W. Lloyd, S.V. Mikhalovsky, S.R. Tennison, A.P. Rawlinson, O.P. Kozynchenko, H.L.H. Owen, J.D.S. Gaylor, J.J. Rouse, J.M. Courtney, Assessing the *in vitro* biocompatibility of a novel carbon device for the treatment of sepsis, Biomaterials 26 (34) (2005) 7124–7131.

[74] D.O. Cooney, Rates of heparin adsorption in hemoperfusion devices, Clin. Toxicol. 15 (3) (1979) 287–291.

[75] J.G. O'Grady, A.E. Gimson, C.J. O'Brien, A. Pucknell, R.D. Hughes, R. Williams, Controlled trials of charcoal hemoperfusion and prognostic factors in fulminant hepatic failure, Gastroenterology 94 (5 Pt 1) (1988) 1186–1192.

[76] G. Hampbel, B. Widdop, R. Goulding, Adsorptive capacities of hemoperfusion devices in clinical use, Artif. Organs 2 (4) (1978) 363–366.

[77] D.S. Terman, G. Buffaloe, C. Mattioli, G. Cook, R. Tillquist, M. Sullivan, J.C. Ayus, Extracorporeal immunoadsorption: initial experience in human systemic lupus erythematosus, Lancet 2 (8147) (1979) 824–827.

[78] V.V. Sarnatskaya, L.A. Yushko, L.A. Sakhno, V.G. Nikolaev, A.V. Nikolaev, D.V. Grinenko, S.V. Mikhalovsky, New approaches to the removal of protein-bound toxins from blood plasma of uremic patients, Artif. Cells Blood Substit. Immobil. Biotechnol. 35 (3) (2007) 287–308.

[79] V.G. Nikolaev, V.V. Sarnatskaya, K.I. Bardakhivskaya, Y.M. Susak, E.A. Snezhkova, V.D. Chornomaz, P.C. Cymbajuk, V.Y. Uvarov, N.M. Gurina, New approaches to adsorptive therapy of liver failure, Efferent Ther. 9 (1) (2003) 26–39.

[80] R.N. Place, A.J. Blackburn, S.R. Tennison, A.P. Rawlinson, B.D. Crittenden, (2003). Method and equipment for removing volatile compounds from air, PCT/GB02/01142.

[81] S.R. Sandeman, C.A. Howell, S.V. Mikhalovsky, G.J. Phillips, A.W. Lloyd, J.G. Davies, S.R. Tennison, A.P. Rawlinson, O.P. Kozynchenko, Inflammatory cytokine removal by an activated carbon device in a flowing system, Biomaterials 29 (11) (2008) 1638–1644.

[82] S.R. Sandeman, C.A. Howell, G.J. Phillips, S.V. Mikhalovsky, S.R. Tennison, O.P. Kozynchenko, O. Boyd, S. Holt, Characterising advanced nanoporous activated MAST carbons for the treatment of kidney disease, Int. J. Artif. Organs 34 (8) (2011) 607.

[83] S.R. Ash, D.E. Blake, D.J. Carr, C. Cater, T. Howard, L. Makowka, Clinical effects of a sorbent suspension dialysis system in treatment of hepatic coma (the BioLogic-DT), Int. J. Artif. Organs 15 (3) (1992) 151–161.

[84] B. Carpentier, S.R. Ash, Sorbent-based artificial liver devices: principles of operation, chemical effects and clinical results, Expert. Rev. Med. Devices 4 (6) (2007) 839–861.
[85] S.R. Ash, C.D. Carr, Characterizing activated carbon for sorbent therapy, ASAIO J. 52 (2) (2006) 11A.
[86] S.R. Mitzner, J. Stange, S. Klammt, P. Peszynski, R. Schmidt, G. Noldge-Schomburg, Extracorporeal detoxification using the molecular adsorbent recirculating system for critically ill patients with liver failure, J. Am. Soc. Nephrol. 12 (Suppl. 17) (2001) S75–S82.
[87] J.W. Agar, Review: understanding sorbent dialysis systems, Nephrology (Carlton) 15 (4) (2010) 406–411.
[88] V. Gura, C. Ronco, A. Davenport, The wearable artificial kidney, why and how: from holy grail to reality, Semin Dial 22 (1) (2009) 13–17.
[89] One Team, One Mission, Securing Our Homeland. (2008). U.S. Department of Homeland Security Strategic Plan. Fiscal Years 2008–2013.
[90] A.L. Melnick, Biological, Chemical, and Radiological Terrorism. Emergency Preparedness and Response for the Primary Care Physician, Springer, 2008.
[91] V.G. Nikolaev, L.B. Pinchuk, M.A. Umansky, V.G. Pinchuk, T.N. Burushkina, S.V. Petrenko, N.K. Rodionova, E.A. Snezhkova, N.P. Bichkova, Early experimental studies on hemoperfusion as a treatment modality for acute radiation disease, J Artif Organs 17 (5) (1993) 362–365.
[92] M. Akashi, T. Tamura, T. Tominaga, K. Abe, M. Hachiya, F. Nakayama, Radiation damage in the human body: acute radiation syndrome and multiple organ failure, Nuclear Eng. Technol. 38 (2006) 231–238.
[93] L.P. Kindzelsky, A.I. Pozmogov, V.A. Chernichenko, V.A. Lisetskĭ, B.K. Nikishin, T.V. Sycheva, N.A. Tomilina, Half-body irradiation in the combined therapy of disseminated lymphomas (in Russian), Meditsinskaya Radiologiya (Moscow) 35 (1990) 9–13.
[94] L.P. Kindzelsky, E.A. Dyomina, Clinical and hematological features of acute radiation sickness in early post-Chernobyl period (in Russian), Likars'ka Sprava 3 (1998) 7–11.
[95] I.K. Dedenko, A.V. Starikov, V.V. Strelko, Efferent Methods of Treatment of Radiation Related Diseases (in Russian), ISPE Publ., Kiev, Ukraine, 1996.
[96] J.R. Hanna, J.A. Giacopelli, A review of wound healing and wound dressing products, J. Foot Ankle Surg. 36 (1) (1997) 2–14. discussion 79.
[97] T. Abdelrahman, H. Newton, Wound dressings: principles and practice, Surgery 29 (10) (2011) 491–495.
[98] A. Goossens, M.B. Cleenewerck, New wound dressings: classification, tolerance, Eur. J. Dermatol. 20 (1) (2010) 24–26.
[99] L. Ovington, Advances in wound dressings, Clin. Dermatol. 25 (2007) 33–38.
[100] J. Chen, C.M. Han, G.L. Su, Z.J. Tang, S.J. Su, X.W. Lin, Randomized controlled trial of the absorbency of four dressings and their effects on the evaporation of burn wounds, Chin Med J (Engl) 120 (20) (2007) 1788–1791.
[101] C. Bouza, A. Munoz, J.M. Amate, Efficacy of modern dressings in the treatment of leg ulcers: a systematic review, Wound Repair Regen. 13 (3) (2005) 218–229.
[102] P.C. Molan, The role of honey in the management of wounds, J. Wound Care 8 (8) (1999) 415–418.
[103] S.S. Brennan, M.E. Foster, D.J. Leaper, Antiseptic toxicity in wounds healing by secondary intention, J. Hosp. Infect. 8 (3) (1986) 263–267.
[104] S.S. Brennan, I.R. Pickford, M. Evans, A.V. Pollock, The prophylaxis of wound infection after abdominal operations: is one dose of antibiotic enough? J. Hosp. Infect. 3 (4) (1982) 351–356.

[105] C.M. da Costa Santos, C.A. de Mattos Pimenta, M.R. Nobre, A systematic review of topical treatments to control the odor of malignant fungating wounds, J. Pain Symptom Manage. 39 (6) (2010) 1065–1076.

[106] T.S. Ustinova, R.I. Kaem, A.V. Yanshevskii, Y.S. Lopatto, Use of activated charcoal fabric in composite dressings, Bull. Exp. Biol. Med. 104 (2) (1987) 1166–1168.

[107] V.K. Sologub, R.I. Kaem, V.V. Pavlova, T.S. Ustinova, Y.S. Lopatto, A.V. Yanshevski, Morphological study of healing of burns covered with activated charcoal cloth, Bull. Exp. Biol. Med. 107 (3) (1989) 401–404.

[108] J. Laszlo, A.L. McLeod, (1986). Wound dressings. Worldwide patent 8605970.

[109] J. Laszlo, A.L. McLeod, (1987). Wound dressings. USP 4715857.

[110] R.D. Langston, F.C. Webb, S.C. McBeath, (1990). Wound dressing. USP 4928681.

[111] K.S.W. Sing, F.G.R. Gimblett, J.J. Freeman, (1987). Manufacture of fibrous activated carbons. USP 4699896.

[112] J.E. Wright, J.J. Freeman, K.S.W. Sing, S.W. Jackson, R.J.M. Smith, (1992). Wound dressing with activated carbon. EU patent 0311364 B1.

[113] B. Griffiths, E. Jacques, S. Bishop, (2002). Multilayered wound dressing. USP 6348423.

[114] J.C. Kerihuel, Effect of activated charcoal dressings on healing outcomes of chronic wounds, J. Wound Care 19(5) (208) (2010) 210–212. 214-5.

[115] M. Rai, A. Yadav, A. Gade, Silver nanoparticles as a new generation of antimicrobials, Biotechnol. Adv. 27 (1) (2009) 76–83.

[116] A.D. Russell, W.B. Hugo, Antimicrobial activity and action of silver, Prog. Med. Chem. 31 (1994) 351–370.

[117] J.R. Furr, A.D. Russell, T.D. Turner, A. Andrews, Antibacterial activity of Actisorb Plus, Actisorb and silver nitrate, J. Hosp. Infect. 27 (3) (1994) 201–208.

[118] M.E. Innes, N. Umraw, J.S. Fish, M. Gomez, R.C. Cartotto, The use of silver coated dressings on donor site wounds: a prospective, controlled matched pair study, Burns 27 (6) (2001) 621–627.

[119] M.A. O'Neill, G.J. Vine, A.E. Beezer, A.H. Bishop, J. Hadgraft, C. Labetoulle, M. Walker, P.G. Bowler, Antimicrobial properties of silver-containing wound dressings: a microcalorimetric study, Int. J. Pharm. 263 (1-2) (2003) 61–68.

[120] G.R. Newman, M. Walker, J.A. Hobot, P.G. Bowler, Visualisation of bacterial sequestration and bactericidal activity within hydrating hydrofiber wound dressings, Biomaterials 27 (7) (2006) 1129–1139.

[121] R. Khundkar, C. Malic, T. Burge, Use of acticoat dressings in burns: what is the evidence? Burns 36 (6) (2010) 751–758.

[122] J.C. Kerihuel, Charcoal combined with silver for the treatment of chronic wounds, Wounds UK 5 (3) (2009) 87–93.

[123] S.M. Bergin, P. Wraight, Silver based wound dressings and topical agents for treating diabetic foot ulcers, Cochrane Database Syst. Rev. 1 (2006) CD005082.

[124] M.J. Carter, K. Tingley-Kelley, R.A. Warriner 3rd, Silver treatments and silver-impregnated dressings for the healing of leg wounds and ulcers: a systematic review and meta-analysis, J. Am. Acad. Dermatol. 63 (4) (2010) 668–679.

[125] J.J. Freeman, F.G.R. Gimblett, R.A. Roberts, K.S.W. Sing, Studies of activated charcoal cloth. III. Mesopore development induced by phosphate impregnants, Carbon 26 (1) (1988) 7–11.

Index

Note: Page numbers with "f" denote figures; "t" tables.

Symbols

π–π dispersion interaction mechanism, 480–481
π-electron system of basal planes, 189, 192f, 447, 461–462, 472

A

A20, 304, 304f, 316–317
Acetylene, 299–301
Acid-doped activated carbon catalysts, 633
Acidity, 173–174, 252–253
Acrylic acid, 365, 376
Actisorb, 659–661
Activated carbon (AC), 4, 245, 473, 503–505
  in hemoperfusion, 645–655
    combined extracorporeal devices, 652–655
    complications of, 650–651
    devices, 651–652
    history of, 645–647
    principles of, 647–650
  in poisoning, 640–645
  scanning electron microscopy of, 504f
  for wound dressings, 655–657
Activated carbon, nonenvironment industrial applications of, 605–638
  in catalysis, 629–634
    acid-doped activated carbon catalysts, 633
    basic activated carbon catalysts, 632–633
    catalytic hydrogenations, 631–632
    C–C coupling reactions, 632
    combination reactions, 633–634
    oxidation reactions, 630–631
  as a decolorizer and a purifier, 607–614
    alcoholic beverage industries, 610–611
    chemical industry, 611–612
    in pharmaceutical industry, 612–613
    for purifying electrolytic baths, 613–614
    oils and fats industry, 609
    sugar and sweetener industries, 607–609
  in gas chromatography, 617–618
  for human body protection, 621–624
  as ion exchanger, 618–619
  in nuclear technology, 614–616
  as porous electrodes, 619–620
  for recovery, 624–628
    of gold, 625–626
    of organic solvents, 624–625
    of sulfur, 626–628
  in refining of mineral oil and gasoline, 620–621
  in vacuum technology, 617
  in wastewater treatment, 628–629
Activated carbon fiber (ACF), 120, 151–152, 404–405, 473–474, 487, 564–565, 614, 616, 627
Active surface for adsorption (ASA) method, 86–87
Adsorba® 300C, 647f
Adsorba® C cartridges, 651–652
Adsorbate, 108–109
Adsorbate molecules and surface groups
  geometry of, 70, 71t
Adsorbed phase, cluster analysis of, 51–53
Adsorption, 4, 376–380
  capacity, of carbon materials, 480
  by carbon materials, 500–501
  energetics, 176
  gas adsorption, 377–380
    CO_2 capture on HTC carbons, 377–378, 378t
    hydrogen storage on HTC carbons, 379–380, 380t
    methane storage on HTC carbons, 378–379, 379t
  hysteresis phenomena in, 117
  isotherms of, 152

Adsorption (*Continued*)
liquid phase, 376–377
by means of pressure, 584–585
by means of temperature, 584–585
by means of vacuum, 584–585
mechanism, 116–119
by phosphorus-containing carbons, 247–263
ammonia, adsorption of, 248–249, 249f
protons, adsorption of, 250–254
water (hydrophilicity), adsorption of, 247–248
Adsorption–catalytic process (ACP), 227
Advanced oxidation processes (AOPs), 510–511, 523
Aerogels, 208–209, 211
Aging, 208
Alcoholic beverage industries
decolorization, activated carbon applications in, 610–611
Allient Sorbent Hemodialysis System, 654
Alpha-s method, 124
Amide, 11–12
Amination/ammoxidation of carbons above 500 K, 185
Ammonia
adsorption of, 248–249, 249f, 460–461
decomposition, 587
gaseous, 185
treatment of oxidized carbons with, 619
Ammoxidation of carbons above 500 K, 185
Anhydroglucose units (AGU), 356–357
Anilines, 185
Anticoagulation solutions, 646
Applications of carbons, 179–180
Aptamer-modified graphene field effect (G-FET) transistor, 20
Aqueous media, environmental catalysis and oxidation reactions in, 510–512
Argon
adsorption, 40, 41f, 82, 111–113, 112f, 116
and nitrogen isotherms, 134–137, 136f
Aromatic compounds
adsorption, on novel carbon materials, 509–510
in water and wastewater, 479–480
AST-120Ї, 642–643
Atomistic calculation method, 409–410
Azomethine ylides, 338–340, 340f

B

Barret, Joyner, and Halenda (BJH) method, 77–78, 80, 122–123, 125–126, 217–219, 218f
Basic activated carbon, 181–183, 632–633
Basicity of carbons, 173–204
6 Ws of, 174–180
applications of carbons, 179–180
basic sites on carbons, 177–179
of basal plane and graphene edges, 189–193
identification of, 175
carbon material, 174
carbon surface, 175
definition of, 174
determination of, 175–177
heteroatom functional groups, 184–189
nitrogen, 184–188
phosphorus and sulfur, 188–189
inorganic (mineral) matter, 193–196
measurements, 180–184
oxygen-containing basic groups, 180–184
Beer manufacturing, activated carbon applications in, 610
Benzalkonium chloride (BKC), 21–22
Benzo[a]anthracene, 226–227
Benzo[a]pyrene, 226–227
Benzo[b]fluoranthene, 226–227
Benzoyl peroxide decomposition, activated carbon catalysts for, 634
BET surface area, 120, 122, 304–305, 393
average micropore diameter and reality of, 85–87
Bhattacharya and Gubbins (BG) method, 67–68, 68f, 69f, 77
Bicyclic pyrones, 181
Biochar, 374–375
BioLogic-DTPF, 652–654
Biomass-derived HTC, 357
Biomedical applications of carbon adsorbents, 639–669
hemoperfusion, activated carbon in, 645–655
complications of, 650–651
extracorporeal devices, 652–655
hemoperfusion devices containing AC as adsorbent, 651–652
history of, 645–647
principles of, 647–650
oral activated carbon in poisoning and beyond, 640–645
for wound dressings, 655–657

Biomorphic carbide
ceramics, 270
chlorination of, 276
Boehm titration method, 180
Boron-doped carbons (BDCs), 573
Boron-doped diamonds (BDDs), 573
Boyle, Robert, 173–174
Brandies manufacturing, activated carbon applications in, 611
Breadth-first search algorithm, 52
"Brick-and-mortar" self-assembly, 329–330
Broekhoff–de Boer (BDB) methods, 122–123
Brunauer classification, 281
Brunauer–Emmett–Teller (BET) method, 119, 213, 285

C

^{13}C solid-state NMR, 357
C_π sites, basicity of, 191
C_{60}/ZnO composites, 538
Calcium-decorated graphene, 458
Calorimetric techniques, 176
Capillary condensation, 117
Carbide-derived carbons (CDCs), porous, 269–294
characterization of, 277–283
degree of structural order, 277–279
mechanical properties, 283
oxidation behavior (TGA), 279–280
pore structure, 280–283
chlorination, in CDC synthesis, 270
with enhanced porosity by post-synthesis treatment in CO_2, 284–290
post-synthesis treatment, 286–290
thermogravimetric analysis (TGA), 285
nitrogen adsorption isotherm of, 137f, 138
processing of, 271–276, 272f
etching of carbide templates with chlorine, 271–276
processing of fibrous carbide templates, 271
Carboflex, 659–660
Carbohydrate-based HTCs, 353–356
Carbon
aerogels, characteristics of, 162t
allotropes, 5–6
atoms removal from building unit, 307–308
basicity. *See* Basicity of carbons
definition of, 4
Carbon blacks (CB), 329–330, 505
for fuel cells, 551–554, 552t
Carbon cryogel microspheres (CCM), 225
adsorption of dyes from water on, 226f
Carbon cryogels, hydrophobicity of, 157
Carbon dioxide
adsorption capacity, 458
capture, 456–459
on HTC carbons, 377–378
technologies, 583–584
porous carbon materials as, 586–598. *See* Porous carbon materials as CO2 adsorbents
uptake, on carbon nanostructures, 24–26
Carbon gels, 207–212, 475–476
as adsorbents for hydrogen storage, 221
doping with nitrogen/cobalt nanoparticles, 221
hydrophobicity of, 222–223
and nitrogen adsorption, 213
nonfunctionalized, 235
porosity of, 210–211
as porous hosts for storing ammonia borane, 221
preparation of, 208–210
skeletal density of, 212
structure of, 210–212
Carbon gels, adsorption by, 207–244
in gas phase, 219–225
condensable vapors and detection, 222–225
permanent gases, separation, and storage, 220–222
in liquid phase, 225–237
energy applications framework, 233–237
halides, 229–230, 230f
metals, 230–233
organic compounds, 225–228
for porosity characterization, 212–219
Carbon ion exchangers, 619
Carbon molecular sieves (CMSs), 41–42, 593
for CO_2 capture, 593–594
Carbon monolith, 328
Carbon nanocages, 329
Carbon nanocoils (CNCs), 562f, 567–568
Carbon nanofibers (CNFs), 474–475, 506–507, 564–565, 565f
Carbon nanohorns (CNHs), 567, 568f
Carbon nanomaterials
adsorption on, 9–16
gas adsorption on graphene and graphitic nanoribbons, 13–16

Carbon nanomaterials (*Continued*)
gas and vapors adsorbed on carbon nanotubes, 9–13
biological systems adsorbed on, 16–20
carbon nanotubes, 19–20
fullerenes, 16–19
graphene, 20
carbon dioxide uptake on, 24–26
in fuel cell applications, 561–568
carbon nanofibers and activated carbon fibers, 564–565
carbon nanotubes, 561–564
graphene, carbon nanohorn, and carbon nanocoils, 565–568
heavy-metals adsorption on, 21–24
chromium adsorption, 22
lead adsorption, 21–22
zinc, nickel and copper adsorption, 22–24
wettability of, 150
Carbon nanostructures, general aspects of, 4–7
Carbon nanotubes (CNTs), 19–20, 150, 474–475, 505–506, 514, 561–564
aerogels, 211
CNT/ZnO composites, 538
Carbon onions (COs), 329–330
Carbonaceous nanocomposites, controlled synthesis of, 367
Carbonaceous nanofibers (CNFs), 370, 381–382
Carbon-coating TiO_2 particles, 529
Carbon-doped TiO_2, 526–527
Carbonet, 659–660
Carbonization technique, 352–353
Carbonization temperature, 366–367
Carbon–oxygen surface groups, 609, 629–630
Carbon/TiO_2 photocatalysts, 526–532
Carbon/ZnO photocatalysts, 535–539
Carbosieve, pore-size distribution of, 47, 48f, 49f
Carboxylic acid, 11–12
Carboxylic groups, 157–159
Carboxylic/imidazole-containing HTC materials, 364
Catalysis, activated carbon application in, 629–634
acid-doped activated carbon catalysts, 633
basic activated carbon catalysts, 632–633
catalytic hydrogenations, 631–632
C–C coupling reactions, 632
combination reactions, 633–634
oxidation reactions, 630–631
Catalyst-assisted chlorination of metal carbides, 275–276
Catalyst-loading techniques, 558
Catalytic carbon gasification, 476–477
Catalytic hydrodechlorination, 512–513
Catalytic oxidation processes, 510
Catalytic wet air oxidation (CWAO), 510–511
Cavitation-induced evaporation, 119
C–C coupling reactions, activated carbon catalysts for, 632
Chahine' rule, 221
Characterization of carbons
argon adsorption, 116
general comments, 109–110
helium adsorption, 116
hydrogen adsorption, 113
by low-temperature adsorption, 39–42
nitrogen adsorption, 115–116, 115f
nonane preadsorption, 115
standard adsorptives for, 111–113
water adsorption, 115–116
Charcoal, 640–641
Chemical doping, 6–7, 9–11
Chemical functionalization, 6–7
Chemical industry
decolorization and clarification, activated carbon applications in, 611–612
Chemical reduction and thermal exfoliation, 440–442
Chemical vapor deposition (CVD) method, 15, 298–301, 402
Chemical vapor infiltration and reaction (CVI-R) technique, 271
Chemically activated carbons, 473–474
Chemviron Carbon, 659–660
Chloride removal from water, 230
Chlorination
of biomorphic carbides, 276
in CDC synthesis, 270, 278
of metal carbides, 270, 275–276
Chromium adsorption
on carbon materials, 491–493, 492f
by carbon nanotubes, 22
Ciprofloxacin (CPX) adsorption, 227, 489–491, 490f
Clinisorb, 659–660
Cluster analysis of adsorbed phase, 51–53
CMK-1 materials, 476
CMK-3 materials, 476

CO_2/CH_4 separation, 594–595
CO_2/N_2 separation, 594–595
Colloidal carbon spheres (CCSs), 537
Colloidal crystals, 325
Colloidal silica, 325, 327–328
Colloidal templating, 334–335
Colloid-templated carbons, 327–328
Combination reactions, activated carbon for, 633–634
Completely reversible water uptake isotherms, 151–152
Computer simulation method, 126–127
Confocal laser scanning microscopy (CLSM), 394–395
Conformal transformation process, 270
Continuous countercurrent method, 608–609
Controlled carbon deposition, 476–477
Copper adsorption by carbon nanotubes, 22–24
Cryogels, 209, 220
Cryogenic distillation, 584–585
Curing, 208
Cyclic amines and imines, 185
Cyclic voltammetry, 235, 235f
Cyanogen chloride, activated carbon application for, 622–623

D

Dahlia-flower-type SWCNH, 404–405
D'Arcy–Watt isotherm, 154–155
D-band (defect band), 403
Decolorization, activated carbons for, 607–614
  in alcoholic beverage industries, 610–611
  in chemical industry, 611–612
  in electrolytic baths purification, 613–614
  in oils and fats industry, 609
  in pharmaceutical industry, 612–613
  in sugar and sweetener industries, 607–609
Deep eutectic solvents (DESs), polycondensation of, 586–587
Defect generation of graphitic surface, 6–7
Dehydrochlorination, activated carbon for, 631
Denitrification of water, activated carbon for, 632
Density functional theory (DFT), 66–67, 108–109, 122–123, 125–130, 217–218, 218f, 304
  calculations of cadmium atom, 12–13
Density of states (DOS) of graphene, 460–461, 461f
Deodorization, activated carbon filters for, 614
Depyrogenation, activated carbon application in, 613
Desorption mechanism, 118–119, 118f
Desulfurization, using activated carbons, 630–631
Dibenzothiophene (DBT), 263
  adsorption, from n-hexadecane (n-HD), 228f
Diels–Alder reactions, 355
Diethyl carbonate, 630
4,6-Dimethyldibenzothiophene (DMDBT), 228, 263
Direct methanol fuel cells (DMFCs), 237
Dispersive interaction, 480–481
Do equation, 155
Dollimore and Heal (DH) method, 216–219, 218f
"Dry methodologies" of preparing carbon materials, 195–196
Dubinin–Astakhov (DA) model, 75–76, 83–85, 88, 154–155
Dubinin–Kaganer method, 120–122
Dubinin–Radushkevich (DR) equation, 123–124, 217–219, 360
Dubinin–Radushkevich–Stoeckli (DRS) model, 80–81, 80f
Dubinin–Serpinsky (DS) equation, 154–155
Dubinin's isotherm, 80–81

E

Electrical double-layer capacitors (EDLC), 233–234, 316–317
  application for, 316–317
Electroadsorption of ions, 260–262
Electrochemical advanced oxidation processes (EAOP), 512
Electrochemical capacitors. *See* Supercapacitors (SCs)
Electrochemical measurements, 177
Electro-Fenton (EF) process, 512
Electrolytic baths purification, activated carbon applications in, 613–614
Electron donor–acceptor (EDA) interactions, 189
Electron paramagnetic resonance (EPR), 531
Electron transfer materials
  activated carbons as, 619–620
Electroplating, 613–614
Electrosorption, 230
Electrostatic interaction, 480–481

"Emerging pollutants", 508–509
Empirical parameters, 87–88
Energy storage, 386–394
 fuel cells, 393–394
 Li/Na ion batteries, 386–390
 supercapacitors, 390–393
Enterosorbent, in poisoning, 640–645
Environmental catalysis and oxidation reactions in aqueous media, 510–512
Epitaxial graphene (EG), production and characterization of, 437, 438f
Epitaxial growth and chemical vapor deposition technique, 437–438
Epitaxially grown graphite, 437
Ethylene glycol (PEG), 11–12
Ethylenediamine triacetic acid (EDTA), 20
Evaporation-induced self-assembly (EISA) method, 340–342
Exfoliated graphene, 441
Extracorporeal blood cleansing, 645t
Extracorporeal hemoperfusion circuit, 646f

F

Few-layer graphene, 442
FexOy@C spheres, fabrication of, 381
Feynman–Hibbs (FH) effective potential, 423–424
Fibrous carbide templates
 etching of, with chlorine, 271–276
 processing of, 271, 272f
Fibrous TiC ceramic, SEM images of, 273f
Field-effect transistor (FET), 20
First-order kinetics, 164
Flexible silver@cross-linked PVA coaxial nanocables, 372
 SEM and TEM images of, 373f
Flow calorimetric measurements, 190
Flue gas, 583–584, 626
 purification of, 626–627
Fluidized-bed method, 608–609
Fly-ash carbon, 588–589
Formaldehyde, polymerization of, 325–326, 327f
Formic acid (FA), 353–354
Forsyth vs Eli Lilly, 640–641
Fredholm integral equation, 77
Freeze-drying aquagels, 209
Fuel cell, 393–394, 549–581
 boron-doped carbons (BDCs), 573
 carbon nanomaterials, 561–568
 conventional carbons, 551–558
  carbon black and electrochemical corrosion of carbon, 551–554
  catalyst-loading techniques, 558
  graphitized carbons and graphite, 554–556
  surface-modified carbons, 556–558
 importance, 549–551
 mesoporous carbons, 559–561
 nitrogen-doped carbons, 569–572
 novel surface treatments of carbons, 558t
 perspectives, 573–574
Fullerenes, 4–5, 16–19
Furfuryl alcohol (FA), 295–296, 337–338

G

Gambro Lundia AB, 651–652
Gas adsorption, 377–380
 HTC carbons
  CO_2 capture on, 377–378, 378t
  hydrogen storage on, 379–380, 380t
  methane storage on, 378–379, 379t
Gas chromatography, activated carbon in, 617–618
Gas sensor, 447–448
Gases' and liquids' accessibility in carbons, 37–60
 atomistic modeling, 51–54
  cluster analysis of adsorbed phase, 51–53
  transition state theory (TST), 53–54
 experimental background, 39–46
 open loop hysteresis, 54–57
 percolation theory-based view, 47–51
Gasoline refining, activated carbon application in, 620–621
Gas–solid chromatography, 176
Gaussian curvature, 6
G-band of SWCNH, 404–405
Genbox procedure, 73–74
Generalized adsorption equation (GAE), 128–129
Generalized adsorption integral (GAI), 39
Global adsorption isotherm (GAI) equation, 66–67
Glucose, 353–355
Glucose-derived HTC materials, 357
Glyoxal, 332–333
Glyoxal–phloroglucinol carbon, 332–333
"Golden fleece" nanocomposites, 370
Gold recovery, activated carbon for, 625–626
Gold-supported activated carbon catalysts, 630

Grand canonical Monte Carlo (GCMC), 66–67, 70–73
Graphene, 20, 26, 312–313, 503, 565–568
Graphene aerogels, 211
Graphene and graphene-related materials, 435–468, 532
 adsorption behaviors, 447–462
  ammonia adsorption, 460–461
  carbon dioxide capture, 456–459
  gas sensor, 447–448
  hydrogen storage, 448–456
  surface functionalization, 461–462
 ammonia on, 461t
 chemically modified preparation methods through
 density of states (DOS) of, 460–461, 461f
 preparation methods, 437–442
  chemical reduction and thermal exfoliation, 440–442
  epitaxial growth and chemical vapor deposition technique, 437–438
  few-layer graphene, 442
  micromechanical exfoliation technique, 439–440
 schematic representation of, 436f
 SEM image of, 439f
 single-layer graphene, 440f
  physical properties of, 436t
Graphene hydrogenation, 461–462
 scheme of, 462f
Graphene nanoribbons, 5–6, 447–448, 449f
Graphene oxide
 chemical structure of, 440f
 hydrogen storage behavior of, 455, 456f
 specific surface area and pore volumes of, 454, 455t
Graphene sheets in ZTC, 306–308
Graphene-like nanosheets, 451
Graphite, 4–6, 502–503, 554–556
Graphite nanofibers (GNFs), 506–507
Graphite nanosheets (GNSs), 183–184
Graphitized carbons and graphite, 554–556
Gravimetric method for adsorption measurements, 109–110
Gurvitsch rule, 123

H

Halides, adsorption of, 229–230, 230f
Hard-templating synthesis of mesoporous carbons, 324f, 327–333
Helium, 116
 adsorption of, 116
Hemicellulose, 356–357
Hemoperfusion (HP), activated carbon in, 645–655
 AC containing combined extracorporeal devices, 652–655
 AC containing hemoperfusion devices, 651–652
 complications of, 650–651
 history of, 645–647
 principles of, 647–650
Heteroatom functional groups, 184–189
 nitrogen, 184–188
 phosphorus and sulfur, 188–189
Heteroatom surface groups, 189
 in carbon skeleton, 156
Heteroatom-doped carbons, 568–573
 boron-doped carbons (BDCs), 573
  boron-doped diamonds (BDDs), 573
 nitrogen-doped carbons, 569–572
  metal-free nitrogen-doped carbons, 572
  nitrogen-doped carbon nanocoils (NCNCs), 570
  nitrogen-doped carbon nanotubes (NCNTs), 569–570
  nitrogen-doped mesoporous carbons (NMCs), 570
Heterogeneous catalysis, 380–383
Heterogeneous photocatalysis, 523–525
n-Hexadecane (*n*-HD), dibenzothiophene adsorption from, 228f
Hexamethoxymethylmelamine (HMMM), 333–334
Hierarchical porous carbons (HPCs), 316
High surface area graphites (HSAGs), 502–503
 scanning and transition electron microscopy of, 502f
Highest occupied molecular orbital (HUMO), 460–461
Highly ordered pyrolytic graphite (HOPG), 502–503
"Highly pure" carbons, 189
High-pressure adsorption, 595–597
High-resolution transmission electron microscopy (HRTEM), 63–64
High-surface-area carbons, 342–343
High-temperature fuel cells, 550
High-temperature gas adsorption
 based on low-temperature characterization, 42–46
l-Histidine, 329
Harkins-Jura (HK) pore-size distribution, 134
Honey-based dressings, 658

Horvath–Kawazoe (HK) method, 77–78, 124–125, 217
HSGD (HaemoSorbent Granulated Deliganding), 652
Human body protection, activated carbon application for, 621–624
against toxic gases, 623
Hybrid QSDFT method, 138
Hydrogen
adsorption, 113
capacity, 453–454
and carbon cap, interaction between, 409
storage, 448–456
application for, 314–315
behavior, of graphene oxide, 455, 456f
on HTC carbons, 379–380
mechanism, on metal/carbon surfaces, 456, 457f
Hydrogen peroxide decomposition, activated carbon catalysts for, 634
Hydrogenation, catalytic
activated carbon application in, 631–632
Hydrogenation reactions, 512–513
Hydrogenation–hydrodechlorination methods, 513
Hydrophilicity, 148
Hydrophobic poly(ethylene oxide)-b-poly(methyl methacrylate) (PEO-b-PMMA) diblock copolymer, 329
Hydrophobicity, 148–150
of carbon cryogels, 157
of carbon surface, 163
Hydrothermal carbons (HTCs), 351–399
with acidic groups, 383
advantage of, 383
applications of, 373–395
adsorption, 376–380
agricultural applications, 374–375
biological applications, 394–395
CO_2 capture on, 377–378, 378t
energy storage, 386–394
fuel cells, 393–394
gas adsorption, 377–380
heterogeneous catalysis, 380–383
hydrogen storage, 379–380, 380t
Li/Na ion batteries, 386–390
liquid phase, 376–377
magnetic HTCs, 383–385
methane storage, 378–379, 379t
photocatalysis, 385–386
supercapacitors, 390–393
carboxylic rich HTC
SEM micrograph of, 365–366, 365f
cellulose conversion into, 356f
formation mechanism and final chemical structure of, 352–357
carbohydrate-based HTCs, 353–356
lignocellulosic-biomass-derived HTCs, 356–357
functionalization of, 362–367
in situ functionalization, 364–367
postmodification, 363–364
in situ functionalization of, 366–367
nanocomposites, 367–373
in situ HTC nanocomposites, 370–373
loading inorganic nanostructures onto HTC materials, 370
preformed nanostructures, coating of, 367–370
porous materials, 357–362
chemical activation, 358–361
soft-templating, 362
SEM micrographs of, 354f
Hydroxymethylfurfural (HMF), 353–354
(N-(4-Hydroxyphenyl)ethanamide), 508–509
Hysteresis phenomena, 117

I

Imidazole-containing porous HTC materials, 382–383
Immunoglobulin E (IgE) aptamers, 20
In situ functionalization of hydrothermal carbons, 366–367
In situ hydrothermal carbon nanocomposites, 370–373
Indoxyl sulfate (IS), 643
Inorganic compounds
adsorption of, 491–493
on carbons, 194–195
Insensitive nuclei-enhanced by polarization transfer (INEPT), 354–355
Integral adsorption equation. *See* Generalized adsorption equation (GAE)
Inverse gas chromatography, 176
Ion exchange mechanism, 254–258
Ion-exchange activated carbons, 618–619
Ions, electroadsorption of, 260–262

J

Jaroniec–Choma (JCh) model, 80–81, 80f

K

Kelvin equation, 125–126
Kremazin, 643
Kubas interaction, 456

L

Langmuir equation, adsorption constants of, 225
Langmuir isotherm model, 57, 376
Large-scale greenhouse gas emissions, 583–584
Laundry wastewater circulation, activated carbon filters for, 616
Lead adsorption by carbon nanotubes, 21–22
Lennard–Jones (LJ) centers, 70
Levulinic acid (LA), 353–354
Lewis basicity type of interaction, 190–191, 197
Light energy, use of
  in nonbiological photochemical processes, 522
Lignin, 356–357
Lignocellulosic-biomass-derived HTCs, 356–357
Lipid peroxidation, 657
Liquids' accessibility in carbons. *See* Gases and liquids accessibility in carbons
Lithium-doped pillared graphene oxide (PGO), 451–452
  gravimetric hydrogen storage capacity of, 452f
Lithium/sodium ion batteries, 386–390
Liver Dialysis Unit, 652–654
Lobry de Bruyn–Alberta van Ekstein rearrangement (BER), 353–354
Lowest unoccupied molecular orbital (LUMO), 18–19, 460–461
Low-temperature adsorption
  characterization of carbons by, 39–42
Low-temperature autoclaving (LTA) route, 340–342
Low-temperature characterization
  high-temperature gas adsorption based on, 42–46
Low-temperature fuel cells, 550

M

Macropore volumes of activated carbons, 473–474
Maggs' results, 39
Magnetic HTCs, 383–385
Magnetite-doped SWCNH, 426–427, 427f
  magnetization curves of, 428f
MCM-48 silica template, 476
Mechanical grinding, 413
Megalopores, 209
Mercury removal from water, 232f
Mesopore analysis, 117, 125–126
Mesopore volumes of activated carbons, 473–474
Mesoporous carbons (MCs), 329–331, 559–561
  surface modification of, 333–340
    functionalization during synthesis, 333–338
    postsynthesis modification, 338–340
Mesoporous melamine–formaldehyde carbon aerogel (MFCAG)
  water adsorption isotherms of, 161
Metal carbides
  catalyst-assisted chlorination of, 275–276
Metal ions, adsorption of, 254–262
  electroadsorption of ions, 260–262
  ion exchange, 254–258
  reduction–oxidation adsorption, 258–260
Metal/carbon surfaces
  hydrogen storage mechanism on, 456, 457f
Metal-containing carbons, 336–337
Metal-doped carbon gels, 237
Metal-free nitrogen-doped carbons, 572
Metallic impurities
  removal by carbon ion exchangers, 619
Metals, adsorption of, 230–233
Methane, 222
  self-diffusivity of, 222f
  storage, on HTC carbons, 378–379
((RS)-2-(4-(2-Methylpropyl)phenyl) propanoic acid), 508–509
Microhoneycomb structures, 209
Micromechanical exfoliation technique, 439–440
Micro–mesoporous carbon materials, pore-size characterization of, 130–138
Micropollutants, 508–509
Micropore analysis, 123–125
Micropore-filling process, 116
Micropores
  physical adsorption in, 110
  volume, of activated carbons, 473–474
Microporous-activated carbon, 661–662
  nitrogen sorption isotherm for, 134, 135f
Microporous carbons, 37–38

Mineral oil, refining of
activated carbon application in, 620–621
Metal-organic framework materials (MOFs), 595–597
Molecular Adsorbent Recirculating System (MARS) System, 652–654
Molecular dynamics, 73–74
simulation, 122–123
Molecular simulations techniques, 61, 88–95
Monte Carlo (MC) simulation, 51, 122–123
MSC30, 304–305, 304f
Multilayer adsorbed film, 125–126
Multiwall carbon nanotubes (MWCNTs), 9–11, 21–23, 148–150, 505–506, 531–532, 538
self-photo-activity of, 535
SEM and TEM micrographs of, 505f

N

Nanodiamond particles, 442
Nanoporosity, 407–415
accessible sites of SWCNH assembly, 411–413
donation of nanowindows to, 410–411
interaction potential view of single SWCNH particle, 407–410
surface chemistry of SWCNH, 414–415
Nanoporous carbons, 38, 107–145
adsorption mechanism, 116–119
density functional theory (DFT), 126–130
experimental aspects, 109–116
argon adsorption, 116
general comments, 109–110
helium adsorption, 116
hydrogen adsorption, 113
nitrogen adsorption, 115–116, 115f
nonane preadsorption, 115
standard adsorptives for carbon characterization, 111–113
water adsorption, 115–116
mesopore analysis, 125–126
micro–mesoporous carbon materials, pore-size characterization of, 130–138
micropore analysis, 123–125
pore size and porosity analysis, 122–138
general comments, 122–123
surface area assessment, aspects of, 119–122
Nanoscale materials growth fields, 426–430
Nanowindow-donated SWCNH, 412
Natural gas, 583–584
Nguyen and Do (ND) model, 77–79
Nickel adsorption by carbon nanotubes, 22–24
Nickel-OMC composites, ferromagnetic properties of, 336–337
Nitrogen, 184–188
adsorption isotherms of, 128, 129f
and argon isotherms, 134–137, 136f
in carbon matrix, 479
doping, 476
and oxygen surface groups on carbon, 473f
physisorption, analysis by, 304–305
pore-filling pressures for, 125–126, 126f
sorption isotherm, for microporous-activated carbon, 134, 135f
in water vapor uptake, 160–161
Nitrogen adsorption, 39, 111–112, 112f
comparison of NLDFT and QSDFT methods for, 131f
on CMK-3, 130–134, 132f
and carbon gels, 213
and desorption isotherms, 213–214, 214f, 215f
isotherm
of carbide-derived carbon, 137f, 138
of typical carbide-derived carbon, 138
studies, 505–506
and water adsorption, 115–116, 115f
Nitrogen-based surface functionalities, 586–587, 599
Nitrogen-containing functional groups, 159–161, 159f
Nitrogen-containing ordered mesoporous polymers, 333–334
Nitrogen-doped carbons, in fuel cell applications, 569–572
metal-free nitrogen-doped carbons, 572
nitrogen-doped carbon nanocoils (NCNCs), 570
nitrogen-doped carbon nanotubes (NCNTs), 569–570
nitrogen-doped mesoporous carbons (NMCs), 570
Nitrogen-doped graphene (N-graphene), 572
Nitrogen-doped multi-walled carbon nanotubes (CN_X-MWNTs), 9–13
Noble metals
as catalysts in fuel cells, 550
nanoparticles, TEM images of, 370, 371f
nanostructures, 370
Nonane preadsorption, 115

Nonlocal density functional theory (NLDFT), 122–123, 127, 129–130
for pore-size characterization of disordered carbon materials, 127
Non-noble metals, 571t
Novel carbon materials
aromatic compounds' adsorption on, 509–510
for fuel cells
carbon nanomaterials, 561–568
mesoporous carbons, 559–561
Novolac, 333–334
Nuclear technology, activated carbon adsorption in, 614–616

O

Octyl-phenol-ethoxylate (TX-100), 21–22
Oils and fats industry
decolorization, activated carbon applications in, 609
One-dimensional (1D) systems, 4–5
Open loop hysteresis, 54–57
Open-SWCNH, 415–419, 422–423, 426–427
magnetite nanoparticles-doped
TEM images of, 427f
pore volume of, 416–419
vapor water adsorption isotherms of, 418f
Oral activated carbon in poisoning, 640–645
Ordered mesoporous carbons (OMCs), 130–134, 313–314, 327–328, 336–337, 476, 559, 560f, 570
synthesis of, 324–325
hard-templating synthesis, 324f, 327–333
with magnetic properties, 336
scheme, 313f
soft-templating synthesis, 326f, 333–340
Ordered mesoporous silica (OMS), 324–325, 327–328
Organic aromatic pollutants in wastewater, 508, 508t
Organic compounds, adsorption of, 225–228, 262–263, 479–491
surface chemistry, effect of, 481–491
Organic gels, 208
pyrolysis of, 212–213, 238
Organic solvents recovery, activated carbon for, 624–625
Oxidation reactions, activated carbon application in, 630–631
Oxidation treatments, of carbon materials, 476–477
Oxidized multi-walled carbon nanotubes (Ox-MWCNTs), 21–23
Oxygen
and carbon basicity, relationship between, 181
functional groups, 310
functionalities, 157–159
Oxygen-containing basic groups, 180–184
Oxygen-containing functional groups (OFGs), 364, 483–485

P

Partial charge density, isosurface plots of, 448, 450f
Path-integral grand canonical Monte Carlo (PI-GCMC) simulation, 423–424
Pd@C core–shell nanoparticles, 372
Pd-open-SWCNH
TEM image of multiwall carbon nanotubes in, 429f
Pd-SWCNH, TEM images of, 429f, 430
Percolation theory, 38, 47–49
Perfluorohexane (PFH), 163
Periodic organosilicas (PMOs), 337
Peroxidase enzyme, 227
Persistent organic pollutants, 226–227
Pharmaceutical industry
activated carbon applications in, 612–613
Phenol, adsorption of, 225
Phenol–formaldehyde-based carbon, 248, 250, 644f
Phloroglucinol, polymerization of, 325–326, 327f
Phosphorus and sulphur-containing functional groups, 188–189
Phosphorus-containing carbons, 245–267
adsorption by, 247–263
ammonia, adsorption of, 248–249, 249f
protons, adsorption of, 250–254
water (hydrophilicity), adsorption of, 247–248
cation-exchange capacity of, 252
metal ions, adsorption of, 254–262
electroadsorption of ions, 260–262
ion exchange, 254–258
reduction–oxidation adsorption, 258–260
from nitrogen-containing copolymer, 251
organics, adsorption of, 262–263
phosphorus heteroatoms, chemical state of, 246–247

Phosphorus–nitrogen co-dopants, 9–11
Photocatalysis, 385–386, 525–539
carbon/TiO2 photocatalysts, 526–532
carbon-doped TiO_2, 526–527
porous carbon/TiO_2 composites
carbon/ZnO photocatalysts, 535–539
photocatalyst, carbon as, 532–535
Photochemical behavior of carbon adsorbents, 521–548
photocatalysis, carbon's role on, 525–539
carbon/TiO_2 photocatalysts, 526–532
carbon/ZnO photocatalysts, 535–539
photocatalyst, carbon as, 532–535
semiconductor photocatalysis, fundamentals of, 523–525
Physisorption, 108–109
Pillared graphene oxide (PGO), 451
Plasma-enhanced chemical vapor deposition (PECVD), 163
Pluronic F127, 333–334
Pluronic® amphiphilic block copolymers, 362
Polar surface groups, 154
Polyacrylamide (PAAM)-grafted MWNTs (MWCNTs/PAAM), 21–22
Polyacrylic acid (PAA), 23
Polyacrylonitrile, 296–297
Poly-(m-amino benzene sulfonic acid), 9–11
Polycyclic aromatic hydrocarbons (PAHs), 226–227
Poly(ethylene oxide)–poly(propylene oxide)–poly(ethylene oxide) (PEO–PPO–PEO) triblock copolymers, 325–326
Polyfuranic intermediate, 357
Poly(furfuryl alcohol) (PFA)–silica nanocomposites, 337–338
Polymer electrolyte membrane fuel cells (PEMFCs), 237
Polymer-based carbons, characteristics of, 161t
Polymer-templated carbons, 338–340
Polypyrrole, 9–11
Polystyrene, 642–643
Polystyrene-block-poly(4-vinylpyridine), 325–326
Poly(vinyl alcohol) (PVA)-assisted synthesis of coaxial nanocables, 372
Polyvinylpyridine, 642–643
Pore
blocking, 118–119
condensation, 117
diameter, 85
morphology, 108–109
network connectivity, 47, 48f
size and porosity analysis, 122–138
general comments, 122–123
topology, 108–109
Pore-size distribution (PSD), 49, 62, 82–83, 127, 130f, 327–328
of Carbosieve, 47, 48f, 49f
determination methods
of gases having permanent electric moments, 82–85
of noble gases, 76–81
Porosity characterization, adsorption for, 212–219
Porous carbon materials, as CO_2 adsorbents, 586–598
adsorption behavior, improving, 587–588
high-pressure performance, 595–597
porous structure, effect of, 589–593
regeneration, 597–598
selectivity, 593–595
surface chemistry, 586–589
Porous carbons, 148, 161–162, 324–325
Porous hydrothermal carbon materials, 357–362, 363f
chemical activation, 358–361
soft-templating, 362
Porous texture versus surface chemistry, 471–498
carbon materials, 472–479
strategies to modify textural and surface properties of, 476–479
structure and properties, 472–476
inorganic compounds, adsorption of, 491–493
organic compounds, adsorption of, 479–491
surface chemistry, effect of, 481–491
Potassium iodide-doped SWCNH, 427–430
Prato's reaction, 338–340
Pratt–Chandler theory, 94–95
Precombustion gas mixtures, 583–584
Prismaflex® HP system, 651–652
Prometheus®, 652–654
Propylene, 299–300
Protective clothing, 622
Protein, 366–367
Proton exchange membrane fuel cells (PEMFCs), 553–554
Proton sinks
basic carbons as, 177
carbons' behaviour of, 175

Protons, adsorption of, 250–254
Pseudocapacitance effect, 236
"Pure" carbons as proton sinks, 189
Purification, activated carbons for, 607–614
Pyridone-like structures, 186–188
Pyrolysis of organic gels, 212–213
Pyrone groups on carbon surfaces, 182–183
Pyrone-like structures, 183, 186–187
Pyrone oxygen, 183
Pyrrole-like structures, 186–188
Pyrrole nitrogen, carbons with, 476

Q

Quantum fluid, adsorption of, 423–426
Quenched solid density functional theory (QSDFT), 109, 128–134

R

Radial breathing mode (RBM) vibrations, 306–307, 403
Radioactive methyl iodide, 615
Radioiodine, 615
Radiological dispersal device (RDD), 655
Radiotoxins, 656
Raman spectroscopy, 277–278
  SWCNT, 403, 403f, 405f
Ramaprabhu's CO_2 adsorption capacity, 458
Rare-earth-doped gadolinium oxide, 385
Recovery, activated carbons for, 624–628
Rectified spirit, activated carbon applications in, 610–611
Reduced graphite oxides (rGOs), 184
Reduction–oxidation adsorption, 258–260
REDY system, 654
Regenerability, 597–598
Resin–TEOS–block copolymer, 331–332
Resol copolymer, 325–326
Resorcinol–formaldehyde resins, 209–210, 210f
Resorcinol–formaldehyde aqueous gels, 475
Resorcinol–furfural oligomers, 333–334
Respirator filters, 623
Reverse Monte Carlo (RMC) simulation method, 62

S

SBA-15 silica templates, 476
Scanning electron microscope (SEM), 211f
Schwarzite, 307–308
SCN oral therapy, 642–643
Selective catalytic reduction (SCR), 180
Semiconductor photocatalysis, fundamentals of, 523–525
SiC-CDC, 279t, 280, 282f, 287–288
  bending strength of, 284f
  cumulative pore volume of, 289f
  and decline of SSA, 281
  micropore volume fraction of, 271
  pore structure of, 289, 290t
  Raman spectra of, 278
  treatment in carbon dioxide, 286
Silica
  interpenetration of, 327–328
  scaffold etching, 325
Silver-carbonaceous nanocables, synthesis of, 372
Silver-doped carbon aerogels, 229–230
Single-layer graphene, physical properties of, 436t
Single-stranded DNA (ssDNA), 20
Single-wall carbon nanohorn (SWCNH), 401–434
  accessible sites of SWCNH assembly, 411–413
  adsorbed density of, 421–422
  adsorption of molecules on, 415–423
    adsorption of supercritical gases, 419–423
    adsorption of vapors, 416–419
  adsorption of quantum fluid, 423–426
  carbon components of, 410–411
  CH_4 adsorption isotherms of, 419f, 420f
  cone-shaped cap on, 409
  donation of nanowindows to, 410–411
  HR-TEM images of, 402f
  interaction potential view of, 407–410
  mechanical grinding, 413
  nanoporosity, 407–415
  nanoscale materials growth fields, 426–430
  nanowindow-donated, 412
  as one of carbon nanotubilites, 402–407
  primary SWCNH particle, 404–405, 404f
  Raman spectra of, 403, 403f, 405f
  SERS spectra of, 405f
  surface chemistry of, 414–415
  SWNCT, interaction potential of
    external, 407–409
    internal, 407–409
  vapor water adsorption isotherms of, 418f
Single-wall carbon nanotube (SWCNT), 9–12, 12f, 13f, 14f, 17f, 18f, 22–23, 25, 316, 505–506
Single-walled carbons, 407–409

SiO_2@C nanocomposites, 367–370
SKN oral therapy, 642–643
Slit-like carbon pore model, 77
Small-angle neutron scattering (SANS), 165–166, 213
Small-angle X-ray scattering (SAXS), 148, 165–166, 213
Sodium dodecylbenzene sulfonate (SDBS), 21–22
Sodium ion batteries, 388
Soft-templated carbons, adsorption by, 323–350
  macroscopic morphologies of, 340–342
  mesoporous carbon materials, surface modification of, 333–340
    functionalization during synthesis, 333–338
    postsynthesis modification, 338–340
  ordered mesoporous carbons (OMCs), synthesis of, 324–325
    hard-templating synthesis of mesoporous carbons, 324f, 327–333
    soft-templating synthesis of mesoporous carbons, 326f, 333–340
  soft-templated carbons, macroscopic morphologies of, 340–342
  tailoring porosity of carbon adsorbents, 327–333
Soft-templating synthesis, 362
  of mesoporous carbons, 326f, 333–340
Sol–gel transition, 208, 209f
Solid electrolyte interface (SEI), 389–390
Solid-phase extraction (SPE) technique, 226–227
Sorbent Management for Advanced Renal Replacement Therapy (SMARRT®) technology, 654
Sorption performance, 148
Specific surface area (SSA), 280
  and hydrogen storage capacity, linear relationship between, 453–454, 453f
Standard hydrogen electrode (SHE), 524
Stationary-bed method, 608–609
Stoeckli and Morell's experimental correlation, 76
Sugar and sweetener industries
  decolorization, activated carbon applications in, 607–609
Sulfur atoms, in carbons, 7
Sulfur-containing chemicals, 227–228
Sulfur–oxygen functional groups, 188–189
Sulfur recovery, activated carbon for, 626–628
Supercapacitors (SCs), 233–234, 390–393
Supercritical carbon dioxide ($scCO_2$), 232
Supercritical fluid-aided deposition, 232
Supercritical gases, adsorption of, 419–423
Surface area assessment, aspects of, 119–122
Surface characterization methodologies, 178
Surface chemistry, effect of, on adsorption of organic compounds, 481–491
Surface functionalization, 461–462
Surface hydrophobicity, 148–150
Surface modification, 585
Surface-enhanced laser desorption ionization time-of-flight mass spectrometry (SELDI-TOF-MS), 20
Surface-enhanced Raman scattering (SERS) technique, 405–406
Surface-modified carbons, for fuel cells, 556–558
Sustainable energy, 549–550
Synergistic factor, 528
Systagenix Wound Management, 659–660

T

Tailoring porosity of carbon adsorbents, 327–333
Te@carbonaceous composite nanocables
  SEM and TEM images of, 369f
Tellurium nanowires, 367
Temperature-programmed desorption (TPD), 310, 477–479
Template carbonization method, 295–296
Templated carbons. *See* Ordered mesoporous carbons (OMCs)
Templated mesoporous carbons (TMCs), 316
Tetraethyl orthosilicate (TEOS), 335–337
Thermogravimetric analysis (TGA), 279, 285
Thermal treatments, of carbon materials, 476–477
Thermoresponsive polymer (N-isopropylacrlyamine), 363–364
Thiolated ssDNA, 20
Three-dimensional (3D) systems, 4–5
  carbon nanomaterials, 7
Three-dimensionally ordered mesoporous (3DOm) carbons, 134
Thrower–Stone–Wales (TSW) defect, 405–406

Tin-nanoparticles encapsulated in elastic hollow carbon spheres (TNHCs), 367–370
Titania-containing mesostructures, 335–336
Titanium-anchored graphene oxide, 454, 454f
Titanium carbide (TiC), 270
TiC-CDC composite, 273–274, 278, 279t, 284–287
bending strength of, 284f
micropore volume fraction of, 271
nitrogen adsorption isotherms of, 282f
pore structure of, 289, 290t
SEM images of, 273f, 290f, 291
specific surface area (SSA) of, 280, 280f, 283f
Titanium oxide, 385, 522–523, 527
immobilization of, 527–528
as a photoactive material, 526
Toluene, adsorption of, 224
Toxic substancesÏ removal of, 641–642
Toxic/hazardous gases, protection against, 621
TPD spectra for AC samples, 481, 483f
T-plot method, 124, 217
Transition state theory (TST), 53–54
Triethyl orthoacetate, 333–334
Triethylene diamine (TEDA), 615
Tris(3-trimethoxysilylpropyl)isocyanurate, 337
Turbomolecular pump, 110
Two-dimensional (2D) structures, 4–5
Typical carbide-derived carbon
nitrogen adsorption isotherm of, 138

U

Ultra-large-pore mesoporous carbon, formation of, 329, 330f
Ultramicroporosity, 153
Ultrathin epitaxial graphite, 437
Urea–phenol–formaldehyde system, 333–334
Uremia, treatment of, 646
UV–vis diffuse reflectance spectroscopy, 386

V

Vacuum technology, activated carbon in, 617
Vapors, adsorption of, 416–419
Vega ZZ package, 86
Vertically aligned nitrogen-containing carbon nanotubes (VANCNTs), 569–570, 572
Virtual porous carbons (VPCs), 61–104
models, 63–68
virtual oxidation, 64–66, 65f
VPC absolute (geometric) porosity, 66–68
results, 74–95
examples of VPC application, 76–95
reality of studied VPC models, 74–76
simulation details, 69–74
grand canonical Monte Carlo (GCMC), 70–73
molecular dynamics, 73–74
Vodka manufacturing, activated carbon applications in, 610
Vulcan XC-72, 552–553, 553f

W

Wastewater treatments, 500
activated carbon application in, 628–629
Water, 227–228
adsorption of, 82, 116, 147–172, 247–248
accepted mechanism of, 154
coadsorption of water in multicomponent vapor systems, 164–166
isotherms, 151–155, 151f, 165f
kinetics of, 164
of mesoporous melamine–formaldehyde carbon aerogel, 161
and nitrogen adsorption, 115–116, 115f
surface chemistry, 155–163
chloride removal from, 230
general pollutants of, 508–509
interaction of carbon materials with, 148
isotherms, 116
mercury removal from, 232f
Water–carbon interaction, 175–176
Water-solved aromatic compounds, catalytic removal of, 174, 472–479, 499–520
carbon materials' application, for removing organic aromatic compounds in polluted waters, 507–514
aromatic compounds' adsorption on novel carbon materials, 509–510
environmental catalysis and oxidation reactions in aqueous media, 510–512
general pollutants of water, 508–509
hydrogenation reactions, 512–513
photocatalysts, 513–514
carbon materials' application, in heterogeneous catalysis, 501–507

Water-solved aromatic compounds, catalytic removal of (*Continued*)
activated carbon and carbon blacks, 503–505
carbon nanofibers, 506–507
carbon nanotubes, 505–506
graphite materials, 502–503
strategies to modify textural and surface properties of, 476–479
structure and properties of, 472–476
Water vapor adsorption, 151–153, 151f, 222–223
Water vapor uptake, nitrogen in, 160–161
Wearable artificial kidney (WAK), 654–655
Wet gels, 208
Wettability of carbon nanoparticles, 150
Wide-angle X-ray scattering (WAXS) measurements, 153
Wines manufacturing, activated carbon applications in, 611
Wound dressings, activated carbon materials for, 655–657

X

X-ray photoelectron spectroscopy (XPS), 414
Xerogels, 208–209, 220, 227

Z

Zeolite, 297–299
cation type, 298–299
pore structure, 298
Si/Al ratio, 298
Zeolite micropores, carbon introduction into, 299–301
CVD method, 299–301
solid-phase carbonization, 299
two-step method, 301
Zeolite-templated carbon, 295–322
electrical double-layer capacitors (EDLC), application for, 316–317
hydrogen adsorption isotherms of, 315f
hydrogen storage, application for, 314–315
molecular structure of, 305–314
construction of possible molecular model, 307–311
information obtained from various kinds of analysis, 305–307
ordered mesoporous carbons (OMCs), comparison with, 313–314
origin of extremely large surface area of ZTC, 311–313
SEM images of, 302f, 305–306
structure of, 301–305
bulk shape, 301
ordered structure, 301–303
textural properties, 304–305
synthesis method, 296f, 297–301
carbon introduction, 299–301
choice of zeolite template, 297–299
transmission electron microscopy (TEM) images of, 301–302, 302f, 305–306
XRD patterns of, 302–303, 303f, 305–306
Zhao's method, 325–326
Zinc adsorption by carbon nanotubes, 22–24
Zinc oxide, 536
nanoparticles (ZnO-NP), 22–23
Zorflex®Medical cloth, 659–662, 661f

FSC
www.fsc.org
MIX
Papier | Fördert
gute Waldnutzung
FSC® C083411

Kolibri 360 GmbH
Ferdinand-Jühlke-Straße 7
99095 Erfurt, Deutschland
produktsicherheit@kolibri360.de